Mitigation Strategies and Tolerance of Plants to Abiotic Stresses

Mitigation Strategies and Tolerance of Plants to Abiotic Stresses

Guest Editors

Geovani Soares de Lima
Lauriane Almeida dos Anjos Soares
Francisco Vanies Da Silva Sá

MDPI

Basel • Beijing • Wuhan • Barcelona • Belgrade • Novi Sad • Cluj • Manchester

Guest Editors
Geovani Soares de Lima
Academic Unit of
Agricultural Sciences
Federal University of
Campina Grande
Pombal
Brazil

Lauriane Almeida
dos Anjos Soares
Academic Unit of
Agricultural Sciences
Federal University of
Campina Grande
Pombal
Brazil

Francisco Vanies Da Silva Sá
Department of Agrarian and
Exact Sciences
State University of Paraíba
Catolé do Rocha
Brazil

Editorial Office
MDPI AG
Grosspeteranlage 5
4052 Basel, Switzerland

This is a reprint of the Special Issue, published open access by the journal *Plants* (ISSN 2223-7747), freely accessible at: www.mdpi.com/journal/plants/special_issues/D91C7VZLRL.

For citation purposes, cite each article independently as indicated on the article page online and using the guide below:

Lastname, A.A.; Lastname, B.B. Article Title. *Journal Name* **Year**, *Volume Number*, Page Range.

ISBN 978-3-7258-2734-3 (Hbk)
ISBN 978-3-7258-2733-6 (PDF)
https://doi.org/10.3390/books978-3-7258-2733-6

Contents

About the Editors

Geovani Soares de Lima

Geovani Soares de Lima holds a degree in agronomy (2012), a Master's degree (2013), and a Ph.D. (2014) in agricultural engineering with a concentration in irrigation and drainage from the Federal University of Campina Grande, Campina Grande Campus (PB). He completed a postdoctoral internship at the National Postdoctoral Program CAPES/PPGEA/UFCG. He is currently an adjunct professor, Class A Level I, at the Federal University of Campina Grande (UFCG), assigned to the Academic Unit of Agricultural Sciences (UAGRA/CCTA). He participates as a permanent member in the Postgraduate Program in Agricultural Engineering (PPGEA/UFCG) in the area of irrigation and drainage and as a collaborator in the Postgraduate Program in Tropical Horticulture (PPGHT/UFCG) in the line of water and soil management in vegetable production systems. He is the editor of the *Brazilian Journal of Agricultural and Environmental Engineering* (ISSN 1807-1929) and *Agriculture* (EISSN 2077-0472). He is the leader of the research group "Water, soil, and plant management in irrigated production systems" (MASP/CNPq) and a researcher at the National Institute of Science and Technology in Sustainable Agriculture in the Tropical Semi-Arid (INCT AgriS), working in the area of agricultural engineering with an emphasis on irrigation and drainage and developing research on the following topics: crops in hydroponic systems, management and recovery of soils affected by salts, strategies for mitigating abiotic stresses (saline and water), physiology, and management of irrigated crops.

Lauriane Almeida dos Anjos Soares

Lauriane Almeida dos Anjos Soares holds a degree in agronomy from the Federal University of Campina Grande, Center for Agri-Food Science and Technology (2012), a Master's degree (2014), and a Ph.D. (2016) in agricultural engineering from UFCG in the area of concentration of irrigation and drainage. She held a Junior Postdoctoral Scholarship, CNPq (2017–2018). She is currently an adjunct professor I of the Agronomy Course at the UAGRA/CCTA/Federal University of Campina Grande (UFCG) Campus de Pombal (PB). She participates as a permanent professor in the Postgraduate Program in Natural Resources Engineering and Management (PPGEGRN/UFCG) and the Postgraduate Program in Tropical Horticulture (PPGHT/UFCG) and is a collaborator in the Postgraduate Program in Agroindustrial Systems (PPGSA/UFCG). She has experience in agricultural engineering, with an emphasis on irrigation, drainage, and salinity. Working mainly with the physiology and production of irrigated crops, saline and water stress, soil–water–plant management, vegetables, large crops, and oilseeds.

Francisco Vanies Da Silva Sá

Francisco Vanies Da Silva Sá has a Ph.D. in agricultural engineering and a Ph.D. in agronomy. He also holds a Master's in soil and water management, a degree in agronomy, and has completed a technical course in agriculture. He has a postdoc in agricultural engineering and a postdoc in soil and water management. He is currently a professor at the UEPB, working in the Postgraduate Program in Agrarian Science, and is also a colaborating professor at the UFERSA working in the Postgraduate Program in Soil and Water Management. He also has experience in agricultural engineering and agronomy.

Preface

Semiarid and arid areas worldwide are subject to qualitative and quantitative scarcity of water resources. In these regions, water scarcity (water deficit) and water sources with high salt concentrations limit agricultural production. Due to high temperatures and light intensity, climate change intensifies abiotic stresses in plants, especially cross-stresses. These changes have led to soil water restriction, soil salinization, accumulation of heavy metals, and decreased plant growth and food production. Thus, there is an urgent need to establish strategies that mitigate abiotic stresses to facilitate crop sustainability and meet the growing needs of food production.

In arid and semiarid regions, abiotic stresses cause genetic alterations, changes in phytohormone levels, oxidative stress, changes in leaf gas exchange and photosynthetic efficiency, and nutritional disorders that decrease plant growth and production. Genomic, biochemical, physiological, and agronomic responses depend on plant tolerance or sensitivity to abiotic stresses. This Special Issue titled "Mitigation Strategies and Plant Tolerance to Abiotic Stresses" presents original research results on the effects of abiotic stresses, tolerance mechanisms, crop tolerance associated with mitigation strategies, and biomolecular mechanisms.

This Special Issue of *Plants* comprises 24 articles covering various species, including fruit, cereal, vegetable, oilseed, and energy crops. The included literature shows mechanisms for inducing tolerance to abiotic stresses such as drought, salinity, excess metals in the soil, and high temperatures. The results achieved show advances in irrigation and fertilizer management (N, P, K, Ca, Se, and organic residues), exogenous application of organic acids (phytohormones, proline, ascorbic acid, and pyruvate), use of plant growth-promoting microorganisms, and tolerant cultivars and genotype selection. However, due to climate change, many knowledge gaps regarding the tolerance of plants to abiotic stresses require further investigation.

In summary, this collection reflects the efforts of multiple researchers in the field of plant sciences, who have contributed different insights to investigate mitigation strategies and tolerance of plants to abiotic stresses. Therefore, this Special Issue will contribute to instigating new studies that allow the incorporation of scientific and technological knowledge in production processes in the face of abiotic stresses.

Geovani Soares de Lima, Lauriane Almeida dos Anjos Soares, and Francisco Vanies Da Silva Sá
Guest Editors

plants MDPI

Article

The Effects of Irrigation Water Salinity on the Synthesis of Photosynthetic Pigments, Gas Exchange, and Photochemical Efficiency of Sour Passion Fruit Genotypes

Francisco Jean da Silva Paiva [1], Geovani Soares de Lima [1,*], Vera Lúcia Antunes de Lima [1], Weslley Bruno Belo de Souza [1], Lauriane Almeida dos Anjos Soares [2], Rafaela Aparecida Frazão Torres [1], Hans Raj Gheyi [1], Luderlândio de Andrade Silva [1], Francisco Vanies da Silva Sá [3], Valeska Karolini Nunes Oliveira de Sá [1], Smyth Trotsk de Araújo Silva [1], Reynaldo Teodoro de Fátima [1], Pedro Dantas Fernandes [1] and Alan Keis Chaves de Almeida [2]

1 Academic Unit of Agricultural Engineering, Federal University of Campina Grande, Campina Grande 58430-380, PB, Brazil; francisco.jean@estudante.ufcg.edu.br (F.J.d.S.P.); vera.lucia@professor.ufcg.edu.br (V.L.A.d.L.); weslley.bruno@estudante.ufcg.edu.br (W.B.B.d.S.); rafaela.aparecida@estudante.ufcg.edu.br (R.A.F.T.); hans.gheyi@ufcg.edu.br (H.R.G.); luderlandio.andrade@estudante.ufcg.edu.br (L.d.A.S.); valeska.karolini@estudante.ufcg.edu.br (V.K.N.O.d.S.); smyth.trotsk@estudante.ufcg.edu.br (S.T.d.A.S.); reynaldo.teodoro@estudante.ufcg.edu.br (R.T.d.F.); pedro.dantas@professor.ufcg.edu.br (P.D.F.)

2 Academic Unit of Agrarian Sciences, Federal University of Campina Grande, Pombal 58840-000, PB, Brazil; lauriane.almeida@professor.ufcg.edu.br (L.A.d.A.S.); alankeisc@gmail.com (A.K.C.d.A.)

3 Agricultural Sciences Center, State University of Paraíba, Catolé do Rocha 58884-000, PB, Brazil; vanies@ufersa.edu.br

* Correspondence: geovani.soares@professor.ufcg.edu.br; Tel.: +55-83-99945-9864

Abstract: The objective of this study was to evaluate the synthesis of photosynthetic pigments, gas exchange, and photochemical efficiency of sour passion fruit genotypes irrigated with saline water under the conditions of the semi-arid region of Paraíba state, Brazil. The experiment was conducted at the experimental farm in São Domingos, PB. A randomized block design was adopted, in a 5 × 3 factorial scheme, with five levels of electrical conductivity of irrigation water—ECw (0.3, 1.1, 1.9, 2.7, and 3.5 dS m^{-1})—and three genotypes of sour passion fruit (Gigante Amarelo—'BRS GA1'; Sol do Cerrado—'BRS SC1'; and Catarina—'SCS 437'. The increase in the electrical conductivity of irrigation water negatively affected most of the physiological characteristics of the sour passion fruit at 154 days after transplanting. Significant differences were observed between sour passion fruit genotypes when its tolerance was subjected to the salinity of irrigation water. There was an increase in the percentage of damage to the cell membrane with the increase in the electrical conductivity of irrigation water, with maximum values of 70.63, 60.86, and 80.35% for the genotypes 'BRS GA1', 'BRS SC1', and SCS 437', respectively, when irrigated with water of 3.5 dS m^{-1}. The genotype 'BRS Sol do Cerrado' showed an increase in the synthesis of photosynthetic pigments when irrigated with water of 3.5 dS m^{-1}, with maximum values estimated at 1439.23 μg mL^{-1} (Chl *a*); 290.96 μg mL^{-1} (Chl *b*); 1730.19 μg mL−1 (Chl *t*); and 365.84 μg mL^{-1} (carotenoids). An increase in photosynthetic efficiency parameters (F_0, Fm, and Fv) of the genotype 'BRS Gigante Amarelo' was observed when cultivated with water with high electrical conductivity (3.5 dS m^{-1}).

Keywords: abiotic stress; genetic variability; *Passiflora edulis Sims*; Brazilian semi-arid region

check for updates

Citation: Paiva, F.J.d.S.; Lima, G.S.d.; Lima, V.L.A.d.; Souza, W.B.B.d.; Soares, L.A.d.A.; Torres, R.A.F.; Gheyi, H.R.; Silva, L.d.A.; Sá, F.V.d.S.; Sá, V.K.N.O.d.; et al. The Effects of Irrigation Water Salinity on the Synthesis of Photosynthetic Pigments, Gas Exchange, and Photochemical Efficiency of Sour Passion Fruit Genotypes. *Plants* **2023**, *12*, 3894. https://doi.org/10.3390/plants12223894

Academic Editor: Alexandra S. Dubrovina

Received: 11 September 2023
Revised: 10 November 2023
Accepted: 15 November 2023
Published: 18 November 2023

1. Introduction

Sour passion fruit (*Passiflora edulis* Sims) is a crop of great social and economic importance and can be used both in the form of food and for medicinal purposes, with fruits rich in several nutrients, vitamins, amino acids, and dietary fiber [1]. The global demand for sour passion fruit juice is growing, with an estimated increase of 15% to 20% per year [2].

In 2021, Brazil produced 683,993 tons of fruits, with the northeastern state standing out among the main producing regions, accounting for 69.6% of the national production [3]. Among the northeastern states, Bahia (207,448 t) and Ceará (177,291 t) stand out as the largest producers, while the state of Paraíba produced only 10,444 t in that year. Although the northeastern region is the largest producer, the average yield per hectare is only 14,758 kg ha^{-1}, which is below the Brazilian average (15,259 kg ha^{-1}).

Factors that may be related to the low yield in the northeastern state include the edaphoclimatic conditions of this region. High evapotranspiration rates and low rainfall, distributed irregularly in the atmosphere, are intrinsic conditions that force producers to use groundwater, which contains high levels of dissolved salts [4].

Salt stress creates conditions of imbalance in plants by limiting physiological and biochemical processes [5,6]. Under salt stress conditions, the first effect to occur is osmotic, characterized by the restriction in water availability to plants, due to the reduction in the osmotic potential of the soil solution [7]. Then, the excessive accumulation of ions in plant tissues leads to ionic stress, which induces morphological, physiological, biochemical, and molecular changes in cell metabolism [8].

In general, fruit crops have greater sensitivity to salt stress than perennial crops, and their success in agricultural production depends on the use of alternative practices, including the use of salt-tolerant species [9,10]. Salinity tolerance mechanism in plants depends on other factors, such as the capacity for transport and accumulation of toxic ions, activation of the antioxidant defense system, and biosynthesis of compatible solutes [11,12].

Studies with various crops, such as sesame [13], cotton [14], cowpea [15], and sunflower [11], have been carried out to evaluate the tolerance of different genotypes to salt stress, with genotypic variations being observed regarding the tolerance of the crops to the salinity of irrigation water. However, research with fruit species is scarce in the literature, especially with the sour passion fruit crop under the semi-arid conditions of the northeastern Brazilian state. In this context, the hypothesis which was tested was that the effects of saline stress differ in sour passion fruit genotypes, with various impacts on the synthesis of photosynthetic pigments, gas exchange, and photochemical efficiency of the plants resulting from their genetic variability. Therefore, the identification of genotypes tolerant to salt stress is important for the adoption of appropriate agronomic strategies for agricultural production in conditions where only waters with higher salt contents are available. In view of the above, the objective of this study was to evaluate the synthesis of photosynthetic pigments, gas exchange, and photochemical efficiency of sour passion fruit genotypes irrigated with saline water under the conditions of the semi-arid region of Paraíba, Brazil.

2. Results

2.1. *Analysis of Principal Components and Multivariate Variance of the Effects of Salt Stress on the Physiology of Sour Passion Fruit Genotypes*

The multidimensional space of the original variables was reduced to two principal components (PC1 and PC2) with eigenvalues greater than $\lambda \geq 1.0$, as highlighted by [16]. The eigenvalues and percentage of variance explained for each component are shown in Table 1. The two components together represented 76.30% of the total variation. PC1 explained 49.59% of the total variance, which is represented by most variables, and PC2 explained 26.71% of the remaining variance.

There was a significant effect ($p \leq 0.01$) of the interaction between the electrical conductivity of water (ECw) and sour passion fruit genotypes (GEN) for PC1 and PC2 (Table 1). When analyzed separately, significant effects ($p \leq 0.01$) were observed for ECw and GEN in the two principal components (PC1 and PC2).

Table 1. Eigenvalues, percentage of total variance explained in the multivariate analysis of variance (MANOVA), probability of significance using the Hotelling test ($p \leq 0.05$) for the factors of the electrical conductivity of irrigation water (ECw), sour passion fruit genotypes (GEN), ECw × GEN interaction, and correlation coefficients (r) between original variables and principal components.

	Principal Components (PCs)	
	PC1	PC2
Eigenvalues (λ)	6.44	3.47
Percentage of total variance (S^2%)	49.59	26.71
Hotelling test (T^2) for electrical conductivity (ECw)	0.01	0.01
Hotelling test (T^2) for genotypes (GEN)	0.01	0.01
Hotelling test (T^2) for ECw × GEN interaction	0.01	0.01

PCs	Correlation coefficient												
	RWC	Chl *a*	Chl *b*	Chl *t*	Car	Ci	E	gs	A	WUEi	CEi	ETR	PAR
PC1	−0.75	−0.60	−0.70	−0.62	−0.60	−0.39	0.72	0.81	0.87	0.72	0.67	−0.76	−0.78
PC2	−0.19	0.78	0.63	0.76	0.76	−0.66	0.17	0.18	0.40	0.41	0.63	−0.08	−0.02

TREAT	Mean values												
	RWC	Chl *a*	Chl *b*	Chl *t*	Car	Ci	E	gs	A	WUEi	CEi	ETR	PAR
EC1G1	83.98 ± 0.59	537.66 ± 77.61	106.76 ± 2.03	644.43 ± 79.64	106.01 ± 6.11	232.00 ± 9.24	3.60 ± 0.16	0.40 ± 0.01	31.71 ± 0.56	8.87 ± 0.53	0.14 ± 0.01	25.83 ± 2.93	95.00 ± 8.08
EC1G2	89.49 ± 2.67	672.01 ± 61.11	225.95 ± 19.64	897.97 ± 54.03	221.51 ± 5.79	228.00 ± 2.89	2.43 ± 2.89	0.32 ± 0.05	15.64 ± 2.25	6.40 ± 0.76	0.07 ± 0.01	37.40 ± 2.48	153.33 ± 5.49
EC1G3	79.02 ± 4.12	723.31 ± 18.04	154.94 ± 5.19	878.25 ± 23.96	192.45 ± 14.29	140.00 ± 23.09	3.42 ± 23.09	0.34 ± 0.01	38.38 ± 7.03	11.44 ± 2.54	0.31 ± 0.11	20.30 ± 1.96	92.00 ± 1.15
EC2G1	76.84 ± 3.27	289.36 ± 44.45	31.85 ± 3.95	321.21 ± 42.61	52.4 ± 8.04	209.50 ± 2.60	3.77 ± 0.14	0.45 ± 0.04	30.94 ± 2.25	8.29 ± 0.91	0.15 ± 0.01	32.25 ± 1.01	113.33 ± 6.64
EC2G2	81.68 ± 1.18	299.27 ± 10.94	88.37 ± 1.45	387.64 ± 9.48	185.08 ± 46.99	239.00 ± 5.51	2.94 ± 0.20	0.52 ± 0.03	29.21 ± 2.28	9.98 ± 0.78	0.12 ± 0.01	39.40 ± 2.37	174.00 ± 8.08
EC2G3	85.47 ± 2.83	201.6 ± 1.51	43.9 ± 0.21	245.49 ± 1.29	59.6 ± 1.80	230.00 ± 4.62	2.93 ± 0.39	0.17 ± 0.02	15.79 ± 0.12	5.61 ± 0.86	0.07 ± 0.01	39.35 ± 4.24	155.33 ± 7.80
EC3G1	88.16 ± 3.37	1386.47 ± 137.18	295.34 ± 52.70	1681.81 ± 189.88	337.28 ± 22.75	206.00 ± 3.46	2.81 ± 0.31	0.12 ± 0.04	17.87 ± 0.83	6.48 ± 0.58	0.09 ± 0.01	77.55 ± 6.90	477.00 ± 58.31
EC3G2	88.74 ± 4.14	593.98 ± 22.22	215.94 ± 8.632	809.92 ± 30.85	188.32 ± 24.68	345.50 ± 10.68	1.66 ± 0.04	0.10 ± 0.01	9.51 ± 0.14	5.75 ± 0.12	0.03 ± 0.01	43.90 ± 1.91	321.00 ± 15.01
EC3G3	85.96 ± 2.91	1155.18 ± 103.71	275.06 ± 35.36	1430.23 ± 139.07	280.77 ± 33.62	292.00 ± 11.26	2.38 ± 0.12	0.12 ± 0.01	9.11 ± 0.46	3.74 ± 0.33	0.04 ± 0.01	55.50 ± 3.70	260.33 ± 26.85
EC4G1	90.28 ± 0.61	159.14 ± 12.50	31.13 ± 1.07	190.27 ± 10.79	46.63 ± 3.83	209.00 ± 12.12	3.62 ± 0.25	0.24 ± 0.03	15.08 ± 1.05	4.18 ± 0.27	0.07 ± 0.01	58.43 ± 5.82	191.00 ± 42.15
EC4G2	86.53 ± 2.15	138.27 ± 13.18	32.35 ± 4.12	170.62 ± 17.29	45.55 ± 3.21	258.50 ± 4.33	3.10 ± 0.03	0.34 ± 0.02	22.69 ± 0.14	7.33 ± 0.02	0.09 ± 0.01	48.50 ± 1.56	250.33 ± 11.26
EC4G3	87.7 ± 1.35	487.55 ± 0.36	48.54 ± 0.09	536.09 ± 0.44	126.45 ± 1.93	214.00 ± 2.02	3.21 ± 0.04	0.31 ± 0.03	25.18 ± 0.85	7.86 ± 0.25	0.12 ± 0.01	51.80 ± 3.15	287.67 ± 25.12
EC5G1	77.76 ± 5.33	218.43 ± 29.02	49.47 ± 5.88	267.9 ± 35.49	86.51 ± 6.34	274.50 ± 1.44	3.66 ± 0.24	0.39 ± 0.01	25.81 ± 2.38	7.03 ± 0.18	0.09 ± 0.01	24.40 ± 2.31	146.00 ± 6.35
EC5G2	86.52 ± 2.52	1439.23 ± 76.14	290.96 ± 25.16	1730.19 ± 101.30	365.84 ± 20.52	150.50 ± 7.79	3.48 ± 0.22	0.38 ± 0.02	22.13 ± 0.38	6.40 ± 0.43	0.15 ± 0.01	40.30 ± 0.98	167.00 ± 6.35
EC5G3	82.33 ± 1.91	642.39 ± 47.18	138.9 ± 6.57	781.28 ± 53.75	168.52 ± 12,59	231.50 ± 10.10	3.69 ± 0.34	0.37 ± 0.01	24.26 ± 0.08	6.70 ± 0.65	0.11 ± 0.01	40.20 ± 0.75	166.33 ± 7.22

ECw—electrical conductivity of water: EC1 (0.3 dS m^{-1}); EC2 (1.1 dS m^{-1}); EC3 (1.9 dS m^{-1}); EC4 (2.7 dS m^{-1}); EC5 (3.5 dS m^{-1}); GEN—sour passion fruit genotypes: G1 ('BRS GA1'—Gigante Amarelo); G2 ('BRS SC1'—Sol do Cerrado); G3 ('SCS 437'—Catarina); RWC—relative water content (%); Chl *a*—chlorophyll *a* (μg mL^{-1}); Chl *b*—chlorophyll *b* (μg mL^{-1}); Chl *t*—total chlorophyll (μg mL^{-1}); Car—carotenoids (μg mL^{-1}); Ci (intercellular CO_2 concentration—μmol CO_2 m^{-2} s^{-1}); E—transpiration (mmol H_2O m^{-2} s^{-1}); gs—stomatal conductance (mol H_2O m^{-2} s^{-1}); A—CO_2 assimilation rate (μmol CO_2 m^{-2} s^{-1}); WUEi (water use efficiency—[(μmol CO_2 m^{-2} s^{-1}) (mmol H_2O m^{-2} s^{-1})$^{-1}$]; CEi (intrinsic carboxylation efficiency—[(μmol CO_2 m^{-2} s^{-1}) (μmol CO_2 m^{-2} s^{-1})$^{-1}$]; ETR (electron transport rate); PAR (photosynthetically active radiation). Cells with stronger green and red shades, respectively, represent the highest and lowest average values observed for variables depending on the treatments.

Effects of treatments and the variables are expressed in Figure 1A,B, referring to the first and second principal components (PC1 and PC2). In the first principal component (PC1), there is a possible interaction between the levels of electrical conductivity of irrigation water and the sour passion fruit genotypes (ECw × GEN). Correlation coefficients were higher than 0.70 for relative water content, chlorophyll *b*, transpiration, stomatal conductance, CO_2 assimilation rate, water use efficiency, intrinsic carboxylation efficiency, electron transport rate, and photosynthetically active radiation.

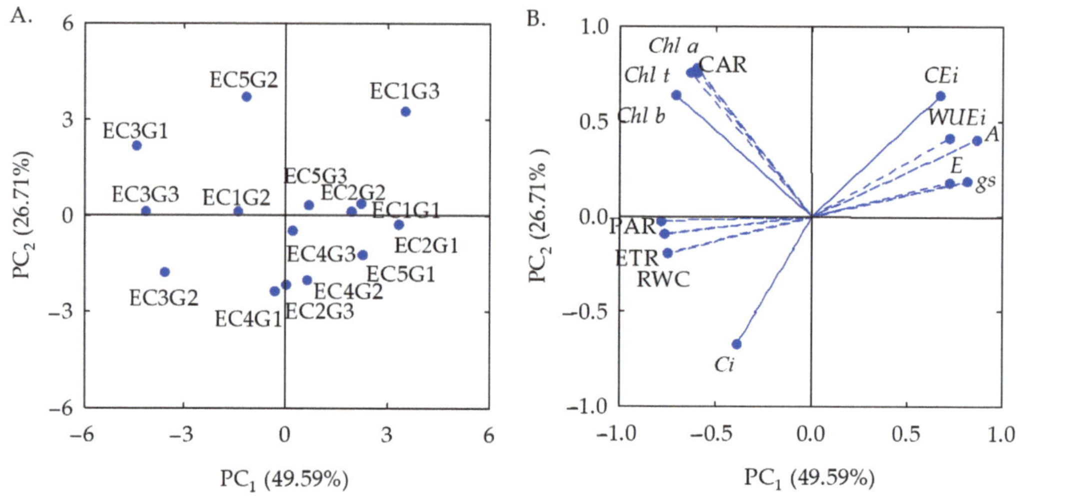

Figure 1. Two-dimensional projection of the scores of the principal components for the factors levels of electrical conductivity of irrigation water (ECw) and genotypes of sour passion fruit (**A**) and of the variables analyzed (**B**) in the two principal components (PC1 and PC2). Electrical conductivity (EC) of water: EC1 (0.3 dS m^{-1}); EC2 (1.1 dS m^{-1}); EC3 (1.9 dS m^{-1}); EC4 (2.7 dS m^{-1}); EC5 (3.5 dS m^{-1}); G—sour passion fruit genotypes: G1 ('BRS GA1'—Gigante Amarelo); G2 ('BRS SC1'—Sol do Cerrado); G3 ('SCS 437'—Catarina).

The highest relative water content (90.28%) was observed in the genotype G1 under irrigation water of 2.7 dS m^{-1} (EC4G1), which was equivalent to a 14.89% increase in RWC compared to the lowest value (76.84%) obtained in plants of the same genotype subjected to ECw of 1.1 dS m^{-1} (EC2G1) (Table 1).

The highest contents of chlorophyll *b* (295.34 µg mL^{-1}) were obtained in the genotype G1 subjected to ECw of 1.9 dS m^{-1} (EC3G1), which was 827.28% higher (264.1 µg mL^{-1}) than that obtained in plants of the same genotype grown under ECw of 2.7 dS m^{-1} (31.13 µg mL^{-1}) (EC4G1).

Regarding transpiration (E), the genotype G1 irrigated with water of 1.1 dS m^{-1} (EC2G1) reached the highest value (3.77 mmol H_2O m^{-2} s^{-1}), which was 127.10% (2.11 mmol H_2O m^{-2} s^{-1}) higher than that obtained in plants irrigated with water of 1.9 dS m^{-1} of the genotype G2, which reached the lowest value (1.66 mmol H_2O m^{-2} s^{-1}).

The genotype G2 irrigated with water of 1.1 dS m^{-1} (EC2G2) showed the highest stomatal conductance (0.52 mol H_2O m^{-2} s^{-1}). When compared to the lowest value observed (0.10 mol H_2O m^{-2} s^{-1}) referring to the same genotype (G2) irrigated with water of 1.9 dS m^{-1} (EC3G2), a reduction of 0.42 mol H_2O m^{-2} s^{-1} was observed.

The highest CO_2 assimilation rate (38.38 µmol CO_2 m^{-2} s^{-1}) was observed in the genotype G3 irrigated with ECw of 0.3 dS m^{-1} (EC1G3), which was 76.25% (29.26 µmol CO_2 m^{-2} s^{-1}) higher than that obtained under water salinity of 1.9 dS m^{-1} (EC3G3), equal to 9.11 µmol CO_2 m^{-2} s^{-1}.

Water use efficiency (WUEi) showed a behavior similar to that obtained for CO_2 assimilation rate (Table 1), with the highest value obtained with the genotype G3 irrigated with water of 0.3 dS m^{-1} (EC1G3) and equal to 11.44 [(µmol CO_2 m^{-2} s^{-1}) (mol H_2O m^{-2} s^{-1}) $^{-1}$], which was 205.88% (7.69 [(µmol m^{-2} s^{-1}) (mol H_2O m^{-2} s^{-1})$^{-1}$]) higher than that obtained by the same genotype irrigated with water of 1.9 dS m^{-1} (EC3G3).

For intrinsic carboxylation efficiency (CEi), the genotype G3 irrigated with water of 0.3 dS m^{-1} (EC1G3) obtained the highest value (0.31 [(µmol CO_2 m^{-2} s^{-1}) (µmol CO_2 m^{-2} s^{-1}) $^{-1}$], which was an increase of 933.33% (0.28 [(µmol CO_2 m^{-2} s^{-1}) (µmol CO_2

m^{-2} s^{-1}) $^{-1}$] compared to the lowest value observed (0.03 [(μmol CO_2 m^{-2} s^{-1}) (μmol CO_2 m^{-2} s^{-1})$^{-1}$], referring to the genotype G2 irrigated with water of 1.9 dS m^{-1} (EC3G2).

For the electron transport rate (ETR), the genotype G1 irrigated with water of 1.9 dS m^{-1} (EC3G1) obtained the highest value (77.55), which was 282.01% (57.25) higher than that obtained by the genotype G3 (20.30) irrigated with water of 0.3 dS m^{-1} (EC1G3). Similarly, a reduction of 80.71% (385.00 μmol m^{-2} s^{-1}) was observed in the PAR of the genotype G1 irrigated with water of 1.9 dS m^{-1} (EC3G1) compared to the genotype G3 cultivated with water of 0.3 dS m^{-1}.

Regarding the results obtained in the principal component 2, the eigenvalues were higher than 0.66 for the following variables: chlorophyll *a*, total chlorophyll, carotenoids, and intercellular CO_2 concentration. The highest content of chlorophyll *a* (Chl *a*) was observed in the genotype G2 (1439.23 μg mL^{-1}) irrigated with water of 3.5 dS m^{-1} (EC5G2), which was 940.88% (1300.96 μg mL^{-1}) higher than that obtained by the genotype G2 (138.27 μg mL^{-1}) irrigated with water of 2.7 dS m^{-1} (EC4G2).

Total chlorophyll (Chl *t*) contents showed a behavior similar to that observed for Chl *a* contents (Table 1), for which the highest value was obtained by the genotype G2 (1730.19 μg mL^{-1}) irrigated with water of 3.5 dS m^{-1} (EC5G2). This value was 914.06% (1559.57 μg mL^{-1}) higher than that obtained by the same genotype irrigated with water of 2.7 dS m^{-1} (EC5G2) and equal to 170.62 μg mL^{-1}.

The behavior of carotenoid (Car) contents was also similar to that of Chl *a* contents (Table 1), with the highest value obtained by the genotype G2 (365.84 μg mL^{-1}) irrigated with water of 3.5 dS m^{-1} (EC5G2), which was 703,16% higher than that obtained by the same genotype irrigated with water of 2.7 dS m^{-1} (45.55 μg mL^{-1}).

For intercellular CO_2 concentration (Table 1), the highest value (345.50 μmol CO_2 m^{-2} s^{-1}) was obtained by the genotype G2 irrigated with water of 1.9 dS m^{-1} (EC3G2). On the other hand, the lowest value (140.00 μmol CO_2 m^{-2} s^{-1}) was observed in the genotype G3, under irrigation with water of 0.3 dS m^{-1}.

Changes in the behavior of physiological variables for each genotype are presented in Pearson's correlation matrix (Figure 2), with all genotypes showing a strong positive correlation between the CO_2 assimilation rate and stomatal conductance (>0.80), as well as with the intrinsic efficiencies of water use (>0.85) and carboxylation (>0.80). Behavior was similar to that obtained for photosynthetic pigments, which maintained a correlation greater than 0.90 between themselves, with a weak correlation to the CO_2 assimilation rate.

In G1 (Figure 2A), differing from the other genotypes, transpiration presented a strong negative correlation with photosynthetic pigments (<−0.90), similar to that observed for stomatal conductance, which presented a correlation in the range of −0.70, and differing from photosynthetically active radiation, which maintained a positive correlation greater than 0.88 with these variables.

The electron transport rate presented different responses between genotypes, showing the least influence in G2 (Figure 2B), with correlations lower than 0.50 in most variables, except for the negative correlations observed in chlorophyll *b* (−0.57) and carotenoids (−0.68). ETR presented a strong positive correlation (>0.70) with PAR in all genotypes.

Relative water content was negatively correlated with gas exchange variables in all genotypes, except for Ci in G3, which showed a positive correlation of 0.68 (Figure 2C). Furthermore, RWC presented a strong positive correlation with ETR and PAR in genotypes G1 and G3, with no effect on G2.

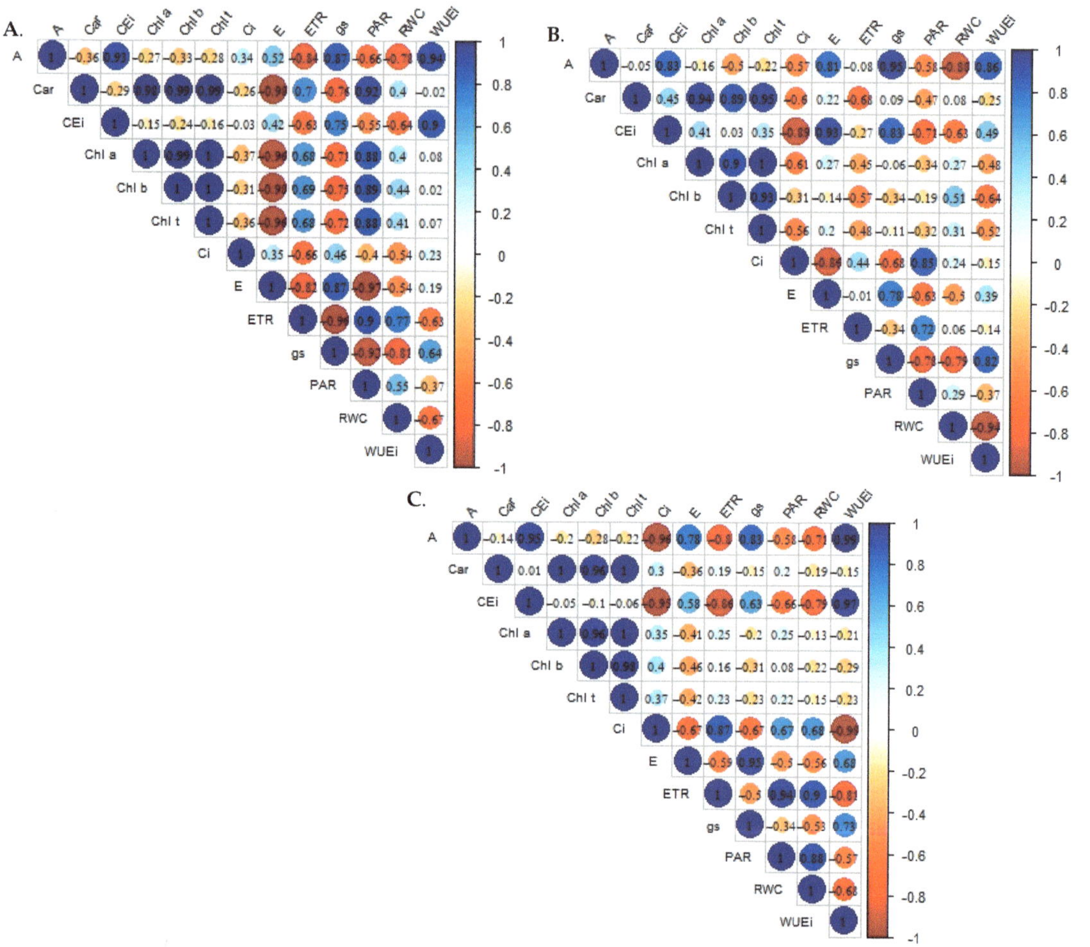

Figure 2. Pearson's correlation matrix for the variables analyzed within the electrical conductivity levels in the genotypes 'BRS GA1'—Gigante Amarelo—G1 (**A**); 'BRS SC1'—Sol do Cerrado—G2 (**B**); and 'SCS 437'—Catarina—G3 (**C**) at 154 days after transplantation. RWC—relative water content (%); Chl *a*—chlorophyll *a* (μg mL^{-1}); Chl *b*—chlorophyll *b* (μg mL^{-1}); Chl *t*—total chlorophyll (μg mL^{-1}); Car—carotenoids (μg mL^{-1}); Ci—intercellular CO_2 concentration (μmol CO_2 m^{-2} s^{-1}); E—transpiration (mmol H_2O m^{-2} s^{-1}); gs—stomatal conductance (mol H_2O m^{-2} s^{-1}); A—CO_2 assimilation rate (μmol CO_2 m^{-2} s^{-1}); WUEi—water use efficiency [(μmol CO_2 m^{-2} s^{-1}) (mmol H_2O m^{-2} s^{-1}) $^{-1}$]; CEi—intrinsic carboxylation efficiency [(μmol CO_2 m^{-2} s^{-1}) (μmol CO_2 m^{-2} s^{-1})$^{-1}$]; ETR—electron transport rate; PAR—photosynthetically active radiation.

2.2. *Cellular Damage and Chlorophyll a Fluorescence in Sour Passion Fruit Genotypes Irrigated with Saline Water*

There was an interaction between the ECw × GEN factors for all variables analyzed (Table 2). The electrical conductivity of irrigation water (ECw) significantly affected all variables measured, except for initial fluorescence (F_0). Electrolyte leakage (%EL) was statistically influenced by sour passion fruit genotypes at 154 DAT.

Table 2. Summary of the analysis of variances for electrolyte leakage (%EL) in the leaf blade, initial fluorescence (F_0), maximum fluorescence (Fm), variable fluorescence (Fv), and quantum efficiency of photosystem II (Fv/Fm) of sour passion fruit genotypes grown under irrigation water salinity at 154 days after sowing.

Sources of Variation	DF	Mean Squares				
		%EL	F_0	Fm	Fv	Fv/Fm
Electrical conductivity (ECw)	4	1716.72 **	1312.57 ns	96,165.81 **	65,375.66 **	0.02 **
Linear Regression	1	6809.31 **	664.22 ns	62,357.34 *	2230.04 ns	0.01 ns
Quadratic Regression	1	3.56 ns	4165.87 *	58,933.53 *	24,9600.50 **	0.04 **
Genotypes (GEN)	2	1860.60 **	1857.05 ns	8633.95 ns	13,440.15 ns	0.00 ns
Interaction (ECw × GEN)	8	371.57 **	1713.15 *	87,023.17 **	17,940.35 *	0.00 ns
Blocks	2	7.24	21,774.15	85,552.15	102,639.62	0.00
Residual	28	25.69	581.50	11,812.66	6728.00	0.00
CV (%)		9.57	5.94	6.72	6.76	4.33

DF—degrees of freedom; CV (%)—coefficient of variation; * significant at 0.05 probability level; ** significant at 0.01 probability level; ns not significant with Tukey's test ($p \leq 0.05$).

The increase in the salinity levels of irrigation water promoted increments in %EL of 148.21, 6.12, and 36.57% per unit increase in ECw in the genotypes G1, G2, and G3, respectively (Figure 3A). The genotypes G1, G2, and G3 were subjected to ECw of 3.5 dS m^{-1} and showed increments of 220.60%, 22.06%, and 133.17% in %EL, respectively, compared to those subjected to the lowest ECw level (0.3 dS m^{-1}).

When the genotype factor was analyzed as a function of each salinity level (Figure 3A), significant differences in %EL were observed at all ECw levels. When cultivated with waters of 0.3, 1.1, and 1.9 dS m^{-1}, the genotype G1 had the lowest %EL (22.03, 27.09, and 28.24%), being statistically inferior to G3 and G2. Under irrigation with water of 2.7 and 3.5 dS m^{-1}, the lowest values were obtained by the genotype G2, being statistically inferior to G3 at both ECw levels. It is worth pointing out that, when irrigated with water of 0.3 dS m^{-1}, plants did not suffer cell damage because the %EL must be higher than 50% for this to be the case—a behavior not observed under these conditions.

For initial fluorescence (F_0) (Figure 3B), the genotype G3 obtained the lowest estimated value (377.56) under ECw of 1.7 dS m^{-1} and the highest value (445.75) under ECw of 3.5 dS m^{-1}. For the genotypes G1 and G2, the F_0 data were not satisfactorily described with the tested models ($R^2 < 0.6$). A significant difference was observed between the genotypes only when they were subjected to irrigation with ECw from 2.7 dS m^{-1}. The genotype G1 obtained the highest mean value (439), being statistically superior to G2 (387) (Figure 3B). When the genotypes were subjected to ECw of 3.5 dS m^{-1}, the highest value (459) was obtained with G3, which did not differ from G1, and these were statistically superior to G2.

The maximum fluorescence (Fm) of plants of the genotype G1 irrigated with ECw of 3.5 dS m^{-1} reached the highest estimated value (1803.64) (Figure 3C), whereas the lowest estimated value was obtained under ECw of 1.4 dS m^{-1} (1417.96). For the genotype G3, there was a linear increase of 4.70% per unit increase in ECw. When comparing the Fm of the genotype G3 in plants subjected to ECw of 3.5 dS m^{-1} to the value of those that received 0.3 dS m^{-1}, an increase of 17.02% (259) was observed.

For Fm, genotypic variations were observed as a function of ECw levels (Figure 3C). Under ECw of 0.3 dS m^{-1}, the genotype G1 was statistically superior to the others, with increments of 26% and 17.72% compared to G2 and G3, respectively. Under ECw of 1.1 and 1.9 dS m^{-1}, the genotype G2 stood out in comparison to the others, and its values were 26.24% and 15.5% higher than those of G1 and G3, respectively, when irrigated with water of 1.1 dS m^{-1}, and 17.26% higher than that of G1 when irrigated with water of 1.9 dS m^{-1}. When the plants were subjected to ECw of 2.7 dS m^{-1}, the genotype G3 obtained the highest mean value (1813), which was 23.01% higher than that of G2. There was no statistical differences among the genotypes when irrigated with water of 3.5 dS m^{-1}.

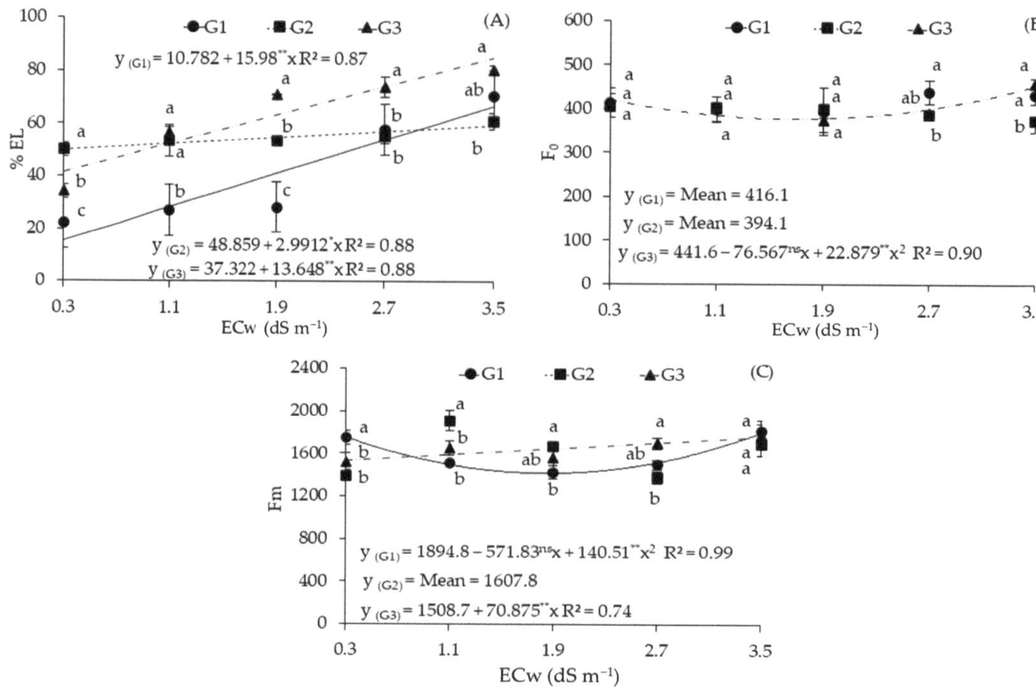

Figure 3. Electrolyte leakage (%EL) (**A**) in the leaf blade; initial fluorescence (F_0) (**B**); and maximum fluorescence (Fm) (**C**) of sour passion fruit genotypes as a function of the electrical conductivity of irrigation water (ECw) at 154 days after transplantation. Bars with the same letter indicate no significant difference between the means of the sour passion fruit genotypes with Tukey's test ($p \leq 0.05$). Sour passion fruit genotypes: G1 ('BRS GA1'—Yellow Giant); G2 ('BRS SC1'—Sol do Cerrado); G3 ('SCS 437'—Catarina). ns, **, and * represent, respectively, non-significant, significant at $p \leq 0.01$, and $p \leq 0.05$ with the F test. The vertical lines represent the standard error of the mean (n = 3).

For variable fluorescence (Fv), the results obtained with the genotypes G1 and G3 were described using the quadratic regression model (Figure 4A), with the maximum and minimum estimated values of 1384.92 and 1058.75 obtained in plants subjected to ECw of 3.5 dS m^{-1} and 1.8 dS m^{-1}, respectively. Similar behavior can be observed for the genotype G3, as the maximum and minimum values were estimated at ECw of 0.3 dS m^{-1} and 2.0 dS m^{-1}, which are equivalent to 1353.94 and 1161.94, respectively.

In the decomposition of the interaction for Fv, a significant difference was observed between the genotypes only when they were subjected to ECw of 1.9 dS m^{-1}; G2 obtained the highest value (1223), which was equal to the value obtained with G3 and statistically higher than that obtained with G1 (1030), a value that was 15.78% lower than that of G2 (Figure 4A).

The quantum efficiency of photosystem II (Fv/Fm) of the genotypes G2 and G3 were described using the quadratic regression model (Figure 4B), with estimated maximum values of 0.827 and 0.884, respectively, under irrigation with water of 0.3 dS m^{-1}. Decomposition of the interaction showed a significant difference between the genotypes only when they were subjected to ECw of 0.3 dS m^{-1} and when G3 obtained the highest value (0.9), which was equal to the value obtained with G2 and statistically higher than that of G1 (0.75), which was 16.66% lower than that obtained with G3 (Figure 4B).

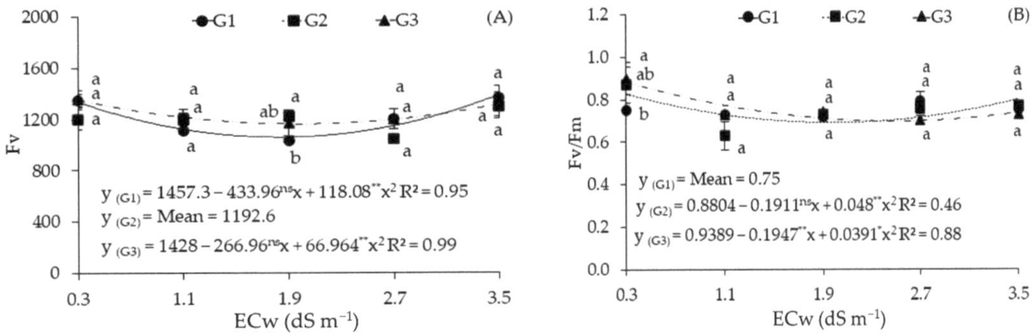

Figure 4. Variable fluorescence (Fv) (**A**) and quantum efficiency of photosystem II (Fv/Fm) (**B**) of sour passion fruit genotypes as a function of the electrical conductivity of irrigation water (ECw) at 154 days after transplantation. Bars with the same letter indicate no significant difference between the means of the sour passion fruit genotypes with Tukey's test ($p \leq 0.05$). Sour passion fruit genotypes: G1 ('BRS GA1'—Yellow Giant); G2 ('BRS SC1'—Sol do Cerrado); G3 ('SCS 437'—Catarina). ns, **, and * represent, respectively, non-significant, significant at $p \leq 0.01$, and $p \leq 0.05$ with the F test. The vertical lines represent the standard error of the mean (n = 3).

There was a significant effect of the interaction between the ECw × GEN factors for all variables evaluated, except for leaf temperature (LTemp) (Table 3). The ECw levels influenced all variables, except for LTemp. For the genotypes, a significant effect was observed only on the leaf temperature of sour passion fruit plants at 154 DAT.

Table 3. Summary of the analysis of variance for initial fluorescence before saturation pulse (Fs), maximum fluorescence after adaptation to saturating light (Fms), potential quantum efficiency of photosystem II (Y), and leaf temperature (LTemp) of sour passion fruit genotypes grown under irrigation water salinity at 224 days after sowing.

Sources of Variation	DF	Mean Squares			
		Fs	Fms	Y	LTemp
Electrical Conductivity (ECw)	4	2357.88 **	21,234.08 **	0.03 **	1.31 ns
Linear Regression	1	3074.17 **	67,240.00 **	0.03 **	0.18 ns
Quadratic Regression	1	528.28 ns	12,520.12 **	0.10 **	0.58 ns
Genotypes (GEN)	2	1912.08 **	9881.31 **	0.05 **	0.20 ns
Interaction (ECw × GEN)	8	2146.42 **	4205.16 **	0.00 *	0.08 ns
Blocks	2	401.42	6175.31	0.02	5.22
Residual	28	150.18	914.99	0.00	0.16
CV (%)		10.03	10.55	10.05	1.30

DF—degrees of freedom; CV (%)—coefficient of variation; * significant at 0.05 probability level; ** significant at 0.01 probability level; ns—not significant.

The increase in the electrical conductivity of irrigation water caused a linear reduction in the initial fluorescence before the saturation pulse (Fs) of the sour passion fruit genotype G2, which was equal to 26.55% per unit increase of ECw (Figure 5A). The data referring to the genotypes G1 and G3 were not satisfactorily described with the regression models studied.

When the genotype factor was analyzed as a function of each ECw level, it was observed that, in plants irrigated with water of 0.3 dS m^{-1}, the highest mean value of Fs was obtained with G2, which was equal to 149 and statistically higher than that of G1 (120), which was 19.46% lower than that obtained with G2 (Figure 5A). When subjected to ECw of 1.9 dS m^{-1}, the genotype G3 obtained the highest value (209), being statistically superior to G1 and G2, with increments of 91.74% and 81.73%, respectively. There was no significant difference in Fs when the genotypes were subjected to the other ECw levels.

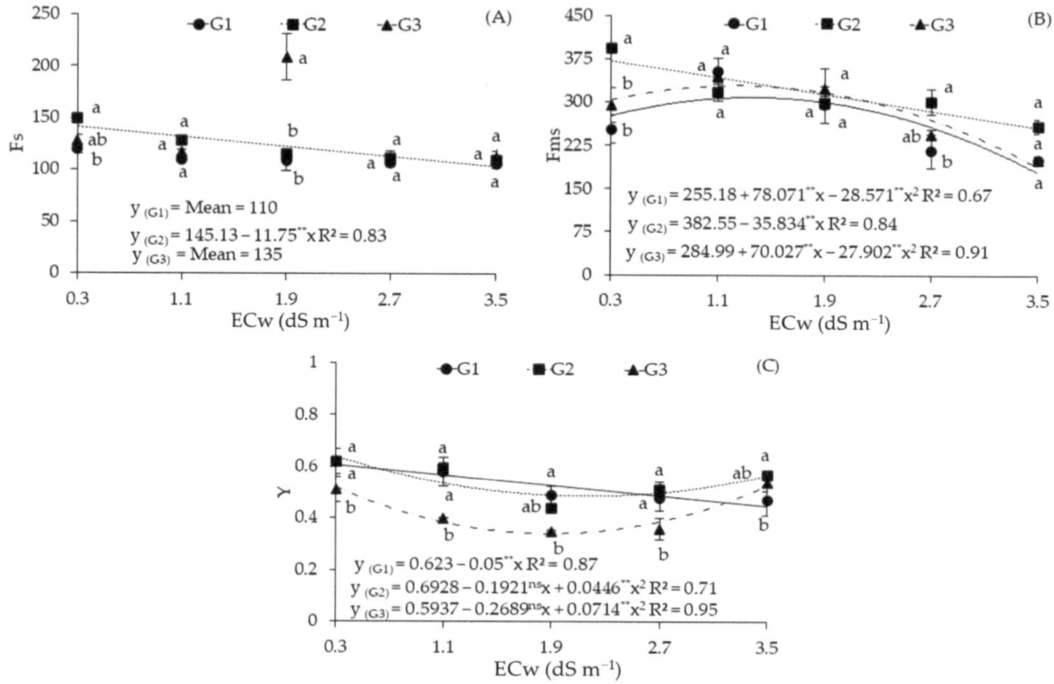

Figure 5. Initial fluorescence before saturation pulse (Fs) (**A**), maximum fluorescence after adaptation to saturating light (Fms) (**B**), and potential quantum efficiency of photosystem II (Y) (**C**) of sour passion fruit genotypes as a function of the electrical conductivity of irrigation water (ECw) at 154 days after transplantation. Bars with the same letter indicate no significant difference between the means of the sour passion fruit genotypes with Tukey's test ($p \leq 0.05$). Sour passion fruit genotypes: G1 ('BRS GA1'—Yellow Giant); G2 ('BRS SC1'—Sol do Cerrado); G3 ('SCS 437'—Catarina). ns and ** represent, respectively, non-significant, significant at $p \leq 0.01$ with the F test. The vertical lines represent the standard error of the mean (n = 3).

A reduction in Fms was observed in the sour passion fruit genotypes with an increase in ECw (Figure 5B). The genotypes G1 and G3 were described using the quadratic regression model, with maximum values estimated at 308.48 and 328.87, respectively, under irrigation water salinity of 1.3 dS m^{-1}. The genotype G2 showed a decreasing linear behavior with an estimated reduction of 30.84% in Fms when comparing the lowest ECw (0.3 dS m^{-1}) to the highest ECw (3.5 dS m^{-1}), which corresponded to a decrease of 9.36% per unit increase in ECw.

When the genotype factor was analyzed, considering each ECw level, it was observed that the highest value of Fms (394) was obtained for the genotype G2 for plants irrigated with water of 0.3 dS m^{-1}, being statistically superior to G1 and G3, with increments of 55.73% and 34.01%, respectively (Figure 5B). When subjected to ECw of 2.7 dS m^{-1}, G2 also obtained the highest value (301.33), but it was statistically superior only to G1. There was no significant difference in Fs when the genotypes were subjected to Ecw of 1.1, 1.9, and 3.5 dS m^{-1}.

For the potential quantum efficiency of photosystem II (Y), the genotypes G2 and G3 were described using the quadratic regression model, with maximum values estimated at 0.639 and 0.527 under irrigation of 0.3 dS m^{-1} and 3.5 dS m^{-1}, respectively (Figure 5C). The data of the genotype G1 were described using the decreasing linear regression model, with an estimated reduction of 26.32% in Y when comparing the lowest ECw (0.3 dS m^{-1}) with the highest ECw (3.5 dS m^{-1}), being equivalent to a reduction of 8.02% per unit increase in ECw.

Decomposition of the interaction of genotypes as a function of ECw levels for Y showed that under ECw of 0.3, 1.1, and 2.7 dS m^{-1}, the genotypes G1 and G2 were statistically superior to G3, with reductions of 17.74% and 17.72% at ECw of 0.3 dS m^{-1}, 31.03%, and 32.20% at ECw of 1.1 dS m^{-1}, and 25% and 29.41% at ECw of 2.7 dS m^{-1}, respectively (Figure 5C). When subjected to ECw of 1.9 dS m^{-1}, the genotype G1 obtained the highest value (0.49), being statistically superior to G3, but not differing from G2. Under ECw of 3.5 dS m^{-1}, the highest value (0.57) was obtained with G2, which was statistically superior to G1, but there were no significant differences from the genotype G3.

3. Discussion

In plants grown under salt stress conditions, the first physiological response of accumulation of salts in the soil is a reduction of stomatal conductance, this occurs as a way to compensate for the water restriction caused by the osmotic imbalance in the soil, leading to a reduction in leaf transpiration—CO_2 entry into the substomatal chamber which leads to the inhibition of photosynthesis by the plant [17,18]. This is a behavior that corroborates with the positive correlation between gs, E, and A in this research.

However, as observed in the main components, the inverse relationship between Ci and A may be related to the carbon consumption for the functioning of the Calvin cycle at low salinities. However, this effect is more significant in genotypes 2 and 3, which show a strong correlation between carbon consumption and photosynthesis. Unlike genotype 1, which showed a weak positive correlation between photosynthesis and carbon input, this behavior comes from the high values of gas exchange in the initial salinity, which, although consumes carbon through the activity of Rubisco, keeps the atmospheric carbon input high through the stomata. This process is affected by the accumulation of salts in the soil [8,10].

Similar results were also reported by Lima et al. [19], who evaluated the effects of saline water irrigation management strategies (1.3 and 4.0 dS m^{-1}) at different phenological stages of plants and potassium fertilization and found that ECw of 4.0 dS m^{-1} caused reductions in the chlorophyll's synthesis, transpiration, and instantaneous carboxylation efficiency of 'BRS GA1' sour passion fruit plants, regardless of their development stage. A reduction in the photosynthetic efficiency in sour passion fruit plants with increasing salinity levels of irrigation water (0.5 to 4.5 dS m^{-1}) was also observed by Souto et al. [20].

In this study, different effects were observed regarding the gas exchange of sour passion fruit genotypes. This is related to the genetic variability present in the genotypes studied, since the intensity of salt stress effects depends on factors such as developmental stage, duration of exposure to stress, irrigation management, edaphoclimatic conditions, and genotypes used, among others [21].

Furthermore, in glycophytes, excess salts inhibit protein formation and increase the activity of the chlorophyllase enzyme, which is responsible for the degradation of photosynthetic pigments (besides inducing oxidative stress through the accumulation of reactive oxygen species), negatively affecting photosynthesis [22,23].

This is a behavior that was not observed in the photosynthetic pigments of the passion fruit, with an increase being observed in the contents of chlorophyll *a*, total chlorophyll, and carotenoids and found at the highest electrical conductivities of irrigation water (3.5 dS m^{-1}). This fact explains the weak negative correlation of photosynthetic pigments with gas exchange, demonstrating a lower propensity for loss in the plant's photochemical apparatus concerning the biochemistry of photosynthesis under conditions of saline stress. This may be related to the intrinsic characteristics of the genotype 'BRS Sol do Cerrado' and vary according to the tolerance level of the crop. Divergent results regarding the synthesis of chlorophylls in plants subjected to salt stress have been reported by several authors; while some record reductions, others report increments [24]. Paiva et al. [25], while studying the effects of salinity (0.3 and 3.5 dS m^{-1}) of irrigation water with distinct cationic natures on 'BRS Rubi do Cerrado' sour passion fruit, observed that the use of water with an electrical conductivity of 3.5 dS m^{-1} and sodic nature (Na^+) promoted the greatest synthesis of chlorophyll *a*, chlorophyll *b*, and carotenoids in the plants at 180 DAT.

Under severe salt stress conditions, the increase in Ci is indicative of the deterioration of the photosynthetic structure, since the damage caused to the structures responsible for CO_2 fixation is also due to the accumulation of salts in the leaves. The increase in *Ci* observed in plants irrigated with the highest level of electrical conductivity (3.5 dS m^{-1}) is an indication that the carbon absorbed by the leaves was not being metabolized by the photosynthetic system due to the low activity of the enzyme ribulose- 1,5-biphostat carboxylase oxygenase (RuBisCO) [26] and due to the effects of salt stress. This due to the osmotic effect, which limits water absorption, and due to which plants tend to close their stomata to reduce water loss to the atmosphere.

This is a behavior that helps explain the increase in %EL, since the change in the soil water potential was caused by salinity results in a condition of physiological drought, nutritional and hormonal imbalances, formation of reactive oxygen species, with these factors leading to the destabilization of the cell membrane, among other effects [27]. The degradation of lipids, caused by the accumulation of reactive oxygen species (ROS), damaged the cell membrane and caused oxidative stress because of an excess of energy being directed to oxygen and thus generating several ROS, including superoxide, hydrogen peroxide, hydroxyl radicals, and singlet oxygen [28].

The decrease in the potential quantum efficiency of photosystem II observed in plants under salt stress indicates that a lower amount of excitation energy was used for the synthesis of ATP and NADPH in chloroplasts [29]. This reduction is often attributed to an imbalance in the electron transport rate and a reduction in ATP and NADPH consumption during the CO_2 assimilation process in plants [30,31].

In this work, it is noted that there was a considerable variation in photosynthetic variables depending on the treatments. This is possibly related to the variability observed in photosynthetically active radiation (PAR), which may have been caused due to external factors, such as the climatic conditions imposed during the evaluations. This may have occurred since the experiment was carried out in the field and was subject to the action of external factors such as the presence of dust particles under the leaf surface, variations in luminosity caused by the presence of passing clouds, temperature, humidity, and characteristics of edaphoclimatic factors of the region where the study was carried out.

4. Materials and Methods

The experiment was conducted from November 2021 to July 2022 in the fruit growing sector, located at the 'Rolando Enrique Rivas Castellón' experimental farm, belonging to the Center of Sciences and Agri-Food Technology of the Federal University of Campina Grande, in the municipality of São Domingos, PB, Brazil, whose coordinates are 06°48′50″ S latitude and 37°56′31″ W longitude, at an altitude of 190 m.

A randomized block design was used, in a 5 × 3 factorial scheme, whose treatments were obtained by combining 2 factors: 5 levels of electrical conductivity of irrigation water—ECw (0.3, 1.1, 1.9, 2.7, and 3.5 dS m^{-1})—and 3 genotypes of sour passion fruit (Gigante Amarelo—'BRS GA1'; Sol do Cerrado—'BRS SC1'; and Catarina—'SCS 437') with 3 replicates, totaling 45 experimental units. The ECw levels used were adopted in accordance with a study conducted by Lima et al. [32].

The genotypes 'BRS Gigante Amarelo' ('BRS GA1') and 'BRS Sol do Cerrado' ('BRS SC1') were released by the Brazilian Agricultural Research Corporation (EMBRAPA) in 2008. They are characterized by producing fruits with a yellow color and oblong shape, weighing from 120 to 350 g, with a pulp yield of 40% and 38%, and soluble solids contents of 13 to 15° Brix, respectively. 'BRS SC1' is tolerant to anthracnose, bacteriosis, and various viruses, and the two genotypes bloom throughout the year, with a greater concentration of them being seen in the driest periods [33,34].

The genotype 'SCS 437' stands out for its tolerance to bacteriosis and anthracnose, resistance to transport, and excellent visual appearance. Its fruits are large, with a mean weight of 315 g, a juice yield of 33 to 50%, soluble solids content of 9 to 14.5 °Brix, a yellow color, and an orange pulp [35].

The seedlings were formed under protected environmental conditions, arranged on benches located near the experimental area. Polyethylene bags with dimensions of 15 × 20 cm were used as containers and filled with substrate composed of a mixture of soil and aged cattle manure at a ratio of 2:1 (on a volume basis), respectively.

Three seeds were sown per bag at a 0.5 cm depth. After the emergence of seedlings, thinning was carried out, leaving only one plant per bag. During the formation of the seedlings, irrigation was performed using public-supply water from the municipal network (0.3 dS m^{-1}), which is considered the lowest level of electrical conductivity control.

The experiment was conducted under field conditions using 100 L pots adapted as drainage lysimeters, filled with approximately 110 kg of a Neossolo Flúvico Ta Eutrófico típico (Fluvent) of loamy sand texture from a private property located near the experimental area, whose physical-hydraulic, chemical, and salinity characteristics were determined according to the methodology recommended by Teixeira et al. [36]: Ca^{2+}, Mg^{2+}, Na^+, K^+, $Al^{3+} + H^+$ = 3.00; 2.44; 0.05; 0.12; 0.69 $cmol_c\ kg^{-1}$, respectively; pH (soil:water, 1:2.5) = 6.01; ECse = 0.71 dS m^{-1}; organic matter = 0.21 dag kg^{-1}; sand, silt, and clay = 75.65, 20.21, and 4.34 dag kg^{-1}, respectively; p = 0.53 mg kg^{-1}; SAR = 0.61 $(mmol\ L^{-1})^{0.5}$; ESP = 0.8%; CEC = 6.25 $cmol_c\ kg^{-1}$.

Plants were trained on a vertical trellis system, built with smooth galvanized steel wire No. 12, fixed at 1.2 m height from the lysimeter soil surface, and guided by nylon strings. When they exceeded 10 cm above the trellis, their apical bud was pruned to stimulate the growth of secondary branches, which were guided in opposite directions until reaching 1.50 m in length. After the secondary branches reached the pre-established length, the apical bud was pruned to stimulate the growth of tertiary branches, which were grown until they were 30 cm away from the soil surface of the experimental area. This was performed in order to avoid possible contamination through the contact with the soil. The spacing adopted was 3.0 m between plants and 2.5 m between rows.

Seedlings were transplanted to the lysimeters 60 days after sowing, a period characterized by the beginning of tendril growth. During the first 30 days after transplantation (DAT), in order to allow acclimatization under field conditions, all plants were irrigated with water of 0.3 dS m^{-1} and, from 31 DAT, irrigation began to be performed using water with the respective levels of electrical conductivity.

Fertilization with NPK was performed using urea (45% N), single superphosphate (20% P_2O_5), and potassium sulfate (51.5% K_2O) as sources of nitrogen, phosphorus, and potassium, respectively. Fertilization with nitrogen and potassium was split and applied monthly in equal doses throughout the cycle of sour passion fruit, with ratios of 1/1 (N/K) in the flowering stage, 1/2 at the beginning of harvest, and 1/3 until the end of harvest, whereas phosphorus was applied all at once as a basal dose, incorporated into the soil as the lysimeters were filled, as recommended by Costa et al. [37].

Application of micronutrients was performed biweekly using Dripsol® micro (Mg^{2+} = 1.1%; B = 0.85%; Cu (Cu-EDTA) = 0.5%; Fe (FeEDTA) = 3.4%; Mn (Mn-EDTA) = 3.2%; Mo = 0.05%; Zn = 4.2%; with 70% of EDTA chelating agent) at a concentration of 1 g L^{-1}, with foliar spraying. During the applications of micronutrients, Haiten® adhesive spreader was used at a concentration of 0.15 mL L^{-1}.

Irrigation water used in the treatment with the lowest electrical conductivity level (0.3 dS m^{-1}) came from a well located in the experimental area of the CCTA/UFCG, and its chemical composition was Ca^{+2}, Mg^{+2}, K^+, SO_4^{2-}, HCO_3^-, CO_3^{2-}-, Cl^- = 0.17; 0.61; 1.41; 0.29; 0.18; 0.81; 0.00; and 1.26 $mmol_c\ L^{-1}$, respectively; electrical conductivity (EC) = 0.30 dS m^{-1}; pH = 7.10; sodium adsorption ratio (SAR) = 2.25 $(mmol\ L^{-1})^{0.5}$.

When necessary, to obtain the higher levels of electrical conductivity, sodium chloride (NaCl) was added to the water from the wells, adjusting the concentrations of the available water considering the relationship between ECw and salt concentration [38].

Irrigation was performed using a drip system, with two pressure-compensating drippers with a flow rate of 10 L h^{-1} for each plant. Irrigations were performed daily from 7:00 a.m., applying (in each container) the volume corresponding to that obtained by the water balance, which was determined using Equation (1) as follows:

$$VI = \frac{(Va - Vd)}{(1 - LF)} \tag{1}$$

where VI = volume of water to be applied in the irrigation event (L); Va = volume applied in the previous irrigation event (L); Vd = volume drained (L); and LF = leaching fraction of 0.2, applied every 15 days to reduce the accumulation of salts in the root zone.

Weed control was performed via manual weeding between the rows of drainage lysimeters and around the plant collar. Control of pests and diseases was carried out preventively using chemical products recommended for the crop with doses established by the manufacturers.

Effects of the different treatments were evaluated at 154 DAT, when the plants were in the phenological stage of fruiting, based on photosynthetic pigments, water relations, gas exchange, and chlorophyll *a* fluorescence in the light and dark phase. Contents of photosynthetic pigments were determined according to the methodology of [39].

Gas exchange was evaluated with the CO_2 assimilation rate—A; transpiration—E; internal CO_2 concentration—Ci; and stomatal conductance—gs. All of which were determined on the third leaf counted from the apex of the tertiary branch using the LCPro+ portable photosynthesis meter (IRGA) from ADC BioScientific, UK. Ltd (Hoddesdon, England). These data were then used to quantify the intrinsic water use efficiency—WUEi (A/gs)—and instantaneous carboxylation efficiency—CEi (A/Ci). These determinations were performed between 6:30 and 10:00 a.m. on the third fully expanded leaf counted from the apical bud under natural conditions of air temperature, CO_2 concentration and using an artificial radiation source of 1200 μmol m^{-2} s^{-1}.

Chlorophyll *a* fluorescence was measured based on initial fluorescence (F_0), maximum fluorescence (Fm), variable fluorescence (Fv), and quantum efficiency of photosystem II (Fv/Fm) in leaves preadapted to the dark using leaf clips for 30 min, between 6:00 and 9:00 a.m., in the median leaf of the intermediate productive branch of the plant, using an OS5p pulse-modulated fluorometer from Opti Science.

Evaluations under conditions using light followed the yield protocol, which were performed using an actinic light source with a multi-flash saturation pulse, coupled to a photosynthetically active radiation determination clip (PAR-Clip), which determined the initial fluorescence before the saturation pulse (Fs), maximum fluorescence after adaptation to the saturating light (Fms), electron transport rate (ETR), potential quantum efficiency of photosystem II (Y), photosynthetically active radiation (PAR), and leaf temperature (LTemp.).

Relative water content (RWC) was determined according to [40] and electrolyte leakage in the leaf blade (%EL) was measured according to [41].

The multivariate structure of the results was evaluated using principal component analysis (PCA), synthesizing the amount of relevant information contained in the original data set in a smaller number of dimensions. This resulted from linear combinations of the original variables generated from the eigenvalues ($\lambda \geq 1.0$) in the correlation matrix, which explained a percentage greater than 10% of the total variance [42].

From the reduction of the dimensions, the original data of the variables of each component were subjected to multivariate analysis of variance (MANOVA) with the Hotelling test [43] at a 0.05 probability level for the levels of electrical conductivity of irrigation water and genotypes of sour passion fruit, as well as for the interaction between them. Only variables with a correlation coefficient greater than or equal to 0.65 were maintained in each principal component (PC) [44]. Statistical analyses were performed using Statistica v. 7.0 software [45].

The data that did not obtain a correlation coefficient above 0.6 were subjected to the Shapiro–Wilk normality test to check for normality. Analysis of variance was applied with the F test ($p \leq 0.05$) and, when significant, linear and quadratic polynomial regression analysis was performed for the factor of electrical conductivity of irrigation water, while a means comparison test (Tukey's test, $p \leq 0.05$) was performed for the sour passion fruit genotypes, using the statistical program SISVAR—ESAL version 5.7 [46]. To verify the relationship between the variables analyzed within the levels of electrical conductivity in the passion fruit genotypes ('BRS GA1'—Gigante Amarelo—G1; 'BRS SC1'—Sol do Cerrado—G2; and 'SCS 437'—Catarina), bivariate analyses were carried out using Pearson's correlation matrix for each genotype.

5. Conclusions

According to the results obtained, it appears that the increase in the electrical conductivity of irrigation water negatively affected most of the physiological characteristics of the sour passion fruit at 154 days after transplanting, causing damage to pigment synthesis, gas exchange, and the photosynthetic efficiency of the studied genotypes. Furthermore, the hypothesis of there being a variation in the degree of tolerance of sour passion fruit genotypes concerning salt stress was confirmed. An increase in the synthesis of photosynthetic pigments (chlorophyll *a*, *b*, *t* and carotenoids) of the 'BRS Sol do Cerrado' genotype was found when it was irrigated with water of 3.5 dS m^{-1}. Furthermore, there was also an increase in the photosynthetic efficiency of the genotype 'BRS Gigante Amarelo' when it was irrigated with water of higher electrical conductivity. Thus, we suggest that new research should be carried out to investigate the effect of the salinity of irrigation water, testing these and other genotypes of sour passion fruit. Trials should preferably be carried out under field conditions with a longer period of exposure of plants to this type of stress.

Author Contributions: Conceptualization, F.J.d.S.P., G.S.d.L. and V.L.A.d.L.; methodology, W.B.B.d.S., R.A.F.T., L.d.A.S. and S.T.d.A.S.; software, F.V.d.S.S. and R.T.d.F.; validation, F.J.d.S.P., G.S.d.L. and L.A.d.A.S.; formal analysis, V.K.N.O.d.S. and L.A.d.A.S.; investigation, G.S.d.L. and F.J.d.S.P.; resources, G.S.d.L. and L.A.d.A.S.; data collection, A.K.C.d.A.; writing—original draft preparation, F.J.d.S.P. and G.S.d.L.; writing—review and editing, G.S.d.L., H.R.G. and L.A.d.A.S.; supervision, H.R.G. and P.D.F.; project administration, G.S.d.L., R.T.d.F., P.D.F. and L.A.d.A.S. All authors have read and agreed to the published version of the manuscript.

Funding: Funding was acquired from the National Council for Scientific and Technological Development—CNPq (Proc. 429732/2018-0), Coordination for the Improvement of Higher Education Personnel—CAPES (financial code—001).

Data Availability Statement: Data are contained within the article.

Acknowledgments: The authors thank the Graduate Program in Agricultural Engineering of the Federal University of Campina Grande, for the support in carrying out this research.

Conflicts of Interest: The authors declare no conflict of interest.

References

1. Fonseca, A.M.A.; Geraldi, M.V.; Maróstica Junior, M.R.; Silvestre, A.J.D.; Rocha, S.M. Purple passion fruit (*Passiflora edulis* f. *edulis*): A comprehensive review on the nutritional value, phytochemical profile and associated health effects. *Food Res. Int.* **2022**, *160*, e111665. [CrossRef] [PubMed]
2. Zhao, L.; Wu, L.; Li, L.; Zhu, J.; Chem, X.; Zhang, S.; Lee, L.; Yan, J.-K. Physicochemical, structural, and rheological characteristics of pectic polysaccharides from fresh passion fruit (*Passiflora edulis f. flavicarpa* L.) peel. *Food Hydrocoll.* **2023**, *136*, 108301. [CrossRef]
3. Instituto Brasileiro de Geografia e Estatística. *Área Plantada e Quantidade Produzida*; Instituto Brasileiro de Geografia e Estatística: Rio de Janeiro, Brazil, 2021. Available online: https://sidra.ibge.gov.br/pesquisa/pam/tabelas (accessed on 30 April 2023).
4. Lima, G.S.; Fernandes, C.G.J.; dos Soares, L.A.; Gheyi, H.R.; Fernandes, P.D. Gas exchange, chloroplast pigments and growth of passion fruit cultivated with saline water and potassium fertilization. *Rev. Caatinga* **2020**, *33*, 184–194. [CrossRef]

5. Oliveira, F.F.M.; Morais, M.B.; Silva, M.E.S.; Saraiva, Y.K.F.; Arruda, M.V.M.; Silva, J.N.C.; Albuquerque, C.C. Ecophysiological response of *Lippia gracilis* (Verbanaceae) to duration of salt stress. *Ecotoxicol. Environ. Saf.* **2019**, *178*, 202–210. [CrossRef]
6. Da Lima, A.F.; da Luz, L.N.; dos Santos, M.F.; da Silva Filho, F.V.; de Gouveia, F.A.L.; de Casemiro, J.A.O. Eficiência fisiológica e desempenho do amendoim sob estresse salino e inoculado com *Bradyrhizobium*. *Water Resour. Irrigat. Manag.* **2022**, *11*, 22–35. [CrossRef]
7. Stavi, I.; Thevs, N.; Priori, S. Soil salinity and sodicity in drylands: A review of causes, effects, monitoring, and restoration measures. *Front. Environ. Sci.* **2021**, *9*, 712831. [CrossRef]
8. Gupta, B.; Huang, B. Mechanism of salinity tolerance in plants: Physiological, biochemical, and molecular characterization. *Int. J. Genom.* **2014**, *2014*, 701596. [CrossRef] [PubMed]
9. Vangelisti, A.; Zambrano, L.S.; Caruso, G.; Macheda, D.; Bernardi, R.; Usai, G.; Giordani, F.M.T.; Gucci, R.; Cavallini, A.; Natali, L. How an ancient, salt-tolerant fruit crop, *Ficus carica* L., copes with salinity: A transcriptome analysis. *Sci. Rep.* **2019**, *9*, 109037. [CrossRef]
10. De Azevedo Neto, A.D.; Mota, K.N.A.B.; Silva, P.C.C.; Cova, A.M.W.; Ribas, R.F.; Gheyi, H.R. Selection of sunflower genotypes for salt stress and mechanisms of salt tolerance in contrasting genotypes. *Ciênc. Agrotec.* **2020**, *44*, e020120. [CrossRef]
11. Liang, W.; Ma, X.; Wan, P.; Liu, L. Plant salt-tolerance mechanism: A review. *Biochem. Biophys. Res. Commun.* **2018**, *495*, 286–291. [CrossRef]
12. Isayenkov, S.V.; Maathuis, F.J.M. Plant salinity stress: Many unanswered questions remain. *Front. Plant Sci.* **2019**, *10*, 80. [CrossRef] [PubMed]
13. De Lima, G.S.; de Lacerda, C.N.; dos Soares, L.A.; Gheyi, H.R.; Rocha, R.H.C. Production characteristics of sesame genotypes under different strategies of saline water application. *Rev. Caatinga* **2020**, *33*, 490–499. [CrossRef]
14. Veloso, L.L.S.A.; Azevedo, C.A.V.; Nobre, R.G.; de Lima, G.S.; Silva, I.J.; Lacerda, C.N. Hydrogen peroxide in the acclimation of colored-fiber cotton genotypes to salt stress. *Rev. Caatinga* **2023**, *36*, 414–423. [CrossRef]
15. Araújo Neto, A.C.; Nunes, R.T.C.; de Costa, R.Q.; Moreira, G.L.P.; de Silva, R.A.; São José, A.R. Germinação e crescimento inicial de *Vigna unguiculata* (L). sob estresse salino. *Rev. Ciências Agrárias* **2020**, *43*, 283–292. [CrossRef]
16. Kaiser, H.F. The application of electronic computers to factor analysis. *Educ. Psychol. Meas.* **1960**, *20*, 141–151. [CrossRef]
17. Da Prazeres, S.S.; de Lacerda, C.F.; Barbosa, F.E.L.; Amorim, A.V.; Araújo, I.C.S.; Cavalcante, L.F. Crescimento e trocas gasosas de plantas de feijão-caupi sob irrigação salina e doses de potássio. *Rev. Agro@Mbiente* **2015**, *9*, 111–118. [CrossRef]
18. Silva, E.M.; Lima, G.S.; Gheyi, H.R.; Nobre, R.G.; da Sá, F.V.S.; de Souza, L.P. Growth and gas exchanges in soursop under irrigation with saline water and nitrogen sources. *Rev. Bras. Eng. Agríc. Ambient.* **2018**, *22*, 776–781. [CrossRef]
19. De Lima, G.S.; Pinheiro, F.W.A.; de Souza, W.B.B.; dos Soares, L.A.A.; Gheyi, H.R.; Nobre, R.G.; de Queiroga, R.C.F.; Fernandes, P.D. Physiological indices of sour passion fruit under brackish water irrigation strategies and potassium fertilization. *Rev. Bras. Eng. Agríc. Ambient.* **2023**, *27*, 383–392. [CrossRef]
20. De Souto, A.D.L.; Cavalcante, L.F.; de Melo, E.N.; Cavalcante, Í.H.L.; da Oliveira, R.Í.L.S.; de Mesquita, E.F.; Mendonça, R.M.N. Gas exchange and yield of grafted yellow passion fruit under salt stress and plastic mulching. *Rev. Bras. Eng. Agríc. Ambient.* **2022**, *26*, 823–830. [CrossRef]
21. Alvarenga, C.F.S.; Silva, E.M.; Nobre, R.G.; Gheyi, H.R.; de Lima, G.S.; de Silva, L.A. Morfofisiologia de aceroleira irrigada com águas salinas sob combinações de doses de nitrogênio e potássio. *Rev. Ciências Agrárias* **2019**, *42*, 194–205. [CrossRef]
22. Salimi, F.; Shekari, F.M.R.; Azimi, E.Z. Role of methyl jasmonate on improving salt resistance through some physiological characters in German chamomile (*Matricaria chamomilla* L.). *Iran. J. Med. Aromat. Plants.* **2012**, *27*, 700–711. [CrossRef]
23. Khaliq, A.; Haq, M.Z.U.; Aslam, F.; Matloob, A.; Saddam, H. Salinity tolerance in wheat cultivars is related to enhanced activities of enzymatic antioxidants and reduced lipid peroxidation. *Clean Soil Air Water* **2015**, *43*, 248–1258. [CrossRef]
24. Do Sacramento, B.L.; Cruz, T.S.; Silva, L.L.; Mota, K.N.A.B.; de Azevedo Neto, A.D. Pigmentos e teores de solutos orgânicos em plantas de aguapé sob estresse salino. *Encic. Biosf.* **2014**, *10*, 33–54.
25. Da Paiva, F.J.S.; de Lima, G.S.; de Lima, V.L.A.; Ramos, J.G.; Gheyi, H.R.; de Farias, M.S.S.; Fernandes, P.D.; de Azevedo, C.A.V. Growth, photosynthetic pigments, and photochemical efficiency of sour passion fruit as a function of the cationic nature of water. *Semina Ciênc. Agrár.* **2021**, *42*, 583–598. [CrossRef]
26. Habermann, G.; Machado, E.C.; Rodrigues, J.D.; Medina, C.L. CO_2 assimilation, photosynthetic light response curves, and water relations of 'Pêra' sweet orange plants infected with Xylella fastidious. *Braz. J. Plant Physiol.* **2003**, *15*, 79–87. [CrossRef]
27. Aycan, M.; Baslam, M.; Asiloglu, R.; Mitsui, T.; Yildiz, M. Development of new high-salt tolerant bread wheat (*Triticum aestivum* L.) genotypes and insight into the tolerance mechanisms. *Plant Physiol. Biochem.* **2021**, *166*, 314–327. [CrossRef] [PubMed]
28. Wu, W.; Zhang, Q.; Ervin, E.H.; Yang, Z.; Zhang, X. Physiological mechanism of enhancing salt stress tolerance of perennial ryegrass by 24-epibrassinolide. *Front. Plant Sci.* **2017**, *8*, 1017. [CrossRef] [PubMed]
29. Taiz, L.; Zeiger, E.; Myller, I.M.; Murphy, A. *Fisiologia e Desenvolvimento Vegetal*; ARTMED: Porto Alegre, Brazil, 2017; 858p.
30. Silva, E.N.; Ribeiro, R.V.; Silva-Ferreira, S.L.; Vieira, S.A.; Ponte, L.F.A.; Silveira, A.G. Coordinate changes in photosynthesis, sugar accumulation and anti-oxidative enzymes improve the performance of *Jatropha curcas* plants under drought stress. *Biomass Bioenergy* **2012**, *45*, 270–279. [CrossRef]
31. Pompeiano, A.; Landi, M.; Meloni, G.; Vita, F.; Guglielminetti, L.; Guidi, L. Allocation pattern, ion partitioning, and chlorophyll *a* fluorescence in *Arundo donax* L. in responses to salinity stress. *Plant Biosyst.* **2016**, *151*, 613–622. [CrossRef]

32. De Lima, G.S.; de Souza, W.B.B.; dos Soares, L.A.A.; Pinheiro, F.W.A.; Gheyi, H.R.; Oliveira, V.K.N. Dano celular e pigmentos fotossintéticos do maracujazeiro-azedo em função da natureza catiônica da água. *Irriga* **2020**, *25*, 663–669. [CrossRef]
33. Empresa Brasileira de Pesquisa Agropecuária—EMBRAPA. *Híbrido de Maracujazeiro-Azedo de Alta Produtividade: BRS Gigante Amarelo*; Empresa Brasileira de Pesquisa Agropecuária—EMBRAPA: Brasília, Brazil, 2014; 2p.
34. Empresa Brasileira de Pesquisa Agropecuária—EMBRAPA. *Híbrido de Maracujazeiro-Azedo para mesa e Indústria: BRS Sol do Cerrado*; Empresa Brasileira de Pesquisa Agropecuária—EMBRAPA: Brasília, Brazil, 2014; 2p.
35. Petry, H.B.; Bruna, E.D.; Moreto, A.L.; Brancher, A.; Sônego, M. 'SCS437 Catarina': Maracujá-azedo de alta qualidade para o mercado de mesa. *Agropec. Catarin.* **2019**, *32*, 49–52. [CrossRef]
36. Teixeira, P.C.; Donagemma, G.K.; Fontana, A.; Teixeira, W.G. *Manual de Métodos de Análise de Solo*, 3rd ed.; Embrapa: Brasília, Brazil, 2017; 573p.
37. De Costa, A.F.S.; Costa, A.N.; Ventura, J.A.; Fanton, C.J.; de Lima, I.M.; Caetano, L.C.S.; de Santana, E.N. *Recomendações Técnicas para o Cultivo do Maracujazeiro*; (Incaper. Documentos, 162); Incaper: Vitória, Brazil, 2008; 56p.
38. Richards, L.A. *Diagnosis and Improvement of Saline and Alkali Soils*; Department of Agriculture: Washington, DC, USA, 1954; 160p.
39. Arnon, D.I. Copper enzymes in isolated cloroplasts: Polyphenoloxidases in *Beta vulgaris*. *Plant Physiol.* **1949**, *24*, 1–15. [CrossRef] [PubMed]
40. Weatherley, P.E. Studies in the water relations of the cotton plant. I- The field easurements of water deficits in leaves. *New Phytol.* **1950**, *49*, 81–97. [CrossRef]
41. Scotti-Campos, P.; Pham-Thi, A.-T.; Semedo, J.N.; Pais, I.P.; Ramalho, J.C.; Matos, M.C. Physiological responses and membrane integrity in three Vigna genotypes with contrasting drought tolerance. *Emir. J. Food Agric.* **2013**, *25*, 1002–1013. [CrossRef]
42. Govaerts, B.; Sayre, K.D.; Lichter, K.; Dendooven, L.; Deckers, J. Influence of permanent raised bed planting and residue management on physical and chemical soil quality in rain fed maize/wheat systems. *Plant Soil* **2007**, *291*, 39–54. [CrossRef]
43. Hotelling, H.; Eisenhart, C.; Hastay, M.W.; Wallis, W.A. *Multivariate Quality Control*; Techniques of statistical analysis; John Wiley & Sons: New York, NY, USA, 1947; 73p.
44. Hair, F.J.; Black, W.C.; Babin, B.J.; Anderson, R.E.; Tatham, R.L. *Análise Multivariada de Dados, 6th ed*; Tradução Adonai Schlup Sant'Anna; Bookman: Porto Alegre, Brazil, 2009; 688p.
45. Statsoft, I.N.C. *Programa Computacional Statistica 7.0*; The European Association of Urology: Arnhem, The Netherlands, 2004.
46. Ferreira, D.F. SISVAR: A computer analysis system to fixed effects split plot type designs. *Rev. Bras. Biom.* **2019**, *37*, 529–535. [CrossRef]

plants

MDPI

Article

Potential of Ca-Complexed in Amino Acid in Attenuating Salt Stress in Sour Passion Fruit Seedlings

Antônio Gustavo de Luna Souto [1,*], Angela Maria dos Santos Pessoa [1], Sarah Alencar de Sá [1], Nayana Rodrigues de Sousa [1], Emerson Serafim Barros [1], Francimar Maik da Silva Morais [1], Fagner Nogueira Ferreira [1], Wedson Aleff Oliveira da Silva [1], Rafael Oliveira Batista [1], Daniel Valadão Silva [1], Rita Magally Oliveira da Silva Marcelino [1], Hans Raj Gheyi [2], Geovani Soares de Lima [2], Rosa Maria dos Santos Pessoa [3] and Mailson Monteiro do Rêgo [3]

1 Universidade Federal Rural do Semi-Árido, Mossoro 59625-900, RN, Brazil; angelapessoapb@gmail.com (A.M.d.S.P.); sarahalencardesa28@gmail.com (S.A.d.S.); nayanasousa12@hotmail.com (N.R.d.S.); emerson.serafim.barros@gmail.com (E.S.B.); francimar.morais@alunos.ufersa.edu.br (F.M.d.S.M.); fagnernf@gmail.com (F.N.F.); wedson.silva@alunos.ufersa.edu.br (W.A.O.d.S.); rafaelbatista@ufersa.edu.br (R.O.B.); daniel.valadao@ufersa.edu.br (D.V.S.); rm.magally@gmail.com (R.M.O.d.S.M.)

2 Programa de Pós-Graduação em Engenharia Agrícola, Universidade Federal de Campina Grande, Campina Grande 58429-900, PB, Brazil; hans.gheyi@ufcg.edu.br (H.R.G.); geovani.soares@professor.ufcg.edu.br (G.S.d.L.)

3 Centro de Ciências Agrárias, Universidade Federal da Paraíba, Areia 58397-000, PB, Brazil; rosapessoaa@gmail.com (R.M.d.S.P.); mailson@cca.ufpb.br (M.M.d.R.)

* Correspondence: gustavo.luna@ufersa.edu.br; Tel.: +55-83-99981-7898

Abstract: Salt stress results in physiological changes that inhibit plant growth and development. Ca-complex sources are used as a potential salt stress attenuator. This study was carried out with the aim of verifying the effects of Ca-complex sources in reducing the effects of saline water stress on the physiological aspects of sour passion fruit seedlings. The experiment was carried out in a randomized block design with a 2 × 2 × 3 factorial scheme, consisting of two cultivars of sour passion fruit (BRS GA1 and BRS SC1), two levels of water salinity (electrical conductivity of 0.5 and 4.0 dS m^{-1}) and three sources of Ca-complex (without, organic acids and amino acids). The traits measured at 60 days after sowing were gas exchange, chlorophyll indices, chlorophyll fluorescence, electrolyte leakage, and relative water content in the leaf limb. Under moderate water salinity, the application of Ca-complex in amino acids promoted increases of 49.84% and 43.71%, respectively, in the efficiency of water use and carboxylation. The application of complex sources increased the stability of cell membranes, reducing electrolyte leakage, providing higher relative water content in seedlings irrigated with moderately saline water. From the results, we conclude that Ca-complex sources have potential as modulators of moderately saline water stress in sour passion fruit seedlings.

Keywords: *Passiflora edulis* Sims; sustainability; water salinity; chelated calcium; gas exchange

Citation: Souto, A.G.d.L.; Pessoa, A.M.d.S.; Sá, S.A.d.; Sousa, N.R.d.; Barros, E.S.; Morais, F.M.d.S.; Ferreira, F.N.; Silva, W.A.O.d.; Batista, R.O.; Silva, D.V.; et al. Potential of Ca-Complexed in Amino Acid in Attenuating Salt Stress in Sour Passion Fruit Seedlings. *Plants* **2024**, *13*, 2912. https://doi.org/10.3390/plants13202912

Academic Editor: Juan Barceló

Received: 22 September 2024
Revised: 13 October 2024
Accepted: 15 October 2024
Published: 17 October 2024

1. Introduction

Increased soil salinity due to irrigation with water containing high concentrations of salts is one of the main obstacles to agricultural expansion and food production in various regions of the world [1]. The accumulation of salts in the topsoil near plant roots causes loss of fertility and changes in the soil's structural properties, resulting in significant negative impacts on crop yields [2]. This problem is particularly relevant in arid and semi-arid climate regions, where low rainfall, high temperatures and high evaporation rates exacerbate the effects of salinization [3].

In Brazil, the occurrence of saltwater is a persistent challenge, especially in the semi-arid areas of the northeast [4,5]. Many farmers in these regions depend exclusively on this type of water for their agricultural activities [6]. The high concentration of salts creates a pressure gradient in the soil, which limits the absorption of water by plants and implies deleterious effects on physiological functioning and nutritional status, mainly due to the presence of toxic ions (Na^+ and Cl^-) in plant tissues that reduce the capacity for plant growth and development [7,8].

The sour passion fruit (*Passiflora edulis* Sims), a species classified as sensitive to salt stress [9], shows a reduction in chlorophyll fluorescence, gas exchange and increased cell damage through greater electrolyte leakage and lower relative water content, affecting the growth and quality of seedlings during the vegetative phase [4,10]. However, the intensity with which salt stress affects plants depends on a number of factors, such as the variation in tolerance levels between passion fruit cultivars, soil and climate conditions, cultural treatments and irrigation and fertilization management [4,6].

Calcium (Ca^{2+}), a secondary macro nutrient for plants, acts in metabolism through molecular signaling and plays an active role in mineral nutrition and plant growth, as well as being a structural constituent of the cell wall [11]. Ca^{2+} is also considered crucial in altering selective ion uptake and increasing plant tolerance to salt stress [8,12]. The role of this macro nutrient in salt stress tolerance is related to the accumulation of the element in cells or intracellular release as a way of signaling plants to adjust their physiological functions under stress conditions [13,14].

Calcium is an immobile element in the phloem, being transported exclusively via the xylem by transpiratory flow. It acts in the rhizosphere, reducing water availability, affecting leaf transpiration due to the stimulus of the partial closure of stomata, which can aggravate calcium deficiency in plant tissues [15]. The use of chelating or complexing agents in agriculture aims to protect metal or non-metal ions from undesirable chemical reactions and improve their availability for plant root absorption, including calcium ions [16,17]. Among natural complexing agents, such as amino acids and organic acids, each act differently on the bioavailability of nutrients for plant absorption [17–19].

The sustainability of agriculture and food security depends on research efforts aimed at selecting varieties that are better adapted to saline growing conditions and developing management practices that minimize the effects of salt stress on plants [20]. Therefore, our hypothesis is that the use of Ca-complexed sources can increase calcium mobility in plants, promoting the modulation of the physiological mechanisms of acclimatization of sour passion fruit seedlings via the supply of Ca^{2+} as a secondary messenger under conditions of salt stress and that this response may vary between different genetic materials of sour passion fruit. Therefore, the aim of this study was to verify the potential of complexed calcium sources in reducing the effects of water salinity on physiological aspects in seedlings of sour passion fruit cultivars in a semi-arid area.

2. Results

2.1. Gas Exchange and Photosynthetic Efficiency

According to the results in Table 1, the interaction among cultivars × water salinity × calcium sources influenced the rate of net photosynthesis and the water use efficiency and instantaneous carboxylation efficiency. Passion fruit leaf transpiration responded to the interaction cultivars × water salinity and cultivars × calcium sources, while stomatal conductance was influenced by the interaction between water salinity and calcium sources. In isolation, water salinity interfered with internal CO_2 concentration and instantaneous water use efficiency, while complex calcium sources had a significant effect on the leaf Ci of seedlings.

The rate of net photosynthesis showed a significant difference between the cultivars, with BRS GA1 standing out, which was superior in the treatments under irrigation with 4.0 dS m^{-1} water, especially when associated with the application of calcium complexed with amino acids (Figure 1A). Irrigation with 4.0 dS m^{-1} water reduced the photosynthetic rate of the seedlings in most of the treatments with different calcium sources. However, an exception was observed in the BRS GA1 cultivar treated with calcium complexed with amino acids, where there was no significant difference between the salinities of the irrigation water.

The application of calcium complexed with amino acids (Ca-AA) promoted a higher rate of photosynthesis in the BRS GA1 cultivar irrigated with moderately saline water, showing an increase of 37.96 and 55.07%, respectively, compared to the treatments without calcium and with calcium complexed in organic acids (Figure 1A).

The water use efficiency (WUE) values of seedlings did not differ significantly between the BRS GA1 and BRS SC1 cultivars when irrigated with saline water and the application of Ca-complex sources (Figure 1B). In the BRS GA1 cultivar, WUE decreased when irrigated with moderately saline water, except in those treated with Ca-complex in amino acids, which did not differ statistically from the mean value of seedlings irrigated with low salinity water. On the other hand, cv. BRS SC1 showed a reduction in WUE when irrigated with moderately saline water, regardless of the application of complexed calcium sources. The losses were 23.98, 34.88 and 36.59%, respectively, for plants in the treatments without calcium (W-Ca), calcium complexed with organic acids (Ca-OA) and calcium complexed with amino acids (Ca-AA). The application of calcium sources had a significant effect on water use efficiency in BRS GA1 seedlings when irrigated with moderately saline water. The Ca-AA treatment stood out, showing an increase of 49.84% compared to the control treatment (Figure 1B). Calcium sources did not cause significant effects on WUE in BRS SC1 seedlings irrigated with moderately saline water.

Table 1. Summary of analysis of variance for gas exchange variables in sour passion fruit cultivars (Cul) irrigated with saline water (Sal), and application of complex calcium sources (Cal) as salinity attenuators at 60 days after sowing.

S.V.	D.F.	Mean Squares			
		E	gs	A	
Cul	2	0.007 ^{ns}	32.832 ^{ns}	1.782 ^{ns}	
Sal	1	0.390 **	1638.610 **	82.344 **	
Cal	2	0.084 *	10.737 ^{ns}	0.196 ^{ns}	
Cul × Sal	1	0.110 *	38.274 ^{ns}	1.288 ^{ns}	
Cul × Cal	2	0.092 *	69.151 ^{ns}	0.104 ^{ns}	
Sal × Cal	2	0.002 ^{ns}	100.330 *	2.079 ^{ns}	
Cul × Sal × Cal	2	0.005 ^{ns}	95.102 ^{ns}	4.367 *	
Block	2	0.138	19. 543	1.369	
Residue	22	0.023	28.303	0.913	
Total	35	-	-	-	
CV (%)	-	8.61	4.83	6.69	
		Ci	WUE	iWUE	iCE
Cul	2	15.853 ^{ns}	0.213 ^{ns}	0.001 ^{ns}	0.001 ^{ns}
Sal	1	13,599.058 *	15.510 **	0.006 **	0.002 ^{ns}
Cal	2	1854.504 *	0.374 ^{ns}	0.001 ^{ns}	0.001 ^{ns}
Cul × Sal	1	108.889 ^{ns}	0.002 ^{ns}	0.001 ^{ns}	0.001 ^{ns}
Cul × Cal	2	413.560 ^{ns}	0.381 ^{ns}	0.001 ^{ns}	0.001 ^{ns}
Sal × Cal	2	290.317 ^{ns}	0.514 ^{ns}	0.001 ^{ns}	0.001 ^{ns}
Cul × Sal × Cal	2	786.346 ^{ns}	1.203 *	0.003 ^{ns}	0.001 *
Block	2	1528.473	0.992	0.002	0.001
Residue	22	243.847	0.318	0.001	0.000
Total	35		-	-	-
CV (%)	-	6.08	15.81	14.40	24.00

SV = source of variation; DF = degree of freedom; CV = coefficient of variation; ^{ns}, * and ** = not significant, significant at 0.05 and 0.01 probability, respectively. E—leaf transpiration; gs—stomatal conductance; A—net photosynthesis rate; Ci—internal CO_2 concentration, WUE—water use efficiency; iWUE—intrinsic water use efficiency; iCE—instantaneous carboxylation efficiency.

The instantaneous carboxylation efficiency (iCE) did not vary between the passion fruit cultivars, regardless of irrigation with saline water or the application of Ca-complex sources, as shown in Figure 1C. In cv. BRS GA1, irrigation with moderately saline water resulted in a reduction in iCE in the treatments without calcium and with Ca-complexed with organic acids, but there was no significant difference between plants irrigated with low and moderate salinity water when applying Ca-complexed with amino acids (Figure 1C). On the other hand, in the BRS SC1 cultivar treated with Ca-AA, the iCE was reduced by 35% when irrigated with moderately saline water. In the W-Ca and Ca-OA treatments, there was no significant difference in iCE between the seedlings irrigated with moderate and low salinity water. The instantaneous carboxylation efficiency in the seedlings of the sour passion fruit cultivars did not differ significantly for the treatments without and with the application of complex calcium sources when irrigated with low salinity water (Figure 1C). However, in the BRS GA1 cultivar irrigated with moderately saline water, the iCE was higher in the Ca-AA treatments, being 43.71% higher than W-Ca, while in the BRS SC1 cultivar, there was no difference between the calcium sources.

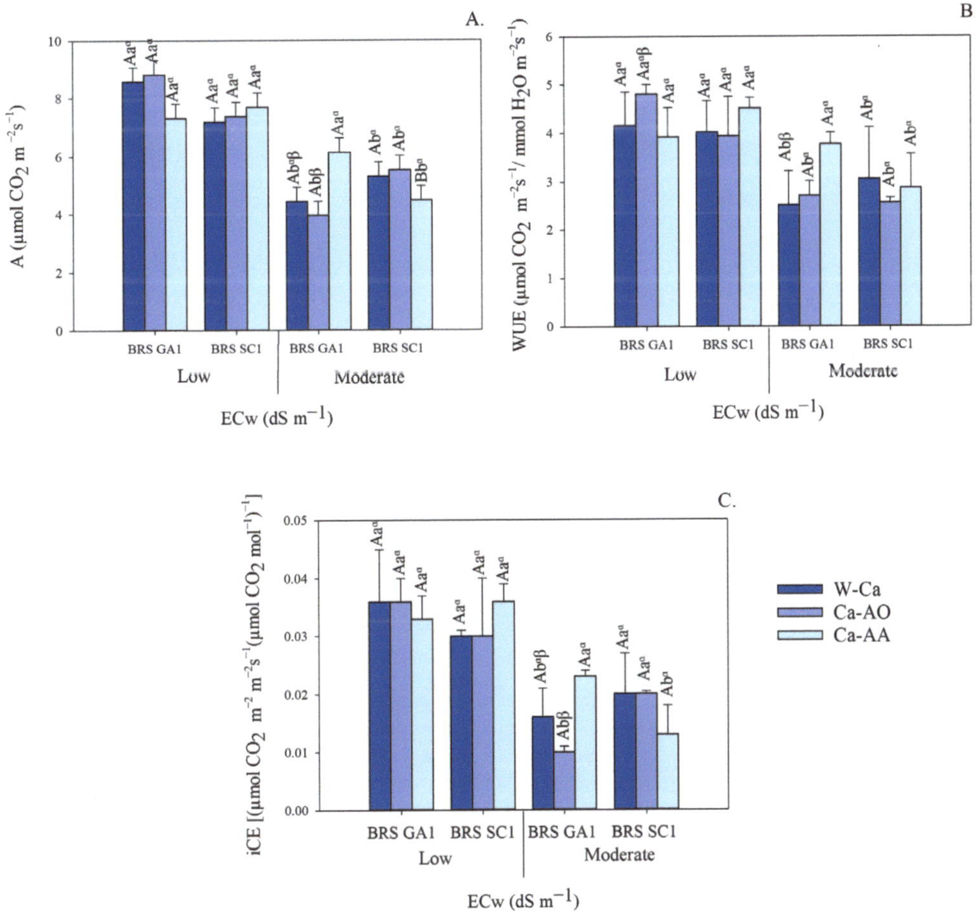

Figure 1. Net assimilation rate—A (**A**), water use efficiency—WUE (**B**) and instantaneous carboxylation efficiency—iCE (**C**) of seedlings of sour passion fruit cultivars irrigated with low and moderate salinity water under application of Ca-complexed sources as attenuators. Means followed by same uppercase letters do not differ for cultivars within each irrigation water salinity and application of calcium sources, according to F test ($p \leq 0.05$). Means followed by same lowercase letters do not differ for irrigation water salinity within each cultivar and calcium sources, by F test ($p \leq 0.05$). Means followed by same Greek letters do not differ for calcium sources within each cultivar and irrigation water salinity, using Tukey test ($p \leq 0.05$). Scatter above bar represents average standard deviation ($n = 3$).

These results indicate that the effect of moderately saline water varies between passion fruit cultivars and the calcium sources tested. In this condition, the seedlings of the BRS GA1 cultivar showed positive responses in photosynthetic rate and in water use and carboxylation efficiencies with the application of Ca-AA, while there were no differences in these variables in the seedlings of the BRS SC1 cultivar.

As for leaf transpiration, a reduction of 16.46% was observed in sour passion fruit seedlings of the BRS GA1 cultivar due to irrigation with saline water, as shown in Figure 2A. In contrast, no significant differences were found in the transpiration rate between the cultivars, regardless of the salinity of the irrigation water. These results highlight how the response of cultivars to saline water irrigation and the application of complex calcium can vary, affecting important variables such as carboxylation efficiency and leaf transpiration in different ways.

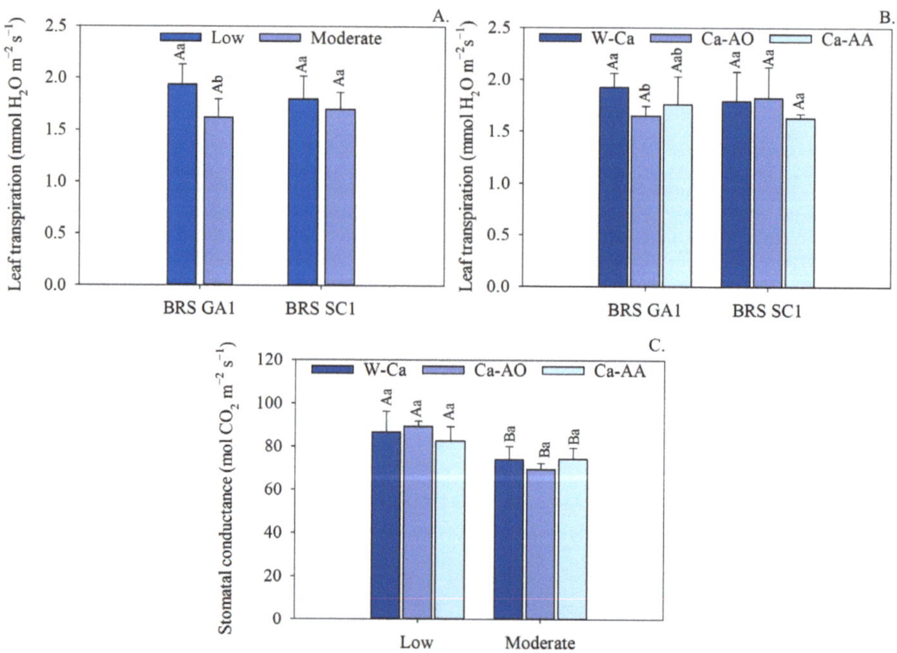

Figure 2. Leaf transpiration in seedlings of sour passion fruit cultivars irrigated with low and moderate salinity water (**A**), application of complex calcium sources (**B**) and internal CO_2 concentration of sour passion fruit seedlings irrigated with low and moderate salinity water and application of complex calcium sources (**C**). (**A**) Means followed by same uppercase letters do not differ for cultivars within each irrigation water salinity by F test ($p \leq 0.05$) and means followed by same lowercase letters do not differ for irrigation water salinity within each cultivar by F test ($p \leq 0.05$). (**B**) Means followed by same uppercase letters do not differ for cultivars within each calcium source by F test ($p \leq 0.05$) and means followed by same lowercase letters do not differ for calcium sources within each passion fruit cultivar by Tukey test ($p \leq 0.05$). (**C**) Means followed by same uppercase letters do not differ from each other for salinity of irrigation water within each calcium source by F test ($p \leq 0.05$) and means followed by same lowercase letters do not differ from each other for calcium sources within each salinity of irrigation water by Tukey test ($p \leq 0.05$). Scatter above bar represents average standard deviation (n = 3).

Leaf transpiration showed no significant differences between the sour passion fruit cultivars in the treatments with the different calcium sources (W-Ca, Ca-OA and Ca-AA), as illustrated in Figure 2B. In the BRS GA1 cultivar, it was observed that the plants that did not receive calcium showed higher leaf transpiration, while those treated with Ca complexed in organic acids showed lower leaf transpiration values. On the other hand, the application of calcium sources had no significant effect on the leaf transpiration of sour passion fruit seedlings cv. BRS SC1.

Irrigation with moderately saline water resulted in a reduction in the stomatal conductance of seedlings, regardless of the calcium source applied, with losses of 14.49, 22.03 and 9.96%, respectively, in the treatments without calcium, with calcium complexed in organic acids and with Ca-complexed in amino acids (Figure 2C). However, there were no significant differences in stomatal conductance between calcium sources in seedlings irrigated with saline water (low and moderate salinity).

Irrigation with moderately saline water increased internal CO_2 concentration by 16.36% (Figure 3A), while reducing instantaneous water use efficiency by 29.67% (Figure 3B). It was observed that the application of Ca-AA resulted in a reduction in the CO_2 concentration of seedlings, followed by the treatment with Ca-OA. On the other hand, the plants in the treatments without calcium had 18.80% higher internal CO_2 concentrations than those treated with Ca-AA (Figure 3C).

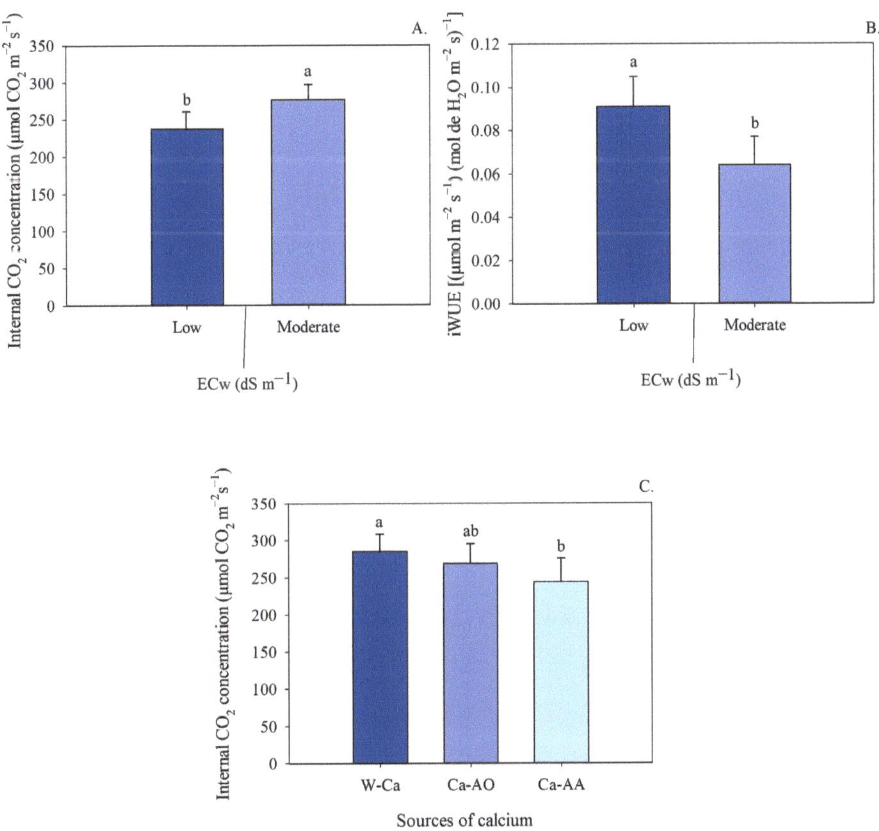

Figure 3. Leaf transpiration in seedlings of sour passion fruit cultivars irrigated with low and moderate salinity water (**A**), application of complex calcium sources (**B**) and stomatal conductance of sour passion fruit seedlings irrigated with low and moderate salinity water and application of complex calcium sources (**C**). (**A**,**B**) Means followed by same lowercase letters do not differ for irrigation water salinity according to F test ($p \leq 0.05$). (**C**) Means followed by same lowercase letters do not differ for calcium sources using Tukey test ($p \leq 0.05$). Scatter above bar represents average standard deviation (n = 3).

2.2. Chlorophyll Indices and Chlorophyll a Fluorescence

The leaf chlorophyll indices of sour passion fruit seedlings responded to the interaction between cultivars, water salinity and sources of complex calcium, as shown in Table 2. The variables related to chlorophyll *a* fluorescence were not affected by the treatments.

Table 2. Summary of analysis of variance, referring to the variables leaf chlorophyll index and chlorophyll fluorescence in the cultivars of sour passion fruit (Cul) irrigated with saline water (Sal) and application of Ca-complexed sources (Cal) as salinity attenuators at 60 days after sowing.

S.V.	D.F.	Mean Squares			
		ICHl*a*	IChl*b*	IChl*t*	
Cul	2	76.038 **	27.825 **	195.860 **	
Sal	1	9.261 ^{ns}	2.068 ^{ns}	20.085 ^{ns}	
Cal	2	62.779 **	4.145 ^{ns}	97.511 **	
Cul × Sal	1	94.997 **	7.084 *	153.966 **	
Cul × Cal	2	2.203 ^{ns}	1.815 ^{ns}	4.704 ^{ns}	
Sal × Cal	2	39.151 **	4.689 *	70.693 **	
Cul × Sal × Cal	2	171.36 **	9.689 *	262.611 **	
Block	2	14.086	4.492	34.096	
Residue	22	3.393	1.323	6.376	
Total	35	-	-	-	
CV (%)	-	4.82	15.86	5.55	
		F_0	F_m	F_V	RQPII
Cul	2	380.250 ^{ns}	1980.250 ^{ns}	245.444 ^{ns}	0.002 ^{ns}
Sal	1	406.694 ^{ns}	4074.694 ^{ns}	1089.000 ^{ns}	0.001 ^{ns}
Cal	2	88.583 ^{ns}	238.083 ^{ns}	149.527 ^{ns}	0.009 ^{ns}
Cul × Sal	1	72.250 ^{ns}	793.361 ^{ns}	81.000 ^{ns}	0.002 ^{ns}
Cul × Cal	2	424.083 ^{ns}	1629.250 ^{ns}	440.527 ^{ns}	0.001 ^{ns}
Sal × Cal	2	105.194 ^{ns}	189.194 ^{ns}	207.750 ^{ns}	0.002 ^{ns}
Cul × Sal × Cal	2	217.583 ^{ns}	12,097.694 ^{ns}	1170.083 ^{ns}	0.010 ^{ns}
Block	2	991.083	1160.583	138.861	0.001
Residue	22	179.659	2204.765	1720.830	0.005
Total	35	-	-	-	-
CV (%)	-	18.30	14.04	15.79	9.95

SV = source of variation; DF = degree of freedom; CV = coefficient of variation; ^{ns}, * and ** = not significant, significant at 0.05 and 0.01 probability, respectively. IChl*a*—chlorophyll *a* index; IChl*b*—chlorophyll *b* index; ICh*t*—chlorophyll total index; F_0—initial fluorescence; F_m—maximum fluorescence; F_v—variable fluorescence; RQPII—quantum yield of photosystem II.

In the seedlings irrigated with low salinity water, there was no difference in the chlorophyll *a* index (IChl*a*) between the cultivars in the treatment without calcium application (Figure 4A). Under irrigation with moderately saline water, sour passion fruit cv. BRS SC1 showed increases of 17.50% and 45.40% in the treatments without and with calcium complexed in organic acids, respectively. Irrigation with moderately saline water reduced the chlorophyll *a* indices of the cv. BRS GA1 in the treatment without calcium and with Ca-OA, while it increased the IChl*a* values when Ca-AA was applied (Figure 4A). For seedlings of the BRS SC1 cultivar, irrigation with moderately saline water increased IChl*a* by 29.60% in the treatments with calcium complexed in organic acids.

The seedlings of the cv. BRS SC1, irrigated with low salinity water and with the application of Ca-complexed with amino acids, showed a higher IChl*b* compared to the seedlings of cv. BRS GA1 (Figure 4B). Under conditions of irrigation with moderately saline water, the seedlings of cv. BRS SC1 treated with Ca-OA showed an IChl*b* value 114.96% higher than those from the other cultivar.

For the treatments without and with the application of Ca-OA, respectively, the seedlings of the cv. BRS GA1 showed a reduction in IChl*b* when irrigated with moderately saline water (Figure 4B). However, there was an increase in this variable with the application of Ca-AA, resulting in a 25.30% increase compared to the control treatment. The application of Ca-OA also increased the chlorophyll b index of BRS SC1 seedlings under moderately saline water conditions.

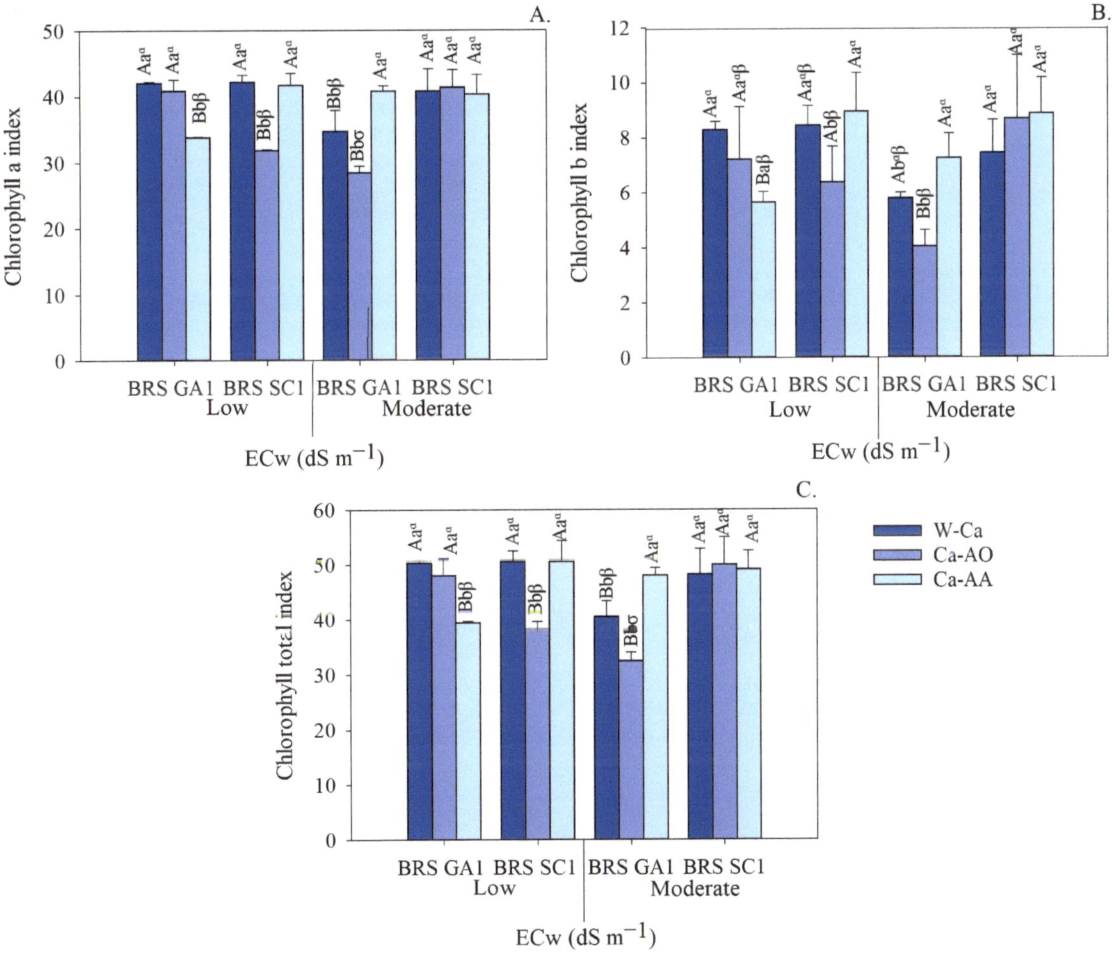

Figure 4. Chlorophyll *a* (**A**), chlorophyll *b* (**B**) and total chlorophyll (**C**) indices of seedlings of sour passion fruit cultivars irrigated with low and moderate salinity water under application of Ca-complex sources as attenuators. Means followed by same uppercase letters do not differ for cultivars within each irrigation water salinity and application of calcium sources, according to F test ($p \leq 0.05$). Means followed by same lowercase letters do not differ for irrigation water salinity within each cultivar and calcium sources, by F test ($p \leq 0.05$). Means followed by same Greek letters do not differ for calcium sources within each cultivar and irrigation water salinity, using Tukey test ($p \leq 0.05$). Scatter above bar represents average standard deviation ($n = 3$).

Under irrigation with moderately saline water, sour passion fruit seedlings cv. BRS SC1 showed higher IChl*t* compared to seedlings of cv. BRS GA1 in the treatments without calcium and with Ca-complexed with organic acids (Figure 4C). The total chlorophyll index of the cv. BRS GA1 was reduced in the treatments without calcium and with Ca-OA when irrigated with moderately saline water, with losses of 19.56% and 32.49%, respectively. The application of Ca-complexed with AA prompted an increment of 21.75% in the total IChl*t* index of BRS GA1 seedlings exposed to the moderately saline treatment. For seedlings of the BRS SC1 cultivar, there was no significant difference in the treatments without calcium and with Ca-AA at both salinity levels. However, in the treatments with Ca-OA, there was a 30.64% increase in IChl*t* when the seedlings were irrigated with moderately saline water.

The application of Ca-complexed with amino acids resulted in a higher IChl*t* in the cv. BRS GA1 irrigated with moderately saline water, with an increase of 18.7% compared to the treatment without calcium. There were no significant differences between calcium sources in the sour passion fruit cv. BRS SC1 when irrigated with moderately saline water.

Similar to Figure 1, the results indicate that the leaf chlorophyll indices in sour passion fruit seedlings irrigated with moderately saline water were higher with the application of calcium complexed in amino acids for the BRS GA1 cultivar, but there was no difference between the calcium sources in the BRS SC1 cultivar.

2.3. Electrolyte Leakage and Relative Water Content

The interaction cultivars × water salinity × complexed calcium sources had a significant effect on electrolyte leakage and relative water content in seedlings (Table 3).

Table 3. Summary of analysis of variance for electrolyte leakage and relative water content in sour passion fruit cultivars (Cul) irrigated with saline water (Sal) and application of Ca-complex sources (Cal) as salinity attenuators at 60 days after sowing.

S.V.	D.F.	Mean Squares	
		EL	RWC
Cul	2	41.667 ^{ns}	0.043 ^{ns}
Sal	1	18.94.135 **	588.790 **
Cal	2	74.488 *	17.423
Cul × Sal	1	25.050 ^{ns}	5.405
Cul × Cal	2	119.695 **	81.170 *
Sal × Cal	2	30.544 ^{ns}	57.721 **
Cul × Sal × Cal	2	49.778 *	98.228 **
Block	2	24.809	22.508
Residue	22	11.285	10.002
Total	35	-	-
CV	-	13.60	3.78

SV = source of variation; DF = degree of freedom; CV = coefficient of variation; ^{ns}, * and ** = not significant, significant at 0.05 and 0.01 probability, respectively. EL—electrolyte leakage; RWC—relative water content.

Irrigation with moderately saline water increased electrolyte leakage, regardless of the application of complex calcium sources (Figure 5A). The highest EL values were observed in BRS SC1 seedlings irrigated with moderately saline water and Ca-OA (41.11%), while the lowest values were found in BRS GA1 seedlings irrigated with low salinity water and Ca-AA (12.98%).

Under the conditions of the substrate complexed without calcium, the BRS SC1 cultivar showed a lower value of electrolyte leakage compared to the BRS GA1 seedlings (Figure 5A). Under moderately saline water conditions, the application of complex calcium to BRS GA1 seedlings resulted in the lowest EL values (29.68%—Ca-OA and 28.61%—Ca-AA). In BRS SC1 seedlings, electrolyte leakage was reduced with the application of Ca-AA (26.29%), although it did not differ statistically from the treatment without calcium (29.19%).

Figure 5B shows that, regardless of the calcium sources, there was no significant difference in relative water content (RWC) between the passion fruit cultivars irrigated with low salinity water. However, under irrigation with moderately saline water, it was observed that in the treatments without the application of calcium sources, the seedlings of the cv. BRS GA1 (95.74%) had a higher RWC compared to cultivar BRS SC1 (82.84%).

Irrigation with moderately saline water increased the relative water content in most treatments, especially in the treatments with calcium complexed in amino acids in both cultivars. There was an increase from 81.29% to 87.44% in cv. BRS GA1 and from 79.61%

to 91.40% in cv. BRS SC1. Under low salinity conditions, no statistical differences were observed between the calcium sources.

However, the cultivars behaved differently when irrigated with moderately saline water. In the BRS GA1 seedlings, the highest RWC values were found in the treatments without calcium application, with an increase of 8.98% over the Ca-AA treatments. On the other hand, in the seedlings of the BRS SC1 cultivar, the highest RWC values were observed in the treatments with Ca-AA (91.40%), representing an increase of 10.32% over the treatment without calcium.

In summary, the results indicate that electrolyte leakage from seedlings irrigated with moderately saline water increased in both cultivars when Ca-AA was applied as a stress attenuator.

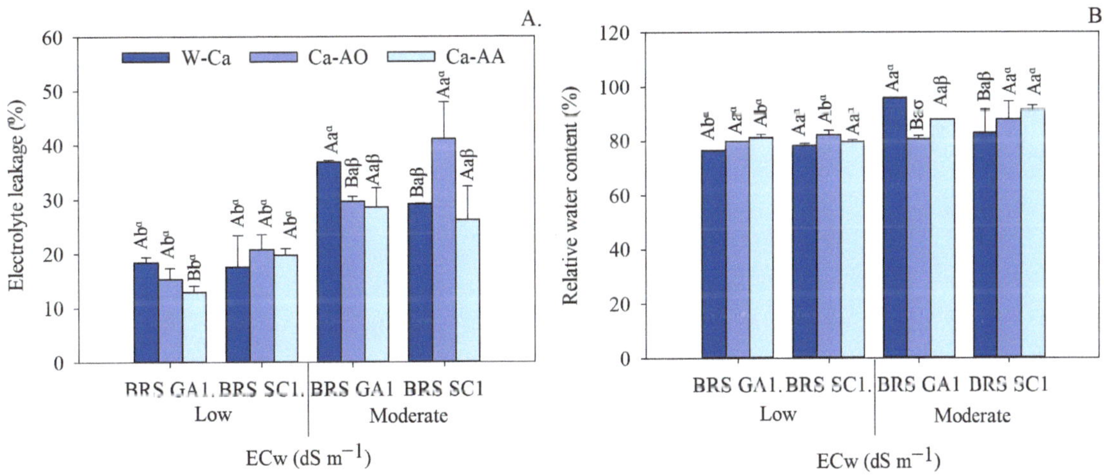

Figure 5. Electrolyte leakage (**A**) and relative water content (**B**) of seedlings of sour passion fruit cultivars irrigated with low and moderate salinity water under application of calcium complex sources as attenuators. Means followed by same uppercase letters do not differ for cultivars within each irrigation water salinity and application of calcium sources by F test ($p \leq 0.05$). Means followed by same lowercase letters do not differ for irrigation water salinity within each cultivar and calcium sources by F test ($p \leq 0.05$). Means followed by same Greek letters do not differ for calcium sources within each cultivar and irrigation water salinity by Tukey test ($p \leq 0.05$). Scatter above bar represents mean standard deviation ($n = 3$).

2.4. Multivariate Analysis of Principal Components

Based on the principal component analysis (PCA) of the physiological traits of the sour passion fruit cv. BRS GA1 (Figure 6), it was observed that the PC1 and PC2 vectors explained 52.8% and 19% of the variance, respectively, totaling 71.8% of the total variance of the data.

Irrigation with moderately saline water had a significant impact on the variables of relative water content, electrolyte leakage and internal CO_2 concentration of sour passion fruit cv. BRS GA1, especially in the treatments without and with Ca-complexed with organic acids. In contrast, seedlings irrigated with low salinity water showed increases in gas exchange variables (stomatal conductance—gs, transpiration rate—E, CO_2 assimilation rate—A, water use efficiency—EUA, instantaneous water use efficiency—iWUE and instantaneous carboxylation efficiency—iCE), and chlorophyll indices (IChl*a*, IChl*b* and IChl*t*), regardless of the application of calcium sources.

When analyzing the principal components of the physiological variables of the BRS SC1 seedlings (Figure 7), it was observed that the eigenvalues accounted for 37.6% (PC1) and 21% (PC2) of the variance, totaling 58.6% of the total variance of the data.

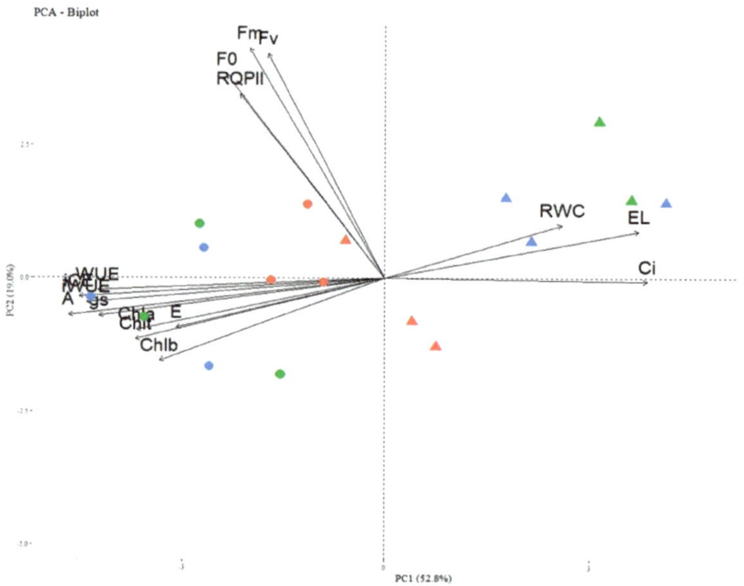

Figure 6. Principal component analysis (PCA) for the physiological traits of sour passion fruit seedlings cv. BRS GA1 irrigated with low and moderate salinity water under sources of complexed calcium as a mitigator ($n = 3$).

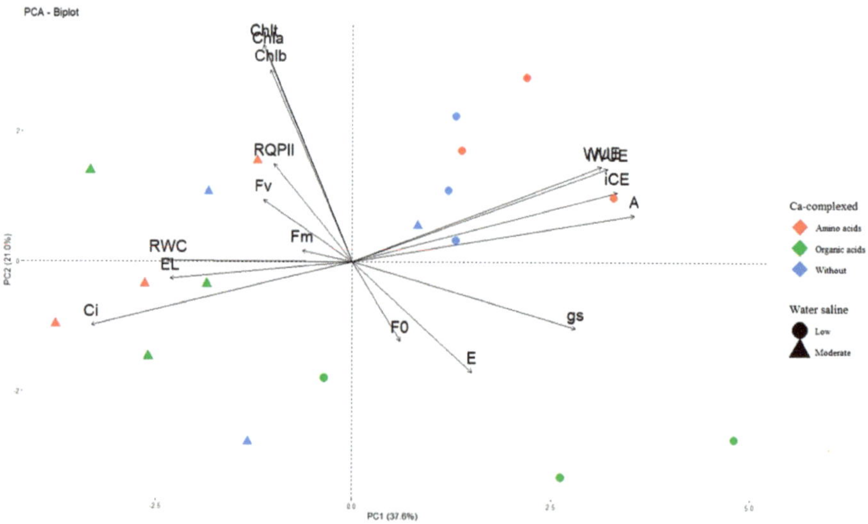

Figure 7. Principal component analysis (PCA) for the physiological traits of sour passion fruit seedlings cv. BRS SC1 irrigated with low and moderate salinity water under sources of calcium complexed as an attenuator ($n = 3$).

This analysis indicates that there was a significant distribution between the main components, reflecting the diversity in the physiological responses of the plants to the different treatments applied, such as irrigation with low and moderate salinity water and the application of complex calcium.

Irrigation with low salinity water associated with the application of Ca-complexed with amino acids promoted higher values of CO_2 assimilation rate (A), water use efficiency (WUE), intrinsic water use efficiency (iWUE) and instantaneous carboxylation efficiency (iCE) in the seedlings of the cv. BRS SC1. These variables showed a negative correlation with internal CO_2 concentration (Ci), transpiration rate (E) and relative leaf water content (RWC), which responded more strongly to irrigation with moderately saline water and the application of complex calcium sources.

Initial fluorescence (F_0), stomatal conductance (gs) and leaf transpiration (E) were most influenced by irrigation with low salinity water associated with the application of calcium complexed with organic acids (Ca-OA). On the other hand, irrigation with moderately saline water had a high influence on the maximum chlorophyll fluorescence (F_m), variable fluorescence (F_v) and quantum efficiency of photosystem II (RQPII) of sour passion fruit cv. BRS SC1. The seedlings' leaf chlorophyll indices were little influenced by the application of the treatments.

3. Materials and Methods

3.1. Description of the Experiment Site

The experiment was conducted between October and December 2023, in a protected environment (green house) located in the Department of Agronomic and Forestry Sciences (DAFS) belonging to the Federal Rural University of the Semi-Arid (UFERSA), which is located in the municipality of Mossoró, Rio Grande do Norte, Brazil. The protected environment is georeferenced by the coordinates 5°12′12″ S and 37°19′26″ W at an altitude of 17 m. It is characterized by transparent low-density polyethylene cover, treated against the action of ultraviolet rays, in the shape of an arch, with dimensions of 6.5 m × 12 m. The sides and front walls were made with anti-aphid screens (50%) and a 0.24 m masonry plinth.

Sour passion fruit seedlings were installed on benches. During the assay, the average daily temperatures and relative humidity of air inside the protected environment were recorded using an HTC-1 Clock/Humidity Digital LCD temperature and humidity meter (Beijing, China), as presented in Figure 8.

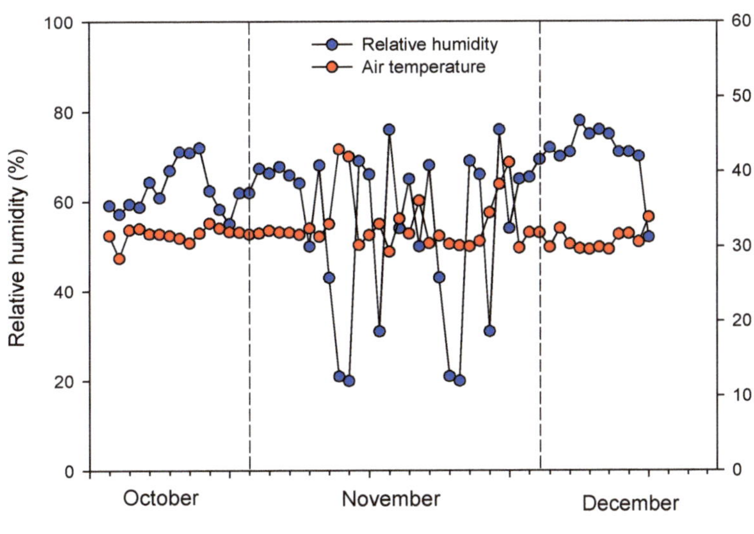

Figure 8. Temperature and relative humidity of air inside the protected environment during the experimental period (18 October to 18 December 2024).

3.2. Experimental Design and Treatments

The experiment was carried out in a randomized block design, in a 2 × 2 × 3 factorial scheme, with three replications and two plants per plot. The treatments referred to two cultivars of sour passion fruit (BRS Gigante Amarelo—BRS GA1 and BRS Sol do Cerrado—BRS SC1), irrigated with low salinity and moderately saline water (electrical conductivity of 0.5 and 4.0 dS m^{-1}, respectively) and the application of three sources of Ca complexed as a water salinity attenuator (control—without calcium [W-Ca], calcium complexed in organic acids [Ca-OA] and calcium complexed in amino acids [Ca-AA]).

The F1 seeds of the passion fruit cultivars BRS GA1 and BRS SC1 were obtained from mother plants in Embrapa Cerrado, Brasilia, Brazil. The low salinity water (0.5 dS m^{-1}) came from the local municipal water supply, while the moderately saline water (4.0 dS m^{-1}) was obtained by diluting the stock solution (8.0 dS m^{-1}), prepared with P.A. sodium chloride (NaCl; PM = 58.44), hydrated calcium chloride ($CaCl_2 \cdot 2H_2O$; PM = 147.01) and magnesium chloride hexahydrate ($MgCl_2 \cdot 6H_2O$; PM = 203.30) in an equivalent ratio of 7:2:1, in supply water (0.5 dS m^{-1}), following the relationship between ECw and its concentration ($mmol_c\ L^{-1} \approx EC \times 10$), according to the methodology contained in Silva [21]. This cation ratio (7:2:1) represents the chemical composition of most surface and underground saline water reservoirs in the Brazilian northeast. The calcium sources used were the commercial products Codasal® (Sustainable Agro Solution, Almacelles, Spain) and Hendosar® (Adriatica S.p.A., Loreo, Italy), respectively, as a source of Ca-complexed in organic acids and amino acids at a concentration of 2% (volume basis).

3.3. Conducting the Experiment

The substrate used to grow the seedlings was made up of a mixture of soil collected from a depth of 0–0.20 m and well-decomposed bovine manure in a ratio of 2:1 (*v/v*), the chemical composition of which was analyzed according to the methodologies presented by Silva [22]. The results of the analysis are shown in Table 4.

Table 4. Chemical attributes of the substrate used in the experiment.

pH	OM	P	K^+	Ca^{2+}	Mg^{2+}	Na^+	Al^{3+}	$H^+ + Al^{3+}$	BS	CEC	ESP	ECse
H_2O	%	mg dm^{-3}		---------------------- ($cmol_c\ dm^{-3}$) ----------------------							%	dS m^{-1}
7.7	33.6	135.40	1.99	10.63	8.43	0.46	0.00	0.70	21.51	22.21	2.07	1.45

pH = hydrogen potential; OM = organic matter (wet oxidation—Walkey–Black); P = phosphorus (Mehlich 1), K^+ = potassium (Mehlich 1), Ca^{2+} = calcium (KCl 1 M); Mg^{2+} = magnesium (KCl 1 M), Na^+ = sodium (KCl 1 M); Al^{3+} = exchangeable acidity (KCl 1 M); $H^+ + Al^{3+}$ = potential acidity (calcium acetate 0.5 N); BS = sum of bases; CEC = cation exchange capacity; ESP = exchangeable sodium percentage; ECse = electrical conductivity of saturation extract.

Subsequently, the substrate (1.7 kg) was placed in black polyethylene bags with a capacity of 1.5 dm^3; three seeds were placed in each bag at a depth of 1 cm, after which they were covered with a thin layer of substrate. Emergence began at 7 days after sowing (DAS) and stabilized at 28 DAS, counting as normal seedlings those that had fully emerged from the substrate and had fully expanded cotyledons. Ten days after emergence had stabilized, the seedlings were thinned out, leaving the most vigorous one per polyethylene bag.

The plants were irrigated daily with the respective irrigation water using the weighing method, replenishing the volume of water corresponding to the evapotranspiration of the plants of the previous day (24 h), so as to leave the moisture content of the substrate at 90% of its field capacity [23]. To do this, before setting up the experiment, ten polyethylene bags containing substrate were selected at random, weighed and then volumes of water were added gradually until all the pore spaces in the substrate were filled and the water began to drain. Immediately, the volume of water drained was measured in a measuring cylinder and the polyethylene bags were weighed to determine the weight of the bags at field capacity. The amount of water to be applied corresponds to the difference between the weight of the polyethylene bag at 90% field capacity and the weight on the day of irrigation. When the weight of the bag with the substrate was the same as that at 90% field capacity, the sour passion fruit seedlings were not irrigated.

The attenuators (Codasal® and Hendosar®) were applied at 12 and 32 days after emergence in a volume equivalent to 50 mL per plant (solution diluted to 2% in water). The solution was prepared by diluting the attenuators in low salinity water (0.5 dS m^{-1}), according to the manufacturer's recommendations. The solution was applied gradually to the surface of the substrate until maximum infiltration was achieved. Codasal® is a dark-colored liquid organo-mineral fertilizer that is Ca-complexed with organic acids (lignosulphonate) and contains a 6.0% N, 8.7% CaO and 14.7% lignosulphonate complexing agent and a salt index of 40.74%. Hendosar® is a light-red liquid mineral fertilizer containing Ca-complexed with amino acids, with 9.0% N, 6.0% K_2O, 7.5% Ca, 1.2% Mg and a salt index of 49.52%.

3.4. *Traits Analysed*

3.4.1. Chlorophyll a Fluorescence

At 60 DAS, the chlorophyll a fluorescence of the leaves was measured using the following variables: initial fluorescence (F_0), maximum fluorescence (F_m), variable fluorescence (F_v) and photosystem II quantum yield (RQPII) using a PEAITM pulse-modulated Plant Efficiency Analyzer (Hansatech Instruments Co., King's Lynn, UK). For this, three plants per treatment were selected for evaluation, in which the third pair of fully expanded leaves was selected, counting from the apex to the base of the plants. Measurements were taken between 08:00 and 11:00 a.m. after the leaves were adapted to the dark for at least 30 min.

3.4.2. Leaf Chlorophyll Indices

The leaf chlorophyll indices of the sour passion fruit seedlings were measured on the same leaves selected for chlorophyll *a* fluorescence analysis at 60 DAS. The chlorophyll *a* index (IChl*a*), chlorophyll *b* index (IChl*b*) and chlorophyll total index (IChl*t*) were measured using a Falker® ChlorofiLOG portable meter (Porto Alegre, Brazil). On each plant, three readings were taken on the basal, median and apical part of the leaf and the average was calculated from these data.

3.4.3. Gas Exchange

At 60 DAS, the third or fourth pair of leaves on the seedlings was selected to analyze gas exchange using the following variables: stomatal conductance (gs—mol H_2O m^{-2} s^{-1}); net photosynthesis rate (A—μmol CO_2 m^{-2} s^{-1}); internal CO_2 concentration (Ci—μmol CO_2 m^{-2} s^{-1}); and leaf transpiration (E—mmol H_2O m^{-2} s^{-1}), measured with a portable infrared carbon dioxide analyzer (IRGA), model LCorp-SD from BioScientific® (Hoddesdon, UK), adjusted to an air temperature of 25 °C, irradiance of 1. 200 μmol m^{-2} s^{-1} and air flow of 200 mL min^{-1} [23]. The readings were taken in the morning, between 08:00 and 11:00 a.m. From these data, water use efficiency (WUE) was determined by the A/E ratio and intrinsic water use efficiency (iWUE) by the A/gs ratio.

The relative water content (RWC) was determined according to Beltrano and Ronco [24], adapted. Leaf disks of 0.7 cm in diameter were removed from the central part of the leaf vein located in the middle leaves of the seedlings to determine the RWC. After determining the mass of fresh matter (MFM), the disks were placed in distilled water for 4 h. The turgid samples were quickly dried with paper towel to determine the turgid weight (WT). The disks were then placed in an oven at 80 °C for 24 h to determine the dry matter weight (DMW). The following equation was used for this: RWC (%) = (MFM − DMW)/(WT − DMW) × 100.

3.4.4. Electrolyte Leakage

Electrolyte leakage (EL) was determined by removing leaf disks with diameters of 0.7 cm, according to the methodology proposed by Beltrano and Ronco [24], with small adaptations. After being washed three times in distilled water for 3 min, 10 leaf disks were placed in 25 mL of deionized water in Falcon tubes, and the initial EC of the contents was read after 90 min. Once this was carried out, the Falcon tubes were taken to a water bath at 85 °C for 4 h, then cooled to a temperature of 25 °C; the final EC was read and the EL was calculated using the following formula: EL = (L1/L2) × 100, where L1—initial EC reading; L2—final EC reading.

3.5. Statistical Analysis

Before the statistical analysis, the data were evaluated for normality and homogeneity of variance using the Shapiro–Wilk test. Subsequently, the analysis of variance was carried out using the F test ($p \leq 0.05$). The means for the cultivars, electrical conductivities of the water and calcium sources were compared using the Tukey test at 0.05 probability. The statistical software Sisvar 5.6 [25] was used to analyze the data. Figures were drawn using Sigmaplot software version 12.5. Multivariate analysis was carried out using principal component analysis (PCA) with the statistical analysis packages available in R Studio version 4.4.0 [26].

4. Discussion

Irrigation with moderately saline water resulted in a reduction in photosynthetic rate, water use efficiency and instantaneous carboxylation efficiency in seedlings (Figure 1). Photosynthesis is the crucial physicochemical process in plant metabolism and is highly sensitive to environmental changes under salt stress; its efficiency is often reduced, either by stomatal and/or non-stomatal limitations [7]. However, the photochemical efficiency of PSII was not affected by irrigation with moderately saline water; the seedlings' temporary stress may have adjusted their photo protection mechanisms (physiological plasticity) without significantly compromising the photochemical efficiency of the plants [27].

The concentrations of sodium and chloride salts in water can significantly alter the physiological and biochemical functions of plants, restricting the growth and development of crops [28], including during the vegetative phase of the sour passion fruit [9], a fact observed in this study during the initial growth of the plants. Exposure to salinity can reduce stomatal conductance due to the partial closure of stomata and decreases in the influx of carbon dioxide and, consequently, the photosynthetic rate, which is limited by the reduced availability of CO_2 for the Calvin–Benson cycle [5,28].

The application of Ca-complexed with amino acids reduced the damage caused by toxic salts in the photosynthetic activity and efficiency of seedlings (Figure 1). In addition, studies have shown that the application of calcium in the form of calcium pyruvate also attenuated the adverse effects of salinity on Redondo Amarelo passion fruit seedlings [10]. These effects are attributed to calcium, which plays a crucial role in maintaining the structural and functional integrity of cell membranes and the cell wall, as well as regulating ion transport and the enzymatic activity of the cell wall [9,10,13]. In addition, the increase in Ca^{2+} in the substrate improves K^+/Na^+ selectivity, altering the absorption rate in favor of K^+ to the detriment of Na^+, reducing Na^+ toxicity in the plant [13,29].

The application of amino acids as a complexing agent may also improve the process of plant acclimatization to abiotic stress conditions, such as salinity, by acting as osmolytes, facilitating the transport of ions, promoting protein synthesis and maintaining the integrity of biomembranes [17]. Furthermore, the source of Ca-complexed with amino acids had in its composition higher percentages of nitrogen, potassium and magnesium, which are involved in the production of proline (osmoregulator and osmoprotector), stomatal regulation, photosynthetic efficiency, protein synthesis and maintenance of the water status of plants, reducing the effect of saline stress [30]. Concerning studies with broccoli var. Italica, Haghighi et al. [31] showed that foliar application of calcium complexed with tryptophan increased the photosynthetic rate of plants under salt stress, highlighting the potential of calcium–amino acid complexes in mitigating the adverse effects of salinity on the crop.

Irrigation with moderately saline water reduced the leaf transpiration of sour passion fruit cv. BRS GA1 (Figure 2A), due in part to the partial closure of the stomata [25,28,32], as seen in Figure 2C. This effect occurs due to osmotic stress, which inhibits the absorption of water by plants, limiting the flow of water vapor into the atmosphere, as described by Lima et al. [6] in studies with seedlings of three cultivars of sour passion fruit irrigated with moderately saline water (3.5 dS m^{-1}). Nevertheless, Paiva et al. [33] found that there is variability in physiological responses among sour passion fruit cultivars (BRS GA1, BRS S1 and SCS 437 Catarina) irrigated with moderately saline water (3.5 dS m^{-1}) at 154 days after transplanting, in which the BRS GA1 cultivar showed higher transpiration and stomatal conductance rates compared to the other cultivars.

Calcium lignosulfonate, calcium complexed in organic acids, as mentioned by Thye et al. [19], disassociates into Ca^{2+} ions and lignosulfonate, thus improving the bioavailability of calcium to plants. These calcium ions can be absorbed and transported by plant cells via calcium ion channels. Under conditions of water restriction due to the high salinity of the substrate, this calcium may contribute to stomatal closure, reducing seedlings' transpiration rate (Figure 2B).

The increase in internal CO_2 concentration seen in seedlings irrigated with moderately saline water (Figure 3A) might be due to the progressive accumulation of salts in the photosynthetically active mesophyll tissues, which leads to an inhibition of CO_2 assimilation, mainly affecting the photosynthetic processes in the chloroplast, also known as non-stomatal limitations [20], as well as due to stomatal constraints (Figure 2C) that may limit CO_2 efflux. These limitations involve suppression of photosynthetic enzymes of the Calvin cycle, disruption of chlorophyll biosynthesis, reduced operational efficiency and structural integrity of the photosynthetic apparatus and the membranes of the thylakoids [33,34].

The seedlings reduced their intrinsic water use efficiency (iWUE) when irrigated with moderately saline water (Figure 3B). Intrinsic water use efficiency refers to the amount of carbon dioxide fixed by the plant for each unit of H_2O that is lost in the fixation process and under conditions of water–salt stress. Under this situation, plants reduce their absorption of water from the soil and, as a way of avoiding loss to the external environment, they acclimatize by reducing the opening of the sub-stomatal chambers and restricting the influx of CO_2 [35]. Similar behavior was observed by Lima et al. [36] in sour passion fruit cv. BRS GA1 under different irrigation strategies with saline water and potassium fertilization, when they found that regardless of the phenological stage, iWUE was reduced when the plants were irrigated with moderately saline water.

The use of Ca-complex reduced the internal concentration of CO_2 in the cells of seedlings, especially with Ca-complexed in amino acids (Figure 3C). The application of complexing or chelating agents protects the calcium applied to the substrate from various harmful reactions, such as fixation, precipitate formation and leaching, resulting in greater efficiency of the attenuator and a better rate of absorption and use by the plants [17]. In addition, amino acids can promote the development of faster and more effective nutrient absorption channels or mechanisms in root cells, optimizing the transport and translocation of calcium by plant cell membrane transporters [18].

Adequate levels of Ca^{2+} ions in the chloroplast regulate the photosynthetic pathway [37]. In light-dependent reactions, calcium acts as a cofactor in the formation of the active site in photosystem II and facilitates the oxidation of water to form ATP [38]. In addition, calcium increases CO_2 consumption by regulating enzymes involved in the Calvin cycle, increasing the plant's photosynthetic capacity and organic compound formation [19].

In general, the leaf chlorophyll indices of seedlings were reduced by irrigation with moderately saline water (Figure 4). High levels of salinity are known to affect plant photosynthesis through non-stomatal limitations, including changes in chlorophyll concentration. This effect can be attributed to an increase in the activity of chlorophyllase, an enzyme responsible for degrading chlorophyll [8].

In Figure 4, it is observed that the application of complexed calcium (organic acids and amino acids) reduced the effects of salinity on chlorophyll indices. Similarly, Bezerra et al. [39] observed increases in leaf chlorophyll indices in adult passion fruit plants irrigated with calcium doses. The authors point out that although calcium is not part of the chlorophyll molecule, it has a synergistic effect with nitrogen, making it possible to increase chlorophyll synthesis. In addition to calcium, the complexing agents in the form of organic acids or amino acids present in these attenuating agents generally contain nitrogen in various forms. These compounds contribute to osmoregulation, chlorophyll biosynthesis, and leaf expansion [17]. Despite not responding to the treatments (Table 3), the F_0, F_v, F_m and RQPII variables showed mean values of 72.25, 262.72, 334.42 and 0.77, respectively. Similar results were observed by Bezerra et al. [40], who found no influence of water salinity and calcium application on chlorophyll fluorescence variables. Values for the maximum photochemical yield of PSII ranging from 0.80 to 0.86 indicate that the plants are not under stress [41]. Therefore, as the values obtained in our research were below this range (0.77), it is possible that the stress was not strong enough to significantly alter PSII photochemistry.

Irrigation with moderately saline water causes cell damage in seedlings, increasing intracellular electrolyte leakage (Figure 5A). Plasma membranes are the primary sites of damage from salinity-specific ions (Na^+ and Cl^-) [42,43]. According to the same authors, maintaining cell membrane stability under conditions of salt stress is one of the most important selection criteria for identifying salt-tolerant species or cultivars. The sour passion fruit cultivars showed greater cell membrane stability with the application of complexed calcium sources (Figure 5A). The calcium chelates (Ca-OA or Ca-AA) present in these sources can influence the distribution of Na and Cl, reducing the phytotoxic effects on plant metabolism, significantly protecting against lipid peroxidation and increasing the integrity of cell membranes under salt stress [17]. In addition, Ca^{2+} is a particularly important nutrient for plants exposed to salt stress (NaCl), due to its role in reducing Na^+ absorption and increasing the absorption of K^+ and Ca^{2+} itself, which contributes to membrane stability and plant growth [13].

A higher RWC was observed in the sour passion fruit seedlings of cv. BRS GA1 compared to cv. BRS SC1 (Figure 5B). Normally, the presence of salts in the root zone can trigger osmotic stress, hindering the plant's ability to extract water from the soil, reducing the conductivity of the water in the root and, consequently, the relative water content in the cells [8]. Thus, maintaining a high RWC under conditions of salt stress can be a criterion for identifying and selecting materials that are more tolerant to salinity [6]. Sour passion fruit cultivars that have the ability to adjust more effectively to the unfavorable conditions of low water availability in the substrate are considered to be more tolerant to salinity.

The sources of complex calcium increased the ability of seedlings to maintain high levels of cellular RWC (Figure 5B). The addition of calcium-complexing agents (such as organic acids and amino acids) as Ca^{2+} stabilizers promotes the absorption of other organic and inorganic compounds (betaine, glycine, acids, calcium and potassium ions) by the plant, facilitating osmotic adjustment and strengthening the cell wall, which provides greater plant resistance to water stress [17]. Similar behavior was observed in barley cultivars (*Hordeum vulgare* L.) under salt stress (NaCl levels up to 300 mM), where the application of calcium complexed with lignosulfonate helped to mitigate the effects of salinity in water with NaCl levels of 200 mM [19].

5. Conclusions

The use of Ca-complexed sources has the potential to mitigate the deleterious effects of irrigation with moderately saline water (electrical conductivity of 4.0 dS m^{-1}) on the physiological parameters of the seedlings of sour passion fruit cultivars at 60 days after sowing. The cultivar BRS SC1 proved to be more tolerant to irrigation with moderately saline water during seedling formation. This includes the promotion of better photosynthetic efficiency, higher chlorophyll indices and cell membrane stability, demonstrated by lower leakage and higher relative water content. Based on these results, it is recommended that future research focus on investigating the role of complex Ca^{2+} sources in modulating the biochemical activity and enzymatic defenses of sour passion fruit under conditions of moderate to severe salinity, as a strategy to increase salt tolerance and enable the sustainable insertion of agricultural areas marginalized by salinity problems.

Author Contributions: Conceptualization, A.G.d.L.S.; methodology, A.G.d.L.S. and A.M.d.S.P.; investigation and validation, A.G.d.L.S., A.M.d.S.P., S.A.d.S., N.R.d.S., E.S.B., F.M.d.S.M., F.N.F., R.M.O.d.S.M., R.M.d.S.P. and W.A.O.d.S.; writing—original draft preparation, A.G.d.L.S.; writing—review and editing, A.G.d.L.S., A.M.d.S.P., R.O.B., D.V.S., H.R.G., G.S.d.L. and M.M.d.R. All authors have read and agreed to the published version of the manuscript.

Funding: CNPq (National Council for Scientific and Technological Development), CAPES (Coordination for the Improvement of Higher Education Personnel) and UFERSA (Universidade Federal Rural do Semi-Árido).

Data Availability Statement: Data are contained within the article.

Conflicts of Interest: The authors declare no conflicts of interest.

References

1. Singh, A. Soil Salinity: A Global Threat to Sustainable Development. *Soil Use Manag.* **2022**, *38*, 39–67. [CrossRef]
2. Mustafa, G.; Akhtar, M.S.; Abdullah, R. Global Zoncern for Salinity on Various Agro-Ecosystems. In *Salt Stress, Microbes, and Plant Interactions: Causes and Solution: Volume 1*; Akhtar, M.S., Ed.; Springer: Singapore, 2019; pp. 1–19. ISBN 978-9-81138-801-9.
3. Fu, Z.; Wang, P.; Sun, J.; Lu, Z.; Yang, H.; Liu, J.; Xia, J.; Li, T. Composition, Seasonal Variation, and Salinization Characteristics of Soil Salinity in the Chenier Island of the Yellow River Delta. *Glob. Ecol. Conserv.* **2020**, *24*, e01318. [CrossRef]
4. Bezerra, J.D.; Pereira, W.E.; Silva, J.M.D.; Raposo, R.W.C. Crescimento de Dois Genótipos de Maracujazeiro-Amarelo Sob Condições de Salinidade. *Rev. Ceres* **2016**, *63*, 502–508. [CrossRef]
5. Fatima, R.T.; Lima, G.S.; Soares, L.a.A.; Veloso, L.L.S.A.; Da Silva, A.A.R.; Lacerda, C.N.; Silva, F.A.; Nobrega, J.S.; Ferreira, J.T.A.; Pereira, W.E. Salicylic Acid Concentrations and Forms of Application Mitigate Water Stress in Sour Passion Fruit Seedlings. *Braz. J. Biol.* **2023**, *83*, e270865. [CrossRef]
6. Lima, G.S.D.; Souza, W.B.B.D.; Paiva, F.J.D.S.; Soares, L.A.D.A.; Torres, R.A.F.; Silva, S.T.D.A.; Gheyi, H.R.; Lopes, K.P. Tolerance of Sour Passion Fruit Cultivars to Salt Stress in a Semi-Arid Region. *Rev. Bras. Eng. Agríc. Ambient.* **2023**, *27*, 785–794. [CrossRef]
7. Arif, Y.; Singh, P.; Siddiqui, H.; Bajguz, A.; Hayat, S. Salinity Induced Physiological and Biochemical Changes in Plants: An Omic Approach towards Salt Stress Tolerance. *Plant Physiol. Biochem.* **2020**, *156*, 64–77. [CrossRef]
8. Shahid, M.A.; Sarkhosh, A.; Khan, N.; Balal, R.M.; Ali, S.; Rossi, L.; Gómez, C.; Mattson, N.; Nasim, W.; Garcia-Sanchez, F. Insights into the Physiological and Biochemical Impacts of Salt Stress on Plant Growth and Development. *Agronomy* **2020**, *10*, 938. [CrossRef]
9. Santos, L.F.S.; De Lima, G.S.; De Lima, V.L.A.; Da Silva, A.A.R.; De Fátima, R.T.; Arruda, T.F.d.L.; Soares, L.A.d.A.; Capitulino, J.D. Proline on the Induction of Tolerance of Sour Passion Fruit Seedlings to Salt Stress. *Rev. Caatinga* **2024**, *37*, e12048. [CrossRef]
10. Da Silva, F.D.A.; Dias, M.D.S.; Fernandes, P.D.; De Lacerda, C.N.; Da Silva, A.A.R.; Marcelino, A.D.A.d.L.; De Lima, A.M.; Barbosa, D.D. Calcium Pyruvate as a Salt Stress Mitigator in Yellow Passion Fruit Seedlings. *Rev. Bras. Eng. Agríc. Ambient.* **2022**, *26*, 763–770. [CrossRef]
11. El-Hady, E.S.; Merwad, M.A.; Shahin, M.F.M.; Hagagg, L.F. Influence of Foliar Spray with Some Calcium Sources on Flowering, Fruit Set, Yield and Fruit Quality of Olive Kalmata and Manzanillo Cultivars under Salt Stress. *Bull. Natl. Res. Cent.* **2020**, *44*, 197. [CrossRef]
12. Wang, G.; Bi, A.; Amombo, E.; Li, H.; Zhang, L.; Cheng, C.; Hu, T.; Fu, J. Exogenous Calcium Enhances the Photosystem II Photochemistry Response in Salt Stressed Tall Fescue. *Front. Plant Sci.* **2017**, *8*, 2032. [CrossRef]
13. Hadi, M.R.; Karimi, N. The Role of Calcium in Plants' Salt Tolerance. *J. Plant Nutr.* **2012**, *35*, 2037–2054. [CrossRef]
14. Elsawy, H.I.A.; Alharbi, K.; Mohamed, A.M.M.; Ueda, A.; AlKahtani, M.; AlHusnain, L.; Attia, K.A.; Abdelaal, K.; Shahein, A.M.E.A. Calcium Lignosulfonate Can Mitigate the Impact of Salt Stress on Growth, Physiological, and Yield Characteristics of Two Barley Cultivars (*Hordeum vulgare* L.). *Agriculture* **2022**, *12*, 1459. [CrossRef]
15. Alikhani, T.T.; Tabatabaei, S.J.; Mohammadi Torkashvand, A.; Khalighi, A.; Talei, D. Effects of Silica Nanoparticles and Calcium Chelate on the Morphological, Physiological and Biochemical Characteristics of Gerbera (*Gerbera jamesonii* L.) under Hydroponic Condition. *J. Plant Nutr.* **2021**, *44*, 1039–1053. [CrossRef]
16. Saeedi, R.; Etemadi, N.; Nikbakht, A.; Khoshgoftarmanesh, A.H.; Sabzalian, M.R. Calcium Chelated with Amino Acids Improves Quality and Postharvest Life of Lisianthus (*Eustoma grandiflorum* Cv. Cinderella Lime). *HortScience* **2015**, *50*, 1394–1398. [CrossRef]
17. Souri, M.K.; Hatamian, M. Aminochelates in Plant Nutrition: A Review. *J. Plant Nutr.* **2019**, *42*, 67–78. [CrossRef]

18. Areche, F.O.; Aguilar, S.V.; López, J.M.M.; Chirre, E.T.C.; Sumarriva-Bustinza, L.A.; Pacovilca-Alejo, O.V.; Córdova, Y.F.C.; Montesinos, C.C.Z.; Astete, J.A.Q.; Quispe-Vidalon, D.; et al. Recent and Historical Developments in Chelated Fertilizers as Plant Nutritional Sources, Their Usage Efficiency, and Application Methods. *Braz. J. Biol.* **2023**, *83*, e271055.
19. Thye, K.-L.; Wan Abdullah, W.M.A.N.; Ong-Abdullah, J.; Lamasudin, D.U.; Wee, C.-Y.; Mohd Yusoff, M.H.Y.; Loh, J.-Y.; Cheng, W.-H.; Lai, K.-S. Calcium Lignosulfonate Modulates Physiological and Biochemical Responses to Enhance Shoot Multiplication in Vanilla Planifolia Andrews. *Physiol. Mol. Biol. Plants* **2023**, *29*, 377–392. [CrossRef]
20. Pan, T.; Liu, M.; Kreslavski, V.D.; Zharmukhamedov, S.K.; Nie, C.; Yu, M.; Kuznetsov, V.V.; Allakhverdiev, S.I.; Shabala, S. Non-Stomatal Limitation of Photosynthesis by Soil Salinity. *Crit. Rev. Environ. Sci. Technol.* **2021**, *51*, 791–825. [CrossRef]
21. Silva, E.M.D.; Gheyi, H.R.; Nobre, R.G.; Soares, L.A.D.A.; Bonifácio, B.F. Saline Waters and Nitrogen/Potassium Fertilization Combinations on Physiological Aspects and Production of West Indian Cherry. *Rev. Ambiente Água* **2021**, *16*, 1–14. [CrossRef]
22. Silva, F.C. Manual de análises químicas de solos, plantas e fertilizantes. In *Portal Embrapa*, 2nd ed.; Embrapa: Brasília, DF, Brazil, 2009; ISBN 978-85-7383-430-7.
23. Guedes, L.R.; Cavalcante, L.F.; Souto, A.G.D.L.; Carvalho, L.H.M.; Cavalcante, Í.H.L.; Diniz Neto, M.A.; Lima, G.S.D.; Melo, T.D.S.; Henrique, J.C.G.D.S. Liquid Fertilizers on Photochemical Efficiency and Gas Exchange in Yellow Passion Fruit under Saline Stress. *Rev. Bras. Eng. Agríc. Ambient.* **2023**, *27*, 839–847. [CrossRef]
24. Beltrano, J.; Ronco, M.G. Improved Tolerance of Wheat Plants (*Triticum aestivum* L.) to Drought Stress and Rewatering by the Arbuscular Mycorrhizal Fungus Glomus Claroideum: Effect on Growth and Cell Membrane Stability. *Braz. J. Plant Physiol.* **2008**, *20*, 29–37. [CrossRef]
25. Ferreira, D.F. Sisvar: A Computer Analysis System to Fixed Effects Split Plot Type Designs: Sisvar. *Rev. Bras. Biom.* **2019**, *37*, 529–535. [CrossRef]
26. R Team. Available online: https://www.r-project.org/ (accessed on 1 July 2024).
27. Zivcak, M.; Kalaji, H.M.; Shao, H.-B.; Olsovska, K.; Brestic, M. Photosynthetic Proton and Electron Transport in Wheat Leaves under Prolonged Moderate Drought Stress. *J. Photochem. Photobiol. B Biol.* **2014**, *137*, 107–115. [CrossRef]
28. Santos, T.B.; Ribas, A.F.; De Souza, S.G.H.; Budzinski, I.G.F.; Domingues, D.S. Physiological Responses to Drought, Salinity, and Heat Stress in Plants: A Review. *Stresses* **2022**, *2*, 113–135. [CrossRef]
29. Gupta, S.; Kaur, N.; Kant, K.; Jindal, P.; Ali, A.; Naeem, M. Calcium: A Master Regulator of Stress Tolerance in Plants. *S. Afr. J. Bot.* **2023**, *163*, 580–594. [CrossRef]
30. Nawaz, F.; Shehzad, M.A.; Majeed, S.; Ahmad, K.S.; Aqib, M.; Usmani, M.M.; Shabbir, R.N. Role of Mineral Nutrition in Improving Drought and Salinity Tolerance in Field Crops. In *Agronomic Crops: Volume 3: Stress Responses and Tolerance*; Hasanuzzaman, M., Ed.; Springer: Singapore, 2020; pp. 129–147. ISBN 978-9-81150-025-1.
31. Haghighi, M.; Khosravi, S.; Sehar, S.; Shamsi, I.H. Foliar-Sprayed Calcium-Tryptophan Mediated Improvement in Physio-Biochemical Attributes and Nutritional Profile of Salt Stressed *Brassica oleracea* Var. Italica. *Sci. Hortic.* **2023**, *307*, 111529. [CrossRef]
32. Zhao, C.; Zhang, H.; Song, C.; Zhu, J.-K.; Shabala, S. Mechanisms of Plant Responses and Adaptation to Soil Salinity. *Innovation* **2020**, *1*, 100017. [CrossRef]
33. Paiva, F.J.d.S.; De Lima, G.S.; De Lima, V.L.A.; De Souza, W.B.B.; Soares, L.A.d.A.; Torres, R.A.F.; Gheyi, H.R.; Silva, L.D.A.; Sá, F.V.d.S.; De Sá, V.K.N.O.; et al. The Effects of Irrigation Water Salinity on the Synthesis of Photosynthetic Pigments, Gas Exchange, and Photochemical Efficiency of Sour Passion Fruit Genotypes. *Plants* **2023**, *12*, 3894. [CrossRef]
34. Sharma, S.; Joshi, J.; Kataria, S.; Verma, S.K.; Chatterjee, S.; Jain, M.; Pathak, K.; Rastogi, A.; Brestic, M. Chapter 27—Regulation of the Calvin Cycle under Abiotic Stresses: An Overview. In *Plant Life Under Changing Environment*; Tripathi, D.K., Pratap Singh, V., Chauhan, D.K., Sharma, S., Prasad, S.M., Dubey, N.K., Ramawat, N., Eds.; Academic Press: Cambridge, MA, USA, 2020; pp. 681–717. ISBN 978-0-12-818204-8.
35. Hameed, A.; Ahmed, M.Z.; Hussain, T.; Aziz, I.; Ahmad, N.; Gul, B.; Nielsen, B.L. Effects of Salinity Stress on Chloroplast Structure and Function. *Cells* **2021**, *10*, 2023. [CrossRef]
36. Lima, G.S.D.; Pinheiro, F.W.A.; Souza, W.B.B.D.; Soares, L.A.D.A.; Gheyi, H.R.; Nobre, R.G.; Queiroga, R.C.F.D.; Fernandes, P.D. Physiological Indices of Sour Passion Fruit under Brackish Water Irrigation Strategies and Potassium Fertilization. *Rev. Bras. Eng. Agríc. Ambient.* **2023**, *27*, 383–392. [CrossRef]
37. Lima, G.S.D.; Pinheiro, F.W.A.; Gheyi, H.R.; Soares, L.A.D.A.; Sousa, P.F.D.N.; Fernandes, P.D. Saline Water Irrigation Strategies and Potassium Fertilization on Physiology and Fruit Production of Yellow Passion Fruit. *Rev. Bras. Eng. Agríc. Ambient.* **2022**, *26*, 180–189. [CrossRef]
38. Liu, Y.-F.; Zhang, G.-X.; Qi, M.-F.; Li, T.-L. Effects of Calcium on Photosynthesis, Antioxidant System, and Chloroplast Ultrastructure in Tomato Leaves under Low Night Temperature Stress. *J. Plant Growth Regul.* **2015**, *34*, 263–273. [CrossRef]
39. Wang, Q.; Yang, S.; Wan, S.; Li, X. The Significance of Calcium in Photosynthesis. *Int. J. Mol. Sci.* **2019**, *20*, 1353. [CrossRef]
40. Bezerra, M.A.F.; Cavalcante, L.F.; Bezerra, F.T.C.; Silva, A.R.; Oliveira, F.F.; Medeiros, S.A.S. Saline Water, Pit Coating and Calcium Fertilization on Chlorophyll, Fluorescence, Gas Exchange and Production in Passion Fruit. *J. Agric. Sci.* **2019**, *11*, 319. [CrossRef]
41. Roháček, K. Chlorophyll Fluorescence Parameters: The Definitions, Photosynthetic Meaning, and Mutual Relationships. *Photosynthetica* **2002**, *40*, 13–29. [CrossRef]

42. Aftab, T.; Khan, M.M.; Idrees, M.; Naeem, M.; Hashmi, N.; Moinuddin, A.S. Effect of Salt Stress on Growth, Membrane Damage, Antioxidant Metabolism and Artemisinin Accumulation in *Artemisia annua* L. *Plant Stress* **2010**, *4*, 36–43.
43. Hniličková, H.; Hnilička, F.; Orsák, M.; Hejnák, V. Effect of Salt Stress on Growth, Electrolyte Leakage, Na^+ and K^+ Content in Selected Plant Species. *Plant Soil Environ.* **2019**, *65*, 90–96. [CrossRef]

plants

Article

Application Techniques and Concentrations of Ascorbic Acid to Reduce Saline Stress in Passion Fruit

Edmilson Júnio Medeiros Caetano [1], André Alisson Rodrigues da Silva [1,*], Geovani Soares de Lima [1,*], Carlos Alberto Vieira de Azevedo [1], Luana Lucas de Sá Almeida Veloso [1], Thiago Filipe de Lima Arruda [1], Allesson Ramos de Souza [1], Lauriane Almeida dos Anjos Soares [2], Hans Raj Gheyi [1], Mirandy dos Santos Dias [1], Lucyelly Dâmela Araújo Borborema [1], Vitória Dantas de Sousa [1] and Pedro Dantas Fernandes [1]

[1] Academic Unit of Agricultural Engineering, Federal University of Campina Grande, Campina Grande 58430-380, PB, Brazil; edmilson.junio@estudante.ufcg.edu.br (E.J.M.C.); carlos.azevedo@professor.ufcg.edu.br (C.A.V.d.A.); luana.lucas@estudante.ufcg.edu.br (L.L.d.S.A.V.); filipe@estudante.ufcg.edu.br (T.F.d.L.A.); allesson.ramos@estudante.ufcg.edu.br (A.R.d.S.); hans.gheyi@ufcg.edu.br (H.R.G.); mirandy.santos@estudante.ufcg.edu.br (M.d.S.D.); lucyelly.damela@estudante.ufcg.edu.br (L.D.A.B.); vitoria.dantas@estudante.ufcg.edu.br (V.D.d.S.); pedro.dantas@professor.ufcg.edu.br (P.D.F.)

[2] Academic Unit of Agrarian Sciences, Federal University of Campina Grande, Pombal 58840-000, PB, Brazil; lauriane.almeida@professor.edu.br

* Correspondence: andrealissonufcg16@gmail.com (A.A.R.d.S.); geovani.soares@professor.ufcg.edu.br (G.S.d.L.); Tel.: +55-83-986898061 (A.A.R.d.S.)

Abstract: Salinity restricts the growth of irrigated fruit crops in semi-arid areas, making it crucial to find ways to reduce salt stress. One effective strategy is using eliciting substances like ascorbic acid. In this context, the objective of this study was to evaluate the effects of application methods and concentrations of ascorbic acid on the morphophysiology and production of sour passion fruit irrigated with saline water. The experiment was organized using a factorial randomized block design ($3 \times 3 \times 2$) with three application methods (soaking, spraying, and soaking and spraying), three concentrations of ascorbic acid (0, 0.8, and 1.6 mM) and two levels of electrical conductivity of irrigation water—ECw (0.8 and 3.8 dS m^{-1}). Foliar spraying of ascorbic acid at a concentration of 0.8 mM mitigated the effects of salt stress on the relative water content of leaves, the synthesis of photosynthetic pigments, gas exchange, and total production of sour passion fruit when irrigated with ECw of 3.8 dS m^{-1}. Plants grown with water of 0.8 dS m^{-1} and under foliar application of 0.8 mM of ascorbic acid achieved the maximum growth in stem diameter and the greatest volume of pulp in the fruits.

Keywords: fruit growing; non-enzymatic compound; *Passiflora edulis* sims; salt stress

Citation: Caetano, E.J.M.; Silva, A.A.R.d.; Lima, G.S.d.; Azevedo, C.A.V.d.; Veloso, L.L.d.S.A.; Arruda, T.F.d.L.; Souza, A.R.d.; Soares, L.A.d.A.; Gheyi, H.R.; Dias, M.d.S.; et al. Application Techniques and Concentrations of Ascorbic Acid to Reduce Saline Stress in Passion Fruit. *Plants* **2024**, *13*, 2718. https://doi.org/10.3390/plants13192718

Academic Editor: Pedro Diaz-Vivancos

Received: 11 July 2024
Revised: 12 September 2024
Accepted: 26 September 2024
Published: 28 September 2024

1. Introduction

Belonging to the Passifloraceae family, sour passion fruit (*Passiflora edulis* Sims) is a fruit crop grown in tropical and subtropical countries. In 2022, Brazil produced 697,859 tons in an area of 45,602 hectares, with the Northeast having the largest planted area, however, the average productivity in this region is only 14,765 kg per hectare, which is lower than the yield in the southern region of the country (21,290 kg per hectare) [1].

From an economic and commercial point of view, the cultivation of sour passion fruit plays a role of considerable importance in Brazil, since this species has a significant impact on the generation of employment and income throughout the year, especially for small- and medium-sized producers living in the semi-arid regions of Brazil [2].

The occurrence of irregular rainfall throughout the year, combined with intense evaporation, favors the accumulation of salts in the water sources of this region. This increase in salinity generates negative impacts on plants, leading to restrictions in the absorption of water and nutrients due to osmotic and ionic stress, thus affecting physiological processes, growth, and production components [3,4].

Using water with high concentrations of salts in irrigation can impair the metabolic and biochemical functions of salt stress-sensitive plants such as sour passion fruit, resulting in the suppression of their production potential through the partial closure of stomata, reductions in photosynthesis, protein synthesis and enzymatic activities, and the degradation of photosynthetic pigments [5]. Salinity also has the potential to modify electron transport, affecting the activity of photosystem II, which plays a crucial role in the oxidation of water molecules to generate electrons [6], resulting in changes in oxidative homeostasis.

Recent studies have shown that irrigation with saline water can negatively affect passion fruit cultivation; for example, Lima et al. [7] highlighted that irrigation with water of ECw of 0.3 dS m^{-1} or higher reduces the relative water content, gas exchange, number of fruits, and production per plant, and increases electrolyte leakage in the leaf blades of passion fruit plants. Galvão Sobrinho et al. [8] evaluated passion fruit production and obtained a threshold water salinity of 0.8 dS m^{-1}, with a reduction of 14.78% per unit increase in ECw above this level. Andrade et al. [9] highlighted that water salinity above 0.7 dS m^{-1} reduces the physical and chemical quality of passion fruit.

Several strategies have been adopted to reduce the effects of salt stress on plants, including the application of ascorbic acid (AsA) [10]. AsA can reduce the excessive generation of reactive oxygen species (ROS) by collaborating with the enzyme ascorbate peroxidase (APX), aiding plant acclimatization during exposure to salinity [11]. AsA also acts to protect proteins and lipids from oxidative stressors such as salinity and drought. This compound contributes to plant growth, photosynthesis rate, transpiration, oxidative defense potential, and photochemical pigments, which can improve tolerance to abiotic stresses [10,12]

In the literature, there are reports on the application of AsA as an alternative to reduce the adverse effects of salt stress on different crops such as radish [13], cowpea [14], barley [15], and sugar beet [16]. However, studies addressing the effects of application methods and concentrations of AsA in fruit species such as sour passion fruit subjected to irrigation with brackish water under semi-arid conditions are incipient.

Considering the importance of this fruit crop in the social and economic context of agribusiness, it is essential to develop strategies capable of mitigating the effects of salt stress on plants and enabling the expansion of irrigated agriculture in semi-arid regions. In this context, this study is based on the hypothesis that the deleterious effects caused by salt stress on the growth, physiology, and production of passion fruit can be mitigated by the use of AsA through the concentration and application method most appropriate for the crop.

In this context, the objective of this study was to evaluate the effects of application methods and concentrations of AsA on the morphophysiology and production of sour passion fruit irrigated with saline waters in a semi-arid area.

2. Results

The original multidimensional space of variables was condensed into two principal components (PC1 and PC2) with eigenvalues of $\lambda \geq 1.0$ [17]. These components, as detailed in Table 1, accounted for 86.1% of the total variance. PC1, which explained 67.52% of the variance, included most of the variables except for average fruit weight (AFW), fruit polar diameter (FPD), and fruit equatorial diameter (FED). PC2 accounted for 18.58% of the remaining variance and comprised the variables AFW, FPD, and FED.

The interaction between application methods, AsA concentrations, and electrical conductivity levels of irrigation water (ECw) had a significant effect ($p \leq 0.05$) on the two principal components (Table 1). Additionally, each individual factor significantly affected all variables ($p \leq 0.01$).

Table 1. Eigenvalues, the percentage of total variance explained in the multivariate analysis of variance, and the correlation coefficients (r) between the original variables and the principal components.

	Principal Components (PCs)	
	PC1	**PC2**
Eigenvalues (λ)	10.80	2.97
Percentage of total variance (S^2%)	67.52	18.58
Hotelling test (T^2) for interaction (ECw × M)	0.01	0.01
Hotelling test (T^2) for interaction (ECw × C)	0.01	0.01
Hotelling test (T^2) for interaction (M × C)	0.01	0.01
Hotelling test (T^2) for interaction (ECw × M × C)	0.03	0.05

The analysis of the correlations between the variables (Table 2) shows that the principal component 1 was affected by the following variables: relative water content (RWC), electrolyte leakage (EL), photosynthetic pigments (Chl *a*, Chl *b*, Chl *t* and Car), internal CO_2 concentration (*Ci*), CO_2 assimilation rate, transpiration (*E*), stomatal conductance (*gs*), instantaneous carboxylation efficiency (*CEi*), number of fruits per plant (NFP), and total production per plant (TPP), with correlation coefficients greater than 0.80. On the other hand, fruit polar diameter (FPD), fruit equatorial diameter (FED), and average fruit weight (AFW) showed a significant effect for principal component 2, with correlation coefficients higher than 0.85.

Table 2. Correlation coefficients (r) between the original variables and the principal components.

PCs	**Correlation Coefficients (r)**							
	RWC	**EL**	**Chl *a***	**Chl *b***	**Chl *t***	**Car**	***Ci***	***A***
PC1	−0.929 *	0.909 *	−0.939 *	−0.937 *	−0.938 *	−0.966 *	0.819 *	−0.947 *
PC2	0.187	−0.308	−0.146	−0.147	−0.149	−0.068	0.417	0.139
PCs	**Correlation Coefficients (r)**							
	E	***gs***	***CEi***	**FPD**	**FED**	**AFW**	**NFP**	**TPP**
PC1	−0.928 *	−0.929 *	−0.937 *	0.219	−0.209	−0.169	−0.809 *	−0.849 *
PC2	0.098	0.078	0.219	−0.898 *	−0.859 *	−0.958 *	−0.119	−0.169

PC1—Principal component 1; PC2—Principal Component 2. *—Variable with Pearson's correlation coefficient ($r \geq 0.80$) considered for PC. Correlation coefficient determined considering the three repetitions (n = 3).

The two-dimensional projections showing the effects of treatments and variables on the first and second principal components are illustrated in Figure 1A,B. The first principal component highlights a process potentially influenced by the interaction between application methods, AsA concentrations, and ECw levels.

When analyzing PC1, it was observed that plants grown under ECw of 0.8 dS m^{-1}, with AsA application by the spraying method and at concentration of 0.8 mM (S1M2C2), stood out with the highest values (Table 3) of RWC (89.51%), Chl *a* (503.33 μg mL^{-1}), Chl *b* (302 μg mL^{-1}), Chl *t* (805.32 μg mL^{-1}), Car (105.70 μg mL^{-1}), *A* (6.18 μmol CO_2 m^{-2} s^{-1}), *E* (1.27 mmol H_2O m^{-2} s^{-1}), *gs* (0.77 mol H_2O m^{-2} s^{-1}), *CEi* [(0.045 μmol CO_2 m^{-2} s^{-1}) (μmol CO_2 m^{-2} s^{-1})$^{-1}$], NFP (15 fruits per plant), and TPP (2011.33 g per plant).

In principal component 2 (PC2), it was observed that plants under ECw of 0.8 dS m^{-1}, with AsA application by the spraying method and at concentration of 1.6 mM (S1M2C3), obtained the maximum values (Table 3) of FPD (78.95 mm), FED (69.95 mm), and AFW (159.5 g per fruit).

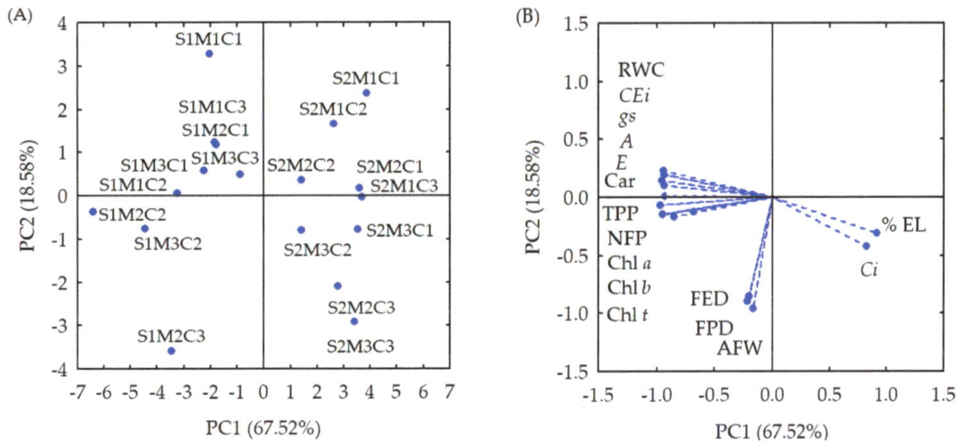

Figure 1. Plots of first two principal components for the three factors interactions (**A**) and analyzed variables (**B**) in sour passion fruit. S—Electrical conductivity of irrigation water (ECw), S1 (0.8 dS m^{-1}); S2 (3.8 dS m^{-1}); M—Application methods, M1 (Soaking); M2 (Spraying); M3 (Spraying + Soaking); C—Ascorbic acid concentration, C1 (0 mM); C2 (0.8 mM); C3 (1.6 mM); Chl *a* (Chlorophyll *a*—μg mL^{-1}); Chl *b* (Chlorophyll *b*—μg mL^{-1}); Chl *t* (Total chlorophyll—μg mL^{-1}); Car (Carotenoids—μg mL^{-1}); RWC (relative water content—%); % EL (percentage of electrolyte leakage); Ci (internal CO_2 concentration—μmol CO_2 m^{-2} s^{-1}); A (CO_2 assimilation rate—μmol CO_2 m^{-2} s^{-1}); E (perspiration—mmol H_2O m^{-2} s^{-1}); gs (stomatal conductance—mol H2O m^{-2} s^{-1}); CEi (instantaneous carboxylation efficiency—[(μmol CO_2 m^{-2} s^{-1}) (μmol CO_2 m^{-2} s^{-1})$^{-1}$]. FPD (Fruit polar diameter—mm); FED (Fruit equatorial diameter—mm); AFW (Average fruit weight—g per fruit); NFP (Number of fruits per plant); TPP (Total production per plant—g per plant).

When comparing the results between plants subjected to the S1M2C2 and S1M2C3 treatments with those cultivated with the S1M2C1 treatment, significant increments of 54.01% (176.51 μg mL^{-1}), 54.01% (105.91 μg mL^{-1}), 54.01% (282.4 μg mL^{-1}), 28.57% (30.2 μg mL^{-1}), 3.2%, 16.82% (0.89 μmol CO_2 m^{-2} s^{-1}), 28.30% (0.28 mmol H_2O m^{-2} s^{-1}), 63.82% (0.03 mol H_2O m^{-2} s^{-1}) and 40.63% [(0.013 μmol CO_2 m^{-2} s^{-1}) (μmol CO_2 mol^{-1})$^{-1}$], 45.21% (4.67 fruits per plant), 57.83% (737.03 g per plant), 15.25% (10.45 mm), 27.90% (15.26 mm), and 78.21% (70.0 g per fruit) were observed for Chl *a*, Chl *b*, Chl *t*, Car, RWC, *A*, *E*, *gs*, *CEi*, NFP, TPP, FPD, FED, and AFW, respectively. In addition, there were reductions of 4.14% (11.14) in electrolyte leakage and 16.46% (27 μmol CO_2 m^{-2} s^{-1}) in internal CO_2 concentration.

In this study, it was also observed that the increase in the electrical conductivity of irrigation water impaired the gas exchange, relative water content, contents of photosynthetic pigments, and production of sour passion fruit. When analyzing PC1, it was observed that sour passion fruit plants grown under ECw of 3.8 dS m^{-1}, without application of ascorbic acid (S2M1C1), stood out with the lowest values (Table 3) of RWC (67.58%), *A* (3.18 μmol CO_2 m^{-2} s^{-1}), *E* (0.67 mmol H_2O m^{-2} s^{-1}), *gs* (0.020 mol H_2O m^{-2} s^{-1}), *CEi* (0.016 [(μmol CO_2 m^{-2} s^{-1}) (μmol CO_2 m^{-2} s^{-1}]$^{-1}$), FPD (63.12 mm), FED (34.64 mm), and AFW (67.44 g per fruit). Lower values of Chl *a* (228.78 μg mL^{-1}), Chl *b* (137.27 μg mL^{-1}), Chl *t* (366.04 μg mL^{-1}), Car (48.04 μg mL^{-1}), and NFP (four fruits) were observed in the S2M2C1 treatment. The lowest TPP values (442.33 g per plant) were obtained in the S2M3C1 treatment.

Table 3. Mean values of variables under the effect of studied treatments.

Treatments	Variables							
	RWC	EL	Chl *a*	Chl *b*	Chl *t*	Car	*Ci*	*A*
S1M1C1	86.48 ± 0.5 a	36.11 ± 1.8 a	341.11 ± 11.8 a	204.67 ± 11.8 a	545.77 ± 9.8 a	78.80 ± 1.8 a	160 ± 1.8 a	5.12 ± 0.9 a
S1M2C1	86.31 ± 0.9 a	37.15 ± 1.5 a	326.82 ± 13.2 a	196.09 ± 5.8 a	522.92 ± 8.7 a	75.50 ± 1.6 a	164 ± 2.9 a	5.29 ± 1.3 a
S1M3C1	83.45 ± 1.1 a	35.94 ± 1.8 a	349.25 ± 12.5 a	209.55 ± 9.2 a	558.80 ± 6.8 a	80.68 ± 2.4 a	173 ± 3.1 a	5.32 ± 1.1 a
S1M1C2	79.66 ± 1.1 a	35.15 ± 1.6 a	402.66 ± 22.6 a	241.60 ± 12.8 a	644.26 ± 7.9 a	84.56 ± 3.1 a	144 ± 4.2 a	5.21 ± 0.9 a
S1M2C2	89.51 ± 2.2 a	33.01 ± 2.1 a	503.33 ± 30.1 a	302.00 ± 14.8 a	805.32 ± 9.8 a	105.70 ± 2.8 a	137 ± 4.9 a	6.18 ± 1.2 a
S1M3C2	83.89 ± 1.4 a	35.54 ± 1.2 a	427.83 ± 26.3 a	256.70 ± 19.5 a	684.53 ± 7.8 a	89.84 ± 3.9 a	146 ± 3.8 a	5.87 ± 1.1 a
S1M1C3	83.50 ± 1.4 a	40.05 ± 5.1 a	342.26 ± 22.1 a	205.36 ± 22.8 a	547.62 ± 9.2 a	71.88 ± 4.1 a	156 ± 7.1 a	5.35 ± 0.8 a
S1M2C3	83.28 ± 0.9 a	41.36 ± 4.3 a	427.83 ± 14.4 a	256.70 ± 29.9 a	684.53 ± 8.6 a	89.84 ± 5.7 a	191 ± 6.4 a	5.01 ± 1.5 a
S1M3C3	82.69 ± 0.8 a	41.72 ± 4.3 a	363.65 ± 13.7 a	218.19 ± 14.7 a	581.85 ± 6.5 a	76.37 ± 3.2 a	196 ± 5.7 a	4.66 ± 0.8 a
S2M1C1	67.58 ± 2.6 a	48.82 ± 6.8 a	238.78 ± 14.7 a	143.27 ± 19.7 a	382.04 ± 6.5 a	50.14 ± 4.5 a	198 ± 4.9 a	3.18 ± 1.2 a
S2M2C1	70.77 ± 1.3 a	50.90 ± 4.2 a	228.78 ± 21.0 a	137.27 ± 14.4 a	366.04 ± 7.1 a	48.04 ± 3.8 a	210 ± 4.3 a	3.68 ± 1.3 a
S2M3C1	68.43 ± 1.1 a	51.11 ± 3.2 a	244.47 ± 19.7 a	146.68 ± 10.2 a	391.16 ± 8.6 a	51.34 ± 4.2 a	227 ± 3.5 a	3.60 ± 1.2 a
S2M1C2	69.54 ± 1.7 a	51.43 ± 1.8 a	281.86 ± 13.7 a	169.12 ± 21.9 a	450.98 ± 8.7 a	59.19 ± 3.1 a	228 ± 9.8 a	4.29 ± 0.9 a
S2M2C2	73.40 ± 1.2 a	52.55 ± 1.7 a	352.33 ± 14.4 a	211.40 ± 24.5 a	563.73 ± 7.6 a	73.99 ± 8.6 a	233 ± 9.7 a	3.69 ± 0.8 a
S2M3C2	69.81 ± 2.2 a	52.98 ± 1.5 a	299.48 ± 16.8 a	179.69 ± 19.6 a	479.17 ± 4.9 a	62.89 ± 9.1 a	257 ± 8.7 a	4.43 ± 0.7 a
S2M1C3	68.47 ± 2.7 a	53.36 ± 1.7 a	239.58 ± 17.9 a	143.75 ± 14.5 a	383.33 ± 3.9 a	50.31 ± 2.4 a	259 ± 6.5 a	3.55 ± 0.7 a
S2M2C3	68.29 ± 2.1 a	55.67 ± 1.6 a	299.48 ± 18.5 a	179.69 ± 17.9 a	479.17 ± 7.8 a	62.89 ± 3.9 a	283 ± 5.4 a	3.41 ± 0.9 a
S2M3C3	67.81 ± 1.8 a	58.56 ± 2.1 a	254.56 ± 16.8 a	152.73 ± 17.1 a	407.29 ± 9.8 a	53.46 ± 8.1 a	288 ± 6.1 a	3.21 ± 0.5 a
Treatments	**Variables**							
	E	*gs*	*CEi*	FPD	FED	AFW	NFP	TPP
S1M1C1	0.96 ± 0.3 a	0.057 ± 0.02 a	0.032 ± 0.01 a	63.97 ± 1.7 a	41.31 ± 2.4 a	67.25 ± 3.1 a	11.33 ± 0.7 a	1500.0 ± 32.2 a
S1M2C1	0.99 ± 0.2 a	0.047 ± 0.06 a	0.032 ± 0.01 a	68.50 ± 1.8 a	54.69 ± 3.4 a	89.50 ± 3.4 a	10.33 ±0.9 a	1274.3 ± 54.1 a
S1M3C1	0.96 ± 0.3 a	0.053 ± 0.04 a	0.031 ± 0.02 a	67.96 ± 1.5 a	56.82 ± 2.3 a	102.17 ± 3.4 a	12.00 ± 1.2 a	1396.7 ± 38.9 a
S1M1C2	1.02 ± 0.2 a	0.070 ± 0.02 a	0.036 ± 0.03 a	72.53 ± 1.1 a	58.58 ± 2.2 a	102.33 ± 2.9 a	8.67 ± 0.9 a	1272.2 ± 41.4 a
S1M2C2	1.27 ± 0.1 a	0.077 ± 0.01 a	0.045 ± 0.02 a	75.90 ± 1.1 a	52.36 ± 3.1 a	120.13 ± 5.1 a	15.00 ± 1.1a	2011.3 ± 81.9 a
S1M3C2	1.13 ± 0.1 a	0.067 ± 0.02 a	0.040 ± 0.01 a	74.74 ± 1.4 a	62.26 ± 4.9 a	118.33 ± 3.4a	9.67 ± 0.5 a	1418.3 ± 56.4 a
S1M1C3	0.93 ± 0.3 a	0.050 ± 0.02 a	0.034 ± 0.02 a	67.57 ± 2.3 a	55.97 ± 3.7 a	84.17 ± 2.8 a	11.67 ± 0.3 a	1284.0 ± 46.1 a
S1M2C3	0.88 ± 0.4 a	0.047 ± 0.03 a	0.026 ± 0.01 a	78.95 ± 1.2 a	69.95 ± 2.7 a	159.50 ± 3.2 a	8.33 ± 0.4 a	1409.0 ± 57.9 a
S1M3C3	0.89 ± 0.3 a	0.043 ± 0.02 a	0.024 ± 0.01 a	72.24 ± 1.8 a	43.23 ± 3.7 a	102.00 ± 4.1a	7.00 ± 0.8 a	1218.0 ± 68.7 a
S2M1C1	0.67 ± 0.5 a	0.020 ± 0.02 a	0.016 ± 0.02 a	63.12 ± 2.1 a	34.64 ± 4.8 a	67.44 ± 2.7 a	5.67 ± 0.2 a	567.7 ± 29.8 a
S2M2C1	0.72 ± 0.3 a	0.023 ± 0.03 a	0.018 ± 0.01 a	72.18 ± 1.9 a	49.77 ± 3.4 a	105.00 ± 3.1 a	4.00 ± 0.4 a	475.3 ± 29.5 a
S2M3C1	0.72 ± 0.4 a	0.021 ± 0.02 a	0.016 ± 0.01 a	72.32 ± 1.6 a	59.15 ± 2.2 a	114.00 ± 2.9a	4.67 ± 0.6 a	442.3 ± 31.8 a
S2M1C2	0.77 ± 0.2 a	0.037 ± 0.02 a	0.019 ± 0.02 a	68.07 ± 2.1 a	39.18 ± 3.1 a	76.50 ± 4.2 a	4.00 ± 0.7 a	456.0 ± 41.2 a
S2M2C2	0.95 ± 0.8 a	0.032 ± 0.01 a	0.016 ± 0.01 a	67.95 ± 1.8 a	45.55 ± 2.4 a	104.50 ± 3.7 a	4.00 ± 0.8 a	477.3 ± 35.6 a
S2M3C2	0.85 ± 0.6 a	0.037 ± 0.02 a	0.017 ± 0.02 a	69.30 ± 2.2 a	58.05 ± 1.8 a	110.00 ± 2.8 a	11.00 ± 1.2 a	1085.3 ± 58.9 a
S2M1C3	0.70 ± 0.2 a	0.027 ± 0.01 a	0.014 ± 0.01 a	68.06 ± 3.4 a	50.36 ± 3.2 a	109.00 ± 3.1 a	6.00 ± 0.9 a	824.0 ± 54.6 a
S2M2C3	0.72 ± 0.1 a	0.040 ± 0.02 a	0.012 ± 0.01 a	76.05 ± 2.1 a	60.07 ± 1.4 a	119.00 ± 2.4 a	4.00 ± 1.1 a	745.0 ± 41.6 a
S2M3C3	0.71 ± 0.3 a	0.035 ± 0.03 a	0.011 ± 0.01 a	77.49 ± 2.1 a	64.70 ± 1.9 a	133.88 ± 5.8 a	8.00 ± 1.4 a	845.7 ± 31.4 a

S—Electrical conductivity of irrigation water (ECw), S1 (0.8 dS m^{-1}); S2 (3.8 dS m^{-1}); M—Application methods, M1 (Soaking); M2 (spraying); M3 (Spraying + Soaking); C—Ascorbic acid concentration, C1 (0 mM); C2 (0.8 mM); C3 (1.6 mM); RWC (relative water content—%); EL (percentage of electrolyte leakage); Chl *a* (Chlorophyll *a*—μg mL^{-1}); Chl *b* (Chlorophyll *b*—μg mL^{-1}); Chl *t* (Chlorophyll total—μg mL^{-1}); Car (Carotenoids—μg mL^{-1}); Ci (internal CO_2 concentration—μmol CO_2 m^{-2} s^{-1}); A (CO_2 assimilation rate—μmol CO_2 m^{-2} s^{-1}); E (transpiration—mmol H_2O m^{-2} s^{-1}); gs (stomatal conductance—mol H2O m^{-2} s^{-1}); CEi (instantaneous carboxylation efficiency—[(μmol CO_2 m^{-2} s^{-1}) (μmol CO_2 m^{-2} s^{-1})$^{-1}$]. FPD (Fruit polar diameter—mm); FED (Fruit equatorial diameter—mm); AFW (Average fruit weight—g per fruit); NFP (Number of fruits per plant); TPP (Total production per plant—g per plant). Averages are followed by standard deviation of the mean (n = 3); means followed by the same letter do not indicate significant differences between treatments according to the Scott–Knott test ($p > 0.05$).

There was a significant effect of the interaction between ascorbic acid (AsA) concentrations and water salinity (ECw) levels only on relative water content (RWC) ($p \leq 0.01$) (Table 4). No significant effect ($p > 0.05$) of the M × AsA × ECw interaction was observed on any of the variables analyzed in sour passion fruit. In isolation, irrigation water salinity levels significantly influenced ($p \leq 0.01$) electrolyte leakage, chlorophyll *a*, total chlorophyll, and carotenoids. AsA concentrations showed a significant effect ($p \leq 0.01$) on Chl *a* and Chl *b* contents. Chlorophyll *b* was also significantly affected ($p \leq 0.01$) by the application methods.

The relative water content of passion fruit plants differed significantly between water salinity levels (Figure 2A). In passion fruit plants grown under ECw of 0.8 dS m^{-1}, the application of AsA at a concentration of 0 mM promoted greater RWC (87.48%) compared to that observed in plants grown under 0.8 and 1.6 mM AsA. However, in plants irrigated with ECw of 3.8 dS m^{-1}, the application of AsA at concentrations of 0.8 and 1.6 mM promoted an increase in RWC in relation to plants subjected to 0 mM concentration. There was also no significant difference between the concentrations of 0.8 and 1.6 mM AsA. The salinity of the irrigation water significantly increased the electrolyte extravasation of sour passion fruit plants grown under ECw of 3.8 dS m^{-1} compared to those that received 0.8 dS m^{-1} (Figure 2B). Plants subjected to irrigation with water of 3.8 dS m^{-1} showed an increase of 15.48% compared to those that received water of 0.8 dS m^{-1}.

Table 4. The analysis of variance for relative water content (RWC), electrolyte leakage (EL), chlorophyll *a* (Chl *a*), chlorophyll *b* (Chl *b*), total chlorophyll (Chl *t*), and carotenoids (Car) at 150 days after transplanting, of sour passion fruit grown under irrigation with water of different levels of electrical conductivity, application methods, and concentrations of ascorbic acid.

Source of Variation	DF	Mean Squares					
		RWC	EL	Chl *a*	Chl *b*	Chl *t*	Car
Application methods (M)	2	1.99 ^{ns}	34.67 ^{ns}	26,737.19 ^{ns}	31,952.81 **	90,892.32 ^{ns}	791.90 ^{ns}
Ascorbic acid (AsA)	2	0.75 ^{ns}	12.72 ^{ns}	5039.03 **	54,984.37 **	77,516.81 ^{ns}	423.37 ^{ns}
Electrical conductivity (ECw)	1	7.05 ^{ns}	2904.00 **	65,022.94 **	2331.82 ^{ns}	42,727.78 **	245.37 **
Interaction (M × AsA × ECw)	4	3.29 ^{ns}	31.69 ^{ns}	15,780.78 ^{ns}	18,571.82 ^{ns}	31,527.80 ^{ns}	538.27 ^{ns}
Interaction (M × AsA)	4	13.61 ^{ns}	27.64 ^{ns}	13,309.78 ^{ns}	21,702.56 ^{ns}	19,106.71 ^{ns}	2565.72 ^{ns}
Interaction (M × ECw)	2	35.77 ^{ns}	20.22 ^{ns}	5038.72 ^{ns}	1607.52 ^{ns}	2205.72 ^{ns}	452.72 ^{ns}
Interaction (AsA × ECw)	2	339.09 **	15.05 ^{ns}	87,585.05 ^{ns}	1268.05 ^{ns}	29,568.05 ^{ns}	421.05 ^{ns}
Blocks	2	26.70 ^{ns}	44.05 ^{ns}	481.17 ^{ns}	217.35 ^{ns}	284.74 ^{ns}	38.27 ^{ns}
Residual	34	15.11	51.25	856.39	1432.39	2910.96	27.65
CV (%)		4.68	15.75	12.67	18.32	12.33	10.51

^{ns,} ** respectively non-significant and significant at $p \leq 0.01$. CV—coefficient of variation, DF—degrees of freedom.

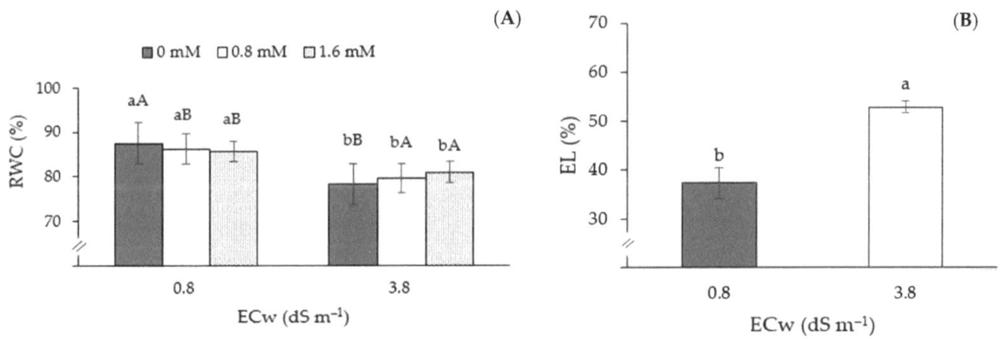

Figure 2. Relative water content of sour passion fruit as a function of the interaction between AsA concentrations and ECw levels (**A**), and electrolyte leakage from sour passion fruit plants as a function of irrigation water salinity (**B**), 150 days after transplanting. Different letters indicate significant differences between AsA concentrations for each electrical conductivity of irrigation water, and different uppercase letters indicate significant differences in AsA concentrations between electrical conductivities of irrigation water according to the Tukey test. ($p \leq 0.05$). The error bars represent standard error (n = 3).

Chlorophyll *a*, total chlorophyll, and carotenoid contents were affected negatively when passion fruit plants were subjected to irrigation with ECw of 3.8 dS m^{-1} (Figure 3A,E,F), with reductions of 30.0, 29.0, and 32.0% in Chl *a*, Chl *t*, and Car, respectively, compared to plants irrigated with 0.8 dS m^{-1} water. Plants that received AsA application at a concentration of 0.8 mM presented the highest values of Chl *a* (377.92 µg mL^{-1}) and Chl *b* (226.75 µg mL^{-1}), differing statistically from plants that received concentrations of 0 and 0.8 mM (Figure 3B,D). It was also observed that the application of AsA through spraying promoted the highest values of Chl *b* (213.86 µg mL^{-1}), standing out in relation to the other application methods (Figure 3C).

The gas exchange variables of passion fruit were not significantly influenced by the interactions between AsA application methods, AsA concentrations, and irrigation water conductivity levels (Table 5). However, ECw levels significantly affected ($p \leq 0.01$) all gas exchange variables. Internal CO_2 concentration and stomatal conductance were also influenced ($p \leq 0.01$) by AsA concentrations (Table 5).

The internal CO_2 concentration increased significantly as a function of irrigation with water, with a CEa of 3.8 dS m^{-1} (Figure 4A). Plants irrigated with the highest CEa (3.8 dS m^{-1}) showed an increase of 48.81% (79.56 µmol CO_2 m^{-2} s^{-1}) compared to those subjected to the lowest CEa (0.8 dS m^{-1}). On the other hand, the CO_2 assimilation rate

(Figure 3B), transpiration (Figure 3C), stomatal conductance (Figure 3D), and instantaneous carboxylation efficiency (Figure 3F) were reduced by irrigation with ECw of 3.8 dS m^{-1}, with reductions of 31.2% (1.66 µmol CO_2 m^{-2} s^{-1}), 24.6% (0.247 mmol H_2O m^{-2} s^{-1}), 46.8% (0.027 mol H_2O m^{-2} s^{-1}), and 53.7% (0.018 [(µmol CO_2 m^{-2} s^{-1}) (µmol CO_2 m^{-2} s^{-1}]$^{-1}$) in *A*, *E*, *gs*, and *CEi*, respectively, when compared to plants irrigated with ECw of 0.8 dS m^{-1}. It is also observed that the stomatal conductance of passion fruit plants was also influenced by AsA concentrations (Figure 3E), with an emphasis on the concentration of 0.8 mM, which promoted the highest gs value (0.053 mol H_2O m^{-2} s^{-1}), differing statistically from the other concentrations (0 and 1.6 mM).

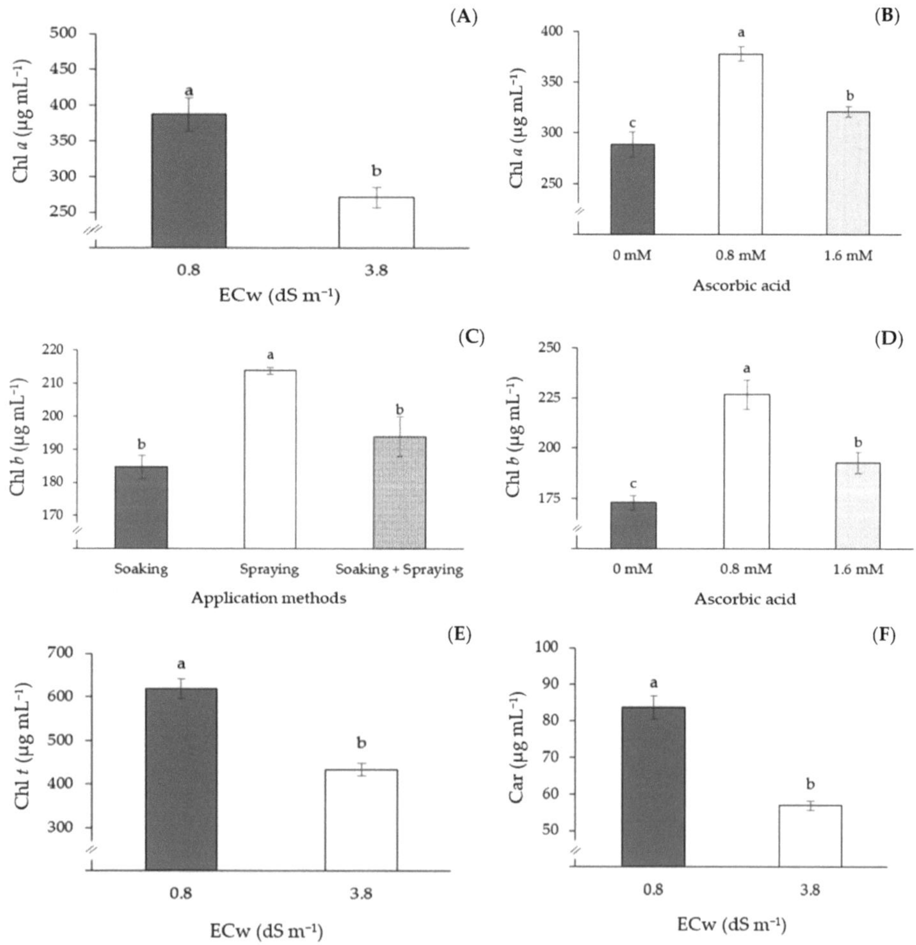

Figure 3. Chlorophyll *a*—Chl *a* (**A**), total chlorophyll—Chl *t* (**E**), and carotenoids—Car (**F**) depending on the electrical conductivity levels of the water (ECw); chlorophyll *a*—Chl *a* (**B**) and chlorophyll *b*—Chl *b* (**D**) depending on AsA concentrations, and Chl *b* of the sour passion fruit plant as a function of the application methods (**C**), 150 days after transplanting. Means followed by the same letter did not differ statistically from each other according to the Tukey test ($p \leq 0.05$). The error bars represent standard error (n = 3).

Table 5. The analysis of variance for internal CO_2 concentration (*Ci*), CO_2 assimilation rate (*A*), transpiration (*E*), stomatal conductance (*gs*), and instantaneous carboxylation efficiency (*CEi*), at 150 days after transplanting, of sour passion fruit grown under irrigation with waters of different electrical conductivity levels, application methods, and ascorbic acid concentrations.

Source of Variation	DF	Mean Squares				
		Ci	*A*	*E*	*gs*	*CEi*
Application methods (M)	2	2521.13 ^{ns}	1.0 ^{ns}	0.16 ^{ns}	0.0003 ^{ns}	0.0002 ^{ns}
Ascorbic acid (AsA)	2	17,420.35 **	2.76 ^{ns}	0.05 ^{ns}	0.01 **	0.0002 ^{ns}
Electrical conductivity (ECw)	1	34,858.96 **	49.06**	1.29 **	0.007 **	0.004 **
Interaction (M × AsA × ECw)	4	3486.60 ^{ns}	0.37 ^{ns}	0.01 ^{ns}	0.0001 ^{ns}	0.0002 ^{ns}
Interaction (M × AsA)	4	1453.51 ^{ns}	3.41 ^{ns}	0.11 ^{ns}	0.0001 ^{ns}	0.0002 ^{ns}
Interaction (M × ECw)	2	3141.51 ^{ns}	1.01 ^{ns}	0.04 ^{ns}	0.0002 ^{ns}	0.0002 ^{ns}
Interaction (AsA × ECw)	2	3486.51 ^{ns}	3.66 ^{ns}	0.01 ^{ns}	0.0002 ^{ns}	0.0002 ^{ns}
Blocks	2	41.51 ^{ns}	2.33 ^{ns}	0.01 ^{ns}	0.00007 ^{ns}	0.0001 ^{ns}
Residual	34	471.34	1.89	0.02	0.0001	0.00001
CV (%)		10.72	21.31	18.86	17.25	19.47

$^{ns,\ **}$ respectively non-significant and significant at $p \leq 0.01$. CV—coefficient of variation, DF—degrees of freedom.

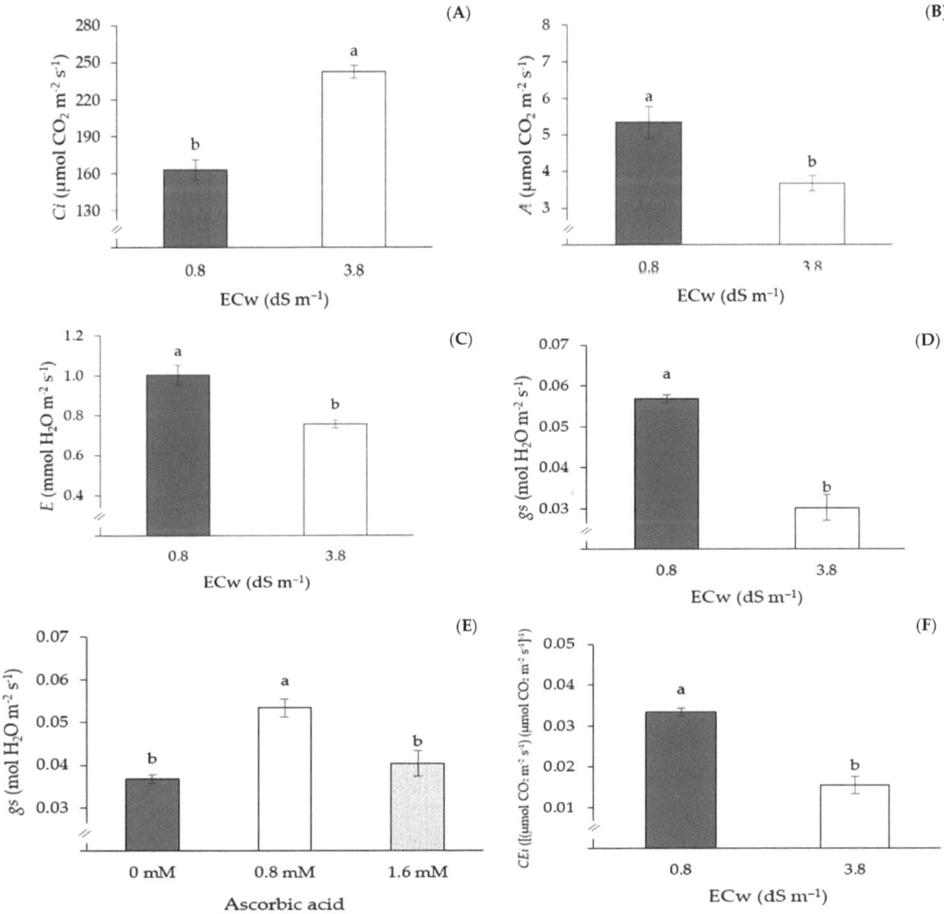

Figure 4. Internal CO_2 concentration—*Ci* (**A**), CO_2 assimilation rate—*A* (**B**), transpiration—*E* (**C**), stomatal conductance—*gs* (**D**), and instantaneous carboxylation efficiency—*CEi* (**F**) as a function of water electrical conductivity levels (ECw), and *gs* (**E**) as a function of AsA concentrations, 150 days after transplantation. Means followed by the same letter did not differ statistically from each other according to the Tukey test ($p \leq 0.05$). The error bars represent standard error (n = 3).

There were significant effects from the interaction between ascorbic acid (AsA) concentrations and water salinity levels (ECw) only for on-stem diameter (SD) ($p \leq 0.05$) (Table 6). There was no significant effect ($p > 0.05$) of the M × AsA × ECw interaction on any of the variables analyzed in sour passion fruit. On the other hand, the salinity levels of the irrigation water significantly influenced all the chlorophyll *a* fluorescence variables.

Table 6. The analysis of variance for initial fluorescence (F_0), variable fluorescence (Fv), quantum efficiency of photosystem II—(Fv/Fm), maximum fluorescence (Fm), and stem diameter (SD), at 150 days after transplanting, of sour passion fruit grown under irrigation with water of different levels of electrical conductivity, application methods, and concentrations of ascorbic acid.

Source of Variation	DF	Mean Squares				
		F_0	Fv	Fv/Fm	Fm	SD
Application methods (M)	2	238.9 ^{ns}	6414.7 ^{ns}	0.000088 ^{ns}	6191.90 ^{ns}	0.348 ^{ns}
Ascorbic acid (AsA)	2	1629 ^{ns}	24,939 ^{ns}	0.000089 ^{ns}	75,752.7 **	4.870 **
Electrical conductivity (ECw)	1	9414 **	42,056 *	0.001645 *	158,437 **	30.31 **
Interaction (M × AsA × ECw)	4	1027 ^{ns}	43,150 ^{ns}	0.000664 ^{ns}	2238.27 ^{ns}	0.905 ^{ns}
Interaction (M × AsA)	4	514.4 ^{ns}	2198.4 ^{ns}	0.00019 ^{ns}	19,700.7 **	1.308 ^{ns}
Interaction (M × ECw)	2	1094 ^{ns}	3404.7 ^{ns}	0.000252 ^{ns}	8583.72 ^{ns}	1.054 ^{ns}
Interaction (AsA × ECw)	2	2086 ^{ns}	3230.2 ^{ns}	0.0000006 ^{ns}	11,268.05 ^{ns}	1.499 *
Blocks	2	225.6	11,128	0.00083	2238.27	0.082
Residual	34	601.7	6975.09	0.00040	3343.65	0.375
CV (%)		5.10	5.70	2.58	3.05	4.02

^{ns}, *, ** respectively non-significant and significant at $p \leq 0.05$ and $p \leq 0.01$. CV—coefficient of variation, DF—degrees of freedom.

The salinity of irrigation water significantly increased the initial fluorescence of sour passion fruit plants grown under ECw of 3.8, compared to those that received 0.8 dS m^{-1} (Figure 2A). Plants subjected to irrigation with water of 3.8 dS m^{-1} showed a 5.64% increase in F_0 (26.4) compared to those that received water of 0.8 dS m^{-1}. Different from the result observed in the initial fluorescence (Figure 5A), sour passion fruit plants irrigated with ECw of 3.8 dS m^{-1} showed a significant reduction in their variable fluorescence compared to those irrigated with a salinity level of 0.8 dS m^{-1} (Figure 5B). When comparing the Fv of plants grown under ECw of 3.8 dS m^{-1} to that of plants subjected to water salinity of 0.8 dS m^{-1}, a reduction of 5.36% was observed (80.11).

Quantum efficiency of photosystem II was also reduced by the increase in the electrical conductivity of water (Figure 5C). The Fv/Fm of plants grown under ECw of 3.8 dS m^{-1} differed significantly from those of plants that received the lowest level of water salinity (0.8 dS m^{-1}). Plants irrigated with water of 3.8 dS m^{-1} had a reduction in maximum fluorescence of 5.56% (108.34) (Figure 5D) compared to those cultivated with water with the lowest salinity (0.8 dS m^{-1}).

Sour passion fruit plants subjected to AsA concentrations of 0.8 and 1.6 mM obtained a higher maximum fluorescence compared to those under 0 mM of AsA (Figure 5E). When comparing the Fm of plants that received AsA concentrations of 0.8 and 1.6 mM, compared to that of plants in the control treatment (0 mM), increments of 7.10 and 3.44% were observed, respectively.

The stem diameter of sour passion fruit plants differed significantly in their water salinity levels (Figure 6). In sour passion fruit plants grown under ECw of 0.8 dS m^{-1}, the application of AsA at a concentration of 0.8 mM promoted a greater growth in stem diameter compared to that observed in plants grown under 1.6 mM of AsA. However, at this water salinity level, there were no significant differences in the SD of plants subjected to AsA concentrations of 0 and 1.6 mM. On the other hand, in plants irrigated with ECw of 3.8 dS m^{-1}, the application of 1.6 mM of AsA resulted in greater SD growth compared to those that received concentrations of 0 and 1.8 mM.

There was no significant effect ($p > 0.05$) of the M × AsA × ECw interaction on any of the production variables analyzed in sour passion fruit (Table 7). However, a significant effect was found when the sources of variation were analyzed in isolation. The application methods significantly influenced ($p \leq 0.01$) the polar and equatorial diameter of the fruits, the average fruit weight, and the pulp volume. AsA concentrations significantly affected ($p \leq 0.01$) polar diameter, average fruit weight, and pulp volume. The salinity levels of irrigation water exhibited a significant effect ($p \leq 0.01$) on the number of fruits and total production per plant. The pulp volume was also influenced by the M × AsA and AsA × ECw interactions (Table 7).

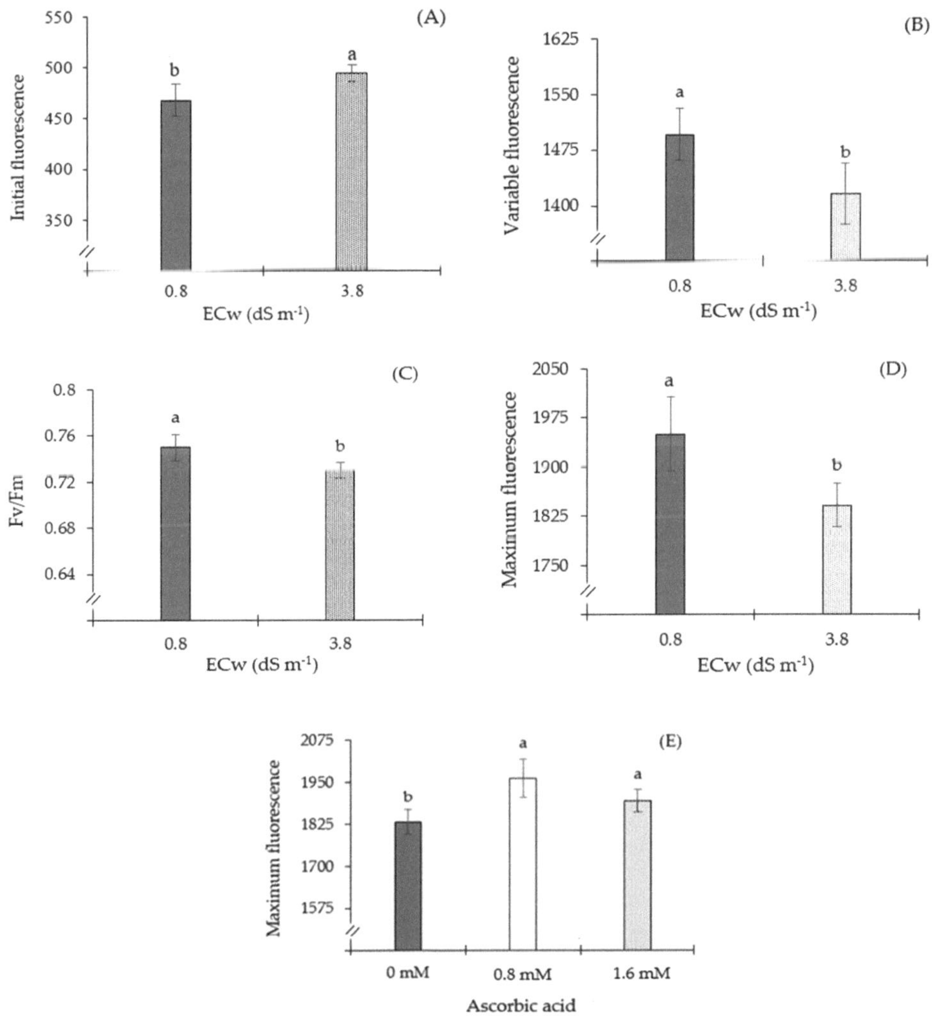

Figure 5. Initial fluorescence—F_0 (**A**), variable fluorescence—Fv (**B**), quantum efficiency of photosystem II—Fv/Fm (**C**), and maximum fluorescence—Fm (**D**) of sour passion fruit cultivated under irrigation with water of different levels of electrical conductivity (ECw), and Fm as a function of ascorbic acid concentrations (**E**), at 150 days after transplanting. Means followed by the same letter did not differ statistically from each other according to the Tukey test ($p \leq 0.05$). The error bars represent standard error (n = 3).

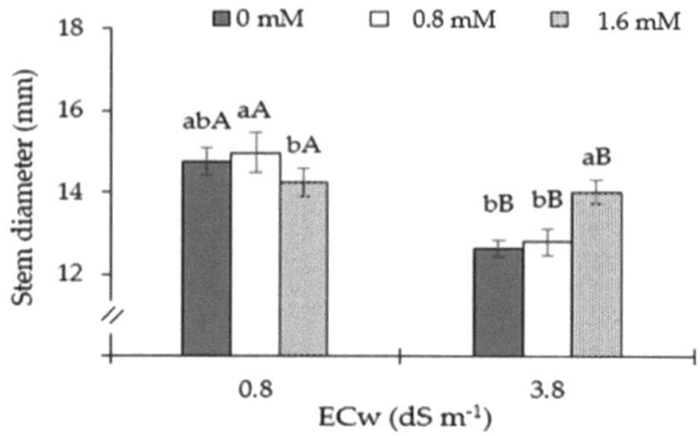

Figure 6. Stem diameter of sour passion fruit cultivated under irrigation with water of different levels of electrical conductivity—ECw, application methods and ascorbic acid concentrations, at 150 days after transplanting. Different letters indicate significant differences between AsA concentrations for each electrical conductivity of irrigation water, and different uppercase letters indicate significant differences in AsA concentrations between electrical conductivities of irrigation water, according to the Tukey test ($p \leq 0.05$). The error bars represent standard error (n = 3).

Table 7. The analysis of variance for fruit polar diameter (FPD), fruit equatorial diameter (FED), average fruit weight (AFW), number of fruits per plant (NFP), total production per plant (TPP), and pulp volume (PV), at 180 days after transplanting, of sour passion fruit grown under irrigation with water of different levels of electrical conductivity, application methods, and concentrations of ascorbic acid.

Source of Variation	DF	Mean Squares					
		FPD	FED	AFW	NFP	TPP	PV
Application methods (M)	2	190.39 **	583.18 **	5576.99 **	6.01 ns	40,933.38 ns	288.06 **
Ascorbic acid (AsA)	2	133.62 **	289.99 ns	3292.81 **	4.57 ns	207,195.71 ns	1362 **
Electrical conductivity (ECw)	1	10.17 ns	189.01 ns	6.11 ns	303.41 **	7,855,116.07 **	46.29 ns
Interaction (M × AsA × ECw)	4	17.93 ns	369.24 ns	722.61 ns	35.91 ns	225,935.91 ns	71.49 ns
Interaction (M × AsA)	4	89.57 ns	167.07 ns	871.78 ns	17.79 ns	59,213.39 ns	382.37 **
Interaction (M × ECw)	2	27.66 ns	284.65 ns	482.68 ns	18.37 ns	238,400.23 ns	323.97 ns
Interaction (AsA × ECw)	2	12.07 ns	98.75 ns	939.10 ns	18.21 ns	150,123.95 ns	314.75 **
Blocks	2	23.28 ns	86.47 ns	28.55 ns	0.90 ns	18,694.25 ns	188.56
Residual	34	28.07	120.27	225.01	15.69	126,741.20	54.242
CV (%)		7.47	10.63	14.33	19.06	14.26	12.77

ns, ** respectively non-significant and significant at $p \leq 0.01$. CV—coefficient of variation, DF—degrees of freedom.

The polar (Figure 7A) and equatorial diameters of passion fruit fruits (Figure 7C) of the plants subjected to the spraying, and soaking and spraying, methods differed significantly from the plants subjected to soaking alone. The highest FPD (73.26 mm) value was obtained in plants subjected to spraying and the highest FED (57.37 mm) value was recorded in plants that received AsA treatment through soaking and spraying. The FPD was also influenced by AsA concentrations (Figure 7B); it can be seen that the 0.8 and 1.6 mM concentrations did not differ from each other, but they differed significantly from the 0 mM concentration. An increase of 5.01 and 7.91% in the FPD of plants subjected to concentrations of 0.8 and 1.6 mM, respectively, was observed in relation to plants that received a concentration of 0 mM.

The average fruit weight of passion fruit plants subjected to the spraying, and soaking and spraying, methods did not differ from each other (Figure 8A). However, there was a significant difference in relation to the plants that were subjected to the soaking treatment, with an increase of 37.68% (31.82 g per fruit) for spraying and 34.28% (28.95 g per fruit) for soaking and spraying, in relation to plants that received only soaking. The average fruit weight was also influenced by AsA concentrations (Figure 8B). It can be seen that the plants subjected to a concentration of 1.6 mM stood out in relation to the other concentrations, presenting the highest AFW value (117.93 g), corresponding to an increase of 29.75%

(27.04 g per fruit) in relation to plants submitted to a concentration of 0 mM and 12.01% (12.64 g per fruit) in relation to plants that received a concentration of 0.8 mM of AsA.

The number of fruits per plant and total production per plant were negatively affected when passion fruit plants were subjected to irrigation with ECw of 3.8 dS m^{-1} (Figure 8C,D), with reductions of 45.38% (4.74 fruits per plant) and 53.70% (762.8 g per plant), respectively, compared to plants irrigated with water of 0.8 dS m^{-1}.

Plants grown under ECw of 0.8 dS m^{-1} obtained a higher pulp volume compared to those that were subjected to water salinity of 3.8 dS m^{-1}, only when they were subjected to an AsA concentration of 0.8 mM (Figure 9A). However, plants grown under AsA spraying achieved the highest pulp volume at the water salinity levels of 0.8 and 3.8 dS m^{-1}, and at the AsA concentration of 1.6 mM. The increase in pulp volume in plants that received AsA at a concentration of 1.6 mM at the two water salinity levels, compared to the results of plants that did not receive AsA, was equal to 38.61% (17.72 mL) and 34.54% (17 mL), respectively. However, plants that did not receive AsA (0 mM) obtained the lowest values of pulp volume (45.89 and 49.22 mL) under water salinity of 0.8 and 3.8 dS m^{-1}, respectively.

The interaction between the application methods (M) and the concentrations of ascorbic acid (Figure 9B) significantly influenced pulp volume, with the highest value (70.83 mL) obtained in plants that received AsA by the spraying method at a concentration of 1.6 mM, however, there was no significant difference in PV between plants that received AsA by spraying, and soaking and spraying. The increase in pulp volume in plants treated with AsA through the spraying, and soaking and spraying, methods was 21.95% (12.75 mL) and 16.22% (9.42 mL), respectively, compared to plants that received AsA only by soaking.

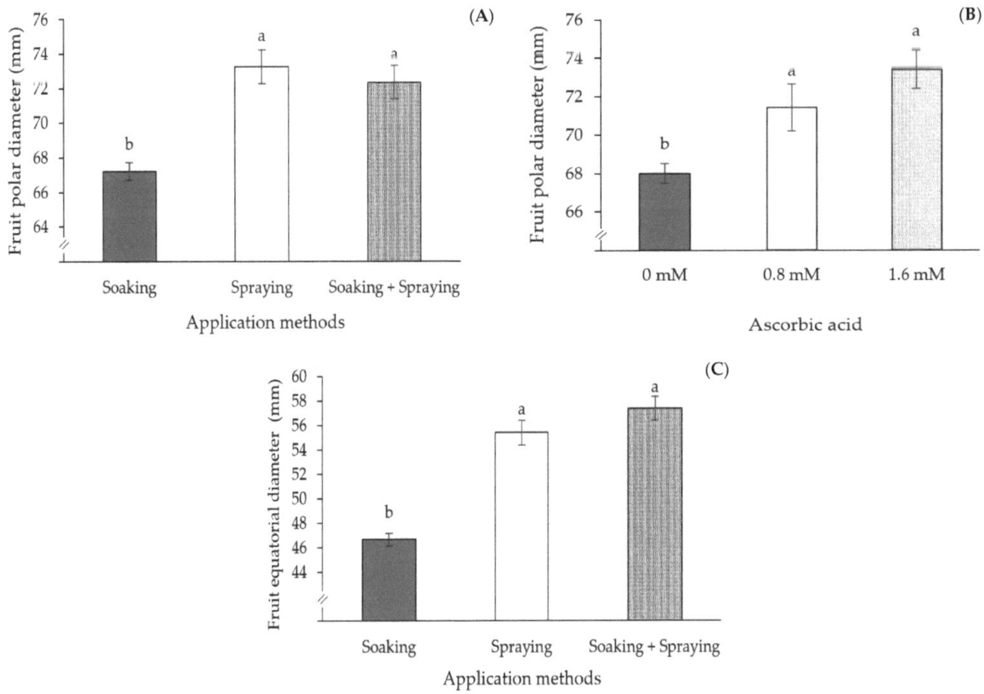

Figure 7. Fruit polar diameter—FPD (**A**) and fruit equatorial diameter—FED (**C**), a function of AsA application methods, and fruit polar diameter as a function of ascorbic acid concentrations (**B**), at 180 days after transplanting. Means followed by the same letter did not differ statistically from each other according to the Tukey test ($p \leq 0.05$). The error bars represent standard error (n = 3).

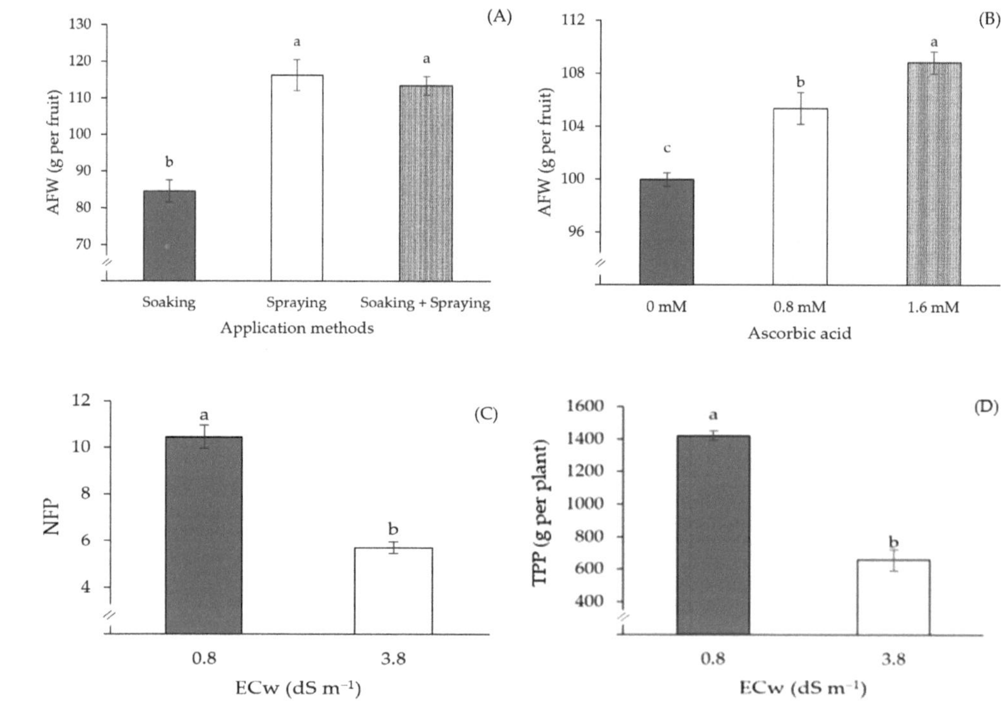

Figure 8. Average fruit weight—AFW as a function of the application methods (**A**) and of AsA concentrations (**B**), number of fruits per plant—NFP (**C**), and total production per plant—TPP (**D**) as a function of water electrical conductivity levels (ECw), at 180 days after transplanting. Means followed by the same letter did not differ statistically from each other according to the Tukey test ($p \leq 0.05$). The error bars represent standard error (n = 3).

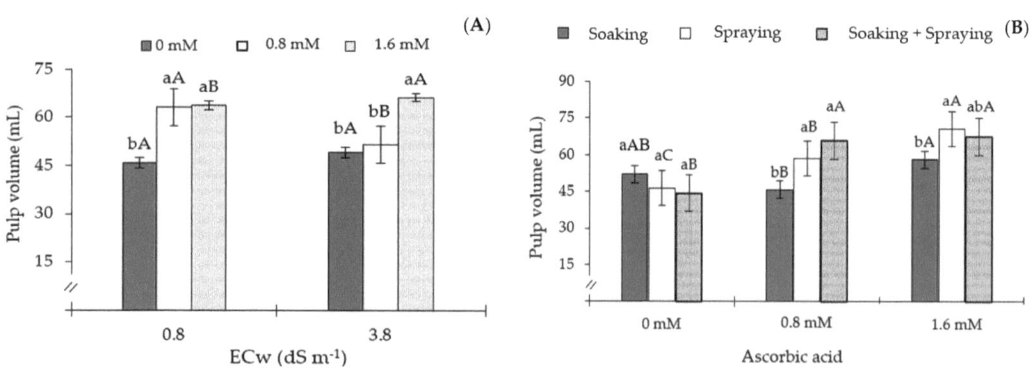

Figure 9. Pulp volume (PV) of passion fruit as a function of the interaction between AsA concentrations and ECw levels (**A**) and as a function of the interaction between application methods and AsA concentration (**B**), at 180 days after transplanting. Different lowercase letters indicate significant differences between AsA concentrations for each electrical conductivity of irrigation water, and different uppercase letters indicate significant differences in AsA concentrations between electrical conductivities of irrigation water, according to the Tukey test. ($p \leq 0.05$). The error bars represent standard errors (n = 3).

The changes in morphophysiology and production for the application methods and concentrations of ascorbic acid in the mitigation of saline stress in passion fruit can be observed in the Pearson correlation matrix (Figure 10). A significant correlation was observed for most variables, with the exception of instantaneous carboxylation efficiency, polar fruit diameter, equatorial fruit diameter, average fruit weight, pulp volume, and variable fluorescence.

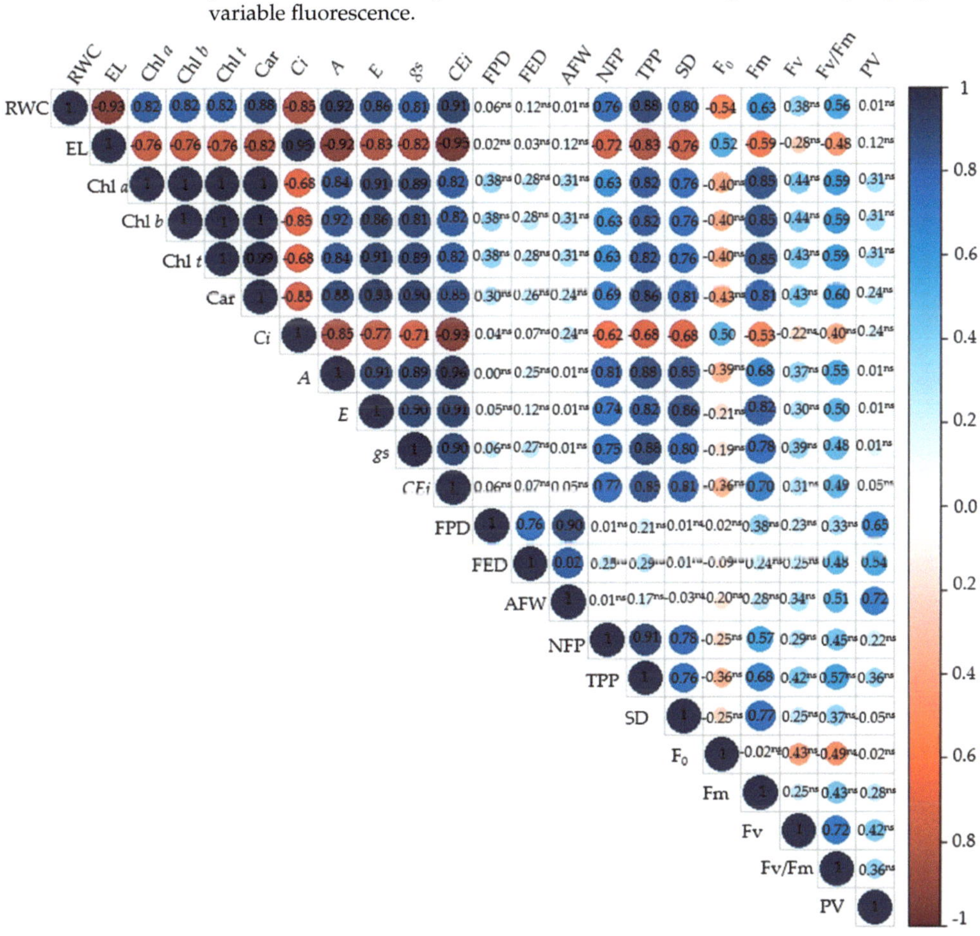

Figure 10. Correlation matrix obtained from the variables analyzed and the treatments applied. ns—non-significant.

It was observed that the relative water content correlated positively with all variables, with the exception of electrolyte leakage (−0.93), internal carbon concentration (−0.85), and initial fluorescence (−0.54). This result is possibly linked to the respective variables being inversely proportional to the RWC. Sour passion fruit plants irrigated with ECw of 3.8 dS m^{-1} showed an increase in internal carbon concentration, which may be related to the fact that the CO_2 absorbed in the substomatal chamber was not being properly assimilated during the photosynthetic process.

Electrolyte leakage showed negative correlations with all variables, except for internal CO_2 concentration (0.95) and initial fluorescence (0.52). This negative correlation in most variables may be associated with the instability of cell membranes, possibly resulting in lipid peroxidation of the membrane. Photosynthetic pigments correlated positively (≥0.80) with all photosynthetic pigment variables, with the exception of internal carbon concentration, which presented a negative correlation of −0.68, −0.85, −0.68, and −0.85 for Chlorophyll *a*, chlorophyll b, total chlorophyll, and carotenoids, respectively.

A positive correlation was also observed between gas exchange parameters, the number of fruits per plant, and the total production per plant. Stomatal conductance, transpiration, the CO_2 assimilation rate, and instantaneous carboxylation efficiency presented a high correlation of $\geq$0.70 in relation to the number of fruits and total fruit production. This result highlights a possible effect of ascorbic acid in mitigating saline stress in passion fruit plants. Since production is a reflection of photosynthesis, for this to occur, it directly requires efficiency in gas exchange.

3. Discussion

The approach with foliar application of AsA showed positive impacts on physiological activity in the current investigation, which may have improved the physiological performance and yield of passion fruit plants under stress. In comparison to other application methods, Nachtigall and Nava [18] claimed that when a product is applied by spraying, the response is nearly instantaneous and more effective in the later phases of growth, when there is a preferential absorption for fruit production compared to other application methods.

Ascorbic acid application by spraying in tomato plants exposed to salt stress increased Chl *t* (8.78%), Car (7.81%), and NFP (24%) [19]. Exogenous administration of AsA also led to an increase in RWC, indicating improved K^+ absorption, in sorghum (*Sorghum vulgare* Pers.) and common beans (*Phaseolus vulgaris* L.) under salt stress, as reported by [20,21]. El-Sayed et al. [22] found that the foliar application of AsA also enhanced the length and diameter of olive fruits under salt stress, as well as the leaf area, total chlorophyll content in leaves, fruit and pulp percentage, and thickness.

As a result of the application of AsA at a concentration of 0.8 mM and using the spraying method, the plants under S1M2C2 treatment had the lowest internal CO_2 concentration (137 μmol CO_2 m^{-2} s^{-1}) and the lowest electrolyte leakage (33.01%) in the leaf blades. The current investigation found that, in plants treated with AsA at a salinity level of 0.8 dS m^{-1}, electrolyte leakage did not signal damage to leaf tissues.

A cell is deemed damaged when the percentage of damage (electrolyte leakage) surpasses 50%, according to Sullivan [23]. Furthermore, a faster rate of CO_2 assimilation was discovered as a result of the decreased internal CO_2 concentration. This happens as a result of AsA's promotion of biomass growth and continuous plant development, which speeds up cell division and growth while strengthening membrane integrity, and lowering ion leakage and internal CO_2 concentration. Since AsA increases the activity of many enzymes, including RuBisCO, these activities are linked to the physiological role of AsA [21,24].

The capacity of roots to absorb water and nutrients from the soil solution is limited by excess salts present in irrigation water, particularly sodium (Na^+) and chloride (Cl^-) ions [25,26]. This leads to osmotic stress and ionic imbalance, which in turn causes stomatal closure and reductions in transpiration, CO_2 assimilation rate, instantaneous carboxylation efficiency, and water use efficiency [3,27]. Other investigations involving various fruit crops, including mandarin [28], West Indian cherry [29], soursop [30], guava [31], and cashew [32], have also noted the decrease in gas exchange as measured by gs, E, and A, with an increase in the electrical conductivity of irrigation water. However, sour passion fruit plants' decreased ability to synthesize photosynthetic pigments as a result of salt stress was also observed by [33].

In research assessing the effects of water salinity on the sour passion fruit crop, the 'Guinezinho' accession, Andrade et al. [34] found that an increase in ECw from 0.7 dS m^{-1} resulted in a lower number of fruits and a lower mean fruit weight. In a different investigation, Lima et al. [33] discovered that sour passion fruit under salt stress had smaller polar and equatorial diameters. According to Souza et al. [35], electrical conductivity of

0.3 dS m^{-1} decreased the sour passion fruit plants' RWC, chlorophyll *a* and *b* levels, CO_2 assimilation rate, and instantaneous water use efficiency.

The increase in F0 in the present study (Figure 5A) indicates that salt stress may have damaged the photosynthesis system, leading to a reduction in light energy. This condition is indicative of a state of oxidation of quinone (the major electron receptor) in the reaction center (P680), which prevents photosystem II from transferring energy [29].

Conversely, Fv (Figure 5B) was decreased through irrigation using 3.8 dS m^{-1} ECw. This decrease indicates a restriction in the activation of the electron transport chain, which plays a crucial role in the synthesis of energy in the form of ATP and NADPH for the Calvin cycle. This reduction is related to the active potential energy in the photosystem. As a result, a plant's ability to photosynthesize is lowered [36,37]. In research with soursop (*Annona muricata* L.) plants irrigated with salty water (ECw ranging from 0.8 to 4.0 dS m^{-1}), Silva et al. [30] discovered that, at 480 days after transplanting, there was a drop in variable fluorescence of 5.21% per unit increase of ECw, starting at 0.8 dS m^{-1}.

The ratio Fv/Fm showed a similar impact to that seen in Fv (Figure 5C), i.e., a decrease promoted by irrigation with water that had a higher salinity (3.8 dS m^{-1}). This scenario shows that sour passion fruit plants can sustain photochemical damage. A portion of the light energy in the thylakoid membrane, linked to the metabolic consequences of salt stress, increases the generation of reactive oxygen species (ROS), which in turn degrades photosynthetic pigments in the reaction center [38,39]. When precocious dwarf cashew seedlings were irrigated with saline waters (ECw varying from 0.4 to 3.6 dS m^{-1}), Silva et al. [32] observed a 19.05% decrease in photosystem II quantum efficiency in the cashew plants.

The results of the present study also revealed that passion fruit plants that were irrigated with ECw of 3.8 dS m^{-1} had less maximum fluorescence (Figure 5D). Salt stress most likely limited the amount of energy absorbed in the light reaction centers because it causes an imbalance in the plant's metabolic activities by causing an excessive build up of harmful ions. Reactive oxygen species are produced as a result, which may restrict energy efficiency [39]. In a study conducted by Silva et al. [31] using soursop (*Annona muricata* L.) cv. Morada Nova plants irrigated with salty water (ECw from 0.8 to 4.0 dS m^{-1}), it was discovered that an increase in ECw levels caused a reduction of 10.9% per unit increment of electrical conductivity in maximum fluorescence.

Conversely, applying AsA at 0.8 and 1.6 mM concentrations promoted an increase in Fm (Figure 5E). Ascorbic acid is responsible for the rise in maximum fluorescence because it acts as a cofactor of violaxanthin de-epoxidase, an enzyme that is essential in the xanthophyll cycle. This cycle represents a mechanism for dissipating excessive light energy that could cause damage to chloroplasts. In the xanthophyll cycle, there is a reversible conversion of violaxanthin to antheraxanthin and zeaxanthin, pigments with the ability to dissipate excess energy in the form of heat, i.e., AsA plays a crucial role in assisting plants in regulating the flow of energy in chloroplasts, thus preventing photoinhibition [40].

Excessive ion accumulation lowers the free energy potential of water in the soil, which makes it more difficult for plants to absorb nutrients and water. Genes that are responsible for the changeover in the synthesis of lignin, suberin, and cell wall polysaccharides may change as a result of this circumstance. These modifications negatively affect the rates of elongation and cell division in tissues, and consequently growth [41,42]. This effect may have contributed to the reduction in the stem diameter of passion fruit plants (Figure 6).

Restrictions in water absorption and the harmful effects of ions (Na^+ and Cl^-), which accumulate in stem tissues during plant development and might change osmotic and ionic balance, may also be linked to the inhibition of plant growth [43,44]. The restrictions in the rate of CO_2 assimilation of plants irrigated with water having an electrical conductivity of 3.8 dS m^{-1}, which modifies the partitioning of photoassimilates, are the cause of the inhibition in the growth of sour passion fruit plants. Furthermore, one factor that leads to growth limitation in plants grown under stress is the energy required to sustain metabolic activities. In a study carried out by Lima et al. [45] with passion fruit irrigated with saline water (ECw of 0 and 150 mM), the authors concluded that the higher salinity level of irrigation water resulted in the inhibition of the growth of the stem diameter.

The pulp volume of passion fruit fruits was likewise observed to be decreased by irrigation with an ECw of 3.8 dS m^{-1} in the current study (Figure 9A). As a protective measure against excessive water loss, plants partially close their stomata due to an increase in the concentration of salts in the soil. This restricts the entry of CO_2 into the substomatal chamber, which in turn affects carboxylation efficiency and the efficient use of water [46]. Ramos et al. [47], in a study with the sour passion fruit cv. BRS Rubi do Cerrado under irrigation with saline water (ECw from 0.6 to 3.0 dS m^{-1}), observed a decrease of 49.84% in the pulp yield of fruits of plants subjected to the higher ECw.

Despite the reduction in pulp volume due to the salinity of the irrigation water (Figure 9A), it was observed that spraying AsA at a concentration of 1.6 mM was able to promote an increase in pulp volume (Figure 9B). This increase in pulp volume may be associated with the role of AsA as an antioxidant, since AsA has the ability to react directly with several reactive oxygen species, resulting in an improvement of the physiological functions of plants [48]. In addition, the application of AsA by spraying induces a practically instantaneous and more efficient response during the advanced stages of growth [18].

In general, the results obtained in this study point to the potential benefits of AsA at concentrations of 0.8 and 1.6 mM, in conjunction with the foliar spraying method, as a means of reducing the negative effects of salt stress on sour passion fruit BRS GA1 plants irrigated with moderately saline waters (ECw of 0.8 and 3.8 dS m^{-1}).

4. Materials and Methods

4.1. Experimental Site

The experiment was conducted from January to October 2022 in a controlled environment at the Academic Unit of Agricultural Engineering, Federal University of Campina Grande, in Campina Grande, Paraíba, Brazil (7°15′18″ S, 35°52′28″ W, at an average altitude of 550 m). The climate in Campina Grande is classified as type As′, characterized by warm and humid conditions with rain in winter and autumn. Figure 11 presents the maximum and minimum air temperatures, as well as the average relative humidity recorded during the study.

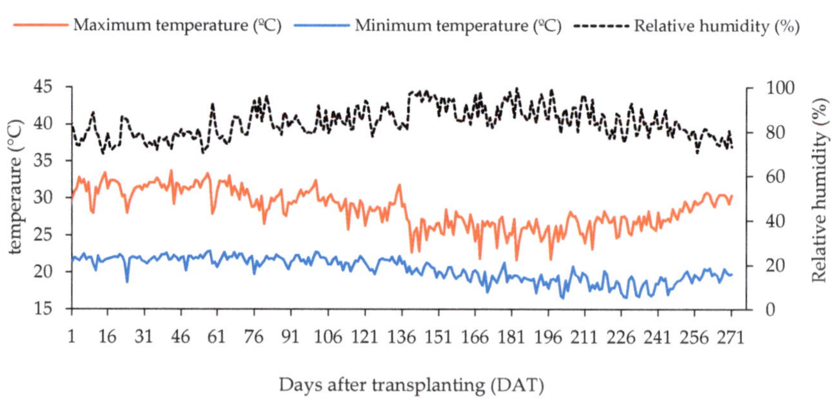

Figure 11. Maximum and minimum temperatures, and average relative humidity, observed in the indoor area of the greenhouse during the experimental period (5 January to 3 October 2022).

4.2. Experimental Design and Treatments

The experimental design was randomized blocks, arranged in a 3 × 3 × 2 factorial scheme, corresponding to three methods of application of ascorbic acid—M (soaking, spraying, and soaking and spraying), three concentrations of ascorbic acid—AsA (0, 0.8, and 1.6 mM) and two electrical conductivities of the irrigation water—ECw (0.8 and 3.8 dS m^{-1}) (Table 8), with three replicates.

Table 8. Description of the treatments used.

ECw	Ascorbic Acid Application Methods (M)			Concentration
	M1 (Soaking)	M2 (Spraying)	M3 (Soaking + Spraying)	
S1 ($0.8\ dS\ m^{-1}$)	S1M1C1 S1M1C2 S1M1C3	S1M2C1 S1M2C2 S1M2C3	S1M3C1 S1M3C2 S1M3C3	C1—0 mM C2—0.8 mM C3—1.6 mM
S2 ($3.8\ dS\ m^{-1}$)	S2M1C1 S2M1C2 S2M1C3	S2M2C1 S2M2C2 S2M2C3	S2M3C1 S2M3C2 S2M3C3	C1—0 mM C2—0.8 mM C3—1.6 mM

ECw—electrical conductivity of irrigation water.

Ascorbic acid (AsA) concentrations of 0.8 and 1.6 mM were based on a study conducted by Fatah and Sadek [16], adjusted to millimolar (mM).

4.3. Plant Material and Seedling Formation

The cultivar BRS GA1 was used in this study. The fruit exhibits a yellow color and has an oblong shape, with a slightly flattened base and apex, and an average weight ranging from 120 to 350 g, while the pulp yield is approximately 40% [49]. Sour passion fruit seedlings were produced in plastic bags with dimensions of 10 × 20 cm using 2.6 kg of substrate consisting of 2.18 kg of soil (84%), 0.39 kg of sand (15%), and 0.03 kg of humus (1%). Prior to sowing, sour passion fruit seeds were soaked in ascorbic acid solutions (0, 0.8, and 1.6 mM) for a period of 24 h in the dark.

Ascorbic acid concentrations were prepared from the dissolution of ascorbic acid in distilled water and stored in a dark environment [50]. Subsequently, three seeds were sown in each plastic bag at 3 cm depth, spaced equidistantly. After 40 days of sowing, thinning was carried out, keeping only the most vigorous plant in each bag. Before sowing, the soil moisture content was brought to field capacity using water according to the treatment.

Irrigation was carried out daily, applying a volume of water to each bag to keep the substrate moisture close to field capacity. The volume applied was determined based on the water needs of the plants, calculated by the water balance, considering the volume drained from the previous irrigation. Every 20 days, the volume to be applied was incremented by a leaching fraction of 0.10 to avoid excessive accumulation of salts in the root zone [51].

4.4. Experiment Setup and Conduction

From the moment the plants began to produce tendrils (at 70 days after sowing—DAS), they were transplanted into pots adapted as drainage lysimeters, with a capacity of 120 L. The lysimeters were perforated at the base to allow drainage, and connected to a 16-mm-diameter transparent drain (green color). The tip of the drain inside the lysimeter was wrapped with a non-woven geotextile (Bidim OP 30, Geomembrana®, Guarulhos—SP, Brazil) to prevent clogging by soil material.

A plastic bottle was placed below each drain to collect drained water and determine water consumption by the plant. The lysimeters were filled with a 1.0 kg layer of crushed stone followed by 160 kg of *Neossolo Regolítico* (Psamment) with sandy loam texture from the municipality of Lagoa Seca, PB, Brazil (7°10′8″ South latitude, 35°51′20″ West longitude), collected at 0–30 cm depth. Before starting the experiment, chemical and physical–hydraulic attributes of the soil were determined according to the methodology proposed by Teixeira et al. [52]: pH (H_2O) = 4.93; organic matter = 9.3 g dm^{-3}; P = 10.7 g dm^{-3}; K^+ = 0.23 $cmol_c\ kg^{-1}$; Na^+ = 0.51 $cmol_c\ kg^{-1}$; Ca^{2+} = 1.77 $cmol_c\ kg^{-1}$; Mg^{2+} = 1.60 $cmol_c\ kg^{-1}$; Al^{3+} = 2.64 $cmol_c\ kg^{-1}$; H^+ = 0.51 $cmol_c\ kg^{-1}$. ECse = 1.15 dS m^{-1}; CEC = 7.23 $cmol_c\ kg^{-1}$; SAR = 0.38 mmol L^{-1}; ESP = 7.05%; particle-size fraction: sand = 760.9 g kg^{-1}; silt = 164.5 g kg^{-1}; clay = 74.6 g kg^{-1}; moisture at 33.42 KPa = 13.07 dag kg^{-1}; and moisture at 1519.5 KPa = 5.26 dag kg^{-1}.

4.5. Preparation of Waters and Irrigation Management

Saline solutions were prepared by adding NaCl, $CaCl_2{\cdot}2H_2O$, and $MgSO_4{\cdot}7H_2O$ salts to the local water supply in Campina Grande (ECw = 0.38 dS m^{-1}), maintaining a ratio of 7:2:1, which reflects the typical composition of water in the north-eastern semi-arid region [53]. Irrigation with these saline waters started 33 days after transplanting (DAT), with applications every 3 days. Each lysimeter received water according to the specific treatments, aiming to keep soil moisture near field capacity. The volume of water applied was based on the plants' water needs, as determined by the water balance.

4.6. Preparation and Application of Ascorbic Acid Solutions

Only distilled water was used to represent the concentration of 0 mM. Prior to sowing, the sour passion fruit seeds corresponding to the soaking, and soaking and spraying, treatments were soaked in ascorbic acid solutions (0, 0.8, and 1.6 mM) for a period of 24 h in the dark. This period was determined by a test conducted before the experiment, with testing times of 8, 12, and 24 h of soaking, and by evaluating the characteristics of germination and seedling quality.

Foliar applications of AsA began at 30 DAT and were later performed at 30-day intervals, totaling 4 applications, carried out with a Jacto XP-12 knapsack sprayer (Jacto®, Pompeia—SP, Brazil) and using a pump with working pressure (maximum) of 6 bar, with a JD-12 nozzle (Jacto®, Pompeia—SP, Brazil) and flow rate of approximately 770 mL min^{-1}. Spraying with AsA was performed on the abaxial and adaxial sides of the leaves between 5:00 p.m. and 5:45 p.m. until the plants were fully wetted. The average volume of spray applied during the experiment was 583 mL per plant.

4.7. Formative Pruning Management

The rows and plants were spaced by 1.5 m using a vertical trellis system. Formative pruning commenced 15 days after transplanting, removing all lateral shoots and leaving only the main stem.

When the plants reached a height of 10 cm above the trellis (2.20 m), the apical bud was pruned to encourage two secondary branches, which were then trained in opposite directions until they grew to 0.75 m. Once this length was achieved, the apical buds of the secondary branches were pruned to promote the growth of tertiary branches, which were trained down to 30 cm from the ground, forming a curtain for inflorescence development. Throughout the experiment, tendrils and unwanted branches were removed to ensure optimal crop development (Figure 12).

Figure 12. Arrangement of sour passion fruit plants in the experimental area in the different phenological stages: vegetative stage (**A**), flowering stage (**B**) and fruiting stage (**C**).

4.8. Fertilization Management

Fertilization with nitrogen, phosphorus, and potassium was based on the recommendation of Cavalcante et al. [54] for the sour passion fruit crop, using 160 g of N, 140 g of P_2O_5, and 480 g of K_2O per plant per year. N and K were applied as top-dressing divided into 18 equal portions and applied at 15-day intervals; P was applied at once, as basal. Urea, single superphosphate, and potassium chloride were used as sources of N, P, and K, respectively.

Micronutrient fertilization was carried out by foliar sprays with a solution containing 1 g L^{-1} of Dripsol® [Mg (1.1%); Zn (4.2%); B (0.85%); Fe (3.4%); Mn (3.2%); Cu (0.5%); Mo (0.05%)] at 30, 60, 90, 120, and 150 DAT.

4.9. Variables Analyzed

4.9.1. Physiological Variables

At 150 days after transplanting (DAT), treatment effects were evaluated through relative water content, electrolyte leakage, gas exchange, chlorophyll *a* fluorescence, photosynthetic pigments, and growth in stem diameter.

Relative water content (RWC) was determined according to the methodology of Weatherley [55]. Electrolyte leakage from the leaf blades was determined according to Scotti-Campos et al. [56].

Gas Exchange

Gas exchange was determined by stomatal conductance—*gs* (mol H_2O m^{-2} s^{-1}), transpiration—*E* (mmol H_2O m^{-2} s^{-1}), internal CO_2 concentration—*Ci* (μmol CO_2 m^{-2} s^{-1}), CO_2 assimilation rate—*A* (μmol CO_2 m^{-2} s^{-1}), instantaneous water use efficiency (WUEi) (A/E) [(μmol CO_2 m^{-2} s^{-1}) (mmol H_2O m^{-2} s^{-1})$^{-1}$], and instantaneous carboxylation efficiency—*CEi* (A/Ci) ([(μmol CO_2 m^{-2} s^{-1}) (μmol CO_2 mol^{-1})$^{-1}$]), between 06:00 and 09:00 a.m. on fully expanded leaves located in the upper third, using a portable infrared carbon dioxide analyzer (IRGA), model LCPro+ Portable Photosynthesis System® (ADC BioScientific Limited, Hoddesdon, United Kingdom), with temperature control at 25 °C, irradiation of 1200 μmol photons m^{-2} s^{-1}, and airflow of 200 mL min^{-1}.

Chlorophyll *a* Fluorescence

Chlorophyll *a* fluorescence was measured by initial fluorescence (F_0), maximum fluorescence (Fm), variable fluorescence (Fv), and quantum efficiency of photosystem II (Fv/Fm), determined by a pulse-modulated fluorometer, model OS5p from Opti Science (Hudson, NH, USA), between 06:00 and 09:00 a.m. on fully expanded leaves located in the upper third. Leaf clips were placed on these leaves and, after a period of 30 min of adaptation to the dark, chlorophyll *a* fluorescence was determined.

Photosynthetic Pigments

Photosynthetic pigments were quantified according to Arnon [57], using extracts obtained from disc samples of the third mature leaf at the apex of the plant. These extracts were used to quantify the contents of chlorophyll *a*, chlorophyll *b*, total chlorophyll, and carotenoids in the solutions with a spectrophotometer at absorbance wavelengths (ABS) of 470, 646, and 663 nm.

4.9.2. Growth Variable

The stem diameter (SD) of sour passion fruit plants was measured close to the plant collar (5.0 cm from the ground) with a digital caliper.

4.9.3. Production Variable

Fruit harvesting began at 150 days after transplanting, and was completed at 180 DAT. The fruits were collected by individually removing the plant, cutting the peduncle when the fruit color changed from green to partially yellow, and before detaching from the mother plant [58,59]. Even when they were not yet fully ripe, the fruits of each plant were individually identified with the numbers corresponding to the pots. Harvests were carried out daily in the morning, followed by the transport of the fruits to the Irrigation and Drainage Laboratory (LEID) for physical characterization.

Sour passion fruit production was evaluated by determining the number of fruits per plant (NFP), obtained by directly counting the fruits that reached the full maturity stage. The average fruit weight (AFW) was obtained by the ratio between the production per plant, the total number of fruits, and the total production per plant (TPP) using a benchtop scale, fruit polar diameter (FPD), fruit equatorial diameter (FED), and digital caliper, and measuring as the fruits were harvested, with the results expressed in millimeters (mm). Pulp volume (PV) was also measured, using a 250 mL beaker.

4.10. Statistical Analysis

The multivariate structure of the results was analyzed using the principal component analysis technique. This analysis synthesizes the relevant information contained in the original dataset into fewer dimensions. The new dimensions are generated from linear combinations of the original variables, based on the eigenvalues ($\lambda \geq 1.0$) found in the correlation matrix. These dimensions explain a significant percentage greater than 10% of the total variance of the data [60].

Once the dimensions were reduced, the original data of the variables of each component were subjected to a multivariate analysis of variance (MANOVA) using the Hotelling test [61] at 0.05 probability level. This was done for the factors' electrical conductivity of irrigation water, ascorbic acid concentrations, and ascorbic acid application methods, and to check the interaction between these factors. Only the variables that showed a correlation coefficient greater than or equal to 0.6 were maintained in each principal component (PC) [62]. Variables that did not show a correlation coefficient greater than or equal to 0.6 were subjected to analysis of variance through the F test at 0.05 and 0.01 probability levels and, in cases of significance, the means were compared using the Tukey test at 0.05 probability level, and using the statistical software SISVAR v.5.6 [63].

5. Conclusions

Foliar spraying of ascorbic acid at a concentration of 0.8 mM mitigates the effects of salt stress on electrolyte leakage, relative water content in leaves, photosynthetic pigments, gas exchange, and the total fruit production of sour passion fruit at 150 days after transplanting. An increase in the electrical conductivity of water above 3.8 dS m^{-1} negatively affects chlorophyll *a* fluorescence in plants that did not receive ascorbic acid treatment. Sour passion fruit plants grown with water of 0.8 dS m^{-1} and foliar application of 0.8 mM of ascorbic acid achieved the highest pulp volume and the highest growth in stem diameter.

The results obtained in this study confirm the hypothesis that ascorbic acid, when applied in the appropriate concentration and method, can act as a signaling molecule and mitigate the effect of the salinity of the irrigation water on the sour passion fruit. However, further studies are needed to analyze the effects of water salinity on the post-harvest quality of the fruits and to understand how ascorbic acid acts in the signaling of salt stress through enzymatic analyses, in addition to validating the results in field research.

Author Contributions: Conceptualization, E.J.M.C., G.S.d.L. and C.A.V.d.A.; methodology, A.A.R.d.S., L.L.d.S.A.V., V.D.d.S. and L.D.A.B.; software, A.A.R.d.S., T.F.d.L.A. and M.d.S.D.; validation, A.R.d.S., G.S.d.L. and L.A.d.A.S.; formal analysis, C.A.V.d.A. and L.A.d.A.S.; investigation, E.J.M.C., G.S.d.L. and L.A.d.A.S.; resources, G.S.d.L. and H.R.G.; data collection, E.J.M.C.; writing—original draft preparation, E.J.M.C. and G.S.d.L.; writing—review and editing, A.A.R.d.S., G.S.d.L., C.A.V.d.A. and P.D.F.; supervision, G.S.d.L., H.R.G. and L.A.d.A.S.; project administration, G.S.d.L. and V.D.d.S. All authors have read and agreed to the published version of the manuscript.

Funding: CNPq (National Council for Scientific and Technological Development) (Proc. 164427/2020-1), CAPES (Coordination for the Improvement of Higher Education Personnel) financial code—001, and UFCG (Universidade Federal de Campina Grande).

Data Availability Statement: Data are contained within the article. No supplemental data are provided.

Conflicts of Interest: The authors declare no conflicts of interest.

References

1. IBGE—Instituto Brasileiro de Geografia e Estatística. Produção Agrícola Municipal. 2022. Available online: https://www.ibge.gov.br/estatisticas/economicas/agricultura-e-pecuaria/9117-producao-agricola-municipal-culturas-temporarias-e-permanentes.html (accessed on 12 July 2023).
2. Aguiar, A.V.M.; Cavalcante, L.F.; Silva, R.M.; Dantas, T.A.G.; Santos, E.C. Effect of biofertilization on yellow passion fruit production and fruit quality. *Rev. Caatinga* **2017**, *30*, 136–148. [CrossRef]
3. Behdad, A.; Mohsenzadeh, S.; Azizi, M. Growth, leaf gas exchange and physiological parameters of two *Glycyrrhiza glabra* L. populations subjected to salt stress condition. *Rhizosphere* **2021**, *17*, e100319. [CrossRef]
4. Pereira, K.T.O.; Torres, S.B.; Paiva, E.P.d.; Sousa, G.D.d.; Sousa Leite, M.d.; Albuquerque, C.C.d.; Sá, F.V.d.S. Priming cycles with salinity tolerance elicitors in seeds of *Mimosa caesalpiniifolia* and *Pityrocarpa moniliformis*. *Biologia* **2024**, *79*, 411–424. [CrossRef]
5. Liang, W.; Ma, X.; Wan, P.; Liu, L. Plant salt-tolerance mechanism: A review. *Biochem. Biophys. Res. Commun.* **2018**, *495*, 286–291. [CrossRef] [PubMed]
6. Nájar, R.; Aydi, S.; Sassi-Aydi, S.; Zarai, A.; Abdelly, C. Effect of salt stress on photosynthesis and chlorophyll fluorescence in *Medicago truncatula*. *Plant Biosyst.* **2019**, *153*, 88–97. [CrossRef]
7. Lima, G.S.d.; Souza, W.B.d.; Paiva, F.J.d.S.; Soares, L.A.d.A.; Torres, R.A.; Silva, S.T.d.A.; Gheyi, H.R.; Lopes, K.P. Tolerance of sour passion fruit cultivars to salt stress in a semi-arid region. *Rev. Bras. Eng. Agríc. Ambient.* **2023**, *27*, 785–794. [CrossRef]
8. Galvão Sobrinho, T.; Silva, A.A.R.d.; Lima, G.S.d.; Lima, V.L.A.d.; Borges, V.E.; Nunes, K.G.; Soares, L.A.d.A.; Saboya, L.M.F.; Gheyi, H.R.; Gomes, J.P.; et al. Foliar applications of salicylic acid on boosting salt stress tolerance in sour passion fruit in two cropping cycles. *Plants* **2023**, *12*, e2023. [CrossRef]
9. Andrade, E.M.G.; Lima, G.S.d.; Lima, V.L.A.d.; Silva, S.S.d.; Gheyi, H.R.; Araújo, A.C.; Gomes, J.P.; Soares, L.A.D.A. Production and postharvest quality of yellow passion fruit cultivated with saline water and hydrogen peroxide. *AIMS Agric. Food* **2019**, *4*, 907–920. [CrossRef]
10. Naz, H.; Akram, N.A.; Ashraf, M. Impact of ascorbic acid on growth and some physiological attributes of cucumber (*Cucumis sativus*) plants under water-deficit conditions. *Pak. J. Bot.* **2016**, *3*, 877–883.
11. Hameed, A.; Gulzar, S.; Aziz, I.; Hussain, T.; Gul, B.; Khan, M.A. Effects of salinity and ascorbic acid on growth, water status and antioxidant system in a perennial halophyte. *AoB Plants* **2015**, *7*, plv004. [CrossRef]
12. Wang, Y.H.; Zhang, G.; Chen, Y.; Gao, J.; Sun, Y.R.; Sun, M.F.; Chen, J.P. Exogenous application of gibberellic acid and ascorbic acid improved tolerance of okra seedlings to NaCl stress. *Acta Physiol. Plant.* **2019**, *41*, 93. [CrossRef]
13. Leal, Y.H.; Sousa, L.V.d.; Gomes, M.J.; Sousa, B.A.G.; Melo Filho, J.S.; Silva, M.G.T.; Carlos, A.; Dias, T.J. Agronomic performance and gaseous exchanges of the radish under saline stress and ascorbic acid application. *Rev. Colomb. Cienc. Hortíc.* **2019**, *13*, 89–98. [CrossRef]
14. Nunes, L.R.d.L.; Pinheiro, P.R.; Cabral, F.A.S.; Silva, J.B.d.; Dutra, A.S. Ascorbic acid of cowpea seeds under saline stress. *J. Seed Sci.* **2019**, *41*, 441–451. [CrossRef]
15. Noreen, S.; Sultan, M.; Akhter, M.S.; Shah, K.H.; Ummara, U.; Manzoor, H.; Ahmad, P.A. Foliar fertigation of ascorbic acid and zinc improves growth, antioxidant enzyme activity and harvest index in barley (*Hordeum vulgare* L.) grown under salt stress. *Plant Physiol. Biochem.* **2021**, *158*, 244–254. [CrossRef]
16. Fatah, E.M.A.; Sadek, K.A. Impact of different application methods and concentrations of ascorbic acid on sugar beet under salinity stress conditions. *J. Agric. Sci.* **2020**, *65*, 31–44. [CrossRef]
17. Kaiser, H.F. The application of electronic computers to factor analysis. *Educ. Psychol. Meas.* **1960**, *20*, 141–151. [CrossRef]
18. Nachtigall, G.R.; Nava, G. Adubação foliar: Fatos e mitos. *Rev. Agropec. Catarinense* **2010**, *23*, 87–97.
19. Alves, R.d.C.; Oliveira, K.R.; Lúcio, J.C.B.; Silva, J.d.S.; Carrega, W.C.; Queiroz, S.F.; Gratão, P.L. Exogenous foliar ascorbic acid applications enhance salt-stress tolerance in peanut plants throughout an increase in the activity of major antioxidant enzymes. *S. Afr. J. Bot.* **2022**, *150*, 759–767. [CrossRef]
20. Azeem, M.; Sultana, R.; Mahmood, A.; Qasim, M.; Siddiqui, Z.S.; Mumtaz, S.; Javed, T.; Umar, M.; Adnan, M.Y.; Siddiqui, M.H. Ascorbic and salicylic acids vitalized growth, biochemical responses, antioxidant enzymes, photosynthetic efficiency and ionic regulation for alleviate salinity stress in Sorghum bicolor. *J. Plant Growth Regul.* **2023**, *2*, 5266–5279. [CrossRef]
21. Saeidi-Sar, S.; Abbaspour, H.; Afshari, H.; Yaghoobi, S.R. Effects of ascorbic acid and gibberellin A3 on alleviation of salt stress in common bean (*Phaseolus vulgaris* L.) seedlings. *Acta Physiol. Plant* **2013**, *35*, 667–677. [CrossRef]
22. El-Sayed, O.M.; El-Gammal, O.H.M.; Salama, A.S.M. Effect of ascorbic acid, proline and jasmonic acid foliar spraying on fruit set and yield of Manzanillo olive trees under salt stress. *Sci. Hortic.* **2014**, *176*, 32–37. [CrossRef]
23. Sullivan, C.Y. Mechanisms of heat and drought resistance in grain sorghum and methods of measurement. In *Sorghum in Seventies*; Rao, N.G.P., House, L.R., Eds.; Oxford & IBH Publishing Co.: New Delhi, India, 1972; 247p.
24. Akram, N.A.; Iqbal, M.; Muhammad, A.; Ashraf, M.; Al-Qurainy, F.; Shafiq, S. Aminolevulinic acid and nitric oxide regulate oxidative defense and secondary metabolisms in canola (*Brassica napus* L.) under drought stress. *Protoplasma* **2018**, *255*, 163–174. [CrossRef] [PubMed]

25. Soni, S.; Sharawat, N.; Kumar, A.; Kumar, N.; Lata, C.; Mann, A. Effect of saline irrigation on plant water traits, photosynthesis and ionic balance in durum wheat genotypes. *Saudi J. Biol. Sci.* **2021**, *28*, 2510–2517. [CrossRef]
26. Farouk, S.; Elhindi, K.M.; Alotaibi, M.A. Silicon supplementation mitigates salinity stress on *Ocimum basilicum* L. via improving water balance, ion homeostasis, and antioxidant defense system. *Ecotoxicol. Environ. Saf.* **2020**, *206*, e111396. [CrossRef]
27. Chrysargyris, A.; Xylia, P.; Nicolas, S.; Tzortzakis, N. Physiochemical properties of petunia edible flowers grown under saline conditions and their postharvest performance under modified atmosphere packaging and ethanol application. *J. Sci. Food Agric.* **2019**, *99*, 3644–3652. [CrossRef] [PubMed]
28. Sá, F.V.S.; Brito, M.E.B.; Silva, L.A.; Moreira, R.C.L.; Fernandes, P.D.; Figueiredo, L.C. Physiology of perception of saline stress in 'Common Sunki' mandarin hybrids under saline hydroponic solution. *Comun. Sci.* **2015**, *6*, 463–470. [CrossRef]
29. Dias, A.S.; Lima, G.S.d.; Gheyi, H.R.; Furtado, G.d.F.; Soares, L.A.d.A.; Nobre, R.G.; Fernandes, P.D. Chloroplast pigments and photochemical efficiency of West Indian cherry under salt stress and potassium-phosphorus fertilization. *Semin. Ciênc. Agrár.* **2021**, *42*, 87–104. [CrossRef]
30. Silva, A.A.R.d.; Lima, G.S.d.; Azevedo, C.A.V.d.; Gheyi, H.R.; Soares, L.A.d.A.; Veloso, L.L.d.S.A. Salicylic acid improves physiological indicators of soursop irrigated with saline water. *Rev. Bras. Eng. Agríc. Ambient.* **2022**, *26*, 412–419. [CrossRef]
31. Lacerda, C.N.d.; Soares, G.S.d.; Soares, L.A.d.A.; Fatima, R.T.d.; Gheyi, H.R.; Azevedo, C.A.V.d. Morphophysiology and production of guava as a function of water salinity and salicylic acid. *Rev. Bras. Eng. Agríc. Ambient.* **2022**, *26*, 451–458. [CrossRef]
32. Silva, A.F.; Sousa, V.F.D.O.; Santos, G.L.d.; Júnior, E.S.A.; Silva, S.L.d.; Macedo, C.E.d.; Maia, J.M. Antioxidant protection of photosynthesis in two cashew progenies under salt stress. *J. Agric. Sci.* **2018**, *10*, 388–404. [CrossRef]
33. Lima, G.S.d.; Pinheiro, F.W.A.; Gheyi, H.R.; Soares, L.A.d.A.; Soares, M.D.M.; Silva, F.A.; Azevedo, C.A.V.d.; Lima, V.L.A. Postharvest quality of sour passion fruit under irrigation strategies with brackish water and potassium application in two crop cycles. *Wat. Air Soil Poll.* **2022**, *233*, e452. [CrossRef]
34. Andrade, E.M.G.; Lima, G.S.d.; Lima, V.L.A.d.; Silva, S.S.d.; Gheyi, H.R.; Silva, A.A.R.d. Gas exchanges and growth of passion fruit under saline water irrigation and H_2O_2 application. *Rev. Bras. Eng. Agríc. Ambient.* **2019**, *23*, 945–951. [CrossRef]
35. Souza, W.B.B.d.; Lima, G.S.d.; Paiva, F.J.d.S.; Soares, L.A.d.A.; Fatima, R.T.d.; Silva, A.A.R.d.; Gheyi, H.R.; Fernandes, P.D. Potassium fertilization as salt stress attenuator in sour passion fruit. *Ciênc. Rural.* **2023**, *53*, e20210769. [CrossRef]
36. Kang, S.M.; Shaffique, S.; Injamum-Ul-Hoque, M.; Alomrani, S.O.; Park, Y.S.; Lee, I.J. Foliar treatment with melatonin modulates photosynthetic and antioxidant responses in *Silybum marianum* L. under salt stress. *Sci. Hortic.* **2024**, *325*, e112664. [CrossRef]
37. Lotfi, R.; Ghassemi-Golezani, K.; Pessarakli, M. Salicylic acid regulates photosynthetic electron transfer and stomatal conductance of mung bean (*Vigna radiata* L.) under salinity stress. *Biocatal. Agric. Biotechnol.* **2020**, *26*, e101635. [CrossRef]
38. Abbas, R.; Rasul, S.; Aslam, K.; Baber, M.; Shahid, M.; Mubeen, F.; Naqqash, T. Halotolerant PGPR: A hope for cultivation of saline soils. *J. King Saud Univ. Sci.* **2019**, *31*, 1195–1201. [CrossRef]
39. Saddiq, M.S.; Iqbal, S.; Hafeez, M.B.; Ibrahim, A.M.; Raza, A.; Fatima, E.M.; Ciarmiello, L.F. Effect of salinity stress on physiological changes in winter and spring wheat. *Agronomy* **2021**, *11*, e1193. [CrossRef]
40. Yang, D.Y.; Ma, N.N.; Zhuang, K.Y.; Zhu, S.B.; Liu, Z.M.; Yang, X.H. Overexpression of tomato SlGGP-LIKE gene improves tobacco tolerance to methyl viologen-mediated oxidative stress. *J. Plant Physiol.* **2017**, *209*, 31–41. [CrossRef]
41. Li, H.; Yan, S.; Zhao, L.; Tan, J.; Zhang, Q.; Gao, F.; Wang, P.; Hou, H.; Li, L. Histone acetylation associated up-regulation of the cell wall related genes is involved in salt stress induced maize root swelling. *BMC Plant Biol.* **2014**, *14*, 127–139. [CrossRef]
42. Byrt, C.S.; Munns, R.; Burtonc, R.A.; Gillihama, M.; Wegea, S. Root cell wall solutions for crop plants in saline soils. *Plant Sci.* **2018**, *269*, 47–55. [CrossRef]
43. Silva, E.B.D.; Viana, T.V.D.A.; Sousa, G.G.D.; Sousa, J.; Santos, M.F.D.; Azevedo, B.M.D. Growth and nutrition of peanut crop subjected to saline stress and organomineral fertilization. *Rev. Bras. Eng. Agríc. Ambient.* **2022**, *26*, 495–501. [CrossRef]
44. Abreu, F.D.S.; Viana, T.V.D.A.; Sousa, G.G.D.; Baldé, B.; Lacerda, C.F.D.; Goes, G.F.; Cambissa, P.B. Salt stress and potassium fertilization on the agronomic performance of peanut crop. *Rev. Caatinga* **2024**, *37*, e11996. [CrossRef]
45. Lima, L.K.d.S.; Jesus, O.N.d.; Soares, T.L.; Santos, I.S.d.; Oliveira, E.J.d.; Coelho Filho, M.A. Growth, physiological, anatomical and nutritional responses of two phenotypically distinct passion fruit species (*Passiflora* L.) and their hybrid under saline conditions. *Sci. Hortic.* **2020**, *263*, e109037. [CrossRef]
46. Skider, R.K.; Wang, X.; Zhang, H.; Gui, H.; Dong, Q.; Jin, D.; Song, M. Nitrogen enhances salt tolerance by modulating the antioxidant defense system and osmoregulation substance content in *Gossypium hirsutum*. *Plants* **2020**, *9*, e450. [CrossRef] [PubMed]
47. Ramos, J.G.; Lima, V.L.A.d.; Lima, G.S.d.; Nunes, K.G.; Pereira, M.O.; Paiva, F.J.S. Produção e qualidade pós-colheita do maracujazeiro-azedo irrigado com águas salinas e aplicação exógena de H_2O_2. *Irriga* **2022**, *27*, 540–556. [CrossRef]
48. Siddiqui, M.H.; Alamri, S.; Al-Khaishany, M.Y.; Khan, M.N.; Al-Amri, A.; Ali, H.M.; Alaraidh, I.A.; Alsahli, A.A. Exogenous melatonin counteracts NaCl-induced damage by regulating the antioxidant system, proline and carbohydrate metabolism in tomato seedlings. *Int. J. Mol. Sci.* **2019**, *20*, e353. [CrossRef]
49. EMBRAPA. *BRS Sol do Cerrado: Híbrido de Maracujazeiro azedo para mesa e indústria*, 2nd ed.; Brasília, D.F., Ed.; Embrapa Cerrado: Planaltina, Brazil, 2008; 2p. [CrossRef]
50. Santos, O.F.; Broetto, F.; Oliveira, D.P.F.; Galvão, Í.M.; Souza, M.L.C.; Basílio, J.J.N. Ácido ascórbico, uma alternativa para minimizar os efeitos da deficiência hídrica em rabanete. *Irriga* **2018**, *1*, 79–91. [CrossRef]

51. Ayers, R.S.; Westcot, D.W. *A Qualidade de Água na Agricultura*, 2nd ed.; Estudos Irrigação e Drenagem, 29 revisado 1; UFPB, FAO: Campina Grande, Brazil, 1999; 153p.
52. Teixeira, P.C.; Donagemma, G.K.; Fontana, A.; Teixeira, W.G. *Manual de Métodos de Análise de Solo*, 3rd ed.; Embrapa: Brasília, Brazil, 2017; 573p.
53. Medeiros, J.F.; Lisboa, R.A.; de Oliveira, M.; de Silva Júnior, M.J.; Alves, L.P. Caracterização das águas subterrâneas usadas para irrigação na área produtora de melão da Chapada do Apodi. *Rev. Bras. Eng. Agríc. Ambient.* **2003**, *7*, 469–472. [CrossRef]
54. Cavalcante, F.J.A. *Recomendação de Adubação para o Estado de Pernambuco: 2. Aproximação*, 3rd ed.; Recife, P.E., Ed.; Instituto Agronômico de Pernambuco: Recife, Brazil, 2008; 212p.
55. Weatherley, P.E. Studies in the water relations of the cotton plant. I-The field measurements of water deficits in leaves. *New Phytol.* **1950**, *49*, 81–97. [CrossRef]
56. Scotti-Campos, P.; Pham-Thi, A.T.; Semedo, J.N.; Pais, I.P.; Ramalho, J.C.; Matos, M.C. Physiological responses and membrane integrity in three Vigna genotypes with contrasting drought tolerance. *Emir. J. Food Agric.* **2013**, *5*, 1002–1013. [CrossRef]
57. Arnon, D.I. Copper enzymes in isolated chloroplasts: Polyphenoloxidase in Beta vulgaris. *Plant Physiol.* **1949**, *24*, 1–15. [CrossRef] [PubMed]
58. Costa, A.d.F.S.d.; Costa, A.N.d.; Ventura, J.A.; Fanton, C.J.; Lima, I.d.M.; Caetano, L.C.S.; Santana, E.N.d. *Recomendações Técnicas para o Cultivo do Maracujazeiro*; Incaper: Vitória, Brazil, 2008; 55p.
59. Pinheiro, F.W.; Lima, G.S.d.; Sousa, P.F.d.N.; Souza, W.B.d.; Soares, L.A.d.A.; Gheyi, H.R.; Felix, C.M.; Paiva, F.J.d.S. Potassium fertilization in the cultivation of sour passion fruit under irrigation strategies with brackish water. *Rev. Bras. Eng. Agríc. Ambient.* **2023**, *27*, 42–50. [CrossRef]
60. Govaerts, B.; Sayre, K.D.; Lichter, K.; Dendoover, L.; Decker, J. Influence of permanent raised bed planting and residue management on physical and chemical soil quality in rainfed maize/wheat systems. *Plant Soil* **2007**, *291*, 39–54. [CrossRef]
61. Hotelling, H. Multivariate quality control. In *Techniques of Statistical Analysis*; Eisenhart, C., Hastay, M.W., Wallis, W.A., Eds.; John Wiley & Sons: New York, NY, USA, 1947; 73p.
62. Hair, F.J.; Black, C.W.; Badin, B.J.; Anderson, R.E.; Tathan, R.L. *Análise Multivariada de Dados*, 6th ed.; Sant'Anna, A.S., Translator; Bookman: Porto Alegre, Brazil, 2009; 688p.
63. Ferreira, D.F. Sisvar: A computer analysis system to fixed effects split plot type designs. *Rev. Bras. Biom.* **2019**, *37*, 529–535. [CrossRef]

plants MDPI

Article

Impact of Nickel Toxicity on Growth, Fruit Quality and Antioxidant Response in Zucchini Squash (*Cucurbita pepo* L.)

Oumayma Labidi [1], Rim Kouki [1], Saida Hidouri [2], Hana Bouzahouane [3,4], Isabel Caçador [5], Rosa M. Pérez-Clemente [6] and Noomene Sleimi [1,*]

1 Laboratory RME—Resources, Materials and Ecosystems, Faculty of Sciences of Bizerte, University of Carthage, Jarzouna, Bizerte 7021, Tunisia; oumayma_abidi11@yahoo.fr (O.L.); kouki.rim.kouki@gmail.com (R.K.)
2 LR12SP13, Faculty of Medicine of Monastir, University of Monastir, Av. Avicenne, Monastir 5000, Tunisia; saida.hidouri2020@gmail.com
3 Faculty of Natural and Life Sciences, University of Mohamed Cherif Messaadia, Souk-Ahras 41000, Algeria; hana_microbiologie@yahoo.fr
4 Laboratory of Environmental Biosurveillance, Faculty of Sciences, University of Badji Mokhtar, Annaba 23000, Algeria
5 MARE—Centro de Ciências do Mar e do Ambiente, Faculdade de Ciências, Universidade de Lisboa, Campo Grande, 1749-016 Lisboa, Portugal; micacador@ciencias.ulisboa.pt
6 Department de Biologia, Bioquímica i Ciències Naturals, Universitat Jaume I, Campus Riu Sec, 12071 Castelló de la Plana, Spain; rosa.perez@uji.es
* Correspondence: noomene.sleimi@gmail.com

Abstract: The impact of trace metal elements (TMEs) on plants is one current pollution problem, the severity of which is increasing with industrial development, population growth and inappropriate agricultural practices. The latter can have irreversible effects on ecosystems, including species extinction, trophic chain contamination and altered human health, particularly in the case of consumed plants such as zucchini squash (*Cucurbita pepo* L.). This study aims to investigate the effects of nickel on various physiological and biochemical parameters of zucchini growth, with a particular focus on how this toxic metal impacts the quality of fruit that is consumed by humans. To achieve this, plants aged 45 days were grown for one month on solid media loaded with different concentrations of Ni (0, 100, 300 and 500 µM). The results showed that exposure of plants to Ni resulted in significantly altered growth and higher accumulation of Ni in the shoots (1314 $\mu g \cdot g^{-1}$ DW) than in roots and fruits. Concerning non-enzymatic antioxidants, the results showed that Ni toxicity significantly increased total polyphenols, especially in shoots at 300 µM Ni, while flavonoid content decreased in the roots and shoots in response to Ni treatment. Our results also show that nickel tolerance in *C. pepo* is ensured by a combination of several mechanisms such as an increase in the content of proline. This species can survive and tolerate, to different degrees, toxic cations at concentrations up to 500 µM but with visible symptoms of toxicity such as chlorosis of the leaves. Indeed, based on thresholds of hyperaccumulation, we can qualify *Cucurbita pepo* as a hyperaccumulator species of nickel.

Keywords: *Zucchini*; TME toxicity; non-enzymatic antioxidants; proline

check for updates

Citation: Labidi, O.; Kouki, R.; Hidouri, S.; Bouzahouane, H.; Caçador, I.; Pérez-Clemente, R.M.; Sleimi, N. Impact of Nickel Toxicity on Growth, Fruit Quality and Antioxidant Response in Zucchini Squash (*Cucurbita pepo* L.). *Plants* **2024**, *13*, 2361. https://doi.org/10.3390/plants13172361

Academic Editors: Geovani Soares de Lima, Lauriane Almeida dos Anjos Soares and Francisco Vanies Da Silva Sá

Received: 30 June 2024
Revised: 7 August 2024
Accepted: 20 August 2024
Published: 24 August 2024

1. Introduction

Soil contamination by trace metal elements (TMEs) is one of the biggest environmental problems worldwide [1]. It presents a risk of toxicity for living beings and humans throughout the food chain [2]. It hampers growth and crop viability depending on the degree of contamination. Studies by Sleimi et al. [3] indicate that many plants from polluted soil were contaminated by TMEs. Likewise, Karimi et al. [4] showed that cultivated crops (*Vicia faba* and *Brassica arvensis*) were contaminated with TMEs such as Pb, Cd and Ni. Altogether, soil contamination finally results in inferior crop quality, harmful effects on human health, contamination of water sources and a negative impact on ecosystem and biodiversity [5].

High levels of geogenic nickel may be transferred from soil to staple crops, resulting in elevated nickel concentrations in cereals. Hseu and Lai [6] reported Ni concentrations ranging from 1.50 to 5.54 mg Kg^{-1} in rice grown in soils with geogenic nickel. Soil pollution is considerable when the compounds introduced alter their function or pose a threat to human health [7]. Health problems may appear following ingestion of food grown on the contaminated site or via inhalation of dust [8].

In plants, some TMEs (Cu, Zn, Ni, Fe, Co) are essential to major physiological processes, in particular, respiration, photosynthesis and macronutrient assimilation. Moreover, some of these TMEs are involved in molecular processes such as the control of gene expression, protein biosynthesis, nucleic acids, phytohormones, chlorophyll and secondary metabolites [9,10].

Environmental contamination by TMEs causes toxicity in biological molecules by producing reactive species, blocking functional groups and replacing basic metal ions [11]. These functional changes translate into visible symptoms at the morphological level [12]. Plant development is reduced, the root system is more compact, and leaves are smaller [13]. The presence of TMEs tends to increase the ratio of root biomass to above-ground biomass, indicating that the plant preferentially maintains its root biomass rather than its above-ground biomass [14]. Root browning, leaf necrosis and chlorosis also occur [15]. Chlorosis is more pronounced in young leaves, while necrosis is more noticeable in older leaves, with higher TME content [16]. However, at low doses, it is possible to observe the opposite effect or hormesis [17].

Nickel is an essential trace element, but at low levels. It is present in high concentrations in ferromagnesian source rocks, in which it partially substitutes for iron or magnesium. It is particularly abundant in the minerals of basic and ultrabasic magmatic rocks and in serpentine ores [18]. Nickel exists in soil mainly in the oxidation state Ni(II). It can occur in the following forms: Ni^{2+}, $NiOH^{+}$, $Ni(OH)_2$ (aq), $Ni(OH)_3{}^{-}$, $Ni(OH)_4{}^{2-}$, $Ni(OH)_2$(s), $NiCO_3$(aq) and $NiHCO^{3+}$ [19]. It can be adsorbed by organic and inorganic matter (Al, Mn, Fe oxides and clay compounds), depending on soil pH [20].

Excessive Ni concentrations in plants can cause iron chlorosis. The nickel content of plants depends on its bioavailability in the soil, the plant species and the specific plant parts analysed. The level of the Ni in plants grown on uncontaminated soil is on the order of 0.05 to 5 mg kg^{-1} dry weight (DW) [21]. Some plants can grow on contaminated or serpentine soils. They can then accumulate more than 1000 mg Kg^{-1} DW, particularly in their leaves [22].

At low concentrations (0.01 to 5 mg Kg^{-1} DW), Ni is considered an essential element. It is involved in the structure of urease, which is responsible for urea hydrolysis in leaves [23]. The toxic effects of nickel and other TMEs mainly manifest through growth inhibition, the extent of which depends on dose, duration of treatment, plant species and chemical form [24,25]. At high concentrations, it becomes toxic: over 10 mg Kg^{-1} for sensitive plants and over 50 mg Kg^{-1} for tolerant plants [26]. This phytotoxicity causes chlorosis [27], reduced growth and metabolic disorders.

Zucchini (*Cucurbita pepo*) is the immature comestible crop of *Cucurbita pepo* spp. It is an important food crop, providing an essential source of minerals, vitamins and fibre [28]. Given the high prevalence of TMEs in agricultural soils, preventives studies and analyzes are therefore necessary as regards the marketing of plants cultivated in contaminated areas. The objective of this research was to investigate the capacity of accumulating Ni in *Cucurbita pepo* and to examine the impact of these elements on the plants' growth. Our objectives were accomplished by measuring the endogenous amounts of K, Ca and Mg as well as the protein content. Additionally, the contents of proline, total phenols, and flavonoids in the different tissues of the zucchini plant were determined to better understand the adaptation levels and tolerance limits of *Cucurbita pepo* in response to nickel-induced stress.

2. Material and Methods

2.1. Plant Culture and Treatments

Cucurbita pepo (*Cucurbitaceae*) plants used in the experiments were grown from seeds germinated in a perlite and gravel substrate within plastic pots, as described by Labidi et al. [29]. The cultivation took place in a greenhouse under semi-controlled conditions at the Faculty of Science of Bizerte, Tunisia. The seedlings were irrigated with 200 mL of Hewitt's nutrient solution [30] three times a week. Plants, aged 45 days, were treated for 4 weeks with various Ni concentrations: (a) 0 μM $NiCl_2$ (control); (b) 100 μM $NiCl_2$; (c) 300 μM $NiCl_2$; and (d) 500 μM $NiCl_2$, added to the nutrient solution.

On the harvest day, plants (aged 75 days) from each treatment group were randomly divided into 2 groups of 5 plants. At the first step, shoots were separated from roots and were washed with cold distilled water and stored in liquid nitrogen for later testing.

To obtain fruits, *Cucurbita pepo* seeds were germinated in peat and then moved to 100% perlite as a substrate. After 2 months, the plants were treated for 10 days with the concentrations of Ni as previously described. This culture was carried out in a greenhouse under semi-controlled conditions at the University of Jaume I, Spain.

The fruits were harvested from 8 weeks after seeding until the 12th week. The harvest was performed by cutting the zucchini fruits to separate them from the plant.

2.2. Parameters of Growth

The ratio of shoot/root dry biomass (S/R) and the tolerance index percentage (TI %) [31] were calculated as follows:

S/R = DW shoots/DW roots

TI (%) = DW-treated plants/DW control plants $\times$ 100

The relative growth rate (RGR) was determined following Sleimi and Abdelly [32]:

RGR = (ln (W2) − ln (W1)/(t2 − t1)

W is the fresh matter at the beginning of treatment (W1) and at the end (W2).

(t2 − t1) is the duration of this period.

2.3. Assay of Nickel and Minerals Elements

The extraction of Ni and mineral element contents from the dry matter of the treated plants was achieved using a mixture of acids: HNO_3:H_2SO_4:$HClO_4$ (10:1:0.5; *v/v/v*) at 110 °C for 2 h [29,33]. Afterward, the obtained extracts were diluted with a volume of 50 mL of nitric acid solution (HNO_3) at 0.5% and filtered. The measurements of Ni, Ca, Mg and K contents was carried out by atomic absorption spectrometry (Perkin Elmer PinAAcle 900T, Waltham, MA, USA). The recorded values were the average of three readings. The limit of detection of Ni, Ca, Mg and K were 0.06, 0.092, 0.0078 and 0.043 mg L^{-1}, respectively.

2.4. Assay of Total Phenolics and Flavonoids

Phenolic compounds were quantitatively determined using the Folin–Ciocalteu method, which involves the oxidation of a reagent with blue tungsten–molybdenum oxide; the intensity of the blue colour informs the concentration of polyphenols in the extracts [34]. Absorbance was measured at 765 nm, using spectrophotometry (UV/visible, TOMAS UV-1200) and expressed as mg gallic acid equivalents.

Total flavonoid content was determined using the colorimetric method described by Zhishen et al. [35], which was based on the formation of a complex between flavonoids and aluminium trichloride. The flavonoid content was measured at 430 nm, using spectrophotometry (UV/visible, TOMAS UV-1200). Rutine was used as a standard reference for the quantitative estimation of flavonoids.

2.5. Assay of Proline

Proline determination followed the method by Bates et al. [36]. Freshly frozen material (50 mg) was ground in 3% sulphosalicylic acid (5 mL) and centrifuged. The supernatant was recovered.

Then, 1 mL of the filtrate was mixed with 1 mL of ninhydrin reagent followed by 1 mL of glacial acetic acid. All samples were incubated at 100 °C for 60 min at 100 °C in a water bath, cooled and centrifuged 5 min at 2000 rpm at 4 °C. Absorbance was read at 520 nm with a spectrophotometer (Thermo Spectronic Genesys 10, Waltham, MA, USA). Total quantification was performed according to a calibration curve prepared with commercial-standard proline.

2.6. Assay of Proteins

Frozen zucchini roots and shoots were ground in poly-vinyl-polypyrrolidone (PVP) and homogenized in 50 mM potassium phosphate buffer (pH 7) containing 0.1 mM EDTA (ethylenediaminetetraacetic), 0.1 mM phenyl-methyl-sulphonyl-fluoride (PMSF) acid and 1 mM DTT. The mixture was centrifuged at 12,000× g at 4 °C for 30 min. Supernatants were stored at −20 °C for protein and enzyme analysis. The protein content of individual extracts was calculated by the Bradford reaction, utilizing bovine serum albumin (BSA) as reference standard [37].

2.7. Statistical Analysis

Samples were tested for a minimum of 3 repetitions, and the mean values and standard error (±) are shown as vertical bars in the graphs. For all data, an analysis of variance (ANOVA) and principal component analysis (PCA) were performed using Statistica 8 and Sigmaplot v. 14.0.0.124 software (Systat Software, San Jose, CA, USA).

3. Results

3.1. Mineral Nutrition

The addition of Ni^{2+} to irrigation solutions induced a significant increase in Ca^{2+}, Mg^{2+} and K^+ content of roots and shoots (Table 1). Indeed, a large accumulation of Mg^{2+} and K^+ was noted in the roots with a concentration of 100 μM Ni^{2+} (Table 1).

Table 1. Variation in calcium (Ca), magnesium (Mg) and potassium (K) contents in the roots, shoots and fruits of zucchini with different concentrations of Ni (0, 100, 300 and 500 μM). Mean values of 6 replicates. Values with different letters are significantly different ($p \leq 0.05$) according to Tukey's HSD test.

Treatments (Ni, μM)	Ca^{2+}, mg·g^{-1} DW			Mg^{2+}, mg·g^{-1} DW			K^+, mg·g^{-1} DW		
	Roots	Shoots	Fruits	Roots	Shoots	Fruits	Roots	Shoots	Fruits
0 μM	54.51 ± 1.38 a	55.90 ± 2.19 a	13.58 ± 1.01 a	7.22 ± 0.20 a	11.48 ± 0.40 a	12.66 ± 0.70 a	11.33 ± 0.52 a	17.63 ± 0.66 a	71.92 ± 3.16 a
100 μM	67.67 ± 4.48 b	75.01 ± 2.21 b	11.94 ± 1.11 ab	23.55 ± 0.98 b	17.31 ± 0.51 b	14.01 ± 0.76 a	73.44 ± 2.26 b	26.74 ± 1.00 b	73.86 ± 3.69 a
300 μM	71.15 ± 1.43 b	84.08 ± 2.59 b	15.20 ± 1.57 a	9.16 ± 0.30 a	11.92 ± 1.11 c	12.48 ± 2.02 a	15.47 ± 1.26 a	53.40 ± 2.78 c	58.13 ± 3.53 a
500 μM	71.25 ± 3.13 b	81.26 ± 3.28 b	7.64 ± 0.89 b	7.18 ± 0.38 a	9.15 ± 0.29 a	12.04 ± 1.60 a	10.70 ± 0.64 a	19.35 ± 0.50 a	73.77 ± 6.72 a

Cucurbita pepo fruits irrigated with the nutrient solution supplemented with different doses of Ni^{2+} showed a decrease in Ca^{2+} concentration, but only with 500 μM Ni. On the other hand, no significant variation in Mg^{2+} and K^+ nutrient content was observed in the fruit. Nevertheless, at a concentration of 300 μM Ni, there was a decrease in the K^+ content of the *Cucurbita* fruits (Table 1).

3.2. Nickel Content

The results showed that Ni^{2+} content increased in different parts of the zucchini plants with increasing Ni^{2+} concentrations in the culture medium. The endogenous concentration of Ni^{2+} ions increased more significantly in the roots and shoots compared to the fruits. Maximum accumulation was observed at 500 µM, reaching 1140.07 µg· g^{-1} DW in the roots, 1314 µg·g^{-1} DW in the shoots, and 195 µg·g^{-1} DW in the fruits (Figure 1).

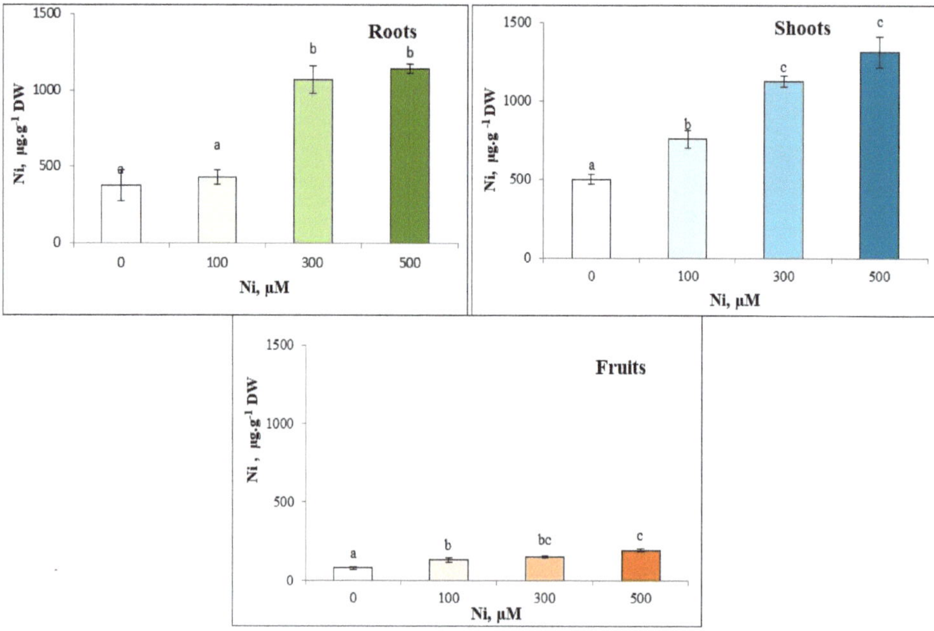

Figure 1. Variation in nickel content in the roots, shoots and fruits of zucchini with different concentrations of Ni (0, 100, 300 and 500 µM). Mean values of 6 replicates. Values with different letters are significantly different ($p \leq 0.05$) according to Tukey's HSD test.

3.3. *Plant Morphology and Growth*

From the 3rd week of treatment, Ni affected plant development and growth. Indeed, chlorosis appeared on *Cucurbita pepo* leaves treated with Ni^{2+} compared to the leaves of control plants (Figure 2a,c). At the end of the treatment, these toxicity symptoms were more pronounced and accompanied by leaf abscission.

Figure 2. Leaf chlorosis (**a**), fruit damage (**b**) in *Cucurbita pepo* exposed to high concentrations of Ni^{2+} (500 µM) after 21 and 60 days of treatment, respectively. (**c**): a leaf of the control plant.

Concerning fruits (60 days of treatment), brown spots were observed on *Cucurbita pepo* fruit treated with Ni^{2+} at 500 µM (Figure 2b).

The presence of Ni^{2+} in the culture medium reduced dry biomass production in *Cucurbita pepo*, with the dry weight decreasing from 186.9 mg to 103.4 mg and from 2699.9 mg to 442.5 mg, respectively, in roots and shoots with 500 µM Ni (Table 2).

Table 2. Variation in dry weight (DW), shoot/root ratio (S/R) and tolerance index (TI) of zucchini with different concentrations of Ni (0, 100, 300 and 500 µM). Mean values of 10 replicates. Values with different letters are significantly different ($p \leq 0.05$) according to Tukey's HSD test.

Treatments (Ni, µM)	DW (g)		S/R	TI (%)
	Roots	Shoots		
0	186.9 ± 9.00 ^{a}	2699.9 ± 173.8 ^{a}	15.1 ± 1.8 ^{a}	
100	204.9 ± 15.2 ^{a}	1901.9 ± 118.0 ^{b}	9.8 ± 1.0 ^{b}	73.9 ± 6
300	127.9 ± 10.5 ^{b}	617.9 ± 49.9 ^{c}	5.1 ± 0.5 ^{c}	22.2 ± 1.6
500	103.4 ± 7.2 ^{b}	442.5 ± 33.5 ^{c}	4.4 ± 0.4 ^{c}	14.6 ± 1.3

The calculation of the S/R ratio shows a significant decrease with Ni^{2+} treatment, indicating that Ni^{2+} affects the shoots more than the roots. This suggests that Ni^{2+} disrupts growth more in the shoots relative to the roots, potentially leading to imbalances in plant development (Table 2).

The estimation of the tolerance index for the whole plant showed a decline as Ni concentrations in the culture medium increased. This decrease indicates that higher levels of Ni reduce the plant's ability to tolerate stress, impacting its overall health and growth (Table 2).

The relative growth rate (RGR) of both shoots and roots shows a notable decrease with increasing Ni^{2+} concentrations in the irrigation solutions (Figure 3). This reduction indicates that higher Ni^{2+} levels significantly hinder the growth of both plant parts.

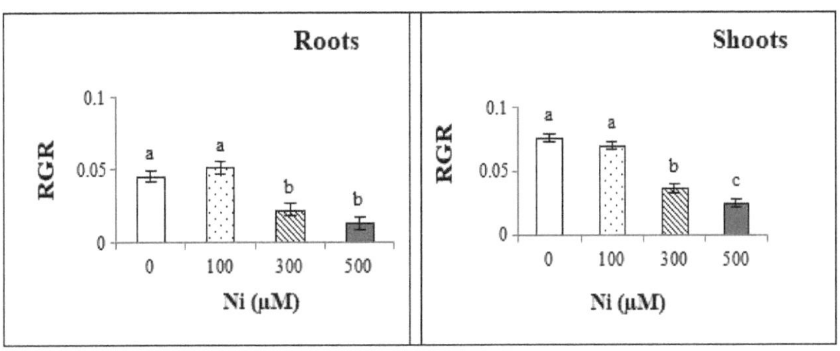

Figure 3. Relative growth rate in the roots and the shoots of *Cucurbita pepo* exposed to different concentrations of Ni^{2+} (0, 100, 300 and 500 µM). Mean values of 10 replicates. Values with different letters are significantly different ($p \leq 0.05$) according to Tukey's HSD test.

3.4. Water Content

The results showed a significant decrease in root water content with increasing Ni^{2+} concentrations, indicating impaired water retention. In contrast, shoot water content increased significantly with higher Ni^{2+} levels, suggesting improved tissue hydration in the shoots. This suggests a shift in water balance within the plant due to Ni^{2+} treatment. (Figure 4).

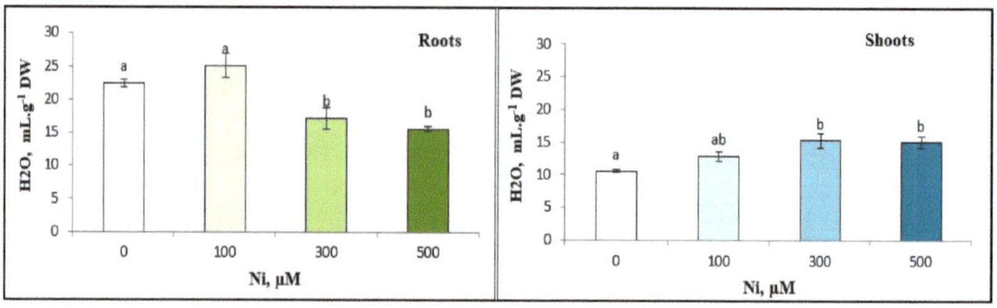

Figure 4. Variation in water content of the roots and the shoots of zucchini with different concentrations of Ni (0, 100, 300 and 500 µM). Mean values of 10 replicates. Values with different letters are significantly different ($p \leq 0.05$) according to Tukey's HSD test.

3.5. Total Polyphenol (TP) Content

The results of the total polyphenol assessment revealed that Ni induced a significant increase in total polyphenol content of all organs of *C. pepo*. In fact, the polyphenol content increased in the roots and shoots according to the Ni^{2+} concentration in the medium. The highest levels of TP in *C. pepo* were found in the shoots of plants exposed to 300 µM Ni^{2+} (34.6 mg·g^{-1} DW) compared with the control (7.98 mg·g^{-1} DW) (Figure 5).

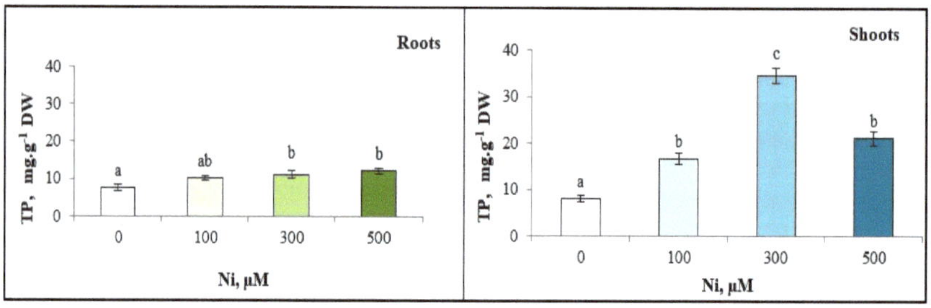

Figure 5. Variation in total polyphenols (TP) content of the roots and the shoots of zucchini with different concentrations of Ni (0, 100, 300 and 500 µM). Mean values of 6 replicates. Values with different letters are significantly different ($p \leq 0.05$) according to Tukey's HSD test.

3.6. Flavonoid Content

Analysis of the variation in flavonoid content of the roots and shoots of *Cucurbita pepo* shows the same behaviour. In fact, these contents decreased as a function of the Ni concentration in the irrigation solutions. However, the higher decrease in flavonoid content was observed in the presence of 500 µM Ni^{2+}, with a 73% decrease in *C. pepo* roots (Figure 6).

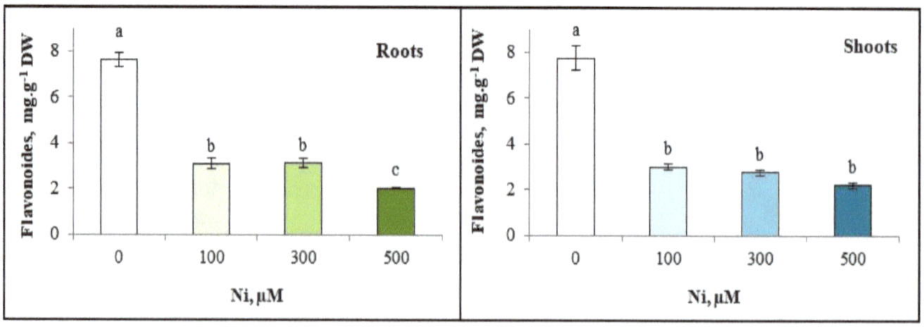

Figure 6. Variation in flavonoid content of the roots and the shoots of zucchini with different concentrations of Ni (0, 100, 300 and 500 µM). Mean values of 6 replicates. Values with different letters are significantly different ($p \leq 0.05$) according to Tukey's HSD test.

3.7. Proline Content

The study of proline content during the vegetative stage in the shoots shows that the addition of Ni^{2+} to the irrigation solutions induces a significant accumulation of proline in *C. pepo* at high doses (500 µM Ni^{2+}). However, Ni-induced stress did not show any significant variations compared with the control in the roots (Figure 7).

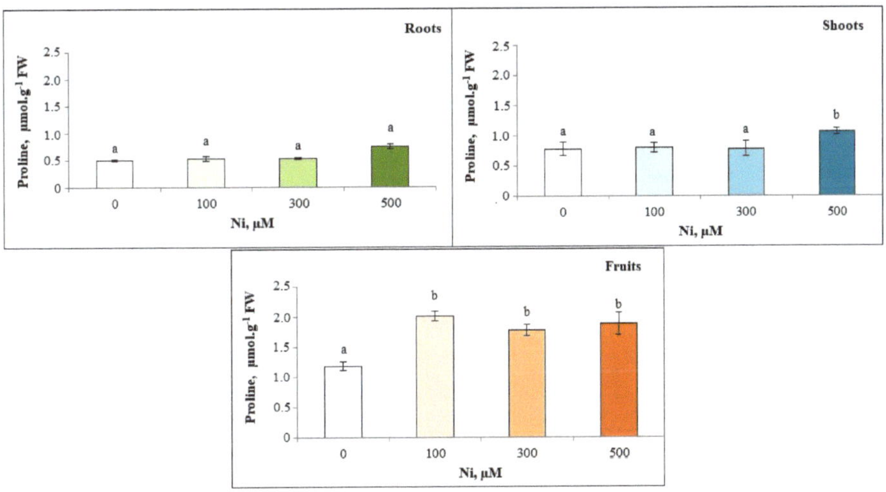

Figure 7. Variation in proline content of the roots, shoots and fruits of zucchini with different concentrations of Ni (0, 100, 300 and 500 µM). Mean values of 6 replicates. Values with different letters are significantly different ($p \leq 0.05$) according to Tukey's HSD test.

3.8. Total Protein Content

The results in Figure 8 show that the addition of the different doses of Ni^{2+} to the irrigation solution did not significantly affect the total protein content of the shoots. However, an accumulation of total protein in the roots was observed with the 100 and 300 µM Ni^{2+} solution (Figure 8).

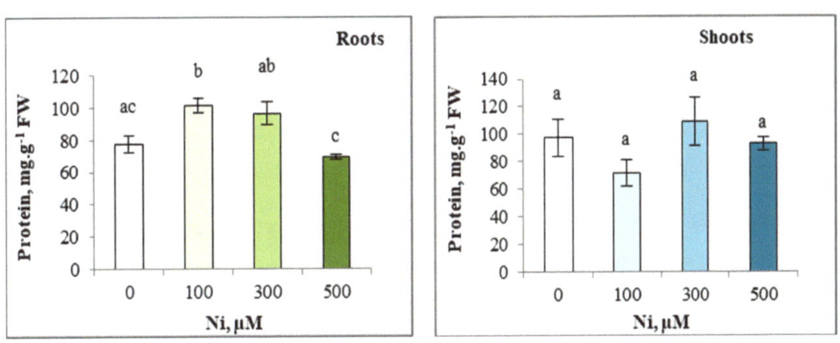

Figure 8. Variation in total protein content of the roots and the shoots of zucchini with different concentrations of Ni (0, 100, 300 and 500 µM). Mean values of 6 replicates. Values with different letters are significantly different ($p \leq 0.05$) according to Tukey's HSD test.

3.9. Principal Component Analysis (PCA)

PCA was carried out in order to study the impact of Ni on the different studied parameters in *C. pepo*, as well as the correlation between these parameters.

Our results showed a high significance; in the shoots of the plants treated with Ni, total variance reached 97.86%, while in the roots, total variance was 95.93%. In the shoots (a), Ni content was negatively correlated with water content, mineral nutrient content, total polyphenols and proline (Figure 9). The data revealed that, in the roots (b), Ca content, total protein and total polyphenols were positively correlated with Ni content (Figure 9). However, DW and Ni content presented negative correlation with flavonoids.

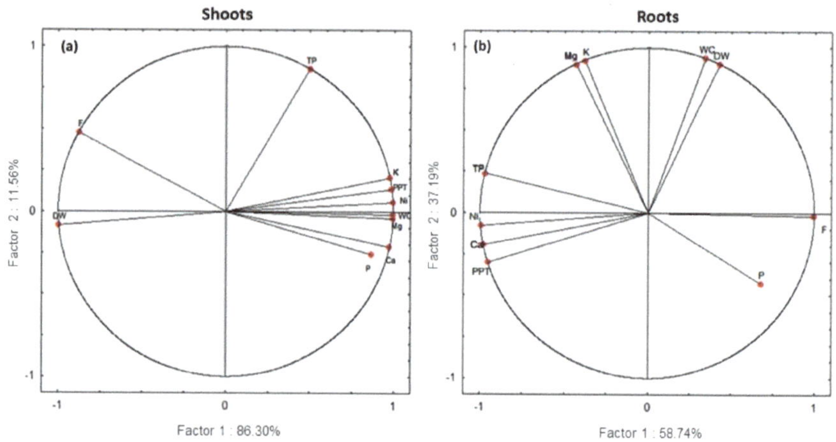

Figure 9. Correlation circle from the PCA of nickel (Ni) content, dry weight (DW), water content (WC), potassium (K), calcium (Ca), magnesium (Mg), proline (P), total protein (TP) total polyphenols (PPT) and flavonoids (F). Data of the shoots (**a**) and the roots (**b**) of *Cucurbita pepo* plants subjected to increasing doses of Ni (0, 100, 300, 500 μM).

4. Discussion

Cultured seedlings exhibit different growth behaviours in response to the presence of TMEs in the culture medium. Our results showed significantly reduced biomass production in *Cucurbita pepo* irrigated with Ni^{2+} supplemented nutrient solution. Indeed, plant growth at the aerial part severely decreased by 84% in *Cucurbita pepo* grown in the presence of 500 μM Ni^{2+}. These seedlings also showed a significant 44% reduction in root growth in the presence of 500 μM Ni. The same results were observed in *Oryza sativa* in the presence of 200 and 400 μM Ni [38]. Similarly, Gajewska and Sklodowska [39] showed that wheat shoot biomass was reduced by 20 and 26% with Ni concentrations of 100 μM and 200 μM, respectively. According to Palacios et al. [40], tomato plant vegetative growth and fruit production decreased significantly with increasing levels of Ni in the nutrient solution. Indeed, plants' Ni uptake reduces shoot and root growth and leaf area [9].

Tolerance index (TI) defines the ability of plants to grow well and tolerate high metal concentrations. Our results show that TI values decreased in *C. pepo* in the presence of Ni in the culture medium. Bouslimi et al. [41] reported a decrease in TI in *Cakile maritima* under the effect of barium.

The depressive effects of Ni on plant growth are due to several factors. Nickel disrupts enzymatic activities [42] and hormonal balances, affecting cell division and increasing stress hormones like abscisic acid [43]. Nickel also displaces essential nutrients like iron and magnesium, causing deficiencies that hinder photosynthesis. Additionally, Ni generates oxidative stress, damaging cellular components. These combined effects significantly disrupt plant growth and development [44].

Disruption of water parameters has also been reported in plants exposed to metal stress [45] (Poschenrieder et al., 1989). Water disruption in plants under metal stress has been documented [45] (Poschenrieder et al., 1989). Our findings show an increase in water content of aerial parts but a decrease in the roots of *C. pepo* under Ni^{2+} stress. Kouki et al. [46] reported similar reductions in water content under metal stress. Barcelo and

Poschenrieder [47] found that metal stress blocks water transport. Ni toxicity reduces water content and transpiration rates, even at low doses [48]. These disturbances cause stomatal closure [49] and inhibit absorptive hair formation, reducing water and nutrient uptake.

In the presence of Ni, the content of mineral nutrients in plant organs may increase, decrease or stay even, and thus may disrupt the uptake and distribution of essential nutrients to different parts of the plant [4,44].

In the present study, the addition of Ni^{2+} to the culture medium induced a significant increase in K^+ levels in the aerial and root parts of *C. pepo*. The improvement in potassium nutrition has been mentioned in several research studies.

Indeed, increased root K^+ levels were observed in tomatoes [40] and maize plants exposed to Ni [50], though some studies reported decreased K^+ under metal stress, possibly due to differences in species and methods [51].

Magnesium, the central atom in chlorophyll, activates many enzymes [52]. Magnesium levels decrease under metal stress as TMEs replace Mg^{2+} in chlorophyll, causing its destruction [53] (Rizwan et al., 2019). Reduced Mg^{2+} has been noted in tomatoes exposed to Ni [40]. TMEs alter nutrient uptake by changing membrane permeability and nutrient transport [54–56]. Conversely, some studies found increased Mg^{2+} under metal stress. Magnesium deficiency could mitigate TME toxicity by maintaining ferric status and increasing iron in Ni^{2+}-treated leaves, preventing chlorosis and restoring chlorophyll [57]. Mg^{2+} may also enhance antioxidant capacity for detoxification and photosynthetic protection.

One of the probable mechanisms for decreasing the uptake of macro- and micronutrients relies on competition for the common binding sites due to the comparable ionic radii of Ni^{2+} and other cations. The decline in nutrient uptake may also result from Ni-induced metabolic disorders that affect the structure and enzyme activities of cell membranes [44].

Calcium nutrition in plants is generally affected by metal stress. Our results showed that the addition of Ni^{2+} to the culture medium increased Ca^{2+} levels in the roots and aerial parts of *C. pepo*. This aligns well with the work of Sghaier et al. [58], who showed that Ca^{2+} uptake was increased in *Tamarix gallica* under the effect of aluminium. Calcium has been reported to attenuate the adverse effects of TMEs on plants [59]. The decrease in Ni^{2+} toxicity by Ca^{2+} may be due to the fact that plasma membrane surfaces are generally negatively charged, and high Ca^{2+} concentrations tend to neutralize them and thus minimize the toxic effect of Ni^{2+}.

Nickel is known to be readily absorbed by most species, as a divalent cation. It can compete with other cations, including Ca^{2+}, Mg^{2+}, Fe^{2+} and Zn^{2+} [60]. Nickel levels in the roots and aerial parts of *C. pepo* increased with increasing Ni concentration in the culture medium. Consequently, Ni phytoavailability depends on both Ni level in solution and treatment time [40]. Moreover, Sleimi et al. [61], show that the increase in TME content in roots and aerial parts is proportional to the increase in TME concentration in the irrigation solution in *Cucumis sativus*. Our results show that *C. pepo* can accumulate 1314 $\mu g \cdot g^{-1}$ DW Ni^{2+} in its aerial parts. These data suggest that *C. pepo* has a strong capacity to accumulate and tolerate these TMEs at the tissue level.

According to Reeves and Baker [62], a species is considered a hyperaccumulator if it accumulates more than 1000 $\mu g \cdot g^{-1}$ DW Ni^{2+} in its aerial parts. However, Ni hyperaccumulators can store more than 1000 $\mu g \cdot g^{-1}$ Ni DW in their tissues when grown on a nickel-rich substrate [10]. Furthermore, we can qualify *C. pepo* as a Ni^{2+} hyperaccumulative species.

In response to abiotic stresses, the biosynthesis of secondary metabolites generally increases in plants. Phenolic compounds confer, in fact, a higher tolerance to plants against various abiotic stresses, such as TME, salinity, drought, temperature, pesticides and UV radiation [63]. Indeed, polyphenol accumulation could play a key role in plant balance and adaptation [64]. This accumulation is more important in the leaves than in the roots of the plant; in this respect, several authors, such as Jonnala et al. [65] and Chaib et al. [66], have shown the richness of cereal leaves with polyphenols, which possess great antioxidant power.

Our results show that the production of total polyphenols increases under Ni^{2+}-induced metal stress in the aerial parts and roots of *C. pepo*. Indeed, a significant threefold increase was observed in plants grown in the presence of 500 µM Ni^{2+} compared with control plants. These results are consistent with Rastgoo et al. [67], who observed an increase in polyphenols under metal stress (Pb, Co, Cd, Ag) in *Aeluropus littoralis*; the same behaviour was reported in lead-stressed *Raphanus sativus* [68].

Stimulating the biosynthesis of phenolic compounds helps protect plants from oxidative stress due to metal exposure [69]. These compounds, with antioxidant properties, scavenge free radicals, reducing cell membrane peroxidation and protecting plant cells. Phenols also enhance membrane stability, contributing to cell rigidity and creating barriers against heavy metals [70]. They chelate metal ions and inhibit the Fenton reaction, promoting plant growth under metal stress [71].

Ma et al. [72] found a direct correlation between TMEs in the culture medium and the expression of phenylalanine ammonia-lyase (PAL), a key enzyme in phenolic compound biosynthesis in response to stress [73]. This explains the accumulation of phenolic acids in TME-stressed plants. Handa et al. [69] showed that increased PAL expression leads to higher levels of total phenols and flavonoids in *Brassica juncea* under Cr stress. Similarly, Pb stimulates PAL activity in soybean [74]. The accumulation of phenolic compounds results from the regulation of phenylpropanoid enzyme biosynthesis, including PAL, chalcone synthase, shikimate dehydrogenase, cinnamyl alcohol dehydrogenase and polyphenol oxidase [75], driven by the modulation of gene transcript levels under metal stress [69].

On the other hand, flavonoid accumulation has been demonstrated in many species and under different abiotic constraints [76,77]. Flavonoid production differs from that of TPP in the aerial parts and roots of *C. pepo* grown in the presence of Ni. Indeed, the flavonoid content of *C. pepo* leaves and roots decreases proportionally with nickel concentration in irrigation solutions; this could be explained by plant resistance to these concentrations [78]. Plant response to metal stress thus varies from one concentration to another [79]. In general, the presence of Ni in irrigation solutions inhibits flavonoid synthesis. Murch et al. [80] reported reduced flavonoid levels in plants exposed to Ni and Cu. Similarly, reduced flavonoid levels were observed in *Atriplex canescens* roots exposed to Cd [81]. However, Bouslimi et al. [77] report an accumulation of flavonoids in different parts of the plant under stress induced by lanthanum. Similarly, Bretzel et al. [82] have shown that flavonoid content increases in *Taraxacum officinale* grown in the presence of Cr, Pb, Cu, Ni or Zn.

Flavonoids are good inhibitors of enzymes responsible for free radical production, such as xanthine oxidase, cyclooxygenase and lipooxygenase [76].

Actually, flavonoids enhance TME chelation, reducing harmful radicals in plant cells [83]. They scavenge H_2O_2 and are crucial in the phenolic/ascorbate peroxidase cycle [84]. Their antioxidant properties come from hydroxyl groups that stabilize radicals [85]. Berni et al. [86] noted that flavonoids act secondarily in plant defence after phenolic acids, conserving energy.

Proline levels in *Cucurbita pepo* increase with Ni concentration more in fruits and leaves than roots, suggesting synthesis in leaves and migration to roots [87]. This is consistent with high proline in *Vaccinium myrtillus* leaves under Zn stress [88], and in *Pinus sylvestris* needles with TMEs [89]. Hadjadj et al. [90] link increased proline to abiotic stress tolerance.

Proline accumulation results from inhibition of its oxidation, increased protein catabolism and/or synthesis of this amino acid. This accumulation protects the cell membrane and is involved in osmotic adjustment, metal chelation and detoxification, enzyme protection, regulation of cytosolic acidity, stabilization of the protein synthesis machinery and scavenging of reactive oxygen species [91].

Overall, the species most sensitive to stress react by accumulating proline more rapidly. On the other hand, those that have proved tolerant show relative stability or low accumulation of their proline content [92]. Brinis and Belkhoudja [93] justified the reduced levels of proline in stressed plants by the fact that the level of stress does not seem to have triggered the proteolysis required to obtain significant quantities of proline, and consequently there is no response to the point of envisaging a possible blockage of the plant's metabolic activity. Thus, Spormann et al. [94] clarify the implications of proline accumulation in plants under abiotic stress induced by soil degradation factors, reinforcing the idea that proline quantification should not be employed as a sole indicator of stress sensitivity or resilience but rather complemented with further biochemical and physiological endpoints. The decrease in proline in plants subjected to TME-induced stress has been examined in several species, *Raphanus sativus* stressed by zinc [95], *Brassica juncea* subjected to high doses of lead and cadmium [96], and *Abelmoschus esculentus* root's subjected to 100 and 200 µM doses of Ba [92].

The results of the present study showed an increase in protein levels in *C. pepo* with increasing the as applied doses of Ni. Dridi et al. [97] reported that Pb application resulted in an increase in nitrogen content of shoots and roots of *Helianthus annuus*. This protein accumulation could be the consequence of the plant's synthesis of defence proteins against this metal stress [98], including proteins involved in maintaining the cell's redox status, such as ascorbate, or in metal sequestration (GSH, PC) [99] (Yadav, 2010). Other studies report the induction of heat shock-specific proteins under metal stress, demonstrating their role in the adaptive response [100].

5. Conclusions

In conclusion, this study demonstrates that Ni concentrations induce notable changes in physiological and biochemical parameters in the roots, shoots, and fruits of zucchini plants (*Cucurbita pepo*). While the doses of Ni used in this study are not highly toxic to zucchini, the plant exhibits significant adaptability, evidenced by its survival and ability to accumulate substantial amounts of Ni in the leaves. This accumulation classifies *C. pepo* as a hyper-accumulator species.

Our findings highlight that increases in proline content play a crucial role in the plant's response to Ni toxicity. Although *Cucurbita pepo* can tolerate Ni levels up to 500 µM, with observable symptoms such as leaf chlorosis, the accumulation of Ni in the fruits poses a risk of contaminating the food chain. Consequently, it is advisable to utilize *C. pepo* primarily for phytoremediation purposes, specifically to clean soils contaminated with nickel, rather than for consumption. This approach can mitigate the potential entry of Ni into the trophic chain and address the broader issue of TMEs in agricultural ecosystems.

Author Contributions: Conceptualization, N.S., I.C. and R.M.P.-C.; methodology, O.L. and R.K.; formal analysis, O.L. and R.K.; writing-original draft preparation, O.L. and H.B.; writing-review and editing, R.M.P.-C. and S.H.; visualization, N.S. and I.C.; supervision, N.S. and R.M.P.-C. All authors have read and agreed to the published version of the manuscript.

Funding: This research received no external funding.

Data Availability Statement: Data are contained within the article.

Acknowledgments: The authors also gratefully acknowledge supported by Universitat Jaume I (Spain) through grant No.UJI6B2019-11 and supported by FCiências.ID | Projeto 1106/Monitorização ativa de metais pesados em Macrófitas.

Conflicts of Interest: The authors declare no conflict of interest.

References

1. Neethu, S.; Midhun, S.J.; Radhakrishnan, E.K.; Jyothis, M. Microbially synthesized nanomaterials for remediation of contaminated soil and water environment. In *Microbe Mediated Remediation of Environmental Contaminants, Series in Food Science, Technology and Nutrition*; Kumar, A., Singh, V.K., Singh, P., Mishra, V.K., Eds.; Woodhead Publishing: Sawston, UK, 2021; pp. 157–176. [CrossRef]
2. Carre, F.; Caudeville, J.; Bonnard, R.; Bert, V.; Boucard, P.; Martine, R. Soil contamination and human health: A major challenge for global soil security. In *Global Soil Security Symposium*; Springer: College Station, TX, USA, 2015; pp. 275–295.
3. Sleimi, N.; Bankaji, I.; Dallai, M.; Kefi, O. Accumulation des éléments traces et tolérance au stress métallique chez les halophytes colonisant les bordures de la lagune de Bizerte. *Rev. Ecol. (Terre Vie)* **2014**, *69*, 49–59. [CrossRef]
4. Karimi, R.; Solhi, S.; Salehi, M.; Solhi, M.; Mollahosaini, H. Effects of Cd, Pb and Ni on growth and macronutrient contents of *Vicia faba* L and *Brassica arvensis* L. *Int. J. Agron. Plant Prod.* **2013**, *4*, 739–744.

5. Cachada, A.; Rocha-Santos, T.; Duarte, A.C. Soil and pollution: An introduction to the main issues. In *Soil Pollution*; Duarte, A.C., Cachada, A., Rocha-Santos, T., Eds.; Academic Press: Cambridge, MA, USA, 2018; pp. 1–28. [CrossRef]
6. Hseu, Z.Y.; Lai, Y.J. Nickel accumulation in paddy rice on serpentine soils containing high geogenic nickel contents in Taiwan. *Environ. Geochem. Health* **2017**, *39*, 1325–1334. [CrossRef]
7. Masciandaro, G.; Macci, C.; Peruzzi, E.; Ceccanti, B.; Doni, S. Organic matter–microorganism–plant in soil bioremediation: A synergic approach. *Rev. Environ. Sci. Biotechnol.* **2013**, *12*, 399–419. [CrossRef]
8. Lupolt, S.N.; Agnew, J.; Burke, T.A.; Kennedy, R.D.; Nachman, K.E. Key Considerations for assessing soil ingestion exposures among agricultural workers. *J. Expo. Sci. Environ. Epidemiol.* **2022**, *32*, 481–492. [CrossRef] [PubMed]
9. Yusuf, M.; Fariduddin, Q.; Hayat, S.; Ahmad, A. Nickel: An overview of uptake, essentiality and toxicity in plants. *Bull. Environ. Contam. Toxicol.* **2011**, *86*, 1–17. [CrossRef] [PubMed]
10. Van der Ent, A.; Baker, A.J.M.; Reeves, R.D.; Pollard, A.J.; Schat, H. Hyperaccumulators of metal and metalloid trace elements: Facts and fiction. *Plant Soil* **2013**, *362*, 319–334. [CrossRef]
11. Mandal, C.; Ghosh, N.; Adak, M.K.; Dey, N. Interaction of polyamine on oxidative stress induced by exogenously applied hydrogen peroxide in *Salvinia natans* Linn. *Theor. Exp. Plant Physiol.* **2013**, *25*, 223–230. [CrossRef]
12. Alloway, B.J. Sources of Heavy Metals and Metalloids in Soils. In *Heavy Metals in Soils: Trace Metals and Metalloids in Soils and Their Bioavailability*; Alloway, B.J., Ed.; Springer: Dordrecht, The Netherlands, 2013; pp. 11–50.
13. Selvam, A.; Wong, J.W.C. Phytochelatin synthesis and cadmium uptake of *Brassica napus*. *Environ. Technol.* **2008**, *29*, 765–773. [CrossRef]
14. Perriguey, J.; Sterckeman, T.; Morel, J.L. Effect of rhizosphere and plant-related factors on the cadmium uptake by maize (*Zea mays* L.). *Environ. Exp. Bot.* **2008**, *63*, 333–341. [CrossRef]
15. Prasad, M.N.V. Cadmium Toxicity and Tolerance in Vascular Plants. *Environ. Exp. Bot.* **1995**, *35*, 525–545. [CrossRef]
16. Carrier, P.; Baryla, A.; Havaux, M. Cadmium distribution and microlocalization in oilseed rape (*Brassica napus*) after long-term growth on cadmium-contaminated soil. *Planta* **2003**, *216*, 939–950. [CrossRef]
17. Poschenrieder, C.; Cabot, C.; Martos, S.; Gallego, B.; Barceló, J. Do Toxic Ions Induce Hormesis in Plants? *Plant Sci.* **2013**, *212*, 15–25. [CrossRef]
18. Baize, D.; Paquereau, H. Teneurs Totales en Éléments Traces dans les Sols Agricoles de Seine-et-Marne. *Etude Gest. Sols* **1997**, *4*, 77–94.
19. WHO. Nickel in Drinking-Water. In *Background Document for Development of WHO: Guidelines for Drinking-Water Quality*; WHO: Geneva, Switzerland, 2005.
20. Mellis, V.E.; da Cruz, M.C.P.; Casagrande, C.J. Nickel adsorption by soils in relation to pH, organic matter, and iron oxides. *Sci. Agric.* **2004**, *61*, 190–195. [CrossRef]
21. Brown, P.H.; Welch, R.M.; Cary, E.E. Nickel: A Micronutrient Essential for Higher Plants. *Plant Physiol.* **1987**, *85*, 801–803. [CrossRef]
22. Reeves, R.D. Tropical hyperaccumulators of metals and their potential for phytoextraction. *Plant Soil* **2003**, *249*, 57–65. [CrossRef]
23. Seregin, I.V.; Kozhevnikova, A.D. Distribution of cadmium, lead, nickel, and strontium in imbibing maize caryopses. *Russ. J. Plant Physiol.* **2005**, *52*, 565–569. [CrossRef]
24. Molas, J.; Baran, S. Relationship between the chemical form of nickel applied to the soil and its uptake and toxicity to barley plants (*Hordeum vulgare* L.). *Geoderma* **2004**, *122*, 247–255. [CrossRef]
25. Markiv, B.; Ruiz-Azcona, L.; Expósito, A.; Santibáñez, M.; Fernández-Olmo, I. Short- and long-term exposure to trace metal(loid)s from the production of ferromanganese alloys by personal sampling and biomarkers. *Environ. Geochem. Health* **2022**, *44*, 4595–4618. [CrossRef]
26. Gerendás, J.; Sattelmacher, B. Influence of Ni supply on growth and nitrogen metabolism of *Brassica napus* L. grown with NH_4NO_3 or urea as N source. *Ann. Bot.* **1999**, *83*, 65–71. [CrossRef]
27. Tipu, M.I.; Ashraf, M.Y.; Sarwar, N.; Akhtar, M.; Shaheen, M.R.; Ali, S.; Damalas, C.A. Growth and physiology of maize (*Zea mays* L.) in a nickel-contaminated soil and phytoremediation efficiency using EDTA. *J. Plant Growth Regul.* **2021**, *40*, 774–786. [CrossRef]
28. Obrero, Á.; Die, J.V.; Román, B.; Gómez, P.; Nadal, S.; González, V.; Clara, I. Selection of reference genes for gene expression studies in zucchini (*Cucurbita pepo*) using qPCR. *J. Agric. Food Chem.* **2011**, *59*, 5402–5411. [CrossRef]
29. Labidi, O.; Vives-Peris, V.; Gómez-Cadenas, A.; Pérez-Clemente, R.M.; Sleimi, N. Assessing of growth, antioxidant enzymes and phytohormone regulation in *Cucurbita pepo* under cadmium stress. *Food Sci. Nutr.* **2021**, *9*, 2021–2031. [CrossRef]
30. Hewitt, E.J. Sand and water culture methods used in the study of plant nutrition. *J. Assoc. Off. Anal. Chem.* **1966**, *49*, 888–889.
31. Sleimi, N.; Abdely, C. Salt-tolerance strategy of two halophytes species: *Spartina alterniflora* and *Suaeda fruticosa*. In *Tasks for Vegetation Science*; Lieth, H., Mochtchenko, M., Eds.; Cash Crop Halophytes: Recent Studies; Kluwer Academic Publishers: Amsterdam, The Netherlands, 2003; Volume 38, pp. 79–85. [CrossRef]
32. Sleimi, N.; Abdelly, C. Growth and mineral nutrition of some halophytes under seawater irrigation. In *Prospects for Saline Agriculture*; Ahmad, R., Malik, K.A., Eds.; Springer: Berlin/Heidelberg, Germany, 2002; pp. 403–410. [CrossRef]
33. Sleimi, N.; Bankaji, I.; Kouki, R.; Dridi, N.; Duarte, B.; Caçador, I. Assessment of extraction methods of trace metallic elements in plants: Approval of a common method. *Sustainability* **2022**, *14*, 1428. [CrossRef]
34. Cicco, N.; Lanorte, M.; Paraggio, M.; Viggiano, M. A reproducible, rapid and inexpensive Folin-Ciocalteu micro-method in determining phenolics of plant methanol extracts. *Microchem. J.* **2009**, *91*, 107–110. [CrossRef]

35. Zhishen, J.; Mengcheng, T.; Jianming, W. The determination of flavonoids contents in mulberry and their scavenging effects on superoxide radicals. *J. Food Chem.* **1999**, *64*, 555–559. [CrossRef]
36. Bates, L.S.; Waldren, R.P.; Teare, I.D. Rapid determination of free proline for water-stress studies. *Plant Soil* **1973**, *39*, 205–207. [CrossRef]
37. Bradford, M.M. A rapid and sensitive method for the quantitation of microgram quantities of protein utilizing the principle of protein-dye binding. *Anal. Biochem.* **1976**, *72*, 248–254. [CrossRef]
38. Maheshwari, R.; Dubey, R.S. Inhibition of ribonuclease and protease activities in germinating rice seeds exposed to nickel. *Acta Physiol. Plant* **2008**, *30*, 863–872. [CrossRef]
39. Gajewska, E.; Sklodowska, M. Effect of nickel on ROS content and antioxidative enzyme activities in wheat leaves. *Biometals* **2007**, *20*, 27–36. [CrossRef]
40. Palacios, G.; Gómez, I.; Carbonell-Barrachina, A.; Pedreño, J.N.; Mataix, J. Effect of nickel concentration on tomato plant nutrition and dry matter yield. *J. Plant Nutr.* **1998**, *21*, 2179–2191. [CrossRef]
41. Bouslimi, H.; Ferreira, R.; Dridi, N.; Brito, P.; Martins-Dias, S.; Caçador, I.; Sleimi, N. Effects of barium stress in *Brassica juncea* and *Cakile maritima*: The indicator role of some antioxidant enzymes and secondary metabolites. *Phyton Int. J. Exp. Bot.* **2021**, *90*, 145–158. [CrossRef]
42. Suwa, R.K.; Jayachandran, N.T.; Nguyen, A.; Boulenouar, K.; Fujita, K.; Saneoka, H. Barium toxicity effects in soybean plants. *Arch. Environ. Contam. Toxicol.* **2008**, *55*, 397–403. [CrossRef] [PubMed]
43. Xu, F.J.; Li, G.; Jin, C.W.; Liu, W.J.; Zhang, S.S.; Zhang, Y.S.; Lin, X.Y. Aluminum-induced changes in reactive oxygen species accumulation, lipid peroxidation and antioxidant capacity in wheat root tips. *Biol. Plant* **2012**, *56*, 89–96. [CrossRef]
44. Bhalerao, S.A.; Sharma, A.S.; Poojari, A.C. Toxicity of Nickel in Plants. *Int. J. Pure Appl. Biosci.* **2015**, *3*, 345–355.
45. Poschenrieder, C.; Gunse, B.; Barcelo, J. Influence of cadmium on water relations, stomatal-resistance, and abscisic-acid content in expanding bean leaves. *Plant Physiol.* **1989**, *90*, 1365–1371. [CrossRef]
46. Kouki, R.; Ayachi, R.; Ferreira, R.; Sleimi, N. Behavior of *Cucumis sativus* L. in presence of aluminum stress: Germination, plant growth, and antioxidant enzymes. *Food Sci. Nutr.* **2021**, *9*, 3280–3288. [CrossRef]
47. Barcelo, J.; Poschenrieder, C.H. Plant water relations as affected by heavy metal stress. *J. Plant Nutr.* **1990**, *13*, 1–37. [CrossRef]
48. Molas, J. Changes in morphological and anatomical structure of cabbage (*Brassica oleracea* L.) outer leaves and in ultrastructure of their chloroplast caused by an in vitro excess of nickel. *Photosynthetica* **1997**, *34*, 513–522. [CrossRef]
49. Brunet, J.; Varrault, G.; Zuily Fodil, Y.; Repellin, A. Accumulation of lead in the roots of grass pea (*Lathyrus sativus* L.) plants triggers systemic variation in gene expression in the shoots. *Chemosphere* **2009**, *77*, 1113–1120. [CrossRef]
50. Amjad, M.; Raza, H.; Murtaza, B.; Abbas, G.; Imran, M.; Shahid, M.; Naeem, M.A.; Zakir, A.; Iqbal, M.M. Nickel toxicity induced changes in nutrient dynamics and antioxidant profiling in two maize (*Zea mays* L.) hybrids. *Plants* **2020**, *9*, 5. [CrossRef]
51. Wang, M.; Zheng, Q.; Shen, Q.; Guo, S. The Critical Role of Potassium in Plant Stress Response. *Int. J. Mol. Sci.* **2013**, *14*, 7370–7390. [CrossRef] [PubMed]
52. Rengel, Z.; Bose, J.; Chen, Q.; Tripathi, B.N. Magnesium alleviates plant toxicity of aluminium and heavy metals. *Crop Pasture Sci.* **2015**, *66*, 1298–1307. [CrossRef]
53. Rizwan, M.; Mostofa, M.G.; Ahmad, M.Z.; Zhou, Y.; Adeel, M.; Mehmood, S.; Ahmad, M.A.; Javed, R.; Imtiaz, M.; Aziz, O.; et al. Hydrogen sulfide enhances rice tolerance to nickel through the prevention of chloroplast damage and the improvement of nitrogen metabolism under excessive nickel. *Plant Physiol. Biochem.* **2019**, *138*, 100–111. [CrossRef]
54. Peralta-Videa, J.R.; Hernandez-Viezcas, J.A.; Zhao, L.J.; Corral-Diaz, B.; Ge, Y.; Priester, J.H.; Holde, P.A.; Gardea-Torresdey, J.L. Cerium dioxide and zinc oxide nanoparticles alter the nutritional value of soil cultivated soybean plants. *Plant Physiol. Biochem.* **2014**, *80*, 128–135. [CrossRef]
55. Hu, H.; Wang, L.; Li, Y.; Sun, J.; Zhou, Q.; Huang, X. Insight into the mechanism of lanthanum (III) induced damage to plant photosynthesis. *Ecotox. Environ. Saf.* **2016**, *127*, 43–50. [CrossRef] [PubMed]
56. Dridi, N.; Romdhane, L.; Ferreira, R.; Sleimi, N. Fertilizer effect of composted sewage sludge and cattle manure on *Pelargonium* growth. *J. Water Sanit. Hygi. Dev.* **2020**, *10*, 1019–1025. [CrossRef]
57. Chaudhry, A.H.; Nayab, S.; Hussain, S.B.; Ali, M.; Pan, Z. Current understandings on magnesium deficiency and future outlooks for sustainable agriculture. *Int. J. Mol. Sci.* **2021**, *22*, 1819. [CrossRef]
58. Sghaier, D.B.; Bankaji, I.; Pedro, S.; Caçador, I.; Sleimi, N. Photosynthetic behaviour and mineral nutrition of *Tamarix gallica* cultivated under aluminum and NaCl combined stress. *Phyton-Int. J. Exp. Bot.* **2019**, *88*, 239–252.
59. Wang, T.; Chen, X.; Ju, C.; Wang, C. Calcium signaling in plant mineral nutrition: From uptake to transport. *Plant Commun.* **2023**, *4*, 100678. [CrossRef]
60. Marschner, H. *Mineral Nutrition of Higher Plants*, 2nd ed.; Academic Press: London, UK, 1995.
61. Sleimi, N.; Kouki, R.; Haj Ammar, M.; Ferreira, R.; Pérez-Clemente, R.M. Barium effect on germination, plant growth, and antioxidant enzymes in *Cucumis sativus* L. plants. *Food Sci. Nutr.* **2021**, *9*, 2086–2094. [CrossRef] [PubMed]
62. Reeves, R.D.; Baker, A.J.M. Metal accumulating plant. In *Phytoremediation of Toxic Metals: Using Plants to Clean up the Environment*; Raskin, I., Ensley, B.D., Eds.; John Wiley and Sons: New York, NY, USA, 2000; pp. 193–230.
63. Naikoo, M.I.; Dar, M.I.; Raghib, F.; Jaleel, H.; Ahmad, B.; Raina, A.; Khan, F.A.; Naushin, F. Role and regulation of plants phenolics in abiotic stress tolerance: An overview. In *Plant Signaling Molecules*; Elsevier: Amsterdam, The Netherlands, 2019; pp. 157–168.

64. Macheix, J.J.; Fleuriet, A.; Jay-Allemand, C. *Les Composés Phénoliques des Végétaux: Un Exemple de Métabolites Secondaires d'Importance économique*; Presses Polytechniques et Universitaires Romandes: Lausanne, Switzerland, 2005; pp. 4–5.
65. Jonnala, R.S.; Irmak, S.; MacRitchie, F.; Bean, S.R. Phenolics in the bran of waxy wheat and triticale. *J. Cereal Sci.* **2010**, *52*, 509–515. [CrossRef]
66. Chaib, G.; Bouchelaleg, H.A.; Talbi, R. Etude phytochimique de quelques variétés de blé tendre (*Triticum aestivum*) et d'orge (*Hordeum vulgare*) et leurs activités biologiques. *Eur. Sci. J.* **2015**, *11*, 1857–7881.
67. Rastgoo, L.; Alemzadeh, A. Biochemical responses of Gouan (*Aeluropus littoralis*) to heavy metals stress. *Aust. J. Crop Sci.* **2011**, *5*, 375–383.
68. Neggaz, N.E.; Reguieg Yssaad, H.A. Effect of lead stress on polyphenols, flavonoids, and proline contents in radish (*Raphanus sativus* L.). *Int. J. Biosci.* **2018**, *12*, 135–144.
69. Handa, N.; Kohli, S.K.; Sharma, A.; Thukral, A.K.; Bhardwaj, R.; Abd_Allah, E.F.; Alqarawi, A.A.; Ahmad, P. Selenium modulates dynamics of antioxidative defence expression, photosynthetic attributes and secondary metabolites to mitigate chromium toxicity in *Brassica juncea* L. plants. *J. Exp. Bot.* **2019**, *161*, 180–192. [CrossRef]
70. Kattab, H. Role of glutathione and polyadenylic acid on the oxidative defense systems of two different cultivars of canola seedlings grown under saline condition. *Aust. J. Basic. Appl. Sci.* **2007**, *1*, 323–334.
71. Dridi, N.; Kouki, R.; Bouslimi, H.; Ferreira, R.; Hidouri, S.; Caçador, I.; Sleimi, N. Assessment of photochemical performance of photosystem II in *Limbarda crithmoides* and *Helianthus annuus* under Pb stress using chlorophyll a fluorescence kinetics. *World. J. Biol. Biotech.* **2023**, *8*, 15–19. [CrossRef]
72. Ma, C.; Liu, H.; Guo, H.; Musante, C.; Coskun, S.H.; Nelson, B.C.; White, J.C.; Xing, B.; Dhankher, O.P. Defense mechanisms and nutrient displacement in *Arabidopsis thaliana* upon exposure to CeO_2 and In_2O_3 nanoparticles. *Environ. Sci. Nano* **2016**, *3*, 1369–1379. [CrossRef]
73. Şen, A. Oxidative stress studies in plant tissue culture, antioxidant enzyme. *Techopen* **2012**, *3*, 59–88.
74. Finger-Teixeira, A.; Ferrarese, M.dL.L.; Soares, A.R.; da Silva, D.; Ferrarese-Filho, O. Cadmium-induced lignification restricts soybean root growth. *Ecotoxicol. Environ. Saf.* **2010**, *73*, 1959–1964. [CrossRef] [PubMed]
75. Chen, S.; Wang, Q.; Lu, H.; Li, J.; Yang, D.; Liu, J.; Yan, C. Phenolic metabolism and related heavy metal tolerance mechanism in *Kandelia Obovata* under Cd and Zn stress. *Ecotoxicol. Environ. Saf.* **2019**, *169*, 134–143. [CrossRef]
76. Shen, N.; Wang, T.; Gan, Q.; Liu, S.; Wang, L.; Jin, B. Plant flavonoids: Classification, distribution, biosynthesis, and antioxidant activity. *Food Chem.* **2022**, *383*, 132531. [CrossRef] [PubMed]
77. Bouslimi, H.; Dridi, N.; Ferreira, R.; Brito, P.; Caçador, I.; Hidouri, S.; Sleimi, N. Appraisal of the physiological response of *Cakile maritima* and *Brassica juncea* for tolerating lanthanum stress. *J. Mar. Sci. Eng.* **2024**, *12*, 65. [CrossRef]
78. Ferreira, I.C.F.R.; Barros, L.; Soares, M.E.; Bastos, M.L.; Pereira, J.A. Antioxidant activity and phenolic contents of *Olea europaea* L. leaves sprayed with different copper formulations. *J. Food Chem.* **2007**, *103*, 188–195. [CrossRef]
79. Khatun, S.; Ali, M.B.; Hahn, E.J.; Paek, K.Y. Copper toxicity in *Withania somnifera*: Growth and antioxidant enzymes responses of in vitro grown plants. *Environ. Exp. Bot.* **2008**, *64*, 279–285. [CrossRef]
80. Murch, S.J.; Haq, K.; Rupasinghe, H.P.V.; Saxena, P.K. Nickel contamination affects growth and secondary metabolite composition of St. John's wort (*Hypericum perforatum* L.). *Environ. Exp. Bot.* **2003**, *49*, 251–257. [CrossRef]
81. Ikram, K.; Abdelhakim, R.Y.H.; Topcuoglu, B.; Badiaa, O.; Houria, T. Accumulation of polyphenols and flavonoids in *Atriplex canescens* (Pursh) Nutt stressed by heavy metals (zinc, lead and cadmium). *Malays. J. Fundam. Appl. Sci.* **2020**, *16*, 334–337. [CrossRef]
82. Bretzel, F.; Benvenuti, S.; Pistelli, L. Metal contamination in urban street sediment in Pisa (Italy) can affect the production of antioxidant metabolites in *Taraxacum officinale* Weber. *Environ. Sci. Pollut. Res.* **2014**, *21*, 2325–2333. [CrossRef]
83. Williams, R.J.; Spencer, J.P.; Rice-Evans, C. Flavonoids: Antioxidants or signalling molecules? *Free Radic. Biol. Med.* **2004**, *36*, 838–849. [CrossRef] [PubMed]
84. Keilig, K.; Ludwig-Mueller, J. Effect of flavonoids on heavy metal tolerance in *Arabidopsis thaliana* seedlings. *Bot. Stud.* **2009**, *50*, 311–318.
85. Heim, K.E.; Tagliaferro, A.R.; Bobilya, D.J. Flavonoid antioxidants: Chemistry, metabolism and structure-activity relationships. *J. Nutr. Biochem.* **2002**, *13*, 572–584. [CrossRef]
86. Berni, R.; Luyckxc, M.; Xud, X.; Legayd, S.; Sergeantd, K.; Hausman, J.F.; Lutts, S.; Cai, G.; Guerriero, G. Reactive oxygen species and heavy metal stress in plants: Impact on the cell wall and secondary metabolism. *Environ. Exp. Bot.* **2019**, *161*, 98–106. [CrossRef]
87. Djerroudi-Zidane, O.; Belkhodja, M.; Bissati, S. Effect du Stress Salin sur l'accumulation de proline chez deux espèces d'*Atriplex Halimus* L. et *Atriplex canescens* (Pursh) Nutt. *Eur. J. Sci. Res.* **2010**, *41*, 248–259.
88. Kandziora-Ciupa, M.; Nadgórska-Socha1, A.; Barczyk, G.; Ciepał, R. Bioaccumulation of heavy metals and ecophysiological responses to heavy metal stress in selected populations of *Vaccinium myrtillus* L. and *Vaccinium vitis-idaea* L. *Ecotoxicology* **2017**, *26*, 966–980. [CrossRef]
89. Kandziora-Ciupa, M.; Ciepał, R.; Nadgórska-Socha, A.; Barczyk, G. Accumulation of heavy metals and antioxidant responses in *Pinus sylvestris* L. needles in polluted and non-polluted sites. *Ecotoxicology* **2016**, *25*, 970–981. [CrossRef]
90. Hadjadj, S.; Djerroudi, O.; Bissati, S. Etude comparative des mécanismes biochimiques de tolérance au stress salin de deux espèces d'*atriplex*: *Atriplex halimus* L. et *Atriplex canescens* (Purch) Nutt. *Alger. J. Arid. Environ.* **2011**, *1*, 3–10.

91. Garg, G.; Neha, P. Plant transcription factors networking of pyrroline-5-carboxylate (P5C) enzyme under stress condition: A review. *Plant Arch.* **2019**, *19*, 562–569.
92. Kouki, R.; Dridi, N.; Vives-Peris, V.; Gómez-Cadenas, A.; Caçador, I.; Pérez-Clemente, R.M.; Sleimi, N. Appraisal of *Abelmoschus esculentus* L. response to aluminum and barium stress. *Plants* **2023**, *12*, 179. [CrossRef]
93. Brinis, A.; Belkhodja, M. Salinity effect on physiological and biochemical parameters of *Atriplex halimus* L. *Sci. Technol. Synthèse* **2015**, *31*, 42–51.
94. Spormann, S.; Nadais, P.; Sousa, F.; Pinto, M.; Martins, M.; Sousa, B.; Fidalgo, F.; Soares, C. Accumulation of proline in plants under contaminated soils-Are we on the same page? *Antioxidants* **2023**, *12*, 666. [CrossRef]
95. Tihana, T.; John, T.H.; Meri, E.; Nada, P.; Vera, C.; Hrvoje, L.; Ivna, Š.; Drago, B. Antioxidative responses in radish (*Raphanus sativus* L.) Plants stressed by copper and lead in nutrient solution and soil. *Acta Biol. Cracoviensia Ser. Bot.* **2008**, *50*, 79–86.
96. John, R.; Ahmad, P.; Gadgil, K.; Sharma, S. Heavy metal toxicity: Effect on plant growth, biochemical parameters and metal accumulation by *Brassica juncea* L. *Int. J. Plant Prod.* **2009**, *3*, 65–75.
97. Dridi, N.; Bouslimi, H.; Caçador, I.; Sleimi, N. Lead tolerance, accumulation and translocation in two *Asteraceae* plants: *Limbarda crithmoides* and *Helianthus annuus*. *S. Afr. J. Bot.* **2022**, *150*, 986–996. [CrossRef]
98. Gupta, D.; Huang, H.; Yang, X.; Razafindrabe, B.; Inouhe, M. The detoxification of lead in *Sedum alfredii H.* is not related to phytochelatins but the glutathione. *J. Hazard. Mater.* **2010**, *177*, 437–444. [CrossRef] [PubMed]
99. Yadav, S.K. Heavy metals toxicity in plants: An overview on the role of glutathione and phytochelatins in heavy metal stress tolerance of plants. *S. Afr. J. Bot.* **2010**, *76*, 167–179. [CrossRef]
100. Wollgieh, R.; Newmann, D. Stress response of tomato cell cultures to toxic metals and heat shock: Differences and similarities. *J. PlantPhysiol.* **1995**, *146*, 736–742.

plants

MDPI

Article

Postponed Application of Phosphorus and Potassium Fertilizers Mitigates the Damage of Late Spring Coldness by Improving Winter Wheat Root Physiology

Hao Fang [1,†], Jinwei Huang [1,†], Xiatong Zhu [1], Muhammad Ahmad Hassan [2], Jin Ren [1], Jingyao Huang [1], Baoqiang Zheng [1], Xiang Chen [1], Feifei Lin [1,*] and Jincai Li [1,3,*]

1 College of Agronomy, Anhui Agricultural University, Hefei 230036, China; fanghao@stu.ahau.edu.cn (H.F.); huangjinwei@stu.ahau.edu.cn (J.H.); zxt200074@stu.ahau.edu.cn (X.Z.); renjin0226@stu.ahau.edu.cn (J.R.); jingyaohuang@stu.ahau.edu.cn (J.H.); zhengbaoqiang@ahau.edu.cn (B.Z.); cx2468@ahau.edu.cn (X.C.)
2 Rice Research Institute, Anhui Academy of Agricultural Sciences, Hefei 230041, China; ahmaduaf93@stu.ahau.edu.cn
3 Jiangsu Collaborative Innovation Centre for Modern Crop Production, Nanjing 210095, China
* Correspondence: lff00261@ahau.edu.cn (F.L.); lijc@ahau.edu.cn (J.L.)
† These authors contributed equally to this work.

Abstract: Late spring coldness (LSC) is the main limiting factor threatening wheat yield and quality stability. Optimal nutrient management is beneficial in mitigating the harms of LSC by improving wheat root physiology. This study proposed a nutrient management strategy that postponed the application of phosphorus (P) and potassium (K), effectively strengthening wheat's defense against LSC. This experiment used the winter cultivar "Yannong19" (YN 19) as plant material for two consecutive years (2021–2022 and 2022–2023). Two fertilizer treatments were used: traditional P and K fertilizers application (R1: base fertilizer: jointing fertilizer = 10:0) and postponed P and K fertilizers application (R2: base fertilizer: jointing fertilizer = 5:5); wheat plants at the anther connective formation stage shifted to temperature-controlled phytotrons for normal (T0, 11 °C/4 h) and low temperatures (T1, 4 °C/4 h; T2, −4 °C/4 h) as treatments of LSC. The results showed that under low temperature (LT) treatment, compared with R1, the R2 treatment increased the concentrations of osmotic adjustment substances (soluble sugars and soluble protein contents by 6.2–8.7% and 3.0–8.9%), enhanced activities of antioxidant enzymes (superoxide dismutase, peroxidase and catalase activities by 2.2–9.1%, 6.2–9.7% and 4.2–8.4%), balanced the hormone concentrations (increased IAA and GA_3 contents by 2.8–17.5% and 10.4–14.1% and decreased ABA contents by 7.2–14.3%), and reduced the toxicity (malondialdehyde, hydrogen peroxide content and $O_2{\cdot}^-$ production rate by 5.7–12.4%, 17.7–22.8% and 19.1–19.1%) of the cellular membranes. Furthermore, the wheat root physiology in R2 significantly improved as the root surface area and dry weight increased by 5.0–6.6% and 4.7–6.6%, and P and K accumulation increased by 7.4–11.3% and 12.2–15.4% compared to R1, respectively. Overall, the postponed application of P and K fertilizers enhanced the physiological function of the root system, maintained root morphology, and promoted the accumulation of wheat nutrients under the stress of LSC.

Keywords: winter wheat; late spring coldness; optimizing phosphorus and potassium application; root physiology; nutrient accumulation

Citation: Fang, H.; Huang, J.; Zhu, X.; Hassan, M.A.; Ren, J.; Huang, J.; Zheng, B.; Chen, X.; Lin, F.; Li, J. Postponed Application of Phosphorus and Potassium Fertilizers Mitigates the Damage of Late Spring Coldness by Improving Winter Wheat Root Physiology. *Plants* **2024**, *13*, 2311. https://doi.org/10.3390/plants13162311

Academic Editors: Geovani Soares de Lima, Lauriane Almeida dos Anjos Soares and Francisco Vanies Da Silva Sá

Received: 23 July 2024
Revised: 15 August 2024
Accepted: 16 August 2024
Published: 20 August 2024

1. Introduction

Wheat (*Triticum aestivum* L.) is the world's most widely cultivated food crop, consumed by more than 35% of the world's population, and is the second-highest grain-producing crop in China [1]. Wheat production is directly related to economic stability and food security; however, with the intensification of global warming in recent years, temperature fluctuations in spring increase the risk of wheat plants suffering from late spring coldness

(LSC) stress [2]. The LSC is an important meteorological factor that adversely affects plant growth and development and causes severe serious economic losses to wheat production in China, the United States, Australia, and Europe [3–5]. It often occurs at the jointing and booting stages during the critical growth periods for the rapid growth and differentiation of young wheat spikelets, which were highly sensitive to low-temperature (LT) stress [6]. When the sensitive growth stage encounters meteorological disasters, yield and grain quality seriously decline [7]. Previous research studies have predominantly concentrated on analyzing the above-ground plant parts, such as leaves, young spikes, and peduncles, with limited attention to exploring the below-ground plant part, i.e., roots.

Roots play a crucial role in the active uptake of nutrients and water while engaging in various symbiotic interactions with the soil [8]. Root systems' spatial and temporal distribution alters below-ground ecological interactions, ultimately impacting plant performance, survival, and productivity [9]. However, the causes of the damage to the root growth under LT stress have still not been explored well. Short-term or continuous LT stress causes morphological, physiological, and biochemical changes that negatively affect the growth and development of plants, resulting in significant yield losses [10]. Buriro et al. [11] found that LT stress reduced root length, fresh stem and root weight, and root dry weight in wheat. Therefore, LT stress limits the growth of roots, which may be related to the physiological damage of roots. Posmyk et al. [12] investigated the changes in antioxidant enzyme activity in soybean roots when exposed to cold. When crops were exposed to LT stress, the activities of SOD, POD, CAT, and MDA contents increased in plant root cells [13]. Additionally, the absorption of water and nutrients and the hormonal balance in roots are severely disturbed [14]. Therefore, it is imperative to improve cultivation techniques to enhance the cold resistance of wheat roots.

Implementing timely preventive measures before LT events is challenging due to their random and unpredictable nature. In this scenario, nutrient management is crucial for enhancing tolerance to extreme temperature stress with P and K playing an important role in mitigating such stresses [15]. The application of P and K fertilizers enhances the adaptation of plants to the external environment by participating in signaling, energy dynamics, and enzyme catalysis, significantly improving the uptake and utilization of water and nutrients in wheat roots [16–18]. Optimizing P fertilizer application increased ascorbic acid, soluble protein, and sugar levels in wheat plants throughout the seedling, tillering, and jointing stages. This enhancement is crucial in improving wheat cold resistance and yield enhancement [19]. K alleviates freeze-induced cell dehydration and regulates the osmotic potential of plants under LT and freezing stress [20]. However, most existing reports on improving cold hardiness in wheat with P and K fertilizers concentrate mainly on the separate effects of each element or on the independent application of P and K fertilizers [21]. Considering that late spring cold primarily occurs during the jointing stage in the Huang Huai Hai wheat region, and conventional fertilization is also conducted at this stage, it is advantageous to concurrently administer nitrogen (N), P, and K fertilizers while ensuring consistent workload and fertilizer application quantities. This study proposed a nutrient management strategy that can effectively mitigate the detrimental impact of low temperatures on wheat. However, the specific mechanism by which postponing the application of P and K fertilizers alleviates the damage caused by LSC in wheat roots remains poorly comprehended.

In the present study, we hypothesized that the postponed application of P and K fertilizers would enhance wheat resilience to LT stress during the anther connective formation stage, improve root physiology, and increase P and K accumulation. Therefore, we performed a two-year pot experiment in which two P and K fertilizer application modes were opted under LSC at the anther connective formation stage. In conjunction with LSC and the treatment involving P and K fertilizers application, the study aimed to (1) investigate the impact of postponed application of P and K fertilizers on the root morphological traits, (2) analyze its effects on the roots physiology, i.e., activities antioxidant enzymes, osmotic adjustments, and hormonal balance; and (3) assess the advantageous outcomes of postponed application of P and K fertilizers on root nutrient accumulation during the anther connective formation stage. This study provided a theoretical basis and extended the technical support for optimizing P and K fertilizers to cope with LSC in the field conditions of China's Huang-Huai-Hai wheat-growing region.

2. Results

2.1. Morphological Traits of Wheat Roots

The application methods of PK fertilizers and LT treatments affected the root surface area (RSA) and root dry weight (RDW) in varying intensities (Figure 1B–E). This study's LT treatment significantly inhibited wheat root growth, resulting in low RSA and RDW (Figure 1A). Specifically, LT treatment reduced the RSA and RDW with a more significant decrease observed at lower temperatures. Compared with T0 treatment, T1 and T2 treatments reduced RSA by 11.9–21.7% and 4.9–16.4% during two growing seasons, respectively (Figure 1B,C). For RDW, the variation was similar to that for the RSA (Figure 1D,E). Compared with R1 treatment, the application of R2 at T1 and T2 enhanced RSA by 5.0–6.6% and RDW by 4.7–6.6%, respectively, during the growing season (2021–2022) (Figure 1B,D). Similar results were observed at the RSA and RDW in 2022–2023 (Figure 1C,E).

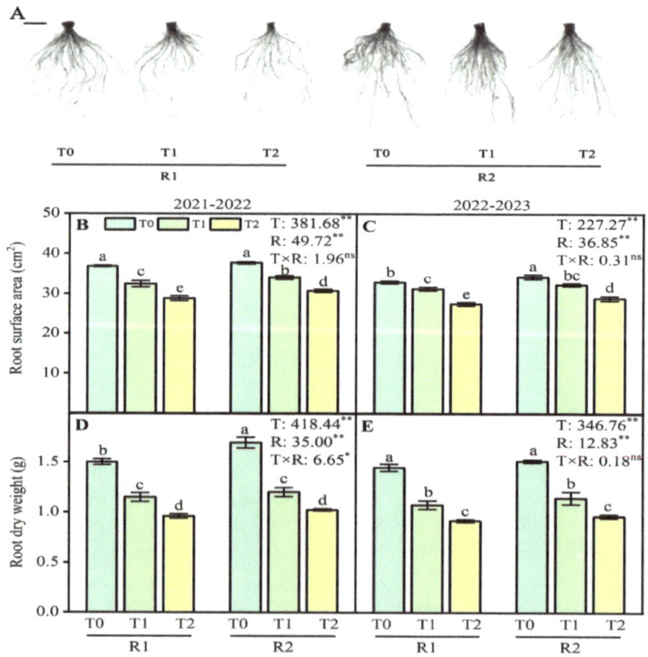

Figure 1. Effects of postponed application of P and K fertilizers on root morphology and dry weight under LT treatments. (**A**) Root morphology of different treatments; (**B**,**C**) difference in the response of root surface area to low temperature and top-dressing PK fertilizers mode at the anther connective formation stage during 2021–2022 and 2022–2023; (**D**,**E**) difference in the response of root dry weight to low temperature and top-dressing PK fertilizers at the anther connective formation stage during 2021–2022 and 2022–2023. T0, 11 °C; T1, 4 °C; T2, −4 °C; R1, traditional P and K fertilizers application; R2, postponed the application of P and K fertilizers; respectively. Different letters represent significant differences for comparing the six groups (LSD–test), *p*-values of multiple comparisons, $p < 0.05$. Analysis of variance with temperature (T) and PK fertilizers application (R) as two factors to analyze effects of single factor and interactions in randomized block experiments, $^{**} p < 0.01$, $^{*} p < 0.05$, and $^{ns} p > 0.05$.

2.2. Soluble Sugar and Soluble Protein Content of Wheat Roots

Wheat roots soluble sugar (SS) content increased, but soluble protein (SP) content decreased following LT treatment at the anther connective formation stage (Figure 2). Compared with the T0 treatment, SS contents increased as the LT decreased; however, the opposite trend was observed for SP contents. Our study showed that compared with the T0 treatment, the LT treatment at T1 and T2 increased SS contents by 58.9–101.7% and 30.7–75.4% during two growing seasons, respectively. Additionally, the contents of SP decreased by 5.5–11.7% and 3.9–11.1% at T1 and T2 under LT treatments in 2021–2022 and 2022–2023, respectively. Compared with the R1 treatment, applying R2 at T1 and T2 enhanced SS contents by 6.2–8.7% and SP contents by 3.0–8.9% during the 2021–2022 growing season (Figure 2A,C). Similar results were observed at the SS and contents in roots during 2022–2023 (Figure 2B,D).

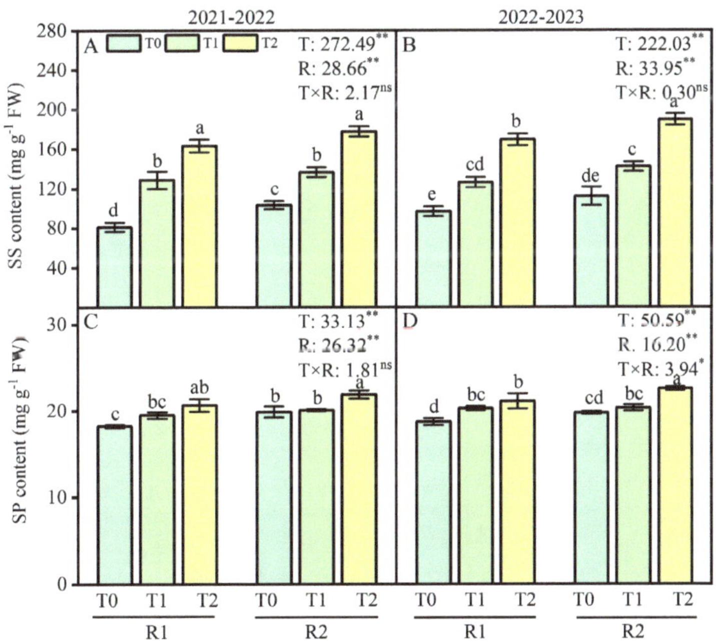

Figure 2. Effects of postponed application of P and K fertilizers on soluble sugar (SS) and soluble protein (SP) contents in wheat root under LT treatments. (**A**,**B**) SS content in root at anther connective formation stage during 2021–2022 and 2022–2023. (**C**,**D**) SP content in root at anther connective formation stage during 2021–2022 and 2022–2023. T0, 11 °C; T1, 4 °C; T2, −4 °C; R1, traditional P and K fertilizers application; R2, postponed the application of P and K fertilizers; respectively. Different letters represent significant differences for comparing the six groups (LSD–test), p values of multiple comparisons, $p < 0.05$. Analysis of variance with temperature (T) and PK fertilizers application (R) as two factors to analyze the effects of single factor and interactions in randomized block experiments, ** $p < 0.01$, * $p < 0.05$, and ^{ns} $p > 0.05$.

2.3. Antioxidant System Enzyme Activities of Wheat Roots

Analysis of variance (ANOVA) revealed that the LSC temperature and PK fertilizers application treatments had a significant effect on the superoxide dismutase (SOD), peroxidase (POD), and catalase (CAT) activity, while their interaction had no significant impact on the activities of SOD, POD, and CAT (Figure 3). In this study, compared with the T0 treatment, the LT treatments at T1 and T2 increased SOD activity by 15.0–28.7% and 17.6–27.6% in 2021–2022 and 2022–2023, respectively. Additionally, the activity of POD increased by 29.7–58.8% and 16.3–33.9% at T1 and T2 under LT treatment in 2021–2022 and 2022–2023 (Figure 3A–D). For the CAT activity of wheat root, the variation was similar to that for the activities of SOD and POD (Figure 3E,F). Compared with R1 treatment, R2 treatment at T1 and T2 enhanced SOD activity by 2.2–9.1%, POD activity by 6.2–9.7%, and CAT activity by 4.2–8.4%, respectively, during 2021–2022 (Figure 3A,C,E). Similar results were observed at the SOD, POD, and CAT activity in 2022–2023 (Figure 3B,D,F).

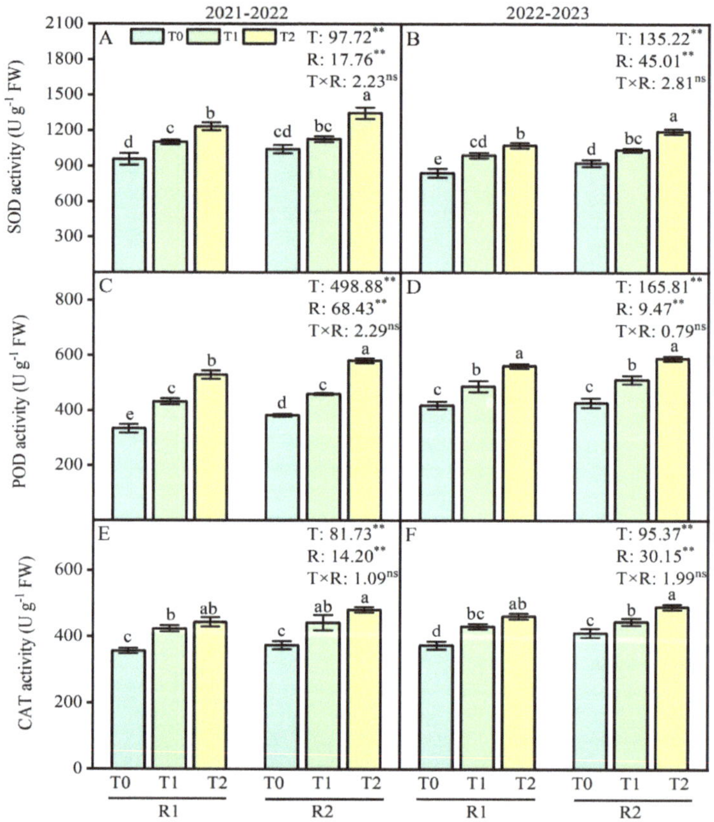

Figure 3. Effects of postponed application of P and K fertilizers on superoxide dismutase (SOD), peroxidase (POD), and catalase (CAT) activities in wheat root under LT treatments. (**A**,**B**) The activity of SOD in root at anther connective formation stage during 2021–2022 and 2022–2023. (**C**,**D**) The activity of POD in root at anther connective formation stage during 2021–2022 and 2022–2023. (**E**,**F**) The activity of CAT in root at anther connective formation stage during 2021–2022 and 2022–2023. T0, 11 °C; T1, 4 °C; T2, −4 °C; R1, traditional P and K fertilizers application; R2, postponed application of P and K fertilizers; respectively. Different letters represent significant differences for comparing the six groups (LSD–test), p values of multiple comparisons, $p < 0.05$. Analysis of variance with temperature (T) and PK fertilizers application (R) as two factors to analyze the effects of single factor and interactions in randomized block experiments, ** $p < 0.01$, and ^{ns} $p > 0.05$.

2.4. ROS Contents of Wheat Roots

ANOVA revealed that the LSC temperature and PK fertilizer application treatments had a significant effect on the malondialdehyde (MDA) content, hydrogen peroxide (H_2O_2) content, and $O_2\cdot^-$ production rate, while their interaction had a significant impact on the MDA content and $O_2\cdot^-$ production rate (Figure 4). In this study, compared with the T0 treatment, the LT treatments at T1 and T2 increased MDA contents by 117.9–242.1% and

73.1–133.7% in 2021–2022 and 2022–2023, respectively. Additionally, the content of H_2O_2 increased by 34.2–81.5% and 46.9–117.5% at T1 and T2 under LT treatments in 2021–2022 and 2022–2023 (Figure 4A–D). The variation for wheat root's $O_2\cdot^-$ production rate was similar to the content of MDA and H_2O_2 (Figure 4E,F). Compared with the R1 treatment, the application of R2 at T1 and T2 decreased the MDA contents by 5.7–12.4%, the contents of H_2O_2 by 17.7–22.8%, and the $O_2\cdot^-$ production rate by 19.1% during the 2021–2022 growing season (Figure 4A,C,E). Similar results were observed for the MDA content, H_2O_2 contents, and $O_2\cdot^-$ production rate in 2022–2023 (Figure 4B,D,F).

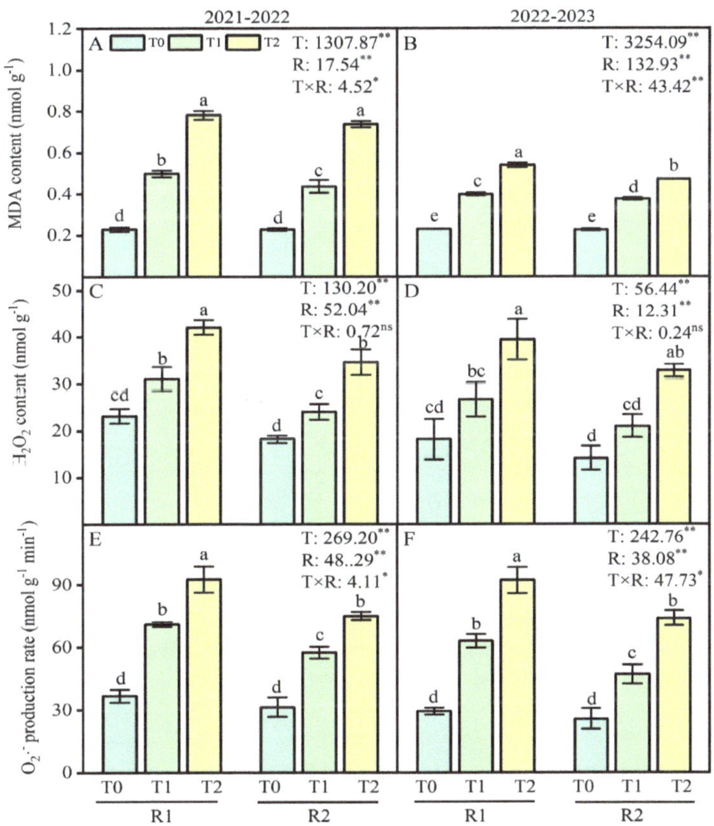

Figure 4. Effects of postponed application of P and K fertilizers on MDA, H_2O_2 content and $O_2\cdot^-$ production rate in wheat root under LT treatments. (**A**,**B**) MDA content in root at anther connective formation stage during 2021–2022 and 2022–2023. (**C**,**D**) H_2O_2 content in root at anther connective formation stage during 2021–2022 and 2022–2023. (**E**,**F**) $O_2\cdot^-$ production rate in root at anther connective formation stage during 2021–2022 and 2022–2023. T0, 11 °C; T1, 4 °C; T2, −4 °C; R1, traditional P and K fertilizers application; R2, postponed the application of P and K fertilizers; respectively. Different letters represent significant differences for comparing the six groups (LSD–test), p values of multiple comparisons, $p < 0.05$. Analysis of variance with temperature (T) and PK fertilizers application (R) as two factors to analyze effects of single factor and interactions in randomized block experiments, ** $p < 0.01$, * $p < 0.05$, and ^{ns} $p > 0.05$.

2.5. Endogenous Hormone Content of Wheat Roots

The ANOVA results revealed that the LSC temperature and PK fertilizer application treatments had a significant effect on the IAA, GA_3, and ABA contents, while their interaction had a significant impact on the IAA content (Figure 5). The ABA content increased in wheat roots, while the IAA and GA_3 content decreased following LT treatments at the anther connective formation stage. The contents of IAA and GA_3 showed a decreasing trend as the LT decreased. However, the opposite trend was observed for ABA contents. In this study, compared with the T0 treatment, the LT treatment at T1 and T2 decreased the IAA content by 7.7–19.9% and 5.8–11.8% during 2021–2022 and 2022–2023, respectively. Additionally, the contents of GA_3 increased by 3.3–20.0% and 8.2–23.8% at T1 and T2 under LT treatment in 2021–2022 and 2022–2023 (Figure 5A–D). However, the opposite trend was observed for ABA contents (Figure 5E,F). Compared with the R1 treatment, the application of R2 at T1 and T2 increased the IAA content by 2.8–17.5% and the content of GA_3 by 10.4–14.1%, respectively, during the 2021–2022 growing season (Figure 5A,C). However, compared with the R1 treatment, the application of R2 at T1 and T2 decreased the ABA content by 7.2–14.3%. Similar results were observed for the IAA, GA_3, and ABA contents in 2022–2023 (Figure 5B,D,F).

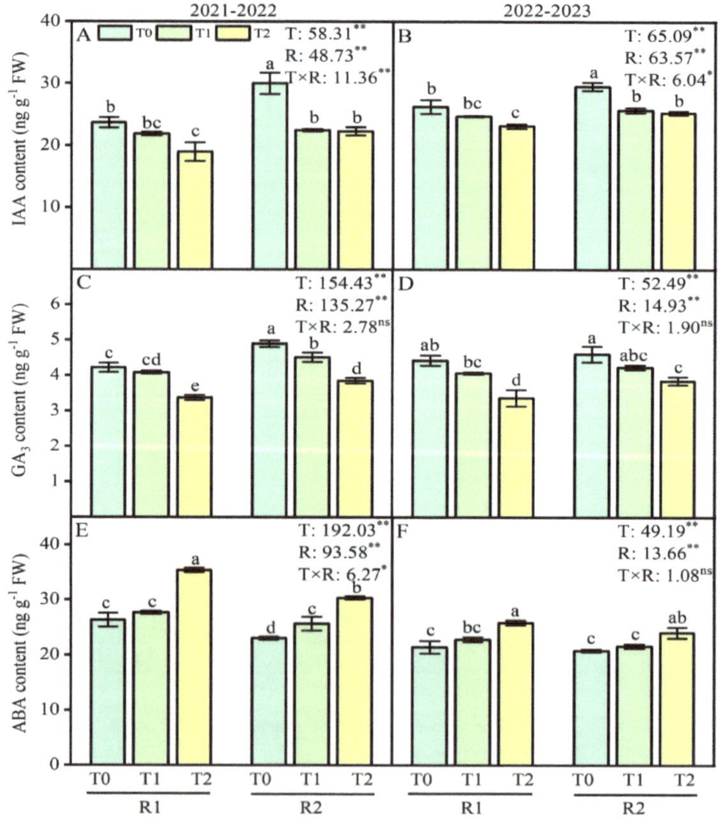

Figure 5. Effects of postponed application of P and K fertilizers on IAA, GA_3, and ABA content in wheat root under LT treatments. (**A**,**B**) IAA content in root at anther connective formation stage during 2021–2022 and 2022–2023. (**C**,**D**) GA_3 content in root at anther connective formation stage during 2021–2022 and 2022–2023. (**E**,**F**) ABA content in root at anther connective formation stage during 2021–2022 and 2022–2023. T0, 11 °C; T1, 4 °C; T2, −4 °C; R1, traditional P and K fertilizers application; R2, postponed application of P and K fertilizers; respectively. Different letters represent significant differences for comparing the six groups (LSD–test), *p* values of multiple comparisons, $p < 0.05$. Analysis of variance with temperature (T) and PK fertilizers application (R) as two factors to analyze effects of single factor and interactions in randomized block experiments, ** $p < 0.01$, * $p < 0.05$, and ^{ns} $p > 0.05$.

2.6. Acid Phosphatase and Alkaline Phosphatase Activities of Wheat Roots

Wheat root acid phosphatase (ACP) and alkaline phosphatase (ALP) activity decreased following LT treatments at the anther connective formation stage (Figure 6). This study showed that compared with the T0 treatment, T1 and T2 treatments decreased ACP activity by 22.2–45.9% and 19.7–27.5%, respectively, during 2021–2022 and 2022–2023. Additionally, the activity of ALP decreased by 36.1–53.3% and 34.1–60.7% at T1 and T2 under LT treatments in 2021–2022 and 2022–2023. Compared with the R1 treatment, the application of R2 at T1 (4 °C) and T2 (−4 °C) enhanced the activity of ACP by 10.1–11.2% and ALP activity by 16.2–20.2% during the 2021–2022 growing season (Figure 6A,C). Similar results were observed for the ACP and ALP activity in 2022–2023 (Figure 6B,D).

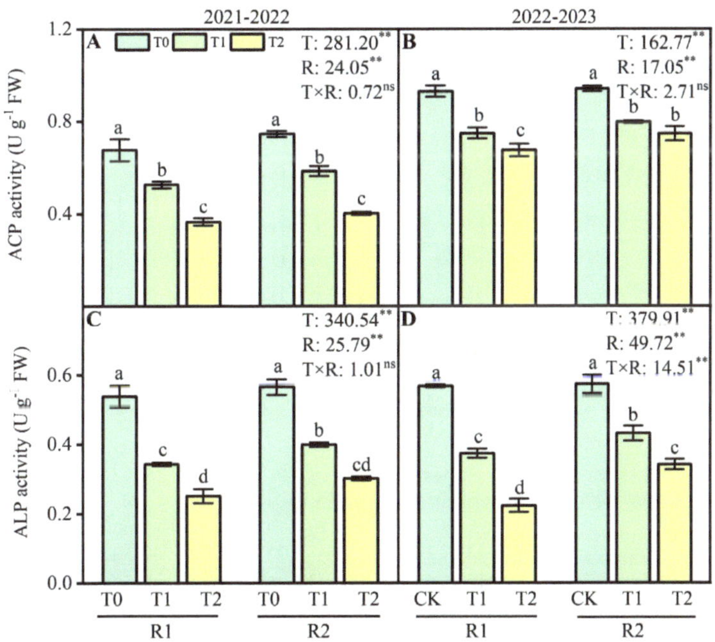

Figure 6. Effects of postponed application of P and K fertilizers on acid phosphatase (ACP) and alkali phosphatase (ALP) activity in wheat root under LT treatments. (**A**,**B**) The activity of ACP in root at anther connective formation stage during 2021–2022 and 2022–2023. (**C**,**D**) The activity of ALP in root at anther connective formation stage during 2021–2022 and 2022–2023. T0, 11 °C; T1, 4 °C; T2, −4 °C; R1, traditional P and K fertilizers application; R2, postponed application of P and K fertilizers; respectively. Different letters represent significant differences for comparing the six groups (LSD -test), p values of multiple comparisons, $p < 0.05$. Analysis of variance with temperature (T) and PK fertilizers application (R) as two factors to analyze effects of single factor and interactions in randomized block experiments, ** $p < 0.01$, and ^{ns} $p > 0.05$.

2.7. Phosphorus Accumulation and Potassium Accumulation Contents of Wheat Roots

Wheat roots' phosphorus accumulation (PA) and potassium accumulation (KA) contents decreased following LT treatments at the anther connective formation stage (Figure 7). This study showed that compared with the T0 treatment, T1 and T2 treatments decreased the PA content by 26.6–44.6% and 29.4–43.4% during the two growing seasons, respectively (Figure 7A,B). Additionally, the content of KA decreased by 28.2–47.0% and 37.4–55.9% at T1 and T2 under LT treatment in 2021–2022 and 2022–2023 (Figure 7C,D). Compared with the R1 treatment, applying R2 at T1 and T2 enhanced the PA content by 7.4–11.3% and KA content by 12.2–15.4% during the 2021–2022 growing season (Figure 7A,C). Similar results were observed in PA and KA in 2022–2023 (Figure 7B,D).

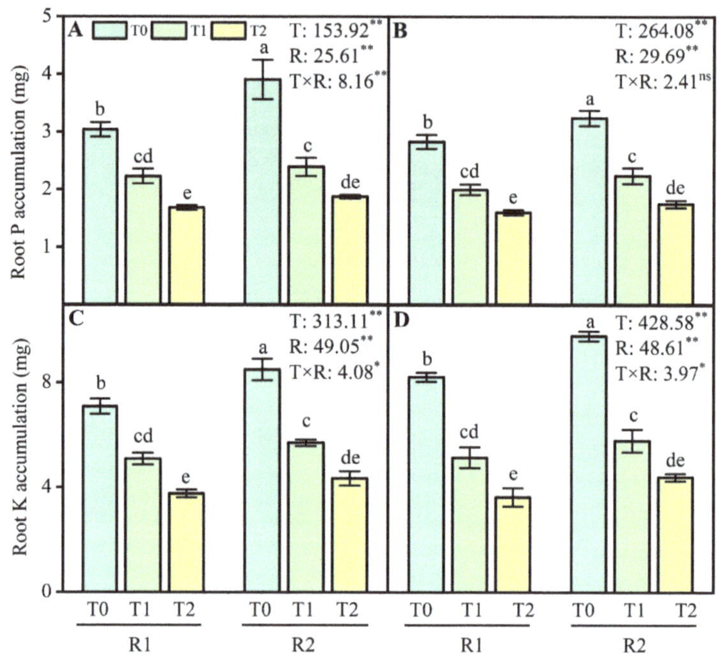

Figure 7. Effects of postponed application of P and K fertilizers on P and K accumulation in wheat root under LT treatments. (**A**,**B**) The content of root P accumulation at anther connective formation stage during 2021–2022 and 2022–2023. (**C**,**D**) The content of root K accumulation at anther connective formation stage during 2021–2022 and 2022–2023. T0, 11 °C; T1, 4 °C; T2, −4 °C; R1, traditional P and K fertilizers application; R2, postponed application of P and K fertilizers; respectively. Different letters represent significant differences for comparing the six groups (LSD-test), *p* values of multiple comparisons, $p < 0.05$. Analysis of variance with temperature (T) and PK fertilizers application (R) as two factors to analyze effects of single factor and interactions in randomized block experiments, $^{**} p < 0.01$, $^{*} p < 0.05$, and $^{ns} p > 0.05$.

2.8. Correlation Coefficients between Root Morphological Traits, Antioxidant Enzymes, Osmotic Adjustment Substance, Endogenous Hormone, Phosphatase Activity, and Nutrient Accumulation

According to Figure 8, the RSA and RDW are positively correlated with the contents of SP, GA3, IAA, and the activities of ACP and ALP. The P and K accumulation are also positively correlated with the contents of SP, GA_3, IAA, and the activities of ACP and ALP. The P accumulation is positively correlated with ACP and ALP activities, which indicates that P metabolizing enzyme activity promotes P accumulation in the root system. However, the P and K accumulation exhibited significant negative correlations with the SOD, POD, CAT, MDA, and H_2O_2 contents as well as the $O_2{\cdot}^-$ production rate. These results indicated that P and K accumulations were primarily influenced by cell membrane and antioxidant activities. However, the IAA and GA_3 contents were positively correlated with root P and K accumulation. The opposite trend was observed for ABA contents. Those results indicated that excessive ABA accumulation suppresses endogenous hormone balance, and the balance between endogenous hormones affects P and K accumulation in the root.

The random forest (RF) analysis demonstrated that POD and GA_3 were the crucial drivers of the variations in RSA (Figure 8B), whereas the RDW was mainly governed by ALP and MDA (Figure 8C). The RF analysis demonstrated that SS and ALP were the crucial drivers of the variations in KA (Figure 8D), whereas the RDW was mainly governed by SS and MDA (Figure 8E).

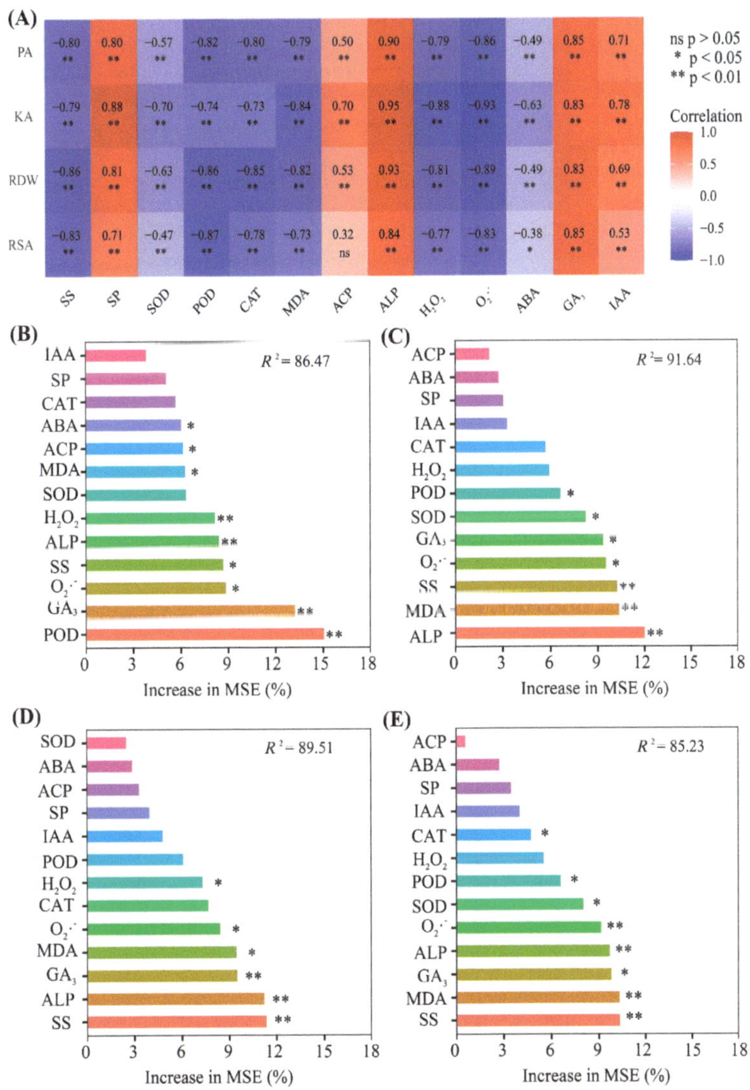

Figure 8. (**A**) The heatmap of correlations among PA, KA, RSA, RDW, and physiological parameters of roots under different treatments. The relative importance (%) of variables for the physiological parameter response to (**B**) RSA, (**C**) RDW, (**D**) KA, and (**E**) PA based on the random forest regression model. PA: phosphorus accumulation; KA: potassium accumulation; RDW, root dry weight; RSA: root surface area; SS: soluble sugar; SP: soluble protein; SOD, superoxide dismutase; POD, peroxidase; CAT, catalase; MDA, malondialdehyde; ACP, acid phosphatase activity; ALP, alkali phosphatase activity; H_2O_2, hydrogen peroxide; $O_2^{\cdot -}$, rate of superoxide anion generation; ABA, abscisic acid; GA_3, gibberellic acid; IAA, auxin under different treatments. ns indicates a non-significant correlation. * and ** represent significant correlations at 5% and 1% probability levels, respectively.

3. Discussion

3.1. Optimizing P and K Application Improved the Growth of Wheat Root Under LT Stress

LT stress affected wheat plants' growth and development with roots typically subjected to more abiotic stress than shoots [11,22]. Hence, it is critical to improve the resistance of wheat roots to LT stress, which is of great significance for disaster prevention and mitigation in wheat production. This study indicated that LT stress decreased the accumulation of wheat root biomass during the anther connective formation stage (Figure 1D,E), which is consistent with previous studies [23]. The decreased wheat root activity slowed root growth and reduced biomass accumulation [10]. The root system is the main organ of wheat that absorbs water and mineral nutrients in the soil, and its growth status will affect the growth and development of the wheat shoot system [24]. This study significantly reduced the root surface area under LT stress (Figure 1B,C). A similar result was reported in rice; cold-sensitive genotypes display a reduced root dry weight and shorter and fewer root hairs associated with a smaller root area [25]. Moreover, chilling stress can decrease the root hydraulic conductance, water uptake, water content, and nutrient uptake [26]. The root system architecture, such as higher root density and more lateral root branches, enables roots to take up more P from the topsoil layer [27,28]. Applying P fertilizer boosts below-ground microbial activity, fostering mutually beneficial conditions for robust root growth and development [29]. Increased microbial activity excites soil temperature to a certain extent and alleviates the adverse impacts of LT stress [21]. Applying P fertilizer increases the root surface area, supporting active nutrient–water uptake [30]. K is engaged in nearly all the plant's physiological processes requiring water [31]. Nutrient management is one of the best options in response to extreme temperature stress tolerance, and among all the nutrients, K plays a significant role in coping with temperature stress [20]. K helps to activate the various physiological and metabolic processes [32]. Several field trials in various crops have also shown a similar phenomenon in which a sufficient K supply can eliminate frost damage [33]. These results indicated that the postponed application of P and K fertilizers alleviated the root morphology of wheat plants after LSC damage, including root surface area and dry weight. So, the postponed application of P and K fertilizers was conducive to maintaining the balance between root and shoot growth by boosting nutrient accumulation and promoting wheat development.

3.2. Optimizing P and K Application Alleviated Root Physiology in Wheat under LT Stress

The MDA content, H_2O_2 concentrations, and $O_2\cdot^-$ production rate in wheat root were decreased under LT stress by the postponed P and K fertilizers application in the present experiment (Figure 4A–F). Optimizing P and K applications have lower levels of MDA [34]. In contrast, producing reactive oxygen species (ROS) during cold leads to damage in roots, as illustrated by higher lipid peroxidation levels and H_2O_2 accumulation in cold-sensitive rice genotypes [35].

Under LT treatments, the antioxidant system of wheat roots was activated, and the antioxidant enzyme activity increased [36]. The induction of antioxidant enzymes in response to cold stress by mineral nutrients facilitates eliminating ROS and enhancing the antioxidant defense against oxidative stress in the root [37]. In this study, antioxidant enzyme activities (SOD, POD, CAT) were significantly increased in wheat roots under LT stress by the postponed application of P and K fertilizer with variations among the different temperature treatments (Figure 3A–F). The previous study found that compared with traditional phosphorus application, the activities of SOD, POD, and CAT of twice-split phosphorus application were significantly increased on the day of LT treatment, respectively [34,38].

Plants have developed various defense systems to respond to cold stress [39]. The regulation of signaling by plant hormones is crucial for antioxidant defense against stressors [40]. In plants, such as ABA, GA, and IAA, endogenous hormones can act as signaling molecules to participate in cold resistance [41]. The results showed that under LT stress, postponed P and K fertilizer application increased the IAA and GA_3 contents in wheat roots while ABA content decreased (Figure 5A–F). LT-induced ABA accumulation in winter wheat at the booting stage altered the activity of enzymes related to sucrose metabolism, which led to sucrose synthesis and accumulation in the young ears, thus causing yield losses [42]. In this study, the IAA and GA_3 concentrations inversely altered to the concentrations of ABA not only after LT but also after the postponed application of P and K fertilizers. In addition, the postponed application of P and K fertilizers maintained lower levels of endogenous ABA and higher GA_3 and IAA levels, which changed the balance of these hormones to adapt to LT stress (Figure 5A–F).

The findings suggest that the postponed application of P and K fertilizers enhanced the cold resistance of wheat roots through a combination of physiological effects, such as antioxidant capacity, membrane lipid protection, and regulation of plant hormones. The equilibrium among these factors reinforces cold resilience.

3.3. Optimizing P and K Application Enhanced Phosphatase Activity and Nutrient Accumulation in Wheat Root under LT Stress

The plant root system is vital for efficient soil water, nutrient uptake, and stress sensing [43]. However, the intake of water and nutrients from the root system is limited in LT environments. LSC is harmful to the growth and development of wheat and the accumulation of nutrients, mainly affecting the metabolism of the root system and nutrient absorption and utilization [17]. Wheat root activity, ACP, and ALP are important in mineral nutrient and water uptake and conductance [44]. The metabolic activities of ACP and ALP in wheat roots were inhibited under the harm of LSC, which led to the decrease in the P absorption capacity of wheat roots and affected the metabolic process of P [34]. In this study, the ACP and ALP activity in wheat root were significantly decreased under LT stress by the postponed application of P and K fertilizers (Figure 6A–D). Under a low root zone temperature, the ability of roots to take up mineral nutrients is significantly decreased because of the reduced activity of enzymes and transporters [25]. LT stress may reduce the absorption efficiency of P in wheat roots, thus reducing the accumulation of P in wheat roots. However, mineral nutrients positively affect the amelioration of cold stress in plants [45]. Optimizing P application increased root ACP activity, promoted nutrient accumulation and transport, and alleviated late spring coldness's harm [34]. The ACP activity in the root system contributes to the activation of P and the symbiotic relationship between beneficial microorganisms [46]. Increasing P activity is plants' primary physiological mechanism of P mobilization [47]. Applying P and K to wheat is an effective nutrient management strategy for nitrogen use efficiency improvement, N losses, and soil N accumulation in a rice–wheat system [48]. Xu et al. [34] pointed out that twice-split phosphorus application increased phosphorus accumulation, and the translocation and partitioning to grains also enhanced after flowering, with the phosphorus harvest index increasing by 2.7–3.1% in the twice-split P application treatments compared with that in the conventional phosphorus application treatments at the low temperature. Various field experiments on different crops have shown that sufficient K supply increases K accumulation and prevents freezing injury [49]. In this study, compared with the R1 treatment, the application of R2 at LT stress enhanced P accumulation by 7.4–11.3% and K accumulation by 12.2–15.4% during the wheat-growing season (Figure 7A–D). These findings suggest that the postponed application of P and K fertilizers protects the phosphatase activity of wheat roots, including ACP and ALP, enhancing the resistance of roots under LT stress and ensuring nutrient accumulation.

4. Materials and Methods

4.1. Plant Growth Conditions

Pot experiments were conducted in the Nongcui Garden of the Anhui Agricultural University (31°8′ N, 117°2′ E) in 2021–2022 and 2022–2023 in Hefei, Anhui Province, China. The subtropical humid monsoon climate zone was selected as the test site. The meteorological data of the wheat growing seasons in this study, including the mean daily temperature and the mean monthly rainfall, are summarized in Figure 9. The potting soil type is yellow–brown and taken from a 0–20 cm cultivation layer. The nutrient contents of the experimental field before sowing are shown in Table 1.

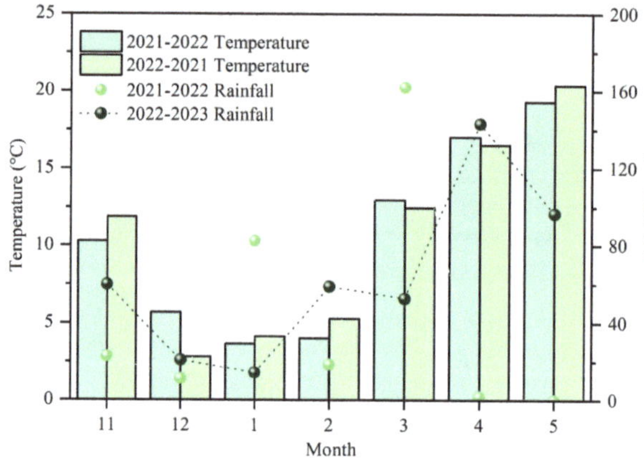

Figure 9. Temperature and precipitation during the wheat growth seasons.

Table 1. The nutrient content of the experimental field before sowing.

Year	Organ Matter ($g \cdot kg^{-1}$)	Total N ($g \cdot kg^{-1}$)	Available N ($mg \cdot kg^{-1}$)	Available P ($mg \cdot kg^{-1}$)	Available K ($mg \cdot kg^{-1}$)
2021–2022	16.5	1.3	112.2	23.0	161.6
2022–2023	16.1	1.0	110.4	26.2	149.5

4.2. Experimental Design and Cropping Management

The high-yield wheat cultivar Yannong 19, which is widely planted in the Huang-Huai-Hai wheat growing area of China, was used in this study. Wheat seeds were sown on 14 November 2021 and 11 November 2022, and harvest time for the two years was 21 May and 18 May. A two-factor experiment with a completely randomized block was designed, including temperature and PK fertilizers application treatment. The pots were 35 cm high and 26 cm in diameter. Each pot was filled with 10 kg of sieved soil, and then 3 cm of soil was covered with wheat seeds after sowing.

Traditionally, field fertilization was applied, and the P and K fertilizers were used as base fertilizers before sowing. Nitrogen was applied to the soil twice: (i) before sowing and (ii) at the jointing stage. The fertilizers used in the experiment were urea (N 46%), superphosphate (P_2O_5 12%), and potassium chloride (K_2O 60%). In this experiment, 1.8 g urea (base application of 1.2 g before sowing + top-dressing application of 0.6 g at the jointing stage) was applied to each pot during the growth period. Before sowing, the traditional P and K fertilizers treatment (R1) was used for all 5.0 g P and K fertilizers applications. The postponed P and K fertilizers treatment (R2) was applied with 2.5 g P and K fertilizers before sowing separately and at the jointing stage. The top dressing during the jointing stage was on March 11 and 15 in 2022 and 2023, respectively.

We observed wheat plants' different growth stages and patterns under a microscope (OLYMPUS SZ2-ILST; Tokyo, Japan). Moreover, all the pots, except control treatments, were shifted to an artificial climate chamber at the anther connective formation stage. To better understand the natural conditions of LT variations, the artificial climate chamber temperature gradually declined from ambient temperature 11 °C (T0) to target LT 4 °C (T1), −4 °C (T2) in 6 h duration, was subjected to LT treatments for the next 4 h, and then increased to the control temperature (Figure 2). Humidity in the chamber was kept at 75%, and light intensity was maintained at 0 μmol m^{-2} s^{-1} over both years (Figure 10). The wheat plants were then kept in the field until maturity. The experiment comprised 6 treatments [(2 fertilization treatments: R1, R2) × (3 temperature treatments: T0, T1, T2) = 6].

Each treatment contained 10 pots and was replicated three times (30 pots for each treatment); there were 180 pots in total. Eighteen seeds were sown in each pot, and nine wheat plants were maintained at the three-leaf stage through thinning.

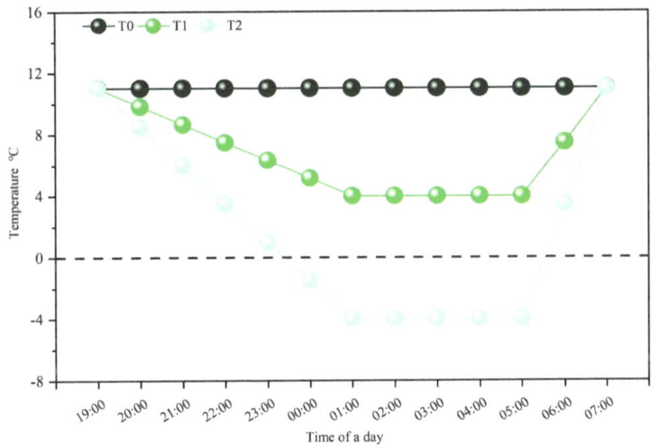

Figure 10. Dynamics of ambient temperature and in the artificial climate chamber during the experiment. T0, 11 °C; T1, 4 °C; T2, −4 °C.

4.3. *Sampling and Measurements*

After LT treatment, three plants were randomly sampled from each pot at the anther connective formation stage. Then, these samples proceeded to the necessary morphological and physiological measurements.

4.3.1. Root Morphological Traits

The roots of sampled wheat plants were washed carefully for scanning using the photo scanner. The root morphological indices were determined by WinRhizo Pro software (v2009, Regent Instrument Inc, Quebec, QC, Canada), according to Armengaud et al. [50]. The scanned roots were placed in an oven at 75 °C until constant weight to determine the dry weight of the roots.

4.3.2. Root Physiology

The wheat root's physiological properties were evaluated under LT treatment at the anther connective formation stage. The wheat root samples were frozen in liquid nitrogen and then sealed in an ultra-low temperature refrigerator with tin foil to protect them from light. The SS and SP content and the activities of SOD, CAT, and POD of wheat root samples were measured according to the assay of Xu et al. [38]. MDA content was determined using the colorimetric thiobarbituric acid method, and the results were expressed as nmol g^{-1} (FW) [51]. H_2O_2 content was determined according to the method of Kamran et al. [52], and the results were expressed as μg g^{-1} (FW). The rate of superoxide anion ($O_2\cdot^-$) production was determined by the hydroxylamine oxidation method [53], and the results were shown as mmol g^{-1} min^{-1} (FW). We analyzed the root activity of ACP and ALP using a plate reader (Multiskan FC, Thermo Fisher Scientific Company, Massachusetts, USA) with kits (Solarbio Science &Technology Company, Beijing, China). Phytohormone concentrations, including ABA, IAA, and GA_3, were quantified using the methodology detailed by Li et al. [54].

4.3.3. Root Nutrient Accumulation

To measure the root's nutrients (P, K) accumulation, wheat root samples were dried and pulverized by a ball mill (MM400, Retsch Company, Arzberg, Germany) and then digested using the H_2SO_4-H_2O_2 method. A fully automated continuous flow analyzer determined the phosphorus content in the digest (AA3, Seal Company, Norderstedt, Germany), and the content of potassium was determined using a flame photometer (FP640, Precision Instrument Science Company, Shanghai, China) [34]. Moreover, the following parameters related to phosphorus and potassium accumulation within the wheat plants during grain filling were calculated:

$$\text{Wheat root phosphorus (potassium) accumulation} = \text{Root dry matter weight} \times \text{phosphorus (potassium) content}$$

4.4. Statistical Analysis

Data were subjected to two-way analyses of variance (ANOVA) using the general linear model to calculate the effects on the measured parameters. Samples were analyzed in triplicate, and mean values were used to compare them. The analysis of variance was performed using the LSD (least significant difference) test in SPSS 26.0 (Statistical Product and Service Solutions, IBM, Armonk, NY, USA). Graphs were drawn using Origin 2021 (OriginLab Corporation, Northampton, MA, USA) and Adobe Photoshop 2021 (Adobe Systems Inc., San Jose, CA, USA).

5. Conclusions

Under LT treatment, wheat roots' antioxidant capacity, active oxygen metabolism, and plant hormones were negatively affected, inhibiting root growth and nutrient accumulation. Our study confirmed that the postponed application of P and K fertilizers enhanced the antioxidant enzyme activity (SOD, POD, and CAT) and reduced membrane damage (MDA, H_2O_2 content, and $O_2{\cdot}^-$ production rate). It improved the balance of wheat root's ABA, IAA, and GA_3. It also increased ACP and ALP activity, enhanced P and K accumulation, and improved the root surface and dry weight (Figure 11). These findings provide effective and practical knowledge and approaches to alleviate the effects of LSC on wheat production. Therefore, this study provides a theoretical basis for stress-resistant cultivation and the efficient utilization of fertilizers.

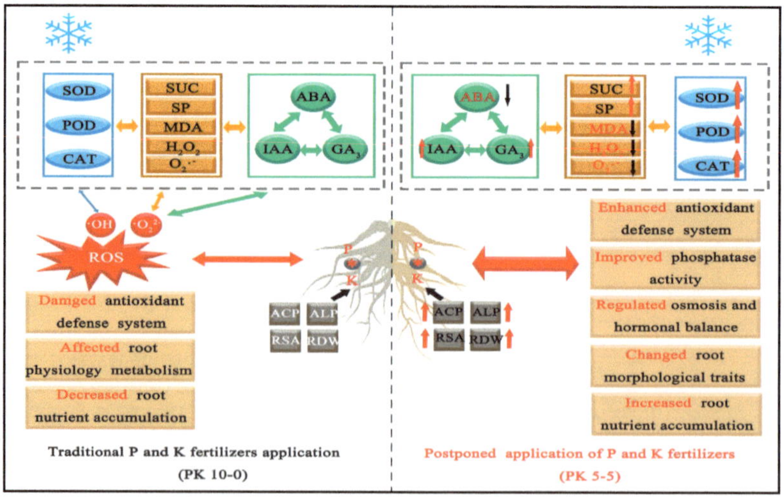

Figure 11. Schematic model of the mechanisms underlying the influence of root characteristics under LT stress by different P and K fertilizers treatment, antioxidant enzymes, osmotic adjustment substance, plant hormone, phosphatase activity, morphological traits, and nutrient accumulation. For LT treatment by postponed application of P and K fertilizers in the wheat root, the black arrows represent downregulation, and the red arrows represent upregulation.

Author Contributions: Conceptualization, H.F., J.H. (Jinwei Huang), F.L. and J.L. conceived and designed the study; H.F., J.H. (Jinwei Huang), X.Z., J.R. and J.H. (Jinwei Huang) participated in the experiments; H.F. and J.H. (Jinwei Huang) performed the data analysis; H.F., F.L. and M.A.H. wrote and edited the manuscript; F.L., J.L., B.Z. and X.C. guided the experiments. Software, J.H. (Jingyao Huang). All authors have read and agreed to the published version of the manuscript.

Funding: This work was supported by the National Key Research and Development Program of China (2023YFD1901005).

Data Availability Statement: The data presented in this study are available upon request from the authors.

Conflicts of Interest: The authors declare no conflicts of interest.

References

1. Wu, M.M.; Xu, J.Y.; Nie, Z.J.; Shi, H.; Liu, H.Y.; Zhang, Y.P.; Li, C.; Zhao, P.; Liu, H.E. Physiological, Biochemical and Transcriptomic Insights into the Mechanisms by Which Molybdenum Mitigates Cadmium Toxicity in *Triticum aestivum* L. *J. Hazard. Mater.* **2024**, *472*, 134516. [CrossRef] [PubMed]
2. Zhang, Y.; Cai, H.M.; Liu, L.Z.; Xu, H.; Chen, X.; Li, J.C. Screening of Varieties Resistant to Late-Spring Coldness in Wheat and Effects of Late-Spring Coldness on the Ultrastructure of Wheat Cells. *Agronomy* **2023**, *13*, 3011. [CrossRef]
3. Jiang, G.; Hassan, M.A.; Muhammad, N.; Arshad, M.; Chen, X.; Xu, Y.H.; Xu, H.; Ni, Q.Q.; Liu, B.B.; Yang, W.K.; et al. Comparative Physiology and Transcriptome Analysis of Young Spikes in Response to Late Spring Coldness in Wheat (*Triticum aestivum* L.). *Front. Plant Sci.* **2022**, *13*, 811884. [CrossRef]
4. Lin, F.F.; Li, C.; Xu, B.; Chen, J.; Chen, A.H.; Hassan, M.A.; Liu, B.B.; Xu, H.; Chen, X.; Sun, J.Q.; et al. Late Spring Cold Reduces Grain Number at Various Spike Positions by Regulating Spike Growth and Assimilate Distribution in Winter Wheat. *Crop J.* **2023**, *11*, 1272–1278. [CrossRef]
5. Xiao, L.J.; Liu, L.L.; Asseng, S.; Xia, Y.; Tang, L.; Liu, B.; Cao, W.X.; Zhu, Y. Estimating Spring Frost and Its Impact on Yield across Winter Wheat in China. *Agric. For. Meteorol.* **2018**, *260–261*, 154–164. [CrossRef]
6. Ji, H.T.; Xiao, L.J.; Xia, Y.M.; Song, H.; Liu, B.; Tang, L.; Cao, W.X.; Zhu, Y.; Liu, L.L. Effects of Jointing and Booting Low Temperature Stresses on Grain Yield and Yield Components in Wheat. *Agric. For. Meteorol.* **2017**, *243*, 33–42. [CrossRef]
7. Liu, L.L.; Song, H.; Shi, K.J.; Liu, B.; Zhang, Y.; Tang, L.; Cao, W.X.; Zhu, Y. Response of Wheat Grain Quality to Low Temperature during Jointing and Booting Stages—On the Importance of Considering Canopy Temperature. *Agric. For. Meteorol.* **2019**, *278*, 107658. [CrossRef]
8. Wang, C.Y.; Li, L.L.; Meiners, S.J.; Kong, C.H. Root Placement Patterns in Allelopathic Plant–Plant Interactions. *New Phytol.* **2023**, *237*, 563–575. [CrossRef]
9. Erktan, A.; McCormack, M.L.; Roumet, C. Frontiers in Root Ecology: Recent Advances and Future Challenges. *Plant Soil* **2018**, *424*, 1–9. [CrossRef]
10. Bhattacharya, A. Effect of Low-Temperature Stress on Germination, Growth, and Phenology of Plants: A Review. In *Physiological Processes in Plants under Low Temperature Stress*; Bhattacharya, A., Ed.; Springer: Singapore, 2022; pp. 1–106.
11. Kul, R.; Ekinci, M.; Turan, M.; Ors, S.; Yildirim, E. How Abiotic Stress Conditions Affects Plant Roots. In *Plant Roots*; Yildirim, E., Turan, M., Ekinci, M., Eds.; IntechOpen: London, UK, 2021; pp. 1–24.
12. Posmyk, M.M.; Bailly, C.; Szafrańska, K.; Janas, K.M.; Corbineau, F. Antioxidant Enzymes and Isoflavonoids in Chilled Soybean (*Glycine max* (L.) Merr.) Seedlings. *J. Plant Physiol.* **2005**, *162*, 403–412. [CrossRef] [PubMed]
13. Ali, M.B.; Hahn, E.J.; Paek, K.Y. Effects of Temperature on Oxidative Stress Defense Systems, Lipid Peroxidation and Lipoxygenase Activity in *Phalaenopsis*. *Plant Physiol. Biochem.* **2005**, *43*, 213–223. [CrossRef]
14. Bhattacharya, A. Plant Growth Hormones in Plants under Low-Temperature Stress: A Review. In *Physiological Processes in Plants under Low Temperature Stress*; Bhattacharya, A., Ed.; Springer: Singapore, 2022; pp. 517–627.
15. Kumari, V.V.; Banerjee, P.; Verma, V.C.; Sukumaran, S.; Chandran, M.A.S.; Gopinath, K.A.; Venkatesh, G.; Yadav, S.K.; Singh, V.K.; Awasthi, N.K. Plant Nutrition: An Effective Way to Alleviate Abiotic Stress in Agricultural Crops. *Int. J. Mol. Sci.* **2022**, *23*, 8915. [CrossRef] [PubMed]
16. Amtmann, A.; Hammond, J.P.; Armengaud, P.; White, P.J. Nutrient Sensing and Signalling in Plants: Potassium and Phosphorus. *Adv. Bot. Res.* **2005**, *43*, 209–257.
17. Hassan, M.A.; Chen, X.; Farooq, M.; Muhammad, N.; Zhang, Y.; Xu, H.; Ke, Y.Y.; Bruno, A.K.; Zhang, L.L.; Li, J.C. Cold Stress in Wheat: Plant Acclimation Responses and Management Strategies. *Front. Plant Sci.* **2021**, *12*, 676884. [CrossRef] [PubMed]
18. Wang, J.P.; Chen, H.Y.; Ma, R.H.; Shao, J.F.; Huang, S.J.; Liu, Y.; Jiang, Y.M.; Cheng, D.D. Novel Water- and Fertilizer-Management Strategy: Nutrient-Water Carrier. *J. Clean. Prod.* **2021**, *291*, 125961. [CrossRef]
19. Nie, Z.J.; Li, S.Y.; Hu, C.X.; Sun, X.C.; Tan, Q.L.; Liu, H.G. Effects of Molybdenum and Phosphorus Fertilizers on Cold Resistance in Winter Wheat. *J. Plant Nutr.* **2015**, *38*, 808–820. [CrossRef]
20. Wang, M.; Zheng, Q.S.; Shen, Q.R.; Guo, S.W. The Critical Role of Potassium in Plant Stress Response. *Int. J. Mol. Sci.* **2013**, *14*, 7370–7390. [CrossRef]

21. Xu, H.; Hassan, M.A.; Sun, D.Y.; Wu, Z.C.; Jiang, G.; Liu, B.B.; Ni, Q.Q.; Yang, W.K.; Fang, H.; Li, J.C.; et al. Effects of Low Temperature Stress on Source–Sink Organs in Wheat and Phosphorus Mitigation Strategies. *Front. Plant Sci.* **2022**, *13*, 807844. [CrossRef]
22. Shi, Y.Y.; Guo, E.J.; Cheng, X.; Wang, L.Z.; Jiang, S.K.; Yang, X.L.; Ma, H.Y.; Zhang, T.Y.; Li, T.; Yang, X.G. Effects of Chilling at Different Growth Stages on Rice Photosynthesis, Plant Growth, and Yield. *Environ. Exp. Bot.* **2022**, *203*, 105045. [CrossRef]
23. Wang, R.; Yu, M.; Xia, J.; Ren, Z.; Xing, J.; Li, C.; Xu, Q.; Cang, J.; Zhang, D. Cold Stress Triggers Freezing Tolerance in Wheat (*Triticum Aestivum* L.) via Hormone Regulation and Transcription of Related Genes. *Plant Biol.* **2022**, *25*, 308–321. [CrossRef]
24. Amirahmadi, E.; Ghorbani, M.; Moudrý, J.; Bernas, J.; Mukosha, C.E.; Hoang, T.N. Environmental Assessment of Dryland and Irrigated Winter Wheat Cultivation under Compost Fertilization Strategies. *Plants* **2024**, *13*, 509. [CrossRef] [PubMed]
25. Zhou, Y.P.; Sommer, M.L.; Hochholdinger, F. Cold Response and Tolerance in Cereal Roots. *J. Exp. Bot.* **2021**, *72*, 7474–7481. [CrossRef]
26. Aroca, R.; Porcel, R.; Ruiz-Lozano, J.M. Regulation of Root Water Uptake under Abiotic Stress Conditions. *J. Exp. Bot.* **2012**, *63*, 43–57. [CrossRef]
27. Lynch, J.P. Root Phenotypes for Improved Nutrient Capture: An Underexploited Opportunity for Global Agriculture. *New Phytol.* **2019**, *223*, 548–564. [CrossRef] [PubMed]
28. Han, Y.; White, P.J.; Cheng, L.Y. Mechanisms for Improving Phosphorus Utilization Efficiency in Plants. *Ann. Bot.* **2022**, *129*, 247–258. [CrossRef]
29. Zhu, S.G.; Cheng, Z.G.; Wang, J.; Gong, D.S.; Ullah, F.; Tao, H.Y.; Zhu, H.; Duan, H.X.; Yang, Y.M.; Xiong, Y.C. Soil Phosphorus Availability and Utilization Are Mediated by Plant Facilitation via Rhizosphere Interactions in an Intercropping System. *Eur. J. Agron.* **2023**, *142*, 126679. [CrossRef]
30. McKay Fletcher, D.M.; Ruiz, S.; Dias, T.; Petroselli, C.; Roose, T. Linking Root Structure to Functionality: The Impact of Root System Architecture on Citrate-Enhanced Phosphate Uptake. *New Phytol.* **2020**, *227*, 376–391. [CrossRef]
31. Shabala, S.; Pottosin, I. Regulation of Potassium Transport in Plants under Hostile Conditions: Implications for Abiotic and Biotic Stress Tolerance. *Physiol. Plant.* **2014**, *151*, 257–279. [CrossRef] [PubMed]
32. Hasanuzzaman, M.; Bhuyan, M.; Nahar, K.; Hossain, M.d.; Mahmud, J.; Hossen, M.d.; Masud, A.; Moumita; Fujita, M. Potassium: A Vital Regulator of Plant Responses and Tolerance to Abiotic Stresses. *Agronomy* **2018**, *8*, 31. [CrossRef]
33. Oosterhuis, D.M.; Loka, D.A.; Raper, T.B. Potassium and Stress Alleviation: Physiological Functions and Management of Cotton. *J. Plant Nutr. Soil Sci.* **2013**, *176*, 331–343. [CrossRef]
34. Xu, H.; Hassan, M.A.; Li, J.C. Twice-Split Phosphorus Application Alleviates Low Temperature Stress by Improving Root Physiology and Phosphorus Accumulation, Translocation, and Partitioning in Wheat. *Agronomy* **2023**, *13*, 2643. [CrossRef]
35. Guo, Z.H.; Cai, L.J.; Liu, C.X.; Chen, Z.Q.; Guan, S.W.; Ma, W.D.; Pan, G.J. Low-Temperature Stress Affects Reactive Oxygen Species, Osmotic Adjustment Substances, and Antioxidants in Rice (*Oryza sativa* L.) at the Reproductive Stage. *Sci. Rep.* **2022**, *12*, 6224. [CrossRef] [PubMed]
36. Wang, W.L.; Wang, X.; Huang, M.; Cai, J.; Zhou, Q.; Dai, T.B.; Jiang, D. Alleviation of Field Low-Temperature Stress in Winter Wheat by Exogenous Application of Salicylic Acid. *J. Plant Growth Regul.* **2021**, *40*, 811–823. [CrossRef]
37. Soualiou, S.; Duan, F.Y.; Li, X.; Zhou, W.B. CROP PRODUCTION UNDER COLD STRESS: An Understanding of Plant Responses, Acclimation Processes, and Management Strategies. *Plant Physiol. Biochem.* **2022**, *190*, 47–61. [CrossRef] [PubMed]
38. Xu, H.; Hou, K.Y.; Fang, H.; Liu, Q.Q.; Wu, Q.; Lin, F.; Deng, R.; Zhang, L.J.; Chen, X.; Li, J.C. Twice-Split Phosphorus Application Alleviates Low-Temperature Impacts on Wheat by Improved Spikelet Development and Setting. *J. Integr. Agric.* **2023**, *22*, 3667–3680. [CrossRef]
39. Adhikari, L.; Baral, R.; Paudel, D.; Min, D.; Makaju, S.O.; Poudel, H.P.; Acharya, J.P.; Missaoui, A.M. Cold Stress in Plants: Strategies to Improve Cold Tolerance in Forage Species. *Plant Stress* **2022**, *4*, 100081. [CrossRef]
40. Raza, A.; Salehi, H.; Rahman, M.A.; Zahid, Z.; Madadkar Haghjou, M.; Najafi-Kakavand, S.; Charagh, S.; Osman, H.S.; Albaqami, M.; Zhuang, Y.; et al. Plant Hormones and Neurotransmitter Interactions Mediate Antioxidant Defenses under Induced Oxidative Stress in Plants. *Front. Plant Sci.* **2022**, *13*, 961872. [CrossRef]
41. Wang, Y.L.; Cui, Y.T.; Hu, G.H.; Wang, X.D.; Chen, H.Z.; Shi, Q.H.; Xiang, J.; Zhang, Y.K.; Zhu, D.F.; Zhang, Y.P. Reduced Bioactive Gibberellin Content in Rice Seeds under Low Temperature Leads to Decreased Sugar Consumption and Low Seed Germination Rates. *Plant Physiol. Biochem.* **2018**, *133*, 1–10. [CrossRef]
42. Zhang, W.J.; Wang, J.Q.; Huang, Z.L.; Mi, L.; Xu, K.F.; Wu, J.J.; Fan, Y.H.; Ma, S.Y.; Jiang, D.G. Effects of Low Temperature at Booting Stage on Sucrose Metabolism and Endogenous Hormone Contents in Winter Wheat Spikelet. *Front. Plant Sci.* **2019**, *10*, 498. [CrossRef]
43. Barzana, G.; Rios, J.J.; Lopez-Zaplana, A.; Nicolas-Espinosa, J.; Yepes-Molina, L.; Garcia-Ibañez, P.; Carvajal, M. Interrelations of Nutrient and Water Transporters in Plants under Abiotic Stress. *Physiol. Plant.* **2021**, *171*, 595–619. [CrossRef]
44. Wang, Q.; Duan, C.J.; Liang, H.X.; Ren, J.W.; Geng, Z.C.; Xu, C.Y. Phosphorus Acquisition Strategies of Wheat Are Related to Biochar Types Added in Cadmium-Contaminated Soil: Evidence from Soil Zymography and Root Morphology. *Sci. Total Environ.* **2023**, *856*, 159033. [CrossRef] [PubMed]
45. Khan, M.I.R.; Nazir, F.; Maheshwari, C.; Chopra, P.; Chhillar, H.; Sreenivasulu, N. Mineral Nutrients in Plants under Changing Environments: A Road to Future Food and Nutrition Security. *Plant Genome* **2023**, *16*, e20362. [CrossRef]

46. Ham, B.K.; Chen, J.Y.; Yan, Y.; Lucas, W.J. Insights into Plant Phosphate Sensing and Signaling. *Curr. Opin. Biotechnol.* **2018**, *49*, 1–9. [CrossRef]
47. Ma, X.M.; Liu, Y.; Shen, W.J.; Kuzyakov, Y. Phosphatase Activity and Acidification in Lupine and Maize Rhizosphere Depend on Phosphorus Availability and Root Properties: Coupling Zymography with Planar Optodes. *Appl. Soil Ecol.* **2021**, *167*, 104029. [CrossRef]
48. Duan, Y.H.; Shi, X.J.; Li, S.L.; Sun, X.F.; He, X.H. Nitrogen Use Efficiency as Affected by Phosphorus and Potassium in Long-Term Rice and Wheat Experiments. *J. Integr. Agric.* **2014**, *13*, 588–596. [CrossRef]
49. Saadati, S.; Baninasab, B.; Mobli, M.; Gholami, M. Foliar Application of Potassium to Improve the Freezing Tolerance of Olive Leaves by Increasing Some Osmolite Compounds and Antioxidant Activity. *Sci. Hortic.* **2021**, *276*, 109765. [CrossRef]
50. Armengaud, P.; Zambaux, K.; Hills, A.; Sulpice, R.; Pattison, R.J.; Blatt, M.R.; Amtmann, A. EZ-Rhizo: Integrated Software for the Fast and Accurate Measurement of Root System Architecture. *Plant J.* **2009**, *57*, 945–956. [CrossRef] [PubMed]
51. Yang, C.; Gao, X.H. Impact of Microplastics from Polyethylene and Biodegradable Mulch Films on Rice (*Oryza sativa* L.). *Sci. Total Environ.* **2022**, *828*, 154579. [CrossRef]
52. Kamran, M.; Wang, D.; Xie, K.Z.; Lu, Y.S.; Shi, C.H.; EL Sabagh, A.; Gu, W.J.; Xu, P.Z. Pre-Sowing Seed Treatment with Kinetin and Calcium Mitigates Salt Induced Inhibition of Seed Germination and Seedling Growth of Choysum (*Brassica Rapa* Var. *Parachinensis*). *Ecotoxicol. Environ. Saf.* **2021**, *227*, 112921. [CrossRef]
53. Xu, X.; Guo, C.; Ma, C.; Li, M.; Chen, Y.; Liu, C.; Chu, J.; Yao, X. Brassinolide Soaking Reduced Nitrite Content and Extended Color Change and Storage Time of *Toona sinensis* Bud during Low Temperature and Near Freezing-Point Temperature Storage. *Int. J. Mol. Sci.* **2022**, *23*, 13110. [CrossRef] [PubMed]
54. Li, C.Y.; Yang, J.; Zhu, M.; Ding, J.F.; Zhu, X.K.; Zhou, G.S.; Guo, W.S. Urea Amendment Alleviated Morphological and Physiological Damages and Yield Loss of Winter Wheat Subjected to Low Temperature Stress at Jointing Stage. *Plant Growth Regul.* **2022**, *98*, 589–598. [CrossRef]

plants MDPI

Article

Genome-Wide Identification and Molecular Evolutionary History of the Whirly Family Genes in *Brassica napus*

Long Wang [1,2,3,4], Zhi Zhao [1,2,3,4], Huaxin Li [1,2,3,4], Damei Pei [1,2,3,4], Qianru Ma [1,2,3,4], Zhen Huang [5], Hongyan Wang [6,*] and Lu Xiao [1,2,3,4,*]

1 Academy of Agricultural and Forestry Sciences, Qinghai University, Xining 810016, China
2 Laboratory for Research and Utilization of Qinghai Tibet Plateau Germplasm Resources, Xining 810016, China
3 Key Laboratory of Spring Rapeseed Genetic Improvement of Qinghai Province, Xining 810016, China
4 Qinghai Spring Rape Engineering Research Center, Xining 810016, China
5 State Key Laboratory of Crop Stress Biology for Arid Areas, College of Agronomy, Northwest A&F University, Yangling 712100, China
6 Laboratory of Plant Epigenetics and Evolution, School of Life Science, Liaoning University, Shenyang 110036, China
* Correspondence: hongyanwang@lnu.edu.cn (H.W.); 2009990025@qhu.edu.cn (L.X.)

Abstract: Whirly transcription factors are unique to plants, playing pivotal roles in managing leaf senescence and DNA repair. While present in various species, their identification in *Brassica napus* L. (*B. napus*) and their differences during hybridization and polyploidy has been elusive. Addressing this, our study delves into the functional and evolutionary aspects of the Whirly gene family during the emergence of *B. napus*, applying bioinformatics and comparative genomics. We identified six *Whirly* genes in *B. napus*. In *Brassica rapa* L. (*B. rapa*), three *Whirly* genes were identified, while four were found in *Brassica oleracea* L. (*B. oleracea*). The results show that the identified *Whirly* genes not only have homology but also share the same chromosomal positions. Phylogenetic analysis indicates that *Whirly* genes in monocots and dicots exhibit high conservation. In the evolutionary process, the Whirly gene family in *B. napus* experienced events of intron/exon loss. Collinearity insights point to intense purifying selection post-duplication. Promoter regions housed diverse cis-acting elements linked to photoresponse, anaerobic initiation, and methyl jasmonate responsiveness. Notably, elements tied to abscisic acid signaling and meristem expression were prominent in diploid ancestors but subdued in tetraploid *B. napus*. Tissue-specific expression unveiled analogous patterns within subfamily genes. Subsequent qRT-PCR analysis spotlighted *BnAWHY1b*'s potential significance in abiotic stress response, particularly drought. These findings can be used as theoretical foundations to understand the functions and effects of the Whirly gene family in *B. napus*, providing references for the molecular mechanism of gene evolution between this species and its diploid ancestors.

Keywords: *Brassica napus* L.; Whirly; evolutionary analysis; expression pattern; abiotic stress

Citation: Wang, L.; Zhao, Z.; Li, H.; Pei, D.; Ma, Q.; Huang, Z.; Wang, H.; Xiao, L. Genome-Wide Identification and Molecular Evolutionary History of the Whirly Family Genes in *Brassica napus*. *Plants* **2024**, *13*, 2243. https://doi.org/10.3390/plants13162243

Academic Editors: Geovani Soares de Lima, Lauriane Almeida dos Anjos Soares and Francisco Vanies Da Silva Sá

Received: 15 July 2024
Revised: 4 August 2024
Accepted: 5 August 2024
Published: 13 August 2024

1. Introduction

Polyploidization, a common event in the diversification and evolution of plants, holds considerable importance for angiosperm speciation [1,2]. This process is often accompanied by genome rearrangement, homologous gene loss, and alterations in gene expression patterns [3]. Concurrently, changes in gene families, such as expansion, may occur in plants [4]. Transcription factors are crucial for plant growth and development, serving key roles in various physiological processes. These proteins have been shown to be integral in responding to both biotic and abiotic stresses, and they are also pivotal in mediating plant hormone signal transduction [5–8]. Whirly proteins constitute a small family of plant-specific transcription factors primarily found in chloroplasts and plastids, with a few localized in mitochondria [9–11]. These proteins consist of three distinct functional domains: the N-terminal domain, which typically contains signal peptides for chloroplasts or mitochondria; the Whirly domain, responsible for binding single-stranded DNA; and the C-terminal variable domain, which regulates this binding activity [12]. Studies reveal that Whirly transcription factors play crucial roles in regulating leaf senescence [13,14], silique development [15,16], plastid and mitochondrial stabilization, and genomic damage repair [17–21]. Moreover, maintaining stable genetic information expression is vital for plant growth and development. Consequently, these multifunctional transcription factors have been extensively studied in various plants, including *Arabidopsis thaliana* L. [9,19], *Oryza sativa* L. [22,23], *Zea mays* L. [15,17], *Solanum tuberosum* L. [24], and *Medicago sativa* L. [25].

Whirly transcription factors were first identified in potatoes. PBF-2 (PR-10a binding factor 2) can combine with the cis-acting element ERE on the promoter of the pathogenesis-related gene (PR-10a) to regulate the expression of defense genes [24]. Three *Whirly* genes (*AtWHY1-3*) were identified in *A. thaliana* by Krause et al., of which *AtWHY1* and *AtWHY3* were mainly localized in chloroplasts or plastids, while *AtWHY2* was localized in mitochondria [9]. The *WHY1* gene plays a dominant role in regulating the leaf senescence and repairing DNA damage. For example, the content of H_2O_2 is remarkably increased when *WHY1* is knocked out from the seeds of *A. thaliana*, eventually resulting in the senescent phenotype in plants [26]. In addition, the deletion of *WHY1/3* significantly inhibits the binding between RNA polymerase and DNA, which diminishes the homologous recombination repair [27]. *WHY1* is also actively involved in the signal transduction of plant hormones, such as salicylic acids and abscisic acids [16,28]. Besides the important effects on the growth and development of plants, this gene can also respond to abiotic stresses. Bonding with the promoter region, *SlWHY1* activates the expression of *SlRbcS1*, making it respond to low-temperature stresses [29]. Furthermore, *SlWHY1* also serves as a heat stress-induced gene to increase the heat resistance of tomatoes by regulating the expression of *SlHSP21.5A* [30]. *WHY1* and *WHY3* can interact with each other to maintain the stability of genomes [20,31]. *WHY2* actively participates in the signal pathway of methyl jasmonate. It is of great significance to repair and stabilize the mitochondrial genomes when acting as a binding protein of DNA/RNA in mitochondria, despite its undefined mechanism of action in chloroplasts or mitochondria [32].

Brassica napus L. (*B. napus*), a characteristic allotetraploid species within the *Brassica* genus, arises from the hybridization and polyploidization of *B. rapa* and *B. oleracea*, and it holds significance as a major oil crop. The Whirly gene family in *B. napus* has not yet been thoroughly examined regarding its number, characteristics, and roles. Consequently, in this study, we carried out in-depth identification and examination of the Whirly gene family in *B. napus* (ZS11), utilizing both bioinformatics and comparative genomics approaches. We examined chromosomal placement, phylogenetic affiliations, and gene configurations, etc. Besides, the transformations that occurred within the Whirly gene family constituents throughout the development of *B. napus* were assessed. This study provides foundational insights into the functions and effects of the Whirly gene family in this plant species, facilitating future research into the molecular mechanisms that drive the evolution of genes in *B. napus* and its diploid ancestors. Our findings enhance understanding of the complex evolutionary dynamics and adaptations in these plants.

2. Methods

2.1. Genome-Wide Identification and Whirly Gene Family Analysis

To pinpoint the Whirly gene family members in *B. napus* and its diploid precursors (*B. rapa* and *B. oleracea*), we sourced the genome and protein reference sequences along with their respective annotation files for the three species from the BRAD database (http://www.brassicadb.cn, accessed on 6 March 2024) [33]. Additionally, *Whirly* sequences and related annotation data for *A. thaliana* were extracted from the TAIR database (https://www.arabidopsis.org, accessed on 7 March 2024) [34]. Our initial step involved securing the hidden Markov model (PF08536) that encodes the Whirly protein domain from the Pfam database (http://pfam-legacy.xfam.org, accessed on 9 March 2024) [35]. Using the hmmsearch tool (version 3.4) under the Linux environment, we searched for *B. napus* protein sequences in our database. Homology comparison was conducted through Blast (E value less than 1×10^{-5}). After excluding sequences with low coverage, we utilized tools such as NCBI CD-search [36] and InterProScan (version 5.32-71.0) [37] to retrieve domains with Whirly protein features. Finally, candidate genes with Whirly domain features were retained for further analysis.

2.2. Protein Physicochemical Properties and Gene Structures Analysis

The physicochemical attributes of the discerned sequences were evaluated utilizing the Expasy website (https://web.expasy.org/protparam/, accessed on 17 March 2024) [38]. Subsequently, WoLF PSORT aided in forecasting the subcellular localization (https://wolfpsort.hgc.jp/, accessed on 18 March 2024) [39]. We explored the Whirly protein sequences' conserved domains and motifs using the NCBI CD-search [36] and MEME Suite 5.5.0 online tools [40]. The architectural analysis of the genes was conducted employing the Gene Structure View feature in TBtools (version 1.120) [41].

2.3. Analysis of the Chromosomal Localization and Phylogeny of the Whirly Gene Family

Based on the genome information about *B. rapa*, *B. oleracea*, and *B. napus*, we plotted the chromosomal localization of *Whirly* genes using the online software MapInspect (version 1.0). The MEGA-X software (version 10.1.8) was used for the repeated alignment of Whirly protein sequences among ten species, i.e., *Z. mays*, *O. sativa*, *A. thaliana*, *M. sativa*, *Glycine max* L., *Brachypodium distachyon* L., *B. rapa*, *Sorghum bicolor* L., *B. oleracea*, and *B. napus* [42]. Then, a phylogenetic tree was constructed by the maximum likelihood (ML) method, with the Bootstrap value set at 1000, and iTOL (https://itol.embl.de/, accessed on 25 March 2024) was applied for visualization [43].

2.4. Protein 3D Structure Prediction and Cis-Acting Element Analysis

Using Phyre2 (http://www.sbg.bio.ic.ac.uk/phyre2/, accessed on 3 April 2024) [44], the three-dimensional protein structures were predicted through homology modeling by comparing the Whirly protein sequences. Then, these structures were visualized using Pymol (version 2.5.4) [45]. The sequence of 2 kb was extracted from the upper stream of the transcription start site in each gene through Perl Script, and then was uploaded to the PlantCARE (http://bioinformatics.psb.ugent.be/webtools/plantcare/html/, accessed on 8 April 2024) database for subsequent analysis on cis-acting elements [46].

2.5. Collinearity Analysis and Protein Interaction Network Prediction of Whirly Gene Family

MCscanX (version 1.1.11) was employed to investigate the gene duplication mechanisms and collinearity within the Whirly gene family in *B. napus* and its diploid progenitors [47]. To assess the rates of synonymous (Ks) and non-synonymous (Ka) substitutions among duplicated gene pairs, along with their ratio (Ka/Ks), we used the Simple Ka/Ks Calculator (NG) package available in TBtools (version 1.120) (Supplementary Table S1) [41]. Based on the protein–protein interaction data from the STRING database (https://cn.string-db.org/, accessed on 13 April 2024) [48], we constructed a network of Whirly protein interactions using the Whirly proteins from *A. thaliana* as a reference. The network analysis and the visualization of the data were conducted using Cytoscape software (version 3.9.1) [49].

2.6. Subcellular Localization Analysis

We first used PCAMBIA2300-GFP as a vector to clone and transform the candidate genes. The successful Agrobacterium cultures were diluted and used to infect *Nicotiana tabacum* L. leaves, marking the infected areas. The infected *N. tabacum* was then dark-cultivated overnight. After 2–3 days of infection, the infected areas were cut, the epidermis was peeled for slide preparation, and finally, the locations affected by the candidate genes were observed under a confocal microscope. GFP represents the green fluorescence field

(488 nm), CHI represents chlorophyll autofluorescence, DAPI represents the DAPI field (358 nm), DIC represents bright field, and Merge represents the combined field.

2.7. Expression Patterns and qRT-PCR Analysis of Whirly Gene Family in B. napus

The Whirly gene family's expression patterns in *B. napus* were visualized and analyzed using the Heatmap package in R (version 4.2.1) based on the data issued by Liu et al. [50] (http://yanglab.hzau.edu.cn/BnIR, accessed on 23 April 2024). These data were concerned with the amount of gene expression (TPM) in *B. napus* under different tissues (root, cotyledon, petal, sepal, filament, pollen, stem epidermis, vegetative rosette, and silique).

To further investigate the potential biological roles of the *Whirly* genes, we conducted qRT-PCR analysis on ZS11 seeds. These seeds were grown in a tissue culture chamber maintained at 25 ± 1 °C, with a 12:12 light/dark cycle. Once the seedlings reached the five-leaf stage, they were subjected to stress treatments, which included irrigation with 200 mmol/L sodium chloride and 10% PEG-6000, along with spraying the adaxial/abaxial leaf surfaces with 200 mmol/L methyl jasmonate. Leaf samples were collected at 0, 3, 6, 9, 12, and 24 h post-treatment, immediately frozen in liquid nitrogen, and stored at −80 °C. Each treatment included three biological replicates.

For RNA extraction, total RNA was isolated from the collected leaf tissues using a specific kit provided by Sangon Biotech Engineering (Shanghai, China). This RNA was then converted to cDNA using the PrimeScriptTM RT reagent kit from TaKaRa (Kusatsu, Japan). The expression levels of the *Whirly* genes were measured using the ABI-7500 fluorescence quantitative system along with Takara's qPCR SYBR Green Master Mix. Specific primers were designed with the assistance of Primer 5 software. BnActin7, with GeneBank ID: GBEQ01027912.1, was used as the internal reference gene, as detailed in Supplementary Table S2. Each reaction was conducted in triplicate to ensure accuracy. The $2^{-\Delta\Delta CT}$ method was used to analyze the expression levels of the *Whirly* genes across various treatments and time points, with the raw data available in Supplementary Table S3.

3. Results

3.1. Identification and Physicochemical Characterization of Whirly Genes

Taking the *Whirly* genes in *A. thaliana* as a model, the homologous alignment was performed in *B. rapa*, *B. oleracea*, and *B. napus* using various tools (Pfam, SMART, CD-search, etc.). Finally, three, four, and six *Whirly* genes were identified in these three species, respectively (Table 1). According to the physicochemical properties of these genes, the number of amino acids ranged between 95 (*BnAWHY3b*) and 269 (*BrWHY3a*); the molecular weight was measured to be 10643.99Da (*BnAWHY3b*)–29404.12Da (*BrWHY3a*); the isoelectric point was 4.91 (*BnAWHY3b*)–10.22 (*BoWHY4*). Most Whirly family members were basic amino acids. In addition, all proteins were unstable, except for *BrWHY1a*, *BoWHY4*, and *BnAWHY1b*, which were defined as stable proteins due to their instability indexes of less than 40 (39.52, 39.28, and 27.58, respectively).

Table 1. Physicochemical properties and basic information of *Whirly* genes in *B. napus* and its diploid ancestors.

Gene Name	Gene ID	Chromosome	Number of Amino Acid (aa)	Molecular Weight (Da)	Isoelectric Point (pI)	Instability Index	Orthologous Gene
BrWHY1a	*BraA06g039670.3C*	A06	261	28,804.77	9.56	39.52	*AT2G02740*
BrWHY2a	*BraA07g035880.3C*	A07	240	26,097.76	9.46	47.03	*AT1G71260*
BrWHY3a	*BraA08g030230.3C*	A08	269	29,404.12	9.30	47.37	*AT1G14410*
BoWHY1a	*BolC06g042680.2J*	C06	238	26,120.87	9.48	43.55	*AT1G71260*
BoWHY2a	*BolC07g032390.2J*	C07	261	28,763.65	9.58	44.89	*AT2G02740*
BoWHY3a	*BolC08g025140.2J*	C08	267	29,185.91	9.30	45.73	*AT1G14410*
BoWHY4	*BolC07g016070.2J*	C07	155	17,246.77	10.22	39.28	*AT1G14410*
BnAWHY1b	*BnaA06G0401800ZS*	chrA06	129	14,290.28	6.10	27.58	*AT2G02740*
BnAWHY2b	*BnaA07G0324700ZS*	chrA07	239	26,033.72	9.46	48.24	*AT1G71260*
BnAWHY3b	*BnaA08G0272100ZS*	chrA08	95	10,643.99	4.91	40.17	*AT1G14410*
BnCWHY1b	*BnaC06G0379900ZS*	chrC06	250	27,481.34	9.18	41.95	*AT1G71260*
BnCWHY2b	*BnaC07G0279600ZS*	chrC07	260	28,635.52	9.58	45.02	*AT2G02740*
BnCWHY3b	*BnaC08G0228600ZS*	chrC08	267	29,225.93	9.30	44.37	*AT1G14410*

3.2. Chromosomal Localization of the Whirly Genes

Using MapInspect, chromosomes were localized based on the genome annotation files concerning *B. rapa*, *B. oleracea*, and *B. napus*. We also found that 13 *Whirly* genes were localized on 12 chromosomes (Figure 1). Three *BrWHY* genes were localized on three chromosomes (Ar06, Ar07, and Ar08) in the genome of *B. rapa*, and four *BoWHY* genes were localized on three chromosomes (Co06, Co07, and Co08) in the genome of *B. oleracea*. Comparing the distribution patterns of *Whirly* genes in A and C subgenomes of *B. napus* and in genomes of *B. rapa* and *B. oleracea*, we found that these genes were homologous, with the same relative locations. Moreover, one gene was lost during the formation of *B. napus* due to the incomplete assembly of the Cn07 chromosome, while the other six (*BrWHY1a* and *BnAWHY1b*, *BrWHY2a* and *BnAWHY2b*, *BrWHY3a* and *BnAWHY3b*, *BoWHY1a* and *BnCWHY1b*, *BoWHY2a* and *BnCWHY2b*, *BoWHY3a* and *BnCWHY3b*) maintained their relative locations on chromosomes in this process. Based on these results, we deduced that there was no significant change in the *Whirly* genes of *B. napus* that originated from its diploid ancestors during hybridization and polyploidization.

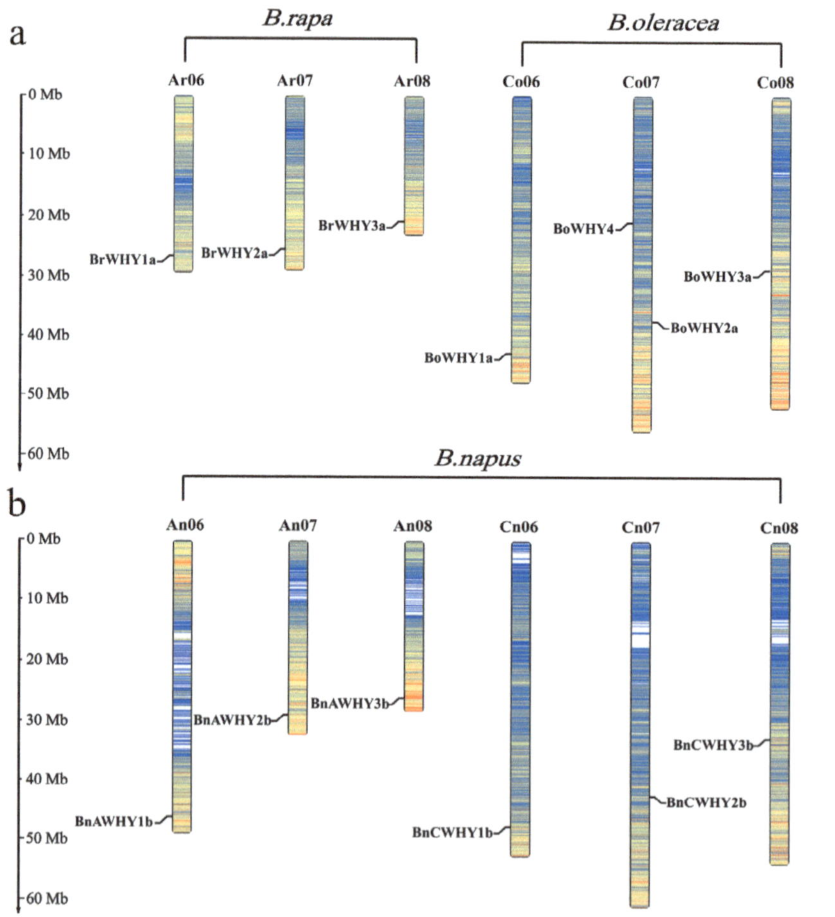

Figure 1. Chromosomal location of the Whirly gene family in *B. rapa*, *B. oleracea* (**a**), and *B. napus* (**b**).

3.3. Phylogenetic and Homology Modeling Analysis of Whirly Proteins

To further analyze the phylogenetic relationship in the Whirly gene family, a phylogenetic tree was constructed by referring to 34 Whirly protein sequences in 10 types of monocots and dicots (Figure 2). The 34 proteins were classified into six subgroups (I~VI). Subgroup IV has the largest number of members, which are *BnAWHY1b*, *BnAWHY3b*, *BnCWHY2b*, and *BnCWHY3b*; subgroup I contains *BnAWHY2b* and *BnCWHY1b*; subgroup III shows a close genetic relationship of members between soybean and alfalfa. In this tree, the Whirly proteins in monocotyledon clustered in subgroups II, V, and VI, while those in dicotyledon clustered in others. The Whirly proteins in *B. rapa* and *B. oleracea* gathered with the homologous ones in *B. napus* in subgroups I and IV, showing a good phylogenetic relationship. These results implied the highly conserved Whirly proteins in monocotyledon and dicotyledon.

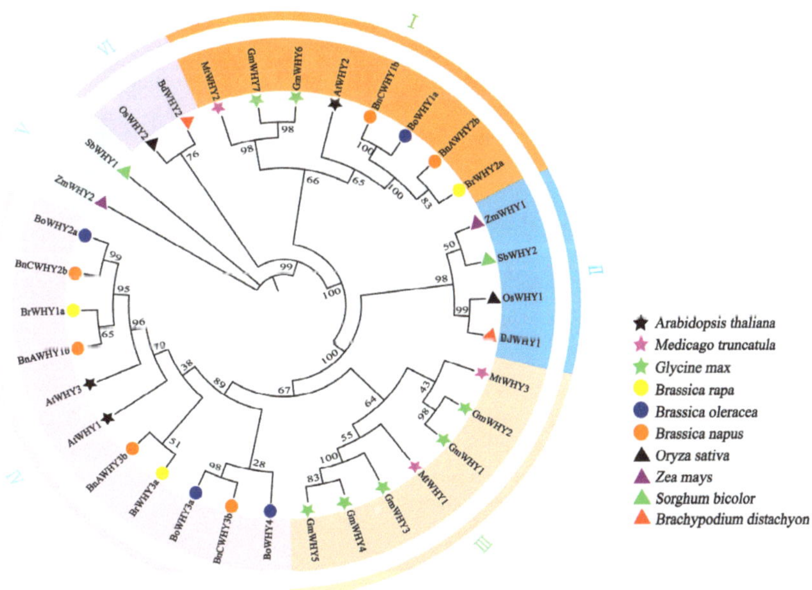

Figure 2. Evolutionary analysis of the Whirly gene family. Utilizing the protein sequences from 10 variations of monocots and dicots, an evolutionary tree for the Whirly gene family was established employing the maximum likelihood (ML) technique, substantiated by 1000 bootstrap replicates.

Based on the homology modeling and prediction of Whirly protein sequences from *A. thaliana*, *B. rapa*, *B. oleracea*, and *B. napus*, we found that the Whirly proteins in *B. napus* and its diploid ancestors, except for BoWHY4, BnAWHY1b, and BnAWHY3b, have a highly similar three-dimensional structure to that of AtWHY proteins (Figure 3) [51]. The Whirly family members are composed of three α-helices (α1–α3) and eight β-sheets (β1–β8), of which groups β1–β4 and β5–β8 are distributed perpendicularly and connected by the α1-helix. In this family, the three-dimensional structures of BoWHY4, BnAWHY1b, and BnAWHY3b are not exactly the same as other members. These similarities and differences can explain the reasons for similar or different functions among Whirly family members.

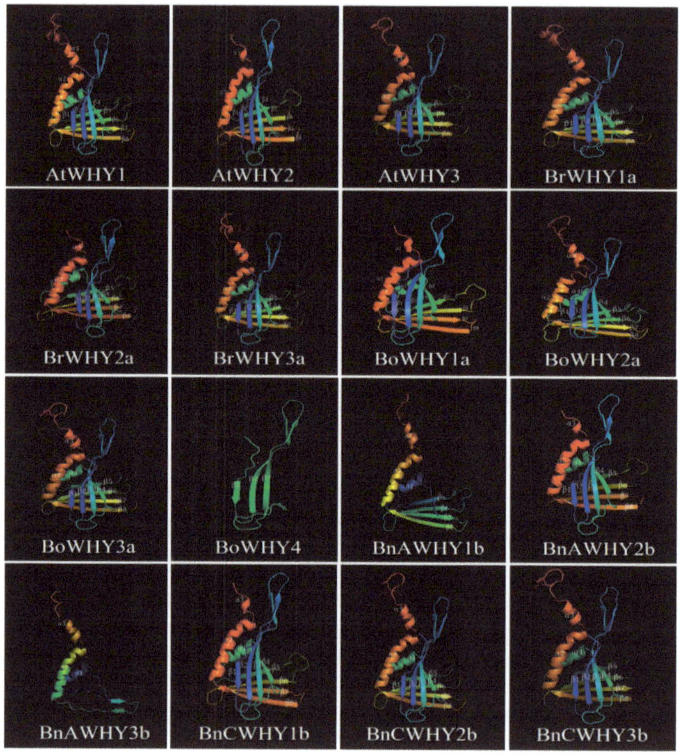

Figure 3. The predicted three-dimensional structure of Whirly proteins in *A. thaliana* (AtWHY1–3), *B. rapa* (BrWHY1a–3a), *B. oleracea* (BoWHY1a–3b), and *B. napus* (BnAWHY1b–3b, BnCWHY1b–3b), which primarily includes α-helices (α1–α3) and β-sheets (β1–β8), with the number of structures indicated numerically.

3.4. *Motif Composition, Conserved Structural Domains, and Gene Structures of the Whirly Gene Family*

The *Whirly* genes were compared in terms of gene structures, motifs, and conserved domains to reveal their phylogenetic relationship in diploid *B. rapa*, *B. oleracea*, and in allotetraploid *B. napus* (Figure 4). Figure 4a illustrates that members of the Whirly family contain between three and nine motifs. Most homologous *Whirly* sequences have similar, but not identical, motif compositions, probably because of the functional diversity and evolutionary mutation of *Whirly* genes. Furthermore, although the Whirly family members are composed of different conserved motifs in *B. napus* and its diploid ancestors, there are three overlapped ones (Motif 2, Motif 3, and Motif 5), indicating that these three motifs are probably of great importance (Figure 4a,d). The analysis on conserved domains shows that the pfam08536 domain, which is symbolic in the Whirly gene family, is contained in *BrWHYs*, *BoWHYs*, and *BnWHYs* (Figure 4b). By comparing conservative motifs and domains, we found that most of the conserved domains in the Whirly gene family contain Motif 1, Motif 4, and Motif 2, suggesting that these three motifs may be unique to the Whirly gene family. Their composition and arrangement order are of significant importance for their functional exercise (Figure 4a,b). Additionally, Motif 3, Motif 5, Motif 6, Motif 7, Motif 8, Motif 9, and Motif 10 are not present in the conserved regions of the Whirly gene family.

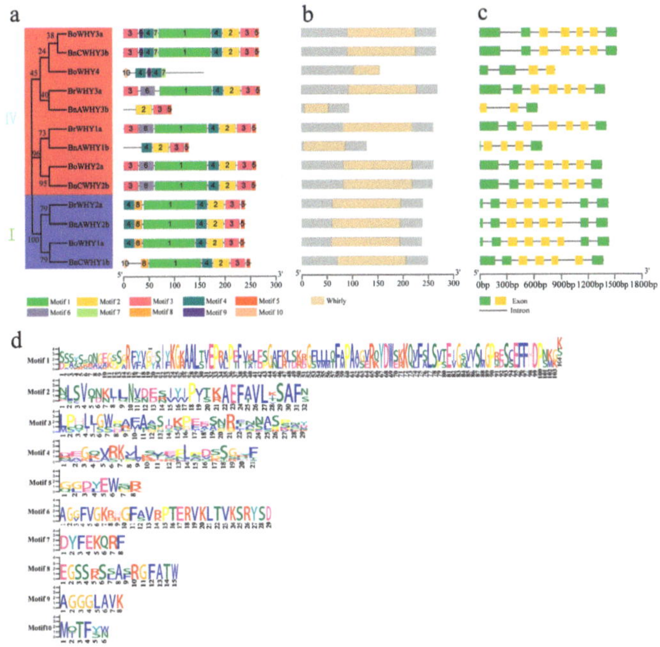

Figure 4. Gene structure and conserved motif analysis of Whirly gene family. (**a**) The 10 conserved motifs have different colors and sequences, and the different motifs correspond to one another in the structure. (**b**) Domains of the Whirly gene family. (**c**) The exon–intron structure of family members. (**d**) Logo of each motif.

A gene family evolves mainly because of its diverse gene structures [52,53]. In this study, we analyzed the intron–exon structure to understand the structural characteristics of identified *Whirly* genes. According to Figure 4c, most homologous *Whirly* genes have similar structures in *B. rapa*, *B. oleracea*, and *B. napus*, while several others are obviously different. This is explicated as follows: compared to *BrWHY3a*, *BnAWHY3b* has four exons and four introns reduced in *B. napus*; compared to *BrWHY1a*, *BnAWHY1b* has two exons and two introns reduced; and compared to *BoWHY1a*, *BnCWHY1b* has one exon and one intron reduced. Besides, despite similar gene structures between *BrWHY2a* and *BnAWHY2b*, the latter's third exon is reduced by 3bp. These findings indicate that some introns/exons may be lost during the evolution of the Whirly gene family in *B. napus*, which can be used to explain the functional difference among different family members. Moreover, in this species, the A subgenome may be more susceptible than the C. Based on the conserved domain characteristics of the Whirly gene family and genome annotation files, we found that in the gene structure, the third to sixth exons of most members of the family are relatively conserved. However, only a few members show a lack of conserved regions (*BoWHY4*, *BnAWHY1b*, *BnAWHY3b*), which may be due to the diversity of *Whirly* gene functions and mutations that occurred during evolution, providing clues for existing functional differences. These conserved exon structures may be more meaningful for their potential biological functions. Furthermore, we also found that compared to subgroup I, members of subgroup IV exhibit instability in conserved regions, which also suggests that different members of the Whirly gene family may have different functions.

3.5. Cis-Acting Elements in the Promoter Region of the Whirly Gene Family

Considering the potential impact of diverse cis-acting elements within the gene promoter on gene functionality, we examined the sequences located 2 kb upstream of the prospective genes. Our analysis led to the identification of 11 distinct categories of cis-acting elements connected with various aspects of plant physiology, as illustrated in Supplementary Figure S1. These elements are implicated in processes such as plant growth, development, photoreactivity, stress response, and hormonal activities. Within this context, *BrWHY3a* and *BnAWHY3b* are distinguished by possessing the highest count of cis-acting elements, totaling 31, while *BoWHY4* is noted for having the least (13). The element (ARE) associated with anaerobic induction was detected in all genes; the two elements (ABRE and CAT-box) participating in the abscisic acid signal transduction and meristem-related gene expression were mostly involved in diploid *B. rapa* and *B. oleracea* but were less involved in the *Whirly* gene promoters of tetraploid *B. napus*. It should be noted that the cis-acting elements identified in most *Whirly* gene promoters are predominantly related to photoresponse (Box4, GT1-motif, and G-box), anaerobic induction (ARE), and methyl jasmonate (CGTCA-motif and TGACG-motif).

3.6. Collinearity Relationship of the Whirly Gene Family

It has been verified that gene duplication is crucial to the expansion of gene families and the evolution of plant genomes [54]. We analyzed the collinearity of *Whirly* genes to explore the duplication events in which *B. napus* and its ancestral species participated (Figure 5). Based on the comparison of collinearity relationship in each species, we detected one pair of collinear genes in each diploid ancestor and seven pairs in *B. napus*. Then, this relationship was compared among these three species. It was found that *B. napus* had 10 pairs of collinear genes with *B. rapa* and *B. oleracea*, respectively, with these genes showing characteristics of segmental duplication. This implied that segmental duplication conduced to the evolution of the Whirly gene family. To better understand whether selection pressure was correlated with *Whirly* genes after duplication events, we calculated the synonymous (Ks) and non-synonymous (Ka) substitution rates of duplicate gene pairs and the ratio (Ka/Ks) between them (Supplementary Table S1). The calculation results showed that the value of Ka/Ks was less than 1 in all pairs, ranging between 0.127 and 0.710, which proved that the *Whirly* genes in *B. napus* and its diploid ancestors underwent strong purifying selection after duplication. Besides, the duplicate gene pairs were duplicated 3.33~28.50 million years ago, indicating that Whirly may be an ancient gene family and was highly conserved during its evolution.

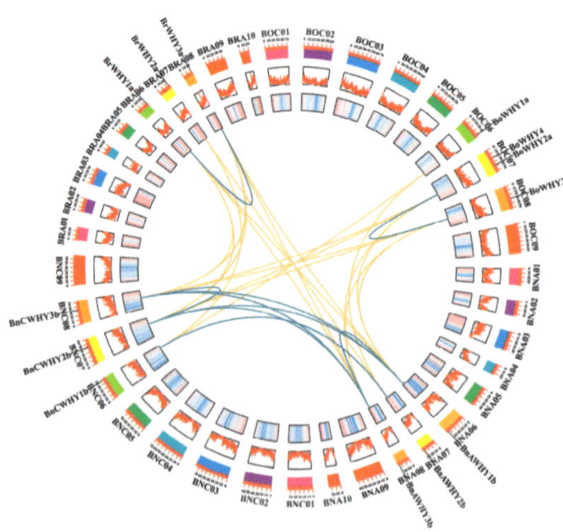

Figure 5. Collinearity analysis of the Whirly gene family. BRA, BOC, and BN represent the chromosomes of *B. rapa*, *B. oleracea*, and *B. napus*, respectively. Members of the Whirly family with gene replication events were all located on corresponding chromosomes. The yellow lines connect collinearity between species, and the blue lines connect collinearity within species.

3.7. *Prediction of Whirly Protein Interactions Network in B. napus*

Using the homologous proteins in *A. thaliana* as a model, we constructed an interaction network for Whirly proteins in *B. napus* in the online database STRING, which was conducive to better understanding the biological functions of these proteins and the dynamic control of various molecules (Figure 6). In this network, the edge number was expected to be 12; the enriched *p*-value was set at 6.66×10^{-6}. As shown in Figure 6, two Whirly family members in *B. napus* were involved in potential interactions. This network contains six proteins maintaining the plastid stability, four DNA repair proteins, four proteins concerning the chloroplast functions, two fructokinases, one galactitol synthetase, one ubiquitin carboxy-terminal hydrolase, one serine carboxypeptidase, and one DNA methyltransferase, which constitutes a total of 104 interactions. BnAWHY1b and BnCWHY3b are essential to the whole network. Furthermore, this network also includes eight key nodal proteins. Therefore, we predicted the direct or indirect interactions among different Whirly family members in *B. napus* and other important interacting proteins in this network.

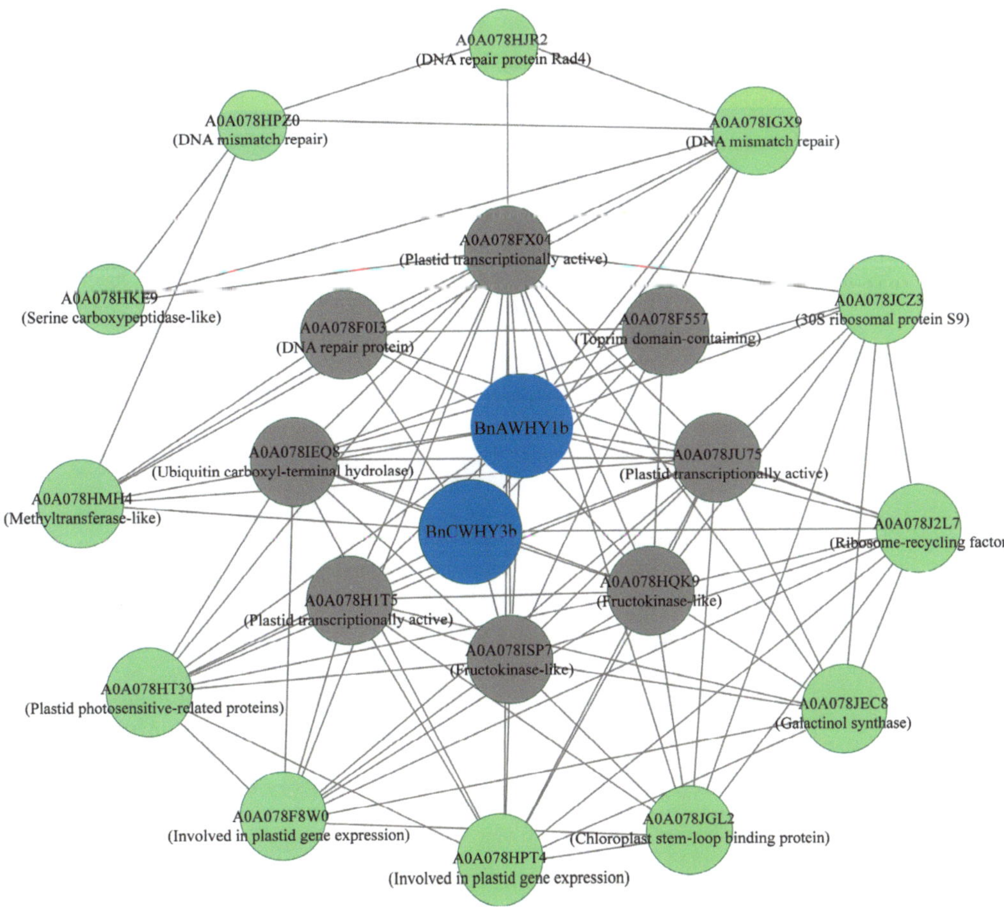

Figure 6. Mapping interactions among Whirly proteins in *B. napus*. Nodes represent different proteins, and lines represent interactions between proteins. In Cytoscape, they are arranged according to the degree algorithm (degree = 1–18), where blue circles represent members of that gene family. Gray circles represent proteins that have strong interactions with it, and green circles represent proteins that may interact with it, albeit with weaker relationships.

3.8. Subcellular Localization of BnAWHY1b and BnCWHY3b

To explore the potential biological functions of BnAWHY1b and BnCWHY3b, we conducted subcellular localization analysis. As shown in Figure 7, the BnAWHY1b protein displayed significant fluorescence signals in both the cell membrane and nucleus, but not in other areas. In contrast, the BnCWHY3b protein showed fluorescence in the chloroplasts. These results suggest that different members of the same family may function in different cellular components. Specifically, BnCWHY3b may play a key role in chloroplasts.

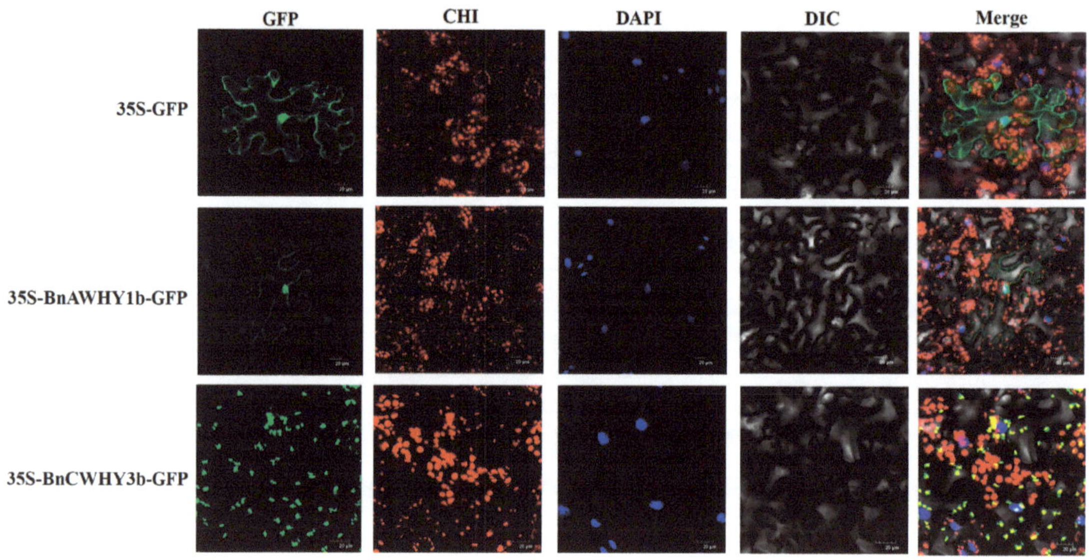

Figure 7. Subcellular localization analysis of BnAWHY1b and BnCWHY3b proteins in *N. tabacum* leaves. The proteins 35S-GFP, 35S-BnAWHY1b-GFP, and 35S-BnCWHY3b-GFP were transiently transformed into *N. tabacum*. GFP represents the green fluorescence field, CHI represents the autofluorescence field of chloroplasts, DAPI represents the DAPI field (a type of nuclear stain), DIC represents the bright field, and Merge represents the composite field. GFP field: 488 nm, DAPI field: 358 nm, CHI field: 488 nm. Note that the excitation light for green fluorescence and chloroplast autofluorescence is the same, but the collected light wavelengths are different. For reference, the scale bar represents a length of 20 μM, facilitating size assessment.

3.9. Tissue Expression Patterns of the Whirly Gene Family in B. napus

To further reveal the expression patterns and potential biological functions of identified *BnWHY* genes, we analyzed the expression of *Whirly* genes in *B. napus* under different tissues based on the RNA-seq data (Figure 8). The results showed that most *Whirly* genes were highly expressed (TPM > 10) in roots, stem epidermises, and rosette leaves. *BnAWHY2b*, *BnCWHY1b*, and *BnCWHY3b* presented similar expression patterns both in these tissues (Figure 8a) and in the development of silique (Figure 8b). Notably, these genes showed higher expressions (TPM > 20) in the initial stage of silique development (2DAF-10DAF). Furthermore, as silique kept growing and developing, *BnCWHY3b* still maintained a higher expression level (TPM > 10), which implied that this gene made great contributions to the growth and development of *B. napus*.

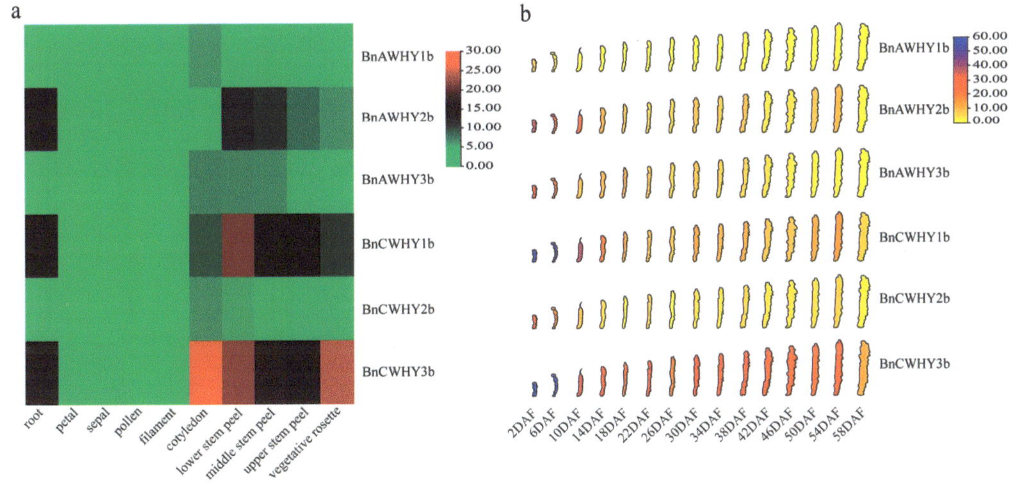

Figure 8. Tissue expression patterns of the Whirly gene family in *B. napus*. (**a**) Tissue expression pattern of the Whirly gene family in *B. napus*. (**b**) Expression patterns of the Whirly gene family in silique at different developmental stages, where DAF refers to days after flowering, and 2DAF refers to 2 days after flowering.

3.10. Expression Pattern Analysis of *BnWHY* Genes by qRT-PCR

Due to the presence of abundant hormone responsive (CGTCA-motif, TGACG-motif, ABRE) and abiotic stress-responsive (TC-rich repeats, ARE) cis-elements in the *Whirly* gene promoter, we analyzed the expression patterns of this family under various hormones and abiotic stresses to further explore the potential functions of the Whirly gene family (Figure 9). The results indicate that under salt stress, all members of the Whirly gene family exhibit an initial upregulation followed by downregulation (Figure 9a). Specifically, compared to the control (0 h), members of subfamily I (*BnAWHY2b*, *BnCWHY1b*) reach a higher expression level 12 h after stress treatment, especially *BnCWHY1b*, which shows the most significant upregulation. Members of subfamily IV (*BnAWHY1b*, *BnAWHY3b*, *BnCWHY2b*, *BnCWHY3b*) similarly reach peak expression levels 6 h post-treatment. Evidently, the *Whirly* genes in *B. napus* exhibit a positive response to salt stress. However, their expression tends to decrease with prolonged treatment. In comparison to salt stress, under PEG stress, members of the Whirly gene family also show an up-then-down expression trend (Figure 9b). Notably, members from subfamilies I and IV reach their peak expression 9 h post-treatment, with *BnAWHY1b* being the most upregulated. This suggests that *BnAWHY1b* from the Whirly gene family might play a vital role in *B. napus*' response to PEG stress. Under MeJA stress, members of subfamily I (*BnAWHY2b*, *BnCWHY1b*) differ markedly in their expression pattern compared to other genes, primarily showing an initial upregulation followed by downregulation, peaking 6 h post-treatment (Figure 9c). However, most members of subfamily IV (*BnAWHY1b*, *BnCWHY2b*, and *BnCWHY3b*) mainly exhibit a downregulation trend. Contrasting with *BnAWHY1b*, *BnCWHY2b*, and *BnCWHY3b* of subfamily IV, *BnAWHY3b* shows an up-then-down trend, peaking 9 h post-treatment. Moreover, the results demonstrate that members of the Whirly gene family in *B. napus* (subfamilies I and IV) have similar response patterns under both salt and PEG stress (Figure 9a,b), primarily characterized by an up-then-down trend. However, under hormone stress, they exhibit different expression patterns (Figure 9c). This variation might be associated with their involvement in distinct biological processes or the execution of different biological functions, hence the specificity.

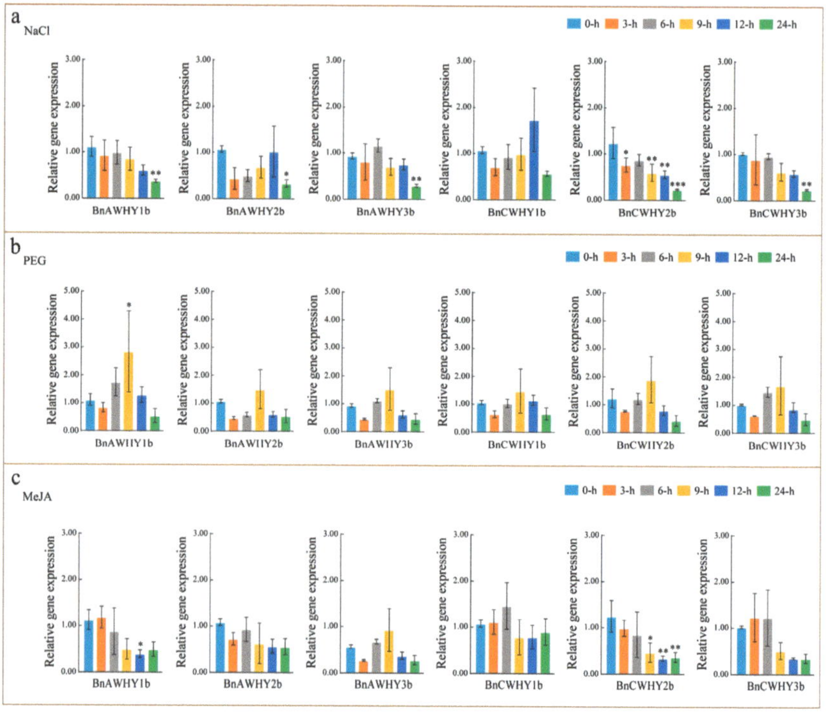

Figure 9. Expression patterns of the Whirly gene family in *B. napus* leaves under (**a**) the NaCl treatment; (**b**) the drought treatment; (**c**) the MeJA treatment. The expression levels of *Whirly* genes are relative to actin. The relative expression levels of each gene were calculated by the $2^{-\Delta\Delta Ct}$ method. The mean (±SE) expression values were calculated from three independent biological replicates and three technical replicates (*, $p < 0.05$; **, $p < 0.01$; ***, $p < 0.001$).

4. Discussion

The process of polyploidization, a recurring event in species evolution, plays a crucial role in plant diversification and the emergence of new species. It not only influences evolutionary trajectories but also enhances the adaptability of species to diverse environmental conditions [55–57]. Plants that have experienced polyploidization will undergo changes in genome size (such as genome rearrangement and homologous gene loss) and gene sequence [3]. The successful sequencing and assembly of genomes in *B. napus* and its diploid ancestors facilitate the genome-wide identification of *Whirly* genes and the exploration of their potential biological functions [58–60]. The Whirly gene family plays a major part in regulating leaf senescence and silique development, stabilizing plastids and mitochondria, and repairing genomic damages. It has been reported in *A. thaliana*, *O. sativa*, *Z. mays*, *S. tuberosum*, and *M. sativa*, but is seldom studied in *Brassica*. Therefore, analyzing the Whirly gene family within the *B. napus* genome and contrasting it with the homologous genes of its diploid predecessors provides deeper insights into the evolutionary trajectory during the emergence of the tetraploid *B. napus*.

In our research, homologous alignment and conserved domain methods led to the discovery of three, four, and six *Whirly* genes within *B. rapa*, *B. oleracea*, and *B. napus*, respectively. Notably, a gene was absent in *B. napus* due to an incomplete assembly of the Cn07 chromosome. The remaining genes in the A and C subgenomes align closely in number and chromosomal distribution with the *Whirly* genes in the genomes of its diploid predecessors, suggesting minimal gene loss in the *B. napus* Whirly gene family. A study of subcellular localization revealed a predominance of these genes in the chloroplasts of the three species, underscoring their significant role, particularly in safeguarding and repairing chloroplast genomes. This finding is consistent with earlier identifications of Whirly proteins, affirming their localization in chloroplasts and involvement in the reparative processes of chloroplast genomes [11]. The phylogenetic relationship among different species indicated that the Whirly gene family in monocotyledon and dicotyledon separately clustered in different branches. This finding aligns with earlier research, suggesting a high level of conservation of this family throughout evolution [25]. A more detailed examination of the conserved motifs and gene structures of this family reveals the presence of Motif 2, Motif 3, and Motif 5 in most *Whirly* genes. Notably, distinctions in *BnAWHY1b*, *BnAWHY3b*, and *BnCWHY1b* between *B. napus* and its diploid forebears are observed, marked by the absence of several conserved motifs. Meanwhile, some introns/exons were also lost in gene structures, which was caused by the functional diversity and evolutionary mutation of *Whirly* genes. In addition, this diversity was promoted by introns/exons due to their varying structures, which helped understand the functional difference among different genes.

The genes, containing various types of cis-acting elements, are different in biological functions [61]. In this investigation, we discerned specific elements within the promoters of *Whirly* genes in *B. rapa*, *B. oleracea*, and *B. napus*. Notably, *B. napus* had fewer cis-acting elements (151) compared to *B. rapa* and *B. oleracea* (176). However, the count of photoresponse-related elements was comparable between *B. napus* (57) and its diploid precursors (58). This discrepancy primarily originated from variations in the cis-acting elements associated with plant growth, development, hormone responses, and stress responses, particularly those linked to abscisic acid signal transduction (ABRE) and meristem (CAT-box). Our findings supported the notion that gene duplication significantly contributes to the proliferation of gene families and the evolution of plant genomes [54]. Our data depicted a segmental duplication pattern characterizing the collinearity between *B. napus* and its diploid forebears. A Ka/Ks ratio below 1 highlighted a robust purifying selection imposed on the *Whirly* genes in these species post-duplication, indicating potential correlations between selection pressure and these genes following duplication. Moreover, since the Ks value in duplicated genes often acts as a molecular clock, uniformity in its values amongst these genes is anticipated over time. Furthermore, we determined that the *Whirly* genes in *B. napus* and its diploid ancestors underwent duplication approximately 3.33 to 28.50 million years ago. This evidence suggests the ancient origins of the Whirly gene family and underscores its significant conservation throughout evolutionary history.

The change in gene expression patterns can directly reflect gene functions. In our research, most of the *Whirly* genes in *B. napus* exhibited expression in roots, stem epidermises, and rosette leaves (TPM >10). However, the organizational structure and expression patterns were fundamentally consistent within the same subgroup. *BnAWHY2b*, *BnCWHY1b*, and *BnCWHY3b* presented similar expression patterns (Figure 8a), which was also the truth during the development of silique (Figure 8b). Moreover, with the continuous growth and development of silique, *BnCWHY3b* was still highly expressed (TPM > 10), which implied that this gene may greatly contribute to the growth and development of *B. napus*. Previous research has reported that plants can reach the stress responses by increasing the transcriptional level of *Whirly* genes, and they can become more tolerant to stresses through the interactions between Whirly family members. By analyzing the expression patterns of the *Whirly* genes in different tissues, it was found that *BnCWHY3b* has a higher expression level in different tissues, especially during the growth and development of silique (Figure 8). This also implies that *BnCWHY3b* may play an essential role during silique maturation. Further analysis of the *Whirly* genes' expression patterns under various abiotic stresses through qRT-PCR revealed that subfamilies I and IV respond similarly under salt and PEG stress (Figure 9), mainly showing an initial upregulation followed by downregulation. However, they display different expression patterns under hormone stress, which might be related to their involvement in different biological processes or performing various biological functions. Additionally, we found that the *Whirly* genes actively respond to drought stress, especially *BnAWHY1b*, suggesting that *BnAWHY1b* might play a significant role in *B. napus* coping with drought stress. This is consistent with previous research results where exogenously applied hydrogen peroxide aids in the accumulation of *WHY1* in chloroplasts [54]. Interestingly, in the analysis of cis-acting elements, we identified numerous antioxidant-related elements (AREs). However, how they function, whether they provide protection to the plant, and the specific mechanisms of action still need further investigation in the future.

5. Conclusions

This study employed bioinformatics and comparative genomics approaches to analyze the *Whirly* genes in allotetraploid *B. napus* and its diploid progenitors. We identified six *Whirly* genes in *B. napus*, three in *B. rapa*, and four in *B. oleracea*. Whirly proteins are categorized into six subgroups based on sequence characteristics and exhibit high conservation across both monocots and dicots. Predictive analyses of cis-acting elements in the promoter regions of the Whirly gene family indicate that many of these elements are linked to light response, anaerobic induction, and methyl jasmonate responsiveness. Elements associated with abscisic acid signal transduction and meristem activity were predominantly found in the diploids *B. rapa* and *B. oleracea* but were less prevalent in the tetraploid *B. napus*. Gene structure analysis indicated a potential loss of introns/exons in the evolutionary history of the Whirly gene family in *B. napus*, likely influenced by segmental duplication as suggested by the collinearity relationships. The *Whirly* genes in *B. napus* and its ancestors appear to have undergone strong purifying selection post-duplication. Additionally, Motif 1, Motif 4, and Motif 2 exhibit high conservation within this gene family. Through protein–protein interaction networks and subcellular localization analysis, BnAWHY1b and BnCWHY3b are identified as potential core members of the Whirly gene family. Further studies on their protein–protein interactions and co-expression relationships could help elucidate their functions. Expression analysis showed that patterns within the same subfamily were consistent, supported by tissue-specific expression data. Further qRT-PCR analysis indicated that the *BnAWHY1b* gene could be particularly important in *B. napus* for responding to abiotic stress, especially under drought conditions. This study, through bioinformatics and comparative genomics analysis of the Whirly gene family in *B. napus* and its diploid ancestors, provides insights into the potential biological functions of this gene family and the occurrence of plant polyploidization events.

Supplementary Materials: The following supporting information can be downloaded at https://www.mdpi.com/article/10.3390/plants13162243/s1: Supplementary Table S1: The Ka, Ks, and Ka/Ks values of duplicated *Whirly* gene pairs; Supplementary Table S2: Specific primers of *Whirly* genes for qRT-PCR; Supplementary Table S3: QRT-PCR raw data; Supplementary Figure S1: Analysis of cis-acting elements in the promoter region of the Whirly gene family.

Author Contributions: L.W. and Z.Z. performed the experiments, conducted the data analysis, and contributed to the initial draft of the manuscript. H.L., D.P., Q.M. and Z.H. assisted with data interpretation. L.X. and H.W. conceptualized and oversaw the study and provided critical revisions to the manuscript. All authors have read and agreed to the published version of the manuscript.

Funding: This work was supported by the Qinghai Province "Handsome Scientist Responsibility System" Project (Grant No. 2022-NK-170), the National Key Research and Development Program (2022YFD1200405), and the China Agriculture Research System (Grant No. CARS-12).

Data Availability Statement: All data created or analyzed during this study are included in this article and its supplementary materials. The sequences discussed are available on Ensembl Plants (http://plants.ensembl.org/index.html, accessed on 6 March 2024).

Acknowledgments: We extend our gratitude to all the members of our research team for their invaluable contributions.

Conflicts of Interest: The authors declare no conflicts of interest. The funders had no role in the design of the study; in the collection, analyses, or interpretation of data; in the writing of the manuscript; or in the decision to publish the results.

References

1. Mayrose, I.; Barker, M.S.; Otto, S.P. Probabilistic models of chromosome number evolution and the inference of polyploidy. *Syst. Biol.* **2010**, *59*, 132–144. [CrossRef] [PubMed]
2. Otto, S.P.; Whitton, J. Polyploid incidence and evolution. *Annu. Rev. Genet.* **2000**, *34*, 401–437. [CrossRef] [PubMed]
3. Adams, K.L.; Wendel, J.F. Polyploidy and genome evolution in plants. *Curr. Opin. Plant Biol.* **2005**, *8*, 135–141. [CrossRef] [PubMed]
4. Yang, Z.; Gong, Q.; Qin, W.; Yang, Z.; Cheng, Y.; Lu, L.; Ge, X.; Zhang, C.; Wu, Z.; Li, F. Genome-wide analysis of WOX genes in upland cotton and their expression pattern under different stresses. *BMC Plant Biol.* **2017**, *17*, 113. [CrossRef] [PubMed]
5. Wang, X.; Niu, Y.; Zheng, Y. Multiple functions of myb transcription factors in abiotic stress responses. *Int. J. Mol. Sci.* **2021**, *22*, 6125. [CrossRef] [PubMed]
6. Liu, T.; Chen, T.; Kan, J.; Yao, Y.; Guo, D.; Yang, Y.; Ling, X.; Wang, J.; Zhang, B. The GhMYB36 transcription factor confers resistance to biotic and abiotic stress by enhancing PR1 gene expression in plants. *Plant Biotechnol. J.* **2022**, *20*, 722–735. [CrossRef] [PubMed]
7. Dudhate, A.; Shinde, H.; Yu, P.; Tsugama, D.; Gupta, S.K.; Liu, S.; Takano, T. Comprehensive analysis of NAC transcription factor family uncovers drought and salinity stress response in pearl millet (*Pennisetum glaucum*). *BMC Genom.* **2021**, *22*, 70. [CrossRef] [PubMed]
8. Wu, Q.; Chen, Y.; Zou, W.; Pan, Y.B.; Lin, P.; Xu, L.; Grisham, M.P.; Ding, Q.; Su, Y.; Que, Y. Genome-wide characterization of sugarcane catalase gene family identifies a ScCAT1 gene associated disease resistance. *Int. J. Biol. Macromol.* **2023**, *232*, 123398. [CrossRef] [PubMed]
9. Krause, K.; Kilbienski, I.; Mulisch, M.; Rödiger, A.; Schäfer, A.; Krupinska, K. DNA-binding proteins of the Whirly family in *Arabidopsis* thaliana are targeted to the organelles. *FEBS Lett.* **2005**, *579*, 3707–3712. [CrossRef]
10. Yoo, H.H.; Kwon, C.; Lee, M.M.; Chung, I.K. Single-stranded DNA binding factor AtWHY1 modulates telomere length homeostasis in *Arabidopsis*. *Plant J.* **2007**, *49*, 442–451. [CrossRef]
11. Ren, Y.; Li, Y.; Jiang, Y.; Wu, B.; Miao, Y. Phosphorylation of WHIRLY1 by CIPK14 shifts its localization and dual functions in *Arabidopsis*. *Mol. Plant* **2017**, *10*, 749–763. [CrossRef] [PubMed]
12. Desveaux, D.; Maréchal, A.; Brisson, N. Whirly transcription factors: Defense gene regulation and beyond. *Trends Plant Sci.* **2005**, *10*, 95–102. [CrossRef] [PubMed]
13. Miao, Y.; Jiang, J.; Ren, Y.; Zhao, Z. The single-stranded DNA-binding protein WHIRLY1 represses WRKY53 expression and delays leaf senescence in a developmental stage-dependent manner in *Arabidopsis*. *Plant Physiol.* **2013**, *163*, 746–756. [CrossRef] [PubMed]
14. Janack, B.; Sosoi, P.; Krupinska, K.; Humbeck, K. Knockdown of WHIRLY1 affects drought stress-induced leaf senescence and histone modifications of the senescence-associated gene *HvS40*. *Plants* **2016**, *5*, 37. [CrossRef] [PubMed]
15. Zhang, Y.F.; Hou, M.M.; Tan, B.C. The requirement of WHIRLY1 for embryogenesis is dependent on genetic background in maize. *PLoS ONE* **2013**, *8*, e67369. [CrossRef] [PubMed]
16. Isemer, R.; Krause, K.; Grabe, N.; Kitahata, N.; Asami, T.; Krupinska, K. Plastid located WHIRLY1 enhances the responsiveness of *Arabidopsis* seedlings toward abscisic acid. *Front. Plant Sci.* **2012**, *3*, 283. [CrossRef]
17. Prikryl, J.; Watkins, K.P.; Friso, G.; van Wijk, K.J.; Barkan, A. A member of the Whirly family is a multifunctional RNA- and DNA-binding protein that is essential for chloroplast biogenesis. *Nucleic Acids Res.* **2008**, *36*, 5152–5165. [CrossRef] [PubMed]
18. Cai, Q.; Guo, L.; Shen, Z.R.; Wang, D.Y.; Zhang, Q.; Sodmergen. Elevation of pollen mitochondrial DNA copy number by WHIRLY2: Altered respiration and pollen tube growth in Arabidopsis. *Plant Physiol.* **2015**, *169*, 660–673. [CrossRef]
19. Maréchal, A.; Parent, J.S.; Véronneau-Lafortune, F.; Joyeux, A.; Lang, B.F.; Brisson, N. Whirly proteins maintain plastid genome stability in *Arabidopsis*. *Proc. Natl. Acad. Sci. USA* **2009**, *106*, 14693–14698. [CrossRef]
20. Maréchal, A.; Parent, J.S.; Sabar, M.; Véronneau-Lafortune, F.; Abou-Rached, C.; Brisson, N. Overexpression of mtDNA-associated AtWhy2 compromises mitochondrial function. *BMC Plant Biol.* **2008**, *8*, 42. [CrossRef]
21. Cappadocia, L.; Maréchal, A.; Parent, J.S.; Lepage, E.; Sygusch, J.; Brisson, N. Crystal structures of DNA-Whirly complexes and their role in *Arabidopsis* organelle genome repair. *Plant Cell* **2010**, *22*, 1849–1867. [CrossRef] [PubMed]
22. Qiu, Z.; Chen, D.; Teng, L.; Guan, P.; Yu, G.; Zhang, P.; Song, J.; Zeng, Q.; Zhu, L. OsWHY1 interacts with OsTRX z and is essential for early chloroplast development in rice. *Rice* **2022**, *15*, 50. [CrossRef] [PubMed]
23. Schwacke, R.; Fischer, K.; Ketelsen, B.; Krupinska, K.; Krause, K. Comparative survey of plastid and mitochondrial targeting properties of transcription factors in *Arabidopsis* and rice. *Mol. Genet. Genom.* **2007**, *277*, 631–646. [CrossRef] [PubMed]
24. Desveaux, D.; Després, C.; Joyeux, A.; Subramaniam, R.; Brisson, N. PBF-2 is a novel single-stranded DNA binding factor implicated in PR-10a gene activation in potato. *Plant Cell* **2000**, *12*, 1477–1489. [CrossRef] [PubMed]
25. Ruan, Q.; Wang, Y.; Xu, H.; Wang, B.; Zhu, X.; Wei, B.; Wei, X. Genome-wide identification, phylogenetic, and expression analysis under abiotic stress conditions of Whirly (WHY) gene family in *Medicago sativa* L. *Sci. Rep.* **2022**, *12*, 18676. [CrossRef] [PubMed]

26. Lin, W.; Huang, D.; Shi, X.; Deng, B.; Ren, Y.; Lin, W.; Miao, Y. H_2O_2 as a feedback signal on dual-located WHIRLY1 associates with leaf senescence in *Arabidopsis*. *Cells* **2019**, *8*, 1585. [CrossRef] [PubMed]
27. Wang, W.; Li, K.; Yang, Z.; Hou, Q.; Zhao, W.W.; Sun, Q. RNase H1C collaborates with ssDNA binding proteins WHY1/3 and recombinase RecA1 to fulfill the DNA damage repair in *Arabidopsis* chloroplasts. *Nucleic Acids Res.* **2021**, *49*, 6771–6787. [CrossRef] [PubMed]
28. Lin, W.; Zhang, H.; Huang, D.; Schenke, D.; Cai, D.; Wu, B.; Miao, Y. Dual-localized WHIRLY1 affects salicylic acid biosynthesis via coordination of ISOCHORISMATE SYNTHASE1, PHENYLALANINE AMMONIA LYASE1, and S-ADENOSYL-L-METHIONINE-DEPENDENT METHYLTRANSFERASE1. *Plant Physiol.* **2020**, *184*, 1884–1899. [CrossRef] [PubMed]
29. Zhuang, K.; Wang, J.; Jiao, B.; Chen, C.; Zhang, J.; Ma, N.; Meng, Q. WHIRLY1 maintains leaf photosynthetic capacity in tomato by regulating the expression of RbcS1 under chilling stress. *J. Exp. Bot.* **2020**, *71*, 3653–3663. [CrossRef]
30. Zhuang, K.; Gao, Y.; Liu, Z.; Diao, P.; Sui, N.; Meng, Q.; Meng, C.; Kong, F. WHIRLY1 regulates HSP21.5A expression to promote thermotolerance in tomato. *Plant Cell Physiol.* **2020**, *61*, 169–177. [CrossRef]
31. Guan, Z.; Wang, W.; Yu, X.; Lin, W.; Miao, Y. Comparative proteomic analysis of coregulation of CIPK14 and WHIRLY1/3 mediated pale yellowing of leaves in *Arabidopsis*. *Int. J. Mol. Sci.* **2018**, *19*, 2231. [CrossRef]
32. Huang, C.; Yu, J.; Cai, Q.; Chen, Y.; Li, Y.; Ren, Y.; Miao, Y. Triple-localized WHIRLY2 influences leaf senescence and silique development via carbon allocation. *Plant Physiol.* **2020**, *184*, 1348–1362. [CrossRef] [PubMed]
33. Cheng, F.; Liu, S.; Wu, J.; Fang, L.; Sun, S.; Liu, B.; Li, P.; Hua, W.; Wang, X. BRAD, the genetics and genomics database for Brassica plants. *BMC Plant Biol.* **2011**, *11*, 136. [CrossRef]
34. Reiser, L.; Subramaniam, S.; Li, D.; Huala, E. Using the *Arabidopsis* information resource (TAIR) to find information about *Arabidopsis* genes. *Curr. Protoc. Bioinform.* **2017**, *60*, e574. [CrossRef]
35. El-Gebali, S.; Mistry, J.; Bateman, A.; Eddy, S.R.; Luciani, A.; Potter, S.C.; Qureshi, M.; Richardson, L.J.; Salazar, G.A.; Smart, A.; et al. The Pfam protein families database in 2019. *Nucleic Acids Res.* **2019**, *47*, D427–D432. [CrossRef]
36. Marchler-Bauer, A.; Bo, Y.; Han, L.; He, J.; Lanczycki, C.J.; Lu, S.; Chitsaz, F.; Derbyshire, M.K.; Geer, R.C.; Gonzales, N.R.; et al. CDD/SPARCLE: Functional classification of proteins via subfamily domain architectures. *Nucleic Acids Res.* **2017**, *45*, D200–D203. [CrossRef] [PubMed]
37. Quevillon, E.; Silventoinen, V.; Pillai, S.; Harte, N.; Mulder, N.; Apweiler, R.; Lopez, R. InterProScan: Protein domains identifier. *Nucleic Acids Res.* **2005**, *33*, W116–W120. [CrossRef]
38. Artimo, P.; Jonnalagedda, M.; Arnold, K.; Baratin, D.; Csardi, G.; de Castro, E.; Duvaud, S.; Flegel, V.; Fortier, A.; Gasteiger, E.; et al. ExPASy: SIB bioinformatics resource portal. *Nucleic Acids Res.* **2012**, *40*, W597–W603. [CrossRef] [PubMed]
39. Horton, P.; Park, K.J.; Obayashi, T.; Fujita, N.; Harada, H.; Adams-Collier, C.J.; Nakai, K. WoLF PSORT: Protein localization predictor. *Nucleic Acids Res.* **2007**, *35*, W585–W587. [CrossRef]
40. Bailey, T.L.; Boden, M.; Buske, F.A.; Frith, M.; Grant, C.E.; Clementi, L.; Ren, J.; Li, W.W.; Noble, W.S. MEME SUITE: Tools for motif discovery and searching. *Nucleic Acids Res.* **2009**, *37*, W202–W208. [CrossRef]
41. Chen, C.; Chen, H.; Zhang, Y.; Thomas, H.R.; Frank, M.H.; He, Y.; Xia, R. TBtools: An integrative toolkit developed for interactive analyses of big biological data. *Mol. Plant* **2020**, *13*, 1194–1202. [CrossRef] [PubMed]
42. Kumar, S.; Stecher, G.; Li, M.; Knyaz, C.; Tamura, K. MEGA X: Molecular evolutionary genetics analysis across computing platforms. *Mol. Biol. Evol.* **2018**, *35*, 1547–1549. [CrossRef] [PubMed]
43. Letunic, I.; Bork, P. Interactive Tree Of Life (iTOL) v5: An online tool for phylogenetic tree display and annotation. *Nucleic Acids Res.* **2021**, *49*, W293–W296. [CrossRef] [PubMed]
44. Kelley, L.A.; Mezulis, S.; Yates, C.M.; Wass, M.N.; Sternberg, M.J. The Phyre2 web portal for protein modeling, prediction and analysis. *Nat. Protoc.* **2015**, *10*, 845–858. [CrossRef] [PubMed]
45. Seeliger, D.; de Groot, B.L. Ligand docking and binding site analysis with PyMOL and Autodock/Vina. *J. Comput. Aided Mol. Des.* **2010**, *24*, 417–422. [CrossRef]
46. Lescot, M.; Déhais, P.; Thijs, G.; Marchal, K.; Moreau, Y.; Van de Peer, Y.; Rouzé, P.; Rombauts, S. PlantCARE, a database of plant cis-acting regulatory elements and a portal to tools for in silico analysis of promoter sequences. *Nucleic Acids Res.* **2002**, *30*, 325–327. [CrossRef] [PubMed]
47. Wang, Y.; Tang, H.; Debarry, J.D.; Tan, X.; Li, J.; Wang, X.; Lee, T.H.; Jin, H.; Marler, B.; Guo, H.; et al. MCScanX: A toolkit for detection and evolutionary analysis of gene synteny and collinearity. *Nucleic Acids Res.* **2012**, *40*, e49. [CrossRef] [PubMed]
48. Franceschini, A.; Szklarczyk, D.; Frankild, S.; Kuhn, M.; Simonovic, M.; Roth, A.; Lin, J.; Minguez, P.; Bork, P.; von Mering, C.; et al. STRING v9.1: Protein-protein interaction networks, with increased coverage and integration. *Nucleic Acids Res.* **2013**, *41*, D808–D815. [CrossRef] [PubMed]
49. Shannon, P.; Markiel, A.; Ozier, O.; Baliga, N.S.; Wang, J.T.; Ramage, D.; Amin, N.; Schwikowski, B.; Ideker, T. Cytoscape: A software environment for integrated models of biomolecular interaction networks. *Genome Res.* **2003**, *13*, 2498–2504. [CrossRef]
50. Liu, D.; Yu, L.; Wei, L.; Yu, P.; Wang, J.; Zhao, H.; Zhang, Y.; Zhang, S.; Yang, Z.; Chen, G.; et al. BnTIR: An online transcriptome platform for exploring RNA-seq libraries for oil crop Brassica napus. *Plant Biotechnol. J.* **2021**, *19*, 1895–1897. [CrossRef]
51. Desveaux, D.; Allard, J.; Brisson, N.; Sygusch, J. A new family of plant transcription factors displays a novel ssDNA-binding surface. *Nat. Struct. Biol.* **2002**, *9*, 512–517. [CrossRef] [PubMed]
52. Pellicer, J.; Hidalgo, O.; Dodsworth, S.; Leitch, I.J. Genome size diversity and its impact on the evolution of land plants. *Genes* **2018**, *9*, 88. [CrossRef] [PubMed]

53. Cao, Y.; Han, Y.; Meng, D.; Li, D.; Jin, Q.; Lin, Y.; Cai, Y. Structural, evolutionary, and functional analysis of the class III peroxidase gene family in Chinese pear (*Pyrus bretschneideri*). *Front. Plant Sci.* **2016**, *7*, 1874. [CrossRef] [PubMed]
54. Lynch, M.; Conery, J.S. The evolutionary fate and consequences of duplicate genes. *Science* **2000**, *290*, 1151–1155. [CrossRef] [PubMed]
55. Burgess, D.J. Evolution: Polyploid gains. *Nat. Rev. Genet.* **2015**, *16*, 196. [CrossRef] [PubMed]
56. Van de Peer, Y.; Maere, S.; Meyer, A. The evolutionary significance of ancient genome duplications. *Nat. Rev. Genet.* **2009**, *10*, 725–732. [CrossRef] [PubMed]
57. Shimizu-Inatsugi, R.; Terada, A.; Hirose, K.; Kudoh, H.; Sese, J.; Shimizu, K.K. Plant adaptive radiation mediated by polyploid plasticity in transcriptomes. *Mol. Ecol.* **2017**, *26*, 193–207. [CrossRef] [PubMed]
58. Wang, X.; Wang, H.; Wang, J.; Sun, R.; Wu, J.; Liu, S.; Bai, Y.; Mun, J.H.; Bancroft, I.; Cheng, F.; et al. The genome of the mesopolyploid crop species *Brassica rapa*. *Nat. Genet.* **2011**, *43*, 1035–1039. [CrossRef] [PubMed]
59. Chalhoub, B.; Denoeud, F.; Liu, S.; Parkin, I.A.; Tang, H.; Wang, X.; Chiquet, J.; Belcram, H.; Tong, C.; Samans, B.; et al. Plant genetics. Early allopolyploid evolution in the post-Neolithic Brassica napus oilseed genome. *Science* **2014**, *345*, 950–953. [CrossRef]
60. Liu, S.; Liu, Y.; Yang, X.; Tong, C.; Edwards, D.; Parkin, I.A.; Zhao, M.; Ma, J.; Yu, J.; Huang, S.; et al. The Brassica oleracea genome reveals the asymmetrical evolution of polyploid genomes. *Nat. Commun.* **2014**, *5*, 3930. [CrossRef]
61. Nuruzzaman, M.; Sharoni, A.M.; Satoh, K.; Kumar, A.; Leung, H.; Kikuchi, S. Comparative transcriptome profiles of the WRKY gene family under control, hormone-treated, and drought conditions in near-isogenic rice lines reveal differential, tissue specific gene activation. *J. Plant Physiol.* **2014**, *171*, 2–13. [CrossRef] [PubMed]

plants MDPI

Article

Fish Waste—A Novel Bio-Fertilizer for Stevia (*Stevia rebaudiana* Bertoni) under Salinity-Induced Stress

Zahra Mahdavi [1], Behrouz Esmailpour [1,*], Rasul Azarmi [1], Sima Panahirad [2], Georgia Ntatsi [3], Gholamreza Gohari [4,5] and Vasileios Fotopoulos [5,*]

[1] Department of Horticulture, Faculty of Agriculture and Natural Resources, Mohaghegh Ardabili University, Ardabil 5619911367, Iran; memahdavi2016@gmail.com (Z.M.); r_azarmi@uma.ac.ir (R.A.)
[2] Department of Horticultural Sciences and Landscape Engineering, Faculty of Agriculture, University of Tabriz, Tabriz 5166616471, Iran; s.panahirad@tabrizu.ac.ir
[3] Laboratory of Vegetable Production, Department of Crop Science, Agricultural University of Athens, Iera Odos 75, 11855 Athens, Greece; ntatsi@aua.gr
[4] Department of Horticultural Sciences, Faculty of Agriculture, University of Maragheh, Maragheh 551877684, Iran; gohari.gh@maragheh.ac.ir
[5] Department of Agricultural Sciences, Biotechnology and Food Science, Cyprus University of Technology, Limassol 3036, Cyprus
* Correspondence: behsmaiel@yahoo.com (B.E.); vassilis.fotopoulos@cut.ac.cy (V.F.)

Abstract: Currently, different strategies, including the application of bio-fertilizers, are used to ameliorate the adverse effects posed by salinity stress as the major global problem in plants. Fish waste is suggested as a novel bio-fertilizer to mitigate the effects of biotic and abiotic stresses. In this investigation, an experiment was conducted to investigate the effects by applying different concentrations (0, 5, 10, and 15% (*v*/*v*)) of fish waste bio-fertilizer on stevia plants grown under salt stress conditions (0, 20, 40, and 60 mM of NaCl). Results showed that salinity negatively affected growth parameters, the photosynthetic pigments, the relative water content, and the chlorophyll fluorescence parameters while increased the activity of antioxidant enzymes, total phenol, hydrogen peroxide (H_2O_2), malondialdehyde (MDA), proline, and total carbohydrates compared with control samples. On the other hand, the application of fish waste bio-fertilizer mitigated the effects of salinity stress by enhancing growth and mitigating stress-relative markers, especially at the highest salinity level (60 mM). Overall, fish waste bio-fertilizer could be considered a sustainable, innovative approach for the alleviation of salinity stress effects in plants and, in addition, fish waste bio-fertilizer did not cause more salinity issues, at least with the applied doses and experiment time, which is an imperative aspect.

Keywords: abiotic stress; physiological attributes; stevia; fish waste; bio-fertilizer

Citation: Mahdavi, Z.; Esmailpour, B.; Azarmi, R.; Panahirad, S.; Ntatsi, G.; Gohari, G.; Fotopoulos, V. Fish Waste—A Novel Bio-Fertilizer for Stevia (*Stevia rebaudiana* Bertoni) under Salinity-Induced Stress. *Plants* **2024**, *13*, 1909. https://doi.org/10.3390/plants13141909

Academic Editors: Geovani Soares de Lima, Lauriane Almeida dos Anjos Soares and Francisco Vanies Da Silva Sá

Received: 3 June 2024
Revised: 24 June 2024
Accepted: 4 July 2024
Published: 11 July 2024

1. Introduction

Stevia rebaudiana Bertoni (stevia) is a perennial medicinal plant belonging to the Asteraceae family. Stevia plants contain many compounds, importantly steviol glycosides (e.g., stevioside and ribosediosides A), in the leaves that are used as natural sweeteners in the food industry [1]. These natural sweeteners are 30 to 400 times sweeter than sucrose and calorie-free. More importantly, since the human body cannot digest these compounds, they are used to prevent diabetes, high blood pressure, fungal diseases, etc. Furthermore, this plant is rich in minerals, proteins, fibers, and essential oils (e.g., caryophyllene oxide and spathulenol as main constituents) [2,3].

Salinity stress, as one of the main abiotic stressors in the world, affects the crop production of most regions in all climate zones [4]. Soluble salts in saline soils are composed of several ions (Na^+, Ca^{2+}, Mg^{2+}, Cl^-, $SO_4{}^{2-}$, and $HCO_3{}^-$), resulting in destructive effects in plants [5]. The adverse effects of salinity on plant growth and performance can be related to the induced osmotic stress [6] and ionic imbalance related to the high presence of sodium (Na^+) and chloride (Cl^-) ions, leading to the reduced uptake of potassium (K^+), calcium (Ca^{2+}), magnesium (Mg^{2+}) and nitrate (NO_3^-) ions [7]. Furthermore, salinity leads to an excessive production of reactive oxygen species (ROS), thereby causing oxidative stress [8–10]. Salinity stress causes an increase in biochemical compounds, antioxidant activity, anti-inflammatory activity, and antimicrobial activity [11]. In addition, salinity has negative impacts on seed germination, photosynthesis, transpiration, chlorophyll, carotenoids, chloroplasts, PSII photosystems, and stomatal conductance [5]. Salinity decreased growth, photosynthetic pigments, and relative water content (RWC) and increased proline, total phenolic content, antioxidant activity, antioxidant enzymatic activities, H_2O_2, MDA, electrolyte leakage (EL), and essential oil content as well as stevioside and rebaudioside A constituents of essential oils in stevia [1]. Plants tolerate salinity by accumulating osmolytes (e.g., proline or sugar), regulating ion homeostasis, and increasing antioxidant system activity. Nevertheless, the responses and defense strategies of plants to survive and also maintain growth are extremely complex and involve multiple pathways [12].

Organic fertilizers (bio fertilizers) reduce chemical fertilizer input and improve the chemical structure and biological activity of soils, leading to increased crop yield by assisting in nutrient uptake and ionic balance, particularly under stressful conditions [13]. Fish waste production is of global concern [14]. Fish waste has a high content of proteins, amino acids, peptides, collagen, minerals, enzymes, and other valuable compounds [15]. Therefore, it can be used as a bio-fertilizer. Fish waste as a bio-fertilizer stimulates plant growth by providing amino acids and a slow release of essential macro- and micro-nutrients, and preventing nutrient leaching [16,17]. Amino acids are key elements needed for plant growth and initiate a number of cellular processes, such as the production of indole acetic acid [18]. Fish waste bio-fertilizer contains N-P-K in a 10-6-2 ratio [19]. Fish waste bio-fertilizer increased the growth of eggplants [16]. Bio-fertilizer foliar application supplied P and K^+ required for plant growth with a poor rooting system under stress conditions [20].

Fish waste production is increasing globally, leading to the disposal of a high content of nutrients and amino acids, and so the collection of the waste and production of liquid bio-fertilizer have become an ambitious project for the fish industry. Accordingly, the purpose of this study was to evaluate the effect of the foliar spraying of liquid fish waste bio-fertilizer, as a food supplement, on key morphophysiological and biochemical characteristics of stevia plants grown under non-stress and salinity stress conditions. To the best of our knowledge, inadequate information is available about the effect of fish waste bio-fertilizer on salinity conditions, likely marking the current study as a step forward toward its effect on plants under stress condition.

2. Results

2.1. Morphological Traits

Salinity stress significantly ($p < 0.05$) decreased plant growth traits (Figure 1A–F). Fish waste bio-fertilizer significantly ($p < 0.05$) increased root length (Figure 2A) and leaf area (Figure 2B) at 10 and 15% concentrations, and branches (Figure 2C) and fresh and dry weight of shoots (Figure 2D,E) at a 15% concentration. According to the results, fish waste bio-fertilizer at a concentration of 15% achieved optimal results, leading to an increase of 14.54% in fresh weight of shoots, 14.44% in dry weight of shoots, 12.64% in leaf area, and 14.27% in root length compared with control samples (Table S1). Figure 3 presents salinity effects on plant morphological traits (Figure 3A) as well as fish waste bio-fertilizer on the traits under 60 mM salinity (Figure 3B).

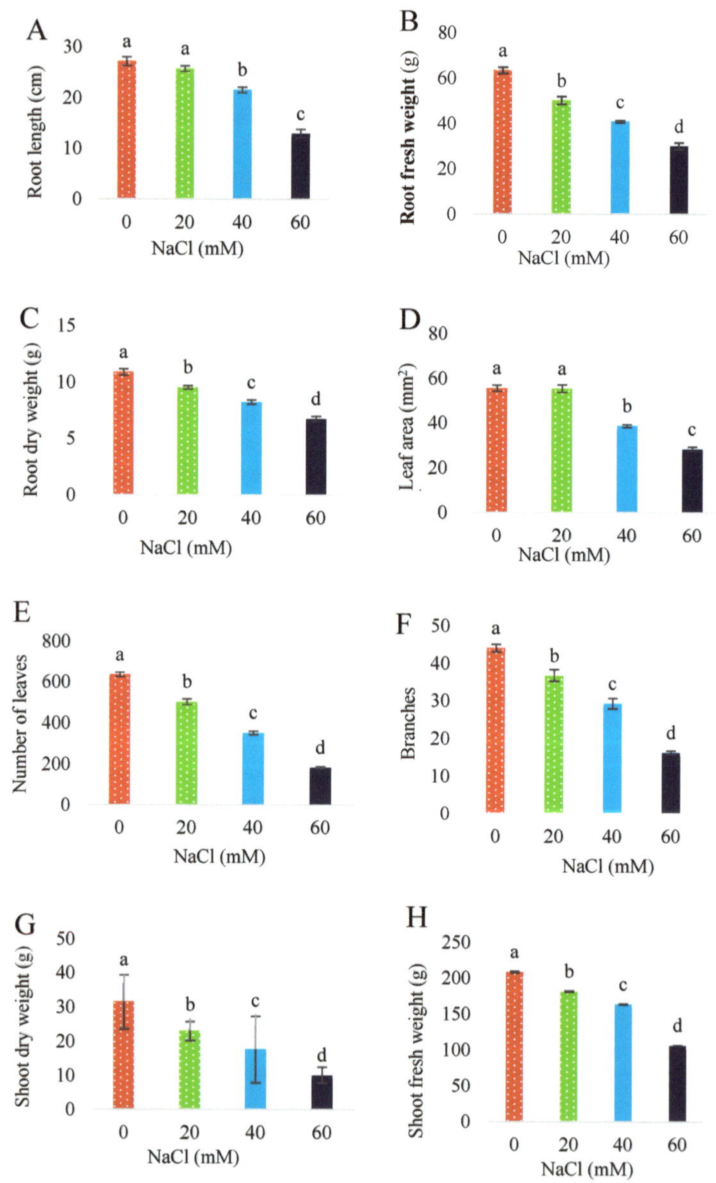

Figure 1. Effect of various concentrations of NaCl on (**A**) root length, (**B**) root fresh weight, (**C**) root dry weight, (**D**) leaf area, (**E**) number of leaves, (**F**) branches, (**G**) shoot dry weight, and (**H**) shoot fresh weight of *Stevia rebaudiana* Bertoni. Same letters indicate no significant differences ($p < 0.05$) based on Duncan's Multiple Range test.

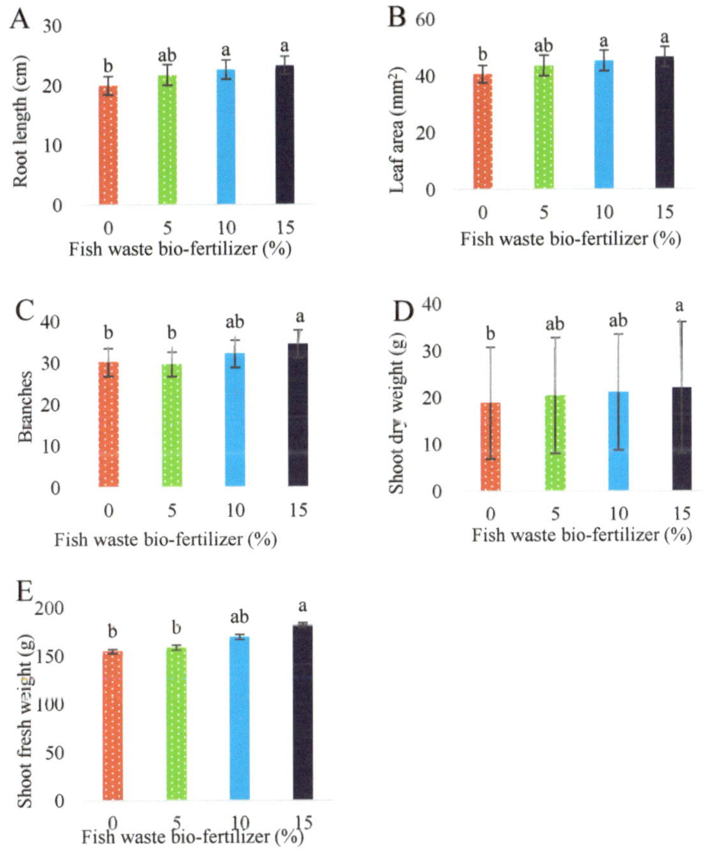

Figure 2. Effect of various concentrations of fish waste bio-fertilizer on (**A**) root length, (**B**) leaf area, (**C**) branches, (**D**) shoot dry weight, and (**E**) shoot fresh weight of *Stevia rebaudiana* Bertoni. Same letters indicate no significant differences ($p < 0.05$) based on Duncan's Multiple Range test.

(**A**)

Figure 3. *Cont.*

(**B**)

Figure 3. (**A**) Effect of various concentrations of NaCl stress (a: 60 mM, b: 40 mM, c: 20 mM, and d: Control) on *Stevia rebaudiana* Bertoni under no fish waste bio-fertilizer treatment. (**B**) Effect of various concentrations of fish waste bio-fertilizer treatment (a: control, b: 5% (*v*/*v*), c: 10% (*v*/*v*), and d: 15% (*v*/*v*)) on *Stevia rebaudiana* Bertoni under 60 mM NaCl.

2.2. Photosynthetic Pigments and $^{Fv}/_{Fm}$

Salinity at a 60 mM concentration caused a decrease of 61.27% in Chl a (Figure 4A), 65.85% in Chl b (Figure 4B), 62.46% in total Chl (Figure 4D), 44.84% in carotenoid (Figure 4C), and 15.78% in $^{Fv}/_{Fm}$ (Figure 4E) in comparison with control. Fish waste bio-fertilizer, at all applied concentrations, enhanced total Chl and Chl a, while $^{Fv}/_{Fm}$ increased using 10 and 15% concentrations under non-stress condition. Under 20 mM NaCl, higher Chl b and carotenoid content was recorded using 10 and 15% bio-fertilizer. Under 40 mM NaCl, all concentrations of fish waste bio-fertilizer increased total Chl and Chl b, while 10 and 15% concentrations enhanced Chl a content. Under 60 mM NaCl, all concentrations of bio-fertilizer enhanced Chl a, Chl b, and $^{Fv}/_{Fm}$, while 10 and 15% increased total Chl and carotenoids (Table S2).

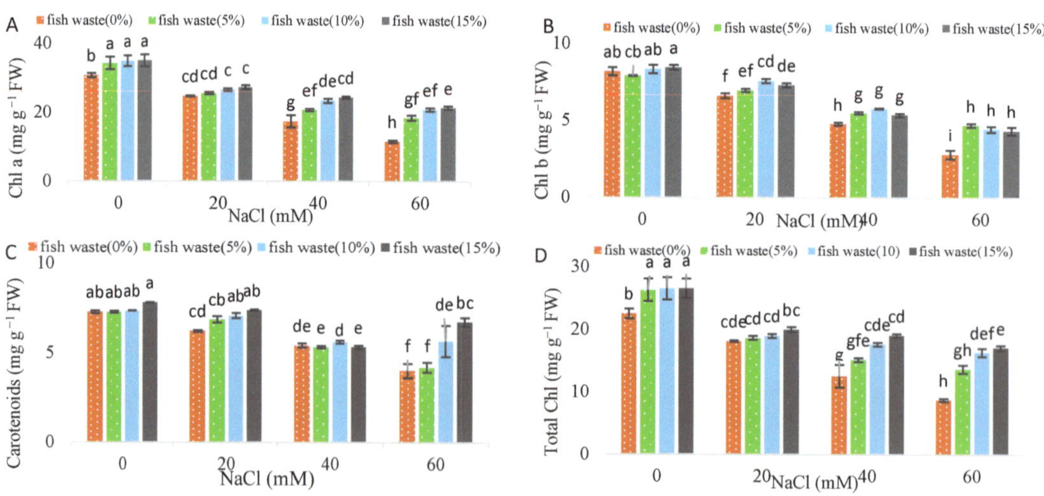

Figure 4. *Cont.*

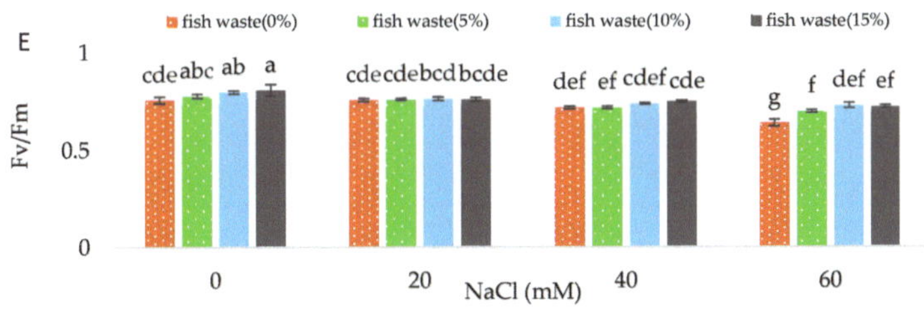

Figure 4. Effect of various concentrations of fish waste bio-fertilizer on photosynthetic parameters ((**A**) Chl a, (**B**) Chl b, (**C**) carotenoids, (**D**) total Chl, and (**E**) $^{Fv}/_{Fm}$) of *Stevia rebaudiana* Bertoni under salinity stress conditions (0, 20, 40, and 60 mM NaCl). Same letters indicate no significant differences ($p < 0.05$) based on Duncan's Multiple Range test.

2.3. Relative Water Content (RWC)

All salinity levels significantly ($p < 0.05$) reduced RWC compared with the control; the higher the salinity concentration, the lower the RWC. Salinity at 60 mM NaCl decreased RWC by 40.52% (Figure 5A). Fish waste bio-fertilizer at 10 and 15% significantly enhanced the RWC of stevia compared with the control (Figure 5B).

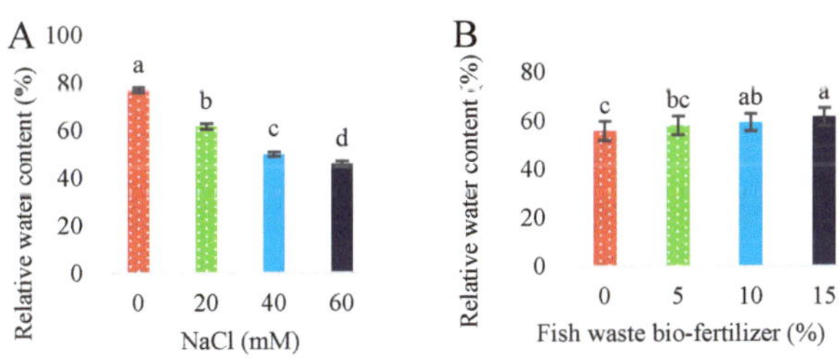

Figure 5. Effect of various concentrations of NaCl (**A**) and fish waste bio-fertilizer (**B**) on relative water content of *Stevia rebaudiana* Bertoni. Same letters indicate no significant differences ($p < 0.05$) based on Duncan's Multiple Range test.

2.4. Total Carbohydrate Content

Salinity at 40 and 60 mM NaCl significantly enhanced total carbohydrate content compared with the control. Contrarily, the application of fish waste bio-fertilizer had no effects on the content under non-stress and 20 and 40 mM NaCl stress conditions. On the other hand, bio-fertilizer application significantly increased total carbohydrate content at 10 and 15% concentrations under 60 mM salinity stress (Figure 6).

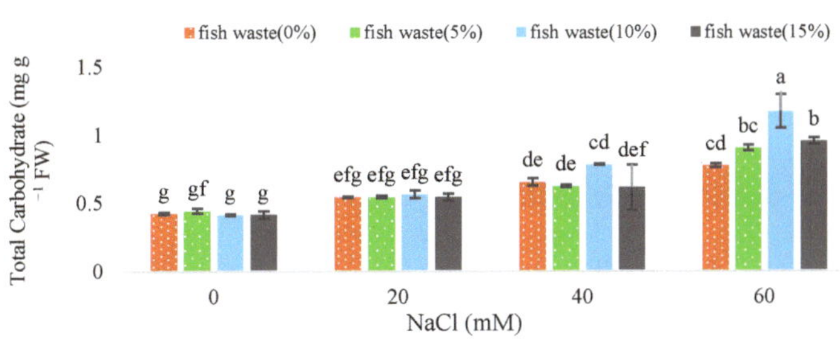

Figure 6. Effect of various concentrations of fish waste bio-fertilizer on total carbohydrates of *Stevia rebaudiana* Bertoni under salinity stress conditions (0, 20, 40, and 60 mM NaCl). Same letters indicate no significant differences ($p < 0.05$) based on Duncan's Multiple Range test.

2.5. Cellular Damage Indicators

Salinity at all applied concentrations caused significant ($p < 0.05$) enhancement in H_2O_2 (Figure 7A), MDA (Figure 7B), and electrolyte leakage (EL) (Figure 7C). The bio-fertilizer had no effect on H_2O_2 and MDA content under non-stress conditions, while 10 and 15% concentrations of fish waste bio-fertilizer led to a significant reduction in electrolyte leakage (EL). Under 20 and 40 mM NaCl, the bio-fertilizer applied at 10 and 15% concentrations significantly decreased H_2O_2, while no effect was observed in MDA content. Under 60 mM NaCl, all applied concentrations of the bio-fertilizer significantly lowered H_2O_2 and MDA content (Figure 7).

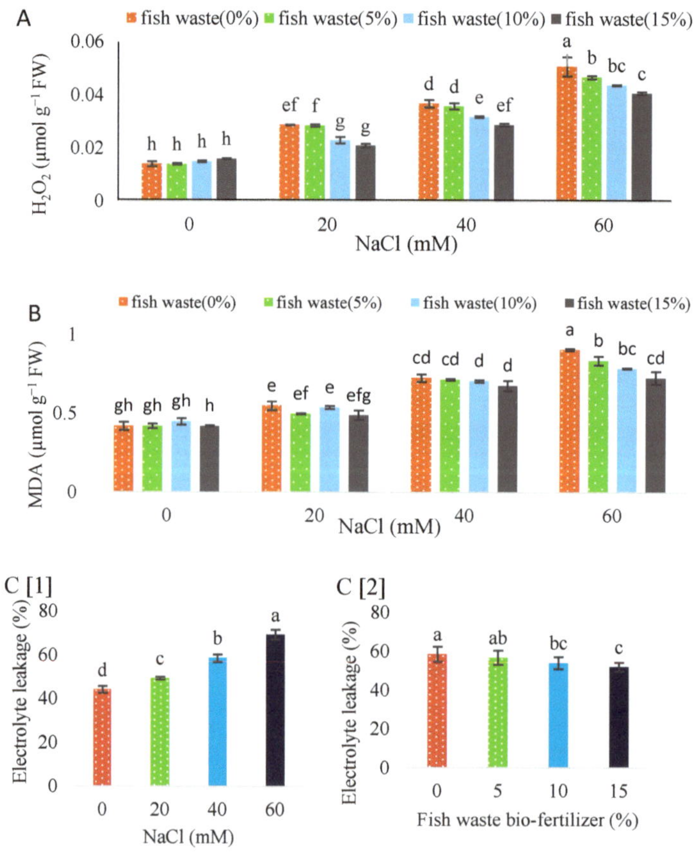

Figure 7. Effect of various concentrations of fish waste bio-fertilizer on (**A**) H_2O_2 and (**B**) MDA of Stevia rebaudiana Bertoni under salinity stress conditions (0, 20, 40, and 60 mM NaCl). Effect of various concentrations of NaCl (**C[1]**) and fish waste bio-fertilizer (**C[2]**) on electrolyte leakage of *Stevia rebaudiana* Bertoni. Same letters indicate no significant differences ($p < 0.05$) based on Duncan's Multiple Range test.

2.6. Proline and Total Phenolic Content

Salinity, at all applied concentrations, increased proline (Figure 8A) and total phenols (Figure 8B) compared with the control. Salinity at 60 mM NaCl increased total phenols by 54.67% and proline by 68.88% compared to unstressed control. Fish waste bio-fertilizer had no effect on the phenol content under non-stress conditions, while a 15% concentration significantly enhanced proline content (Figure 7A). Under 20 and 40 mM NaCl, 15% bio-fertilizer enhanced total phenols, while both 10 and 15% concentrations caused an increase in total phenols under 60 mM NaCl (Figure 8).

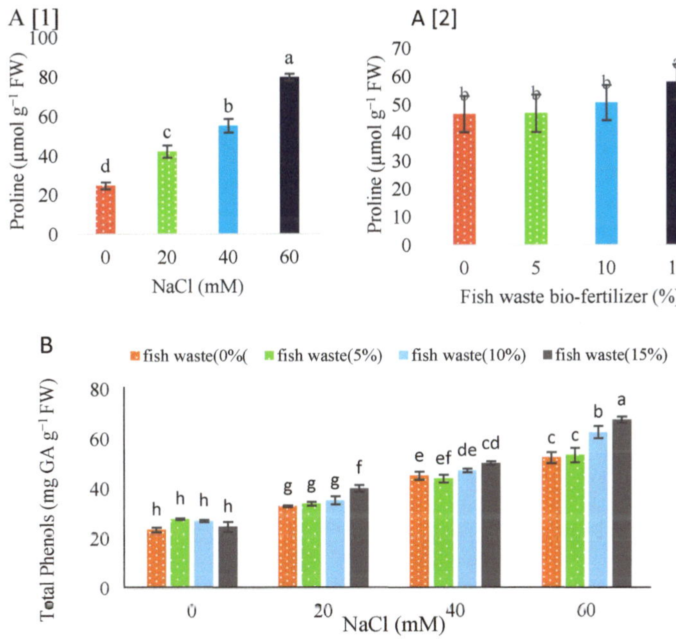

Figure 8. Effect of various concentrations of NaCl (**A[1]**) and fish waste bio-fertilizer (**A[2]**) on proline of *Stevia rebaudiana* Bertoni. Effect of various concentrations of fish waste bio-fertilizer on total phenols (**B**) of *Stevia rebaudiana* Bertoni under salinity stress conditions (0, 20, 40, and 60 mM NaCl). Same letters indicate no significant differences ($p < 0.05$) based on Duncan's Multiple Range test.

2.7. Antioxidant Enzymatic Activities

All salinity levels significantly ($p < 0.05$) enhanced enzymatic activities as compared with to the control (Figure 9). CAT activity in leaf tissues under all NaCl concentrations significantly increased compared to the untreated control. The maximum and minimum activities of CAT were recorded in 60 mM NaCl-treated plants under 10 and 15% fish waste bio-fertilizer concentration application and control samples, respectively (Figure 9B). In regard to APX, the enzymatic activity significantly increased under 20, 40, and 60 mM NaCl compared with the control. Therefore, increasing NaCl levels resulted in increasing APX activity. The highest activity among treatments was achieved at 10% fish waste bio-fertilizer concentrations under 60 mM NaCl, while the lowest was in all bio-fertilizer-treated plants under non-stress conditions (Figure 9A). The highest and lowest POD activities were observed in 10 and 15% bio-fertilizer-treated plants under 60 mM NaCl and non-stress conditions, respectively. Similar to APX and CAT, increasing salinity levels led to increasing POD activity, under no fish waste bio-fertilizer treatment. Fish waste bio-fertilizer treatments increased POD activity under both non-stress and stress conditions, with these increases being higher than non-treated plants at the same conditions (Figure 9C). Overall, fish waste bio-fertilizer at a 10% concentration generally increased antioxidant enzymatic activity under stress conditions compared with plants at similar conditions without receiving any bio-fertilizer treatments.

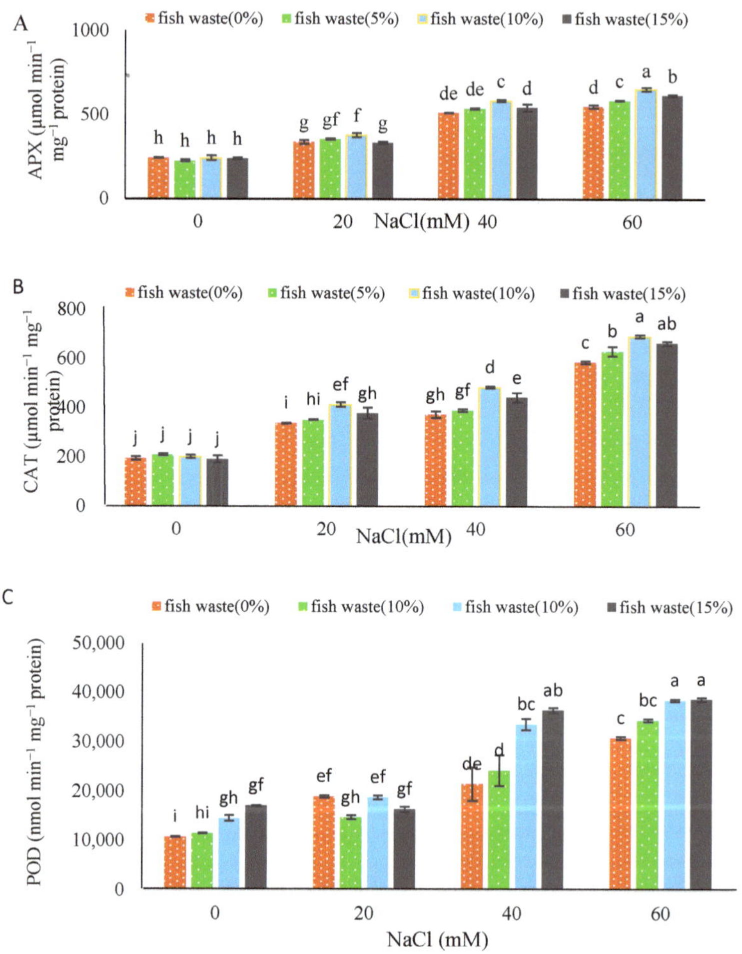

Figure 9. Effect of various concentrations of NaCl and fish waste bio-fertilizer on the enzymatic activity of (**A**) APX, (**B**) CAT, and (**C**) POD of *Stevia rebaudiana* Bertoni under salinity stress conditions (0, 20, 40, and 60 mM NaCl). Same letters indicate no significant differences ($p < 0.05$) based on Duncan's Multiple Range test.

2.8. Elemental Composition of Shoots and Roots

Salinity stress increased the Na^+ content of shoots and roots (Figure 10) and reduced the Ca^{2+} shoots and K^+ roots contents (Figure 10), with increasing NaCl concentrations showing a positive correlation with Na^+ content, and a negative correlation with Ca^{2+} and K^+ contents, respectively. The bio-fertilizer had no effect on Na^+ content in shoots and roots, Ca^{2+} content in shoots, or K^+ content in roots (Figure 10). Salinity stress reduced the Ca^{2+} roots and K^+ shoots contents (Figure 11). Whereas the bio-fertilizer enhanced Ca^{2+} content in roots at all concentrations and K^+ content in shoots at 10 and 15% bio-fertilizer concentrations under non-stress conditions (Figure 11). Under 20 mM NaCl, the bio-fertilizer at all applied concentrations decreased Na^+ content in roots, whereas K^+ content increased in roots following an application at a 15% concentration. Under moderate salinity levels, fish waste bio-fertilizer decreased Na^+ content in shoots at all applied concentrations, while application at a 15% concentration decreased Na^+ content in roots and increased Ca^{2+} content in shoots (Figure 10). Under severe salinity levels, all concentrations of the bio-fertilizer reduced Na^{+} content in shoots and roots and enhanced Ca^{2+} content in shoots and K^{+} content in roots (Table S3).

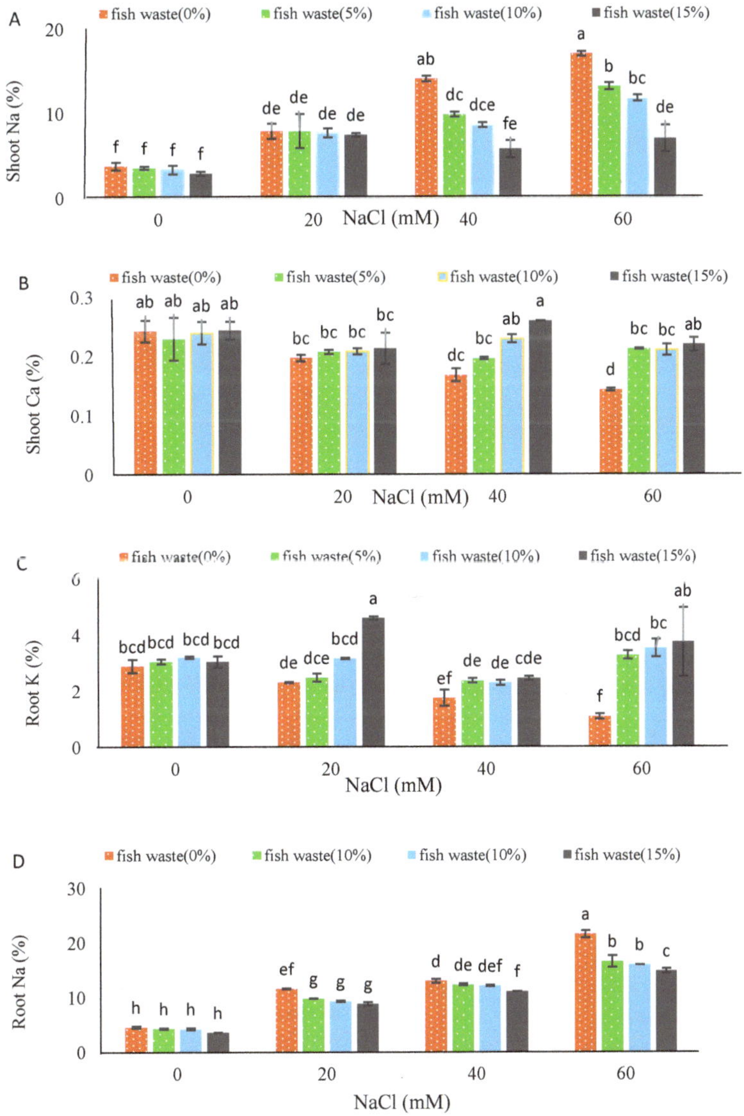

Figure 10. Effect of various concentrations of fish waste bio-fertilizer on essential nutrients: (**A**) shoot Na, (**B**) shoot Ca, (**C**) root K, and (**D**) root Na, of *Stevia rebaudiana* Bertoni under different concentrations of salinity stress (0, 20, 40, and 60 mM NaCl). Same letters indicate no significant differences ($p < 0.05$) based on Duncan's Multiple Range test.

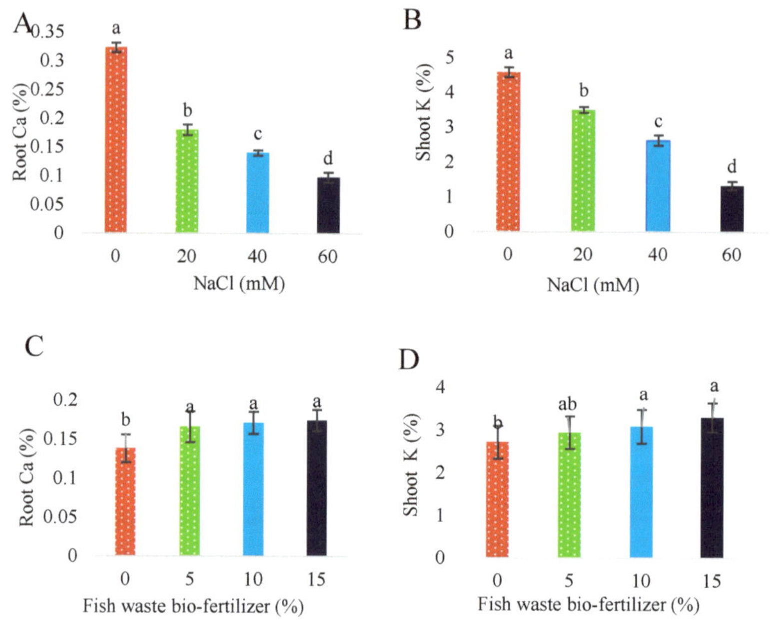

Figure 11. Effect of various concentrations of NaCl on (**A**) root Ca and (**B**) shoot K and effect of various concentrations of fish waste bio-fertilizer on (**C**) root Ca and (**D**) shoot K, of *Stevia rebaudiana* Bertoni. Same letters indicate no significant differences ($p < 0.05$) based on Duncan's Multiple Range test.

2.9. Principal Component Analysis (PCA)

To demonstrate the negative impact of salinity stress in *Stevia rebaudiana* Bertoni plants with the application of fish waste bio-fertilizer, a principal component analysis was performed (Table 1, Figure 12). For the first two components, i.e., PC1 and PC2, cover was about 92.83% of the overall data base. Here, component PC1 contributes about 84.78%, correlating (biochemical and growth traits) significantly mainly with the activity of antioxidant enzymes (APX, POD, and CAT), total phenol content, carbohydrate, proline, malondialdehyde, chlorophyll b, relative water content, electrolyte leakage and growth traits (LN, SDW, RFW, B, LA, RDW, SFW, and RL), Na^+ and Ca^{2+} contents in the roots, and K^+ contents in the shoot. On the other hand, the second component, PC2, contributes 8.05%, correlating (photosynthetic pigments) mainly with the activity of chlorophyll a, total chlorophyll, carotenoids, $^{Fv}/_{Fm}$, plant height, Na^+ and Ca^{2+} contents in the shoots, and K^+ contents in the root. Similarly, the principal component analysis revealed that different physiological and biochemical parameters correlated with different growth traits.

Table 1. Principal components analysis for morphological, physiological, and biochemical traits of *Stevia rebaudiana* Bertoni under salinity stress conditions.

Traits	First Principal Components	Second Principal Components
Total phenol content (TPCs)	−0.967	-
Catalase (CAT)	−0.953	-
Carbohydrate (Carbo)	−0.948	-
Peroxidase (POD)	−0.943	-
Proline (PRO)	−0.941	-
Ascorbate peroxidase (APX)	−0.908	−0.314
Leaf number (LN)	0.903	0.420
Shoot dry weight (SDW)	0.860	0.479

Table 1. *Cont.*

Traits	First Principal Components	Second Principal Components
Root fresh weight (RFW)	0.848	0.513
Shoot K (KS)	0.845	0.525
Branches (B)	0.845	0.493
Leaf area (LA)	0.824	0.485
Root dry weight (RDW)	0.814	0.550
Hydrogen peroxide (H_2O_2)	−0.801	−0.587
Malondialdehyde (MDA)	−0.799	−0.574
Shoot fresh weight (SFW)	0.788	0.522
Root length (RL)	0.786	0.532
Chlorophyll b (ChLb)	0.785	0.584
Relative water content (RWC)	0.777	0.572
Root Na (NaR)	−0.757	−0.629
Root Ca (CaR)	0.755	0.623
Electrolyte leakage (EL)	−0.747	−0.628
Shoot Ca (CaS)		0.859
Root K (KR)	-	0.816
Shoot Na (NaS)	−0.516	−0.808
Chlorophyll a (ChLa)	0.582	0.776
Total chlorophyll (ChLT)	0.646	0.741
Plant height (PH)	0.652	0.734
$^{Fv}/_{Fm}$	0.604	0.729
Carotenoids (Caros)	0.576	0.707
Percentage of variance	840.78%	80.05%

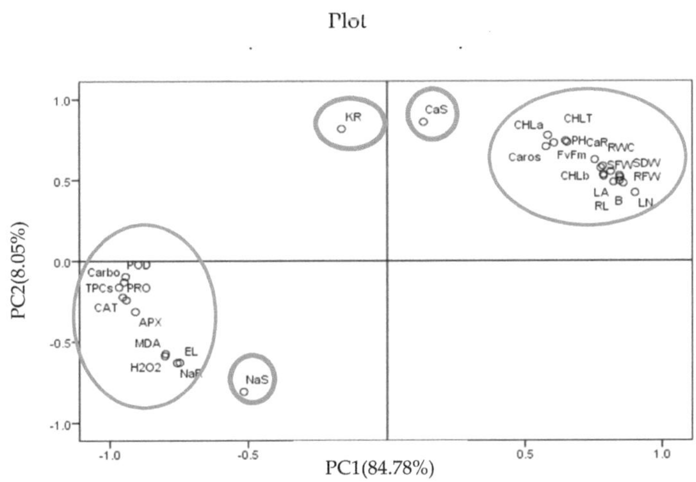

Figure 12. Plot from PCA performed on morphological, physiological, and biochemical traits in various concentrations of NaCl and fish waste bio-fertilizer of *Stevia rebaudiana* Bertoni. Abbreviations: root fresh weight (RFW), shoot fresh weight (SFW), shoot dry weight (SDW), root dry weight (RDW), relative water content (RWC), leaf number (LN), chlorophyll a (Chl_a), chlorophyll b (Chl_*b*), total chlorophyll (Chl_T), total carotenoids (Caros), Na^+ concentration in shoots and roots (Na^+ S; Na^+ R), K^+ concentration in shoots and roots (K^+ S; K^+ R), Ca^{2+} concentration in shoots and roots (Ca^{2+} S; Ca^{2+} R), proline (Pro), malondialdehyde (MDA), hydrogen peroxide (H_2O_2), total phenolic compounds (TPCs), Ascorbate peroxidase (APX), catalase (CAT), Peroxidase (POD), total carbohydrate (carbo), electrolyte leakage (EL), plant height (PH), root length (RL), leaf area (LA), and branches (B).

2.10. Correlation Analysis

Pearson's correlation was performed among different photosynthetic pigments, $^{Fv}/_{Fm}$ and the relative content of water and Na, K, and Ca in roots and shoots in the present work. Chl *a*, Chl *b*, total Chl, carotenoid content, RWC, and $^{Fv}/_{Fm}$ content are positively correlated with K and Ca in roots and shoots and negatively correlated with Na in roots and shoots at a significant level ($p < 0.05$) (Table 2).

Table 2. Pearson correlation coefficients (r) between photosynthetic pigments, $^{Fv}/_{Fm}$ and the relative content of water and sodium, potassium, and calcium in roots and shoots of *Stevia rebaudiana* Bertoni under salinity stress conditions (* $p < 0.05$; ** $p < 0.01$).

	$^{Fv}/_{Fm}$	Chl a	Chl b	Total Chl	Carotenoids	RWC	Shoot Na	Shoot K	Shoot Ca	Root Na	Root K	Root Ca
$^{Fv}/_{Fm}$	1											
Chla	0.927 **	1										
Chlb	0.916 **	0.898 **	1									
Total Chl	0.942 **	0.993 **	0.942 **	1								
Carotenoids	0.858 **	0.856 **	0.865 **	0.857 **	1							
RWC	0.847 **	0.929 **	0.49 **	0.95 **	0.883 **	1						
Shoot Na	−0.883 **	−0.948 **	−0.849 **	−0.942 **	−.846 **	−0.859 **	1					
Shoot K	0.901 **	0.905 **	0.964 **	0.938 *	0.843 **	0.951 **	−0.864 **	1				
Shoot Ca	0.706 **	0.769 **	0.587 *	0.737 **	0.518 *	0.556 *	−.839 **	0.577 **	1			
Root Na	−0.931 **	−0.939 **	−0.960 **	−0.963 **	−0.861 **	−0.949 **	0.904 **	−0.971 **	−0.673 **	1		
Root K	0.457	0.48	0.398	0.468	0.562 *	0.371	−0.464	0.27	0.488	−0.363	1	
Root Ca	0.897 **	0.933 **	0.957 **	0.957 **	0.864 **	0.957 **	−0.837 **	0.967 **	0.616 **	−0.97 **	0.408	1

3. Discussion

In the current study, the effects of fish waste bio-fertilizer were investigated on the growth performance and physiological and biochemical characteristics of stevia plants subjected to salinity stress conditions.

According to the results, salinity stress mostly caused reduction in all morphological parameters whereas fish waste bio-fertilizer particularly at the 15% level increased these parameters, resulting in better plant growth. Salinity stress reduces photosynthesis and chlorophyll pigments, relative water content, and the absorption of important mineral ions including N and K, and simultaneously increases Na^+ accumulation, all of which lead to a decrease in growth [21,22]. In fact, salinity has a negative effect on the growth and performance of plants by creating ionic toxicity, osmotic stress, and nutritional imbalance. In addition, salinity affects the availability, absorption, and transport of nutrients and water, and cell division, resulting in a decrease in growth previously confirmed in some plants [6,22–24], all in line with the current findings. Fish waste bio-fertilizer contains amino acids, organic acids, lactic acids, and acetic acids, all of which activate bacteria such as rhizobium and bacillus with the ability to dissolve P and stabilize N, Ca^{2+}, K^+, Na^+, Mg^{2+}, Zn^{2+}, and Mn^{2+} contents that then increase organic compounds and siderophore photosynthesis and chlorophyll pigments and synthesis for iron (Fe) sequestration. Increased access to the mentioned elements enhances photosynthesis, chlorophyll pigments, and plant growth, and, as a result, increases plant fresh weight [16]. Application of rural slaughterhouse waste, as a bio-fertilizer, improved plants growth when applied after planting (at two and six weeks) [25]. Fish waste bio-fertilizer causes the release of fulvic acid and humic acid, the decomposition of which creates auxin, as well as the increase in P and K and other nutrient absorption. The application of fish waste bio-fertilizer increased the growth of red chili and tomato plants [26], and the root length of *Solanum melongena* [16], onion [27], and *Prunus persica* [28], in line with the current results. Salinity reduced the amount of photosynthetic pigments and $^{Fv}/_{Fm}$ in stevia plants, while fish waste bio-fertilizer improved their amounts probably via preventing pigment decomposition. The chlorophyll fluorescence index is one of the most important factors to evaluate photosynthetic efficiency due to its sensitivity to early plant responses to stress conditions. According to our findings, a decrease in the $^{Fv}/_{Fm}$ index was reported in sweet pepper and wheat under salinity stress [28].

Salinity damages chloroplast membranes and the electron transport chain in photosystem I, and reduces the absorption of Mg and K, resulting in negative feedback on chlorophyll fluorescence parameters [29]. Salinity-induced Cl^- ions have destructive impacts on the reaction center of photosystem II (PSII), the quinone receptor, and the oxygen transport system. This destructive impacts results in a decrease in light energy of the reaction center and photosynthetic pigments and finally decrease photosynthesis process. Salinity-induced Cl^- ions have destructive impacts on the reaction center of photosystem II (PSII), the quinone receptor, and the oxygen transport system, which then results in a decrease in light energy of the reaction center and photosynthetic pigments and process. Salinity causes osmotic tension via producing reactive oxygen species (ROS) through causing a high concentration of Na^+ in plant tissues that decreases CO_2 and finally photosynthetic processes. Moreover, salinity increases the activity of chlorophyllase, the enzyme involved in the decomposition of photosynthetic pigments [30] and damages chloroplasts [31].

There was a negative correlation between the increase of Na^+ ions and the amount of photosynthetic pigments. Lower chl content during salinity stress was identified as a good indicator of the sensitivity of stevia plants to salinity conditions because it connotes damage in the chloroplast membranes and oxygen transport system, and finally a reduction in photosynthesis [32]. The negative effect of salinity on photosynthetic pigments and processes was previously confirmed in several plant species [22,24,30,33].

There is a positive correlation between the increase in Ca^{2+} and K^+ ions and the amount of photosynthetic pigments. Higher chl content during the application of fish waste bio-fertilizer was identified as a good indicator of the fact that the application of the bio-fertilizer is effective under salinity conditions. Ca enhances the number of leaves and maintains the pH of the cell for electron transports in chloroplast membranes, and finally maintains photosynthesis and reduces oxygen species (ROS). K causes an increase in the carbon assimilation rate and chl metabolism, and increases photosynthesis rate [34]. In line with the current findings, the positive effect of K and Ca on photosynthetic pigments was reported in Wheat [34].

Fish waste bio-fertilizer enhances the production of antioxidants, vitamins, proteins, and alkaloids in plants thanks to its amino acid content such as glycine and glutamine, as precursors for Chl biosynthesis [35]. Moreover, this bio-fertilizer enhances nutrient absorption, especially P [26], and metabolic processes of the plant that result in increased leaf surface area and finally photosynthetic pigments and photosynthesis. As mentioned, the bio-fertilizer is rich in Mg and N, two main elements of Chl [16]. The amino acid glutamate in fish waste bio-fertilizer increases vegetative growth by increasing the synthesis of chlorophyll and other amino acids (aspartic acid, serine, alanine, lysine, and proline). This amino acid increased vegetative growth in soybean plant [36]. The amino acid alanine found in fish waste bio-fertilizer has a dual function between carbon and nitrogen metabolism, and, finally, this amino acid is associated with increasing chlorophyll synthesis and photosynthesis [37]. The amino acid glutamine in fish waste fertilizer increases nitrogen uptake in rice plants under cadmium stress by increasing plant growth regulators [38]. Therefore, fish waste bio-fertilizer enhanced the photosynthetic pigments and processes thanks to the mentioned reasons formerly confirmed [16,39]. That could be observed under salinity conditions. In line with the current findings, the positive effect of fish waste bio-fertilizer was reported on the leaf area and photosynthetic pigment biosynthesis of pumpkin [40] and cucumber [40] plants.

Salinity led to a reduction in RWC due to an imbalance in ions and osmotic pressure. In addition, salinity caused an enhancement in proline and sugar. On the other hand, the bio-fertilizer improved RWC as well as the proline and sugar content of stevia plants. Plant relative water content (RWC) is a vital physiological indicator in plants. Salinity stress disturbs the reasonable absorption of water and concomitantly osmotic equilibrium in the root zone, and, as a result, plant RWC decreased as previously reported [41,42]. The fish waste bio-fertilizer has a high K^+ content, an element with positive impacts on the osmotic potential and root growth that leads to increased water absorption [43]. The amino acids glycine and proline in fish waste bio-fertilizer increase the osmotic potential of plant cells, thereby leading to increased water absorption [44], in line with the current findings.

Plants commonly enhance the release of compatible solutes such as proline and sugars in response to stress-induced reduced osmotic potential and cellular water [45]. Proline is an osmolyte capable of neutralizing ROS over stress conditions. Proline and carbohydrates increase the activity of the Rubisco enzyme, the primary carboxylase of photosynthesis which then protects photosynthesis, cell expansion, membrane stability and flexibility, and plant growth under stress conditions [21]. Moreover, the biosynthesis pathways of proline and sugar are closely tied together due to glutamate production, and an increase in proline biosynthesis leads to sugar production and vice versa. Salinity caused an increase in sugar and proline in some plants [22,24,46]. Mostly, an increase in proline synthesis and a decrease in its oxidation are the reasons for proline accumulation under salinity stress [47]. A high concentration of Na^+ reduces the absorption of Ca^{2+}, Mg^{2+}, and K^+, leading to an enhanced Na^+/K^+ ratio. On the contrary, fish waste bio-fertilizer has a high K^+ content that results in a reduced Na^+/K^+ ratio and then better stomatal performance. As a result, stomata open and CO_2 diffusion to plant tissues increases, and then the photosynthesis rate is enhanced, which finally leads to carbohydrate (sugar) production [43], as previously confirmed in spinach plants [19]. In addition, the amino acid leucine in fish waste bio-fertilizer acts as a precursor for alanine synthesis in plants. In plants, alanine is important for the synthesis of pantothenate and subsequently coenzyme A, an essential coenzyme in carbohydrates. The carbohydrate produced in the plant plays a key role in the opening and closing of the stomata [48]. An increase in carbohydrate production by using the amino acids leucine and alanine has been reported in wheat plants [48], in line with the current results.

The enhanced amount of proline after the bio-fertilizer application referred to its high N and P content that helped mitigating the salinity effects [49]. In addition, fish waste fertilizer contains proline, which acts as an osmotic protector in abiotic stress and also as a metal chelator [36]. Proline and glutamine make the tomato plant resistant to salt stress [44], and all mentioned studies and explanations are in accordance with the current findings.

Salinity increased the total phenols and the bio-fertilizer increased the content even more in the stevia plant. Under stress conditions, plants activate enzymatic and non-enzymatic antioxidant systems to lessen the stress impacts. Phenolics are non-enzymatic antioxidants whose amounts are enhanced under salinity to neutralize ROS and prevent the decomposition of hydroperoxide [50]. An increase in phenols was previously reported in some plants under salinity conditions [51,52], as observed in the current study. Fish waste bio-fertilizer contains some phenol- and flavonoid-based compounds that could stimulate phenolic content in the plant [53]. Also, the amino acid phenylalanine in fish waste fertilizer increases the phenolic compounds of the plant and increases the antioxidant capacity through the shikimic acid pathway. The production of phenolic compounds from the amino acid phenylalanine has been observed in grape plants [54], in line with the current findings. Salinity increased electrolyte leakage (EL), MDA, and H_2O_2 and the application of fish waste bio-fertilizer reduced their amounts. Normally, plant cells produce small amounts of H_2O_2 and this creates a defense mechanism in the plant. Salinity causes oxidative stress, leading to ROS and H_2O_2 high production and accumulation which then damages membranes and enhances electrolyte leakage (EL) and MDA [22,24]. An increase in MDA demonstrated the plant's incapability to remove ROS and H_2O_2 since ROS and H_2O_2 accumulation led to the peroxidation of fats, the deactivation of enzymes, damage to nucleic acids, and the destruction of cell membranes. Salinity caused the conversion of superoxide radical (O_2^-) to H_2O_2 inside the cell, which hinders the Calvin cycle and finally the activity of antioxidant enzymes (e.g., CAT and SOD). Next, H_2O_2 prevents sugar biosynthesis in chloroplasts [45]. Salinity increased MDA content in soybean [55] and stevia [2] and H_2O_2 content in basil [56] and wheat [57]. An increase in H_2O_2 resulted in enhanced electrolyte leakage (EL) [24,58]. An increase in salinity resulted in enhanced MDA, which was additionally confirmed in sweet basil [59,60]. P content in the bio-fertilizer increases the production of phospholipids that strengthen the cell membranes and thus reduce MDA and H_2O_2 [26]. Also, glycine and proline in fish waste fertilizer increase the stability of the two layers of the plasma membrane by increasing the activity of catalase and superoxide dismutase enzymes and regulating the peroxidation of lipids and homeostasis of ions. Glycine treatment increased salinity tolerance in safflower, reduced MDA and H_2O_2, and improved homeostasis [61], which could explain the current results. Salinity enhanced antioxidant enzymes (POD, CAT, and APX) and fish waste bio-fertilizer increased their activities further. Under salinity conditions, plants increase several antioxidant enzymes' activities (e.g., POD, CAT, and APX) to neutralize salinity-induced ROS effects [62], as was formerly confirmed [22,24,63,64]. The nutrient content of the bio-fertilizer, especially C, N, and P, as well as the amino acids, increased the antioxidant enzymes' activities [26] like CAT and POD in lettuce and cowpea [13,39]. N present in fish waste bio-fertilizer is used as a precursor for the production of protein and antioxidant enzymes under salinity conditions. Arginine, as an amino acid in the bio-fertilizer, is involved in salinity stress tolerance by producing spermidine (a polyamine) or nitrate reductase (reducing nitrate to nitrite) [65]. Different amino acids, such as the amino acid proline found in fish waste bio-fertilizer, play a key role in the antioxidant defense system and the production of various enzymes in stressful conditions [36]. The positive effects of fish waste bio-fertilizer have been proven in this regard [66], all in line with the current results.

The current results demonstrated that salinity increased Na^+ and reduced K^+ and Ca^{2+} contents of stevia plants, as previously confirmed [1,28,41,67]; on the contrary, the application of fish waste bio-fertilizer reduced Na^+ and increased K^+ and Ca^{2+} contents. Under salinity, H^+-ATPases of the plasma membrane create a H^+ gradient that provides the necessary energy for the secretion of K^+ via the H^+/K^+ antiport and thereby ensures Na re-absorption [68]. Salinity additionally causes osmotic pressure in the rhizosphere soil solution that, as a result, decreases the uptake of water and minerals, (e.g., K^+ and Ca^{2+}) due to the antagonistic effect [69]. It seems that the bio-fertilizer reduced Na^+ accumulation in the aerial parts through Na^+ removal from the xylem [70]. Thanks to the high solubilization and availability of the Ca^{2+} and K^+ contents of the fish waste bio-fertilizer, its application could compensate for the lack of these elements caused by salinity [15]. The bio-fertilizer application increased N, P, K, and Ca in spinach [19], in line with the current results.

The PCA analysis showed that the measured traits were divided into two principal components. The first one, PC1, included biochemical and morphological traits and the second principal component, PC2, contained physiological traits. The plot also showed a high correlation between physiological and morphological traits.

PCA was also identified for the assessment of stress conditions in crop species such as wheat [71]. These principal components were most likely to have the most effect on the overall variation of the data set and could be identified as significant contributors to the salinity response in various plants. In other words, these components demonstrated the strongest association with salinity stress and the modulator of stress. These statistical tools allow the identification of probable components and associations among accessions and traits [71].

4. Materials and Methods

This study was conducted in the research greenhouse of the Faculty of Agriculture, Mohaghegh Ardabili University (38 2514′ N; 48), in a factorial experiment, and was based on a completely randomized design with three replications in the spring-summer of 2021. The first factor was salinity stress at four levels (0, 20, 40, and 60 mM NaCl) and the second factor was liquid fish waste bio-fertilizer (foliar application) at four levels (0, 5, 10, and 15% (v/v)). The fish waste of *Hypophthalmichthys molitrix* (head, tail, and fin) was purchased from local fishmongers in Ardabil city. After washing, the waste was dried in the shade and then ground, while the resulting powder was autoclaved for 20 min. After mixing the dry material with distilled water at a ratio of 1:5, 30 mL protein-hydrolyzing bacteria (*Bacillus subtilis*) and 150 g sugar were added and incubated at 25 °C for two weeks according to [72]. Finally, membrane filters (Filtration-Micro MF) were used for separating the liquid phase, and then free amino acids (Table 3) as well as an elemental analysis (Table 4) of the total liquid fish bio-fertilizer were measured using HPLC.

Table 3. Composition of amino acids in fish waste bio-fertilizer. Content shown as mg Amino Acid (AA)/g sample.

Amino Acid	Free Amino Acid	Total Amino Acid	Unit (mg Amino Acid(AA)/g Sample)
Asp	0.7	1.36	mg AA/g sample
Glu	1.49	3.57	mg AA/g sample
Ser	0.17	0.87	mg AA/g sample
Gly	1.08	3.42	mg AA/g sample
His	0.15	0.48	mg AA/g sample
Arg	0.23	1.02	mg AA/g sample
Thr	0.14	0.84	mg AA/g sample
Ala	1.39	2.77	mg AA/g sample
Pro	0.61	1.86	mg AA/g sample
Tyr	0.48	0.48	mg AA/g sample
Val	0.72	1.63	mg AA/g sample
Met	0.26	0.51	mg AA/g sample
(cys)2	0.00	0.00	mg AA/g sample
Ile	0.47	0.94	mg AA/g sample
Leu	0.98	1.7	mg AA/g sample
Phe	0.41	0.83	mg AA/g sample
Lys	0.28	0.92	mg AA/g sample

Table 4. The composition of major nutrients of fish waste bio-fertilizer.

Ca%	P%	K%	N_{total}%	OC (Organic Carbon)%
1.4	4.1	6.23	5.92	12.5

Stevia seedlings (*Stevia rebaudiana* Bertoni) were purchased from a medicinal plant greenhouse in Shiraz city and planted in 10 L pots containing a mixture of cocopeat and perlite (1:1). The plants were first irrigated with half-strength Hoagland's solution (4 weeks) and then irrigated with full-strength nutrient solution (up to harvest: 400 mL every day). The hydroponic solution contained the following macronutrients (mg/L): nitrate (N) 210, potassium (K) 204, calcium (Ca) 140, sulphur (S) 64, magnesium (Mg) 48, phosphorus (P) 31, and micronutrients (mg/L): iron (Fe) 4, boron (B) 0.5, manganese (Mn) 0.5, copper (Cu) 0.1, zinc (Zn) 0.1, and molybdenum (Mo) 0.05. The half-strength Hoagland's solution had a conductivity of 1.022 dS/m. The pH of the Hoagland's solution was sustained at 5.97 throughout the experiment. Following plant establishment (two weeks after the transferred seedlings to pots), salinity was applied by adding the mentioned concentrations of NaCl (0, 20, 40, and 60 mM) to the nutrient solution. The plants were first irrigated with half NaCl concentration (7 days) and then irrigated with the desired NaCl concentrations (0, 20, 40, and 60 mM) until harvesting the plant. After the initiation of salinity treatments, plants were grown for 8 weeks. The leaves (aerial parts) were sprayed with the mentioned concentrations of fish waste bio-fertilizer (0, 5, 10, and 15% along with Tween 0.1%) two weeks after salinity application 4 times, with an 8-day interval. The control plants were irrigated with the nutrient solution in the same manner up to the harvest and treated with any treatments. The plants were harvested 80 days after planting and samplings were performed through the leaves. Each measurement was performed in triplicate.

4.1. Morphological Parameters

Shoot height, root length, fresh and dry weights of roots and shoots, number of leaves, branches, and leaf area were recorded as morphological parameters. To measure dry weight, the plant materials were kept at 70 °C for 72 h in a drying oven. Ten leaves were selected of random from each replicate to measure leaf area using Pamwin software (PAM 2500).

4.2. Photosynthetic Pigments and Chlorophyll Fluorescence Parameter

Photosynthetic pigments and carotenoids were measured according to the method described in [73]. For this purpose, 0.1 g of fresh leaves was homogenized with 5 mL of acetone (80%) and then the obtained mixture extract was centrifuged at 10,000 rpm for 10 min. Using a UV-V device (Hitachi U-2910, Tokyo, Japan), the absorption of the samples was recorded at wavelengths of 470, 646.8, and 663.2 nm to determine carotenoids and chlorophyll (Chl a and Chl b), respectively, based on the following formulas:

Chl a = (19.3 × A663.2 − 0.86 × A646.8) [V/100 × W]

Chl b = (19.3 × A646.8 − 3.6 × A663.2) [V/100 × W]

Carotenoids = [100(A470) − 3.27 (chl b)]/227

Chl_{Total} = Chl a + Chl b

Where V is the volume of the extract and W is the weight of fresh material.

The chlorophyll fluorescence parameter ($^{Fv}/_{Fm}$) was measured using a fluorometer (model OS-30p) after a 20 min adaption of leaves in the dark.

4.3. Relative Water Content (RWC)

Relative water content was measured according to the method described in [74]. For this purpose, three young and developed leaves from each pot were weighed (FW) after harvesting. The leaves were then placed in water for 24 h, in darkness and in a refrigerator; then, turgid weight (TW) was measured. At that time, the leaves were dried in an oven at 70 °C for 48 h inside paper envelopes to obtain their dry weight (DW). Relative water content was calculated using the following formula:

$$RWC\% = [(FW - DW)/(TW - DW)] \times 100$$

4.4. Soluble Carbohydrates

The amount of soluble carbohydrates was measured using anthrone reagent [75]. For this purpose, 0.1 g of fresh leaves was homogenized with 5 mL ethanol (80%) and placed in a hot water bath (95 °C, 10 min). The resulting suspension was centrifuged at 10,000 rpm for 10 min. Then, 3 mL of anthrone reagent was added to 100 µL of prepared extract and the absorbance was recorded at 620 nm using a UV-V spectrophotometer (Hitachi U-2910, Tokyo, Japan). Finally, glucose concentration was determined using a standard curve.

4.5. Total Protein Content

To assay the concentration of protein in stevia extracts, the wet leaves of samples were homogenized by 3 mL of sodium phosphate buffer (pH = 6.8). The samples were centrifuged at 13,000 rpm for 15 min. The reaction mixture contained 100 µL of enzyme extract and 900 µL of Bradford reagent. The reaction was measured at a wavelength of 595 nm [76].

4.6. Hydrogen Peroxide (H_2O_2)

H_2O_2 content was determined first by homogenizing leaf tissues (0.1 g fresh weight) in 2 mL of trichloroacetic acid (TCA, 0.1% (*w*/*v*)). Then, 500 µL of the extract was added to the reaction mixture containing 500 µL of phosphate buffer (100 mM, pH = 7) and 2 mL of potassium iodide (1 mM). Lastly, the absorbance was measured at 620 nm using a spectrophotometer [77].

4.7. Malondialdehyde (MDA)

To measure MDA, 0.5 g of fresh leaves was homogenized in 6 mL of trichloroacetic acid (1% (*w*/*v*)) and then centrifuged. Next, the supernatant was separated and 2 mL of thiobarbiotic acid solution was added and then centrifuged (10,000 rpm, 10 min). The absorption was recorded at 532 nm and 600 nm and converted to the exact amount [78].

4.8. Electrolyte Leakage (EL)

The method of [79] was used to measure electrolyte leakage (EL). Based on this method, EL was determined using an electrical conductivity meter (Hanna, HI98192, Hanna Instruments, Inc., Woonsocket, RI, USA). The initial electrical conductivity (EC1) was recorded after washing the discs (0.5 cm diameter) of leaves three times with deionized water and incubating them at ambient temperature (24 h). The final electrical conductivity (EC2) was measured after incubating the samples in a water bath (95 °C, 20 min) and cooling down the samples at 25 °C. Lastly, electrolyte leakage EL was calculated from the following equation.

$$EL\ (\%) = (EC1/EC2) \times 100$$

4.9. Proline Content

To measure proline content, 0.1 g of fresh leaf tissues was homogenized in 5 mL of sulfosalicylic acid) 3% (*w*/*v*)) and centrifuged. The reaction mixture included 2 mL of extract, 2 mL of ninhydrin acid, 2 mL of acetic acid, and toluene. The absorbance was measured at 520 nm using a spectrophotometer (Hitachi U-2910, Tokyo, Japan) [80].

4.10. Total Phenolic Content

Phenolic content was determined through homogenizing leaf tissues (0.2 g of fresh weight) in 2 mL of ethanol (70%), kept for 24 h in the dark at 4 °C. Then, to 500 μL of extract, 500 μL of ethanol (96%) and distilled water (1.5 mL) were added; then, Folin–Ciocalteu reagent and sodium carbonate were mixed with the mixture. Finally, the absorbance was measured at 725 nm with the spectrophotometer [81].

4.11. Activity of Antioxidant Enzymes

After homogenizing leaves (0.5 g) with potassium phosphate buffer (pH 6.8, 10 mM), the extract was centrifuged (6000 rpm, 20 min) and the supernatant was used for the assay of the enzymatic activities.

4.11.1. Peroxidase Enzyme (POD)

To assay POD activity, the reaction mixture contained 2.9 mL of sodium phosphate buffer (100 mM), guaiacol (100 mM), and 50 μL of enzyme extract. The reaction was initiated at a wavelength of 470 nm by the addition of H_2O_2 (20 mM) after 60 s. A blank sample was prepared without enzyme extract [82].

4.11.2. Catalase (CAT)

CAT activity was measured according to the method described in [83]. The reaction mixture consisted of 2.5 mL of potassium phosphate buffer (100 mM, pH = 7), H_2O_2 (5 mM), and 60 μL of enzyme extract. The reduction in absorbance was measured by the degradation of H_2O_2 at 240 nm using the spectrophotometer.

4.11.3. Ascorbate peroxidase (APX)

The reaction mixture for the APX assay consisted of potassium phosphate buffer (50 mM, pH = 7), 100 μL of enzyme extract, ascorbic acid (50 μM), and H_2O_2 (1.5 mM), and the absorbance was recorded at 290 nm using the spectrophotometer [84].

4.12. Sodium (Na), Potassium (K), and Calcium (Ca) Content of Shoots and Roots

To measure the amount of macronutrients (potassium, K; calcium, Ca; and sodium, Na), stevia shoots and roots were ashed in an oven at 550 ± 25 °C. White ash was digested in 10 mL of concentrated hydrochloric acid (HCl) and brought up to a volume of 100 mL for the measurement of Na^+, potassium (K), and calcium (Ca). Sodium and potassium concentrations were quantified using a flame photometer. Ca concentration was recorded with an atomic absorption instrument [85].

4.13. Statistical Analysis

Data were analyzed using SAS 9.1 software by analyzing the means using Duncan's multiple range test, with a significant difference level at $p < 0.05$.

5. Conclusions

In spite of the fact that fish waste can be used as a bio-fertilizer, little research has been conducted on its use to lessen the effects of various stress conditions. Accordingly, the current study aimed to shed light on the application of fish waste as a bio-fertilizer on stevia plants under salinity stress. The bio-fertilizer increased morphological parameters, photosynthetic pigments, $^{Fv}/_{Fm}$, RWC, proline, phenols, and antioxidants, and reduced sugar, EL, MDA, and H_2O_2. In fact, the presence of different amino acids and nutritional elements in fish waste bio-fertilizer caused an enhancement in antioxidant enzyme activities and a reduction in the salinity-induced oxidative damages in stevia plants. The bio-fertilizer, particularly at a 15% concentration, could be introduced as the best dose based on the measured parameters to lessen salinity effects. Hence, fish waste bio-fertilizer could be considered an effective bio-fertilizer to apply on plants under different stress conditions to mitigate stress effects through a safe and environmentally friendly method. In addition, fish waste bio-fertilizer did not cause more salinity issues, at least with the applied doses and experiment time, which is an imperative aspect.

Supplementary Materials: The following supporting information can be downloaded at: https://www.mdpi.com/article/10.3390/plants13141909/s1, Table S1: Mean comparisons of salinity stress and fish waste bio-fertilizer on morphological traits of *Stevia rebaudiana* Bertoni. Same letters indicate no significant differences ($p < 0.05$) based on Duncan's test; Table S2: Mean comparisons of salinity stress and fish waste bio-fertilizer on proline, electrolyte leakage (EL) and relative water content (RWC) of *Stevia rebaudiana* Bertoni. Same letters indicate no significant differences ($p < 0.05$) based on Duncan's test; Table S3: Effect of salinity stress and fish waste bio-fertilizer on shoot potassium (K) and root calcium (Ca) of *Stevia rebaudiana* Bertoni. Same letters indicate no significant differences ($p < 0.05$) based on Duncan's test.

Author Contributions: Conceptualization Z.M. and B.E.; methodology, Z.M. and R.A.; software, Z.M. and B.E.; investigation, Z.M.; data curation, Z.M., R.A., G.N., and B.E.; formal analysis, B.E.; writing—original draft preparation, Z.M.; writing—review and editing, Z.M., B.E., R.A., S.P., G.N., G.G. and V.F.; project administration, B.E.; funding acquisition, V.F. All authors have read and agreed to the published version of the manuscript.

Funding: This research received no external funding.

Data Availability Statement: Data will be made available upon request.

Conflicts of Interest: The authors declare no conflicts of interest.

References

1. Sheikhalipour, M.; Esmaielpour, B.; Gohari, G.; Haghighi, M.; Jafari, H.; Farhadi, H.; Kulak, M.; Kalisz, A. Salt Stress Mitigation via the Foliar Application of Chitosan Functionalized Selenium and Anatase Titanium Dioxide Nanoparticles in Stevia (*Stevia rebaudiana* Bertoni). *Molecules* **2021**, *26*, 4090. [CrossRef] [PubMed]
2. Gerami, M.; Majidian, P.; Ghorbanpour, A.; Alipour, Z. *Stevia rebaudiana* Bertoni responses to salt stress and chitosan elicitor. *Physiol. Mol. Biol. Plants* **2020**, *26*, 965–974. [CrossRef] [PubMed]
3. Bahari Saravi, H.; Gholami, A.; Pirdashti, H.; Firouzabadi, M.B.; Asghari, H.; Yaghoubian, Y. Improvement of salt tolerance in *Stevia rebaudiana* by co-application of endophytic fungi and exogenous spermidine. *Ind. Crops Prod.* **2022**, *177*, 114443. [CrossRef]
4. Evelin, H.; Devi, T.S.; Gupta, S.; Rupam Kapoor, R. Mitigation of Salinity Stress in Plants by *Arbuscular Mycorrhizal* Symbiosis: Current Understanding and New Challenges. *Front. Plant Sci.* **2019**, *10*, 470. [CrossRef]
5. Arif, Y.; Singha, P.; Siddiquia, H.; Bajguzb, A.; Hayata, S. Salinity induced physiological and biochemical changes in plants: An omic approach towards salt stress tolerance. *Plant Physiol. Biochem.* **2020**, *156*, 64–77. [CrossRef] [PubMed]
6. Giuffrida, F.; Graziani, G.; Fogliano, V.; Scuderi, D.; Romano, D.; Leonardi, C. Effect of nutrient and NaCl salinity on growth, yield, quality and compotion of pepper growth in soilless closed system. *J. Plant Nutr.* **2014**, *37*, 1455–1474. [CrossRef]
7. Sapre, S.; Gontia-Mishra, I.; Tiwari, S.H. Plant Growth-Promoting Rhizobacteria Ameliorates Salinity Stress in Pea (*Pisum sativum*). *J. Plant Growth Regul.* **2021**, *41*, 647–656. [CrossRef]

8. Ramadan, A.A.; Ebtihal, M.; Elhamid, A.; Mervat, S.h. Comparative study for the effect of arginine and sodium nitroprusside on sunflower plants grown under salinity stress conditions. *Bull. Natl. Res. Center* **2019**, *43*, 118. [CrossRef]
9. Hasanuzzaman, M.; Bhuyan, M.H.M.B.; Zulfiqar, F.; Raza, A.; Mohsin, S.M.; Al Mahmud, J.; Fujita, M.; Fotopoulos, V. Reactive oxygen species and antioxidant metabolism in plants under abiotic stress: Revisiting the crucial role of a universal defense regulator. *Antioxidants* **2020**, *9*, 681. [CrossRef]
10. Raza, A.; Tabassum, J.; Fakhar, A.Z.; Sharif, R.; Chen, H.; Zhang, C.; Ju, L.; Fotopoulos, V.; Siddique, K.H.; Singh, R.K.; et al. Smart reprograming of plants against salinity stress using modern biotechnological tools. *Crit. Rev. Biotechnol.* **2023**, *43*, 1035–1062. [CrossRef]
11. Magangana, T.P.; Stander, M.A.; Masondo, N.A.; Makunga, N.P. Steviol glycoside content and essential oil profles of Stevia rebaudiana Bertoni in response to NaCl and polyethylene glycol as inducers of salinity and drought stress in vitro. *Plant Cell Tissue Organ Cult.* **2021**, *145*, 1–18. [CrossRef]
12. Mahajana, M.; Sharma, S.; Kumar, P.; Pala, P.K. Foliar application of KNO_3 modulates the biomass yield, nutrient uptake and accumulation of secondary metabolites of *Stevia rebaudiana* under saline conditions. *Ind. Crops Prod.* **2020**, *145*, 112102. [CrossRef]
13. Yidirim, E.; Kul, R.; Turan, M.; Ekinci, M.; Alak, G.; Atamanalp, M. Effect of Nitrogen and Fish Manure Fertilization on Growth and Chemical Composition of Lettuce. In Proceedings of the International Conference on Advances in Natural and Applied Sciences 2016, Antalya, Turkey, 21–23 April 2016.
14. Choe, U.; Mustafa, A.M.; Lin, H.; Choe, U.; Sheng, K. Anaerobic co-digestion of fish processing waste with a liquid fraction of hydrothermal carbonization of bamboo residue. *Bioresour. Technol.* **2020**, *297*, 122542. [CrossRef] [PubMed]
15. Ahuja, I.; Dauksas, E.; Remmec, J.F.; Richardsen, R.; Loes, A.K. Fish and fish waste-based fertilizers in organic farming—With status in Norway: A review. *Waste Manag.* **2020**, *115*, 95–112. [CrossRef] [PubMed]
16. Balraj, H.T.; Palani, S.; Arumugam, G. Influence of Gunapaselam, a liquid fermented fish waste on the growth characteristics of *Solanum melongena*. *J. Chem. Pharm. Res.* **2014**, *6*, 58–66.
17. Radziemska, M.; Vaverková, M.D.; Adamcová, D.; Brtnický, M.; Mazur, Z. Valorization of Fish Waste Compost as a Fertilizer for Agricultural Use. *Waste Biomass Valorization* **2019**, *10*, 2537–2545. [CrossRef]
18. Sadak, M.T.; Abdelhamid, M.S.H. Influence of Amino Acids Mixture Application on Some Biochemical Aspects, Antioxidant Enzymes and Endogenous Polyamines of *Vicia faba* Plant Grown under Seawater Salinity Stress. *Gesunde Pflanz.* **2015**, *67*, 119–129. [CrossRef]
19. Ekinci, M.; Atamanalp, M.; Turan, M.; Alak, G.; Kul, R.; Kitir, N.; Yildirim, E. Integrated Use of Nitrogen Fertilizer and Fish Manure: Effects on the Growth and Chemical Composition of Spinach. *Commun. Soil Sci. Plant Anal.* **2019**, *50*, 1580–1590. [CrossRef]
20. Adhikari, B.; Dhungana, S.K.; Kim, I.D.; Shin, D.H. Effect of foliar application of potassium fertilizers on soybean plants under salinity stress. *J. Saudi Soc. Agric. Sci.* **2019**, *19*, 261–269. [CrossRef]
21. Ahanger, M.A.; Qin, C.; Begum, N.; Maodong, Q.; Dong, X.X.; El-Esawi, M.; El-Sheikh, M.A.; Alatar, A.A.; Zhang, L. Nitrogen availability prevents oxidative effects of salinity on wheat growth and photosynthesis by up-regulating the antioxidants and osmolytes metabolism, and secondary metabolite accumulation. *BMC Plant Biol.* **2019**, *19*, 479. [CrossRef]
22. Gohari, G.; Farhadi, H.; Panahirad, S.; Zareei, E.; Labib, p.; Jafari, H.; Mahdavinia, G.; Hassanpouraghdam, M.B.; Ioannou, A.; Kulak, M.; et al. Mitigation of salinity impact in spearmint plants through the application of engineered chitosan-melatonin nanoparticles. *Int. J. Biol. Macromol.* **2023**, *224*, 893–907. [CrossRef] [PubMed]
23. Rostami, G.; Moghaddam, M.; Narimani, R.; Mehdizadeh, L. The effect of different priming treatments on germination, morphophysiological, and biochemical indices and salt tolerance of basil (*Ocimum basilicum* L. cv. Keshkeni Levelou). *Environ. Stresses Crop Sci.* **2018**, *11*, 1107–1123.
24. Gohari, G.; Mohammadi, A.; Akbari, A.; Panahirad, S.; Dadpour, M.R.; Fotopoulos, V.; Kimura, S. Titanium dioxide nanoparticles (tio2 nps) promote growth and ameliorate salinity stress effects on essential oil profile and biochemical attributes of *Dracocephalum moldavica*. *Sci. Rep.* **2020**, *10*, 912. [CrossRef] [PubMed]
25. Roy, M.; Sukalpa, K.; Anupam, D.; Pradip, K.S.; Joydeep, M. Application of rural slaughterhouse waste as an organic fertilizer for pot cultivation of solanaceous vegetables in India. *Int. J. Recycl. Org. Waste Agric.* **2013**, *2*, 6–16. [CrossRef]
26. Metwally, R.; Solimana, S.A.; Hamed, A.A.; Latef, A.; Abdelhamee, R. The individual and interactive role of arbuscular mycorrhizal fungi and Trichoderma viride on growth, protein content, amino acids fractionation, and phosphatases enzyme activities of onion plants amended with fish waste. *Ecotoxicol. Environ. Saf.* **2021**, *214*, 112072. [CrossRef]
27. Baldi, E.; Tosel li, M. Root growth and survivorship in cow manure and compost amended soils. *Plant Soil Environ.* **2013**, *59*, 221–226. [CrossRef]
28. Azami, M.A.; Maleki, M.; Rasouli, F.; Gohari, G. Protective effects of chitosan based salicylic acid nanocomposite (CS-SA NCs) in grape (*Vitis vinifera* cv. 'Sultana') under salinity stress. *Sci. Rep.* **2023**, *13*, 883. [CrossRef] [PubMed]
29. Wright, A.H.; DeLong, J.M.; Lada, R.R.; Prange, R.K. The relationship between water status and chlorophyll a fluorescence in grapes (*Vitis* spp.). *Postharvest Biol. Technol.* **2009**, *51*, 193–199. [CrossRef]
30. Salim Akhter, M.; Noreen, S.; Mahmood, S.; Athar, H.U.R.; Ashraf, M.; Alsahli, A.A.; Ahmad, P. Influence of salinity stress on PSII in barley (*Hordeum vulgare* L.) genotypes, probed by chlorophyll-a fluorescence. *J. King Saud Univ.-Sci.* **2021**, *33*, 101239. [CrossRef]

31. Hameed, A.; Ahmed, M.Z.; Hussain, T.; Aziz, I.; Ahmad, N.; Gul, B.; Nielsen, B. Effects of Salinity Stress on Chloroplast Structure and Function. *Cells* **2021**, *7*, 2023. [CrossRef]
32. Akhtar, N.; Ilyas, N.; Arshad, M.; Meraj, T.A.; Hefft, D.I.; Jan, B.L.; Ahmad, P. The Impact of Calcium, Potassium, and Boron Application on the Growth and Yield Characteristics of Durum Wheat under Drought Conditions. *Agronomy* **2022**, *12*, 1917. [CrossRef]
33. Kamanga, M.; Echigo, K.; Yodoya, K.; Mohammad, A.; Mekawy, M.; Ueda, A. Salinity acclimation ameliorates salt stress in tomato (*Solanum lycopersicum* L.) seedlings by triggering a cascade of physiological processes in the leaves. *Sci. Hortic.* **2020**, *270*, 109434. [CrossRef]
34. Rana, V.; Ram Sewa, R.; Kiran Nehra, S.; Sharma, I. Physiological, biochemical and morphological study in wheat (*Triticum aestivum* L.) RILs population for salinity tolerance. *J. Agric. Sci.* **2015**, *7*, 119–128. [CrossRef]
35. Ishak, N.H.; Sarbon, N.M.A. Review of Protein Hydrolysates and Bioactive Peptides Deriving from Wastes Generated by Fish Processing. *Food Bioprocess Technol.* **2017**, *11*, 2–16. [CrossRef]
36. Teixeira, W.F.; Soares, L.H.; Fagan, E.B.; Costa Mello, S.D.; Reichardt, K.; Dourado Neto, D. Amino Acids as Stress Reducers in Soybean Plant Growth Under Diferent Water Defcit Conditions. *J. Plant Growth Regul.* **2019**, *39*, 905–919. [CrossRef]
37. Alfosea-Simon, M.; Simon-Grao, S.; Zavala-Gonzalez, E.A.; Camara-Zapata, J.M.; Inmaculada Simon, I.; Martinez-Nicolas, J.J.; Lidon, V.; Garcia-Sanchez, F. Physiological, Nutritional and Metabolomic Responses of Tomato Plants After the Foliar Application of Amino Acids Aspartic Acid, Glutamic Acid and Alanine. *Orig. Res.* **2021**, *11*, 5811234. [CrossRef] [PubMed]
38. Jiang, M.; Jianga, J.; Lia, S.H.; Lia, M.; Yuanyuan Tana, Y.; Songa, S.H.; Shua, Q.A.; Huanga, J. Glutamate Alleviates Cadmium Toxicity in Rice via Suppressing Cadmium Uptake and Translocation. *J. Hazard. Mater.* **2019**, *384*, 121319. [CrossRef] [PubMed]
39. Shahsavani, S.; Abaspour, A.; Parsaeeyan, M.; Yonesi, Z. Effect of fish waste, chemical fertilizer and bio-fertilizer on yield and yield components of bean (Vigna sinensis) and some soil properties. *J. Pulses Res.* **2017**, *8*, 45–59.
40. Saifuddin, M.; Sharif Hossain, A.B.M.; Normaniza, O.; Moneruzzaman, K.M. Bract size enlargement and longevity of *Bougainvillea spectabilisas* affected by GA3 and phloemic stress. *Asian J. Plant Sci.* **2009**, *8*, 212–217. [CrossRef]
41. Sheikhalipour, M.; Mohammadi, S.A.; Esmaielpour, B.; Zareei, E.; Kulak, M.; Ali, S.; Nouraein, M.; Bahrami, M.K.; Gohari, G.; Fotopoulos, V. Exogenous melatonin increases salt tolerance in bitter melon by regulating ionic balance, antioxidant system and secondary metabolism-related genes. *BMC Plant Biol.* **2022**, *22*, 380. [CrossRef]
42. El-Beltagi, H.S.; El-Yazied, A.A.; Hany, G.; El-Gawad, A.; Kandeel, M.; Shalaby, T.A.; Abdallah Tageldein Mansour, A.T.; Al-Harbi, N.A.; Al-Qahtani, S.M.; Alkhateeb, A.A.; et al. Synergistic Impact of Melatonin and Putrescine Interaction in Mitigating Salinity Stress in Snap Bean Seedlings: Reduction of Oxidative Damage and Inhibition of Polyamine Catabolism. *Horticulturae* **2023**, *9*, 285. [CrossRef]
43. Shams, H.; Yildirim, E.; Arslan, E.; Agar, G. Salinity induced alteration in DNA methylation pattern, enzyme activity, nutrient uptake and H_2O_2 content in pepper (*Capsicum annuum* L.) cultivars. *Acta Physiol. Plant* **2020**, *42*, 59. [CrossRef]
44. Alfosea-Simon, M.; Zavala-Gonzalez, E.A.; Camara-Zapata, J.M.; Martinez Nicolas, J.J.; Simon, I.; Simon-Grao, S. Effect of foliar application of amino acids on the salinity tolerance of tomato plants cultivated under hydroponic system. *Sci. Hortic.* **2020**, *272*, 109509. [CrossRef]
45. Hnilickova, H.; Kraus, K.; Vachova, P.; Hnilicka, F. Salinity Stress Affects Photosynthesis, Malondialdehyde Formation, and Proline Content in *Portulaca oleracea* L. *Plants* **2021**, *10*, 845. [CrossRef] [PubMed]
46. Alikhani, S.; Mahmudi Zarandi, M. Effect of coinoculation with endomycorrhiza, *Pseudomonas aeroginosa* and *Rhizobium meliloti* on *Medicago sativa* under water stress. *Plant Res.* **2019**, *32*, 155–166.
47. Shafi, A.; Zahoor, I.; Mushtaq, U. Proline Accumulation and Oxidative Stress: Diverse Roles and Mechanism of Tolerance and Adaptation under Salinity Stress. In *Salt Stress, Microbes, and Plant Interactions: Mechanisms and Molecular Approaches*; Akhtar, M., Ed.; Springer: Singapore, 2019; pp. 269–300.
48. Yang, Q.; Zhao, D.; Liu, Q. Connections between amino acid metabolisms in plants: Lysine as an example. *Front. Plant Sci.* **2020**, *11*, 928. [CrossRef] [PubMed]
49. Fageria, N.K. *The Use of Nutrients in Crop Plants*; CRC Press: Boca Raton, FL, USA, 2009; pp. 230–240.
50. Minh, D.T.K.; Ha, P.T.T.; Tuyen, P.T.; Minh, T.N.; Quan, N.V.; Xuan, T. Effects of Salinity Stress on Growth and Phenolics of Rice (*Oryza sativa* L.). *Luong. Int. Lett. Nat. Sci.* **2016**, *57*, 1–10. [CrossRef]
51. Ksouri, R.; Megdiche, W.; Debez, A.; Falleh, H.; Grignon, C.; Abdelly, C. Salinity effects on polyphenol content and antioxidant activities in leaves of the halophyte *Cakile maritima*. *Plant Physiol. Biochem.* **2007**, *45*, 244–249. [CrossRef]
52. Bistgani, Z.E.; Hashemi, M.; DaCosta, M.; Craker, L.; Maggi, F.; Morshedloo, M.R. Effect of salinity stress on the physiological characteristics, phenolic compounds and antioxidant activity of *Thymus vulgaris* L. and *Thymus daenensis Celak*. *Ind. Crops. Prod.* **2019**, *135*, 311–320. [CrossRef]
53. Muscolo, A.; Mauriello, F.; Marra, F.; Calabrò, P.S.; Russo, M.; Ciriminna, R.; Pagliaro, M. AnchoisFert: A New Organic Fertilizer from Fish Processing Waste for Sustainable Agriculture. *Glob. Chall.* **2022**, *6*, 2100141. [CrossRef]
54. Cheng, X.; Wang, X.; Zhang, A.; Wang, P.; Chen, Q.; Ma, T.; Li, W.; Liang, Y.; Sun, X.; Fang, Y. Foliar. Phenylalanine Application Promoted Antioxidant Activities in Cabernet Sauvignon by Regulating Phenolic Biosynthesis. *J. Agric. Food Chem.* **2020**, *68*, 15390–15402. [CrossRef] [PubMed]
55. Othman, M.; Richard, P.; Jacoby, A.; Millar, H.; Nicolas, L.; Taylor, T. Wheat mitochondrial respiration shifts from the tricarboxylic acid cycle to the GABA shunt under salt stress. *New Phytol.* **2020**, *225*, 1047–1048.

56. Gohari, G.; Alavi, Z.; Esfandiari, E.; Panahirad, S.; Hajihoseinlou, S.; Fotopoulos, V. Interaction between hydrogen peroxide and sodium nitroprusside following chemical priming of *Ocimum basilicum* L. against salt stress. *Physiol. Plant.* **2019**, *168*, 361–373. [CrossRef]
57. Ali, Q.; Daud, M.K.; Haider, M.Z.; Ali, S.; Aslam, N.; Noman, A.; Iqbal, N.; Shahzad, F.; Rizwan, M.; Deeba, F.; et al. Seed priming by sodium nitroprusside improves salt tolerance in wheat (*Triticum aestivum* L.) by enhancing physiological and biochemical parameters. *Plant Physiol. Biochem.* **2017**, *119*, 50–58. [CrossRef]
58. Safikhan, S.; Chaichi, M.R.; Khoshbakht, K.; Amini, A.; Motesharezadeh, B. Application of nanomaterial graphene oxide on biochemical traits of milk thistle (*Silybum marianum* L.) under salinity stress. *Aust. J. Crop Sci.* **2018**, *12*, 931. [CrossRef]
59. Ahmad, P.; Ahanger, M.A.; Alam, P.; Alyemeni, M.N.; Wijaya, L.; Ali, S.; Ashraf, M. Silicon (Si) supplementation alleviates NaCl toxicity in mung bean *Vigna radiata* L. *through the modifications of physio-biochemical attributes and key antioxidant enzymes. J. Plant Growth Regul.* **2019**, *38*, 70–82. [CrossRef]
60. Gohari, G.; Panahirad, S.; Sadeghi, M.; Akbari, A.; Zareei, E.; Zahedi, S.M.; Fotopoulos, V. Putrescine-functionalized carbon quantum dot (put-CQD) nanoparticles effectively prime grapevine (*Vitis vinifera* cv. 'Sultana') against salt stress. *BMC Plant Biol.* **2021**, *21*, 1–15.
61. Alasvandyari, F.; Mahdavi, B.; Hosseini, S.M. Glycine betaine affects the antioxidant system and ion accumulation and reduces salinity-induced damage in safflower seedlings. *Arch. Biol. Sci.* **2017**, *69*, 139–147. [CrossRef]
62. Ali, M.; Kamran, M.; Abbasi, G.H.; Saleem, M.H.; Ahmad, S.; Parveen, A.; Malik, A.; Afza, S.; Ahmar, S.; Dawar, K.M.; et al. Melatonin-Induced Salinity Tolerance by Ameliorating Osmotic and Oxidative Stress in the Seedlings of Two Tomato (*Solanum lycopersicum* L.) Cultivars. *J. Plant Growth Regul.* **2020**, *40*, 2236–2248. [CrossRef]
63. Ramesh Kannan, P.; Deepa, S.; Kanth, S.V.; Rengasamy, R. Growth, osmolyte concentration and antioxidant enzymes in the leaves of *Sesuvium portulacastrum* L. under salinity stress. *Appl. Biochem. Biotech.* **2013**, *171*, 1925–1932. [CrossRef]
64. Moghaddam, M.; Nasrin, F.; Panjtandoust, M.; Ghanati, F. Seed germination, antioxidant enzymes activity and proline content in medicinal plant *Tagetes minuta* under salinity stress. *Plant Biosyst.* **2020**, *154*, 835–842. [CrossRef]
65. Batista-Silva, W.; Heinemann, B.; Rugen, N.; Nunes-Nesi, A.; Araújo, W.L.; Braun, H.P.; Hildebrandt, T.M. The role of amino acid metabolism during abiotic stress release. *Plant Cell Environ.* **2019**, *42*, 1630–1644. [CrossRef] [PubMed]
66. Sahu, B.; Barik, N.K.; Paikaray, A.; Agnibesh, A.; Mohapatra, S.; Jayasankar, P. Fish Waste Bio-Refinery Products: Its application in Organic Farming B. *Int. J. Environ. Agric. Biotechnol.* **2016**, *1*, 4. [CrossRef]
67. Alnusairi, G.; Mazrou, Y.; Qari, S.; Elkelish, A.; Soliman, M.; Eweis, M.; Abdelaal, K.; El-Samad, G.A.; Ibrahim, M.; ElNahhas, N. Exogenous Nitric Oxide Reinforces Photosynthetic Efficiency, Osmolyte, Mineral Uptake, Antioxidant, Expression of Stress-Responsive Genes and Ameliorates the Effects of Salinity Stress in Wheat. *Plants* **2022**, *11*, 576. [CrossRef] [PubMed]
68. Ketehouli, T.; Idrice Carther, K.F.; Noman, M.; Wang, F.W.; Li, X.W.; Hai-Yan Li, H.Y. Adaptation of Plants to Salt Stress: Characterization of Na^+ and K^+ Transporters and Role of CBL Gene Family in Regulating Salt Stress Response. *Agronomy* **2019**, *9*, 687. [CrossRef]
69. Lindberg, S.; Abdul Kader, M.D.; Yemelyanov, V. Calcium Signalling in Plant Cells under Environmental Stress. In *Environmental Adaptations and Stress Tolerance of Plants in the Era of Climate Change*; Springer: New York, NY, USA, 2012.
70. Adem, G.D.; Roy, S.J.; Zhou, M.; Bowman, J.P.; Shabala, S. Evaluating contribution of ionic, osmotic and oxidative stress components towards salinity tolerance in barley. *BMC Plant Biol.* **2014**, *14*, 113. [CrossRef] [PubMed]
71. Reyes, J.A.O.; Carpentero, A.S.; Santos, P.J.A.; Delfin, E.F. Santos and Evelyn F. Effects of Water Regime, Genotype, and Formative Stages on the Agro-Physiological Response of Sugarcane (*Saccharum officinarum* L.) to Drought. *Plants* **2020**, *9*, 661. [CrossRef] [PubMed]
72. Whiting, D.; Wilson, C.; Card, A. Organic Fertilizers. In *Colorado Master Gardener*; Colorado State University: Denver, CO, USA, 2005; pp. 1–5.
73. Arnon, D. Copper enzymes in isolation chloroplast phenoloxidase in *Beta vulgaris*. *Plant Physiol.* **1949**, *24*, 1–15. [CrossRef] [PubMed]
74. Smart, R.; Bingham, G.E. Rapid estimates of relative water content. *Plant Physiol.* **1974**, *53*, 258–260. [CrossRef] [PubMed]
75. Omokolo, N.D.; Tsala, N.G.; Kanmegne, G.; Balange, A.P. In vitro induction of multiple shoots, plant regeneration and tuberization from tips of cocoyam. *C. R. Acad. Sci.* **1992**, *318*, 773–778.
76. Bradford, M.M. A rapid and sensitive method for the quantitation of microgram quantities of protein utilizing the principle of protein-dye binding. *Anal. Biochem.* **1976**, *72*, 248–254. [CrossRef] [PubMed]
77. Sergiev, I.; Alexieva, V.; Karanov, E. Effect of spermine, atrazine and combination between them on some endogenous protective systems and stress markers in plants. *C. R. Acad. Sci.* **1997**, *51*, 121–124.
78. Heath, R.L.; Packer, I. Photoperoxidation in isolated chloroplasts: I. Kinetics and stoichiometry of fatty acid peroxidation. *Arch. Biochem. Biophys.* **1968**, *125*, 189–198. [CrossRef] [PubMed]
79. Redman, R.; Haraldson, J.; Gusta, L. Leakage of UV- absorbing substances as a measure of salt injury in leaf tissue of woody spicies. *Physiol. Plant.* **1986**, *67*, 87–91. [CrossRef]
80. Bates, L.; Waldren, R.; Teare, I. Rapid determination of free proline for water-stress studies. *Plant Soil* **1973**, *39*, 205–207. [CrossRef]
81. Xu, C.; Zhang, Y.; Cao, L.; Lu, J. Phenolic compounds and antioxidant properties of different grape cultivars grown in China. *Food. Chem.* **2010**, *119*, 1557–1565. [CrossRef]

82. Hemeda, H.M.; Klein, B.P. Effects of naturally occurring antioxidants onperoxidase activity of vegetable extracts. *J. Food Sci.* **1990**, *55*, 184–185. [CrossRef]
83. Kar, M.; Mishra, D. Polyphenol oxidase activities during rice leaf senescence. *Plant Physiol.* **1976**, *57*, 315–319. [CrossRef] [PubMed]
84. Nakano, Y.; Asada, K. Hydrogen peroxide is scavenged by ascorbate specific peroxidase in spinach chloroplasts. *Plant Cell Physiol.* **1981**, *22*, 867–888.
85. Chapman, H.D.; Pratt, D.F. *Methods of Analysis for Soil, Plant and Water*; University of California Agricultural Science: Davis, CA, USA, 1961; pp. 60–62.

plants MDPI

Article

Use of Proline to Induce Salt Stress Tolerance in Guava

Smyth Trotsk de Araújo Silva [1], Geovani Soares de Lima [1], Vera Lúcia Antunes de Lima [1], Jackson Silva Nóbrega [1,*], Saulo Soares da Silva [2], Jean Telvio Andrade Ferreira [1], Maila Vieira Dantas [1], Iara Almeida Roque [1], Lauriane Almeida dos Anjos Soares [3], Rafaela Aparecida Frazão Torres [1], Cassiano Nogueira de Lacerda [1], Hans Raj Gheyi [1], Luderlândio de Andrade Silva [2] and Valéria Fernandes de Oliveira Sousa [3]

1 Academic Unit of Agricultural Engineering, Federal University of Campina Grande, Campina Grande 58430-380, PB, Brazil; smythtrotsk18@gmail.com (S.T.d.A.S.); geovani.soares@professor.ufcg.edu.br (G.S.d.L.); vera.lucia@professor.ufcg.edu.br (V.L.A.d.L.); jeantelvioagronomo@gmail.com (J.T.A.F.); maila.vieira02@gmail.com (M.V.D.); yara.roque.sb@gmail.com (I.A.R.); rafaela.aparecida@estudante.ufcg.edu.br (R.A.F.T.); cassianonogueiraagro@gmail.com (C.N.d.L.); hans.gheyi@ufcg.edu.br (H.R.G.)
2 Postgraduate Program in Agroindustrial Systems, Federal University of Campina Grande, Pombal 58840-000, PB, Brazil; saulosoares90@gmail.com (S.S.d.S.); luderlandioandrade@gmail.com (L.d.A.S.)
3 Academic Unit of Agrarian Sciences, Federal University of Campina Grande, Pombal 58840-000, PB, Brazil; lauriane.almeida@professor.ufcg.edu.br (L.A.d.A.S.); valeriafernandesbds@gmail.com (V.F.d.O.S.)
* Correspondence: jackson.nobrega@ufcg.edu.br; Tel.: +55-83-99974-9357

Abstract: Guava is a fruit tree with high potential in the semi-arid region of northeast Brazil. However, qualitative and quantitative water scarcity is a limiting factor for the expansion of irrigated agriculture. Thus, it is necessary to use techniques to mitigate the effects of salt stress, such as foliar application of proline. The objective of this study was to evaluate the effect of foliar application of proline as a mitigator of salt stress effects on the morphophysiology of guava cv. Paluma. The experiment was carried out under field conditions at the 'Rolando Enrique Rivas Castellón' Experimental Farm in São Domingos, PB, Brazil, using a randomized block design in a 5 × 4 factorial scheme referring to five levels of electrical conductivity of irrigation water, ECw (0.8, 1.5, 2.2, 2.9, and 3.5 dS m^{-1}) and four concentrations of proline (0, 8, 16, and 24 mM). Salinity above 0.8 dS m^{-1} compromised gas exchange, photosynthetic pigment synthesis, photochemical efficiency, and growth of guava plants at 360 days after transplanting. Foliar application of proline at a concentration of 24 mM mitigated the effect of salt stress on the relative water content, stomatal conductance, and carotenoid contents in plants irrigated with 3.6 dS m^{-1} water. Meanwhile, a proline concentration of up to 18 mM resulted in higher transpiration, CO_2 assimilation rate, instantaneous carboxylation efficiency, and absolute growth rate in stem diameter under ECw of 0.8 dS m^{-1}. Proline concentration of up to 24 mM increased the biosynthesis of photosynthetic pigments and the relative growth rate in stem diameter of guava in the period from 190 to 360 days after transplanting.

Keywords: *Psidium guajava* L.; water scarcity; amino acid

Citation: Silva, S.T.d.A.; Lima, G.S.d.; Lima, V.L.A.d.; Nóbrega, J.S.; da Silva, S.S.; Ferreira, J.T.A.; Dantas, M.V.; Roque, I.A.; Soares, L.A.d.A.; Torres, R.A.F.; et al. Use of Proline to Induce Salt Stress Tolerance in Guava. *Plants* **2024**, *13*, 1887. https://doi.org/10.3390/plants13141887

Academic Editor: Chien Van Ha

Received: 15 June 2024
Revised: 28 June 2024
Accepted: 5 July 2024
Published: 9 July 2024

1. Introduction

Guava (*Psidium guajava* L.) is a species belonging to the family Myrtaceae, whose origin is located between the tropical and subtropical regions of Mexico, Central, and South America [1]. Guava fruit has great economic importance in Brazil, being consumed fresh and/or in the agro-industrial segment as pulp, ice cream, puree, and jam. In addition, it has nutritional value as a source of iron, calcium, phosphorus, potassium, ascorbic acid, and vitamin A and B vitamins, besides being rich in fiber [2].

Brazil ranks third in guava production in the world, with a total of 567,764 tons being produced in the 2022 season in a harvested area of 22,684 hectares, obtaining an average production per hectare of almost 25 thousand tons. Pernambuco and São Paulo stand out among the largest guava-producing states [3]. The expansion of guava cultivation areas in the northeast occurred mainly through the quick financial return of the capital invested and due to it being a versatile fruit in its forms of use; as such, guava represents a real alternative in the development of the fruit-growing sector of the northeast [4].

In 2022, the state of Paraíba produced only 2557 tons in an area of 338 hectares, and its yield was lower than those of the other guava-producing states [3]. The limitation of guava production is related to irregular rainfall associated with high temperatures and evapotranspiration rates, making the use of irrigation indispensable for safe production [5]. The water sources available for irrigation in the Brazilian semi-arid region, with some exceptions, contain high levels of adsorbed salts, predominantly chloride and sodium [6].

High concentrations of salts in water and/or soil represent one of the abiotic stresses that most limit crop production [7] due to the osmotic and ionic effects, which hinder the absorption of water and nutrients and cause specific toxicity due to the high concentration of phytotoxic ions (Na^+ and Cl^-) in the cytoplasm, nutritional imbalance, and physiological and metabolic changes, such as oxidative stress, resulting in changes in the cell structure and the rupture of membranes [8,9].

However, the intensity of salt stress effects on plants varies depending on the cationic nature of the water, the time of exposure to stress, the genotype, the developmental stage, irrigation management, fertilization, climatic conditions, and the application of exogenous substances [10,11]. The guava cultivar Paluma in its initial stage of development is sensitive to irrigation water salinity, with a threshold ECw of 0.3 dS m^{-1} [12] and 2.15 dS m^{-1} in the post-grafting stage [13].

Considering the socioeconomic importance of guava cultivation in the northeast region of Brazil, it is necessary to use strategies capable of mitigating the harmful effects of salt stress on this fruit crop. Several strategies have been used to mitigate the effects of salt stress on plants, including foliar application of proline [14–16]. Under conditions of abiotic stress, proline can act as a signaling and/or regulatory osmolyte that activates multiple physiological, biochemical, and molecular responses [17], and it can be used in the stabilization of membranes and proteins, the scavenging of reactive oxygen species (ROS), the maintenance of intracellular pH, acting on molecular signaling, oxidation reduction balancing, and the induction of gene expression [18].

Several studies have been conducted to evaluate the effects of foliar application of proline in mitigating salt stress effects on plants, such as guava in the seedling formation stage. Veloso [19] concluded that foliar application of proline reduced the harmful effects of water salinity on seedling growth. Gohari [20], in grapevines, found that foliar application of proline improved growth, photosynthetic pigment synthesis and antioxidant activity under saline conditions. However, studies evaluating the effects of foliar application of proline in guava under irrigation with saline waters throughout the development cycle are scarce in the literature. Thus, it is extremely important to conduct research with the purpose of evaluating mitigation strategies with foliar application of proline in guava under irrigation with waters of different salinity levels.

In this context, the hypothesis of this study is that foliar application of proline increases the tolerance of guava plants through the elimination of ROS and by aiding in the accumulation of osmoprotective substances, thus attenuating the deleterious effects of salt stress on gas exchange, water status, pigment synthesis, and photochemical efficiency supporting the growth of guava cv. Paluma.

The objective of this study was to evaluate the effect of foliar application of proline as a mitigating factor of salt stress on the morphophysiology of guava cv. Paluma.

2. Results

2.1. Relative Water Content, Electrolyte Leakage in the Leaf Blade, and Water Saturation Deficit

According to the summary of the analysis of variance (Table 1), there was a significant effect ($p \leq 0.01$) of the interaction between the electrical conductivity of irrigation water (ECw) and proline concentrations (PROL) on the relative water content (RWC), water saturation deficit (WSD), and electrolyte leakage (EL%) in the leaf blade of guava plants cv. Paluma at 360 days after transplantation (DAT).

Table 1. Summary of the analysis of variance for the relative water content (RWC), water saturation deficit (WSD), and electrolyte leakage (EL%) in the leaf blade of guava plants cv. Paluma cultivated under different electrical conductivities of irrigation water (ECw) and proline concentrations (PROL) at 360 days after transplanting.

Source of Variation	DF	Mean Squares		
		RWC	WSD	EL%
Salinity levels (ECw)	4	1022 **	1033 **	746 **
Linear regression	1	3899 **	3934 **	2886 **
Quadratic regression	1	138 ^{ns}	144 ^{ns}	73 *
Proline (PROL)	3	998 **	992 **	112 **
Linear regression	1	2500 **	2473 **	16 ^{ns}
Quadratic regression	1	491 **	500 **	164 **
Interaction (ECw × PROL)	12	126 **	127 **	63 **
Blocks	2	90 ^{ns}	89 ^{ns}	14 ^{ns}
Residual	38	45	45	17
CV (%)		8.51	31.80	14.58

DF—Degrees of freedom; CV (%)—coefficient of variation; ** significant at 0.01 probability level; * significant at 0.05% probability level; ^{ns} not significant.

Foliar application of proline mitigated the effect of salt stress on the relative water content (RWC) in the leaf blade, with the highest estimated value (100%) obtained in plants subjected to ECw of 0.8 dS m^{-1} and a proline concentration of 24 mM (Figure 1A). In contrast, the lowest value (63.6%) occurred in plants cultivated under an estimated water salinity of 3.0 dS m^{-1} and that did not receive foliar application of proline.

Water saturation deficit (WSD) in the leaf blade was higher (37.17%) in plants subjected to ECw of 3.6 dS m^{-1} and foliar application of proline at a concentration of 4.5 mM (Figure 1B). However, foliar application of proline reduced WSD in plants irrigated using water with electrical conductivity of 0.8 dS m^{-1}, and a value of 0.79% was observed in plants that received proline application at a concentration of 24 mM.

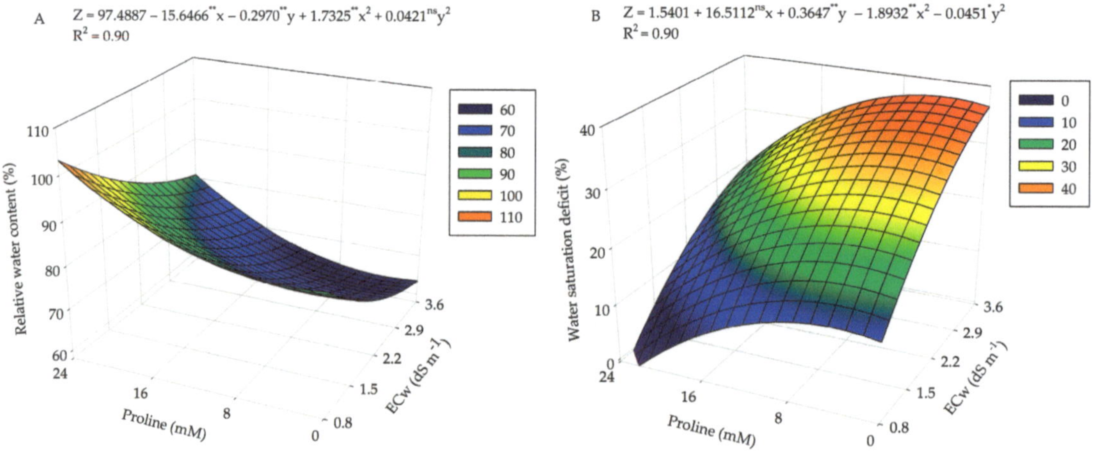

Figure 1. *Cont.*

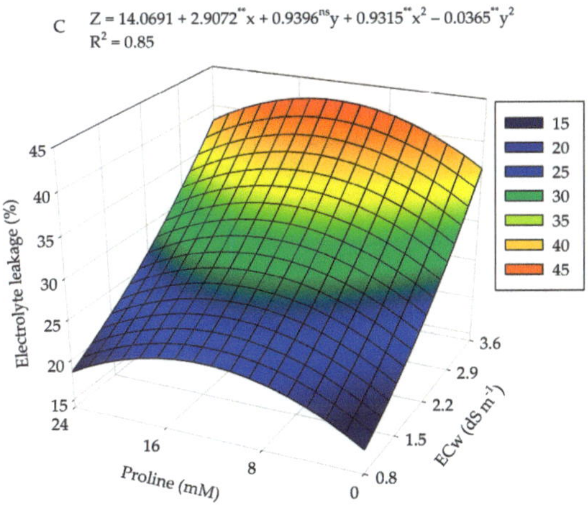

Figure 1. Relative water content (RWC) (**A**), water saturation deficit (WSD) (**B**), and electrolyte leakage (EL%) (**C**) in the leaf blade of guava plants as a function of the interaction between electrical conductivity levels (ECw) and proline concentrations at 360 days after transplanting. X and Y: levels of electrical conductivity of water (ECw) and proline concentrations (PROL), respectively. ** and * significant at $p \leq 0.01$ and $p \leq 0.05$ and ^{ns} not significant ($p > 0.05$) according to the F test, respectively.

Electrolyte leakage (EL%) in the leaf blade of guava increased with the increase in irrigation water salinity, with the highest value (42.6%) reached in plants subjected to ECw of 3.6 dS m^{-1} and under a proline concentration of 13.5 mM (Figure 1C). Conversely, the lowest EL% value (17%) occurred in plants cultivated under the lowest ECw level (0.8 dS m^{-1}) and in plants that did not receive foliar application of proline (0 mM), indicating that the increase in salt stress promoted a 25.6% increase when comparing the maximum value obtained at ECw of 3.6 dS m^{-1} to that obtained at the lowest salinity level (0.8 dS m^{-1}).

2.2. Gas Exchange

There was a significant effect of the interaction between the factors electrical conductivity of irrigation water (ECw) and proline concentrations (PROL) on the stomatal conductance, transpiration, internal CO_2 concentration, CO_2 assimilation rate, instantaneous carboxylation efficiency, and instantaneous water use efficiency of guava plants (Table 2).

For stomatal conductance (Figure 2A), plants grown under a proline concentration of 24 mM and irrigation water electrical conductivity of 0.8 and 3.6 dS m^{-1} obtained higher values (0.24 and 0.11 mol H_2O m^{-2} s^{-1}), respectively. The lowest value of 0.05 mol H_2O m^{-2} s^{-1} was reached in plants subjected to ECw of 2.9 dS m^{-1} without proline application (0 mM).

For the transpiration (E) of guava plants (Figure 2B), the highest value obtained (19.73 mmol H_2O m^{-2} s^{-1}) occurred in plants subjected to ECw of 0.8 dS m^{-1} and under foliar application of proline at a concentration of 18 mM. On the other hand, the lowest value (6.04 mMol H_2O m^{-2} s^{-1}) of E was observed in plants subjected to the highest salinity (3.6 dS m^{-1}) and that did not receive proline application, which resulted in a decrease of 69.3% when comparing the maximum and minimum values obtained.

Regarding internal CO_2 concentration, the maximum value of 252.27 μmol CO_2 m^{-2} s^{-1} was observed in guava plants irrigated using water with electrical conductivity of 3.6 dS m^{-1} and foliar application of proline at a concentration of 24 mM (Figure 2D). In contrast, the minimum value of 131.43 μmol CO_2 m^{-2} s^{-1} was observed in plants under ECw of 0.8 dS m^{-1} and an estimated proline concentration of 9.0 mM. When comparing the maximum and minimum values obtained, a decrease of 47.9% was observed.

Regarding the CO_2 assimilation rate (Figure 2D), plants irrigated using water with electrical conductivity of 0.8 dS m^{-1} and foliar application of proline at a concentration of 18 mM obtained the highest value (19.73 μmol CO_2 m^{-2} s^{-1}). The lowest value obtained was 6.04 μmol CO_2 m^{-2} s^{-1}, which was reached in plants irrigated with ECw of 3.6 dS m^{-1} and without proline application (0 mM).

Instantaneous water use efficiency (WUEi) was higher (4.64 [(μmol CO_2 m^{-2} s^{-1}) (mmol H_2O m^{-2} s^{-1})$^{-1}$]) when plants were subjected to ECw of 0.98 dS m^{-1} and a proline concentration of 10.5 mM (Figure 2E). However, the lowest WUEi (3.05 [(μmol CO_2 m^{-2} s^{-1}) (mmol H_2O m^{-2} s^{-1})$^{-1}$]) was observed in plants irrigated with 3.6 dS m^{-1} water and under foliar application of proline at a concentration of 24 mM, which represents a reduction of 34.3% when compared to the highest value obtained.

For instantaneous carboxylation efficiency (Figure 2F), the electrical conductivity of water of 0.8 dS m^{-1} and an estimated proline concentration of 13.5 mM resulted in a maximum value of 0.13 [(μmol CO_2 m^{-2} s^{-1}) (μmol CO_2 m^{-2} s^{-1})$^{-1}$]. The minimum value of 0.02 [(μmol CO_2 m^{-2} s^{-1}) (μmol CO_2 m^{-2} s^{-1})$^{-1}$] was found in plants without proline application (0 mM) and under ECw of 3.6 dS m^{-1}.

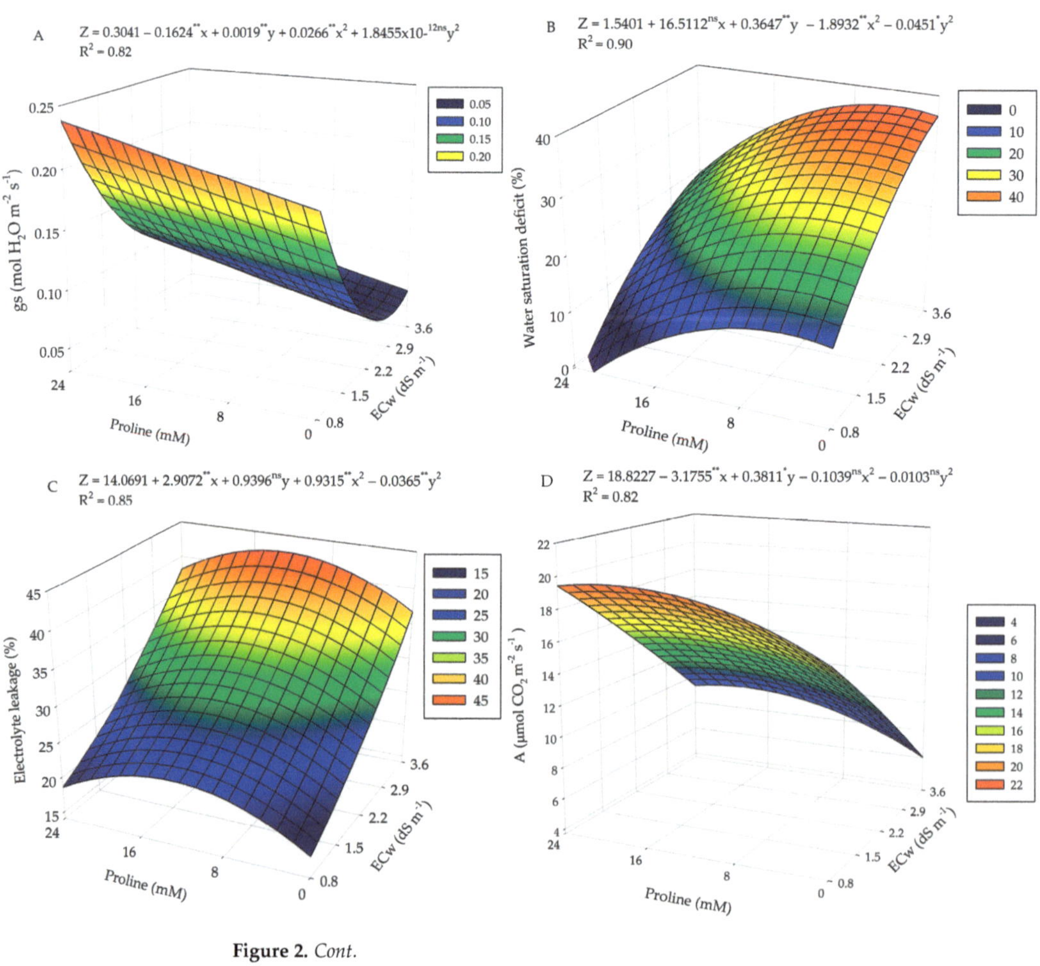

Figure 2. *Cont.*

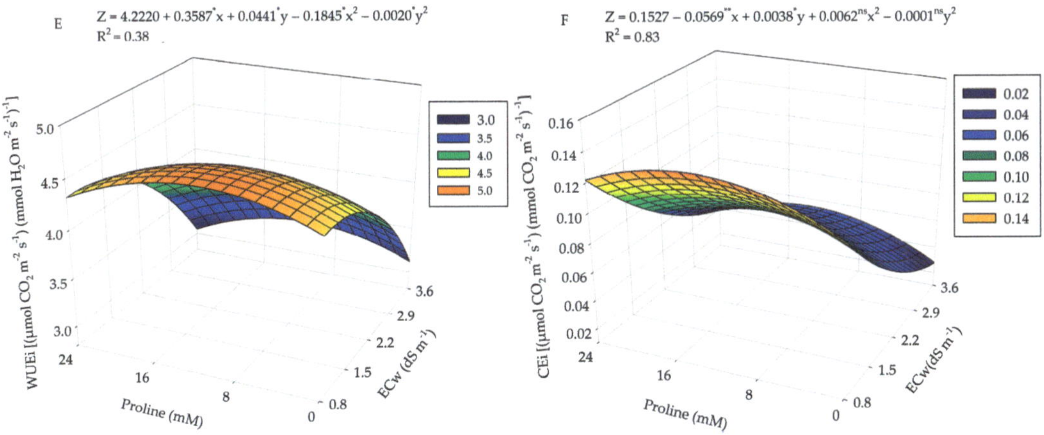

Figure 2. Stomatal conductance (gs) (**A**), transpiration (E) (**B**), internal CO_2 concentration (Ci) (**C**), CO_2 assimilation rate (A) (**D**), instantaneous water use efficiency (WUEi) (**E**), and instantaneous carboxylation efficiency (Cei) (**F**) of guava plants cv. Paluma as a function of the interaction between levels of electrical conductivity of water (ECw) and proline concentrations at 360 days after transplanting. X and Y: levels of electrical conductivity of water (ECw) and proline concentrations, respectively. ** and * significant at $p \leq 0.01$ and $p \leq 0.05$ and ^{ns} not significant ($p > 0.05$) according to the F test, respectively.

Table 2. Summary of the analysis of variance for stomatal conductance (gs), transpiration (E), internal CO_2 concentration (Ci), CO_2 assimilation rate (A), instantaneous water use efficiency (WUEi), and instantaneous carboxylation efficiency (CEi) of guava plants cv. Paluma cultivated under different electrical conductivities of irrigation water (ECw) and proline concentrations (PROL) at 360 days after transplanting.

Source of Variation	DF	Mean Squares					
		gs	*E*	*Ci*	*A*	*WUEi*	*CEi*
Salinity levels (ECw)	4	0.044 **	4.87 **	12106 **	196.55 **	4.67 ^{ns}	0.0141 **
Linear regression	1	0.124 **	17.34 **	46256 **	775.71 **	9.22 ^{ns}	0.0550 **
Quadratic regression	1	0.039 **	1.64 ^{ns}	125 ^{ns}	0.43 ^{ns}	6.94	0.0010 ^{ns}
Proline (PROL)	3	0.005 **	0.42 ^{ns}	6356 **	45.22 *	0.44 ^{ns}	0.0018 ^{ns}
Linear regression	1	0.015 **	1.09 ^{ns}	5534 **	85.86 **	0.30 ^{ns}	0.0008 ^{ns}
Quadratic regression	1	0.0004 ^{ns}	0.17 ^{ns}	9238 **	26.00 ^{ns}	0.61 ^{ns}	0.0045 **
Interaction (ECw × PROL)	12	0.005 **	2.29 *	3295 **	33.62 **	8.38 **	0.0020 *
Blocks	2	0.004 *	1.39 ^{ns}	229 ^{ns}	7.11 ^{ns}	1.32 ^{ns}	0.0006 ^{ns}
Residual	38	0.001	0.87	769	12.09	2.50	0.0008
CV (%)		26.17	27.42	26.17	25.76	14.64	36.12

DF—Degrees of freedom; CV (%)—coefficient of variation; ** significant at 0.01 probability level; * significant at 0.05% probability level; ^{ns} not significant.

2.3. Photosynthetic Pigments

There was a difference between the salinity levels, with a significant effect ($p \leq 0.01$) on the contents of chlorophyll *a*, chlorophyll *b*, total chlorophyll, and carotenoids of guava cv. Paluma (Table 3). Proline concentrations significantly influenced ($p \leq 0.01$) all variables related to photosynthetic pigments. Regarding the interaction between the factors (ECw × PROL), a significant effect was observed only in carotenoids ($p \leq 0.01$) of guava cv. Paluma at 360 days after transplanting

Table 3. Summary of the analysis of variance for chlorophyll *a* (Chl *a*), chlorophyll *b* (Chl *b*), Chl *a*/Chl *b* ratio (Chl *a*/Chl *b*), total chlorophyll (Chl *T*), and carotenoids (Car) of guava plants cv. Paluma cultivated under different electrical conductivities of irrigation water (ECw) and proline concentrations (PROL) at 360 days after transplanting.

Source of Variation	DF	Mean Squares				
		Chl *a*	Chl *b*	Chl *a*/Chl *b*	Chl *T*	Car
Salinity levels (ECw)	4	74.34 **	8.76 **	0.45 ns	132.43 **	6.00 **
Linear regression	1	277.3 **	33.88 **	0.04 ns	504.99 **	23.46 **
Quadratic regression	1	15.16 **	0.02 ns	1.46 *	14.03 *	0.10 ns
Proline (PROL)	3	17.09 **	2.36 **	1.73 **	26.46 **	4.70 **
Linear regression	1	41.21 **	4.60 **	0.03 ns	73.22 **	13.79 **
Quadratic regression	1	5.11 ns	2.47 *	4.48 **	0.47 ns	0.00008 ns
Interaction (ECw × PROL)	12	2.97 ns	0.60 ns	0.89 ns	3.84 ns	1.15 **
Blocks	2	0.21 ns	0.26 ns	0.11 ns	0.02 ns	0.53
Residual	38	2.12	0.46	0.29	3.03	0.40
CV (%)		13.13	15.92	20.50	11.32	13.15

DF—Degrees of freedom; CV (%)—coefficient of variation; ** significant at 0.01 probability level; * significant at 0.05% probability level; ns not significant.

Irrigation with saline water significantly affected the synthesis of photosynthetic pigments, and it was observed for the contents of chlorophyll *a*, chlorophyll *b*, and total chlorophyll (Figure 3A–C, respectively) that the plants had the highest values at ECw of 0.8 dS m^{-1} with the increase in salinity levels, with reductions of 43.0, 39.7, and 42.1%, respectively, when comparing the values obtained in plants irrigated with water of 3.6 dS m^{-1} to those of plants subjected to ECw of 0.8 dS m^{-1}.

Foliar application of proline positively influenced the contents of chlorophyll *a*, chlorophyll *b*, and total chlorophyll of guava cv. Paluma. For Chl *a* and Chl *b* contents, a quadratic response was observed (Figure 3B,D), with maximum values of 12, 04, and 4.88 µg mL^{-1}, respectively. For the total chlorophyll contents (Figure 3F), it was observed that foliar application of proline resulted in a linear increase equal to 0.88% per unit increase in proline concentration. When comparing plants that received the highest proline concentration (24 mM) to those in the control treatment (0 mM), an increase of 2.46% was observed in Chl *T* contents.

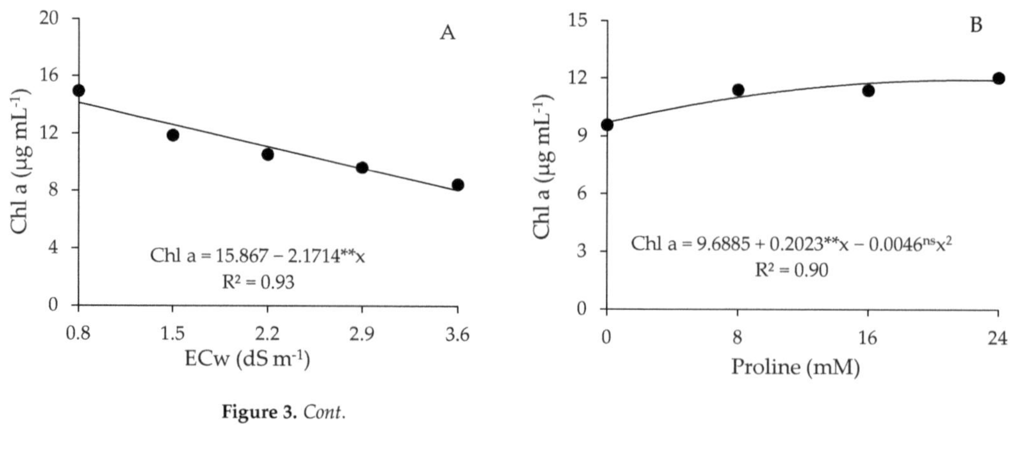

Figure 3. *Cont.*

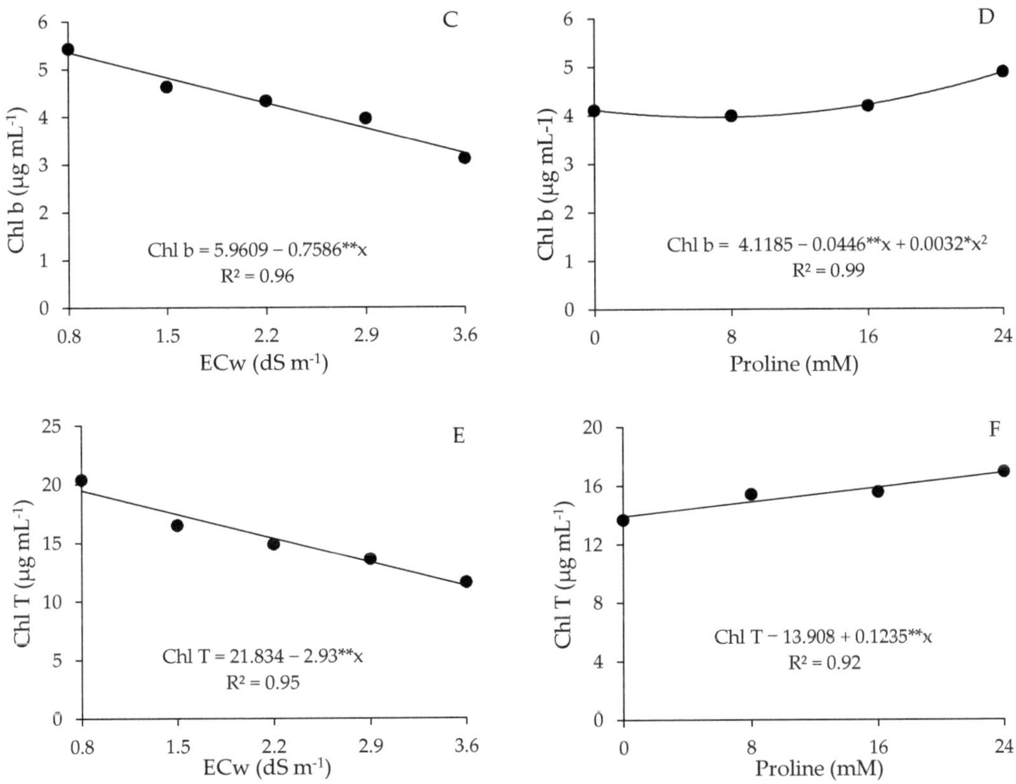

Figure 3. Contents of chlorophyll *a* (Chl a), chlorophyll *b* (Chl b), and total chlorophyll (Chl T) of guava plants cv. Paluma as a function of the levels of electrical conductivity of the water (ECw) (**A**,**C**,**E**) and proline concentrations (**B**,**D**,**F**) at 360 days after transplanting. ** and * significant at $p \leq 0.01$ and $p \leq 0.05$ and ^{ns} not significant ($p > 0.05$) according to the F test, respectively.

The increase in the electrical conductivity of irrigation water and proline concentrations stimulated the synthesis of carotenoids in guava cv. Paluma (Figure 4), with the highest content (6.25 m µg mL) observed in plants subjected to the highest ECw level (3.6 dS m^{-1}) and a proline concentration of 24 mM. On the other hand, the lowest value (3.36 µg mL^{-1}) was found in plants subjected to ECw of 0.8 dS m^{-1} and that did not receive foliar application of proline (0 mM). When comparing the Car contents, it was observed that the increase up to the highest ECw and a proline concentration of 24 mM promoted an increase of 46.2% in comparison to the values obtained at ECw 0.8 dS m^{-1} and a proline concentration of 0 mM.

2.4. Photochemical Efficiency

There was a significant effect ($p \leq 0.05$) of salinity levels (ECw) and proline concentrations (PROL) on the initial fluorescence (F_0) of guava cv. Paluma (Table 4). In contrast, the water salinity levels significantly affected the maximum fluorescence (F_m) and variable fluorescence (F_v) of guava plants cv. Paluma at 360 days after transplanting.

For the initial fluorescence (F_0), the increase in water salinity and proline concentrations promoted an increase in F_0 (Figure 5), with the highest value (172.66) obtained in plants irrigated with ECw of 3.6 dS m^{-1} and under 24 mM of proline. On the other hand, the lowest value (108.21) of F_0 occurred in plants subjected to water salinity of 0.8 dS m^{-1} and a proline concentration of 1.5 mM. When comparing the maximum and minimum values obtained, an increase of 37.3% in the initial fluorescence was observed when plants were subjected to ECw of 3.6 dS m^{-1} and a proline concentration of 24 mM.

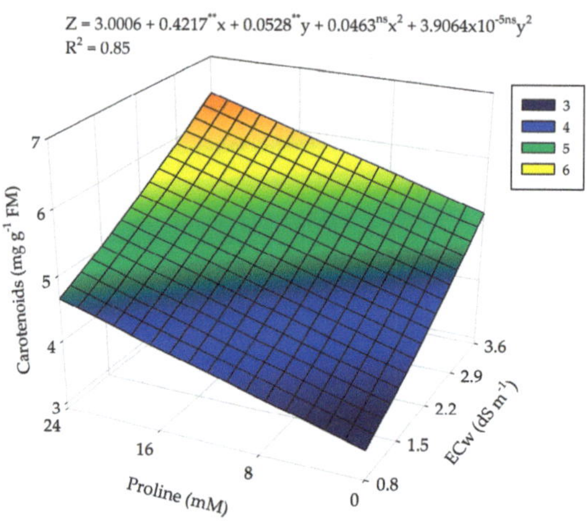

Figure 4. Carotenoid content of guava plants as a function of the interaction between levels of electrical conductivity of water (ECw) and proline concentrations at 360 days after transplanting. X and Y: levels of electrical conductivity of water (ECw) and proline concentrations, respectively; ** and * significant at $p \leq 0.01$ and $p \leq 0.05$ and ns not significant ($p > 0.05$) according to the F test, respectively.

Table 4. Summary of the analysis of variance for initial (F_0), maximum (F_m), variable (F_v) fluorescence, and quantum efficiency of photosystem II (F_v/F_m) of guava plants cv. Paluma cultivated under different electrical conductivities of irrigation water (ECw) and proline concentrations (PROL) at 360 days after transplanting.

Source of Variation	DF	Mean Squares			
		F_0	F_m	F_v	F_v/F_m
Salinity levels (ECw)	4	5073 **	26,326 **	25,561 *	0.014 ns
Linear regression	1	20,254 **	75,075 **	49,227 *	0.006 ns
Quadratic regression	1	3.72 ns	25,271 *	50,596 *	0.051 *
Proline (PROL)	3	502 ns	4951 ns	16,240 ns	0.023 ns
Linear regression	1	1310 *	8883 ns	26,498 ns	0.030 ns
Quadratic regression	1	176 ns	2554 ns	5890 ns	0.005 ns
Interaction (ECw × PROL)	12	581 *	3874 ns	6013 ns	0.009 ns
Blocks	2	383 ns	1649 ns	3257 ns	0.003 ns
Residual	38	256	6021	7314	0.009
CV (%)		11.54	15.55	23.65	13.78

DF—Degrees of freedom; CV (%)—coefficient of variation; ** significant at 0.01 probability level; * significant at 0.05% probability level; ns not significant.

Maximum fluorescence (F_m) showed a quadratic behavior, and the maximum estimated value (536.18) occurred in plants irrigated with 1.5 dS m^{-1} water, decreasing from this salinity level and reaching the lowest value (424.34) under ECw of 3.6 dS m^{-1} (Figure 6A). When comparing the F_m of plants subjected to the highest salinity (3.6 dS m^{-1}) to the value of those cultivated under ECw of 0.8 dS m^{-1}, a reduction of 20.85% was observed.

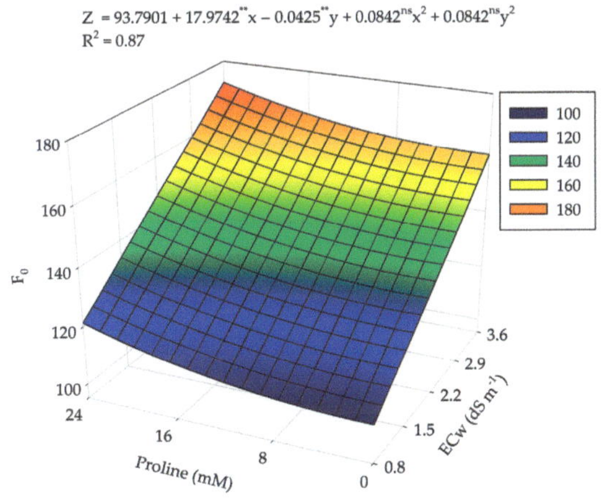

Figure 5. Initial fluorescence (F_0) of guava plants cv. Paluma as a function of the interaction between levels of electrical conductivity of water (ECw) and proline concentrations at 360 days after transplanting. X and Y: levels of electrical conductivity of water (ECw) and proline concentrations, respectively; ** and * significant at $p \leq 0.01$ and $p \leq 0.05$ and ^{ns} not significant ($p > 0.05$) according to the F test, respectively.

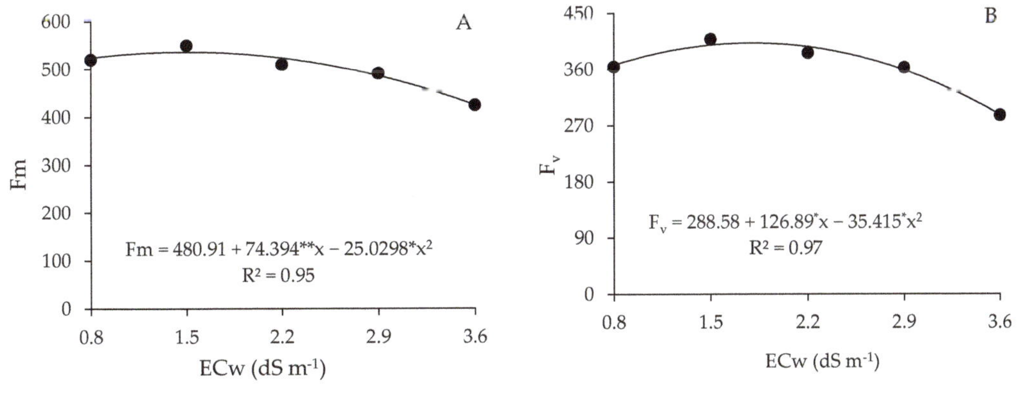

Figure 6. Maximum fluorescence (F_m) (**A**) and variable (F_v) (**B**) of guava plants cv. Paluma as a function of the levels of electrical conductivity of water (ECw) at 190 days after transplanting. ** and * significant at $p \leq 0.01$ and $p \leq 0.05$ according to the F test, respectively.

As observed for F_m (Figure 6A), variable fluorescence (F_v) (Figure 6B) reached the maximum estimated value (402.23) in plants irrigated under water salinity of 1.8 dS m^{-1}. When irrigated with water of 3.6 dS m^{-1}, it resulted in the lowest Fv value (286.40), leading to a decrease of 28.8% when compared to the maximum values obtained in ECw of 1.8 dS m^{-1}.

2.5. Growth Parameters

There was a significant effect between the factors (ECw × PROL) on crown diameter (D_{crown}) and the absolute growth rate in stem diameter of guava plants (AGR_{SD}) (Table 5). The water salinity levels significantly affected the stem diameter (SD), crown volume (V_{crown}), and vegetative vigor index (VVI) at 360 DAT. The proline concentration had a significant effect on the relative growth rate in stem diameter (RGR_{SD}) during the period from 190 to 360 days after transplanting.

Table 5. Summary of the analysis of variance for stem diameter (SD), crown volume (V_{crown}), vegetative vigor index (VVI), and crown diameter (D_{crown}) at 360 days after transplanting (DAT) and absolute growth rate (AGR_{SD}) and relative growth rate (RGR_{SD}) in stem diameter of guava cv. Paluma cultivated under different electrical conductivities of irrigation water (ECw) and proline concentrations (PROL) during the period from 190 to 360 days after transplanting (DAT).

Source of Variation	DF	Mean Squares					
		SD	V_{crown}	VVI	D_{crown}	AGR_{SD}	RGR_{SD}
Salinity levels (ECw)	4	287.47 **	4.75 **	1.11 **	1.47 **	0.0021 **	0.000007 ^{ns}
Linear regression	1	1016.40 **	17.91 **	4.00 **	5.69 **	0.0078 **	0.000003 ^{ns}
Quadratic regression	1	115.46 **	0.48 ^{ns}	0.14 ^{ns}	0.19 ^{ns}	0.0006 ^{ns}	0.000000 ^{ns}
Proline (PROL)	3	8.60 ^{ns}	0.61 ^{ns}	0.07 ^{ns}	0.58 **	0.0011 *	0.000003 *
Linear regression	1	25.28 ^{ns}	1.42 ^{ns}	0.14 ^{ns}	1.69 **	0.0021 *	0.000003 ^{ns}
Quadratic regression	1	0.03 ^{ns}	0.35 ^{ns}	0.0001 ^{ns}	0.01 ^{ns}	0.0012 ^{ns}	0.000004 *
Interaction (ECw × PROL)	12	14.93 ^{ns}	1.40 ^{ns}	0.30 ^{ns}	0.44 **	0.0007 *	0.000002 ^{ns}
Blocks	2	12.85 ^{ns}	0.09 ^{ns}	0.07 ^{ns}	0.05 ^{ns}	0.0002 ^{ns}	0.000002 ^{ns}
Residual	38	8.57	0.83	0.16	0.06 ^{ns}	0.0003	0.000008
CV (%)		11.44	26.14	14.20	15.22	35.58	34.50

DF—Degrees of freedom; CV (%)—coefficient of variation; ** significant at 0.01 probability level; * significant at 0.05% probability level; ^{ns} not significant.

For crown diameter (D_{crown}), the highest value (2.39 m) was observed in plants subjected to irrigation with ECw of 0.8 dS m^{-1} and foliar application of proline at a concentration of 24 mM, promoting an increase of 1.18 m (15.84%) when compared to plants under the control treatment (0.0 mM) and irrigated with the water salinity level of 3.6 dS m^{-1} (Figure 7A). Meanwhile, irrigation with 3.6 dS m^{-1} water and application of proline at a concentration of 0.0 mM resulted in the lowest D_{crown} (1.07 m).

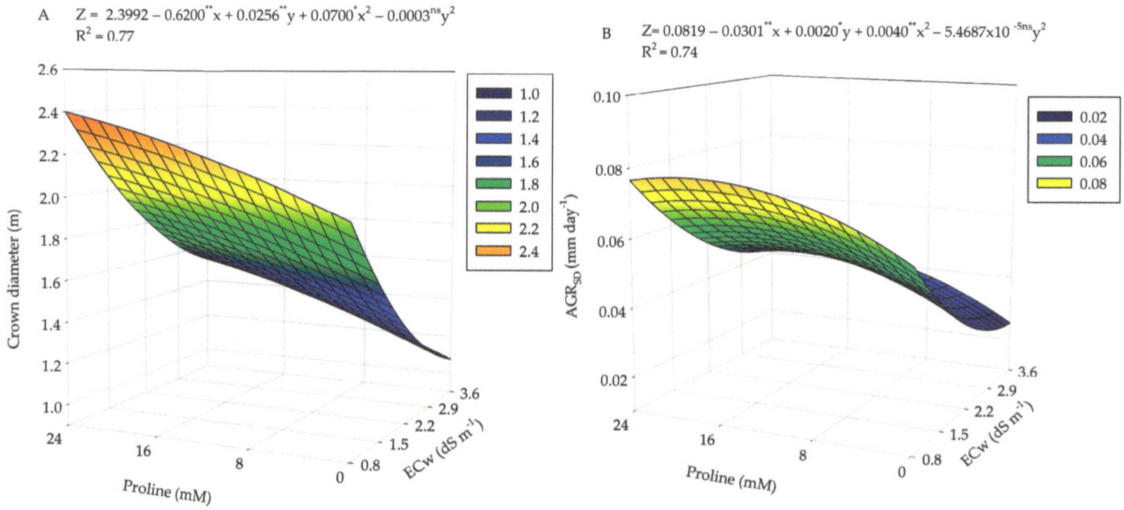

Figure 7. Crown diameter (**A**) and absolute growth rate in stem diameter (AGR_{SD}) (**B**) of guava cv. Paluma as a function of the interaction between levels of electrical conductivity of water (ECw) and proline concentrations at 190 days after transplanting. X and Y: levels of electrical conductivity of water (ECw) and proline concentrations, respectively; ** and * significant at $p \leq 0.01$ and $p \leq 0.05$ and ^{ns} not significant ($p > 0.05$) according to the F test, respectively.

For the absolute growth rate in stem diameter (AGR_{SD}) (Figure 7B), it was observed that irrigation with ECw of 0.8 dS m^{-1} associated with foliar application of proline at the estimated concentration of 18 mM resulted in the highest values (0.0784 mM day^{-1}). When comparing the AGR_{SD} of plants that did not receive foliar application of proline and were irrigated with water of 0.8 dS m^{-1}, whose estimated value was 0.0603, an increase of 23.08% was observed. Meanwhile, plants that did not receive foliar application of proline and were irrigated with ECw of 3.6 dS m^{-1} obtained the lowest value found (0.0253 mM day^{-1}), which indicates a gain of 67.7% when compared with the maximum values obtained.

The increase in irrigation water salinity reduced the growth in stem diameter of guava plants (Figure 8A) by 11.96% per unit increment of ECw. When comparing the SD of plants cultivated with ECw 3.6 dS m^{-1} with that of plants subjected to water salinity of 0.8 dS m^{-1}, a reduction of 37.03% (11.64 mM) was observed.

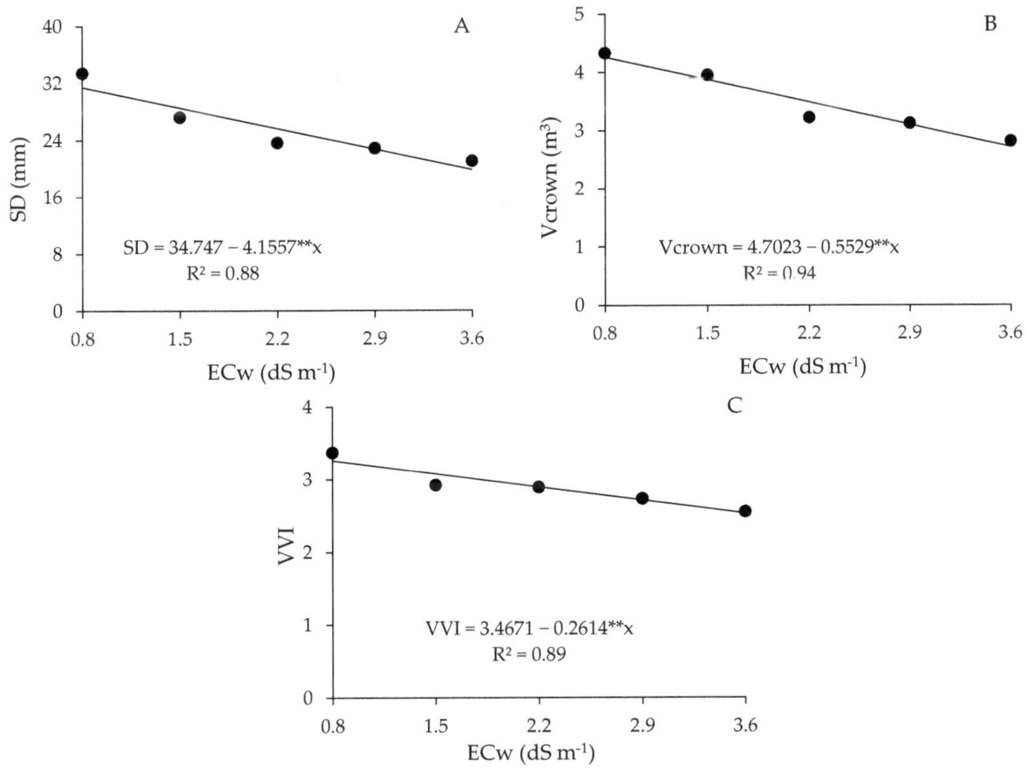

Figure 8. Stem diameter (SD) (**A**), crown volume (V_{crown}) (**B**), and vegetative vigor index (VVI) (**C**) of guava plants cv. Paluma as a function of the levels of electrical conductivity of the water (ECw) at 190 days after transplanting. ** significant at $p \leq 0.01$ according to the F test.

Crown volume (V_{crown}) (Figure 8B) and vegetative vigor index (VVI) (Figure 8C) also decreased linearly with the increase in the salinity levels of irrigation water by 11.76 and 7.54% per unit increase in ECw, respectively. When comparing the V_{crown} and VVI of plants irrigated with the highest salinity (3.6 dS m^{-1}) with those obtained under the lowest ECw (0.8 dS m^{-1}), reductions of 36.3% (1.55 m^3) and 22.6% (0.73) were observed, respectively.

Foliar application of proline favored the relative growth rate in stem diameter (RGR_{SD}) of guava plants (Figure 9), with a maximum estimated value of 0.002625 mM mM^{-1} day^{-1} obtained at a concentration of 12.6 mM, representing an increase of 31.25% compared to plants that did not receive proline (0.0 mM).

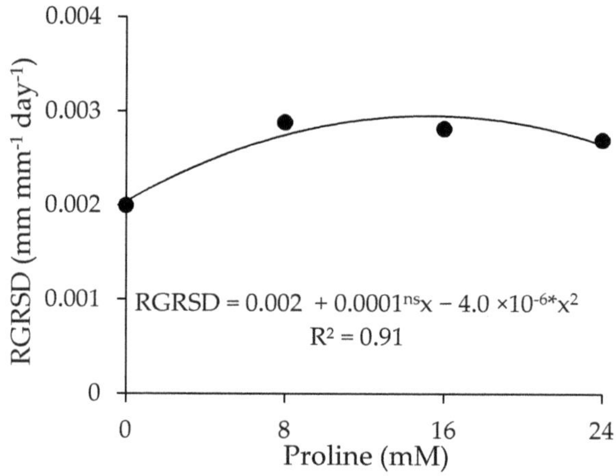

Figure 9. Relative growth rate in stem diameter (RGR_{SD}) of guava cv. Paluma as a function of proline concentrations during the period from 190 to 360 days after transplanting. * significant at $p \leq 0.05$ and ^{ns} not significant ($p > 0.05$) according to the F test, respectively.

2.6. Principal Component Analysis

The correlation between physiological and growth variables with saline levels and proline concentrations presented by principal component analysis (PCA) indicates a representation of 69.9% in the first two components (Figure 10). It is possible to highlight that component 1 (PCA1) is responsible for 53.2%, being associated with treatments S1P1, S1P2, S1P3, and S1P4, which presented scores of −4.0418, −4.0919, −5.0365, and −5.5716, respectively. When associated with the variables evaluated, it is observed that these treatments showed high affinity with Chl a, Chl b, Chl t, and SD, which reached scores of −0.8903, −0.8629, −0.9190, and −0.9311, respectively. Also highlighted is a strong relationship between treatments and gas exchange *gs*, *E*, and A, which obtained scores of −0.7771, −0.6201, and −0.7886, respectively.

It was found that the second component (PCA 2) provided a contribution of 16.7%, with a strong association of Fv, Fm, and Fv/Fm (0.6242, 0.8539, and 0.8448, respectively) with treatments S2P2, S2P1, and S3P3, which have score values of 2.1306, 2.2656, and 1.3930, respectively. It is also worth mentioning the existence of a strong relationship between the S4P4 treatment (−2.2126) and the Car variable (−0.6101). Treatments S5P3 and S5P4 presented a good contribution to PC2, with scores of −3.7088 and −3.3274; however, they demonstrated low affinity with the variables analyzed (Figure 10).

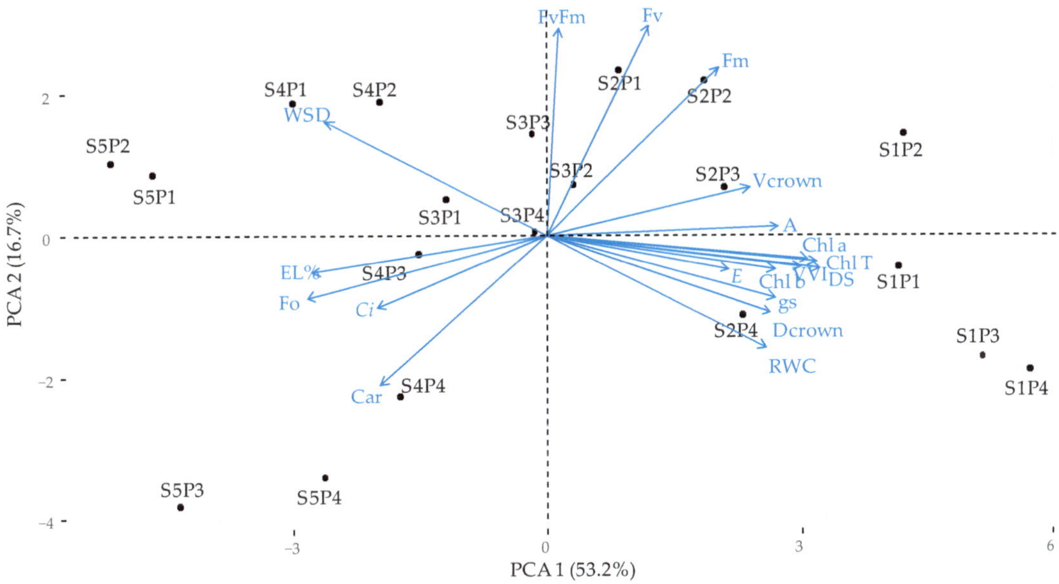

Figure 10. Principal component analysis (PCA) of correlation between physiological and growth variables of guava plants cv. 'Paluma' due to the interaction between the electrical conductivity levels of the irrigation water and proline concentrations at 390 days after transplanting. gs—stomatal conductance, E—transpiration, Ci—internal CO_2 concentration, A—CO_2 assimilation rate, RWC—relative water content; WSD—water saturation deficit; EL%—electrolyte leakage; Chl *a*—chlorophyll *a*; Chl *b*—chlorophyll *b*; Chl t—total chlorophyll; Car—carotenoids; F0—initial fluorescence; Fv—variable fluorescence; F_m—maximum fluorescence; F_v/F_m—quantum efficiency of photosystem II; SD—stem diameter; V_{crown}—crown volume; VVI—vegetative vigor index; D_{crown}—crown diameter; absolute growth rate (AGR_{SD}) and relative growth rate (RGR_{SD}); S1P1—0.8 dS m^{-1} and 0 mM; S1P2—0.8 dS m^{-1} and 8 mM; S1P3—0.8 dS m^{-1} and 16 mM; S1P4—0.8 dS m^{-1} and 24 mM; S2P1—1.5 dS m^{-1} and 0 mM; S2P2—1.5 dS m^{-1} and 8 mM; S2P3—1.5 dS m^{-1} and 16 mM; S2P4—1.5 dS m^{-1} and 24 mM; S3P1—2.2 dS m^{-1} and 0 mM; S3P2—2.2 dS m^{-1} and 8 mM; S3P3—2.2 dS m^{-1} and 16 mM; S3P4—2.2 dS m^{-1} and 24 mM; S4P1—2.9 dS m^{-1} and 0 mM; S4P2—2.9 dS m^{-1} and 8 mM; S4P3—2.9 dS m^{-1} and 16 mM; S4P4—2.9 dS m^{-1} and 24 mM; S5P1—3.6 dS m^{-1} and 0 mM; S5P2—3.6 dS m^{-1} and 8 mM; S5P3—3.6 dS m^{-1} and 16 mM; S5P4—3.6 dS m^{-1} and 24 mM.

3. Discussion

Foliar application of proline was beneficial in maintaining the water status of guava plants, causing an increase in the relative water content, which indicates that its use increases the plant's tolerance to salt stress, thus favoring the absorption of water and nutrients and making the plant able to maintain the turgidity of its tissues. This is due to the fact that proline stimulates greater antioxidant activity, allowing the plant to adjust osmotically to stress conditions [21].

Water saturation deficit (WSD) was reduced by the foliar application of 24 mM of proline up to 1.5 dS m^{-1}, which may be associated with an acclimatization mechanism through the accumulation of organic solutes, such as the accumulation of this amino acid in the vacuole of the plants, promoting osmotic adjustment, which led to a lower water deficit in the leaf blade [22].

It is possible to observe that the increase in salinity up to 3.6 dS m^{-1} compromised the maintenance of water status in the leaf blade of guava plants, with the highest WSD and EL% being observed under this condition. This behavior results from the deleterious effects caused by salt stress induced by the reduction of the plant's ability to absorb water due to the decrease in the osmotic potential in the soil [12].

In the gas exchange, it was observed that the increase in water salinity above 0.8 dS m^{-1} led to partial closure of the stomata, thus reducing *gs* and transpiration and, consequently, the CO_2 assimilation rate of guava plants. This limiting effect caused by salt stress occurs because plants partially close their stomata in order to avoid the loss of water to the atmosphere and reduce the absorption of toxic ions, such as Na^+ and Cl^-, thus avoiding the dehydration of the guard cells [9,23], a mechanism triggered to promote the maintenance of tissue turgidity [24].

Despite the stomatal limitation, there was an increase in the internal CO_2 concentration in the substomatal chamber, highlighting non-stomatal factors, such as the limitation of the enzymatic activity of RuBisCO [25]. This fact did not directly affect the photosynthetic activity, as there was a reduction in CO_2 assimilation with the increase in salinity above 0.8 dS m^{-1}, which may have occurred due to the water deficit caused by the reduction in the osmotic potential of the soil, resulting in damage to PSII and restricting the capacity of CO_2 assimilation [26], which explains why the increase in the internal CO_2 concentration did not lead to an increase in the photosynthesis of guava plants.

This limitation caused by salt stress in gas exchange has also been observed by Nobre [27] in guava, Figueiredo [28] in pomegranate seedlings, Souza [29] in West Indian cherry plants, and Capitulino [30] in soursop.

Under low-salinity conditions (0.8 dS m^{-1}), foliar application of proline improved gas exchange, as concentrations of 18 and 24 mM stimulated *gs*, transpiration, internal CO_2 concentration, the CO_2 assimilation rate, and the instantaneous carboxylation efficiency of guava plants, whereas WUEi was higher in plants under ECw of 0.98 dS m^{-1} and a proline concentration of 10.5 mM. When accumulated in cells, proline acts in plant acclimatization by protecting macromolecules against tissue damage and denaturation and increasing stress tolerance with osmotic adjustment and control of redox potential in energy production [31]. In addition, the application of proline may have a safeguarding effect against oxidative damage to cell organelles, especially leaf mesophylls, thus protecting them from the risks of cytotoxicity in leaf tissues [32] and ultimately mitigating the effects of salinity on guava gas exchange.

The occurrence of this beneficial effect through the foliar application of proline in attenuating the effects of stress conditions was observed by Santos [16] in sour passion fruit (*Passiflora edulis* Sims.), for which the application of up to 6.5 mM of proline stimulated gas exchange in plants under salt stress. In sugar apple (*Annona squamosa* L.), Torres [33] reported that the application of 10 mM of proline promoted an increase in gas exchange in plants subjected to water deficit. Zahedi [34] observed that the combination of 100 mM of proline with graphene oxide stimulated the activity of antioxidant enzymes, thus reducing the effect of salt stress on grapevine plants (*Vitis vinifera* L.)

Salt stress compromised the synthesis of photosynthetic pigments in guava plants, possibly due to the excess of salts accumulated in the plant tissues, thus inducing the occurrence of damage to chloroplasts and resulting in protein denaturation and destabilization of membranes due to the excessive accumulation of toxic ions, causing inhibition in the synthesis of photosynthetic pigments [35,36]. In addition, salt stress also stimulates the activity of the enzyme chlorophyllase, which acts on the degradation of chlorophyll pigment molecules [37].

On the other hand, the application of proline was beneficial in the synthesis of chlorophyll, and increases in the contents of chlorophyll *a*, chlorophyll *b*, and total chlorophyll were observed with the increase in proline concentration. The increase in chlorophyll synthesis may be related to the fact that proline is an antioxidant that interacts with several enzymes, preserving and supporting the activity of proteins, as well as being involved in the regulation of genes that act in the biosynthesis of chlorophyll [38,39]. In addition, proline is closely linked to the increase in the synthesis of aminolevulinic acid, a precursor molecule of chlorophyll synthesis [40], and it participates as a constituent of several structuring proteins necessary for the synthesis and activation of chlorophyll [41], thus reducing the activity of the enzyme chlorophyllase, which acts on the degradation of photosynthetic pigments [42].

Similar results have been observed by Santos [16] in sour passion fruit irrigated with saline waters and subjected to proline concentrations, as foliar application of proline increased the synthesis of photosynthetic pigments, and by Zahedi [34] in grapevine plants, for which the combination of proline and graphene oxide promoted an increase in the synthesis of photosynthetic pigments.

In relation to the synthesis of carotenoids, the increase in water salinity and proline concentrations promoted an increase in their synthesis, which may be associated with the fact that carotenoids are responsible for the photoprotection of photosynthetic membranes, acting as an accessory pigment that helps in the dissipation of the excited state of chlorophyll and the neutralization of ROS [43]. Thus, foliar application of proline stimulates the activity of the antioxidant system against oxidative damage due to the ability to eliminate ROS from the cell [44], resulting in a higher carotenoid content in guava plants, even under conditions of salt stress.

The photochemical efficiency of guava plants was reduced by the increase in water salinity levels. However, an increase in F_0 was observed as a function of water salinity, which is indicative of damage to the photosynthetic apparatus of the plants, as it is related to the loss of photochemical energy by chlorophylls in the antennae complexes of the photosystems [45]. In addition, F_v and F_m were reduced by salt stress, which implies the correct functioning of the reaction centers of PSII, which reduce its maximum capacity due to the reduction of quinone (QA) by the electrons transferred from P680, thus indicating a reduction in the activity of chlorophyll *a* [46].

The effect of reduction in gas exchange and photochemical efficiency was reflected in the inhibition of the growth of guava plants under salt stress. This is a consequence of the restriction in the absorption of water and nutrients caused by the decrease in the osmotic potential that induces partial closure of the stomata, leading to a reduction in the production of photoassimilates, and, combined with the accumulation of Na^+ and Cl^-, resulting in the decrease in pectin crosslinking, possibly produced by calcium deficiency. As a result, there is a loss of the turgor pressure of the cell in cell expansion and division, which directly interferes with plant growth [47,48]. In addition, salt stress affects the synthesis of structural constituents of the cell wall, such as suberin, lignin, and polysaccharides, which are related to plant growth [49].

It is worth pointing out that under conditions of low salinity (0.8 dS m^{-1}), foliar application of proline at concentrations of 24 and 18 mM stimulated the growth in stem diameter and the absolute growth rate, which indicates the occurrence of a beneficial effect of proline on the growth of guava plants. The increase in growth may be related to the fundamental physiological and biochemical functions that proline exerts in plants, such as in the oxidative pathway of pentose phosphate, generating $NADP^+$ in the cytosol, in addition to being a primary component of some important proteins in plant growth [21].

The principal component analysis shows that the foliar application of proline provided beneficial effects on the synthesis of photosynthetic pigments and gas exchange in guava, especially under low-salinity conditions (0.8 dS m^{-1}). Still, according to the PCA, it is possible to highlight that the photochemical efficiency, represented by the variable and maximum fluorescence indices and the quantum yield of photosystem II, also benefited from the application of up to 16 mM of proline. It is worth noting that some authors classify guava as moderately tolerant to salinity [50], while others classify it as sensitive [12]. Thus, the foliar application of proline at the tested levels is an alternative option for the cultivation of guava cv. Paluma under conditions of moderate salinity, as our results demonstrated improvements in the physiological parameters evaluated.

4. Materials and Methods

4.1. Experimental Area Location and Characterization

The experiment was conducted from October 2022 to October 2023 under field conditions at the 'Rolando Enrique Rivas Castellón' Experimental Farm, belonging to the Center for Science and Agrifood Technology (CCTA) of the Federal University of Campina Grande (UFCG), located in the municipality of São Domingos, Paraíba, Brazil, located at the coordinates 06°48′50″ latitude (S) and 37°56′31″ longitude (W), at an altitude of 190 m. During the experimental period, relative humidity and temperature (maximum and minimum) data were collected through the weather station of São Gonçalo, as presented in Figure 11.

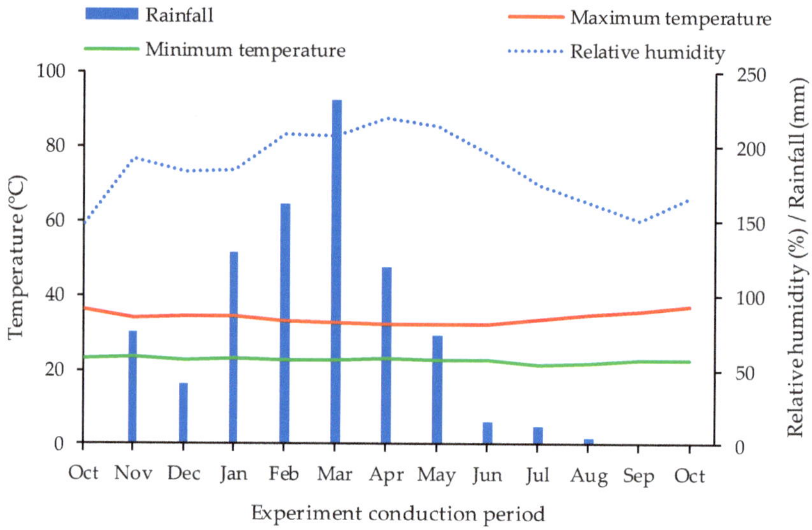

Figure 11. Maximum and minimum temperature and relative humidity data during the experimental period.

4.2. Experimental Design and Treatments

The experiment was set up in a randomized block design, with the treatments arranged in a 5 × 4 factorial scheme, referring to five levels of electrical conductivity of water (ECw) (0.8, 1.5, 2.2, 2.9, and 3.6 dS m^{-1}) and four concentrations of proline (0, 8, 16, and 24 mM) with three replicates, with the experimental plot consisting of one plant, totaling 60 experimental units. ECw levels were established based on a study conducted by Xavier [36] with guava cv. Paluma in the seedling formation stage. Proline concentrations were based on the results of the study conducted by Lima [51] with All Big bell pepper crop.

The guava cultivar Paluma was used in the present study, which stands out for its vigor and earliness, and whose propagation by cuttings results in flowering at 6 or 7 months of age after planting in the definitive location. The fruits are pyriform with smooth skin, and, when thinned, they can reach more than 500 g, especially in the first production cycles [52].

The seedlings were obtained from a commercial nursery accredited by the National Registry of Seeds and Seedlings located in the District of São Gonçalo, Sousa, PB, cultivated in black polyethylene bags with dimensions of 10 × 20 cm and a volumetric capacity of 0.5 L, and produced using the cutting technique.

4.3. Experiment Setup and Conduct

Guava plants were grown in plastic pots adapted as drainage lysimeters with 100 L capacity. Two holes were equidistantly made at the bottom of the lysimeters and connected to two 4 mM diameter plastic hoses, which were used for drainage and leaching of salts, and these drains were connected to 2 L plastic bottles. Inside of the lysimeters, a geotextile and a 0.5 kg layer of crushed stone were placed first, and then the lysimeters were filled with 110 kg of soil.

The soil used in the experiment was a *Neossolo Flúvico Ta Eutrófico típico* (Ultisol), with a loam texture, collected at a 0–30 cm depth from the experimental farm, belonging to the Center for Science and Agrifood Technology (CCTA) in São Domingo, PB. Chemical and physical attributes of the soil (Table 6) were determined according to the methodology of Teixeira [53].

Table 6. Chemical and physical attributes of the soil used during the experiment before the application of the treatments.

Chemical characteristics									
pH (H_2O) 1:2.5	OM dag kg^{-1}	P (mg kg^{-1})	K^+	Na^+	Ca^{2+}	Mg^{2+}	Al^{3+} + H^+	ESP (%)	ECse (dS m^{-1})
					($cmol_c$ kg^{-1})				
7.19	1.40	59.5	0.49	0.07	4.70	3.63	0	0.79	0.58
Physical characteristics									
Particle-size fraction (g kg^{-1})			Textural class	Moisture (kPa)		AW	Total porosity	BD	PD
				33.43 *	1519.8 **				
Sand	Silt	Clay						(kg dm^3)	
				....	dag kg^{-1}	....	%		
735.1	201.4	60.30	SL	15.78	6.41	9.37	55.05	1.20	2.67

OM—Organic Matter: Walkley–Black Wet Digestion; Ca^{2+} and Mg^{2+} extracted with 1 M KCl at pH 7.0; Na^+ and K^+ extracted with 1 M NH_4Oac at pH 7.0; Al^{3+} and H^+ extracted with 0.5 M CaOAc at pH 7.0; ESP—exchangeable sodium percentage; ECse—electrical conductivity of saturation extract; SL—sandy loam; AW—available water; BD—bulk density; PD—particle density; *—field capacity; **—wilting point.

The seedlings were transplanted to the lysimeters when the plants reached 150 days after propagation, a height of 32 cm, and a stem diameter of 7 mM. Soil moisture was raised to the level corresponding to field capacity using water with ECw of 0.8 dS m^{-1} up to 40 days after transplanting (DAT) to promote acclimation to the field conditions.

4.4. Preparation of Waters and Irrigation Management

The different levels of electrical conductivity of the irrigation water were prepared by diluting NaCl in water from the treatment of lowest salinity (0.8 dS m^{-1}), from the supply system of the experimental farm, from an artesian well, following the relationship between ECw and the concentration of salts [54] according to Equation (1).

$$Q \cong 10 \times \text{ECw} \quad (1)$$

where

Q = sum of cations ($mmol_c$ L^{-1});

ECw = electrical conductivity after deducting ECw of water from the supply system (dS m^{-1}).

After the acclimatization period of the plants in the lysimeters at 40 days after transplanting (DAT), irrigation began to be performed with the waters of the different salinity levels, and the volume of water was applied according to the water needs of the plants, determined by the water balance, according to Equation (2).

$$VI = \frac{(Va - Vd)}{(1 - LF)} \tag{2}$$

where

VI = volume of water to be used in the irrigation event (mL);
Va = volume of water applied in the previous irrigation event (mL);
Vd = volume of water drained (mL);
LF = leaching fraction (0.10) applied at 15-day intervals.

4.5. Preparation and Application of Proline Concentrations

Proline concentrations were prepared in each application event through dilution in distilled water, and 2 mL of Tween 80® adjuvant solution was added to facilitate the fixation of the solution on the leaves. Applications were carried out monthly using a manual sprayer (Lynus PL-2A) with an adjustable 1 cm conical metal nozzle at an operating pressure of 64 Psi from 5 p.m. due to the lower temperature. A physical structure of polypropylene plastic tarpaulin was used to prevent the drift from affecting plants of another treatment. In total, an average volume of 331.73 mL per plant of the solutions with the different concentrations of proline was applied.

4.6. Crop Management

Fertilization with nitrogen (N), phosphorus (P), and potassium (K) was carried out according to the recommendation of Cavalcanti [55], considering the nutritional requirements of the crop and the contents of the elements in the soil. The sources used were urea (45% N), potassium sulfate (50% K_2O), and monoammonium phosphate (50% P_2O_5 and 11% N). Fertilization started at 15 DAT and was split and applied via fertigation at 10-day intervals. Micronutrient fertilization was carried out weekly through the leaves, starting at 20 DAT, on the adaxial and abaxial surfaces, considering the nutritional requirements of the crop at the concentration of 1 g L^{-1} of Dripsol Micro® (1.2% magnesium, 0.85% boron, 3.4% iron, 4.2% zinc, 3.2% manganese, 0.5% copper, and 0.06% molybdenum).

Guava plants were grown with a single stem until they reached 50 cm in height, and the apical bud was eliminated to stimulate the emergence of lateral bud shoots. From the appearance of the lateral branches, three to four well-located branches with 30 cm lengths, symmetrically and spirally distributed, were left. Subsequently, these branches were pruned when they reached 60 cm in length in order to stimulate the sprouting of secondary branches and to control lateral growth to form the basic structure of the crown and promote adequacy of the plants to the orchard spacing. During pruning, unwanted and poorly located branches were removed, especially those that were directed to the soil and to the interior of the crown.

The plants were trained by staking with a vertical trellis system by directing the branches toward the rows and the interrows. Phytosanitary control was carried out by manually eliminating weeds that appeared on the soil of the lysimeters and by using a hoe in the rows and interrows.

Pest control was carried out preventively upon the appearance of green aphid (*Myzus persicae*) and yellow beetle (*Costalimaita ferruginea vulgata*) through chemical intervention using Actara® 250 WG (neonicotinoid–thiamethoxam) at a dose of 1 g for 10 L of water as recommended according to the package and Provado® 200 SC (neonicotinoid–imidacloprid) at a dose of 2.5 mL for 10 L of water as recommended according to the package, respectively.

4.7. Variables Analyzed

4.7.1. Relative Water Content, Electrolyte Leakage in the Leaf Blade, and Water Saturation Deficit

Analyses of relative water content, electrolyte leakage in the leaf blade, and water saturation deficit were performed at 360 days after transplanting. For the relative water content (RWC), fully expanded leaves were collected from the intermediate third of the branches, located in their terminal portion, and 8 leaf discs with 12 mM diameters were collected and their fresh mass (FM) was determined. Then, the samples were placed in plastic bags, immersed in distilled water, and stored for 24 h; after this period, excess water was removed with paper towels, and the turgid mass (TM) was obtained. Subsequently, the samples were dried in a forced-air circulation oven adjusted to a temperature of 65 °C until reaching constant mass to obtain the dry mass (DM). The samples were weighed on a 0.001 g precision semi-analytical scale.

RWC in the leaf blade was calculated according to Weatherley [56].

Electrolyte leakage in the leaf blade was determined by collecting eight leaf discs located in the middle third of the branch, in the terminal portion of the branch, and in the median region of the crown. After collection, the discs were immediately washed with distilled water in order to remove the contents of cells ruptured during collection and other electrolytes adhered to the leaves. After washing, the discs were dried on absorbent paper and placed in beakers containing 25 mL of distilled water at 25 °C for 24 h, where the initial electrical conductivity (Ci) was measured using an mCA 150 benchtop conductivity meter. Soon after, the beakers with the discs were placed in an incubator at 90 °C for 2 h, and then, with the temperature equilibrium, the final electrical conductivity (Cf) was measured. Electrolyte leakage (EL%) in the leaf blade was measured according to the methodology of Scotti-Campos [57].

Water saturation deficit is an indicator of the plant's water balance, as it represents the amount of water it needs to reach saturation. In this context, WSD was determined according to the methodology described by Taiz [58].

4.7.2. Gas Exchange

At 360 DAT, gas exchange was evaluated in the plants between 6:00 a.m. and 9:00 a.m. on the fourth fully expanded leaf from the apex to the base of the branch through determination of transpiration (E—mmol H_2O m^{-2} s^{-1}), stomatal conductance (gs—mol H_2O m^{-2} s^{-1}), CO_2 assimilation rate (A—µmol CO_2 m^{-2} s^{-1}), intracellular CO_2 concentration (Ci—µmol CO_2 m^{-2} s^{-1}), instantaneous carboxylation efficiency (CEi) (A/Ci) [(µmol CO_2 m^{-2} s^{-1}) (µmol CO_2 mol^{-1}]$^{-1}$, and instantaneous water use efficiency (WUEi) [(µmol CO_2 m^{-2} s^{-1}) (mol H_2O m^{-2} s^{-1})$^{-1}$].

Readings were performed using a portable infrared carbon dioxide analyzer (IRGA) and a model LCPro+ Portable Photosynthesis System® (ADC BioScientific Limited, United Kingdom) with irradiation of 1200 µmol photons m^{-2} s^{-1}, air flow of 200 mL min^{-1}, and atmospheric CO_2 concentration.

4.7.3. Photosynthetic Pigments

Contents of photosynthetic pigments were quantified at 306 DAT when leaf discs with an area of 3.14 cm^2 were collected, placed in test vials, and wrapped in aluminum foil for protection against light. Then, 5 mL of dimethyl sulfoxide was added per sample. The protocol used for extraction and quantification was that described by Cruz [59]. In the laboratory, the vials were kept in the dark until completing 48 h from the moment of collection, after which the extracts were passed to the quartz cuvette and thus analyzed using the spectrophotometer at absorbances of 665, 649, and 480 nm. Contents of chlorophyll a (Chl a), chlorophyll b (Chl b), total chlorophyll (Chl total), and carotenoids (CAR) were determined by following the methodology of Wellburn [60].

4.7.4. Photochemical Efficiency

Chlorophyll *a* fluorescence was measured through initial fluorescence (F_0), maximum fluorescence (F_m), variable fluorescence (F_v), and quantum yield of PSII (F_v/F_m) in leaves pre-adapted to the dark using leaf clips for 30 min in the median leaf from the intermediate productive branch of the plant so as to ensure that all primary acceptors were oxidized, i.e., the reaction centers were open, using an OS5p pulse-modulated fluorometer from Opti Science.

4.7.5. Growth

Growth was evaluated at 190 and 360 DAT based on stem diameter (SD), measured with a digital caliper at 3 cm height from the plant collar, crown diameter (D_{crown}), obtained through the average of crown diameter observed in the row direction (RD) and interrow direction (IRD), crown volume (V_{crown}), defined according to plant height (H), RD, and IRD, and the vegetative vigor index (VVI), obtained according to Portella [61].

The absolute (AGR_{SD}) and relative (RGR_{SD}) growth rates in stem diameter were determined according to the methodology of Benincasa [62].

4.8. Statistical Analysis

The data were analyzed for normality (Shapiro–Wilk test) and then subjected to analysis of variance (F test) at 0.05 and 0.01 probability levels; in cases of significance, linear and quadratic regression analysis was performed for the levels of electrical conductivity of irrigation water and proline concentrations using the statistical software SISVAR—ESAL version 5.7 [63]. To evaluate the correlations between physiological and growth parameters and the salinity levels of the irrigation water and proline concentrations, a principal component analysis (PCA) was performed.

5. Conclusions

Irrigation water salinity compromises the growth and development of guava plants, being one of the main problems faced by producers in the semi-arid region of northeast Brazil. In this study, exogenous application of proline proved to be a viable and low-cost alternative to reducing the deleterious effects of salt stress, thus promoting improvements in gas exchange, pigment synthesis, and growth of guava plants at 360 days after transplanting. However, further research is still needed in order to elucidate the biochemical effects that proline provides as an attenuator of the harmful effects of salt stress. This demonstrates that its use can become an important tool for producers in semi-arid regions, such as the Brazilian northeast.

Author Contributions: Conceptualization, L.A.d.A.S., G.S.d.L. and V.L.A.d.L.; methodology, S.T.d.A.S., R.A.F.T., L.d.A.S. and C.N.d.L.; software, I.A.R., J.T.A.F. and V.F.d.O.S.; validation, J.S.N., G.S.d.L. and L.A.d.A.S.; formal analysis, J.S.N. and S.S.d.S.; investigation, S.T.d.A.S., L.d.A.S., C.N.d.L., J.T.A.F. and M.V.D.; resources, G.S.d.L. and L.A.d.A.S.; data collection, I.A.R., R.A.F.T. and M.V.D.; writing—original draft preparation, J.S.N., S.S.d.S. and G.S.d.L.; writing—review and editing, G.S.d.L., J.S.N. and H.R.G.; supervision, H.R.G. and V.L.A.d.L.; project administration, G.S.d.L. and V.L.A.d.L. All authors have read and agreed to the published version of the manuscript.

Funding: Funding was acquired from the National Council for Scientific and Technological Development (CNPq), Coordination for the Improvement of Higher Education Personnel (CAPES).

Data Availability Statement: Data are contained within the article.

Acknowledgments: The authors thank the Graduate Program in Agricultural Engineering of the Federal University of Campina Grande for the support in carrying out this research.

Conflicts of Interest: The authors declare no conflicts of interest.

References

1. Diaz-Garcia, L.; Padilla-Ramírez, J.S. Development of single nucleotide polymorphism markers and genetic diversity in guava (*Psidium guajava* L.). *Plants People Planet* **2023**, *5*, 58–69. [CrossRef]
2. Arango, V.A.V.; Martínez, A.A.B.; Santos, L.E.O.; Zapata, J.L.H. Caracterización del epicarpio de guayaba (*Psidium guajava* L.) como alternativa natural para uso en productos alimenticios processados. *Biotecnol. Sect. Agropecu. Agroindustrial* **2020**, *18*, 26–36. [CrossRef]
3. IBGE. Instituto Brasileiro de Geografia e Estatística. Produção Agrícola—Lavoura Permanente, Safra—2022. Available online: https://cidades.ibge.gov.br/brasil/pesquisa/15/11954 (accessed on 18 January 2024).
4. Castro, J.M.C.; Ribeiro, J.M. *Pesquisa e Desenvolvimento para a Cultura da Goiabeira: A Contribuição da Embrapa Semiárido*, 1st ed.; Embrapa Semiárido: Petrolina, Brazil, 2020; 82p.
5. Melo, E.N.; Nobre, R.G.; Pinheiro, F.W.A.; Souza, L.P.; Lima, G.S.; Gheyi, H.R.; Elias, J.J.; Silva, W.L. Evaluation of West Indian cherry (*Malpighia emarginata*) rootstock under saline water irrigation and nitrogen fertilization. *Aust. J. Crop Sci.* **2018**, *12*, 1034–1040. [CrossRef]
6. Nunes, K.G.; Costa, R.N.T.; Cavalcante, I.N.; Gondim, R.S.; Lima, S.C.R.V.; Mateos, L. Groundwater resources for agricultural purposes in the Brazilian semi-arid region. *Rev. Bras. Eng. Agrícola Ambient.* **2022**, *26*, 915–923. [CrossRef]
7. Ondrasek, G.; Rathod, S.; Manohara, K.K.; Gireesh, C.; Anantha, M.S.; Sakhare, A.S.; Parmar, B.; Yadav, B.K.; Bandumula, N.; Raihan, F.; et al. Salt stress in plants and mitigation approaches. *Plants* **2022**, *11*, 717. [CrossRef]
8. Silva, E.M.; Gheyi, H.R.; Nobre, R.G.; Lima, G.S.; Soares, L.A.A.; Bonifácio, B.F. Saline waters and nitrogen/potassium fertilization combinations on physiological aspects and production of West Indian cherry. *Rev. Ambiente Água* **2021**, *16*, e2780. [CrossRef]
9. Nóbrega, J.S.; Gomes, V.R.; Soares, L.A.A.; Lima, G.S.; Silva, A.A.R.; Gheyi, H.R.; Torres, R.A.F.; Silva, F.J.L.; Silva, T.I.; Costa, F.B.; et al. Hydrogen peroxide alleviates salt stress effects on gas exchange, growth, and production of naturally colored cotton. *Plants* **2024**, *13*, 390. [CrossRef]
10. Negrão, S.; Schmöckel, S.M.; Tester, M. Evaluating physiological responses of plants to salinity stress. *Ann. Bot.* **2017**, *119*, 1–11. [CrossRef]
11. Kumar, N.; Haldarb, S.; Saikia, R. Root exudation as a strategy for plants to deal with salt stress: An updated review. *Environ. Exp. Bot.* **2023**, *216*, 105518. [CrossRef]
12. Ferreira, J.T.A.; Lima, G.S.; Silva, S.S.; Soares, L.A.A.; Fátima, R.T.; Nóbrega, J.S.; Gheiy, H.R.; Almeira, F.A.; Mendonça, A.J.T. Hydrogen peroxide in the induction of tolerance of guava seedlings to salt stress. *Semin. Ciências Agrárias* **2023**, *44*, 739–754. [CrossRef]
13. Bezerra, I.L.; Gheyi, H.R.; Nobre, R.G.; Barbosa, J.L.; Fátima, R.T.; Elias, J.J.; Azevedo, F.L. Physiological alterations and production of guava under water salinity and nitrogen fertilizer application. *Semin. Ciências Agrárias* **2018**, *39*, 1945–1956. [CrossRef]
14. Orsini, F.; Pennisi, G.; Mancarella, S.; Al-Nayef, M.; Sanoubar, R.; Nicola, S.; Gianquinto, G. Hydroponic lettuce yields are improved under salt stress by utilizing white plastic film and exogenous applications of proline. *Sci. Hortic.* **2018**, *233*, 283–293. [CrossRef]
15. Noreen, S.; Faiz, S.; Akhter, M.S.; Shah, K.H. Influence of foliar application of osmoprotectants to ameliorate salt stress in sunflower (*Helianthus annuus* L.). *Sarhad J. Agric.* **2019**, *35*, 1316–1325. [CrossRef]
16. Santos, L.F.S.; Lima, G.S.; Lima, V.L.A.; Silva, A.A.R.; Silva, A.A.R.; Nóbrega, J.S.; Capitulino, J.D.; Borborema, L.D.A.; Souza, A.R.; Caetano, E.J.M. Foliar application of proline on the mitigation of salt stress in the physiological indices of sour passion fruit. *Semin. Ciências Agrárias* **2023**, *44*, 1637–1660. [CrossRef]
17. Nounjan, N.; Chansongkrow, P.; Charoensawan, V.; Siangliw, J.L.; Toojinda, T.; Chadchawan, S.; Theerakulpisut, P. High performance of photosynthesis and osmotic adjustment are associated with salt tolerance ability in rice carrying drought tolerance QTL: Physiological and co-expression network analysis. *Front. Plant Sci.* **2018**, *9*, e1135. [CrossRef]
18. Mansour, M.M.F.; Ali, E.F. Glycinebetaine in saline conditions: An assessment of the current state of knowledge. *Acta Phys. Plant.* **2017**, *39*, 56. [CrossRef]
19. Veloso, L.L.S.A.; Nobre, R.G.; Souza, C.M.A.; Fátima, R.T.; Souza, L.P.; Elias, J.J.; Azevêdo, F.L.; Santos, J.B. Morphophysiology of guava cv. Paluma with water of different salt concentrations and proline doses. *Semin. Ciências Agrárias* **2018**, *39*, 1877–1886. [CrossRef]
20. Gohari, G.; Panahirad, S.; Sepehri, N.; Akbari, A.; Zahedi, S.M.; Jafari, H.; Dadpour, M.R.; Fotopoulos, V. Enhanced tolerance to salinity stress in grapevine plants through application of carbon quantum dots functionalized by proline. *Environ. Sci. Pollut. Res.* **2021**, *28*, 42877–42890. [CrossRef]
21. El-Moukhtari, A.; Cabassa-Hourton, C.; Farissi, M.; Savouré, A. How does proline treatment promote salt stress tolerance during crop plant development? *Front. Plant Sci.* **2020**, *11*, e1127. [CrossRef]
22. Silva, F.A.; Pereira, F.H.F.; Campos Júnior, J.E.; Nóbrega, J.S.; Dias, M.S. Aplicação foliar de prolina no crescimento e fisiologia do milho verde cultivado em solo salinizado. *Colloq. Agrar.* **2020**, *16*, 23–34. [CrossRef]
23. Lu, Y.; Fricke, W. Salt Stress—Regulation of root water uptake in a whole-plant and diurnal context. *Int. J. Mol. Sci.* **2023**, *24*, 8070. [CrossRef] [PubMed]
24. Lima, G.S.; Pinheiro, F.W.A.; Souza, W.B.B.; Soares, L.A.A.; Gheyi, H.R.; Nobre, R.G.; Queiroga, R.C.F.; Fernandes, P.D. Physiological indices of sour passion fruit under brackish water irrigation strategies and potassium fertilization. *Rev. Bras. Eng. Agrícola Ambient.* **2023**, *27*, 383–392. [CrossRef]

25. Jacinto Júnior, S.G.; Moraes, J.G.L.; Silva, F.D.B.; Silva, B.N.; Sousa, G.G.; Oliveira, L.L.B.; Mesquita, R.O. Respostas fisiológicas de genótipos de fava (*Phaseolus lunatus* L.) submetidas ao estresse hídrico cultivadas no Estado do Ceará. *Rev. Bras. Meteorol.* **2019**, *34*, 413–422. [CrossRef]
26. Sarabi, B.; Fresneau, C.; Ghaderi, N.; Bolandnazar, S.; Streb, P.; Badec, F.W.; Citerne, S.; Tangama, M.; David, A.; Ghashaghaie, J. Stomatal and non-stomatal limitations are responsible in down-regulation of photosynthesis in melon plants grown under the saline condition: Application of carbon isotope discrimination as a reliable proxy. *Plant Phys. Biochem.* **2019**, *141*, 1–19. [CrossRef] [PubMed]
27. Nobre, R.G.; Rodrigues Filho, R.A.; Lima, G.S.; Linhares, E.R.L.; Soares, L.A.A.; Silva, L.A.; Teixeira, A.D.S.; Macumbi, M.J.V. Gas exchange and photochemical efficiency of guava under saline water irrigation and nitrogen-potassium fertilization. *Rev. Bras. Eng. Agrícola Ambient.* **2023**, *27*, 429–437. [CrossRef]
28. Figueiredo, F.R.A.; Fátima, R.T.; Nóbrega, J.S.; Silva, T.I.; Ferreira, J.T.A.; Ribeiro, J.E.S.; Leal, M.P.S.; Soares, L.A.A.; Dias, T.J. Morphophysiology and gas exchange of pomegranate under salt stress and foliar application of nitrogen. *Rev. Bras. Eng. Agrícola Ambient.* **2023**, *27*, 958–965. [CrossRef]
29. Souza, L.P.; Lima, G.S.; Gheyi, H.R.; Fátima, R.T.; Silva, A.A.R.; Nobre, R.G.; Soares, L.A.A.; Lacerda, C.N. Morphophysiology and production of West Indian cherry under salt stress and nitrogen-phosphorus-potassium fertilization. *Rev. Bras. Eng. Agrícola Ambient.* **2023**, *27*, 539–549. [CrossRef]
30. Capitulino, J.D.; Lima, G.S.; Azevedo, C.A.V.; Silva, A.A.R.; Arruda, T.F.L.; Soares, L.A.A.; Gheyi, H.R.; Fernandes, P.D.; Farias, M.S.S.; Silva, F.A.; et al. Influence of foliar application of hydrogen peroxide on gas exchange, photochemical efficiency, and growth of soursop under salt stress. *Plants* **2023**, *12*, 599. [CrossRef]
31. Zheng, J.L.; Zhao, L.Y.; Wu, C.W.; Shen, B.; Zhu, A.Y. Exogenous proline reduces NaCl-induced damage by mediating ionic and osmotic adjustment and enhancing antioxidant defense in *Eurya emarginata*. *Acta Physiol. Plant.* **2015**, *37*, 181. [CrossRef]
32. El-Banna, M.F.; Masa, A. Exogenous application of proline mitigates deteriorative effects of salinity stress in NFT closed-loop system: An ultrastructural and physio-biochemical investigation on hydroponically grown tomato (*Solanum lycopersicon* L.). *Sci. Hortic* **2024**, *330*, 113061. [CrossRef]
33. Torres, R.A.F.; Lima, G.S.; Paiva, F.J.S.; Soares, L.A.A.; Silva, F.A.; Silva, L.A.; Oliveira, V.K.N.; Mendonça, A.J.T.; Roque, I.A.; Silva, S.T.A. Physiology and production of sugar-apple under water stress and application of proline. *Braz. J. Biol.* **2023**, *83*, e273404. [CrossRef] [PubMed]
34. Zahedi, S.M.; Abolhassani, M.; Hadian-Deljou, M.; Feyzi, H.; Akbary, A.; Rasouli, F.; Koçak, M.Z.; Kulak, M.; Gohari, G. Proline-functionalized graphene oxide nanoparticles (GO-Pro NPs): A new engineered nanoparticle to ameliorate salinity stress on grape (*Vitis vinifera* L. cv Sultana). *Plant Stress* **2023**, *7*, 100128. [CrossRef]
35. Shahverdi, M.A.; Omidi, H.; Tabatabaei, S.J. Stevia (*Stevia rebaudiana* Bertoni) responses to NaCl stress: Growth, photosynthetic pigments, diterpene glycosides and ion content in root and shoot. *J. Saudi Soc. Agric. Sci.* **2019**, *18*, 355–360. [CrossRef]
36. Xavier, A.V.O.; Lima, G.S.; Gheyi, H.R.; Silva, A.A.R.; Soares, L.A.A.; Lacerda, C.N. Gas exchange, growth and quality of guava seedlings under salt stress and salicylic acid. *Rev. Ambiente Água* **2022**, *17*, e2816. [CrossRef]
37. Soares, L.A.A.; Oliveira, S.G.; Lima, G.S.; Fernandes, P.D.; Araújo, R.H.C.R.; Fernandes, E.A. Physiological changes of pomegranate seedlings under salt stress and nitrogen fertilization. *Rev. Bras. Eng. Agrícola Ambient.* **2021**, *25*, 453–459. [CrossRef]
38. El-Betalgi, H.S.; Mohamed, H.I.; Sofy, M.R. Role of ascorbic acid, glutathione and proline applied as singly or in sequence combination in improving chickpea plant through physiological change and antioxidant defense under different levels of irrigation intervals. *Molecules* **2020**, *25*, 1702. [CrossRef]
39. El-Shawa, G.M.R.; Rashwan, E.M.; Abdelaal, K.A.A. Mitigating salt stress effects by exogenous application of proline and yeast extract on morphophysiological, biochemical and anatomical characters of calendula plants. *Sci. J. Flowers Ornam. Plants* **2020**, *7*, 461–482. [CrossRef]
40. Merwad, A.R.M.; Desoky, E.S.M.; Rady, M.M. Response of water deficit stressed *Vigna unguiculata* performances to silicon, proline or methionine foliar application. *Sci. Hortic.* **2018**, *228*, 132–144. [CrossRef]
41. Monteiro, J.G.; Cruz, F.J.R.; Nardin, M.B.; Santos, D.M.M. Crescimento e conteúdo de prolina em plântulas de guandu submetidas a estresse osmótico e à putrescina exógena. *Pesqui. Agropecu. Bras.* **2014**, *49*, 18–25. [CrossRef]
42. Butt, M.; Sattar, A.; Abbas, T.; Sher, A.; Ijaz, M.; Ul-Allah, S.; Shaheen, M.R.; Kaleem, F. Foliage applied proline induces salt tolerance in chili genotypes by regulating photosynthetic attributes, ionic homeostasis, and antioxidant defense mechanisms. *Hortic. Environ. Biotechnol.* **2020**, *61*, 693–702. [CrossRef]
43. Stefanov, M.; Yotsova, E.; Gesheva, M.; Dimitrava, D.; Maskovska, L.; Dancheva, S.; Apostolova, E.L. Role of flavonoids and proline in the protection of photosynthetic apparatus in Paulownia under salt stress. *S. Afr. J. Bot.* **2021**, *139*, 246–253. [CrossRef]
44. Menegazzo, R.F.; Rickli, M.E.; Scanavacca, J.; Lima, J.D.; Silveira, A.C.; Menegazzo, A.W.; Souza, S.G.H.; Lopes, A.D. Resposta diferencial a diferentes classes de herbicidas: *Tradescantia pallida* (Rose) D. R. Hunt var. *purpurea* Boom como planta modelo. *Res. Soc. Dev.* **2021**, *10*, e6910111452. [CrossRef]
45. Veloso, L.L.S.A.; Capitulino, J.D.; Lima, G.S.; Azevedo, C.A.V.; Silva, A.A.R.; Gheyi, H.R. Application methods of hydrogen peroxide in soursop seedlings irrigated with saline water. *Comun. Sci.* **2021**, *12*, e3288. [CrossRef]
46. Faseela, P.; Sinisha, A.K.; Brestič, M.; Puthur, J.T. Chlorophyll a fluorescence parameters as indicators of a particular abiotic stress in rice. *Photosynthetica* **2020**, *58*, 293–300. [CrossRef]

47. Lacerda, C.N.; Lima, G.S.; Soares, L.A.A.; Fátima, R.T.; Gheyi, H.R.; Azevedo, C.A. Morphophysiology and production of guava as a function of water salinity and salicylic acid. *Rev. Bras. Eng. Agrícola Ambient.* **2022**, *26*, 451–458. [CrossRef]
48. Queiroz, G.C.M.; Medeiros, J.F.; Silva, R.R.; Morais, F.M.S.; Sousa, L.V.; Souza, M.V.P.; Santos, E.N.; Ferreira, F.N.; Silva, J.M.C.; Clemente, M.I.B.; et al. Growth, solute accumulation, and ion distribution in sweet sorghum under salt and drought stresses in a brazilian potiguar semiarid area. *Agriculture* **2023**, *13*, 803. [CrossRef]
49. Byrt, C.S.; Munns, R.; Burton, R.A.; Gilliham, M.; Wege, S. Root cell wall solutions for crop plants in saline soils. *Plant Sci.* **2018**, *269*, 47–55. [CrossRef]
50. Abrar, M.M.; Sohail, M.; Saqib, M.; Akthar, J.; Abbas, G.; Wahad, H.A.; Mumtaz, M.Z.; Mehmood, K.; Memon, M.S.; Sun, N.; et al. Interactive salinity and water stress severely reduced the growth, stress tolerance, and physiological responses of guava (*Psidium guajava* L.). *Sci. Rep.* **2022**, *12*, 18952. [CrossRef]
51. Lima, G.S.; Santos, J.B.; Soares, L.A.A.; Gheyi, H.R.; Nobre, R.G. Irrigação com águas salinas e aplicação de prolina foliar em cultivo de pimentão 'All Big'. *Comun. Sci.* **2016**, *7*, 513–522. [CrossRef]
52. Pommer, C.V.; Murakami, K.R.N.; Watlington, F. Goiaba no mundo. *Agronômico* **2006**, *58*, 22–26.
53. Teixeira, P.C.; Donagema, G.K.; Fontana, A.; Teixeira, W.G. *Manual de Métodos de Análise de Solo*, 3rd ed.; EMBRAPA: Brasília, Brazil, 2017; 573p.
54. Richards, L.A. *Diagnosis and Improvement of Saline and Alkali Soils*; Department of Agriculture: Washington, DC, USA, 1954; 160p.
55. Cavalcanti, F.D.A.; Santos, J.C.P.; Pereira, J.R.; Leite, J.P.; Silva, M.C.L.; Freira, F.J.; Silva, D.J.; Sousa, A.R.; Messias, A.S.; Farias, C.M.B.; et al. *Recomendações de Adubação para o Estado de Pernambuco: 2. Aproximação*, 3rd ed.; Instituto Agronômico de Pernambuco-IPA: Recife, Brazil, 2008; 212p.
56. Weatherley, P.E. Studies in the water relations of the cotton plant. I. The field measurements of water deficits in leaves. *New Phytol.* **1950**, *49*, 81–97. [CrossRef]
57. Scotti-Campos, P.; Pham-Thi, A.T.; Semedo, J.N.; Pais, I.P.; Ramalho, J.C.; Matos, M.C. Physiological responses and membrane integrity in three *Vigna* genotypes with contrasting drought tolerance. *Emir. J. Food Agric.* **2013**, *25*, 1002–1013. [CrossRef]
58. Taiz, L.; Zeiger, E.; Møller, I.M.; Murphy, A. *Fisiologia e Desenvolvimento Vegetal*; Artmed: Porto Alegre, Brazil, 2017; 888p.
59. Cruz, A.C.F.; Santos, R.P.; Iarema, L.; Fernandes, K.R.G.; Kuki, K.N.; Araújo, R.F.; Otoni, W.C. Métodos comparativos na extração de pigmentos foliares de três híbridos de *Bixa orellana* L. *Rev. Bras. Biociências* **2007**, *5*, 777–779.
60. Wellburn, A.R. The spectral determination of chlorophylls a and b, as well as total carotenoids, using various solvents with spectrophotometers of different resolution. *J. Plant Physiol.* **1994**, *144*, 307–313. [CrossRef]
61. Portella, C.R., Marinho, C.S., Amaral, B.D.; Carvalho, W.S.G.; Campos, G.S.; Silva, M.P.S.; Sousa, M.C. Desempenho de cultivares de citros enxertados sobre o trifoliateiro 'Flying Dragon' e limoeiro 'Cravo' em fase de formação do pomar. *Bragantia* **2016**, *75*, 70–75. [CrossRef]
62. Benincasa, M.M.P. *Análise de Crescimento de Plantas, Noções Básicas*, 2nd ed.; FUNEP: Jaboticabal, Brazil, 2003; 41p.
63. Ferreira, D.F. Sisvar: A computer analysis system to fixed effects split plot type designs. *Braz. J. Biom.* **2019**, *37*, 529–535. [CrossRef]

plants MDPI

Article

Exogenous Sodium Nitroprusside Affects the Redox System of Wheat Roots Differentially Regulating the Activity of Antioxidant Enzymes under Short-Time Osmotic Stress

Alsu Lubyanova * and Chulpan Allagulova

Institute of Biochemistry and Genetics-Subdivision of the Ufa Federal Research Centre of the Russian Academy of Sciences, Prospect Oktyabrya 71, lit.1e, 450054 Ufa, Russia; allagulova-chulpan@rambler.ru
* Correspondence: lubyanova555@mail.ru; Tel.: +7-987-0974257

Abstract: Nitric oxide (NO) is a multifunctional signalling molecule involved in the regulation of plant ontogenesis and adaptation to different adverse environmental factors, in particular to osmotic stress. Understanding NO-induced plant protection is important for the improvement of plant stress tolerance and crop productivity under global climate changes. The root system is crucial for plant survival in a changeable environment. Damages that it experiences under water deficit conditions during the initial developmental periods seriously affect the viability of the plants. This work was devoted to the comparative analysis of the pretreatment of wheat seedlings through the root system with NO donor sodium nitroprusside (SNP) for 24 h on various parameters of redox homeostasis under exposure to osmotic stress (PEG 6000, 12%) over 0.5–24 h. The active and exhausted solutions of SNP, termed as (SNP/+NO) and (SNP/−NO), respectively, were used in this work at a concentration of 2×10^{-4} M. Using biochemistry and light microscopy methods, it has been revealed that osmotic stress caused oxidative damages and the disruption of membrane cell structures in wheat roots. PEG exposure increased the production of superoxide ($O_2^{\bullet-}$), hydrogen peroxide (H_2O_2), malondialdehyde (MDA), and the levels of electrolyte leakage (EL) and lipid peroxidation (LPO). Stress treatment enhanced the activities of superoxide dismutase (SOD), ascorbate peroxidase (APX), catalase (CAT), the excretion of proline, and the rate of cell death and inhibited their division. Pretreatment with (SNP/+NO) decreased PEG-induced root damages by differently regulating the antioxidant enzymes under stress conditions. Thus, (SNP/+NO) pretreatment led to SOD, APX, and CAT inhibition during the first 4 h of stress and stimulated their activity after 24 h of PEG exposure when compared to SNP-untreated or (SNP/−NO)-pretreated and stress-subjected plants. Osmotic stress triggered the intense excretion of proline by roots into the external medium. Pretreatment with (SNP/+NO) in contrast with (SNP/−NO) additionally increased stress-induced proline excretion. Our results indicate that NO is able to mitigate the destructive effects of osmotic stress on the roots of wheat seedlings. However, the mechanisms of NO protective action may be different at certain periods of stress exposure.

Keywords: catalase; electrolyte leakage; hydrogen peroxide; malondialdehyde; ascorbate peroxidase; polyethylene glycol; proline; superoxide dismutase; superoxide anion; *Triticum aestivum* L.

Citation: Lubyanova, A.; Allagulova, C. Exogenous Sodium Nitroprusside Affects the Redox System of Wheat Roots Differentially Regulating the Activity of Antioxidant Enzymes under Short-Time Osmotic Stress. *Plants* **2024**, *13*, 1895. https://doi.org/10.3390/plants13141895

Academic Editors: Geovani Soares de Lima, Lauriane Almeida dos Anjos Soares and Francisco Vanies Da Silva Sá

Received: 4 June 2024
Revised: 28 June 2024
Accepted: 29 June 2024
Published: 9 July 2024

1. Introduction

Plants are periodically affected by unfavourable environmental factors, which have recently increased significantly due to global climate change and human economic activity. Drought, salinity, extreme temperatures, excess of fertilizers, and heavy metals are the most widespread negative factors, all leading to water regime disturbances initiating osmotic and oxidative stress. Water deficit becomes especially damaging in early ontogeny, during germination and seedling emergence, inhibiting further plant growth and ultimately crop productivity. The root system is crucial for plant survival in a changeable environment, providing water and nutrient uptake, anchorage to the substrate, and signal transduction to the shoots or surrounding medium [1]. Damages to a root system under water deficit conditions at the initial periods of vegetation can seriously affect plant viability, increasing the death rate dramatically. The reactions of roots to external factors may be different than in aerial parts of the plants [2–5]. The information related to the cellular and molecular mechanisms of roots development and their stress tolerance is mainly based on the studies of dicot plant species. However, many crops, including wheat (*Triticum aestivum* L.), which is the world's major cereal, are classified as monocots. Therefore, the study of root reactions in wheat plants at the early stages of development under osmotic stress has a fundamental importance.

There are different adaptive mechanisms of plant survival under water deficit, among which the most common one is osmotic adjustment, which involves a diversity of osmoprotective compounds such as polyamines, sugars, and amino acids, including proline (Pro) [6]. Proline also has antioxidant properties, contributing to plant defence reactions against the damaging effects of reactive oxygen species (ROS) such as superoxide anion radical ($O_2^{\bullet-}$), hydrogen peroxide (H_2O_2), hydroxyl radicals ($^{\bullet}OH$), and singlet oxygen ($^1O_2$) [7,8]. Stress-induced oxidative burst leads to the injury of biomolecules including nucleic acids, lipids, or proteins, causing the degradation of cellular structures [9,10]. Membrane damages are coupled with an increase in malondialdehyde (MDA) accumulation and enhanced electrolyte leakage (EL). ROS neutralization and the maintenance of redox homeostasis occur due the action of the plant's antioxidant system (AOS), composed of non-enzymatic components and antioxidant enzymes like superoxide dismutases (SODs), catalases (CATs), and peroxidases (PODs), in particular, ascorbate peroxidase (APX), etc. SOD catalyses the transformation of toxic $O_2^{\bullet-}$ to less toxic H_2O_2, which is ultimately converted to O_2 and H_2O through POD or CAT activity [9,11,12].

Plant defence reactions and the establishment of their stress resistance are regulated by a complex signalling system through interaction with molecular messengers, among which nitric oxide (NO) has received significant attention [10]. NO is a gaseous redox-active signalling molecule which fulfils multiple regulatory functions at all stages of the plant life cycle, including seed germination, vegetative growth, root morphogenesis, flowering, and fruit ripening [13,14]. It plays an important role in the adaptation of plants to numerous stresses, in particular, water deficit. The stress-induced increase in endogenous NO levels is well known, as is the enhancement of plant stress tolerance by the exogenous application of NO in gaseous form or as donors [15]. However, excessive amounts are toxic and can induce nitro-oxidative stress in tissues [13,16]. Therefore, practical use requires scrupulous study regarding the mechanisms of NO action and those of its active derivatives—reactive nitrogen species (RNS). Their biological activity is realized through interaction with various mediators (glutathione, ROS, polyunsaturated fatty acids), with subsequent signal transduction to its own molecular targets [10,14]. For example, when NO interacts with $O_2^{\bullet-}$, peroxynitrite ($NOOO^-$) is formed, which is characterized by less toxic properties than $O_2^{\bullet-}$. Thus, $NOOO^-$ formation can be recognized as a direct antioxidant effect of NO [10]. In addition, $NOOO^-$ acts as a nitrating agent in reactions of tyrosine nitration in NO-dependent post-translational modifications of proteins (NO-PTMs). Besides tyrosine nitration, other types of NO-PTMs are S-nitrosylation and metal nitrosylation, all leading to the structural reorganization of target proteins, thus changing their functional activity [14]. For instance, NO-induced S-nitrosylation of APX1 at cysteine-32 enhances its enzymatic activity in *Arabidopsis* [17]. Potential targets of S-nitrosylation are the protein molecules SOD and CAT, so their enzymatic activity can be regulated by NO-PTMs [18–20], which could contribute significantly to the NO-dependent regulation of the redox homeostasis of plants.

Experimental studies of NO functional activity are commonly conducted using different donors, in particular, sodium nitroprusside (SNP), which is an unstable photosensitive compound with a half-life time of about 12 h [10,13]. Therefore, the assessment of its effects must be performed with parallel additional control, such as exhausted SNP [21]. This work was devoted to a comparative analysis of the effects of active and exhausted SNP, termed as (SNP/+NO) and (SNP/−NO), respectively [21], on various parameters of redox homeostasis in roots of wheat seedlings subjected to osmotic stress. $O_2^{\bullet-}$, H_2O_2 production, antioxidant enzyme activity (APX, SOD, and CAT), MDA accumulation, EL, the level of cell viability, mitotic activity, and the integrity of cellular structures were quantified in the roots of SNP-pretreated and stress-subjected seedlings.

2. Results

2.1. $O_2^{\bullet-}$ Production and SOD Activity in Roots of Wheat Seedlings Pretreated with SNP and Subjected to Osmotic Stress for 4 h

The production of $O_2^{\bullet-}$ by roots was evaluated from its accumulation both in the roots and in the growth medium of SNP-pretreated and osmotic-stress-subjected wheat seedlings. The (SNP/+NO) pretreatment of 5-day-old seedlings had no effect on $O_2^{\bullet-}$ accumulation in roots and reduced its level in the growth medium by 8–29% relative to the control (Figure 1a,b). Osmotic stress induced an enhancement of $O_2^{\bullet-}$ levels in the root tissues as well as in the growth medium. $O_2^{\bullet-}$ content increased in roots by 26–67% relative to the control, with a maximum observed at 1 h of stress exposure (Figure 1a). The most pronounced increase in $O_2^{\bullet-}$ accumulation in the growth medium, exceeding the control by more than two-fold, was also observed after 1 h of stress treatment (Figure 1b). Then, the content of $O_2^{\bullet-}$ in the medium gradually decreased but still elevated in the control in the next 2–4 h of the experiment (Figure 1b). (SNP/+NO) pretreatment decreased, but did not prevent, stress-induced $O_2^{\bullet-}$ production by seedlings. The levels of $O_2^{\bullet-}$ accumulation under (SNP/+NO) pretreatment followed by 4 h of stress exposure exceeded the control by 7–53% in the roots and by 30–60% in the growth medium.

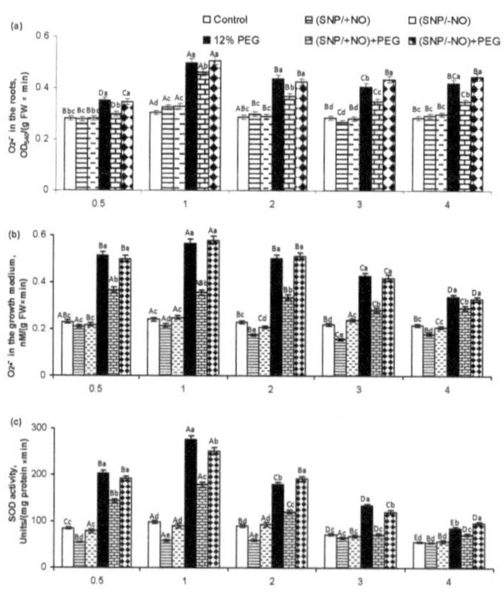

Figure 1. Effect of 2×10^{-4} M sodium nitroprusside (SNP) pretreatment of 5-day-old wheat seedlings for 24 h on superoxide ($O_2^{\bullet-}$) production in roots (**a**), $O_2^{\bullet-}$ accumulation in the growth medium (**b**), and superoxide dismutase (SOD) activity in the roots (**c**) of 6-day-old seedlings subjected to short-term (4 h) osmotic stress, caused by 12% polyethylene glycol 6000 (PEG). (SNP/+NO)—active SNP; (SNP/−NO)—exhausted (light-inactivated) SNP. Data are given as mean values and their standard errors from three biological and four analytical repeats. Lowercase letters above the columns indicate significant differences between variants of treatment at the particular time point at $p < 0.05$ (ANOVA, LSD test). Capital letters indicate significant differences for the same treatment at different time points at $p < 0.05$ (ANOVA, LSD test).

SOD activity in wheat roots after the first 2 h of (SNP/+NO) pretreatment was 33–40% lower than the control, which then gradually increased, reaching control values at 4 h (Figure 1c). Stress treatment caused a sharp increase in this enzyme's activity. A maximum 2.4- to 2.8-fold stimulation of SOD activity was observed in roots after 0.5–1 h of stress exposure, which then gradually decreased. (SNP/+NO) pretreatment decreased the level of stress-induced enzyme activation. In this case, the roots of (SNP/+NO)-pretreated and stress-subjected seedlings differed from control by just a 1.7- to 1.9-fold increase in the SOD activity. No changes were detected in the roots and in the medium for $O_2^{\bullet-}$ accumulation and also for root's SOD activity under pretreatment with a 2×10^{-4} M solution of exhausted NO donor (SNP/−NO) both in normal and stressful conditions.

2.2. H_2O_2 Production and Activity of APX and CAT in Roots of Wheat Seedlings Pretreated with SNP and Subjected to Osmotic Stress for 4 h

The production of hydrogen peroxide by roots was also analysed by its endogenous content in roots and accumulation in the external medium of SNP-pretreated seedlings subjected to PEG-induced osmotic stress, and it had a similar trend with $O_2^{\bullet-}$ (Figure 2).

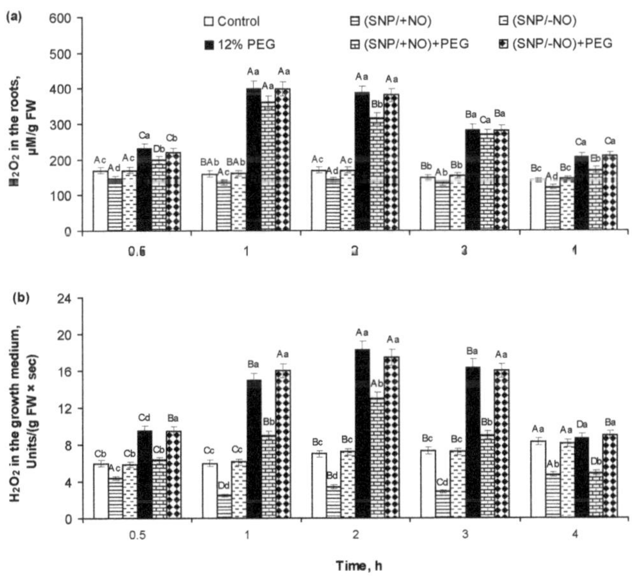

Figure 2. Effect of SNP pretreatment of wheat seedlings on hydrogen peroxide (H_2O_2) accumulation in roots (**a**) and in the growth medium (**b**) of 6-day-old seedlings subjected to short-term (0.5–4 h) osmotic stress, caused by PEG exposure. Data are given as mean values and their standard errors from three experimental and four analytical repeats. Lowercase letters above the columns indicate significant differences between variants of treatment at the particular time point at $p < 0.05$ (ANOVA, LSD test). Capital letters indicate significant differences for the same treatment at different time points at $p < 0.05$ (ANOVA, LSD test).

(SNP/+NO) pretreatment reduced root endogenous levels of H_2O_2 by 10–15% (Figure 2a) and its excretion to the medium by 30–60% (Figure 2b) as compared to control. Osmotic stress resulted in a rapid 1.5–2.5-fold increase in H_2O_2 content in roots and in the growth medium during the first 4 h of PEG exposure. The maximum H_2O_2 accumulation occurred at 1 h of 12% PEG influence. (SNP/+NO) pretreatment decreased the stress-induced accumulation of H_2O_2 both in the roots and in the growth medium. H_2O_2 content increased 1.2–2.3-fold in the root tissues and 1.1–1.9-fold in the growth medium over the control level. The maximum levels of H_2O_2 accumulation were detected in roots after 1–3 h and in the growth medium after 1–2 h of stress influence. The (SNP/-NO) pretreatment showed no effect of active SNP on the root H_2O_2 production and excretion under 12% PEG exposure (Figure 2a,b).

Roots of (SNP/+NO)-pretreated seedlings were characterized by no significant changes in APX activity and by 17–23% reduction in CAT activity as compared to control (Figure 3). The activity of these related to the regulation of H_2O_2 balance enzymes was changed by osmotic stress. APX activity in the roots of PEG-subjected seedlings was 1.6–1.9-fold higher than that in the control. The maximum APX activity occurred at 1–3 h of stress influence (Figure 3a). CAT activity increased 2−6-fold during 4 h of PEG exposure. The peak of CAT activity in roots took place at 3 h of stress impact (Figure 3b). The roots of (SNP/+NO)-pretreated seedlings were characterized by the reduction in the stress-induced activation of APX and CAT relative to NO-untreated samples, which were still higher than the control. (SNP/+NO)-pretreated samples exposed to PEG for 4 h differed from the control by a 21–29% increase in APX activity and a 2.5- to 3-fold increase in CAT activity. In contrast, (SNP/−NO) pretreatment had no effect on APX and CAT activity in wheat roots under normal and stressful conditions (Figure 3).

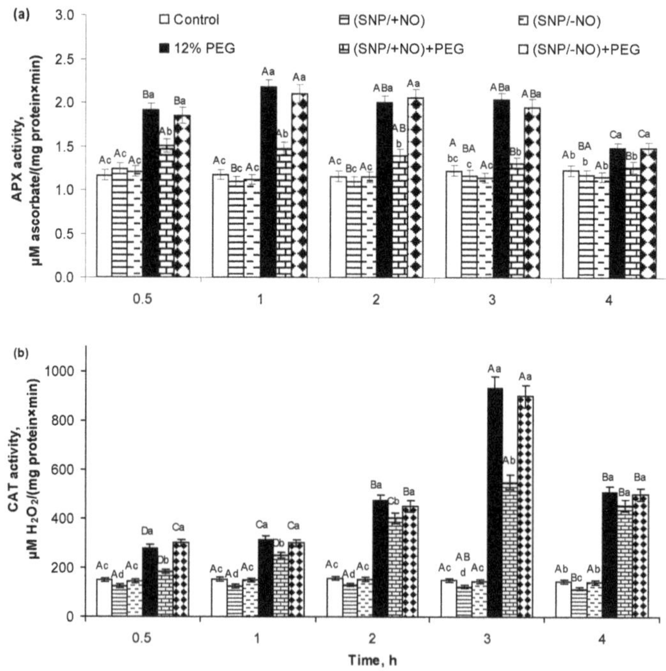

Figure 3. Effect of SNP pretreatment of wheat seedlings on activities of ascorbate peroxidase (APX) (**a**) and catalase (CAT) (**b**) in roots of 6-day-old seedlings subjected to short-term (0.5–4 h) osmotic stress, caused by 12% PEG exposure. Data are given as mean values and their standard errors from three experimental and four analytical repeats. Lowercase letters above the columns indicate significant differences between variants of treatment at the particular time point at $p < 0.05$ (ANOVA, LSD test). Capital letters indicate significant differences for the same treatment at different time points at $p < 0.05$ (ANOVA, LSD test).

2.3. *ROS Content and Antioxidant Enzymes Activity after 24 h of 12% PEG Exposure*

Control and (SNP/+NO)- and (SNP/−NO)-pretreated wheat roots 24 h later after seedlings' treatment did not differ significantly in the $O_2^{\bullet-}$ and H_2O_2 content under non-stressful conditions (Table 1). Exposure to 12% PEG for 24 h resulted in the enhancement of

$O_2^{\bullet-}$ production in wheat roots by 29% as compared to control. (SNP/+NO) pretreatment almost completely prevented PEG-induced $O_2^{\bullet-}$ accumulation in the roots of seedlings subjected to stress during 24 h. The H_2O_2 content in the roots of SNP-untreated and (SNP/-NO)-pretreated plants increased more than 1.5-fold over the control under 24 h of stress exposure. The (SNP/+NO) pretreatment markedly reduced but did not prevent stress-induced H_2O_2 production. It was found that in the roots of (SNP/+NO)-pretreated seedlings exposed to PEG for 24 h, the H_2O_2 production was higher by 13.4% than in the control plants and lower by about 27% relative to NO-untreated and stress-subjected samples.

Table 1. Effect of SNP pretreatment (2×10^{-4} M, 24 h) on the content of $O_2^{\bullet-}$ and H_2O_2 and the activity of SOD, APX, and CAT in the roots of 7-day-old wheat seedlings subjected to osmotic stress (12% PEG) over 24 h.

Treatment	Root $O_2^{\bullet-}$ Content, OD_{540}/(g FW × min)	Root H_2O_2 Content, μM/g FW	SOD Activity, Units/(mg Protein × min)	APX Activity, μM Ascorbate/(mg Protein × min)	CAT Activity, μM H_2O_2/(mg Protein × min)
Control	0.34 ± 0.017 b	153.85 ± 7.60 c	95.9 ± 4.5 d	1.31 ± 0.06 c	121.8 ± 5.9 d
(SNP/+NO)	0.33 ±0.019 b	154.85 ± 7.31 c	107.7 ± 5.2 c	1.23 ± 0.05 c	140.2 ± 6.7 c
(SNP/−NO)	0.34 ± 0.017 b	150.00 ± 7.40 c	92.3 ± 4.1 d	1.25 ± 0.06 c	118.2 ± 5.1 d
12% PEG	0.44 ± 0.020 a	240.00 ± 11.12 a	236.2 ± 11.1 b	1.71 ± 0.08 b	273.0 ± 11.6 b
(SNP/+NO) + PEG	0.37 ± 0.016 b	175.00 ± 8.52 b	292.2 ± 13.9 a	2.14 ± 0.10 a	322.8 ± 14.4 a
(SNP/−NO) + PEG	0.42 ± 0.021 a	235.06 ± 11.56 a	240.0 ± 11.6 b	1.74 ± 0.08 b	268.0 ± 14.3 b

Values are the mean of three replicates ± SE (n = 12). Different letters show significant difference at $p < 0.05$.

An increase in SOD and CAT activity by approximately 12% over the control was detected in the roots 24 h later after (SNP/+NO) pretreatment, while no significant changes in APX activity were detected. The stress exposure for 24 h caused a 2.5-fold increase in SOD activity, 1.95-fold increase in CAT activity, and the enhancement of APX activity by only 31% as compared to control. The roots of (SNP/+NO)-pretreated seedlings, in contrast to untreated ones, were characterized by the additional PEG-induced stimulation of SOD, APX, and CAT activities under 24 h of stress exposure. (SNP/−NO) pretreatment showed no effects of the active SNP on the studied antioxidant enzymes under control and stressful conditions.

2.4. Proline Accumulation in the Growth Medium

The amino acid proline belongs to low-molecular-weight osmolytes, and it is actively synthesized by plants in the response to water deficit conditions. It is characterized by a wide range of protective functions, including osmoprotective and antioxidant activity. Taking into account that exogenous pretreatment with active SNP reduced the level of oxidative stress in PEG-exposed wheat seedlings, it can be expected that Pro is an important component of NO-induced defence reactions. Therefore, we analysed proline excretion by wheat roots, which was determined by its accumulation in the growth medium of 7-day-old seedlings pretreated with (SNP/+NO) or (SNP/−NO) for 24 h and subjected to 12% PEG for 24 h.

No significant changes in Pro content were found under the influence of SNP; its concentrations in the growth medium of (SNP/+NO)-pretreated, (SNP/−NO)-pretreated, and control samples were approximately the same and ranged from 0.42 to 0.44 pM/mg (Figure 4). Osmotic stress tremendously increased Pro production by roots, causing more than 50-fold enhancement of its excretion to the growth medium, where its concentrations reached about 23.8 pM/mg. (SNP/+NO) pretreatment additionally stimulated stress-induced Pro production by roots, increasing its medium concentration by about 1.8-fold relative to SNP-untreated and stress-subjected samples, while (SNP/−NO) pretreatment had no such an effect. The level of Pro accumulation in the growth medium of (SNP/−NO)-pretreated and stress-subjected plants was the same as under 12% PEG exposure only.

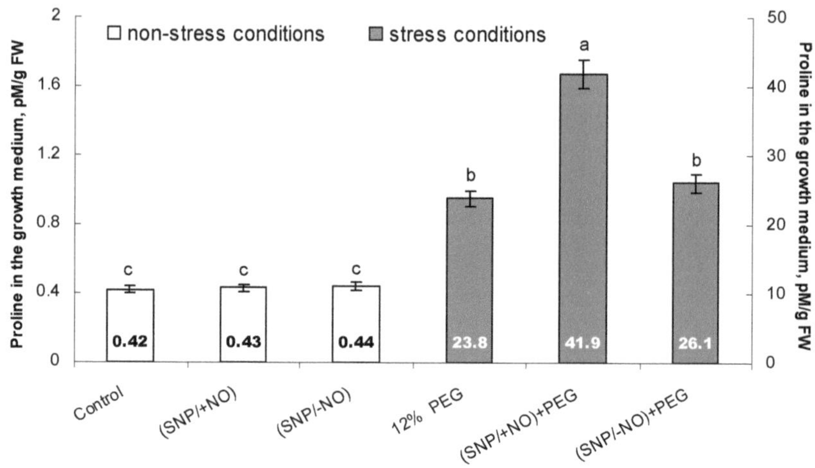

Figure 4. Effect of 2×10^{-4} M SNP pretreatment on proline (Pro) accumulation in the growth medium of wheat seedlings subjected to long-term (24 h) osmotic stress, caused by 12% PEG exposure. Data are given as mean values and their standard errors from three experimental and four analytical repeats. Distinct letters imply that the values differ significantly at $p < 0.05$ (ANOVA, LSD test).

2.5. The Levels of Malondialdehyde Accumulation and Electrolytes Leakage

Malondialdehyde (MDA) and electrolyte leakage (EL) were measured for the determination of lipid peroxidation (LPO) level and membrane structures integrity. Pretreatment with (SNP/+NO) or (SNP/−NO) had no significant effect on any of these parameters in the roots of 7-day-old non-stressed seedlings as compared to control (Figure 5a,b). Osmotic stress (PEG, 12%) during 24 h caused severe damage to membrane structures, which was evidenced by a 68% increase in MDA content and a 4-fold increase in EL level over the control. Exogenous (SNP/+NO) application had a protective effect on membrane structures of the roots of wheat seedlings under osmotic stress. (SNP/+NO)-pretreated samples were characterized by a decrease in stress-induced MDA accumulation and EL level, which exceeded control values by no more than 25% and 62%, respectively. The exhausted NO-donor had no effect on MDA content and EL level in the wheat roots under stress conditions. No significant differences in these two parameters were observed between samples pretreated or untreated with (SNP/−NO) and subjected to 12% PEG (Figure 5a,b).

2.6. Histochemical Detection of Lipid Peroxidation

Quantitative data regarding MDA accumulation and EL levels were confirmed during the histochemical studies using light microscopy and Schiff's reagent, which stains aldehydes produced by lipid peroxidation (LPO) (Figure 5c). Histochemical analysis revealed that LPO predominantly localized in the root tips and central cylinder. The baseline level of LPO was detected in the tissues of control roots. Their tips, which are the zones of active proliferation and cell expansion, as well as the central cylinder, were stained by a purple colour (Figure 5c). Wheat exposure to 12% PEG for 24 h significantly increased the intensity of staining of all plant tissues, indicating the stress-triggered enhancement of LPO in roots in addition to the physiological one. (SNP/+NO) pretreatment decreased lipid peroxidation in the roots of wheat seedlings under osmotic stress as compared with SNP-untreated and stressed plants, while (SNP/−NO) had no such an effect (Figure 5c).

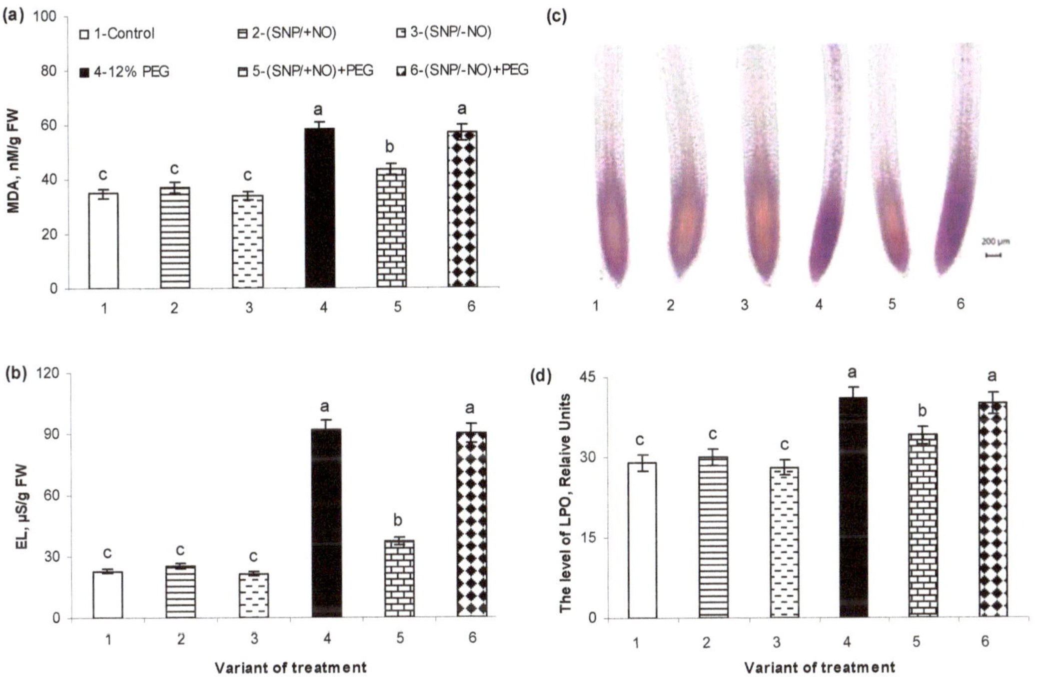

Figure 5. Effect of 2×10^{-4} M SNP pretreatment on cell structures integrity in roots of wheat seedlings subjected to long-term (24 h) osmotic stress, caused by PEG exposure. Accumulation of malondialdehyde (MDA) (**a**), level of electrolyte leakage (EL) (**b**), histochemical detection of lipid peroxidation (**c**), relative quantification of the lipid peroxidation (**d**). Image J software (NIH, Bethesda, MD, USA) was used to calculate relative level of peroxidation of lipids by integrated density. Data are the mean ± SE (n = 10). Distinct letters above the columns imply that the values differ significantly at $p < 0.05$ (ANOVA, LSD test).

Ten representative images were then quantified using Image J software to identify differences in the staining intensity in root tissues of wheat seedlings pretreated or untreated with (SNP/+NO) and subjected to osmotic stress (Figure 5d). The intensity and area of staining were almost identical in the roots of control and (SNP/+NO)- or (SNP/−NO)-treated wheat roots under non-stress conditions. Under 12% PEG exposure the LPO level increased by 35% over the control. (SNP/+NO)-pretreated samples differed from NO-untreated ones by a significantly lower degree of stress-induced LPO, which exceeded the control values by no more than 11%. Under osmotic stress, (SNP/−NO)-pretreated samples were characterized by practically the same LPO level as SNP-untreated ones (Figure 5d).

2.7. The Percent of Cell Death and MI of Meristematic Cells of Wheat Roots

The level of cell death in roots of (SNP/−NO)-pretreated seedlings was not significantly different from control under normal growth conditions (Figure 6a). (SNP/+NO) application reduced the cell death rate by about 8% and 20% relative to the control 4 h and 24 h after of SNP pretreatment, respectively. Osmotic stress had a strong negative effect on the cell survival, increasing their death rate by 44% and 80% over the control after 4 h and 24 h of stress exposure, respectively. An active NO donor increased the viability of wheat cells under stress conditions. (SNP/+NO) pretreatment prevented PEG-induced cell death in roots at 4 h of stress exposure and decreased its level by about 22% at 24 h in comparison to SNP-untreated and stress-subjected samples. (SNP/−NO) pretreatment caused no changes in the stress-induced cell death rate.

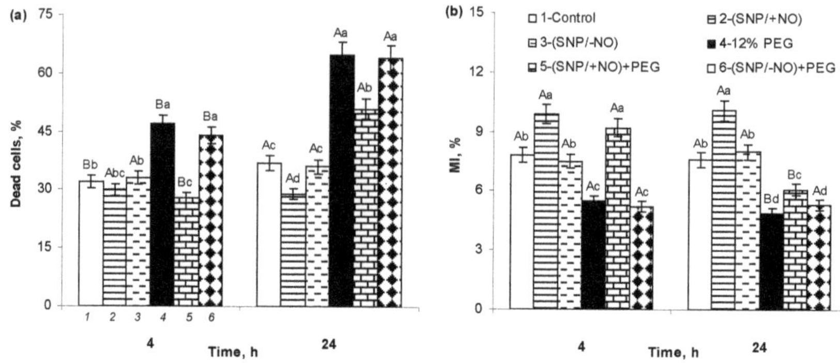

Figure 6. Effect of 2×10^{-4} M SNP pretreatment on the cells death percent (**a**) and mitotic index (MI) of meristematic cells (**b**) in roots of wheat seedlings subjected to short-term (4 h) and long-term (24 h) osmotic stress, caused by 12% PEG exposure. The numbers at the bottom of the columns indicate the variant of treatment. The number of roots in cell death assay: $n = 10$; the cells number in MI assay: $n = 2000$. Distinct lowercase letters above the columns indicate significant differences between variants of treatment at the certain time point at $p < 0.05$ (ANOVA, LSD test). Capital letters indicate significant differences for the same treatment at distinct time points at $p < 0.05$ (ANOVA, LSD test).

An important indicator of the cells' vital activity is the process of their division, which can be assessed using light microscopy by determining in the root apical meristem the parameter of mitotic index (MI), expressed as the percentage of dividing cells to the total number of analysed cells. MI ranged from 7.6 to 7.8% in the roots of 7-day-old control seedlings (Figure 6b). (SNP/+NO) pretreatment stimulated cell division as indicated by a 2.5% increase in MI over the control. Osmotic stress decreased the values of MI by 2.1% and 2.9% at 4 h and 24 h of PEG exposure, respectively, as compared to control. (SNP/+NO) pretreatment prevented the reduction in MI caused by 4 h of stress influence and significantly mitigated its inhibitory effect at 24 h of PEG exposure. Moreover, at the initial stages of stress, the (SNP/+NO)-stimulating effect on MI was still maintained in the root apical meristem, while exhausted SNP solution had no effect on the mitotic activity of the root's cells either under non-challenging or stressful conditions.

2.8. Correlation Matrices

The analysis of correlation matrices revealed that during 12% PEG exposure for 0.5–2 h, almost all the studied indicators had a positive correlation with coefficients more than 0.9 (Figure 7a–c).

During the third hour of stress, the indicators could be grouped into two positive correlation clusters, with coefficients more than 0.9 and 0.773–0.893 (Figure 7d). The parameter of H_2O_2 in wheat roots showed declined correlation with APX activity, H_2O_2 in the growth medium, and $O_2^{\bullet-}$ levels both in roots and in the medium. The parameter of H_2O_2 in roots had the weakest correlation with SOD activity during 3 h of stress exposure of wheat seedlings. H_2O_2 level in the roots demonstrated the reduction in its correlation with $O_2^{\bullet-}$ generation and utilization.

The correlation coefficients between H_2O_2 accumulation in the growth medium and H_2O_2 content and SOD activity in wheat roots as well as $O_2^{\bullet-}$ levels both in roots and in the medium decreased about two-fold during 4 h of PEG exposure (Figure 7e). After 4 h of stress influence, the correlation coefficients between the percent of root cell death and the indicators, illustrating redox status, decreased and reached the values of 0.636–0.895. CAT activity showed the weakest correlation with the percent of dead cells of wheat roots, while APX activity demonstrated the highest correlation with this parameter. Such data indicate the time-varying involvement of studied antioxidant enzymes in the processes of plant survival under osmotic stress. During the stress exposure of wheat seedlings for 4 h and 24 h, the parameter of MI showed a negative correlation with all other investigated indicators (Figure 7e,f). These data point to a switch from active growth to stress adaptation under PEG exposure. It should be noted that under normal conditions, active growth and adaptation processes are also antagonistic.

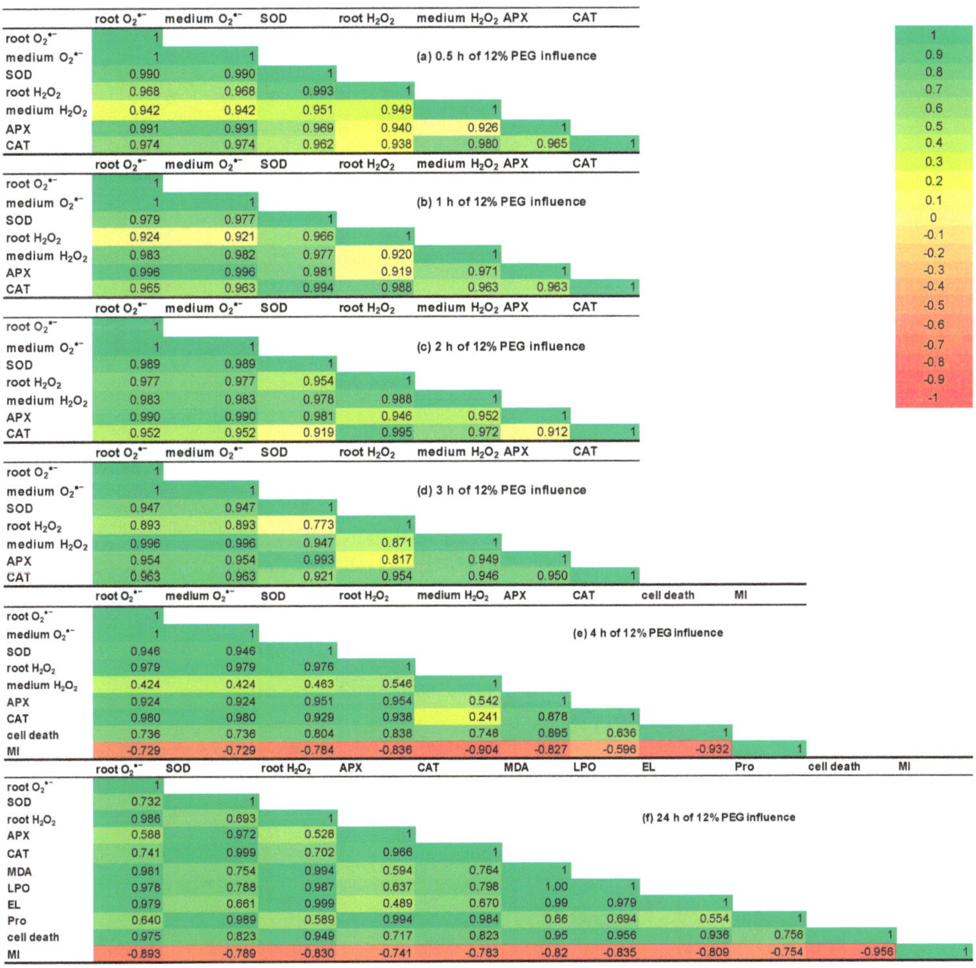

Figure 7. The correlation coefficient matrices for all examined parameters for 6–7-day-old wheat seedlings subjected to SNP pretreatment and osmotic stress, caused by 12% PEG. The matrices are coloured by magnitude and parity (green positive and red negative) for correlations at $p < 0.05$ significance level.

After 24 h of stress exposure, the most investigated parameters had changeable but positive correlation coefficients, except MI (Figure 7f). The levels of $O_2^{\bullet-}$ and H_2O_2 in roots had a significant correlation not only between each other but also with MDA, EL, the percent of cell death, the level of LPO. The correlation between activities of antioxidant enzymes and MDA, EL, the percent of cell death, and the level of LPO declined and reached the less positive values 0.498–0.823. The correlation analysis data additionally confirm the stress-induced oxidative burst and its negative consequences on the plant cells. Pro in the growth medium has significant correlation coefficients with SOD, APX, and CAT activities. The correlations between Pro in the growth medium and ROS generation, MDA, EL, LPO level, and the percent of cell death were weaker (0.554–0.694) as compared to Pro's relationship with the activity of the redox enzymes, indicating that Pro may act as an antioxidant.

3. Discussion

NO is an endogenous signalling molecule with multiple functions in the regulation of the plant's development and its stress responses. Abiotic stresses have a negative impact on the water regime of plants, leading to osmotic and oxidative stress. This, in turn, disrupts the metabolic activity of the cells, limiting the growth and productivity of crop plants, including wheat, the world's major crop [15,21]. The root system of the plants is in direct contact with osmotic stress-inducing agents, and it determines the viability of the whole plant. NO plays a pivotal role in the formation of root architecture, modulating the growth of the primary roots, lateral and adventitious roots, and root hair development [22–25]. NO biosynthesis is increased in roots under abiotic stress conditions, and this can have a dual effect depending on the production level. NO can act as a signalling molecule at low concentrations, or it can be toxic at high concentrations and can provoke nitro-oxidative stress [24,25]. There is considerable evidence demonstrating the alleviation of the negative effects of different stresses via the exogenous treatment of plants with NO donors in proper concentrations through the root system, indicating their potential practical application to improve crop growth and productivity [15,21]. It has been previously shown that SNP pretreatment has stimulatory and protective effects on the growth of shoots and roots of wheat seedlings subjected to salinity or PEG-induced dehydration [15,26]. The beneficial effects of NO treatment were associated with its influence on the hormonal system, water regime, and also with the gene expression of dehydrins, which are well-known plant protective proteins [26]. In this study, we investigated the effect of the 24 h pretreatment of 5-day-old wheat seedlings (*Triticum aestivum* L.) through the root system with the active NO-donor (SNP/+NO) on various parameters of the redox homeostasis, as well as on the production of proline, the integrity of the membrane structures, and the mitotic activity and viability of the root cells under PEG-induced osmotic stress. Experiments were conducted using an exhausted SNP solution, (SNP/−NO), which served as an additional control of particular effects of NO.

The analysis of ROS content in the roots and in the growth medium revealed that the wheat seedlings produce $O_2^{\bullet-}$ and H_2O_2 under control conditions at the basic levels, which were maintained at about the same values during the whole experiment (Figures 1a,b and 2a,b; Table 1). In addition, the roots of the control plants excreted proline (Figure 4). These data are quite understandable, since it is well-known that the plant root systems secrete various biologically active compounds into the medium: sugars, organic acids, amino acids and fatty acids, sterols, enzymes, flavonoids, growth regulators, inorganic ions, RNS including NO, ROS, etc. [1,18,27]. Roots' exudates have a wide physiological significance. For example, NO and ROS can be involved in the plants' interactions with each other or with symbiotic microorganisms, as well as in the regulation of water and nutrient uptake. ROS production during normal metabolic processes occurs constantly. They are generated in cell walls and membrane structures, including plasmalemma, chloroplasts, mitochondria, peroxisomes, and in other cellular compartments, during 1-, 2-, or 3-electron reactions of oxygen reduction and in photoinduced reactions, and they are involved in the regulation of various metabolic processes [20,28,29]. In particular, H_2O_2 is capable of oxidating the thiol groups in protein molecules of enzymes or transcription factors, modulating the rate of the biochemical reactions or the expression of the corresponding genes. Apoplastic H_2O_2, hydroxyl, and superoxide radicals participate in the mediation of important plant processes, such as root cell elongation controlled by indole-3-acetic acid [30,31]. The baseline level of ROS production depends on the plant developmental stage, circadian clock, environmental and physiological conditions, and interactions with its root and leaf microbiomes, and it is under the influence of phytohormones, nutrient and water availability [1,32].

PEG-induced osmotic stress caused a rapid and reversible increase in root $O_2^{\bullet-}$ and H_2O_2 production, as indicated by changes in their contents in the wheat tissues and growing medium (Figure 1a,b and Figure 2a,b). These results are consistent with the literature data. An increase in ROS production has been recorded in the different plant species exposed to adverse factors, inducing water stress in a huge number of studies [4,6,33–36]. Under such conditions, they can be both markers of the stress state of plants and signalling mediators in the adaptive reactions [9]. ROS can serve as short distance signals in plant tissues, acting along membranes between different organelles and/or neighbouring cells [32]. It is known that apoplast and cytoplasm interconnection plasmodesmata are involved in transducing RBOH-mediated cell-to-cell ROS and redox signals in plants [36]. This process is autopropagating and capable of transferring stress-induced ROS and redox signals from cell to cell over long distances [32]. The PEG-induced production of $O_2^{\bullet-}$ and H_2O_2 led to the oxidative stress in the roots of wheat seedlings, as indicated by data on the increasing levels of MDA and EL (Figure 5a,b), damage to the cell membrane structures, and the enhancement of LPO (Figure 5c,d). Moreover, stress treatment has a strong negative effect on the viability of the root cells and their division (Figure 6a,b). The enlargement of stress exposure from 4 to 24 h significantly increased the death rate of the cells, notably decreasing their mitotic activity (Figure 6a,b).

ROS neutralization and plant redox homeostasis are ensured by AOS involving multiple antioxidant enzymes, in particular, SOD, CAT, and APX [37]. A significant increase in SOD activity was detected in the roots after 30 min of PEG exposure and reached a maximum after 1 h of stress, which then gradually declined but was significantly higher than the control after 4 h of the experience (Figure 1c). Then, SOD activity increased again after 24 h of stress exposure, exceeding the control by approximately 2.5-fold (Table 1). In this case, there was a direct connection between the levels of $O_2^{\bullet-}$ content both in the roots and medium and the change in SOD activity in roots (Figure 1a–c; Table 1). An increase in the H_2O_2 content in the roots and medium under osmotic stress was in parallel with the activation of APX and especially CAT (Figure 2, Table 1). It is known that drought stress can activate plant enzymes POD and CAT differentially [18]. Literature data noticed that APX has higher affinity for H_2O_2 and is located in more diverse subcellular compartments than CAT, which works only in plant's peroxisomes [20]. It is necessary to take into account that plants respond to stress-triggered oxidative burst not only through the stimulation of antioxidant enzyme's activities but also through the up-regulation of non-enzymatic antioxidants [27,29,38,39].

It seems promising that the exogenous application of NO donors, including SNP, could increase plant stress resistance and crop productivity [24,25]. However, SNP is an unstable compound decomposing with the release of iron and cyanides, which can have a toxic effect on the plants, negating the positive effects of NO [40]. Meanwhile, compared to other NO donors, the relatively low cost and beneficial effects of its exogenous application were convincingly demonstrated in different plant species. Pretreatment with active or exhausted SNP solutions at a concentration of 2×10^{-4} M did not lead to oxidative damage in the wheat roots under control conditions, as was indicated by data on the production of $O_2^{\bullet-}$, H_2O_2 (Figures 1a,b and 2a,b; Table 1), MDA (Figure 5a), EL level (Figure 5b), histochemical analysis of integrity of cell membrane structures, evaluation of LPO (Figure 5c,d), measurement of MI, and cell viability (Figure 6a,b). Moreover, pretreatment with an active NO donor (SNP/+NO) promoted wheat root growth, as indicated by MI data (Figure 6a), whereas an exhausted NO donor (SNP/−NO) had no such effect. These data are consistent with the previously obtained data on root growth stimulation in different plant species through treatment with SNP in optimal concentrations [15,26,41]. The incubation of the root tips of 3-day-old maize seedlings in the presence of various NO donors, including SNP, stimulated their elongation [41,42]. Wheat seeds' pretreatment with SNP or its presence in the germination medium at the concentrations of 50–200 μM promoted the subsequent increase in the linear sizes of the shoots and roots of 4–7-day-old seedlings [15,26]. NO-deficient mutants of *Arabidopsis thaliana* showed the delayed elongation of primary roots and were characterized by a poorly developed root meristem with abnormal cell division pointing to the pivotal role of NO in the root development [25]. On the other hand, it has been reported that NO is able to suppress root growth, as demonstrated in studies with NO-overproducing mutants or with exogenous NO treatments, indicating its concentration-dependent role in the regulation of root growth [25]. The exogenous NO treatment of tomato or cucumber plants inhibited the growth of the primary roots, although the growth of the lateral and adventitious roots was stimulated. This was accompanied by an increase in the branching of their root systems [22,23,43].

(SNP/+NO) pretreatment decreased the stress-induced accumulation of $O_2^{\bullet -}$ and H_2O_2 in the roots and in the medium both under short-term (4 h) (Figures 1a,b and 2a,b) and long-term (24 h) PEG exposure (Table 1), indicating the mitigation of oxidative stress in the wheat seedlings. At the same time, analysing the activity of antioxidant enzymes, we found that (SNP/+NO) pretreatment decreased the PEG-induced activation of SOD, APX, and CAT in the first 0.5–4 h of stress influence (Figures 1c and 3a,b), whereas (SNP/+NO) pretreatment, conversely, resulted in the additional increase in SOD, APX, and CAT activities in the samples subjected to prolonged (24 h) osmotic stress (Table 1). Usually, the decline in ROS level is associated with the activation of endogenous antioxidant enzymes [28,29,38,39]. For example, SNP treatment increased the activity of antioxidant enzymes and upregulated the expression of APX genes in wheat plants under heat stress [34]. Spraying the soybean plants with 100 μM SNP reduced the levels of H_2O_2 accumulation and LPO and contributed to the additional activation of SOD, CAT, and APX in leaves under PEG-induced drought [6]. However, the SNP treatment of crambe plants decreased not only $O_2^{\bullet -}$, H_2O_2, and MDA accumulation but also SOD, CAT, glutathione reductase, and APX activities under water deficit conditions [44]. The decrease in PEG-induced $O_2^{\bullet -}$ and H_2O_2 production by seedling roots under the influence of (SNP/+NO) during the initial periods of stress (Figures 1a,b and 2a,b) may be due to direct ROS interaction with the NO, leading to $NOOO^-$ formation, which in turn may affect the catalytic activity of antioxidant enzymes [27]. It was shown that NO-triggered peroxynitrite accumulation in *A. thaliana* differentially inhibited the mitochondrial Mg-SOD, peroxisomal Cu/Zn-SOD, and chloroplastic Fe-SOD, which was due to tyrosine nitration [45]. Furthermore, SNP-induced ROS accumulation might be limited by the inhibition of the NADPH oxidase activity [27].

After 24 h of PEG exposure, (SNP/+NO)-pretreated samples exhibited a significant decrease in stress-induced H_2O_2 production and an almost complete prevention of $O_2^{\bullet -}$ production, which was accompanied by an additional increase in the stress-induced SOD, APX, and CAT activity (Table 1). Obviously, during the first hours of stress exposure, antioxidant enzymes were not so active in reducing ROS generation (Figures 1 and 3) as later in stress exposure (Table 1). Plants have both short-term and long-term responses to water deficit [46], and the regulation of SOD, APX, and CAT activity may be the result of distinct factors acting simultaneously. There is evidence that the additional activation of antioxidant enzymes induced by NO donors in wheat plants under salinity requires the coordinated action of abscisic acid (ABA), H_2S, and NO [29,47–49]. The obtained results can indicate that NO may be involved in the regulation of antioxidant enzymes activities depending on the duration of osmotic stress. $O_2^{\bullet -}$ or H_2O_2 were quickly scavenged in tissues under the initial stage of osmotic stress, and the activity of redox enzymes may not be the main factor in ROS regulation during the first hours of stress exposure (Figures 1–3). Meanwhile, at later periods of stress, the ROS neutralization in (SNP/+NO)-pretreated wheat plants can be mediated by the stimulation of antioxidant enzyme activity (Table 1).

Various adverse factors, associated with osmotic stress, induce the accumulation of Pro in plant tissues [21,28,44,50–52]. This heterocyclic amino acid, which is characterized by osmoprotective, chaperone, and antioxidant properties, significantly contributes to the preservation of cellular molecular structures under conditions of water stress [7,8]. We found that a 24 h exposure to osmotic stress caused a nearly 50-fold increase in proline content in the growth medium of wheat seedlings, indicating the very strong magnification of its excretion by roots (Figure 4). Pretreatment with an active or exhausted SNP solution had no significant effect on proline production by the roots under non-challenging growth conditions. Meanwhile, (SNP/+NO)-pretreated samples differed from NO-untreated or (SNP/−NO)-pretreated samples by an additional almost 2-fold increase in stress-induced proline excretion into the medium (Figure 4). Such a significant increase in the proline content in the seedlings' surrounding medium can play an especially important role in maintaining the hydration layer and structural integrity of roots in juvenile developmental stages under water deficit and osmotic stress conditions. These results are in good agreement with the literature data. There is evidence that NO donors usually have no significant effect on proline synthesis by plants under non-stressful conditions [53,54]. Meanwhile, NO pretreatment can enhance proline synthesis under water stress, as has been shown in *Lycopersicon esculentum*, *Ginkgo biloba*, and *Glycine max* [6,53]. The constitutive expression of the mammalian neuronal NO synthase gene (nNOS) in *Oryza sativa* led to an increase in proline accumulation induced by drought or salinity [55]. There is a suggestion that NO-induced proline accumulation probably has a stress-specific character and depends on the stress intensity. Additional studies are required to establish the role of NO in the regulation of proline synthesis, as well as its participation in the NO-dependent protection of plants under osmotic stress [53,54].

High MDA level is a marker of lipid peroxidation and its measurement is widely used for the detection of oxidative stress in plants. The significant increase in MDA production and EL level was observed under PEG exposure, indicating the severe damage of membrane cell structures in roots (Figure 5a,b). Pretreatment with (SNP/+NO) or (SNP/−NO) had no significant influence on the levels of MDA and EL under normal growth conditions, confirming the absence of the destructive effect of an NO donor in the used concentrations. (SNP/+NO)-pretreated samples differed from (SNP/−NO)-pretreated and from NO-untreated samples through a reduction in stress-induced MDA accumulation and EL rate, suggesting the NO-mediated mitigation of oxidative stress in wheat seedlings and also confirming the antioxidant activity of NO.

The SNP-dependent decrease in the activity of antioxidant enzymes during the initial periods of osmotic stress, which was revealed in this study (Figures 1c and 3a,b), did not lead to subsequent oxidative stress in the wheat seedlings. On the contrary, SNP pretreatment contributed to the protection of the cells from oxidative damages, as evidenced by data on MDA content and EL rate (Figure 5a,b), which is consistent with the literature [28,56]. The histochemical analysis of the integrity of cell membrane structures using Schiff's reagent revealed areas of ROS accumulation in the tips and central cylinder of control roots (Figure 5c). These data could be evidence of ROS involvement in the regulation of physiological processes such as cell's division and expansion, as well as the formation of xylem and phloem vessels. Pretreatment with active or exhausted SNPs had no effect on staining with the Schiff reagent in comparison with control, which is consistent with the results of MDA and EL quantification, confirming the absence of a negative effect of SNP on the cell structures (Figure 5a,b). The action of osmotic stress led to the intense staining of the root tips and central cylinder, indicating severe damage to the cellular structures in the roots of wheat seedlings. (SNP/+NO) pretreatment reduced the area and intensity of stained zones in the roots of stress-subjected seedlings, whereas (SNP/−NO) had no such effect. Thus, the data of histochemical analysis provide further evidence of NO participation in the regulation of antioxidant defence in plants.

It is known that an oxidative burst, triggered by environmental stressors, can affect fundamental processes such as the cell cycle, cell division, and expansion. It can also initiate signalling towards cell death [30]. Cellular transitions from phase-to-phase are regulated by complex mechanisms in which the interaction between cyclins (Cyc) and cyclin-dependent kinases (CDK) plays an important role [57]. Redox perturbations affect both the activities and transcript levels of CYCs and CDKs through a specific transcription factor, TEOSINTE BRANCHED1-CYCLOIDEA-PROLIFERATING CELL FACTOR1 (TCP) [58,59]. Under oxidative stress, the interaction between a TCP and its promoter might be inhibited due to the formation of covalently linked TCP dimers via disulfide bonds, which in turn affects cell division [58,59]. It has been found in this work that osmotic stress increased the rate of cell death in roots (Figure 6a) and decreased the percentage of dividing cells in the tips, as indicated by MI data (Figure 6b). This may lead to the inhibition of growth and can ultimately reduce the stress resistance and productivity of wheat plants.

It should be noted that cell death occurs during normal plant ontogenesis, which is also evidenced by our findings. A baseline level of cell death ranging from 33% to 40% was found in the control samples (Figure 6a). Programmed cell death is necessary for the selective elimination of redundant or damaged cells, which is essential for maintaining tissue homeostasis. At the same time, cell death plays an important role in various developmental processes, including the formation of xylem and phloem vessels, as well as parenchyma and mechanical tissues [60]. SNP-pretreated samples showed approximately the same percentage of cell death as the control plants under non-challenging conditions, and a reduction in its stress-induced levels, indicating the ability of NO to participate in the maintenance of cell viability (Figure 6a). The presence of 12% PEG in the growth medium significantly reduced the values of MI, suggesting the inhibition of mitotic activity in the wheat roots (Figure 6b). An active NO donor stimulated cell division in the roots, as evidenced by the increase in MI values at 4 h and 24 h after (SNP/+NO) pretreatment. It is worth noting that the stimulating effect of (SNP/+NO) was retained even under short-term 4 h stress exposure. Furthermore, under prolonged 24 h stress, (SNP/+NO) pretreatment noticeably mitigated the stress-induced inhibition of cell division (Figure 6b). Pretreatment with exhausted donor NO had no significant effect on cell death and the mitotic activity of roots either in non-challenging or stressful conditions (Figure 6a,b). The stimulating and protective effect of (SNP/+NO) on the cell division and mitotic activity of wheat roots may be due to its influence on the gene activity linked with the cell cycle. In particular, the SNP treatment of tomato root segments stimulated the expression of cyclin genes, *CYCD3;1* and *CYCA2;1* [43], which may play a role in the NO-mediated regulation of root growth under both normal and stressful conditions.

4. Materials and Methods

4.1. Plant Material and Treatments

Winter wheat (*Triticum aestivum* L.) of cultivar Scepter was used as the object of the study. The seeds were obtained from the Chishminsky Breeding Station in Bashkortostan, Russia. The seeds were surface-sterilized with 96% ethanol and thoroughly washed with tap water. Then they were sown on Petri dishes containing filter paper soaked in 10 mL of 10% Hoagland–Arnon solution and were grown under illumination of 200 mmol $m^{-2}s^{-1}$ at 16 h photoperiod and ambient temperature of 22–24 °C for 3 days. Twenty 3-day-old plants were transferred into glass jars, containing 25 mL of nutrient solution. Regular changing of the solution (every day) was sufficient. The effects of NO were analysed using sodium nitroprusside (SNP), a donor of NO, at the concentration of 2×10^{-4} M, which, as previously shown, has a stimulating and protective effect on young wheat seedlings [15,26]. The experiments were conducted with the parallel application of active (SNP/+NO) and exhausted (SNP/−NO) solution, which lacks the NO activity and thus can serve as an additional control of active SNP. (SNP/−NO) solution was obtained by illuminating the stock (SNP/+NO) solution for 1 day at elevated temperatures. (SNP/−NO) treatment did not stimulate plant growth, which means it contained nitric oxide in quantities insufficient for a physiological effect on plants.

The 5-day-old seedlings were SNP-pretreated through the roots for 24 h via supplementation with (SNP/+NO) or (SNP/−NO) into 10% Hoagland–Arnon growth medium. Control plants were grown on 10% Hoagland–Arnon solution. Then, the roots of wheat plants were carefully washed by distilled water, and the 6-day-old plant samples were subjected to osmotic stress through treatment with a mixture of 10% Hoagland–Arnon and 12% polyethylene glycol 6000 (PEG, PanReac AppliChem, Barcelona, Spain), which induce the physiological drought. The wheat plants were incubated during 1–4 h (short-term stress) or 24 h (long-term stress) (Figure 8). The roots of 6- or 7-day-old seedlings were used for the analysis of different parameters.

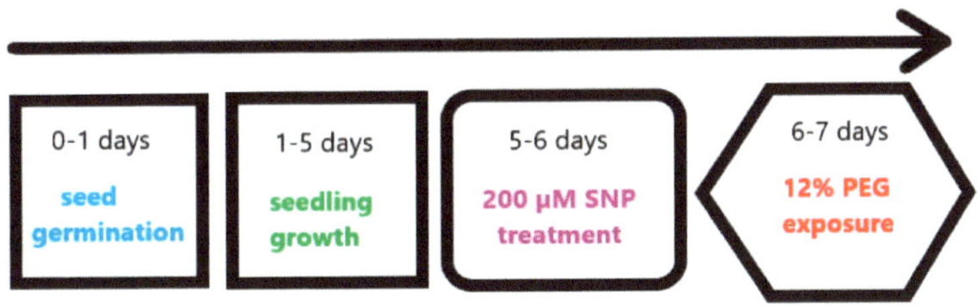

Figure 8. Schematic representation of the treatments in the study of the effects of SNP on wheat seedlings under osmotic stress (PEG, 12%).

4.2. *Assay of Antioxidant Enzymes*

Superoxide dismutase (SOD, EC 1.15.1.1) activity was measured in the wells of a flat bottom plate for immunoassay (Costar, Washington, DC, USA) by spectrophotometer EnSpire 2300 (Perkin Elmer, Waltham, MA, USA) according to Beyer and Fridovich [61] at 540 nm. Enzyme extraction was performed by homogenizing the frozen roots of 5 wheat seedlings (250–300 mg) with extraction solution (0.15 M potassium phosphate buffer at pH 7.0) in a 1:10 ratio (*m*:*v*). The samples were then centrifuged for 10 min at 13,000× *g* and 4 °C. SOD activity was assayed by measuring its ability to inhibit the photochemical reduction of nitro blue tetrazolium (NBT, Merk, Darmstadt, Germany). For this purpose, 10 μL of the enzyme extract was mixed with 200 μL of reagent I (0.15 M potassium phosphate buffer (pH 7.8) containing 0.17 mM EGTA, 6 mM NBT, 3.3 mM phenazine metasulfate) and 10 μL of reagent II (Tris-EDTA buffer (pH 8.0) containing 1 mM NADH). The control was a mixture of 10 μL extraction buffer with 200 μL reagent I and 10 μL reagent II. One unit of the SOD activity (U) was defined as the amount of enzyme required to result in a 50% inhibition of the rate of NBT reduction at 540 nm. Results were given as a specific enzyme activity in Units/(mg protein × min).

Catalase (CAT, EC 1.11.1.6) activity was determined by measuring the decomposition of H_2O_2 [62]. The enzyme extracts were obtained by homogenizing the roots of 5 seedlings in 0.05 M potassium phosphate buffer (pH 6.2) in a ratio of 1:10 (*m*:*v*). The homogenates were centrifuged for 10 min at 13,000× *g* and 4 °C. The supernatants were used in the further analysis. The reaction was started by adding 0.15 mL 0.033% H_2O_2 to the 0.02 mL of enzyme extract in the wells of a flat bottom plate (Costar, USA). After 1 min, the reaction was terminated by adding 0.075 μL 4% $(NH_4)_6Mo_7O_{24} \times 4H_2O$. CAT activity was measured by a Benchmark microplate reader (Bio-Rad, Hercules, CA, USA) as the change in the absorbance at 405 nm, defined as the destruction of complexes of H_2O_2 molecules with molybdenum ions. Results were given as a specific CAT activity in μM H_2O_2/(mg protein×min).

Ascorbate peroxidase (APX, EC: 1.11.1.11) activity was assayed as described by Verma and Dubey [63]. The roots of 5 seedlings were homogenized in extraction solution (50 μM K,Na-phosphate buffer (pH 7.8), 1% polyvinyl pyrrolidone (ABCR, Karlsruhe, Germany), 1 mM ascorbic acid) in a ratio of 1:3 (*m*:*v*) and incubated for 10 min at 4 °C. The extracts were centrifuged for 10 min at 13,000× *g* and 4 °C, and supernatants were used for the enzyme activity assay, which was conducted in the wells of a flat bottom plate (Costar, USA). The reaction was initiated by adding of 0.06% H_2O_2 to the mixture of 1.47 μL 50 mM K,Na-phosphate buffer with 15 μL 17 mM ascorbic acid, 15 μL 5 mM EDTA, and 5 μL extraction solution. H_2O_2 was not added to the control samples. The measurements were carried out for 90 s after the start of the reaction on the spectrophotometer EnSpire 2300 (Perkin Elmer, Waltham, MA, USA) at 290 nm. The APX activity was calculated using the following formula:

$$A = ((OD - OD_{control}) \times X)/(T \times L \times C), \text{ where}$$

OD—optical density in experimental samples;

$OD_{control}$—optical density of control samples;
T—time of reaction, min;
X—terminal dilution of extraction solution in the well;
L—layer thickness, cm;
C—protein content in the probe, mg.

Results were given as a specific activity in μM ascorbate/(mg protein × min). Total protein content in the extracts was quantified by the Bradford method [64].

4.3. Determination of Superoxide Radical in the Growth Medium

The extracellular production of $O_2^{\bullet-}$ was measured by the oxidation of epinephrine (ICN, Costa Mesa, CA, USA) to adrenochrome [65]. A total of 1 mL of 10^{-3} M adrenaline solution was added to the water with the roots of ten intact control or experimental seedlings, which have been pre-incubated in glass jars in 0.025 mM $CaCl_2$ solution for 1 h at 30 °C. The reaction of adrenaline conversion to adrenochrome was stopped by 0.05 n HCl followed by incubation for 15 min. The OD of the solutions was measured by a spectrophotometer Smart Spec Plus (Bio-Rad, Hercules, CA, USA) at 490 nm and expressed as nM $O_2^{\bullet-}$/(g FW × min).

4.4. Estimation of $O_2^{\bullet-}$ Content in the Roots of Wheat Seedlings

The level of $O_2^{\bullet-}$ in root tissues was determined based on the method proposed by Chaitanya and Naithani [66]. The roots of 4 seedlings were homogenized in 0.1 M sodium phosphate buffer at pH 7.2, containing 1 mM sodium diethyldithiocarbamate to inhibit SOD activity, and the extracts were centrifuged at 13,000× *g* for 20 min at 4 °C. The supernatant (600 μL) was added to 1.4 mL of a reaction medium, containing 0.1 M sodium phosphate buffer (pH 7.2), 1 mM sodium diethyldithiocarbamate, and 0.25 mM NBT (Merk, Germany). The superoxide anion concentration was measured by its ability to reduce NBT. The absorbance of the reaction mixture was measured after 1 min at 540 nm using a spectrophotometer Smart Spec Plus (Bio-Rad, Hercules, CA, USA). The results were given as OD_{540}/(g FW × min).

4.5. Extracellular H_2O_2 Content

The content of H_2O_2 in the growth medium was determined by oxidation OPD with hydrogen peroxide according to Khairullin et al. [67]. Six-day-old seedlings were transferred to Petri dishes (ten seedlings per dish) with 0.01 M KCl solution (10 mL) 0.05% OPD. The reaction was initiated by the addition of 1 mL 0.025 M oxalic acid; the reaction mixture was stirred gently, and 0.2 mL aliquots were taken at 2 min intervals. The aliquots were applied to the wells of a flat bottom plate (Costar, USA) preliminarily filled with 0.05 mL 4 M H_2SO_4. The absorption of the samples was measured by a Benchmark microplate reader (Bio-Rad, Hercules, CA, USA) at 492 nm. The OPD oxidizing activity was expressed in absorption Units/(g FW × s).

4.6. Estimation of H_2O_2 Content in the Roots of Wheat Seedlings

H_2O_2 content was determined in the root extracts using the method described by Bellincampi et al. [68]. The roots of 5 seedlings (250–300 mg) were homogenized in 0.01 M potassium phosphate buffer (pH 6.0) in the ratio 1:5 (*m:v*) and incubated for 10–15 min at 4 °C. The extracts were centrifuged at 10,000× *g* for 10 min at 4 °C. One ml of the solution, containing 25 mM $FeSO_4$ and 25 mM $(NH_4)_2SO_4$ dissolved in 2.5 M H_2SO_4, was mixed with 100 mL 125 μM xylenol orange (Merk, Germany) and 100 mM sorbitol. The resulting mixture (100 μL) was used to initiate the reaction by adding supernatant (50 μL). H_2O_2 content was monitored with dye xylenol orange; when hydroperoxides are reduced by ferrous ions in acid solution, they form a ferric product–xylenol orange complex. The reaction was incubated in the dark at 25 °C for 1 h and absorbance were recorded by a spectrophotometer EnSpire 2300 (Perkin Elmer, Waltham, MA, USA) at 570 nm. H_2O_2 content was calculated using a standard curve based on the absorbance of H_2O_2 standards.

4.7. Malondialdehyde (MDA) Accumulation

MDA was determined by measuring by its reaction with the thiobarbituric acid in accordance with Heath and Packer [69] with some modification. The frozen wheat roots (250 mg) were homogenized with 3 volumes of ice cold 10% trichloroacetic acid (TCA) and centrifuged at 10,000× *g* for 15 min. An assay mixture containing 1 mL of the supernatant and 1 mL of 0.5% (*m:v*) 2-thiobarbituric acid in 20% (*m:v*) TCA was heated at 95 °C for 1 h and then rapidly cooled in the ice bath. The absorbance of supernatant was monitored at 532 nm and 600 nm using a spectrophotometer Smart Spec Plus (Bio-Rad, Hercules, CA, USA). MDA content was calculated using 155 $mM^{-1}cm^{-1}$ as a coefficient of absorbance.

4.8. Estimation of Electrolyte Leakage (EL)

EL is considered as the marker of membrane stability. The level of EL in roots of wheat seedlings was determined in accordance with Bari et al. 2019 [35]. After all treatments, the surface of the roots was rinsed with distilled water. Then, the plants were incubated in the glass jars containing 20 mL of distilled water with shaking at 130 rpm on the orbital shaker–incubator ES-20 (Biosan, Riga, Latvia) at 30 °C for 2 h. The electrical conductivity of the incubation solution was recorded by a portable multi-range conductivity/TDS meter HI 8633 (HANNA Instruments, Catania, Italy). Results were given as electrical conductivity µS/g FW.

4.9. Proline Accumulation in the Growth Medium

For the determination of proline in the plant growth medium, 1 mL of growth solution from 10 wheat plants was mixed with 1 mL of glacial acetic acid–ninhydrin reagent and 1 mL of glacial acetic acid in a test tube, and the mixture was incubated in a water bath for 1 h at 100 °C [70]. The absorbance of the cold reaction mixture was measured at 522 nm using a spectrophotometer Smart Spec Plus (Bio-Rad, Hercules, CA, USA). The proline was calculated by comparison to a standard curve of proline, and the results were expressed as pM Pro/FW [71].

4.10. Analysis of Root Cell Death

The percentage of dead cells was analysed by the Evans blue method [35]. The roots of 5 wheat seedlings were kept in 10 mL of 0.25% (*w:v*) Evans blue (Merck, Germany) water solution at room temperature for 20 min. The amount of dye that entered into the cells corresponds to the degree of damage to tissue cells. The roots were washed three times by distilled water with gentle shaking; the trapped Evans blue was released from the plant tissues by homogenizing in 80% ethanol, and then the coloured solution was centrifuged at 12,000 rpm for 10 min. The optical density was monitored at 600 nm using a spectrophotometer Smart Spec Plus (Bio-Rad, Hercules, CA, USA), and the results of cell death in roots were given as percentages of frozen and thawed roots.

4.11. Histochemical Detection of Lipid Peroxidation

The lipid peroxidation of the root tips was determined histochemically [72]. The freshly cut off root tips were stained in Schiff reagent (Merck, Germany) for 1 h in the dark at room temperature until a purple colour appeared and then washed in 0.5% $K_2S_3O_5$ in 0.05 M HCl solution [73]. Each variant of the experiment included 10 seedlings. The red/purple colour, indicating the localization of aldehydes formed during peroxidation, was monitored using a Biozero BZ-8100E microscope (Keyence Co., Osaka, Japan) at a magnification of x4. Then, images were analysed with ImageJ software (NIH, Bethesda, MD, USA) based on the integrated density multiplied by stained area to obtain relative total stained values [74]. For a better presentation of results, relative levels of lipid peroxidation were divided by 10^{10}.

4.12. Mitotic Index (MI) Analysis

MI was determined from changes in the activity of the cells division in the roots' apical meristem [75]. The root tips of wheat seedlings were collected after treatments and fixed for

1 h at room temperature in Carnoy's solution composed of ethanol and glacial acetic acid in the of 3:1. The fixed root tips were kept in 70% ethanol under −4 °C. The slides were prepared by maceration method using 5% pectinase (Merck, Germany) and 5% cellulase (Merck, Germany) and stained by acetocarmine dye. Two thousand cells from the root meristem were used for the analysis of mitosis. Slide evaluations were performed in a light Biozero BZ-8100E microscope (Keyence Co., Osaka, Japan) at a magnification of ×20. Each variant of experiment included not less than 20 seedlings. MI was calculated as percent of the dividing cells.

4.13. *Statistical Analysis*

All experiments were performed in triplicate, and the experimental data were subjected to a one-way analysis of variance (ANOVA) using the SPSS 19.0 software (SPSS Inc., Chicago, IL, USA). Significant differences between values were determined using the LSD test at $p < 0.05$. Data in the figures are presented as mean values and their standard errors (±SE).

5. Conclusions

More and more evidence appears to show that NO is involved in a diversity of plant functions. Much progress has been made in the identification and characterization of enzymes involved in the biosynthesis and metabolism of NO and its derivatives [34,76,77]. NO plays a crucial role in plant signalling under different environmental stresses [21,78]. Under stress conditions, NO has several ameliorative functions as a signalling molecule [77]. The data available and the results of recent work revealed that NO is interconnected with the plant redox system and it's regulation in response to different stress factors [18,39,79]. Osmotic stress triggers the excessive production of ROS of wheat seedling roots at the beginning of stress influence followed by oxidative stress-induced damage, whereas the application of active SNP decreased $O_2^{\bullet -}$, H_2O_2 generation, EL level, and MDA content. It is known that environmental stresses induce short-term and long-term responses in plants [80]. Root-born hydraulic and chemical signals under water deficit conditions consist of two separate phases [81]. To avert the cellular damages, plants tightly regulate ROS production via the recruitment of enzymatic and non-enzymatic antioxidants. The exogenous supply of active SNP before 12% PEG treatment differentially regulated the activities of redox enzymes SOD, APX, and CAT in wheat roots. This regulation depended on time of stress influence, suggesting an important role for SNP during response to osmotic stress. During the first 4 h of osmotic stress, the activity of antioxidant enzymes of SNP-pretreated wheat roots reduced, while after 24 h of 12% PEG exposure their activities increased as compared to SNP-pretreated stressed plants. Under 12% PEG treatment SNP-induced declined of ROS production, which then decreased the damages of oxidative stress were observed in wheat roots. The effects (SNP/+NO) pretreatment on antioxidant enzymes activity demonstrated the dual role of NO donor, as active SNP involved in regulation of ROS signalling and in oxidative stress protection in wheat plants under osmotic stress.

Author Contributions: Conceptualization, A.L.; methodology, A.L.; formal analysis, A.L.; investigation, A.L.; resources, C.A.; data curation, A.L.; writing—original draft preparation, A.L. and C.A.; writing—review and editing, A.L. and C.A.; visualization, A.L.; project administration, A.L.; funding acquisition, A.L. All authors have read and agreed to the published version of the manuscript.

Funding: This research received no external funding.

Data Availability Statement: Data are contained within the article.

Acknowledgments: This research was carried out within the framework of a state assignment of the Ministry of Education and Science of Russia (number of state registration 122041400162-3).

Conflicts of Interest: The authors declare no conflicts of interest.

References

1. Vives-Peris, V.; de Ollas, C.; Gómez-Cadenas, A.; Pérez-Clemente, R.M. Root Exudates: From Plant to Rhizosphere and Beyond. *Plant Cell Rep.* **2020**, *39*, 3–17. [CrossRef] [PubMed]
2. Soares, C.; Rodrigues, F.; Sousa, B.; Pinto, E.; Ferreira, I.M.P.L.V.O.; Pereira, R.; Fidalgo, F. Foliar Application of Sodium Nitroprusside Boosts *Solanum lycopersicum* L. Tolerance to Glyphosate by Preventing Redox Disorders and Stimulating Herbicide Detoxification Pathways. *Plants* **2021**, *10*, 1862. [CrossRef] [PubMed]
3. Skliros, D.; Kalloniati, C.; Karalias, G.; Skaracis, G.N.; Rennenberg, H.; Flemetakis, E. Global Metabolomics Analysis Reveals Distinctive Tolerance Mechanisms in Different Plant Organs of Lentil (*Lens culinaris*) upon Salinity Stress. *Plant Soil* **2018**, *429*, 451–468. [CrossRef]
4. Signorelli, S.; Corpas, F.J.; Borsani, O.; Barroso, J.B.; Monza, J. Water Stress Induces a Differential and Spatially Distributed Nitro-oxidative Stress Response in Roots and Leaves of *Lotus japonicas*. *Plant Sci.* **2013**, *201–202*, 137–146. [CrossRef] [PubMed]
5. Zhang, H.; Zhao, Y.; Zhu, J. Thriving under Stress: How Plants Balance Growth and the Stress Response. *Dev. Cell* **2020**, *55*, 529–543. [CrossRef] [PubMed]
6. Rezayian, M.; Ebrahimzadeh, H.; Niknam, V. Nitric Oxide Stimulates Antioxidant System and Osmotic Adjustment in Soybean Under Drought Stress. *J. Soil Sci. Plant Nutr.* **2020**, *20*, 1122–1132. [CrossRef]
7. Kolupaev, Y.E.; Vainer, A.A.; Yastreb, T.O. Proline: Physiological Functions and Regulation of Content in Plants Under Stress Conditions. *Visn. Khark. Nats. Agrar. Univ. Ser. Biol.* **2014**, *2*, 6–22.
8. Ghosh, U.K.; Islam, M.N.; Siddiqui, M.N.; Cao, X.; Khan, M.A.R. Proline, a Multifaceted Signalling Molecule in Plant Responses to Abiotic Stress: Understanding the Physiological Mechanisms. *Plant Biol.* **2022**, *24*, 227–239. [CrossRef]
9. Kolupaev, Y.E.; Karpets, Y.V. Reactive Oxygen Species and Stress Signaling in Plants. *Ukr. Biochem. J.* **2014**, *86*, 18–35. [CrossRef] [PubMed]
10. Allagulova, C.R.; Lubyanova, A.R.; Avalbaev, A.M. Multiple Ways of Nitric Oxide Production in Plants and Its Functional Activity under Abiotic Stress Conditions. *Int. J. Mol. Sci.* **2023**, *24*, 11637. [CrossRef]
11. Del Río, L.A.; Corpas, F.J.; López-Huertas, E.; Palma, J.M. Plant Superoxide Dismutases: Function Under Abiotic Stress Conditions. In *Antioxidants and Antioxidant Enzymes in Higher Plants*; Gupta, D.K., Palma, J.M., Corpas, F.J., Eds.; Springer International Publishing AG: Cham, Switzerland, 2018; pp. 13–26. [CrossRef]
12. Paul, S.; Roychoudhury, A. Regulation of Physiological Aspects in Plants by Hydrogen Sulfide and Nitric Oxide Under Challenging Environment. *Physiol. Plant.* **2020**, *168*, 374–393. [CrossRef]
13. Roy, S. Role of Nitric Oxide as a Double Edged Sword in Root Growth and Development. In *Rhizobiology: Molecular Physiology of Plant Roots. Signaling and Communication in Plants*; Mukherjee, S., Baluška, F., Eds.; Springer International Publishing AG: Cham, Switzerland, 2021; pp. 167–193. [CrossRef]
14. Corpas, F.J.; Palma, J.M. Functions of NO and H_2S Signal Molecules Against Plant Abiotic Stress. In *Plant Abiotic Stress Signaling. Methods in Molecular Biology*; Couée, I., Ed.; Humana: New York, NY, USA, 2023; Volume 2642, pp. 97–109. [CrossRef]
15. Allagulova, C.; Avalbaev, A.; Lubyanova, A.; Plotnikov, A.; Yuldashev, R.; Lastochkina, O. Nitric Oxide (NO) Improves Wheat Growth under Dehydration Conditions by Regulating Phytohormone Levels and Induction of the Expression of the TADHN Dehydrin Gene. *Plants* **2023**, *12*, 4051. [CrossRef]
16. Corpas, F.J.; Leterrier, M.; Valderrama, R.; Airaki, M.; Chaki, M.; Palma, J.M.; Barroso, J.B. Nitric Oxide Imbalance Provokes a Nitrosative Response in Plants Under Abiotic Stress. *Plant Sci.* **2011**, *181*, 604. [CrossRef]
17. Yang, H.; Mu, J.; Chen, L.; Feng, J.; Hu, J.; Li, L.; Zhou, J.M.; Zuo, J. S-Nitrosylation Positively Regulates Ascorbate Peroxidase Activity During Plant Stress Responses. *Plant Physiol.* **2015**, *167*, 1604–1615. [CrossRef]
18. Laxa, M.; Liebthal, M.; Telman, W.; Chibani, K.; Dietz, K.J. The Role of the Plant Antioxidant System in Drought Tolerance. *Antioxidants* **2019**, *8*, 94. [CrossRef]
19. Correa-Aragunde, C.; Foresi, N.; Lamattina, L. Nitric Oxide is a Ubiquitous Signal for Maintaining Redox Balance in Plant Cells: Regulation of Ascorbate Peroxidase as a Case Study. *J. Exp. Bot.* **2015**, *66*, 2913–2921. [CrossRef]
20. Palma, J.M.; Mateos, R.M.; López-Jaramillo, J.; Rodríguez-Ruiz, M.; González-Gordo, S.; Lechuga-Sancho, A.M.; Corpas, F.J. Plant Catalases as NO and H_2S Targets. *Redox Biol.* **2020**, *34*, 101525. [CrossRef]
21. Lubyanova, A.R.; Bezrukova, M.V.; Shakirova, F.M. Involvement of Nitric Oxide in Methyl Jasmonate-Mediated Regulation of Water Metabolism in Wheat Plants under Drought Stress. *Stresses* **2022**, *2*, 477–492. [CrossRef]
22. Pagnussat, G.C.; Lanteri, M.L.; Lombardo, M.C.; Lamattina, L. Nitric Oxide Mediates the Indole Acetic Acid Induction Activation of a Mitogen-Activated Protein Kinase Cascade Involved in Adventitious Root Development. *Plant Physiol.* **2004**, *135*, 279. [CrossRef]
23. Correa-Aragunde, N.; Graziano, M.; Lamattina, L. Nitric Oxide Plays a Central Role in Determining Lateral Root Development in Tomato. *Planta* **2004**, *218*, 900. [CrossRef]
24. Corpas, F.J.; Barroso, J.B. Functions of Nitric Oxide (NO) in Roots during Development and under Adverse Stress Conditions. *Plants* **2015**, *4*, 240–252. [CrossRef] [PubMed]
25. Wimalasekera, R.; Scherer, G.F.E. Nitric Oxide (NO) and Lateral Root Development in Plants Under Stress. In *Nitric Oxide in Plant Biology: An Ancient Molecule with Emerging Roles*; Singh, V.P., Singh, S., Sandalio, L.M., Eds.; Academic Press: Cambridge, MA, USA, 2022; pp. 319–329. [CrossRef]

26. Maslennikova, D.R.; Allagulova, C.R.; Fedorova, K.A.; Plotnikov, A.A.; Avalbaev, A.M.; Shakirova, F.M. Cytokinins Contribute to Realization of Nitric Oxide Growth-Stimulating and Protective Effects on Wheat Plants. *Russ. J. Plant Physiol.* **2017**, *64*, 665–671. [CrossRef]
27. Groß, F.; Durner, J.; Gaupels, F. Nitric Oxide, Antioxidants and Prooxidants in Plant Defence Responses. *Front. Plant Sci.* **2013**, *4*, 419. [CrossRef] [PubMed]
28. Majeed, S.; Nawaz, F.; Naeem, M.; Ashraf, M.Y.; Ejaz, S.; Ahmad, K.S.; Tauseef, S.; Farid, G.; Khalid, I.; Mehmood, K. Nitric Oxide Regulates Water Status and Associated Enzymatic Pathways to Inhibit Nutrients Imbalance in Maize (*Zea mays* L.) Under Drought Stress. *Plant Physiol. Biochem.* **2020**, *155*, 147–160. [CrossRef] [PubMed]
29. Kaya, C.; Ugurlar, F.; Ashraf, M.; Alam, P.; Ahmad, P. Nitric Oxide and Hydrogen Sulfide Work Together to Improve Tolerance to Salinity Stress in Wheat Plants by Upraising the AsA-GSH Cycle. *Plant Physiol. Biochem.* **2023**, *194*, 651–663. [CrossRef]
30. Mhamdi, A.; Van Breusegem, F. Reactive Oxygen Species in Plant Development. *Development* **2018**, *145*, dev164376. [CrossRef] [PubMed]
31. Ruan, J.; Zhou, Y.; Zhou, M.; Yan, J.; Khurshid, M.; Weng, W.; Cheng, J.; Zhang, K. Jasmonic Acid Signaling Pathway in Plants. *Int. J. Mol. Sci.* **2019**, *20*, 2479. [CrossRef] [PubMed]
32. Mittler, R.; Zandalinas, S.I.; Fichman, Y.; Van Breusegem, F. Reactive Oxygen Species Signalling in Plant Stress Responses. *Nat. Rev. Mol. Cell Biol.* **2022**, *23*, 663–679. [CrossRef] [PubMed]
33. Sharma, A.; Soares, C.; Sousa, B.; Martins, M.; Kumar, V.; Shahzad, B.; Sidhu, G.P.S.; Bali, A.S.; Asgher, M.; Bhardwaj, R.; et al. Nitric Oxide-Mediated Regulation of Oxidative Stress in Plants Under Metal Stress: A Review. *Physiol. Plant.* **2020**, *168*, 318–344. [CrossRef]
34. Iqbal, N.; Sehar, Z.; Fatma, M.; Umar, S.; Sofo, A.; Khan, N.A. Nitric Oxide and Abscisic Acid Mediate Heat Stress Tolerance through Regulation of Osmolytes and Antioxidants to Protect Photosynthesis and Growth in Wheat Plants. *Antioxidants* **2022**, *11*, 372. [CrossRef]
35. Bari, M.A.; Akther, M.S.; Reza, M.A.; Kabir, A.H. Cadmium Tolerance is Associated with the Root-Driven Coordination of Cadmium Sequestration, Iron Regulation, and ROS Scavenging in Rice. *Plant Physiol. Biochem.* **2019**, *136*, 22–33. [CrossRef]
36. Fichman, Y.; Mittler, R. A Systemic Whole-Plant Change in Redox Levels Accompanies the Rapid Systemic Response to Wounding. *Plant Physiol.* **2021**, *186*, 4–8. [CrossRef]
37. Kolupaev, Y.E.; Kokorev, A.I. Antioxidant System and Plant Resistance to Water Deficit. *Fiziol. Rast. Genet.* **2019**, *51*, 28–54. (In Russian) [CrossRef]
38. Cao, X.; Zhu, C.; Zhong, C.; Zhang, J.; Wu, L.; Jin, Q.; Ma, Q. Nitric Oxide Synthase-Mediated Early Nitric Oxide Burst Alleviates Water Stress-Induced Oxidative Damage in Ammonium-Supplied Rice Roots. *BMC Plant Biol.* **2019**, *19*, 108. [CrossRef] [PubMed]
39. Zhang, L.; Zhou, S.; Xuan, Y.; Sun, M.; Zhao, L. Protective Effect of Nitric Oxide Against Oxidative Damage in *Arabidopsis* Leaves Under Ultraviolet-B Irradiation. *J. Plant Biol.* **2009**, *52*, 135–140. [CrossRef]
40. Keisham, M.; Jain, P.; Singh, N.; von Toerne, C.; Bhatla, S.C.; Lindermayr, C. Deciphering The Nitric Oxide, Cyanide and Iron-Mediated Actions of Sodium Nitroprusside in Cotyledons of Salt Stressed Sunflower Seedlings. *Nitric Oxide* **2019**, *88*, 10–26. [CrossRef]
41. Kolbert, Z.; Barroso, J.B.; Brouquisse, R.; Corpas, F.J.; Gupta, K.J.; Lindermayr, C.; Loake, G.J.; Palma, J.M.; Petřivalský, M.; Wendehenne, D.; et al. A Forty Year Journey: The Generation and Roles of NO in Plants. *Nitric Oxide* **2019**, *93*, 53. [CrossRef]
42. Gouvea, C.M.C.P.; Souza, J.F.; Magalhaes, A.C.N.; Martins, I.S. NO•–Releasing Substances That Induce Growth Elongation in Maize Root Segments. *Plant Growth Regul.* **1997**, *21*, 183–187. [CrossRef]
43. Correa-Aragunde, N.; Graziano, M.; Chevalier, C.; Lamattina, L. Nitric Oxide Modulates the Expression of Cell Cycle Regulatory Genes During Lateral Root Formation in Ttomato. *J. Exp. Bot.* **2006**, *57*, 581–588. [CrossRef]
44. Batista, P.F.; Costa, A.C.; Müller, C.; de Oliveira Silva-Filho, R.; da Silva, F.B.; Merchant, A.; Mendes, G.C.; Nascimento, K.J.T. Nitric Oxide Mitigates the Effect of Water Deficit in *Crambe abyssinica*. *Plant Physiol. Biochem.* **2018**, *129*, 310–322. [CrossRef]
45. Holzmeister, C.; Gaupels, F.; Geerlof, A.; Sarioglu, H.; Sattler, M.; Durner, J.; Lindermayr, C. Differential Inhibition of *Arabidopsis* Superoxide Dismutases by Peroxynitrite-Mediated Tyrosine Nitration. *J. Exp. Bot.* **2015**, *66*, 989–999. [CrossRef] [PubMed]
46. Razi, K.; Muneer, S. Drought Stress-Induced Physiological Mechanisms, Signaling Pathways and Molecular Response of Chloroplasts in Common Vegetable Crops. *Crit. Rev. Biotechnol.* **2021**, *41*, 669–691. [CrossRef] [PubMed]
47. Zhang, L.; Shi, X.; Zhang, Y.; Wang, J.; Yang, J.; Ishida, T.; Jiang, W.; Han, X.; Kang, J.; Wang, X.; et al. CLE9 Peptide-Induced Stomatal Closure is Mediated by Abscisic Acid, Hydrogen Peroxide, and Nitric Oxide in *Arabidopsis thaliana*. *Plant Cell Environ.* **2019**, *42*, 1033–1044. [CrossRef] [PubMed]
48. Iqbal, N.; Umar, S.; Khan, N.A.; Corpas, F.J. Nitric Oxide and Hydrogen Sulfide Coordinately Reduce Glucose Sensitivity and Decrease Oxidative Stress via Ascorbate-Glutathione Cycle in Heat-Stressed Wheat (*Triticum aestivum* L.) Plants. *Antioxidants* **2021**, *10*, 108. [CrossRef] [PubMed]
49. Ozfidan-Konakci, C.; Yildiztugay, E.; Elbasan, F.; Kucukoduk, M.; Turkan, I. Hydrogen Sulfide (H_2S) and Nitric Oxide (NO) Alleviate Cobalt Toxicity in Wheat (*Triticum aestivum* L.) by Modulating Photosynthesis, Chloroplastic Redox and Antioxidant Capacity. *J. Hazard. Mater.* **2020**, *388*, 122061. [CrossRef] [PubMed]
50. Reiter, T.A. NO• Chemistry: A Diversity of Targets in the Cell. *Redox Rep.* **2006**, *11*, 194–206. [CrossRef] [PubMed]

51. Khan, M.N.; Mobin, M.; Abbas, Z.K.; Siddiqui, M.H. Nitric Oxide-Induced Synthesis of Hydrogen Sulfide Alleviates Osmotic Stress in Wheat Seedlings Through Sustaining Antioxidant Enzymes, Osmolyte Accumulation and Cysteine Homeostasis. *Nitric Oxide* **2017**, *68*, 91–102. [CrossRef]
52. Ghaffari, H.; Tadayon, M.R.; Bahador, M.; Razmjoo, J. Investigation of the Proline Role in Controlling Traits Related to Sugar and Root Yield of Sugar Beet Under Water Deficit Conditions. *Agric. Water Manag.* **2021**, *243*, 106448. [CrossRef]
53. Santisree, P.; Bhatnagar-Mathur, P.; Sharma, K.K. NO to Drought-Multifunctional Role of Nitric Oxide in Plant Drought: Do We Have All the Answers? *Plant Sci.* **2015**, *239*, 44–55. [CrossRef]
54. Santisree, P.; Sanivarapu, H.; Gundavarapu, S.; Sharma, K.K.; Bhatnagar-Mathur, P. Nitric Oxide as a Signal in Inducing Secondary Metabolites During Plant Stress. In *Co-Evolution of Secondary Metabolites. Reference Series in Phytochemistry*; Merillon, J.M., Ramawat, K.G., Eds.; Springer Nature: Cham, Switzerland, 2019; pp. 593–621. [CrossRef]
55. Cai, W.; Liu, W.; Wang, W.S.; Fu, Z.W.; Han, T.T.; Lu, Y.T. Overexpression of Rat Neurons Nitric Oxide Synthase in Rice Enhances Drought and Salt Tolerance. *PLoS ONE* **2015**, *10*, e0131599. [CrossRef]
56. Bhaskarla, V.; Zinta, G.; Ford, R.; Jain, M.; Varshney, R.K.; Mantri, N. Comparative Root Transcriptomics Provide Insights into Drought Adaptation Strategies in Chickpea (*Cicer arietinum* L.). *Int. J. Mol. Sci.* **2020**, *21*, 1781. [CrossRef] [PubMed]
57. Vysotskaya, L.V. Mitotic Cycle and its Regulation. *Vavilov J. Genet. Breed.* **2014**, *18*, 81–92. (In Russian)
58. Viola, I.L.; Güttlein, L.N.; Gonzalez, D.H. Redox Modulation of Plant Developmental Regulators from the Class I TCP Transcription Factor Family. *Plant Physiol.* **2013**, *162*, 1434–1447. [CrossRef]
59. Viola, I.L.; Alem, A.L.; Jure, R.M.; Gonzalez, D.H. Physiological Roles and Mechanisms of Action of Class I TCP Transcription Factors. *Int. J. Mol. Sci.* **2023**, *24*, 5437. [CrossRef]
60. Lynch, J.P.; Strock, C.F.; Schneider, H.M.; Sidhu, J.S.; Ajmera, I.; Galindo-Castañeda, T.; Klein, S.P.; Hanlon, M.T. Root Anatomy and Soil Resource Capture. *Plant Soil* **2021**, *466*, 21–63. [CrossRef]
61. Beyer, Y.; Fridovich, I. Assaying for Superoxide Dismutase Activity: Some Large Consequences of Minor Changes in Conditions. *Anal. Biochem.* **1987**, *161*, 559–566. [CrossRef]
62. Aebi, H. Catalase in vitro. *Methods Enzymol.* **1984**, *105*, 121–126.
63. Verma, S.; Dubey, R.S. Lead Toxicity Induces Lipid Peroxidation and Alters the Activities of Antioxidant Enzymes in Growing Rice Plants. *Plant Sci.* **2003**, *164*, 645–655. [CrossRef]
64. Bradford, M.M. A Rapid and Sensitive Method for the Quantitation of Microgram Quantities of Protein Utilizing the Principle of Protein-Dye Binding. *Anal. Biochem.* **1976**, *72*, 248–254. [CrossRef]
65. Minibayeva, F.V.; Gordon, L.K.; Kolesnikov, O.P.; Chasov, A.V. Role of Extracellular Peroxidase in the Superoxide Production by Wheat Root Cells. *Protoplasma* **2001**, *217*, 125–128. [CrossRef]
66. Chaitanya, K.K.; Naithani, S.C. Role of Superoxide, Lipid Peroxidation and Superoxide Dismutase in Membrane Perturbation During Loss of Viability in Seeds of *Shorea robusta* Gaertn. f. *New Phytol.* **1994**, *126*, 623–627. [CrossRef]
67. Khairullin, R.M.; Yarullina, L.G.; Troshina, N.B.; Akhmetova, I.E. Chitooligosacharide-Induced Activation of o-Phenylenediamine Oxidation by Wheat Seedlings in the Presence of Oxalic Acid. *Biochemistry* **2001**, *66*, 286–289. [PubMed]
68. Bellincampi, D.; Dipierro, N.; Salvi, G.; Cervone, F.; De Lorenzo, G. Extracellular H_2O_2 Induced by Oligogalacturonides is not Involved in the Inhibition of the Auxin-Regulated *rolB* Gene Expression in Tobacco Leaf Explants. *Plant Physiol.* **2000**, *122*, 1379–1386. [CrossRef] [PubMed]
69. Heath, R.L.; Packer, L. Photoperoxidation in Isolated Chloroplasts: Kinetics and Stoichiometry of Fatty Acid Peroxidation. *Arch. Biochem. Biophys.* **1968**, *125*, 189–198. [CrossRef] [PubMed]
70. Bates, L.S.; Waldran, R.P.; Teare, I.D. Rapid Determination of Free Proline for Water Stress Studies. *Plant Soil* **1973**, *39*, 205–208. [CrossRef]
71. Vives-Peris, V.; Gómez-Cadenas, A.; Pérez-Clemente, R.M. Citrus Plants Exude Proline and Phytohormones Under Abiotic Stress Conditions. *Plant Cell Rep.* **2017**, *36*, 1971–1984. [CrossRef] [PubMed]
72. Arasimowicz-Jelonek, M.; Floryszak-Wieczorek, J.; Kubiś, J. Involvement of Nitric Oxide in Water Stress-Induced Responses of Cucumber Roots. *Plant Sci.* **2009**, *177*, 682–690. [CrossRef]
73. Pompella, A.; Maellaro, E.; Casini, A.F.; Comporti, M. Histochemical Detection of Lipid Peroxidation in the Liver of Bromobenzene-Poisoned Mice. *Am. J. Pathol.* **1987**, *129*, 295–301.
74. Gill, M.B.; Zeng, F.; Shabala, L.; Zhang, G.; Yu, M.; Demidchik, V.; Shabala, S.; Zhou, M. Identification of QTL Related to ROS Formation Under Hypoxia and Their Association With Waterlogging and Salt tolerance in Barley. *Int. J. Mol. Sci.* **2019**, *20*, 699. [CrossRef]
75. Bezrukova, M.; Kildibekova, A.; Shakirova, F. WGA Reduces the Level of Oxidative Stress in Wheat Seedlings Under Salinity. *Plant Growth Regul.* **2008**, *54*, 195–201. [CrossRef]
76. Farnese, F.S.; Menezes-Silva, P.E.; Gusman, G.S.; Oliveira, J.A. When Bad Guys Become Good Ones: The Key Role of Reactive Oxygen Species and Nitric Oxide in the Plant Responses to Abiotic Stress. *Front. Plant Sci.* **2016**, *7*, 471. [CrossRef] [PubMed]
77. Sami, F.; Faizan, M.; Faraz, A.; Siddiqui, H.; Yusuf, M.; Hayat, S. Nitric Oxide-Mediated Integrative Alterations in Plant Metabolism to Confer Abiotic Stress Tolerance, NO Crosstalk with Phytohormones and NO-Mediated Post Translational Modifications in Modulating Diverse Plant Stress. *Nitric Oxide* **2018**, *73*, 22–38. [CrossRef] [PubMed]
78. Allagulova, C.R.; Avalbaev, A.M.; Lubyanova, A.R.; Lastochkina, O.V.; Shakirova, F.M. Current Concepts of the Mechanisms of Nitric Oxide Formation in Plants. *Russ. J. Plant Physiol.* **2022**, *69*, 339–351. [CrossRef]

79. Fancy, N.N.; Bahlmann, A.K.; Loake, G.J. Nitric Oxide Function in Plant Abiotic Stress. *Plant Cell Environ.* **2017**, *40*, 462–472. [CrossRef]
80. Ahluwalia, O.; Singh, P.C.; Bhatia, R. A Review on Drought Stress in Plants: Implications, Mitigation and the Role of Plant Growth Promoting Rhizobacteria. *Resour. Environ. Sustain.* **2021**, *5*, 100032. [CrossRef]
81. Tatar, Ö.; Brück, H.; Asch, F. Atmospheric and Soil Water Deficit Induced Changes in Chemical and Hydraulic Signals in Wheat (*Triticum aestivum* L.). *J. Agron. Crop. Sci.* **2023**, *209*, 242–250. [CrossRef]

plants

MDPI

Article

High Concentrations of Se Inhibited the Growth of Rice Seedlings

Ying Liu, Jiayu Ma, Feng Li, Xiang Zeng, Zhengwei Wu, Yongxiang Huang, Yingbin Xue and Yanyan Wang *

Department of Agronomy, College of Coastal Agricultural Sciences, Guangdong Ocean University, Zhanjiang 524088, China; liuying85168@gdou.edu.cn (Y.L.)
* Correspondence: yanyanwang@gdou.edu.cn

Abstract: Selenium (Se) is crucial for both plants and humans, with plants acting as the main source for human Se intake. In plants, moderate Se enhances growth and increases stress resistance, whereas excessive Se leads to toxicity. The physiological mechanisms by which Se influences rice seedlings' growth are poorly understood and require additional research. In order to study the effects of selenium stress on rice seedlings, plant phenotype analysis, root scanning, metal ion content determination, physiological response index determination, hormone level determination, quantitative PCR (qPCR), and other methods were used. Our findings indicated that sodium selenite had dual effects on rice seedling growth under hydroponic conditions. At low concentrations, Se treatment promotes rice seedling growth by enhancing biomass, root length, and antioxidant capacity. Conversely, high concentrations of sodium selenite impair and damage rice, as evidenced by leaf yellowing, reduced chlorophyll content, decreased biomass, and stunted growth. Elevated Se levels also significantly affect antioxidase activities and the levels of proline, malondialdehyde, metal ions, and various phytohormones and selenium metabolism, ion transport, and antioxidant genes in rice. The adverse effects of high Se concentrations may directly disrupt protein synthesis or indirectly induce oxidative stress by altering the absorption and synthesis of other compounds. This study aims to elucidate the physiological responses of rice to Se toxicity stress and lay the groundwork for the development of Se-enriched rice varieties.

Keywords: rice; selenium toxicity stress; physiological response; roots and leaves

Citation: Liu, Y.; Ma, J.; Li, F.; Zeng, X.; Wu, Z.; Huang, Y.; Xue, Y.; Wang, Y. High Concentrations of Se Inhibited the Growth of Rice Seedlings. *Plants* **2024**, *13*, 1580. https://doi.org/10.3390/plants13111580

Academic Editors: Geovani Soares de Lima, Lauriane Almeida dos Anjos Soares and Francisco Vanies Da Silva Sá

Received: 17 May 2024
Revised: 4 June 2024
Accepted: 5 June 2024
Published: 6 June 2024

1. Introduction

Se (Selenium) is a critical element for organic life [1]. It is essential for various life forms, including humans, as it is a component of the amino acid selenocysteine (SeCys) [1]. Se is important for human and animal health [2]. It plays a vital role in maintaining physical well-being [3], particularly in terms of its antioxidant [4], anticancer [5], cardiovascular disease prevention [6], immune improvement [7], and other good auxiliary effects. The recommended daily intake of Se for humans is 55 μg (MCGs) [8]. As a trace element, Se is crucial for numerous biological activities, such as thyroid hormone metabolism, antioxidant defense, immune system strengthening, and prevention of certain types of cancer [9]. Epidemiological studies have shown a negative correlation between Se intake and mortality from various cancers [10]. Se is primarily obtained from food sources and some Se supplements, including plants, animals, and certain microorganisms [11]. In the human body, Se is metabolized into 25 selenoproteins, each of which serves distinct biological functions, including antioxidant and anticancer effects, as well as improving fertility and reproduction [12]. Furthermore, Se has been found to be beneficial for crop growth and protection against specific biological and abiotic stresses when it is maintained at appropriate concentrations [13].

In soils, Se levels range from 0.01 to 2.0 milligrams per kilogram, with a mean value of 0.4 milligrams per kilogram, except in selenic soils, where the Se content is >10 mg/kg [14].

The difference between the nutritional dose of Se and the maximum safe intake is very narrow. The general intake for the human body is 50~250 μg/d, and high levels of Se have detrimental effects on biological organisms [15]. Chronic Se poisoning occurs when individuals consume excessive amounts of Se over a prolonged period, and common symptoms in humans include hair loss, nail damage, and rashes [1]. Furthermore, studies indicate that excessive Se may contribute to the development of type 2 diabetes [16,17] and severe bowel disease [18]. In plants, superfluous Se interferes with nutrient absorption, inhibits the formation of photosynthetic pigments, causes oxide damage, and induces genotoxic effects [19]. Additionally, elevated Se levels above this threshold impede plant growth and development, thus reducing grain yield [20].

Plants primarily assimilate and transfer Se through phosphate and sulfate transport proteins, which accumulate Se in their roots [19,21]. Se has been proven to play a role in the adjustments to stress response, respiration, photosynthesis, abiotic stress survivability, oxidation resistance, and remission of heavy metal stress [22,23]. Land plants predominantly take up Se through active transport processes, specifically in the form of selenates or selenites, which are primarily absorbed through sulfate carrier proteins [24,25]. Selenites have also been found to enter root cells via phosphate transporters [26]. Generally, young leaves have a greater Se concentration during seedling growth than older leaves [27,28]. Within plant cells, Se primarily accumulates in vacuoles [29] and can be excreted through sulfate transporters present in flat bodies [30].

In plants, Se assimilation results in the production of ROS (reactive oxygen species), which are quickly neutralized by antioxidant substances [31]. Nevertheless, when the ratio of Se assimilability is very high or Se poisoning is present, an imbalance between ROS and antioxidant substances occurs, leading to oxidizing stress that influences different physiological response processes. These processes include chlorophyll deletion, suppression of the tricarboxylic acid cycle (TCA), and decreasing photosynthetic efficiency [31]. It is interesting that some ROS have been found to be involved in regulating genetic expression and influencing Se absorption [31].

Plants employ efficient mechanisms for alleviating stress, such as oxidation resistance systems, to minimize Se toxicity. These systems effectively eliminate excess ROS under Se stress [19]. Optimum Se concentrations stimulate the biosynthesis of antioxidants, thus impacting plants' responses to biotic and abiotic stresses [21].

Plant hormones have a vital role in sustaining optimum cell activity through the regulation of different signaling pathways [32]. They can be produced interiorly in response to special stimuli or applied externally to regulate the growth of plants [32]. Furthermore, the interaction between plant hormones and Se contributes to a range of effects, including the maintenance of a REDOX balance, nutrient absorption, changes in development patterns, and improved crop yield and quality [33,34]. High concentrations of Se, acting as a potent pro-oxidant, promote plant physiological deterioration. The degree of damage is species-specific and concentration-dependent [35]. However, at trace concentrations, Se acts as an antioxidant in several plants. It replenishes metabolic pools, triggers antioxidant mechanisms, enhances the accumulation of nonenzymatic antioxidants, and boosts the activity of antioxidant enzymes [35].

As a crucial cash crop, rice (*Oryza sativa*) is globally recognized for its high yield and extensive cultivation [36]. Moreover, rice serves as an important source of Se for human consumption, and the Se level in rice determines the Se nutritional status of the Chinese population [37,38]. Research has indicated that rice cultivated in high-Se soil accumulates more Se than rice grown in low-Se soil [39]. Although cereals produced from naturally Se-rich soil can meet the daily Se requirements of individuals, approximately 72% of soil in China lacks Se, resulting in insufficient Se content in the plants growing on it [40]. Consequently, the Se content in most rice falls below the required levels, failing to meet the daily Se requirements of individuals [41,42]. In light of this, studies have demonstrated that rice possesses a remarkable ability to enrich Se and serve as a crucial carrier for Se, with the application of Se fertilizer markedly bolstering the Se content in rice [43,44]. Nevertheless, research has also highlighted that low levels of exogenous Se can enhance plant development, whereas high levels can restrain it [25,45]. Several studies have shown that increased Se fertilizer application leads to increased Se content in rice grains [46–48]. Furthermore, researchers are eager to identify suitable rice varieties with robust Se-enrichment capabilities, thus enabling the production of safe Se-rich rice with lower Se supply levels [49].

Current studies on Se in rice predominantly focus on Se accumulation and distribution [46–49]. However, the physiological response mechanism of rice to excessive Se remains unclear [50,51], necessitating further investigation. Consequently, this study aimed to examine the growth phenotype and physiological characteristics of rice subjected to different concentrations of sodium selenite and to explore gene expression in rice. The ultimate goals are to increase our understanding of the molecular regulatory network of Se in rice, elucidate the physiological response mechanism of rice to excessive Se, and provide a foundation for the cultivation of high-quality rice enriched with Se in a targeted manner, thus yielding far-reaching implications.

2. Results

2.1. Influences of Various Concentrations of Se on Rice Development

To assess the effects of Se poisoning stress on rice seedling growth, we measured differences in the plant height and fresh and dry weight of rice plants treated with various levels of Se. The results showed that low Se concentrations promoted rice seedling growth, whereas high Se concentrations significantly impeded growth (Figure 1A–F). Compared with that of the control treatment (0 μM Se), the plant height of the rice plants treated with lower Se concentrations (10 and 20 μM) exhibited insignificant increases of 10.91% and 9.55%, respectively. However, higher Se levels (40, 80, and 100 μM) leaded to a notable reduction in plant height, with reductions of 17.20%, 33.55%, and 60.41%, respectively (Figure 1G). Taking 0 μM of Se treatment as the control (CK), the fresh root weight of rice increased significantly by 61.03% in the 10 μM Se treatment group, whereas there was no significant change in the 20 μM and 40 μM Se treatment groups. Higher concentrations of Se (80 and 100 μM) resulted in significant decreases of 58.09% and 80.15%, respectively, in the fresh root weight of the rice plants (Figure 2A). In terms of rice's root dry weight, the 10 μM Se treatment group showed a significant increase of 54.79%, whereas there was no significant change in the 20 μM and 40 μM Se treatment groups. The rice's root dry weights in the 80 μM and 100 μM Se treatment groups significantly decreased by 43.15% and 67.12%, respectively (Figure 2B).

Compared with that in the CK treatment group, the stem fresh weight in the 10 μM Se treatment group increased by 12.61%, but this increase was not statistically significant. The stem fresh weight in the 20 μM and 40 μM Se treatment groups did not significantly decrease. Nevertheless, higher levels of Se (80 and 100 μM) markedly decreased the fresh weight of the stems by 58.11% and 74.77%, respectively (Figure 2C). The stem weights in the 10 μM and 20 μM treatment groups did not significantly change, whereas those in the other treatment groups significantly decreased by 26.83%, 40.74%, and 72.59%, respectively (Figure 2D). Compared with that in the CK group, the rice leaves' fresh weight from the 10 μM Se processing group did not significantly decrease, whereas there were significant decreases of 32.89%, 39.47%, 74.69%, and 88.59% in the other treatment groups (Figure 2E). The leaf dry weight did not significantly decrease in the 10 μM and 20 μM Se treatment groups compared to that in the CK group, whereas there were evident decreases of 29.88%, 56.59%, and 75.04%, respectively, in the other treatment groups (Figure 2F).

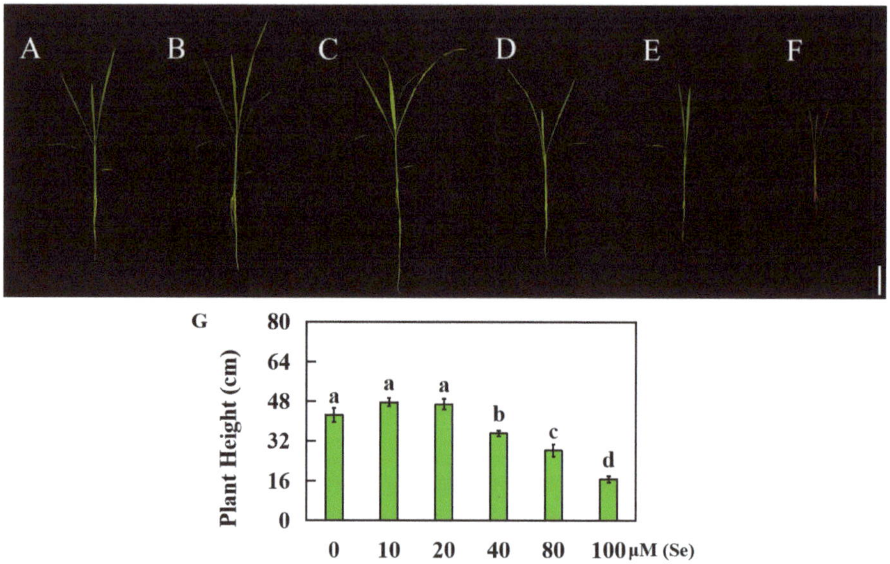

Figure 1. Phenotype of rice treated with different Se concentrations. Rice seedlings with a seedling age of 7 days were used for treatment with (**A**) 0; (**B**) 10; (**C**) 20; (**D**) 40; (**E**) 80; and (**F**) 100 μM Na_2SeO_3 for 20 days (bar = 5 cm). (**G**) The plant height of rice. The results are expressed as the mean and SD (standard deviation) from experiments repeated three times. Duncan's multiple comparison method was used to compare the differences in Se toxicity among all the groups. Different lowercase letters marked on the column diagrams indicate that there were significant differences between these data ($p < 0.05$).

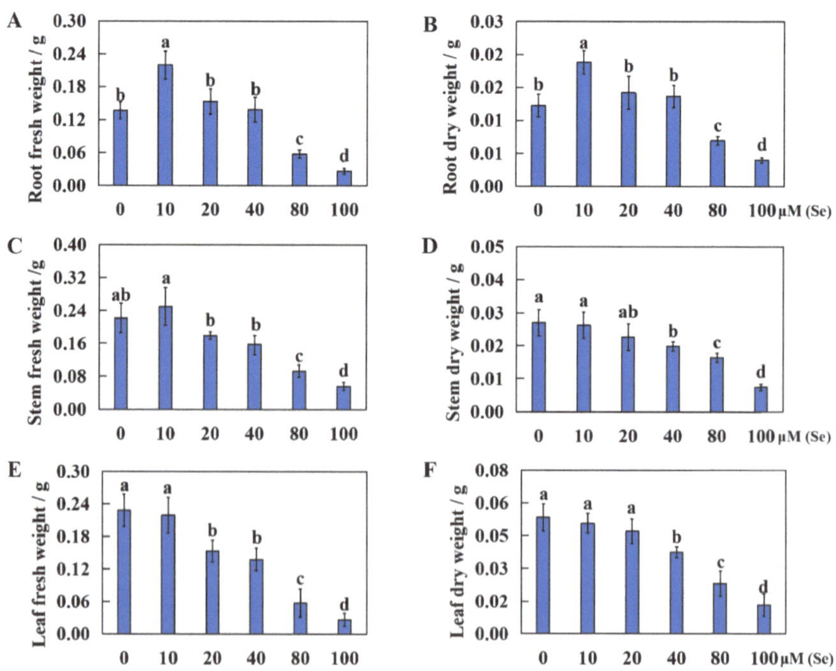

Figure 2. Results for various Se levels on the fresh and dry weight of rice roots, stems, and leaves. After treatment with 0, 10, 20, 40, 80, and 100 μM Se for 20 days, the dry and fresh weights were tested, and

the CK was 0 μM Se. (**A**) Fresh weight of roots; (**B**) dry weight of roots; (**C**) fresh weight of stems; (**D**) dry weight of stems; (**E**) fresh weight of leaves; (**F**) dry weight of leaves. The results are expressed as the mean and SD (standard deviation) from experiments repeated three times. Duncan's multiple comparison method was used to compare the differences in Se toxicity among all the groups. Different lowercase letters marked on the column diagrams indicate that there were significant differences between these data ($p < 0.05$).

2.2. Influences of Various Levels of Se on Root Development in Rice

To analyze the effects of different Se concentrations on rice root development, changes in rice root growth were measured under varying Se concentrations. The CK group was treated with 0 μM of Se. The root surface area of the 10 μM treatment group significantly increased by 53.53% compared to that of the control group. However, there were no significant changes in the root surface area of the plants in the 20 μM and 40 μM Se treatment groups. Conversely, the root surface area in the other treatment groups decreased significantly, by 65.47% and 80.81%, respectively (Figure 3A). Similarly, the root volume in the 10 μM Se treatment group significantly increased by 46.06%, whereas that in the 20 μM and 40 μM Se treatment groups did not significantly change. The root volume in the other treatment groups decreased significantly, by 56.77% and 89.01%, respectively (Figure 3B). In terms of root length, a significant increase of 57.58% was observed under the 10 μM Se treatment. However, there were no significant changes in root length in the 20 μM Se or 40 μM Se treatment groups. On the other hand, the root length decreased significantly by 65.91% and 76.14%, respectively, in the other treatment groups (Figure 3C). The mean root diameter in each treatment group did not markedly change compared with that in the CK group (Figure 3D).

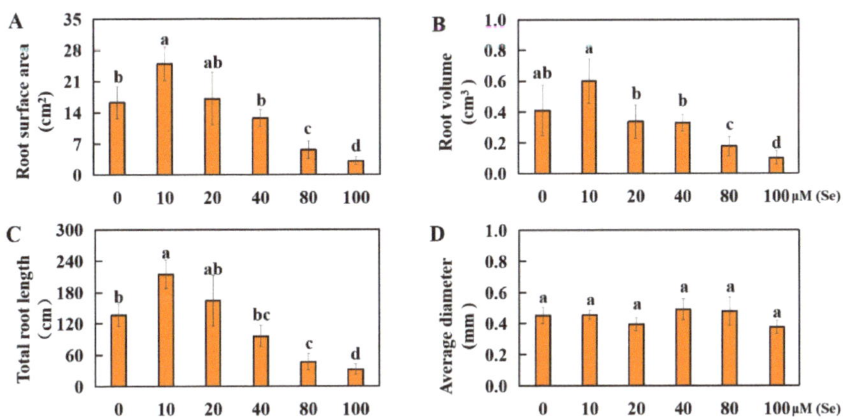

Figure 3. Results of rice root development under different selenium concentrations. (**A**) Surface area of roots; (**B**) volume of roots; (**C**) total root length; (**D**) mean diameter of roots. The results are expressed as the mean and SD (standard deviation) from experiments repeated three times. Duncan's multiple comparison method was used to compare differences in Se toxicity among all the groups. Different lowercase letters marked on the column diagrams indicate that there were significant differences between these data ($p < 0.05$).

2.3. Effects of Se Stress on the Physiological Indicators of Rice

Under high Se stress, the physiological response indices of rice were significantly impacted. Moreover, the rice plants experienced pronounced growth stress under the Se treatment at concentrations of 80 and 100 μM. The severe dwarfing and yellowing observed were a direct result of the excessive growth stress caused by the 100 μM Se treatment. However, this concentration was deemed too high for research purposes, leading to the selection of 80 μM of Se as the stress concentration. A comparison between the control group treated with 0 μM of

Se and the group treated with 80 µM of Se for 20 days revealed the significant influence of Se stress on the physiological indicators of roots, stems, and leaves in rice.

Compared with those in the CK group, the CAT activity in the roots, stems, and leaves increased by 53.33%, 34.21%, and 42.86%, respectively, under Se stress (Figure 4A). Similarly, the superoxide (SOD) activity increased by 113.51%, 96.33%, and 79.44% in the roots, stems, and leaves, respectively, under high Se stress (Figure 4B). Under Se stress, the POD activity in the roots, stems, and leaves significantly increased by 139.67%, 113.34%, and 116.98%, respectively (Figure 4C). The APX activity also significantly increased by 178.26%, 78.99%, and 158.18% in the roots, stems, and leaves, respectively, under stress treatment (Figure 4D). The content of malondialdehyde in the roots, stems, and leaves significantly increased by 160.73%, 137.23%, and 54.29%, respectively, under Se treatment (Figure 4E). Compared with those in the CK group, the proline levels in the leaves did not significantly change under Se stress but increased significantly, by 172.50% and 33.58%, in the roots and stems, respectively (Figure 4F). Compared with those in the CK group, the soluble sugar levels in the leaves, roots, and stems under high Se stress tended to decrease overall, with a significant decrease of 21.95% in the roots (Figure 4G). The levels of soluble protein in the leaves, roots, and stems decreased under Se stress, but not significantly (Figure 4H).

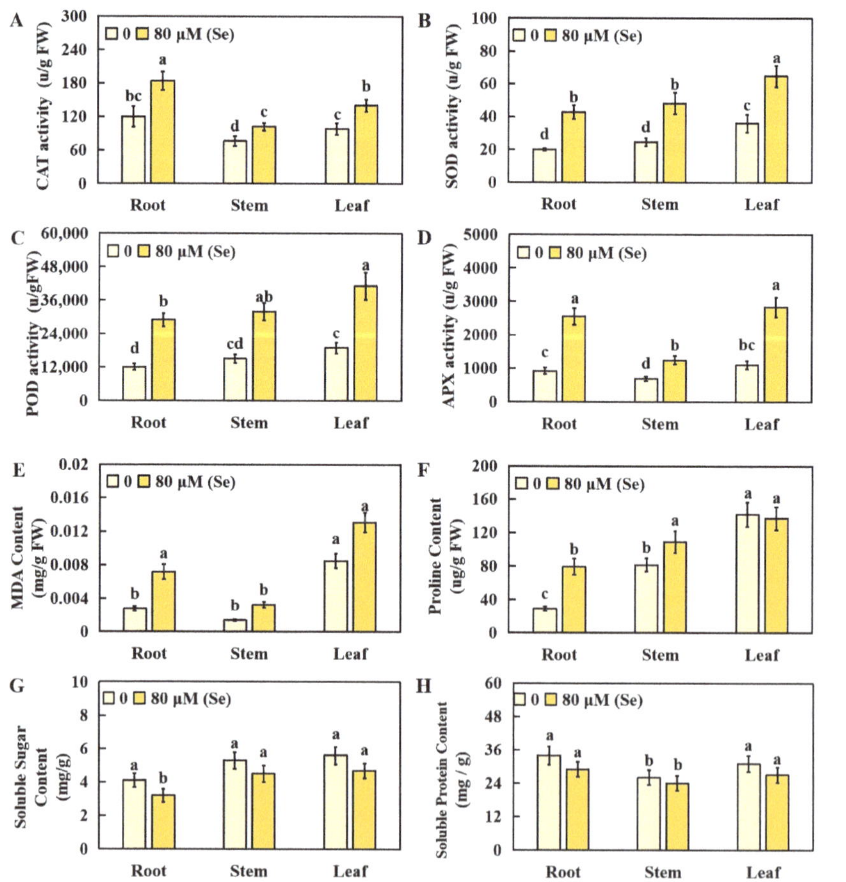

Figure 4. Effects of physiological indices in rice roots, stems, and leaves under selenium poisoning stress. The concentration of selenium toxicity was 80 µM Se, and the control group was 0 µM Se. Eight physiological indices were determined: the activities of (**A**) CAT (catalase), (**B**) SOD (superoxide

dismutase), (**C**) POD (peroxidase), and (**D**) APX (ascorbate peroxidase) and the contents of (**E**) malondialdehyde (MDA), (**F**) proline, (**G**) soluble sugar, and (**H**) soluble protein. The results are expressed as the mean and SD (standard deviation) from experiments repeated three times. Duncan's multiple comparison method was used to compare the differences in Se toxicity among all the groups. Different lowercase letters marked on the column diagrams indicate that there were significant differences between these data ($p < 0.05$).

2.4. *Influences of Se Poisoning Stress on Chlorophyll Levels in Rice*

The Se stress treatment had an impact on the content of photosynthetic pigments in the rice leaves. Compared with that in the CK group treated with 0 µM of Se, the content of chlorophyll in rice decreased by 12.11% under the 80 µM Se treatment, although this change was not significant (Figure 5A). Conversely, the chlorophyll b content significantly decreased by 30.29% under the Se stress treatment (Figure 5B). The overall concentration of total chlorophyll decreased by 17.77%, but this change was not statistically significant (Figure 5C).

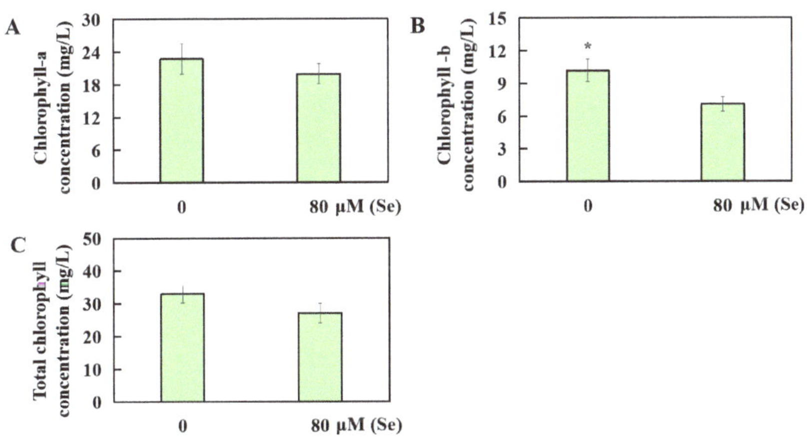

Figure 5. Changes in photosynthetic pigment content in rice leaves under different Se concentrations. The levels of (**A**) chlorophyll a, (**B**) chlorophyll b, and (**C**) total chlorophyll were measured. The results are expressed as the mean and SD (standard deviation) from experiments repeated three times. The *t*-test was used to compare the differences in Se toxicity among all the groups (* $p < 0.05$).

2.5. *Influences of Se Stress on the Accumulation of Several Elements in Rice*

The concentrations of Se, Zn, Cu, Fe, Mn, Ca, K, Al, Mg, and Na were determined after a 20-day treatment with 80 µM of Se. The samples were then dried until a constant weight was achieved. This research aimed to explore the influence of Se toxicity stress on the accumulation of these 10 elements in rice roots and leaves.

In the Se stress treatment group, compared to those in the CK group, the Se concentrations in the roots, stems, and leaves significantly increased by 9003.18%, 14803.87%, and 4645.63%, respectively (Figure 6A). There was no obvious change in the Na levels in the roots, but a significant increase of 290.56% and 146.95% was detected in the stems and leaves, respectively (Figure 6B). The Mg levels in the roots, stems, and leaves decreased significantly by 62.77%, 71.69%, and 16.68%, respectively (Figure 6C). Although there was no obvious change in the Al levels in the rice stems, it significantly decreased in the roots and leaves by 41.49% and 38.95%, respectively (Figure 6D). The K levels in the roots, stems, and leaves decreased significantly, by 84.17%, 77.18%, and 49.39%, respectively (Figure 6E). The Ca contents in the roots did not significantly change, whereas the Ca contents in the stems and leaves significantly decreased by 64.09% and 62.23%, respectively (Figure 6F). The contents of Mn in the roots, stems, and leaves were markedly reduced by 32.37%, 96.68%, and 94.52%, respectively (Figure 6G). The Fe contents did not change significantly in the roots and stems but significantly decreased in the leaves by 15.92% (Figure 6H). The Cu levels in the roots, stems, and leaves decreased significantly, by 20.95%, 81.84%, and 59.19%, respectively (Figure 6I). Furthermore, the Zn contents in the roots, stems, and leaves decreased markedly, by 75.59%, 91.72%, and 51.52%, respectively (Figure 6J).

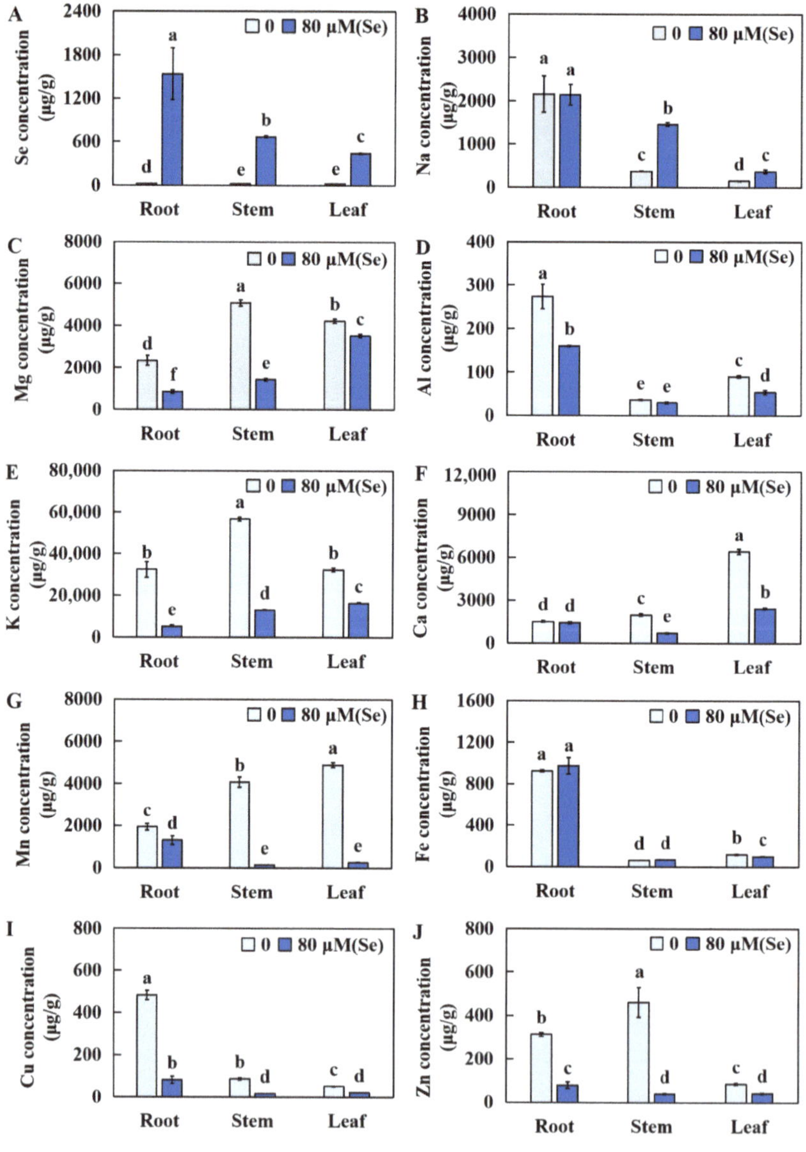

Figure 6. Accumulation of 10 elements in rice roots, stems, and leaves under Se toxicity stress. **(A)** Se (selenium); **(B)** Na (sodium); **(C)** Mg (magnesium); **(D)** Al (aluminum); **(E)** K (potassium); **(F)** Ca (calcium); **(G)** Mn (manganese); **(H)** Fe (iron); **(I)** Cu (copper); **(J)** Zn (zinc). The results were expressed as the mean and SD (standard deviation) from experiments repeated three times. Duncan's multiple comparison method was used to compare differences in Se toxicity among all the groups. Different lowercase letters marked on the column diagrams indicate that there were significant differences between these data ($p < 0.05$).

2.6. Determination of Five Kinds of Hormones from Roots and Leaves in Rice

Under high Se stress, the contents of IAA (indoleacetic acid), SA (salicylic acid), and ABA (abscisic acid) in the roots significantly increased by 29.71%, 463.71%, and 83.91%, respectively. However, the levels of zeatin and JA (jasmonic acid) were significantly reduced by 89.59% and 28.38%, respectively (Figure 7A). Compared with that in the CK group, the level of IAA in the leaves did not significantly change under Se stress. However, the contents of zeatin and JA were significantly reduced by 44.12% and 49.69%, respectively, whereas the contents of SA and ABA were markedly increased by 28.38% and 17.62%, respectively (Figure 7B). The retention time and HPLC chromatograms of zeatin, SA, JA, IAA, and ABA are displayed in Figure S1.

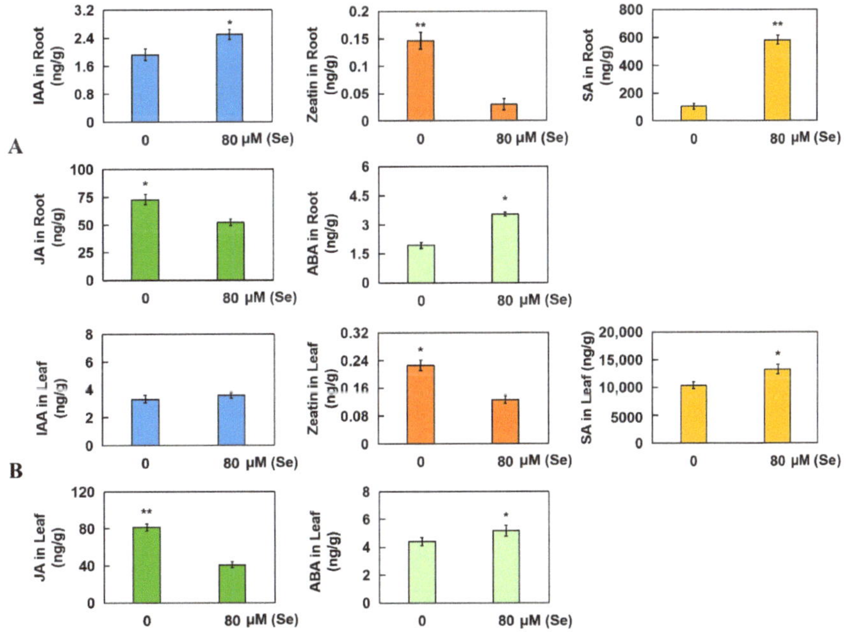

Figure 7. Changes in five hormone levels in rice roots and leaves under Se stress. The five hormones measured in rice roots (**A**) and leaves (**B**) were IAA, zeatin, SA, JA, and ABA. The results are expressed as the mean and SD (standard deviation) from experiments repeated three times. The *t*-test was used to compare differences in Se toxicity among all the groups (* $p < 0.05$, ** $p < 0.01$).

2.7. Effects of Se Poisoning Stress on the Relative Expression of Nine Related Genes in Rice

To investigate the effect of selenium stress on the expression of nine related genes in leaves, the relative transcript levels of three types of genes were measured via real-time fluorescence. These gene categories included genes related to selenium metabolism (*SMT1*, *SMT2*, and *SMT3*), physiological response genes (*Fe-SOD*, *Cu/Zn-SOD*, *CATB*, and *APX6*), and ion transport genes (*NRAMP3* and *NRAMP5*). *SMT1*, *SMT3*, *Fe-SOD*, *Cu/Zn-SOD*, *APX6*, and *NRAMP3* were upregulated in the roots (Figure 8A,C–E,G,H). *CATB*, on the other hand, was significantly downregulated in the roots (Figure 8F), whereas *SMT2* and *NRAMP5* did not significantly change in the roots (Figure 8B,I). In the stem, the significantly upregulated genes were *SMT1*, *SMT2*, *SMT3*, *Fe-SOD*, *Cu/Zn-SOD*, and *NRAMP3* (Figure 8A–E,H). *CATB* and *NRAMP5*, on the other hand, were significantly downregulated in the stem (Figure 8F,I), whereas *APX6* did not significantly change (Figure 8G). In the leaves, *SMT1*, *SMT3*, *Fe-SOD*, *APX6*, and *NRAMP3* were significantly upregulated (Figure 8A,C,D,G,H). Conversely, *CATB* and *NRAMP5* were significantly downregulated (Figure 8F,I), whereas *SMT2* and *Cu/Zn-SOD* did not significantly change (Figure 8B,E).

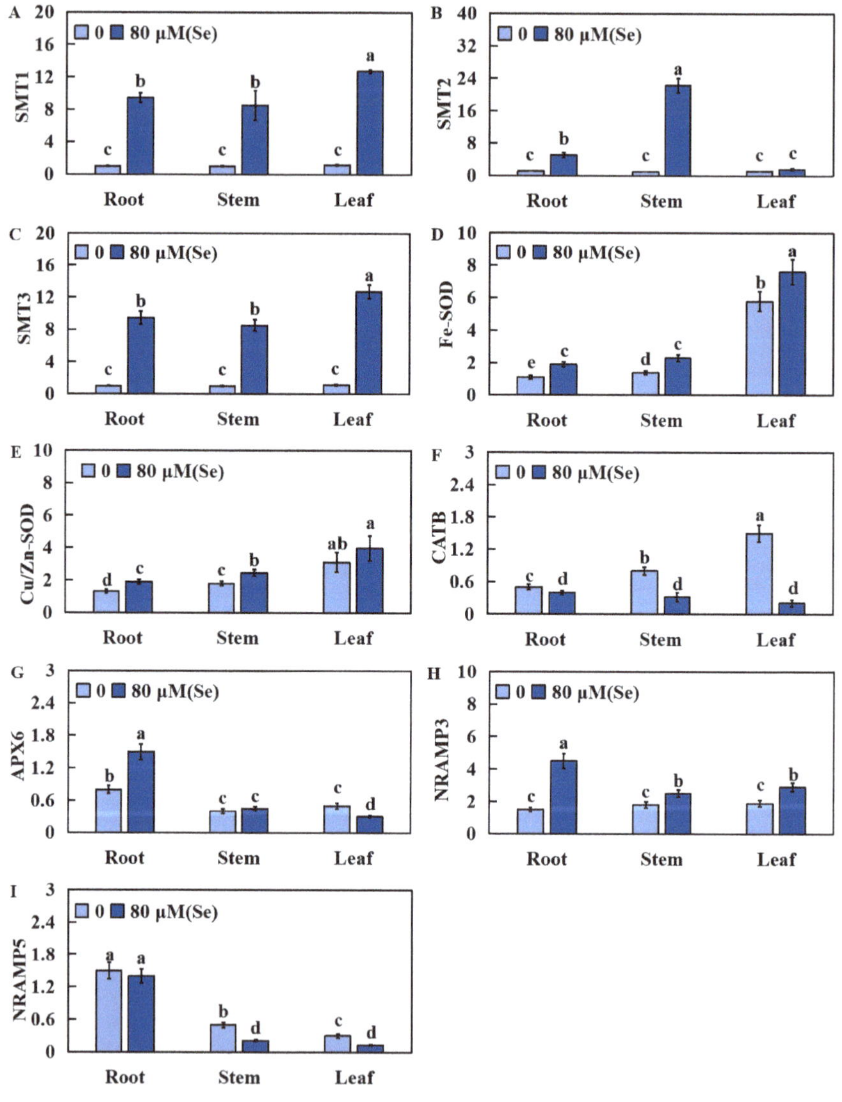

Figure 8. Relative transcript levels of 9 related genes in leaves, stems, and roots of rice under Se stress. (**A**) SMT1; (**B**) SMT2; (**C**) SMT3; (**D**) Fe-SOD; (**E**) Cu/Zn-SOD; (**F**) CATB; (**G**) APX6; (**H**) NRAMP3; (**I**) NRAMP5. The results are expressed as the mean and SD (standard deviation) from experiments repeated three times. Duncan's multiple comparison method was used to compare differences in Se toxicity among all groups. Different lowercase letters marked on the column diagrams indicate that there were significant differences between these data ($p < 0.05$).

3. Discussion

The concentration and availability of Se in crops are affected by soil characteristics and environmental factors [14]. Suitable concentrations of Se can enhance plant development and improve stress resistance, but excess Se can be toxic [14]. Hamilton identified three levels of biological activity for Se. The first level suggests that low Se concentration doses can facilitate plant development, whereas the second level suggests that moderate doses can maintain homeostasis processes. The third level suggests that high doses of Se can have harmful consequences [52,53]. For example, low Se concentrations significantly promote the growth of apple (*Malus pumila*) plants, but once the Se concentration exceeds 12 μM, plant growth is notably inhibited, particularly in the roots [54]. High concentrations of Se in plants can induce phytotoxicity by interfering with metabolism, resulting in chlorosis, cell membrane degradation, senescence, and reduced grain yield [55]. Studies have shown that sodium selenate has dual effects on rice seedling growth and development: low levels facilitate development, whereas high levels restrain development [56]. Similarly, the results of this study demonstrated that Se has a similar effect on rice. Under low Se treatment, rice seedling growth increased, but under high Se treatment, the plant height, root length, root volume, total number of roots, root surface area, and dry and fresh weights of the rice significantly decreased due to high Se stress. Consequently, these changes result in stunted plants, yellow leaves, and hindered growth. These effects are likely due to the strong binding of Se to the cell wall in rice root cells. An excessive Se concentration disrupts the structure and stability of rice root cells, leading to cell deformation, cell death, and diminished nutrient absorption capacity in rice plants.

In addition to its significant effect on human health, Se also has a crucial role in the antioxidant protection of plants [56]. Se has the ability to stimulate growth, resist pathogens, and confer tolerance to oxidative stress [57]. When plants suffer from stress conditions, they create excess ROS, which can result in damage to cell components, the destruction of cell membranes, and the inhibition of plant growth [58]. To combat oxidative stress, plants rely on their antioxidant resistance system and osmotic adjustment system, both of which contribute to their resistance to such stress [59].

When Se is used in conjunction with other nutrients during fertilization, it can activate enzymes such as SOD and catalase (CAT) depending on the dosage. These enzymes are essential for reducing oxidative stress in plant cells and are especially active in the presence of Se [60]. Se in plants can function as a scavenger of ROS at low concentrations by increasing antioxidant metabolism, but at higher concentrations, it can actually enhance oxidative stress [61]. SOD, which is the main antioxidant enzyme responsible for scavenging ROS, helps maintain the balance of reactive oxygen metabolism and protect the integrity of the cell membrane structure [62].

To counter excessive Se accumulation, plants have developed an antioxidant system regulated by enzymes such as SOD, CAT, APX, glutathione reductase (GR), and glutathione (GSH). The rice plants treated with Se exhibited increased SOD and CAT activity, indicating the importance of these antioxidant enzymes [61]. Consequently, it is crucial to measure the antioxidase activities and osmoregulatory components to assess the metabolic activity and overall health of cells [61].

It is widely recognized that Se can increase SOD activity and gene expression levels [61]. In a previous research report, SOD activity was increased by adding 0.5 mg/kg of Se to rice [61]. The control of SOD activity in response to Se is influenced by three main factors in plants, as previously reported [45,57]. The first factor is the stress level [45,57]. Under low stress conditions, the antioxidant capacity of plants is sufficient, whereas under high stress conditions, it may be necessary to increase SOD activity, which can be accomplished by adding Se [45,57]. The second element affecting the activity of SOD is the concentration of Se [45,57]. The third factor is connected with the concentration of SOD helper factors, such as Zn, Cu, Mn, and Fe [45,57]. SOD has an important role as the first defense against antioxidant systems in plants [62]. The family of plant SOD genes can be grouped into three categories on account of their metal helper factors: FeSOD, Cu/ZnSOD, and MnSOD [62]. In oilseed rape (*Brassica napus*), various genes, including *BnFeSOD4*, *BnFeSOD5*, *BnFeSOD6*, *BnMnSOD2*, *BnMnSOD10*, *BnCuSOD1*, *BnCuSOD3*, and *BnCuSOD14*, were markedly upregulated when it was subjected to different abiotic stresses, such as drought, salt, cold, and flooding [63]. In this study, the expression of SOD in the stems, leaves, and roots of rice subjected to Se poisoning stress significantly increased, whereas the levels of Cu and Zn significantly decreased, consistent with previous research results. This suggests that the absorption of Fe, Cu, and Zn in rice might be altered under the influence of Se, which could affect the differential expression of *Cu/ZnSOD* and *FeSOD* in rice as well as the SOD activity in rice leaves and roots. The family of POD genes plays a part in the biosynthesis of oxidoreductases in grapes (*Vitis vinifera*) [64]. The class III plant POD is a plant-specific oxidant and is widely distributed in animals, plants, and microorganisms [65]. Previous studies have demonstrated the involvement of PODs in various processes, such as lignification, plant defense, development, and germination, as well as their mechanisms of action, including substrate oxidation, reactive oxygen species regulation, and free radical formation, particularly in lignification [66]. For instance, PODs (AtPrx72 and PtrPO21) in *Arabidopsis thaliana* and Populus trichocarpa play a crucial role in leaf lignification [67]. The overexpression of *POD* in *A. thaliana* (*AtPrx22*, *AtPrx39*, and *AtPrx69*) has been shown to improve cold tolerance [68]. In this study, Se poisoning induced differential *POD expression*, leading to a significant increase in *POD* expression. These changes could be attributed to the irreversible damage caused by Se to the formation of xylem in leaves, triggering defense stress in rice plants, which affects the capacity of resistance to oxidation in rice as well as the development of leaves and roots. CAT is primarily found in peroxisomes and catalyzes the conversion of H_2O_2 to water [69]. The expression level of *CAT* in some crops is affected by phytohormones, such as ABA, which induces *AtCAT1* expression in *A. thaliana* [69]. CATB, also known as catalase isoenzyme B, works together with APX to clear H_2O_2 and regulate the redox balance. Specifically, CTAB interacts with OsAPX5 and OsAPX6, which are located in the mitochondria [70]. In oilseed rape, various genes, including *BnCAT13*, *BnCAT11*, *BnCAT3*, and *BnCAT1*, were markedly upregulated when it suffered from GA, ABA, salt, and low-temperature treatments [71]. Previous research reports have shown that genes such as *AhAPX4*, *AhAPX17*, and *AhAPX19* in peanuts (*Arachis hypogaea*) were significantly upregulated under abiotic stress and hormone treatments (SA and ABA), suggesting that the synthesis of APX is stimulated by various environmental stresses and that the differential expression of *APX* is related to hormone concentrations in peanut cells [72]. Therefore, the variations observed in APX and CAT activities in the current study might be attributed to the oxidative damage induced by Se stress as well as changes in hormone concentrations in rice cells.

In the current study, compared with those in the control group, the enzymatic activities of APX, CAT, POD, and SOD in the stems, leaves, and roots of the rice plants under Se stress were significantly elevated. The increase in these antioxidase activities suggested that the cell membrane system of the rice was partially damaged by Se poisoning. CAT, APX, SOD, and POD have crucial roles in preserving plant cell membrane systems during Se poisoning [73]. Similar physiological indicators have been surveyed when subjected to other types of abiotic stresses, such as in aluminum-treated soybeans (*Glycine max*) [74].

Furthermore, soluble sugar, proline, and soluble protein contribute to mitigating the effects of high Se stress [75]. This study revealed changes in soluble protein, proline, and soluble sugar levels. Under high Se stress, the soluble protein and soluble sugar levels decreased to varying degrees in the stems, leaves, and roots of rice, whereas the proline content increased. These findings suggest that rice responds collaboratively to Se poisoning by regulating the levels of osmoregulatory constituents such as proline, soluble sugar, and soluble protein. MDA is a final product of unsaturated fatty acid overoxidation, which can injure cell membranes when it accumulates [76]. The MDA content serves as an important indicator of lipid overoxidation in plants, and lipid peroxidation and protein changes induced by ROS are also controlled by high Se concentrations in plants [61]. In this study, the MDA content significantly increased in the stems, leaves, and roots of the plants that suffered high Se stress. This increase may be attributed to the elevated Se concentration and the additional peroxidation damage caused by Se poisoning, resulting in a progressive increase in malondialdehyde level.

In summary, under high Se stress, the oxidation resistance protection system and osmotic regulation system in rice might have important functions, resulting in changes in the levels of APX, POD, SOD, CAT, soluble protein, proline, MDA, and soluble sugars. These changes might serve to protect rice cells and mitigate oxidative damage.

There was a significant increase in the accumulation of Se in the rice stems, leaves, and roots after Se treatment compared with that in the control group. This augmentation might be the leading reason for Se toxicity in rice plants. Previous research reports have demonstrated that the application of selenocysteine to roots leads to a notable increase in extractable Se content in rice leaves [51]. Lin et al. reported that the addition of selenate to soil is the most efficient solution to enhance the Se content in rice grains [77]. It is important to note that in this research, the contents of Mg and K in the rice leaves, stems, and roots decreased distinctly. Past research has shown a synergistic association between Mg and K in plants and that K and Mg have synergistic effects on processes such as photosynthesis, carbohydrate transport and distribution, nitrogen metabolism, and the regulation of expansion [78]. The interaction between Mg and K occurs not only during the root absorption process but also during root-to-bud translocation, distribution, and utilization [79]. Similarly, the findings suggest that Mn plays a synergistic role in the accumulation of Al [80]. Ca is a component of cell wall pectin and affects the mobility of cytomembranes in plants [81]. In the current study, the decrease in the Ca, K, Mg, and Al contents in the rice under Se toxicity stress might have been due to the high level of Se taking up a large quantity of combining sites in the rice cells.

Selenocysteine methyltransferase (SMT) is a crucial enzyme in the plant Se metabolism pathway [82]. Studies on transgenic *Arabidopsis thaliana* plants that overexpress *SMT* have indicated that SMT has a function in the synthesis of MeSeCys and MeCys, suggesting that SMT may catalyze the methylation of selenocysteine and cysteine [82]. LeDuc et al. successfully cloned the SMT gene from Se-enriched Indian mustard (*Brassica juncea*) plants, and the resulting transgenic plants exhibited improved tolerance and accumulation of selenite [83]. These findings imply that SMT has significant potential for regulating the Se tolerance and Se enrichment ability of crops [84]. The results revealed that the SMT-related genes *SMT1*, *SMT2*, and *SMT3* were significantly upregulated under Se toxicity stress in this study. This could be attributed to the application of exogenous Se promoting the expression level of genes connected with the Se metabolism pathway in rice, thus facilitating the absorption of large quantities of Se by individual plants.

The family of NRAMPs is crucial for the uptake and transport of metals in crops, with peanut (*Arachis hypogaea*) *AhNramps* expressed mainly in stamens, immature seeds, and roots [85]. The NRAMPs play a vital role in transporting metallic ions through cytomembranes [86]. For example, the expression levels of *GmNRAMP5* and *GmNRAMP1* in soybeans are upregulated under Cu toxicity stress [86]. Both OsNRAMP3 and OsNRAMP5 are located in the plasma membrane and contribute to the absorption or transport of Mn, with the latter also involved in Fe absorption and transport [87,88]. In this study, *OsNramp5* and *OsNramp3*, members of the rice NRAMP family, exhibited differential expression in response to Se toxicity. These findings suggest that the NRAMP family in rice might have an important function in the resistance of rice to Se poisoning and could impact the absorption or transport of certain ions in rice.

In recent years, several studies have reported the influence of ABA on Se and heavy metal uptake and accumulation in plants [89–91]. ABA is a crucial hormone for plant resistance to metal ion toxicity [92]. Furthermore, ABA primarily plays functions in the elongation and meristem regions of plant roots [92]. In addition, ABA adjusts osmotic pressure equilibrium and stomatal opening and closing, thereby enhancing tolerance to adversity [93]. Similarly, the foliar application of ABA under Se toxicity stress has been found to stimulate the production of stress response phytohormones (such as SA (salicylic acid) and JA (jasmonic acid)), ultimately improving the Se resistance of plants [94]. In the current research, the ABA content in the rice roots markedly increased when suffering from Se poisoning, potentially contributing to Se poisoning in rice. Furthermore, the results might be due to Se-induced genes related to ABA synthesis in rice roots, indicating that increasing the ABA content can enhance the resistance of rice plants to Se toxicity stress.

SA promotes plant stress resistance by regulating photosynthesis and proline and nitrogen metabolism and stimulating the oxidation resistance system [95]. SA has been proven to decrease K outflow and Na afflux, thereby maintain ion balance and saline–alkaline resistance [95]. Adding Se to tomato (*Solanum lycopersicum*) roots and leaves under saline–alkali stress has been found to activate the SA biosynthesis pathway and increase SA levels, consequently improving tomatoes' saline–alkaline resistance [53]. Since both SA and Se facilitate oxidation resistance systems, they may have synergistic effects through feedback loops [53]. SA participates in adjusting various facets of healthy plant development, including protein biosynthesis, signal transduction, material transportation, and so on [96]. Some studies have shown that SA can alleviate selenium toxicity in rice [97]. SA can maintain selenium homeostasis and reduce the oxidative stress induced by selenium stress by increasing the level of ascorbate and the activities of antioxidant enzymes such as superoxide dismutase, catalase, glutathione peroxidase, and glutathione reductase [97]. In addition, SA protects rice from the harmful effects of methylglyoxal by stimulating the glyoxalase activity [97]. In the present research, the SA levels in the rice leaves and roots markedly increased when suffering from Se poisoning. This could be due to Se-induced SA synthesis, leading to elevated SA levels in rice roots and leaves. Consequently, the phytohormone distribution and level in the rice leaves and roots were influenced, resulting in poor rice seedling growth.

JA acquires both biological and abiotic stress survivability by stimulating the oxidation resistance system and interacting with other phytohormones [98]. JA is related to Se tolerance, as Se addition activates JA synthesis and signal transduction, thus maintaining a high antioxidant content in Se superaccumulators [99]. During pathogenic microorganism infections, JA activates a signal pathway that triggers plant self-protection responses, including the synthesis and biological accumulation of plant antitoxins and pathogenic microorganism growth suppressor substances. It also alters the cytoderm structure to inhibit pathogenic microorganism entry [100]. Analogously, Se enhances the biosynthesis of antioxidant substances that can eliminate stress products, thereby improving disease enduring [99]. Jasmonate compounds include JA and its derivative methyl jasmonic acid (MeJA), which are growth regulators and signaling molecules associated with plant damage and can regulate plant growth and stress [101]. In addition, a low concentration of MeJA (0.1–1.0 μM) inhibited the non-reversible toxicity of rice seedlings induced by high selenium by enhancing the antioxidant system, reducing the content of H_2O_2 and MDA, and thereby alleviating the toxicity of high selenium treatment on rice seedlings [101]. Under Se poisoning, the level of JA in the rice roots and leaves significantly decreased. This decrease might be due to an antagonistic interaction between SA and JA, as an increase in the SA level led to a decrease in the JA level.

The stabilization of zeatin contents reflects plants' coping styles in response to environmental changes [102]. Zeatin is primarily created by cells that quickly divide, such as in buds or root tips, and has a crucial role in enhancing root elongation, leaf expansion, the germination of seeds, and cell division and differentiation in plants [103]. Studies have revealed that under NaCl stress treatment, the zeatin concentration in tomato leaves (*Solanum lycopersicum*) decreases significantly within 24–72 h after treatment. Tomatoes respond to this stress by adjusting their zeatin concentration [104]. The zeatin content also significantly reduced the level of suffering from Se poisoning. This decrease might be due to the induction of Se, which affects the expression level of related genes and hinders zeatin biosynthesis. Alternatively, this may be caused by damage to hormone transport in roots due to Se toxicity stress, which inhibits the growth of rice roots.

IAA is a crucial plant hormone essential for various aspects of plant growth and development. It is the most extensively studied auxin in plants [105]. IAA exerts pleiotropic effects in nearly every step of plants' life cycles, such as cell division and differentiation, plant organogenesis, cell development, physiological response to stress, and so on [106]. The metabolic activity and polar apportion of IAA are affected by metallic ions [107]. The results from studies have indicated that IAA has a vital impact on the physiological and biochemical processes of higher plants, regulating plant growth, development, and stress response [108]. In this research, the concentrations of IAA in the rice roots markedly increased when they were subjected to Se poisoning, indicating that Se stress inhibits the growth of rice roots. This increase in IAA may be due to high Se stress triggering the defense mechanism of roots, resulting in substantial production of IAA in the roots to cope with adversity.

Based on the current results of the present study, the physiological regulation mechanisms of rice plants in response to Se poisoning stress were summarized in Figure 9. Firstly, when suffering from Se toxicity stress, the uptake of selenium via rice root cells significantly increased. Under stress conditions, plants generate a large number of reactive oxygen species to cope with stress. Excess Se led to intracellular oxidative damage, and the MDA content served as an important indicator of lipid peroxidation in the cytomembranes of the plants, leading to an increase in the MDA level in various parts of the rice plants (Figure 9I). The rice cells regulated their milieu interne by means of osmotic adjustment substances to maintain an intracellular steady state, as evidenced by the changes in the concentrations of proline, soluble protein, and soluble sugar (Figure 9II). Subsequently, due to the influx of large amounts of Se, the rice plants developed an antioxidant system to balance the increasing amount of ROS attributed to excessive Se accumulation. This led to an improvement in the enzymatic activity of antioxidases (POD, SOD, APX, and CAT) (Figure 9III). Under Se stress, the rice plants mitigated stress by modulating their own hormone levels, resulting in significant variations in the concentrations of phytohormones (JA, SA, Zeatin, IAA, and ABA) in their leaves and roots (Figure 9IV). The changes in the growth of the rice plants under Se stress were jointly determined by genes related to Se metabolism, antioxidase activity, and ion transport (Figure 9V). Consequently, the weight of the rice stems, leaves, and roots decreased, and plant growth was inhibited.

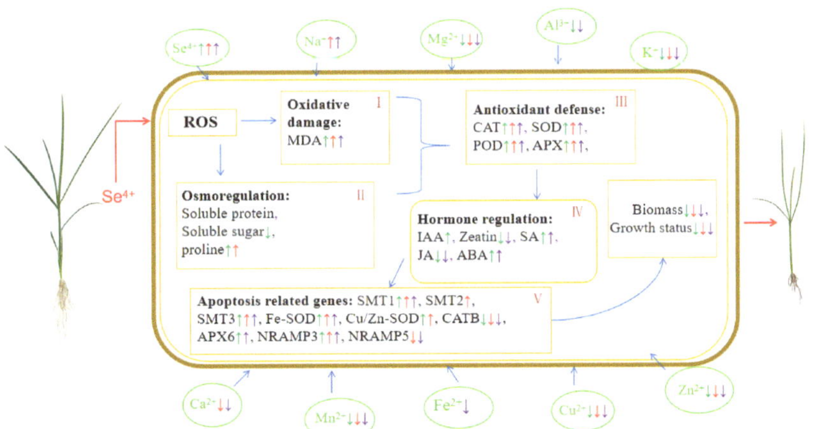

Figure 9. Regulatory pathways of high Se stress in rice. The rice plants on the left are the control group, and the rice plants treated with Se stress are on the right. (**I**) Oxidative damage; (**II**) osmotic adjustment; (**III**) antioxidant defense; (**IV**) hormone regulation; (**V**) expression of related genes. Green arrows indicate significant changes in physiological response indicators or gene expression levels in roots, red arrows indicate significant changes in stems, and purple arrows indicate significant changes in leaves.

4. Materials and Methods

4.1. Plant Materials

The rice variety used in the present study was Haihong 11 (Se sensitivity), which was generously donated by Professor Zhou Hongkai and cultivated at the plantation base of the Coastal Agricultural Sciences College of Guangdong Ocean University (East longitude: 110.30311, Northern latitude: 21.15005).

4.2. Material Handling

Rice seeds with full grains were chosen. The rice seeds were rinsed with tap water and then disinfected with 1.5% sodium hypochlorite (Solarbio, Beijing, China) for 30 min. After the seeds were thoroughly washed with deionized water 5 times, they were placed on moist absorbent paper in a glass culture dish (Devan, Wuxi, Jiangsu, China) with a diameter of 13.5 cm for 7 days of germination culture. After that, the rice seedlings were transplanted to a plastic bucket (5 L) with a 4 L hydroponic nutrient solution for 10 days of strong seedling culture. Finally, rice seedlings were treated by adding 0, 10, 20, 40, 80, or 100 μM sodium selenite (Na_2SeO_3) [50,109,110] (Solarbio, Beijing, China) and keeping them in a hydroponic nutrient solution for 20 days. The rice seedlings were cultivated in a growth chamber at 24–29 °C (daytime)/19–23 °C (night), with a light and dark cycle of 12/12 h, a relative humidity of 70%, a light intensity of 2000 lux, and Kimura Culture B nutrient solution (Table S1) [111]. Then, 8–10 rice seedlings were treated with each concentration. The experiment was repeated three times for each concentration treatment.

4.3. Determination of Fresh and Dry Weight and Plant Height of Rice

The fresh weight and plant height of the rice were measured immediately after harvest. Fresh rice seedlings were accurately weighed using electronic balances (Sartorius, Gottingen, Germany). Prior to dry weight measurement, the samples were moved into a constant-temperature drying oven (Yiheng, Shanghai, China) at 80 °C for 7 days [112].

4.4. Surveying of Root Phenotype Data

A Win RHIZO LA6400XL root scanning system (Regent, Vancouver, BC, Canada) was utilized to scan the root phenotype data of the rice plants, and images of the roots of each individual plant were obtained. Subsequently, Win RHIZO (WinRHIZO 2013e Professional Edition, Regent, Vancouver, BC, Canada) software was adopted to survey root surface area, total root length, number of root tips, and root volume. The measurements were repeated four times [113].

4.5. Surveying of Metal Ion Levels in Rice Stems, Leaves, and Roots

The metal ion levels in the stems, leaves, and roots of rice treated with 0 or 80 μM Se were assessed. Approximately 0.2 g of biologically dry sample was moved into a polytetrafluoroethylene crucible (Yaron Instrument, Shijiazhuang, China), wetted with a little water, and supplemented with 5 mL of 8 M HNO_3 (Solarbio, Beijing, China). The mixture was placed at room temperature for 1 day until the dry sample was fully dissolved in nitric acid. Subsequently, the mixture was transferred to an electric heating plate (DB-XWJ, MITR, Changsha, China) and gradually heated to 180 °C for digestion. Upon reaching a minimal liquid state in the crucible, the crucible was decanted. After slight cooling, 2 mL of 8 M HNO_3 and 2 mL of 30% H_2O_2 (Solarbio, Beijing, China) were supplemented, and the mixture was digested on an electric heating plate (Binzhenghong GS-type, Nanjing, China) at 180 °C. Finally, the mixture was heated and evaporated until nearly dry, after which 1 mL of concentrated nitric acid was added. After slight heating and leaching, all the extract and residue were transferred into 25 mL colorimetric tubes (Tianbo borosilicate glass, Tianjin, China) with distilled water and brought to a constant volume. The contents of Se, Zn, Fe, Ca, Mn, K, Cu, Al, Mg, and Na in the liquid treated with 0 and 80 μM Se were tested by ICP–AES (inductively coupled plasma atomic emission spectrometry) (Hitachi PS7800, Tokyo, Japan). Each index was measured four times [114].

4.6. Determination of Physiological Response Indices and Chlorophyll Content in Rice

The physiological response indices and chlorophyll content of rice were determined as follows: The physiological indices of rice roots subjected to two different concentrations of Se (0 and 80 μM) were assessed. Physiological data, including data on enzymatic antioxidant systems (SOD, CAT, POD, APX activity) and osmoregulatory substances (MDA, soluble protein, soluble sugar, proline), were analyzed. Additionally, the chlorophyll content (chlorophyll a, chlorophyll b, and total chlorophyll) in the rice leaves was measured following Se stress treatment. The chlorophyll measurements were conducted as follows: 0.2 g of fresh leaves was extracted from each treated plant, minced with scissors, and soaked in a 95% ethanol (Solarbio, Beijing, China) solution for 12 h in the absence of light. Subsequently, an ultraviolet spectrophotometer (Yuexi UV5100B, Shanghai, China) was utilized to survey the absorbance at 663 nm, 645 nm, and 652 nm, and the OD value was recorded [115]. The content of soluble protein was tested using the Coomassie brilliant blue method [116]. The level of soluble sugar was surveyed using the anthrone method [116]. The proline level was surveyed by extracting sulfosalicylic acid (Solarbio, Beijing, China) and subsequently measuring its content via the ninhydrin reaction [117]. The MDA content was determined using the TBA method (thiobarbituric acid) [118]. The activity of POD was surveyed using the guaiacol method [119]. The activity of SOD was determined using the nitroblue tetrazolium (NBT) method [120]. CAT activity was measured using a spectrophotometer [121]. The APX activity was surveyed using a spectrophotometer at 290 nm with ascorbic acid (Solarbio, Beijing, China) and a hydrogen peroxide solution (Solarbio, Beijing, China) [122].

4.7. Determination of the Levels of Six Kinds of Hormones in Rice Leaves and Roots

The levels of six hormones in rice leaves and roots treated with 0 and 80 μM Se were determined. Following the reported method [112], the operating solution was prepared, and the hormone standard curve was calculated to extract the hormones from the rice roots and leaves. For this experiment, an HPLC (high-performance liquid chromatography) system (Agilent 1290, Santa Clara, CA, USA) in conjunction with an AB Qtrap6500 mass spectrometer (AB, Madison, WI, USA) was used. The mobile phase was pumped into the system by an HPLC binary pump and mixed, and the sample to be tested was injected into the mobile phase using an automatic sampler before entering the chromatographic column. The components of the sample were separated based on their retention time in the chromatographic column and then entered the ion source sequentially. The levels of the endogenous hormones IAA, ABA, JA, SA, and zeatin in rice roots and leaves were detected using multiple response detection (MRM) mode scanning [123]. The Qtrap6500 mass analyzer consisted of Q1, Q2, and Q3 triple mass spectrometry in series, where Q1 separated the molecular ions based on the mass–charge ratio (m/z), Q2 served as a collision chamber to further split the parent ions into fragment ions, and Q3 functioned as a four-stage bar and a linear ion trap. Internal standard substances, including IAA, ABA, JA, SA, zeatin, and GA_3 (Sigma–Aldrich, St. Louis, MO, USA), were used. External standard substances such as deuterated abscisic acid (D-ABA), deuterated gibberellin (D-GA4), deuterated indole acetic acid (D-IAA), deuterated jasmonic acid (D-JA), deuterated zeatin (D-zeatin), and deuterated salicylic acid (D-SA) (Sigma–Aldrich, St. Louis, MO, USA) were also used. For hormone measurements, a chromatographic column (C18 QuEChERS packing) (Amperex, Shanghai, China) along with chromatographic-grade methanol and acetonitrile (Sigma–Aldrich, St. Louis, MO, USA) were used. The detailed gradient parameters for HPLC (Table S2), mass spectrum parameters (Table S3), and selected monitoring indicators for protonation or deprotonation reactions of phytohormones (Table S4) are displayed in the Supplementary Data. Mobile phase: (methanol/0.1% formic acid): (water/0.1% formic acid).

4.8. cDNA Extraction and qRNA Analysis

Rice samples subjected to different Se treatments (0 and 80 μM) were collected, and RNA was extracted from rice roots, stems, and leaves under Se toxicity stress using an RNA extraction kit (19291ES50, 50 T, Yeasen Biotechnology, Shanghai, China). A reverse transcription kit (11119ES60, 100 T, YEASEN, Shanghai, China) was used to obtain cDNA. A 20 μL reverse transcription system (5 μL of RNA, 7 μL of RNase-free H_2O, 3 μL of 5 × gDNA digester mix, and 5 μL of 4 × Hifair III Super Mix plus) was prepared under the reaction conditions (25 °C for 5 min, followed by reaction at 55 °C for 15 min, and 85 °C for 5 min) [112]. qRT–PCR was implemented using a real-time fluorescent quantitative PCR assay system manufactured by Bio-Rad (Hercules, CA, USA). The PCR system consisted of 20 μL, including 8 μL of ddH_2O, 10 μL of Universal Blue qPCR SYBR Green Master Mix (11184ES08, YEASEN, Shanghai, China), 0.5 μL of reverse primer, 0.5 μL of forward primer, and 1 μL of cDNA. The PCR conditions were as follows: 95 °C for 30 s, followed by 95 °C for 5 s, 58 °C for 60 s, and 72 °C for 30 s. This process was repeated for 45 cycles, and the final reaction was preserved at 16 °C for 20 min [124]. The $2^{-\Delta\Delta CT}$ method, in which *OsActin* (*Os03g0718100*) is used as the internal parameter, was used for data analysis [125]. The primers applied to qPCR in this study are exhibited in Table S5.

4.9. Data Processing

All the data were analyzed using Microsoft Excel 2010 (Microsoft, Redmond, WA, USA) and SPSS version 19.0 software (IBM Corporation, New York, NY, USA). The data were analyzed for significant differences via Duncan's multiple comparison test or the *t*-test [112].

5. Conclusions

This study investigated the physiological regulatory mechanism of selenium stress in rice. Initially, an increase in selenium levels was observed in rice cells due to the presence of a large amount of exogenous selenium during selenium poisoning stress. This increase in selenium activated a cellular defense mechanism in rice, leading to changes in the levels of MDA, APX, CAT, POD, SOD, proline, soluble sugar, and soluble protein in rice. As a result, the rice cells were protected, and oxidative damage was reduced. Furthermore, under selenium toxicity stress, selenium ions might influence the expression levels of genes related to selenium metabolic pathways and metal ion transportation. These inductions resulted in variations in the expression of SMT-related genes (*SMT1*, *SMT2*, and *SMT3*) and metal ion transport genes (*Nramp3* and *Nramp5*), leading to changes in plant hormone levels (IAA, Zeatin, SA, JA, and ABA). Finally, these changes led to symptoms such as a decrease in biomass, growth restriction, and an unnormal aggregation of metallic ions. The findings of this research lay a basis for further study on the molecular and physiological mechanisms of rice's response to selenium stress. Additionally, these results will provide an academic basis for the cultivation of new Se-rich rice varieties through breeding efforts.

Supplementary Materials: The following supporting information can be downloaded at https://www.mdpi.com/article/10.3390/plants13111580/s1: Table S1: Composition and content in hydroponic solution; Table S2: The detailed gradient parameters for HPLC; Table S3: mass spectrum parameters; Table S4: the selected monitoring indicators for protonation or deprotonation reactions of phytohormones; Table S5: the primers applied to qPCR in this study; Figure S1: The retention time and HPLC chromatograms of Zeatin, SA, JA, IAA, and ABA. The retention time and HPLC chromatograms of Zeatin, SA, JA, IAA, and ABA were displayed orderly from top to bottom.

Author Contributions: Y.L. and Y.W. conceived and designed the experiments; Y.L., J.M., F.L., X.Z., Z.W., Y.H., Y.X. and Y.W. performed the experiments; Y.L., J.M., F.L., X.Z., Z.W., Y.H., Y.X. and Y.W. analyzed the data; Y.L., J.M. and Y.W. wrote the manuscript. All authors have read and agreed to the published version of the manuscript.

Funding: This research was funded by the Project of Marine Young Talent Innovation of Zhanjiang City (210818004542097), the Project of Rural Science and Technology Team "One-to-One" Service to Help the Town Full Coverage of Action Funding in Zhanjiang City (A23081), the Postgraduate Education Innovation Project of Guangdong Ocean University–Excellent Dissertation Cultivation Project (040502752325), the Project of Social Services of Rural Science and Technology Specialists of Guangdong Ocean University (080503052212), the Graduate Education Innovation Program of Guangdong Ocean University (202401), and Maintenance Expenses of Key Construction Disciplines in High-level University of Crop Science (080508052201).

Data Availability Statement: Data is contained within the article.

Acknowledgments: We would like to thank Analytical and Testing Center of Guangdong Ocean University for determination of ion content.

Conflicts of Interest: The authors declare no conflicts of interest.

References

1. Richard, C.; Elizabeth, A.H. Selenium transport and metabolism in plants: Phytoremediation and biofortification implications. *J. Hazard. Mater.* **2021**, *404*, 124178.
2. Schiavon, M.; Nardi, S.; Dalla, F.; Ertani, A. Selenium biofortification in the 21st century: Status and challenges for healthy human nutrition. *Plant Soil.* **2020**, *453*, 245–270. [CrossRef] [PubMed]
3. Sun, Y.; Wang, Z.; Gong, P.; Yao, W.; Ba, Q.; Wang, H. Review on the health-promoting effect of adequate selenium status. *Front. Nutr.* **2023**, *10*, 1136458. [CrossRef] [PubMed]
4. Chen, Z.; Li, X.; Liu, T.; Fu, H.; Yuan, X.; Cheng, Q.; Li, X. Strategies for fertilizer management to achieve higher yields and environmental and fertilizer benefits of rice production in China. *Sci. Total Environ.* **2023**, *904*, 166325. [CrossRef] [PubMed]
5. Li, K.; Zhu, Y.; Zhang, S.; Xu, Q.; Guo, Y. Nitrate reductase involves in selenite reduction in *Rahnella aquatilis* HX2 and the characterization and anticancer activity of the biogenic selenium nanoparticles. *J. Trace Elem. Med. Biol.* **2024**, *83*, 127387. [CrossRef]
6. Kuria, A.; Tian, H.; Li, M.; Wang, Y.; Aaseth, J.O.; Zang, J.; Cao, Y. Selenium status in the body and cardiovascular disease: A systematic review and meta-analysis. *Cirt. Rev. Food Sci.* **2021**, *61*, 3616–3625. [CrossRef] [PubMed]
7. Yao, M.; Deng, Y.; Zhao, Z.; Yang, D.; Wan, G.; Xu, X. Selenium nanoparticles based on *Morinda officinalis* polysaccharides: Characterization, anti-cancer activities, and immune-enhancing activities evaluation in vitro. *Molecules* **2023**, *28*, 2426. [CrossRef] [PubMed]
8. Dong, W.; Liu, X.; Ma, L.; Yang, Z.; Ma, C. Association between dietary selenium intake and severe abdominal aortic calcification in the United States: A cross-sectional study. *Food Funct.* **2024**, *15*, 1575–1582. [CrossRef] [PubMed]
9. Padariya, C.; Rutkowska, M.; Konieczka, P. The accessibility, necessity, and significance of certified reference materials for total selenium content and its species to improve food laboratories' performance. *Food Chem.* **2023**, *425*, 136460. [CrossRef]
10. Jabbari, P.; Yazdanpanah, O.; Benjamin, D.J.; Rezazadeh, K.A. Supplement use and increased risks of cancer: Unveiling the other side of the coin. *Cancers* **2024**, *16*, 880. [CrossRef]
11. Chang, C.Y.; Yin, R.S.; Wang, X.; Shao, S.X.; Chen, C.Y.; Zhang, H. Selenium translocation in the soil-rice system in the Enshi seleniferous area, Central China. *Sci. Total Environ.* **2019**, *669*, 83–90. [CrossRef]
12. Schweizer, U.; Fradejas, V.N. The significance of selenoproteins for human health revealed by inborn errors of metabolism. *FASEB J.* **2016**, *30*, 3669–3681. [CrossRef]
13. Yang, H.; Yang, X.; Ning, Z.; Kwon, S.Y.; Li, M.L.; Tack, F.M.; Yin, R. The beneficial and hazardous effects of selenium on the health of the soil-plant-human system: An overview. *J. Hazard. Mater.* **2022**, *422*, 126876. [CrossRef]
14. Ikram, S.; Li, Y.; Lin, C.; Yi, D.; Heng, W.; Li, Q.; Tao, L.; Hong, J.Y.; Wei, J.J. Selenium in plants: A nexus of growth, antioxidants, and phytohormones. *J. Plant Physiol.* **2024**, *21*, 154237. [CrossRef]
15. Zou, J.; Su, S.; Chen, Z.; Liang, F.; Zeng, Y.; Cen, W.; Zhang, X.; Huang, D. Hyaluronic acid-modified selenium nanoparticles for enhancing the therapeutic efficacy of paclitaxel in lung cancer therapy. *Artif. Cells Nanomed. B* **2019**, *47*, 3456–3464. [CrossRef]
16. Cardoso, B.R.; Braat, S.; Graham, R.M. Selenium status is associated with insulin resistance markers in adults: Findings from the 2013 to 2018 National Health and Nutrition Examination Survey (*NHANES*). *Front. Nutr.* **2021**, *8*, 696024. [CrossRef]
17. Casanova, P.; Monleon, D. Role of selenium in type 2 diabetes, insulin resistance and insulin secretion. *World J. Diabetes* **2023**, *14*, 147–158. [CrossRef]
18. Tortelly, V.C.; Melo, D.F.; Matsunaga, A.M. The relevance of selenium to alopecias. *Inter. J. Trichol.* **2018**, *10*, 92–93. [CrossRef]
19. Qu, L.; Xu, J.; Dai, Z.; Elyamine, A.M.; Huang, W.; Han, D.; Jia, W. Selenium in soil-plant system: Transport, detoxification and bioremediation. *J. Hazard. Mater.* **2023**, *452*, 131272. [CrossRef]
20. Tao, J.; Leng, J.; Lei, X.; Wan, C.; Li, D.; Wu, Y.; Gao, J. Effects of selenium (Se) uptake on plant growth and yield in common buckwheat (*Fagopyrum esculentum* Moench). *Field Crops Res.* **2023**, *302*, 109070. [CrossRef]
21. Jia, H.; Song, Z.; Wu, F.; Ma, M.; Li, Y.; Han, D.; Cui, H. Low selenium increases the auxin concentration and enhances tolerance to low phosphorous stress in *tobacco*. *Environ. Exp. Bot.* **2018**, *153*, 127–134. [CrossRef]
22. Regni, L.; Palmerini, C.A.; Businelli, D.D.; Amato, R.; Mairech, H.; Proietti, P. Effects of selenium supplementation on olive under salt stress conditions. *Sci. Hortic.* **2021**, *278*, 109866. [CrossRef]
23. Rady, M.M.; Belal, H.E.; Gadallah, F.M.; Semida, W.M. Selenium application in two methods promotes drought tolerance in *Solanum lycopersicum* plant by inducing the antioxidant defense system. *Sci. Hortic.* **2020**, *266*, 109290. [CrossRef]
24. Shinmachi, F.; Buchner, P.; Stroud, J.L.; Parmar, S.; Zhao, F.J.; McGrath, S.P.; Hawkesford, M.J. Influence of sulfur deficiency on the expression of specific sulfate transporters and the distribution of sulfur, selenium, and molybdenum in wheat. *Plant Physiol.* **2010**, *153*, 327–336. [CrossRef]
25. Terry, N.; Zayed, A.M.; De, S.M.; Tarun, A.S. Selenium in higher plants. *Annu. Rev. Plant Biol.* **2000**, *51*, 401–432. [CrossRef]
26. Zhang, L.; Hu, B.; Li, W.; Che, R.; Deng, K.; Li, H.; Yu, F.; Ling, H.; Li, Y.; Chu, C. OsPT2, a phosphate transporter, is involved in the active uptake of selenite in rice. *New Phytol.* **2014**, *201*, 1183–1191. [CrossRef]

27. Cappa, J.J.; Cappa, P.J.; Mehdawi, A.F.; McAleer, J.M.; Simmons, M.; Pilon-Smits, E.A. Characterization of selenium and sulfur accumulation across the genus Stanleya (*Brassicaceae*): A field survey and common-garden experiment. *Am. J. Bot.* **2014**, *101*, 830–839. [CrossRef]
28. Harris, J.; Schneberg, K.A.; Pilon-Smits, E.A. Sulfur–selenium–molybdenum interactions distinguish selenium hyperaccumulator *Stanleya pinnata* from non-hyperaccumulator *Brassica juncea* (*Brassicaceae*). *Planta* **2014**, *239*, 479–491. [CrossRef]
29. Mazej, D.; Osvald, J.; Stibilj, V. Selenium species in leaves of chicory, dandelion, lamb's lettuce and parsley. *Food Chem.* **2008**, *107*, 175–183. [CrossRef]
30. Gigolashvili, T.; Kopriva, S. Transporters in plant sulfur metabolism. *Front. Plant Sci.* **2014**, *5*, 107212. [CrossRef]
31. Mansoor, S.; Ali, A.; Kour, N.; Bornhorst, J.; AlHarbi, K.; Rinklebe, J.; Chung, Y.S. Heavy metal induced oxidative stress mitigation and ROS scavenging in plants. *Plants* **2023**, *12*, 3003. [CrossRef]
32. Raza, A.; Charagh, S.; Najafi-Kakavand, S.; Abbas, S.; Shoaib, Y.; Anwar, S.; Sharifi, S.; Lu, G.; Siddique, K.H. Role of phytohormones in regulating cold stress tolerance: Physiological and molecular approaches for developing cold-smart crop plants. *Plant Stress* **2023**, *8*, 100152. [CrossRef]
33. Hasanuzzaman, M.; Bhuyan, M.; Raza, A.; Hawrylak-Nowak, B.; Matraszek-Gawron, R.; Mahmud, J.; Nahar, K.; Fujita, M. Selenium in plants: Boon or bane? *Environ. Exp. Bot.* **2020**, *178*, 104170. [CrossRef]
34. Lanza, M.G.; Dos, R.A. Roles of selenium in mineral plant nutrition: ROS scavenging responses against abiotic stresses. *Plant Physiol. Biochem.* **2020**, *164*, 27–43. [CrossRef]
35. Banerjee, A.; Roychoudhury, A. Role of selenium in plants against abiotic stresses: Phenological and molecular aspects. *Mol. Plant Abiotic Stress Biol. Biotechnol.* **2019**, *12*, 123–133.
36. Chen, Z.; Lu, Y.; Dun, X.; Wang, X.; Wang, H. Research progress of selenium-enriched foods. *Nutrients* **2023**, *15*, 41–89. [CrossRef] [PubMed]
37. Hu, Q.H.; Chen, L.C.; Xu, J.; Zhang, Y.L.; Pan, G.X. Determination of selenium concentration in rice and the effect of foliar application of Se-enriched fertiliser or sodium selenite on the selenium content of rice. *J. Sci. Food Agric.* **2002**, *82*, 869–872. [CrossRef]
38. Chen, Q.; Shi, W.; Wang, X. Selenium speciation and distribution characteristics in the rhizosphere soil of rice (*Oryza sativa* L.) Seedlings. *Commun. Soil Sci. Plant Anal.* **2010**, *41*, 1411–1425. [CrossRef]
39. Sun, G.; Liu, X.; Williams, P.; Zhu, Y.G. Distribution and translocation of selenium from soil to grain and its speciation in paddy rice (*Oryza sativa* L.). *Environ. Sci. Technol.* **2010**, *44*, 6706–6711. [CrossRef] [PubMed]
40. Li, B.; Liu, X.; Yu, T.; Lin, K.; Ma, X.; Li, C.; Wang, Y. Environmental selenium and human longevity: An ecogeochemical perspective. *Chemosphere* **2024**, *347*, 140691. [CrossRef]
41. Zhang, B.; Wei, Y.; Yan, S.; Shi, H.; Nie, Y.; Zou, G.; Zhang, X.; Luo, L. Characterization of selenium accumulation of different rice genotypes in Chinese natural seleniferous soil. *Plant Soil Environ.* **2019**, *65*, 15–20. [CrossRef]
42. Chen, L.; Yang, F.; Xu, J.; Hu, Y.; Hu, Q.; Zhang, Y.; Pan, G. Determination of selenium concentration of rice in China and effect of fertilization of selenite and selenate on selenium content of rice. *J. Agric. Food. Chem.* **2002**, *50*, 5128–5130. [CrossRef]
43. Zhang, L.; Shi, W.; Wang, C. Difference in selenium accumulation in shoots of two rice cultivars. *Pedosphere* **2006**, *16*, 646–653. [CrossRef]
44. Song, T.; Su, X.; He, J.; Liang, Y.; Zhou, T.; Liu, C. Selenium (Se) uptake and dynamic changes of Se content in soil-plant systems. *Environ. Sci. Pollut. R.* **2018**, *25*, 34343–34350. [CrossRef]
45. Kong, L.; Wang, M.; Bi, D. Selenium modulates the activities of antioxidant enzymes, osmotic homeostasis and promotes the growth of sorrel seedlings under salt stress. *Plant Growth Regul.* **2005**, *45*, 155–163. [CrossRef]
46. Shen, J.; Jiang, C.; Yan, Y.; Zu, C. Selenium distribution and translocation in rice (*Oryza sativa* L.) under different naturally seleniferous soils. *Sustainability* **2019**, *11*, 520. [CrossRef]
47. Cao, Z.; Wang, X.; Yao, D.; Zhang, X.; Wong, M. Selenium geochemistry of paddy soils in Yangtze River Delta. *Environ. Int.* **2001**, *26*, 335–339. [CrossRef]
48. Lima-Lessa, J.H.; Araujo, A.M.; Ferreira, L.A.; Silva-Júnior, E.C.; Oliveira, C.; Corguinha, A.P.; Martins, F.A.; Carvalho, H.W.; Guilherme, L.R.; Lopes, G. Agronomic biofortification of rice (*Oryza sativa* L.) with selenium and its effect on element distributions in biofortified grains. *Plant Soil.* **2019**, *444*, 331–342. [CrossRef]
49. Schiavon, M.; Pilon-Smits, E.A. The fascinating facets of plant selenium accumulation–biochemistry, physiology, evolution and ecology. *New Phytol.* **2017**, *213*, 1582–1596. [CrossRef]
50. Liang, Y.; Su, Y.; Li, L.; Huang, X.; Panhwar, F.H.; Zheng, T.; Tang, Z.; Ei, H.H.; Farooq, M.U.; Zeng, R.; et al. Quick selenium accumulation in the selenium-rich rice and its physiological responses in changing selenium environments. *BMC Plant Biol.* **2019**, *19*, 559. [CrossRef]
51. Yin, H.; Qi, Z.; Li, M.; Ahammed, G.; Chu, X.; Zhou, J. Selenium forms and methods of application differentially modulate plant growth, photosynthesis, stress tolerance, selenium content and speciation in *Oryza sativa* L. *Ecotoxicol. Environ. Saf.* **2019**, *169*, 911–917.
52. Hamilton, S.; Teven, J. Review of selenium toxicity in the aquatic food chain. *Sci. Total Environ.* **2004**, *326*, 1–31. [CrossRef]
53. Wu, H.; Fan, S.; Gong, H.; Guo, J. Roles of salicylic acid in selenium-enhanced salt tolerance in tomato plants. *Plant Soil.* **2023**, *484*, 569–588. [CrossRef]

54. Liu, C.; Zhou, G.; Qin, H.; Guan, Y.; Wang, T.; Ni, W.; Ge, S. Metabolomics combined with physiology and transcriptomics reveal key metabolic pathway responses in apple plants exposure to different selenium concentrations. *J. Hazard. Mater.* **2024**, *464*, 132953. [CrossRef]
55. Dos-Reis, A.R.; Boleta, E.H.; Alves, C.Z.; Cotrim, M.F.; Barbosa, J.Z.; Silva, V.M.; Porto, R.L.; Lanza, M.G.; Lavres, J.; Gomes, M.H.; et al. Selenium toxicity in upland field-grown rice: Seed physiology responses and nutrient distribution using the μ-XFR technique. *Ecotoxicol. Environ. Saf.* **2020**, *190*, 110147. [CrossRef]
56. Du, B.; Luo, H.; He, L.; Zhang, L.; Liu, Y.; Mo, Z.; Tang, X. Rice seed priming with sodium selenate: Effects on germination, seedling growth, and biochemical attributes. *Sci. Rep.* **2019**, *9*, 4311. [CrossRef]
57. Nedjimi, B. Selenium as a powerful trace element for mitigation of plant salt stress: A review. *J. Trace Elem. Miner.* **2024**, *8*, 100123. [CrossRef]
58. Labanowska, M.; Filek, M.; Koscielniak, J.; Kurdziel, M.; Kulis, E.; Hartikainen, H. The effects of short-term selenium stress on Polish and Finnish wheat seedlings-EPR, enzymatic and fluorescence studies. *Plant Physiol.* **2012**, *169*, 275–284. [CrossRef]
59. Jan, S.; Bhardwaj, R.; Sharma, N.R.; Singh, R. Unraveling the role of plant growth regulators and plant growth promoting rhizobacteria in phytoremediation. *J. Plant Growth Regul.* **2024**, *43*, 1–17. [CrossRef]
60. De-Lima, J.H.; Raymundo, J.F.; Corguinha, A.P.; Martins, F.A.; Araujo, A.M.; Santiago, F.E.; Lopes, G. Strategies for applying selenium for biofortification of rice in tropical soils and their effect on element accumulation and distribution in grains. *J. Cereal Sci.* **2020**, *96*, 103125. [CrossRef]
61. Andrade, F.R.; Silva, G.N.; Guimarães, K.C.; Barreto, H.B.F.; Souza, K.R.D.; Guilherme, L.R.G.; Dos-Reis, A.R. Selenium protects rice plants from water deficit stress. *Ecotoxicol. Environ. Saf.* **2018**, *164*, 562–570. [CrossRef]
62. Dreyer, B.H.; Schippers, J.H.M. Copper-Zinc superoxide dismutases in plants: Evolution, enzymatic properties, and beyond. *Annu. Plant Rev.* **2018**, *2*, 933–968.
63. Su, W.; Raza, A.; Gao, A.; Jia, Z.; Zhang, Y.; Hussain, M.A.; Mehmood, S.S.; Cheng, Y.; Lv, Y.; Zou, X. Genome-wide analysis and expression profile of superoxide dismutase (SOD) gene family in rapeseed (*Brassica napus* L.) under different hormones and abiotic stress conditions. *Antioxidants* **2021**, *10*, 1182. [CrossRef]
64. Xiao, H.L.; Wang, C.P.; Khan, N.; Chen, M.; Fu, W.; Guan, L.; Leng, X. Genome-wide identification of the class III POD gene family and their expression profiling in grapevine (*Vitis vinifera* L.). *BMC Genom.* **2020**, *21*, 444. [CrossRef]
65. Hiraga, S.; Sasaki, K.; Ito, H.; Ohashi, Y.; Matsui, H. A large family of class III plant peroxidases. *Plant Cell Physiol.* **2001**, *42*, 462. [CrossRef]
66. Daudi, A.; Cheng, Z.; O'Brien, J.A.; Mammarella, N.; Khan, S.; Ausubel, F.M.; Bolwell, G.P. The apoplastic oxidative burst peroxidase in *Arabidopsis* is a major component of pattern-triggered immunity. *Plant Cell* **2012**, *24*, 275–287. [CrossRef]
67. Lin, C.Y.; Li, Q.; Tunlaya-Anukit, S.; Shi, R.; Sun, Y.H.; Wang, J.P.; Chiang, V.L. A cell wall-bound anionic peroxidase, PtrPO21, is involved in lignin polymerization in *Populus trichocarpa*. *Tree Genet. Genomes* **2016**, *12*, 22. [CrossRef]
68. Kim, B.H.; Kim, S.Y.; Nam, K.H. Genes encoding plant-specific class III peroxidases are responsible for increased cold tolerance of the *brassinosteroid-insensitive 1* mutant. *Mol. Cells* **2012**, *34*, 539–548. [CrossRef]
69. Zhang, Y.; Zheng, L.J.; Yun, L.; Ji, L.; Li, G.H.; Ji, M.C.; Shi, Y.; Zheng, X. Catalase (CAT) gene family in wheat (*Triticum aestivum* L.): Evolution, expression pattern and function analysis. *Int. J. Mol. Sci.* **2022**, *23*, 542. [CrossRef]
70. Wu, B.; Binbin, W. Comparative analysis of ascorbate peroxidases (APXs) from selected plants with a special focus on *Oryza sativa* employing public databases. *PLoS ONE* **2019**, *14*, 0226543. [CrossRef]
71. Raza, A.; Su, W.; Gao, A.; Mehmood, S.S.; Hussain, M.A.; Nie, W.L.; Lv, Y.; Zhang, X.L. Catalase (CAT) gene family in rapeseed (*Brassica napus* L.): Genome-wide analysis, identification, and expression pattern in response to multiple hormones and abiotic stress conditions. *Int. J. Mol. Sci.* **2021**, *22*, 4281. [CrossRef]
72. Raza, A.; Sharif, Y.; Chen, K.; Wang, L.H.; Fu, H.W.; Zhuang, Y.H.; Chitikineni, A.; Chen, H.; Zhang, C.; Varshney, R.K. Genome-wide characterization of ascorbate peroxidase gene family in peanut (*Arachis hypogea* L.) revealed their crucial role in growth and multiple stress tolerance. *Plant Sci.* **2022**, *13*, 962182. [CrossRef]
73. Ling, J.Y.; Tan, J.H.; Chen, H.; Yang, Z.Q.; Luo, Q.F.; Jia, J. Physiology, transcriptome and root exudates analysis of response to aluminum stress in *Pinus massoniana*. *Forests* **2023**, *14*, 1410. [CrossRef]
74. Shi, J.; Zhao, M.; Zhang, F.; Feng, D.; Yang, S.; Xue, Y.; Liu, Y. Physiological mechanism through which al toxicity inhibits peanut root growth. *Plants* **2024**, *13*, 325. [CrossRef]
75. Pirzadah, T.B.; Malik, B.; Tahir, I.; Rehman, R.U.; Hakeem, K.R.; Alharby, H.F. Aluminium stress modulates the osmolytes and enzyme defense system in Fagopyrum species. *Plant Physiol. Biochem.* **2019**, *144*, 178–186. [CrossRef]
76. Li, F.; Wu, X.J.; Wu, W. Effects of malondialdehyde-induced protein oxidation on the structural characteristics of rice protein. *Int. J. Food Sci. Technol.* **2020**, *55*, 760–768. [CrossRef]
77. Lin, L.; Zhou, W.; Dai, H.; Cao, F.; Zhang, G.; Wu, F. Selenium reduces cadmium uptake and mitigates cadmium toxicity in rice. *J. Hazard. Mater.* **2012**, *235*, 343–351. [CrossRef]
78. Cakmak, I.; Wang, S.; Zhang, F.; Guo, S. Synergistic and antagonistic interactions between potassium and magnesium in higher plants. *Crop J.* **2021**, *9*, 249–256.
79. Senbayram, M.; Gransee, A.; Wahle, V.; Thiel, H. Role of magnesium fertilisers in agriculture: Plant–soil continuum. *Crop Pasture Sci.* **2015**, *66*, 1219–1229. [CrossRef]

80. Taylor, G.J.; Blarney, F.P.; Edwards, D. Antagonistic and synergistic interactions between aluminum and manganese on growth of *Vigna unguiculata* at low ionic strength. *Physiol. Plant.* **1998**, *104*, 183–194. [CrossRef]
81. Zhu, J.K. Abiotic stress signaling and responses in plants. *Cell* **2016**, *167*, 313–324. [CrossRef]
82. Sors, T.G.; Ellis, D.R.; Salt, D.E. Selenium uptake, translocation, assimilation and metabolic fate in plants. *Photosynth. Res.* **2005**, *86*, 373–389. [CrossRef]
83. LeDuc, D.L.; Tarun, A.S.; Montes-Bayon, M.; Meija, J.; Malit, M.F.; Wu, C.P.; AbdelSamie, M.; Chiang, C.Y.; Tagmount, A.; DeSouza, M.; et al. Overexpression of selenocysteine methyltransferase in *Arabidopsis* and Indian mustard increases selenium tolerance and accumulation. *Plant Physiol.* **2004**, *135*, 377–383. [CrossRef]
84. Nikiforova, V.; Freitag, J.; Kempa, S. Transcriptome analysis of sulfur depletion in *Arabidopsis thaliana*: Interlacing of biosynthetic pathways provides response specificity. *Plant J.* **2003**, *33*, 633–650. [CrossRef]
85. Tan, Z.J.; Li, J.X.; Guan, J.H.; Wang, C.H.; Zhang, Z.; Shi, G.R. Genome-wide identification and expression analysis reveals roles of the NRAMP gene family in iron/cadmium interactions in peanut. *Int. J. Mol. Sci.* **2023**, *24*, 1713. [CrossRef] [PubMed]
86. Qin, L.; Han, P.P.; Chen, L.Y.; Walk, T.C.; Li, Y.S.; Hu, X.J.; Xie, L.H.; Liao, H.; Liao, X. Genome-wide identification and expression analysis of NRAMP family genes in soybean (*Glycine max* L.). *Plant Sci.* **2017**, *8*, 1436. [CrossRef]
87. Yamaji, N.; Sasaki, A.; Yokosho, K.; Ma, J.F. A node-based switch for preferential distribution of manganese in rice. *Nat. Commun.* **2013**, *4*, 2442. [CrossRef]
88. Ishimaru, Y.; Takahashi, R.; Bashir, K.; Shimo, H.; Senoura, T.; Sugimoto, K.; Nishizawa, N.K. Characterizing the role of rice NRAMP5 in manganese, iron and cadmium transport. *Sci. Rep.* **2012**, *2*, 1286. [CrossRef]
89. Liu, N.; Jiang, Z.; Li, X.; Liu, H.; Li, N.; Wei, S. Mitigation of rice cadmium (Cd) accumulation by joint application of organic amendments and selenium (Se) in high-Cd-contaminated soils. *Chemosphere* **2020**, *241*, 125106. [CrossRef] [PubMed]
90. Lv, H.; Chen, W.; Zhu, Y.; Yang, J.; Mazhar, S.H.; Zhao, P.; Feng, R. Efficiency and risks of selenite combined with different water conditions in reducing uptake of arsenic and cadmium in paddy rice. *Environ. Pollut.* **2020**, *262*, 114283. [CrossRef] [PubMed]
91. Zhou, X.; Yang, J.; Kronzucker, H.J.; Shi, W. Selenium biofortification and interaction with other elements in plants: A review. *Front. Plant Sci.* **2020**, *11*, 586421. [CrossRef]
92. Tao, Q.; Jupa, R.; Liu, Y.K.; Luo, J.P.; Li, J.X.; Kováč, J.; Li, T.Q. Abscisic acid-mediated modifications of radial apoplastic transport pathway play a key role in cadmium uptake in hyperaccumulator *Sedum alfredii*. *Plant Cell Environ.* **2019**, *42*, 1425–1440. [CrossRef]
93. Chen, K.; Li, G.J.; Bressan, R.A.; Song, C.P.; Zhu, J.K.; Zhao, Y. Abscisic acid dynamics, signaling, and functions in plants. *J. Integr. Plant Biol.* **2020**, *62*, 25–54. [CrossRef]
94. Wang, X.; Lu, W.; Zhao, Z.; Hao, W.; Du, R.; Li, Z.; Lin, L. Abscisic acid promotes selenium absorption, metabolism and toxicity via stress-related phytohormones regulation in *Cyphomandra betacea* Sendt (*Solanum betaceum* Cav). *J. Hazard. Mater.* **2024**, *461*, 132642. [CrossRef]
95. Yang, W.; Zhou, Z.; Chu, Z. Emerging roles of salicylic acid in plant saline stress tolerance. *Int. J. Mol. Sci.* **2023**, *24*, 43388. [CrossRef]
96. Ranjan, A.; Sinha, R.; La, S.K.; Bishi, S.K.; Singh, A.K. Phytohormone signalling and cross-talk to alleviate aluminium toxicity in plants. *Plant Cell Rep.* **2021**, *40*, 1331–1343. [CrossRef] [PubMed]
97. Mostofa, M.G.; Rahman, M.M.; Siddiqui, M.N.; Fujita, M.; Tran, L.S. Salicylic acid antagonizes selenium phytotoxicity in rice: Selenium homeostasis, oxidative stress metabolism and methylglyoxal detoxification. *J. Hazard. Mater.* **2020**, *394*, 122572. [CrossRef]
98. Verma, V.; Ravindran, P.; Kumar, P.P. Plant hormone-mediated regulation of stress responses. *BMC Plant Biol.* **2016**, *16*, 86. [CrossRef]
99. Chen, Z.; Sun, H.; Hu, T.; Wang, Z.; Wu, W.; Liang, Y.; Guo, Y. Sunflower resistance against Sclerotinia sclerotiorum is potentiated by selenium through regulation of redox homeostasis and hormones signaling pathways. *Environ. Sci. Pollut. R.* **2022**, *29*, 38097–38109. [CrossRef] [PubMed]
100. Zhang, N.L.; Zhou, S.; Yang, D.Y.; Fan, Z.J. Revealing shared and distinct genes responding to JA and SA signaling in *Arabidopsis* by meta-analysis. *Plant Sci.* **2020**, *11*, 908.
101. Dai, Z.; Yuan, Y.; Huang, H.; Hossain, M.M.; Xiong, S.; Cao, M.; Ma, L.Q.; Tu, S. Methyl jasmonate mitigates high selenium damage of rice via altering antioxidant capacity, selenium transportation and gene expression. *Sci. Total Environ.* **2021**, *756*, 143848. [CrossRef] [PubMed]
102. Figueiredo, M.D.; Strader, L.C. Intrinsic and extrinsic regulators of Aux/IAA protein degradation dynamics. *Trends Biochem. Sci.* **2022**, *47*, 865–874. [CrossRef] [PubMed]
103. Hwang, I.; Sheen, J.; Müller, B. Cytokinin signaling networks. *Annu. Rev. Plant Biol.* **2012**, *63*, 353–380. [CrossRef] [PubMed]
104. Yürekli, F.; Türkan, I.; Porgali, Z.B.; Topçuoglu, S.F. Indoleacetic acid, gibberellic acid, zeatin, and abscisic acid levels in NaCl-treated tomato species differing in salt tolerance. *Isr. J. Plant Sci.* **2001**, *49*, 269–278. [CrossRef]
105. Kasahara, H. Current aspects of auxin biosynthesis in plants. *Biosci. Biotechnol. Biochem.* **2016**, *80*, 34–42. [CrossRef] [PubMed]
106. Simon, S.; Petrášek, J. Why plants need more than one type of auxin. *Plant Sci.* **2011**, *180*, 454–460. [CrossRef] [PubMed]
107. Mathur, P.; Tripathi, D.K.; Baluška, F.; Mukherjee, S. Auxin-mediated molecular mechanisms of heavy metal and metalloid stress regulation in plants. *Environ. Exp. Bot.* **2022**, *196*, 104796. [CrossRef]

108. Qi, T.; Yang, W.; Hassan, M.J.; Liu, J.; Yang, Y.; Zhou, Q.; Peng, Y. Genome-wide identification of Aux/IAA gene family in white clover (*Trifolium repens* L.) and functional verification of TrIAA18 under different abiotic stress. *BMC Plant Biol.* **2024**, *24*, 346. [CrossRef] [PubMed]
109. Hu, Z.; Cheng, Y.; Suzuki, N.; Guo, X.; Guo, H.; Ogra, Y. Speciation of selenium in brown rice fertilized with selenite and effects of selenium fertilization on rice proteins. *Int. J. Mol. Sci.* **2018**, *19*, 3494. [CrossRef]
110. He, L.X.; Zheng, A.X.; Du, B.; Luo, H.W.; Lu, R.H.; Chen, Y.L.; Zhang, T.T.; Lai, R.F.; Tang, X.R. Low-concentration sodium selenite applications improve oxidation resistance of filling-stage rice. *Appl. Ecol. Environ. Res.* **2019**, *17*, 989–998. [CrossRef]
111. Zhang, W.; Guan, M.; Chen, M.; Lin, X.; Xu, P.; Cao, Z. Mutation of OsNRAMP5 reduces cadmium xylem and phloem transport in rice plants and its physiological mechanism. *Environ. Pollut.* **2024**, *341*, 122928. [CrossRef]
112. Liu, Y.; Pan, Y.H.; Li, J.Y.; Chen, J.Y.; Yang, S.X.; Zhao, M.; Xue, Y.B. Transcriptome sequencing analysis of root in soybean responding to Mn poisoning. *Int. J. Mol. Sci.* **2023**, *24*, 12727. [CrossRef]
113. Pornaro, C.; Macolino, S.; Menegon, A.; Richardson, M. WinRHIZO technology for measuring morphological traits of bermudagrass stolons. *Agron. J.* **2017**, *109*, 3007–3010. [CrossRef]
114. Kalogiouri, N.P.; Manousi, N.; Zachariadis, G.A. Determination of the toxic and nutrient element content of almonds, walnuts, hazelnuts and pistachios by ICP-AES. *Separations* **2021**, *8*, 28. [CrossRef]
115. Li, J.; Zhou, X.; Zhou, J.; Shang, R.; Wang, Y. Comparative study on several determination methods of chlorophyll content in plants. *IOP Conf. Ser. Mater. Sci. Eng.* **2020**, *730*, 012066. [CrossRef]
116. Cao, K.L.; Yin, Y.; Li, S.S.; Li, M.Q.; Tajdar, A.; Tan, S.Q.; Shi, W.P. Chemical composition and amino acids of *Locusta migratoria* (*Orthoptera Acrididae*) fed wheat seedlings (*Triticum aestivum*) or rape leaves (*Brassica napus*). *J. Insects Food Feed* **2024**, *1*, 1–10. [CrossRef]
117. Ghaffari, H.; Tadayon, M.R.; Nadeem, M.; Cheema, M.; Razmjoo, J. Proline-mediated changes in antioxidant enzymatic activities and the physiology of sugar beet under drought stress. *Acta Physiol. Plant.* **2019**, *41*, 23. [CrossRef]
118. Mas-Bargues, C.; Escriva, C.; Dromant, M.; Borras, C.; Vina, J. Lipid peroxidation as measured by chromatographic determination of malondialdehyde. Human plasma reference values in health and disease. *Arch. Biochem. Biophys.* **2021**, *709*, 108641. [CrossRef]
119. Yang, Z.; Qiu, D.; Jiang, K.; Du, M.; Li, H. Abatement effects of different soil amendments on continuous crop of *Codonopsis pilosula*. *Plant Sci.* **2024**, *15*, 1376362.
120. Sadak, M.S.; El-Hameid, A.R.; Zaki, F.S.; Dawood, M.G.; El-Awadi, M.E. Physiological and biochemical responses of soybean (*Glycine max* L.) to cysteine application under sea salt stress. *Bull. Nat. Res. Cent.* **2020**, *44*, 1.
121. Hadwan, M.H.; Hussein, M.J.; Mohammed, R.M.; Hadwan, A.M.; Saad Al-Kawaz, H.; Al-Obaidy, S.S.; Al-Talebi, Z.A. An improved method for measuring catalase activity in biological samples. *Biol. Methods Protoc.* **2024**, *9*, 15. [CrossRef] [PubMed]
122. Hosseini, F.; Mosaddeghi, M.R.; Dexter, A.R.; Sepehri, M. Maize water status and physiological traits as affected by root endophytic fungus *Piriformospora indica* under combined drought and mechanical stresses. *Planta* **2018**, *247*, 1229–1245. [CrossRef] [PubMed]
123. Khew, C.Y.; Mori, I.C.; Matsuura, T.; Hirayama, T.; Harikrishna, J.A.; Lau, E.T.; Mercer, Z.J.A.; Hwang, S.S. Hormonal and transcriptional analyses of fruit development and ripening in different varieties of black pepper (*Piper nigrum*). *J. Plant Res.* **2020**, *133*, 73–94. [CrossRef] [PubMed]
124. Xue, Y.B.; Zhuang, Q.L.; Zhu, S.N.; Xiao, B.X.; Liang, C.Y.; Liao, H.; Tian, J. Genome wide transcriptome analysis reveals complex regulatory mechanisms underlying phosphate homeostasis in soybean nodules. *Int. J. Mol. Sci.* **2018**, *19*, 2924. [CrossRef]
125. Zhuang, Q.L.; Xue, Y.B.; Yao, Z.F.; Zhu, S.N.; Liang, C.Y.; Liao, H.; Tian, J. Phosphate starvation responsive GmSPX5 mediates nodule growth through interaction with GmNF-YC4 in soybean (*Glycine max*). *Plant J.* **2021**, *108*, 1422–1438. [CrossRef]

plants

Review

The Role of Plant Ubiquitin-like Modifiers in the Formation of Salt Stress Tolerance

Siarhei A. Dabravolski [1] and Stanislav V. Isayenkov [2,3,4,*]

1 Department of Biotechnology Engineering, Braude Academic College of Engineering, Snunit 51, Karmiel 2161002, Israel; sergedobrowolski@gmail.com
2 International Research Centre for Environmental Membrane Biology, Foshan University, Foshan 528000, China
3 Institute of Agricultural and Nutritional Sciences, Martin Luther University Halle-Wittenberg, Betty-Heimann-Strasse 3, 06120 Halle, Germany
4 Department of Plant Food Products and Biofortification, Institute of Food Biotechnology and Genomics, The National Academy of Sciences of Ukraine, Baidi-Vyshneveckogo Str. 2a, 04123 Kyiv, Ukraine
* Correspondence: stan.isayenkov@gmail.com

Abstract: The climate-driven challenges facing Earth necessitate a comprehensive understanding of the mechanisms facilitating plant resilience to environmental stressors. This review delves into the crucial role of ubiquitin-like modifiers, particularly focusing on ATG8-mediated autophagy, in bolstering plant tolerance to salt stress. Synthesising recent research, we unveil the multifaceted contributions of ATG8 to plant adaptation mechanisms amidst salt stress conditions, including stomatal regulation, photosynthetic efficiency, osmotic adjustment, and antioxidant defence. Furthermore, we elucidate the interconnectedness of autophagy with key phytohormone signalling pathways, advocating for further exploration into their molecular mechanisms. Our findings underscore the significance of understanding molecular mechanisms underlying ubiquitin-based protein degradation systems and autophagy in salt stress tolerance, offering valuable insights for designing innovative strategies to improve crop productivity and ensure global food security amidst increasing soil salinisation. By harnessing the potential of autophagy and other molecular mechanisms, we can foster sustainable agricultural practices and develop stress-tolerant crops resilient to salt stress.

Keywords: salt stress; ubiquitin; SUMOylation; autophagy; phytohormones

Citation: Dabravolski, S.A.; Isayenkov, S.V. The Role of Plant Ubiquitin-like Modifiers in the Formation of Salt Stress Tolerance. *Plants* **2024**, *13*, 1468. https://doi.org/10.3390/plants13111468

Academic Editors: Geovani Soares de Lima, Lauriane Almeida dos Anjos Soares and Francisco Vanies Da Silva Sá

Received: 9 May 2024
Revised: 22 May 2024
Accepted: 23 May 2024
Published: 25 May 2024

1. Ubiquitin-Mediated Protein Degradation in Plant Adaptation to Salt Stress

The climate of Earth is undergoing constant changes, which are often characterised by the increased number of frequent and severe adverse weather events such as inundations, hurricanes, drought, desertification, and others, making plants, more so than ever, dependent on effective mechanisms to withstand these environmental fluctuations [1,2]. In such challenging environmental conditions, for plants, given their stationary lifestyle, the effective removal and utilisation of malfunctioning and damaged proteins is crucial for the proper functioning of signal transduction systems and timely adaptation to environmental disturbances. Ubiquitination, a process of post-translational protein modification, plays a crucial role in almost every aspect of plant growth, development, and response to various stresses [3,4]. Thus, ubiquitin is the major component orchestrating protein degradation and turnover by ubiquitination [5]. The major ubiquitin-based protein degradation systems are ubiquitin (Ub)-26S proteasome system (UPS), small ubiquitin-like modifier (SUMO), and ubiquitin-like modifier autophagy-related protein 8 (ATG8) (Figure 1).

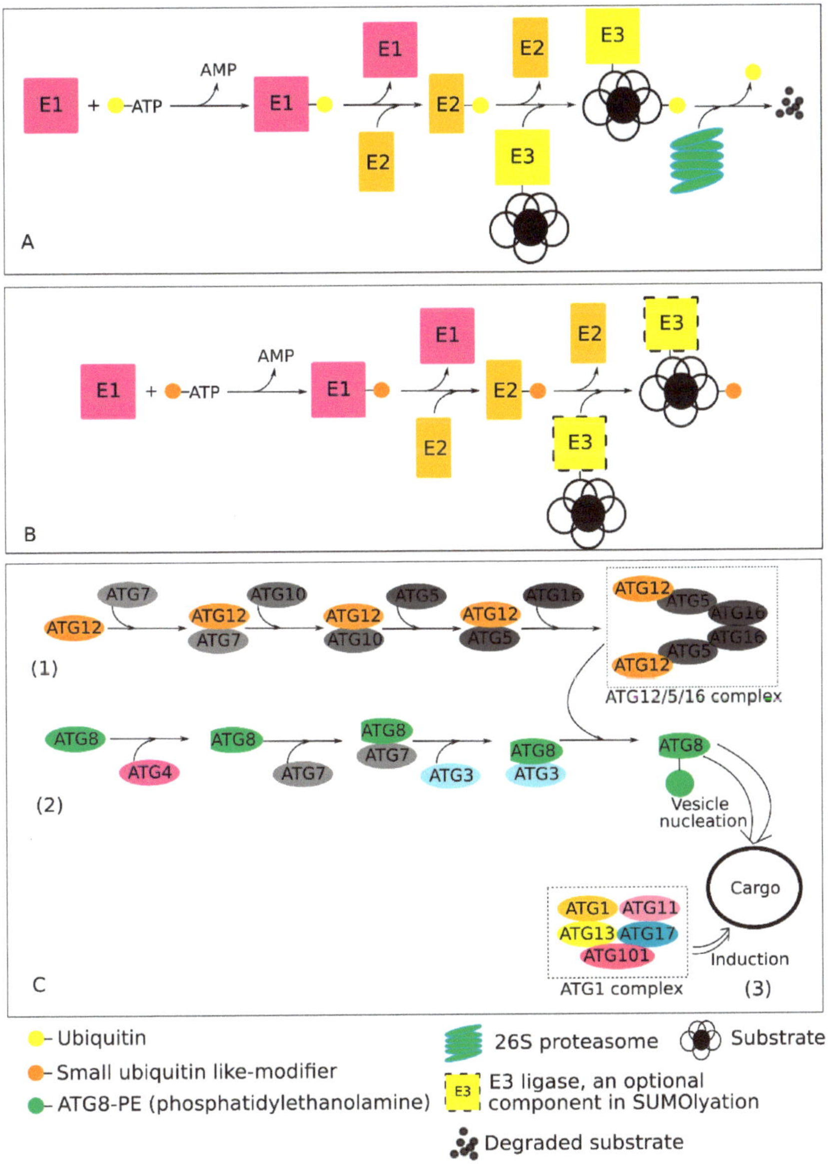

Figure 1. Schematic representation of the functioning mechanics of the ubiquitin (Ub), small ubiquitin-like-modifier (SUMO), and ubiquitin-like modifier autophagy-related protein 8 (ATG8). (**A**,**B**) The Ub and SUMOylation processes engaged similar cascade reactions with different enzymes (E1, E2, and E3) in an ATP-dependent way to add Ub or SUMO moieties, respectively, onto a target protein. In the UPS (**A**), marked proteins are recognised and degraded by the 26S proteasomes, which return back to the cytosol and the nucleus recycled Ub and short peptides. These short peptides then further degraded into free amino acids by various cellular peptidases. SUMOylation (**B**) may occur also in the absence of E3. (**C**) The autophagosome formation is mediated by various ATG proteins with three distinct processes: (1) formation of ATG12-ATG5-ATG16 complex; (2) sequential ATG8 cleavage and lipidation to form ATG8-PE (phosphatidylethanolamine); and (3) cargo, induced by the ATG1 complex, engulfed through interaction with ATG8 to form autophagosome.

Salt stress is one of the most common adverse environmental factors affecting plant growth and productivity around the world. Improper irrigation practices, misuse of fertilisers, and global climate change in general gradually increase the severity of global soil salinisation [6]. Salt stress affects plant growth and productivity in various ways, such as osmotic and water stress, ion toxicity, nutrition disorders, dysregulation of biochemical, metabolic and physiological processes, membrane disorganisation, genotoxicity, repressed cell division, and others [7,8]. The stress-associated accumulation of reactive oxygen species (ROS) plays a crucial role in salt stress responses by causing oxidative damage to biomolecules and by initiating various cell signalling pathways [9]. Also, plants have developed several sophisticated defence mechanisms to alleviate salt stress, such as ion homeostasis, compartmentalisation and export, production of osmoprotectants and ROS scavengers, modulation of cytoskeleton dynamics and the cell wall composition, phytohormones-mediated reactions, and others [8,10]. Additionally, ROS accumulation is a crucial factor linking autophagy and salt stress responses [11]. The UPS and SUMO systems have been extensively reviewed in several recent reviews [12–18]; thus they will be introduced only briefly. Further, in this review we discuss the recent research deciphering the role of autophagy in salt stress tolerance, with a particular focus on ATG8 as the key player in selective autophagy.

2. Ubiquitin (Ub)-26S Proteasome System (UPS) and Its Role in Salinity Stress Tolerance

Ub is extremely conserved across eukaryotic kingdoms, with only minor amino acid variations between such distant organisms as humans and plants [19]. In the process of protein ubiquitylation, a 76-amino acid Ub polypeptide covalently attached to substrate lysine residues through the sequential activation by the E1 Ub-activating enzyme, resulting in conjugation onto an E2 Ub-conjugating enzyme, and subsequent catalysis by a specific E3 Ub ligase [20]. Such ubiquitylated proteins are recognised by the 26S proteasome for degradation [21,22]. Furthermore, the attachment of a single Ub (monoubiquitylation) can serve as a traffic signal for substrate transport within the cell or targeting for degradation in vacuoles or lysosomes, while the topology of polyubiquitylation can regulate protein activity and localisation, or target it for degradation by the 26S proteasome [23]. A diverse array of E3 ubiquitin ligases encoded in plant genomes plays the primary role in regulating the specificity of the UPS in targeting polyubiquitylated proteins for degradation [24]. Based on the structural compositions and the mechanics of the conjugation process of activated Ub moieties, plant E3 ligases can be classified into three main types: really interesting new gene (RING)-containing E3 ligases, homologous to the E6-AP carboxyl terminus (HECT), and plant U-box (PUB) (Figure 1) [25–27].

Considering specifically salt stress tolerance, UPS was recently shown to play a crucial role in various salt stress tolerance-related processes. Thus, the expression of the *Oryza sativa* (L.) *salt-induced RING Finger Protein 1* (*OsSIRP1*) gene was up-regulated in roots after salt, heat, drought, and ABA treatments. Furthermore, transgenic Arabidopsis plants expressing *OsSIRP1* showed lower salt stress tolerance during seed germination and root growth, suggesting that *OsSIRP1* acted through the ubiquitin 26S proteasome system to negatively regulate rice salt stress tolerance [28]. Another rice RING H2-type E3 ligase, OsSIRH2-14, was characterised as a positive regulator of salinity tolerance. Mechanically, OsSIRH2-14 interacted and regulated the stability of the HKT-type Na^+ transporter OsHKT2;1 through the 26S proteasome system only in high salt but not in normal conditions. Accordingly, *OsSIRH2-14*-overexpressing rice plants showed a higher tolerance to salt stress and reduced Na^+ accumulation in the aerial shoot and root tissues compared to wild-type plants [29]. Similarly, wheat (*Triticum. aestivum* L.) E3 ubiquitin ligase TaDIS1 (an orthologous of the *Oryza sativa* gene *drought-induced SINA protein 1*, a C3HC4 RING finger E3 ligase) was shown to interact and regulate the stability of the salt-tolerant protein (TaSTP) via the 26S proteasome pathway. Additionally, the expression level of TaSTP was up-regulated by salt, drought stresses, and ABA treatment, thus suggesting its involvement in different abiotic stress responses [30].

3. Small Ubiquitin Like-Modifiers (SUMO) and The Role in Salinity Stress Tolerance

Similarly to Ub, SUMO resulted in mono- or multi-SUMOylation of a substrate protein by attaching to the lysine residues (Figure 1) [31]. Interestingly, SUMOylation sites often have a consensus pattern (such hydrophobic/KxD/E motif, with x = any amino acid) recognised by a SUMO conjugation enzyme, while in some cases SUMO can be ligated onto its substrate even in the absence of a SUMO E3 ligase [18]. Recent proteome analysis demonstrated the significant enrichment of SUMOylated proteins involved in chromatin remodelling/repair, transcription, RNA metabolism, protein trafficking, and other vital functions [32,33], including also involvement in the response to various biotic and abiotic stresses [34–37]. However, it is important to note that SUMOylation and ubiquitylation intersect in multiple pathways and interact with various hormonal signalling pathways leading to antagonistic, synergistic, and even more complex outcomes [38]. In the case of abiotic stress tolerance, the ABA signalling pathway may be considered the primary target, which is modulated and regulated by Ub-SUMO systems [39–41].

Accordingly, the SUMO system was shown to play a crucial role in plant salt stress tolerance. Thus, overexpressing the rice Ubiquitin-like-specific protease 1D OVERLY TOLERANT TO SALT 1 (OsOTS1) increased the root biomass, reduced the levels of SUMOylated proteins, and increased salt stress tolerance. Also, salt treatment of rice plants with RNAi silenced the OsOTS1/2 increased accumulation of SUMO-conjugated proteins and greatly reduced salt stress tolerance, while in normal conditions these plants were not different from wild-type controls [42]. Similarly, the expression of rice SUMO E3 ligase SIZ1 (SAP AND MIZ1 DOMAIN CONTAINING LIGASE1) in Arabidopsis increased tolerance to heat, drought, and salt stresses and increased seed yields in stressed conditions. Also, *SIZ1-expressing* plants produced more osmoprotectant proline, accumulated more potassium ions, and excluded sodium ions more effectively in comparison to wild-type plants [43]. Furthermore, the heterologous expression of *SIZ1a/b/c* genes from sweet potato (*Ipomoea batatas* (L.) Lam.) in Arabidopsis *atsiz1* plants rescued the defective germination and growth phenotype, while *SIZ1a/b/c* expression in wild-type Arabidopsis plants enhanced tolerance to salt and drought stress [44]. Similarly, the expression of the SUMO-conjugating enzyme SCE1e from *Zea mays* (L.) in tobacco (*Nicotiana tabacum* L.) plants increased levels of SUMO conjugates and enhanced tolerances to salt and drought stress. Additionally, *ZmSCE1e*-expressing transgenic plants demonstrated a higher activity of antioxidant enzymes (SOD, CAT, APX, and GPX), lower H_2O_2 and malondialdehyde (MDA) accumulations, and higher expression of stress-responsive genes *LEA5*, *ERD10C*, *CDPK2*, and *AREB* under salt and drought stress treatments [45]. These results demonstrated that SUMO genes play a crucial role in salt stress tolerance by modulating protein sumoylation levels, antioxidant amount and activities, and the expression of general stress defence genes.

4. Overview of the Core Plant Autophagy Components

Autophagy is a crucial regulatory system to maintain homeostasis; it is based on the degradation and recycling of damaged and malfunctioning cellular components during the normal plant's life cycle and in response to environmental biotic and abiotic stresses [46]. Adverse environmental conditions like starvation, abiotic (such as salt, drought, cold, and heat) and biotic (such as pathogens) stresses are known to activate autophagy [47]. Depending on the target substrate and involved regulatory mechanisms, several forms of autophagy have been identified (macro-autophagy, micro-autophagy, organelle-specific, and chaperone-mediated autophagy) [48,49]. Despite some variation among plant species, different forms of autophagy have well-conserved core regulatory components identified in species ranging from lower photosynthetic algae to major agricultural species [50,51]. Originally, more than 40 *autophagy-related genes* (*ATGs*) were identified in yeast, while their homologues were further searched in *Arabidopsis thaliana* (L.) Heynh. and *Oryza sativa*

L.—the major model species to study autophagy in plants. Among the core *ATG* identified in yeast, Arabidopsis has 3 *ATG1* homologues and 2 *ATG13*, while rice has 4 *ATG1* and 2 *ATG1* homologues. ATG11 acted as a scaffold protein to facilitate the formation of a kinase complex between ATG1 and ATG13 under starvation conditions [52].

Two sensor kinases identified in Arabidopsis and rice, TARGET OF RAPAMYCIN (TOR) and SUCROSE NONFERMENTING1-RELATED PROTEIN KINASE1 (SnRK1), have been found to cooperatively perceive nutritional status and to activate or inhibit autophagy by regulating ATG1–ATG13 kinase complex states [53,54]. Additionally, an ATG1-independent autophagy pathway has been identified in Arabidopsis under prolonged carbon starvation, where the SnRK1 catalytic SNF1 KINASE HOMOLOG 10 (KIN10) subunit directly phosphorylates the ATG6 subunit to trigger autophagy [55]. These data confirmed that, although the core autophagy mechanism is conserved in all eukaryotic cells, plants employ alternative pathways to activate autophagy to ensure an appropriate response to dynamic changes in the environment.

The Arabidopsis homologue of yeast ATG14 was shown to regulate autophagosome assembly, while it was not yet discovered in rice [56]. ATG9 cycles between the ER and the cytoplasmic pool, regulating the trafficking of ATG18 on the autophagosomal membrane in a PI3P-dependent manner, thus ensuring efficient autophagosome budding from the initial autophagosome formed from the ER membrane [57,58].

Two essential ubiquitin-like conjugation systems, ATG8-PE (phosphatidylethanolamine) and ATG12-ATG5-ATG16, are crucial for the progression of autophagy. Thus, the formation of the ATG8-PE lipid complex requires cleavage by the cysteine protease ATG4, and the ATG12-ATG5 complex promotes ATG8 lipidation, while the specific function for ATG16 homologue in plants is still unknown [59–61]. ATG8 plays a central role in selective autophagy, because almost all currently known autophagy-related receptors/adaptors have an ATG8 interaction motif (AIM) or ubiquitin-interacting motif (UIM)-like sequences [62,63]. Subsequently, the autophagosome becomes laden with cargoes. The mature autophagosome subsequently fuses with and delivers its contents to the vacuole, where various hydrolases break down the contents into carbohydrates, amino acids, and lipids. These components are then returned to the cytosol and recycled to synthesize new products, or they are reused for other purposes.

4.1. The TOR-Dependent Autophagy Regulation and Its Interplay with Plant Phytohormones

Through TOR kinase phytohormones, ROS, nutrients, amino acids, and various small molecules regulate the functions of ATG proteins and autophagy in plants [64]. Among amino acids, the role of cysteine has been extensively studied, with sulphur and carbon/nitrogen (C/N) precursors sensed by TOR kinase and GENERAL CONTROL NON-DEPRESSIBLE2 (GCN2) kinase, respectively. Further research demonstrated that elevated sulphur levels led to an increase in glucose levels, activating TOR, while low glucose levels during sulphur deficiency or impaired photosynthesis inhibit TOR and activate autophagy [65]. Prolonged carbon starvation and phosphate deficit resulted in autophagy through the induction of ER stress [55,66]. Nitrogen depletion inhibited TOR and activated autophagy; however, it also led to sugar accumulation (including glucose), which activated TOR and inhibited autophagy [67]. The deficit of many metals (such as zinc, iron, and manganese) also can induce autophagy, although the underlying mechanisms are poorly understood [68,69].

Phytohormones are crucial regulators of the TOR kinase and, subsequently, TOR-dependent autophagy (Figure 2). For example, **auxin** stimulated TOR activity thus inhibiting stress-induced autophagy [70]. Treatment with **salicylic acid** (**SA**) up-regulated autophagy during methyl jasmonate-induced leaf senescence [71]. Also, the accumulation of **SA** and ROS, associated with pathogens that attack and plant immunity, has been considered as autophagy-stimulating factors [72,73]. TOR inhibited **abscisic acid** (**ABA**) signalling under unstressed environmental conditions, while under stressed conditions ABA repressed TOR activity and activated autophagy [74]. The complex interplay between **brassinosteroids** (BRs) regulates plant growth, stress responses and nitrogen starvation.

Thus, the blockage of BR signalling or biosynthesis up-regulated autophagy, while the increased activity of the BR pathway down-regulated autophagy. The brassinazole-resistant 1 (BZR1) transcription factor, brassinosteroid-insensitive 2 (BIN2) kinase, and regulatory-associated protein of TOR 1B (RAPTOR1B) have been defined as the major players coordinating interaction between BR and TOR [75,76]. The levels of active forms of **gibberellins** (GAs) and **cytokinin** (trans-zeatin) were lower in the anthers of autophagy-defective rice mutant *Osatg7–1*, while the exact role of these hormones in autophagy regulation remains unclear [77].

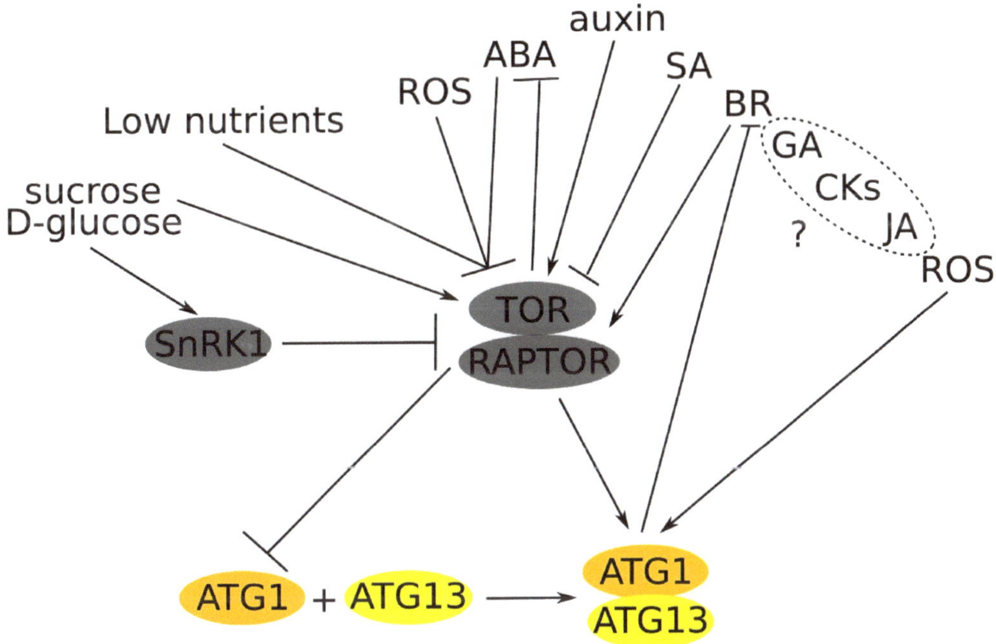

Figure 2. Autophagy sensors and regulators. The deficiency of nutrients activates autophagy through TOR inhibition. SnRK1 kinase senses sugars upstream of TOR. Glucose activates TOR, while glucose deficiency inhibits the TOR kinase, thus activating autophagy. ROS can activate autophagy by inhibiting TOR or acting directly through ATG downstream of the TOR kinase. ABA inhibits TOR kinase activity under stress, while in favourable conditions TOR suppresses ABA signalling. Auxin stimulates TOR, thus inhibiting the autophagy process. SA interacts with ROS to stimulate autophagy during senescence and plant immune response. BR inhibits autophagy through TOR activation, while autophagy regulates BR signalling by degrading its components. GA, CKs, and JA have been shown to regulate the autophagy process, although the exact mechanisms are still unknown (depicted with "?" and dashed circle). Arrows represent positive regulation and blunt lines—negative. TOR—target of rapamycin; RAPTOR—regulatory-associated protein of mTOR; SnRK1—sucrose non-fermenting related kinase 1; ROS—reactive oxygen species; ATG—autophagy related; ABA—abscisic acid; SA—salicylic acid; BR—brassinosteroids; CKs—cytokinins; GA—gibberellic acid; and JA—jasmonic acid.

4.2. Plant Autophagy-Related Ubiquitin-like Modifier (ATG8) and Its Role in the Responses to Environmental Stresses

ATG8 is the key component of the autophagy pathway, it conjugates with phospholipids to form a double-membrane vesicle (autophagosome), which further degrades intracellular components in selective or non-selective manners [78] (Figure 1). Unlike UPS, autophagosome has a much larger volume and is involved in bulk and selective recycling of huge protein complexes and aggregates, and malfunctioning and damaged organelles by delivering them to the vacuole. Since organelles (chloroplasts in particular) are extremely sensitive to various stresses, organelles' recycling is considered as one of the primary targets in autophagy-mediated stress response regulation [reviewed in [79]]. Under normal conditions, the autophagy pathway engages at a basal level to maintain cellular homeostasis and facilitate plant development, but in the case of stress, it is further activated to provide necessary relief from a certain stress factor and to secure plant survival. Thus, ATG8 is tightly regulated by external and internal factors via plant hormones and diverse transcription factors, which, in total, contribute to plant survival during stress conditions [11,80].

ATG8, a protein crucial for the autophagic process, is normally activated through post-translational cleavage by the ATG4 protease at the C-terminus. ROS generated under stressed conditions inactivates ATG4 and guarantees a high level of ATG8 lipidation and, subsequently, autophagy progression. Additionally, the particular ATG4 homologues tend to interact with the preferred ATG8 isoforms, thus fine-tuning the specific and efficient induction of autophagy [59,81]. Furthermore, ATG8 can interact with several receptors and adaptor proteins, such as:

(1) Neighbour of BRCA1 (NBR1) and the ATG8-interacting protein 1 (ATI1) and ATI2 proteins, which target plastid proteins for degradation upon carbon starvation [82];
(2) Arabidopsis Orosomucoid (ORM) proteins 1 and 2, which are involved in plant immunity regulation through the degradation of Flagellin-Sensing 2 (FLS2) receptor kinase [83];
(3) Tryptophan-rich sensory protein/translocator (TSPO), which is involved in the abiotic stress response through the regulation of the levels of PIP2;7 aquaporin on the cell surface [84];
(4) DOMINANT SUPPRESSOR OF KAR 2 (DSK2), targeting the brassinosteroid pathway regulator BES1 for degradation [85];
(5) A dehydrin, *Medicago truncatula (Gaertn.)* MtCAS31 (cold acclimation-specific 31), which under drought stress facilitated the autophagic degradation of MtPIP2;7, thus reducing water loss and improving drought tolerance [86];
(6) S-nitrosoglutathione reductase 1 (GSNOR1) in hypoxic conditions [87];
(7) ROOT HAIR DEFECTIVE3 (RHD3), an atlastin GTPases family member involved in ER stress response, was recently shown to have two AIM, through which it acted as a receptor for the selective autophagy of the ER (ER-phagy) during nutrient starvation. Thus, *rhd3* plants were defective in ER-phagy under ER stress. Furthermore, the RHD3–ATG8e interaction was enhanced under ER stress, but not *RHD3(ΔAIM1)* plants [88];
(8) E3 enzyme Ufm1-protein Ligase 1 (Ufl1), one of the proteins regulating Ufmylation, a novel PTM characterised by multiple functions in diverse cellular processes, was shown to interact with ATG8e under salt stress. Furthermore, the core components in the Ufmylation cascade (Ufl1 and ubiquitin-fold modifier 1 (Ufm1)) interacted also with ATG1 and ATG6. Also, *Ufl1* plants showed salt hypersensitive phenotype and abnormal ER morphology, and the expression of Ufmylation cascade components was up-regulated by salt stress. These results suggested that autophagy machinery acted together with the Ufmylation cascade to maintain ER homeostasis under salt stress by regulating ER-phagy [89], and many others.

However, given the number and diversity of ATG8 isoforms, it is likely that particular ATG8s may be induced by different stress factors and interact with different effectors, while it is difficult to identify the role of each isoform because of the functional redundancy between isoforms and certain overlap in their inducers and interaction partners, which makes the analysis of single knock-out mutants ineffective [62,63,90,91].

4.3. Role of the Neighbour of BRCA1 (NBR1) Protein in the Regulation of Plant Stress Response

NBR1, one of the best-characterised plant autophagy receptors, is a structural homologue and functional hybrid of mammalian autophagy receptors NBR1 and p62. In plants,

NBR1 contains five domains (an N-terminal PB1 (Phox and Bem1p), a central ZZ-type zinc finger and NBR1_like, a two C-terminal UBA (ubiquitin-associated)), and an LC3-interacting region (LIR motif) (also known as AIM motif in yeast and plants) [92]. The biological functions of NBR1 have been investigated through an analysis of *nbr1* mutants or transgenic silencing lines. Arabidopsis *nbr1* plants showed normal growth and development under normal conditions and had no deviation in general and selective (peroxisomes, mitochondria, or the endoplasmic reticulum (ER)) autophagy [93]. Furthermore, NBR1 is not essential for age- and darkness-induced senescence but may modulate growth or senescence in short-day growth conditions or under mineral deficiency [94,95].

The role of NBR1 in plant abiotic stress tolerance is mediated by selective autophagy and, depending on its interaction with ATG8, it is also associated with the clearance of aggregation-prone misfolded proteins and protein aggregates [93]. Also, NBR1 was shown to participate in the regulation of plant tolerance to heat [96], oxidative [97], salt [98], and drought [99] stresses, and regulates heat stress memory [100] and resistance to some bacterial (like, for example, *Pseudomonas syringae* [101]) and viral (like cauliflower mosaic virus [102]) pathogens. In addition, Arabidopsis NBR1 was shown to interact with several crucial proteins of the ABA signalling pathway (such as ABSCISIC ACID INSENSITIVE (ABI3/4/5) [103], which suggested a wide role of autophagy in general stress-related processes.

The role of NBR1 in salt stress tolerance is mediated through several mechanisms. As it was recently shown, the transgenic poplar hybrid plants' (*Populus alba* × *P. tremula* var. *glandulosa*) overexpression *PagNBR1* had higher rates of enzyme activities of antioxidants (SOD, POD, and CAT) and a higher expression of ROS scavengers (*APX, MDHAR, DHAR*, and *GR*), resulting in lower malonaldehyde (MDA) concentration, electrolyte leakage, and membrane lipid peroxidation. Also, *PagNBR1* overexpression prevented a decrease in photosynthetic parameters (PSII conversion efficiency and maximum optical quantum yield), thus providing higher net photosynthesis rates under salt stress. Also, transgenic plants maintained a higher K^+/Na^+ ratio in the roots, which, at least partially, was mediated through the up-regulated expression of *SOS1*. Furthermore, PagNBR1 was localised in the autophagosome and interacted with ATG8, whose expression was also up-regulated in transgenic plants. Finally, transgenic plants formed more autophagosomes, resulting in a lower accumulation of insoluble ubiquitinated protein aggregates compared to wild-type plants under salt stress. In total, these results suggested that PagNBR1 enhanced salt tolerance by up-regulating antioxidant defence and ROS scavengers, protecting photosynthetic apparatus, maintaining the K^+/Na^+ ratio, and promoting the degradation of insoluble ubiquitinated protein in transgenic poplar plants under salt stress [104].

Therefore, NBR1-mediated selective autophagy plays a critical role in response and adaptation to both biotic- and abiotic stresses. Specific to the role of NBR1 in salt stress, it would be interesting to determine the possible regulatory function of NBR1 in other stress-related hormonal signalling pathways (such as ABA and SA), and in the degradation of ABA pathway regulatory proteins ABSCISIC ACID INSENSITIVE (ABI3, ABI4, and ABI5).

5. Universal Role of ATGs in the Regulation of Plant Salinity Stress

5.1. Stress Protective Role of ATGs in Microalgae

The unicellular alga *Chlamydomonas reinhardtii* (P.A.Dang.) has responded to salt stress through the activation of antioxidant enzymes, cell damage, autophagy, and nitric oxide metabolism processes. The application of comparative proteomics and physiological approaches helped to identify 74 proteins, such as antioxidant enzymes (Monodehydroascorbate reductase (MDHAR), Dehydroascorbate reductase (DHAR), Ascorbate peroxidase (APX), and glutathione reductase (GR)), nitric oxide metabolism-related proteins (S-Nitrosoglutathione reductase (GSNOR)), DNA damage (DNA damage checkpoint protein, DNA repair protein, DNA damage-inducible protein, RAN-binding proteins RANBP1, RAD 51, and REX1-B, and putative defender against death (DAD)) and autophagy (VPS30, ATP-dependent Clp protease proteolytic subunit, vacuolar trafficking protein, ATG3, and ATG8)), the accumulation of which was altered by short-term salt stress.

These data suggested that ROS and GSNOR are involved in the activation of salt stress-induced autophagy and DNA damage [105]. Similarly, the cross-species meta-analysis on the transcriptomes of two microalgae (*Dunaliella salina* and *Dunaliella tertiolecta*) species under salt stress demonstrated that chaperone-mediated autophagy, lipid and nitrogen metabolism, photosynthesis, and ROS detoxification were core systems engaged in the *Dunaliella* salt stress response [106].

5.2. ATGs as Crucial Players in Adaptation to Various Abiotic Stresses

Recent genome-wide research has identified autophagy-related genes and their activation by different stresses in Foxtail millet (*Setaria italica* (L.) P. Beauvois), which is generally known for its strong resistance to various environmental stresses. From the total of 37 *SiATG* genes, the expression of 31 *SiATG* genes was induced by phytohormones (GA, methyl jasmonate (MeJA), SA, and ABA), 26 *SiATG* genes by cold, salt, and drought stresses, 24 *SiATG* genes by darkness, and 25 *SiATG* genes by nitrogen starvation [107]. In parallel, 49 *ATG* genes were identified in the genome of Tartary buckwheat (*Fagopyrum tataricum* (L.) Gaertn.). Among them, the eight *FtATG8* genes showed a high expression level under both salt and drought stresses. Also, all *FtATG8* genes contained MYB-family transcription factors binding and light-responsive cis-acting elements, and strongly correlated with other stress-associated transcription factors, suggesting their vital role in stress resistance in buckwheat plants [108].

In total, 29 *ATG* genes were identified in pepper (*Capsicum annuum* L.) another important agricultural species. Various abiotic stresses (heat, drought, cold, salt, and carbohydrate starvation) affected the expression of *CaATG* genes in a stress type-dependent pattern, thus suggesting the connection between autophagy and pepper tolerance to abiotic stresses. In particular, salt stress up-regulated eight *CaATGs* (*2*, *7*, *8a*, *8e*, *12*, *13a*, *18b*, and *18g*) and *VPS34*, with no *ATG* genes that were down-regulated. Stem salt stress up-regulated 15 *CaATG* genes (especially *5*, *8a*, *8c*, and *15*), while 4 *CaATG* genes were down-regulated (especially *2* and *4*) [109].

Also, nine well-expressed *ATG8* genes were identified in pear (*Pyrus* × *bretschneideri* Rehder). While *PbrATG8a/c/g/i* were up-regulated under salt stress conditions, only the *PbrATG8c* gene was up-regulated by all tested stresses (salt, drought, and *Botryosphaeria dothidea* infection) in pear [110]. Similarly, the expression of 35 *ATG* genes identified in sweet orange (*Citrus* × *sinensis* (L.) Osbeck) was affected by different stresses (salt, drought, cold, heat, and mannitol, and surplus manganese, copper, and cadmium). Under salt stress treatment, *CsATG8a*, *CsATG18e*, and *CsVTI12c* were up-regulated, while only *CsATG1a* was down-regulated in the seedlings. Furthermore, Arabidopsis plants transformed with *CsATG18a* and *CsATG18b* showed increased tolerance to salt, drought, cold, and osmotic stresses compared to control plants [111].

In potato (*Solanum tuberosum* L.), SA treatment, heat, and wounding stresses up-regulated the expression of *StATG3*, *StATG9*, *StATG11*, and *StATG13a*, while the expression of *StATG101* was up-regulated after wounding stress, but down-regulated by heat stress. Also, *StATG11* and *StATG101* expressions were up-regulated by salt stress. Furthermore, *StATG8* isoforms responded differently to various stresses, with *StATG8-2.1*, *StATG8-2.2*, *StATG8-3.2*, and *StATG8-4* up-regulated, while the expression of *StATG8-3.1* was down-regulated by heat stress, and *StATG8-1.1*, *StATG8-2.1*, and *StATG8-3.2* were up-regulated by wounding stress. Among the *StATG8* isoforms, only *StATG8-2.1* was up-regulated by SA treatment, and the expression of *StATG8-1.1*, *StATG8-2.1*, and *StATG8-3.2* was up-regulated after NaCl treatment [112].

In total, these results suggested that *ATG* genes are crucial players in response and adaptation to various abiotic stresses, which, at least partially, were mediated through interactions with different plant hormones. Further functional characterisations of *ATG* genes can provide promising new genetic resources for the development of novel stress-resistant varieties of crop species.

5.3. Functional Characterisation of ATG8 Isoforms in Adaptation to Salt Stress

Experiments on marker GFP-ATG8a overexpressing lines demonstrated that autophagy is activated by salt stress, peaking at 30 min after stress exposure to alleviate the negative effects of salt stress on plants and to establish new homeostasis. In particular, autophagy removed more oxidised proteins, facilitated the accumulation of proline and soluble sugars, and promoted Na^+ sequestration in the central vacuole of root cortex cells of the *ATG8* overexpressing plants. Such autophagy induction was absent in autophagy-defective plants (*atg2*, *atg5*, *atg7*, *atg9*, and *atg10*). Thus, *atg* mutants were hypersensitive to both osmotic and salt stress, while *ATG8* overexpressing plants showed stress-tolerant phenotypes, suggesting that a functional autophagy mechanism is absolutely required for salt stress tolerance in Arabidopsis [113].

In total, 77 *ATG* genes were defined in wheat (*Triticum aestivum* L.) and its di- and tetraploid-related species (51 in *T. turgidum* ssp. *dicoccoides* (wild emmer), 29 in *T. urartu*, and 30 in *Aegilops tauschii*). Further characterisations showed that transgenic wheat and Arabidopsis plants expressing *TaVAMP727* have higher resistance to cold, drought, and salt stresses [114]. Furthermore, 29 *ATG8* homologues were identified in allotetraploid rapeseed (AACC, *Brassica napus* L.). Among them, 13 *BnaATG8* genes were up-regulated in shoots, while 3 genes (*BnaA3.ATG8E*, *BnaA7.ATG8E*, and *BnaC3.ATG8E*) were down-regulated in the roots after salt treatment. Additionally, transgenic plants overexpressing the *BnaA8.ATG8F* gene showed reduced tolerance to salt stress, while the transcription level of *BnaA8.ATG8F* in the transgenic plants was about 30% lower in comparison to the wild type. Also, the total Na^+ concentration was higher in the wild type than in *bnaa8.atg8f* plants, and the ratio of shoot/root Na^+ accumulation in *bnaa8.atg8f* plants was almost three times that in wild type. These data suggested that under salt stress *BnaA8.ATG8F* expression is down-regulated and causes more Na^+ translocation to shoots, thus reducing plant tolerance to salt [115].

In the poplar tree (*Populus. alba* × *P. tremula* var. *glandulosa*)), 48 *ATG* genes were identified and characterised. Thus, ABA treatment up-regulated *ATG1c/f*, *ATG2b*, *ATG4b*, *ATG5*, *ATG8a/g/j/k*, *ATG10*, and *ATG8a/c*, while *ATG1b/d*, *ATG2a*, *ATG3a/b*, *ATG7*, *ATG8c/d/e/f/h/i*, and *ATG13b/c/d* were down-regulated. The BR treatment, on the contrary, up-regulated *ATG7*, *ATG8a/b/j*, and *ATG18c/g/h*, while the *ATG1* subfamily was mostly down-regulated after BR treatment. Interestingly, in the *ATG18* subfamily salt treatment up-regulated mostly *ATG18a/c* members. Furthermore, *PagATG18a* overexpression enhanced salt stress tolerance and improved photosynthesis and antioxidant activities in salt-stressed poplar plants [116].

These results provide valuable information for further research on the functional characteristics of *ATG8* isoforms and other *ATG* genes in various agricultural and model species under biotic and abiotic stresses, interaction with hormonal signalling pathways and various stress response-related mechanisms, and prepare the necessary theoretical background for the future breeding of salt stress-tolerant varieties.

6. Conclusions

In conclusion, in this review we discuss the role of ubiquitin-like modifiers in enhancing plant tolerance to salt stress, particularly focusing on the ATG8-mediated autophagy under salt stress. We have unveiled the multifaceted contributions of ATG8 to plant salt stress adaptation mechanisms. Specifically, ATG8 positively influences various physiological processes crucial for salt adaptation, including stomatal regulation, photosynthetic efficiency, osmotic adjustment, Na^+ sequestration, antioxidant defence, and metabolic activities (Figure 3). Notably, *ATG8*-overexpressing plants exhibit enhanced photosynthetic capacity, improved osmotic adjustment, and superior growth performance compared to wild-type plants under salt stress. These results highlight the potential of autophagy modulation as a promising strategy for enhancing plant resilience to salt stress conditions, while minimising trade-offs between growth and stress tolerance.

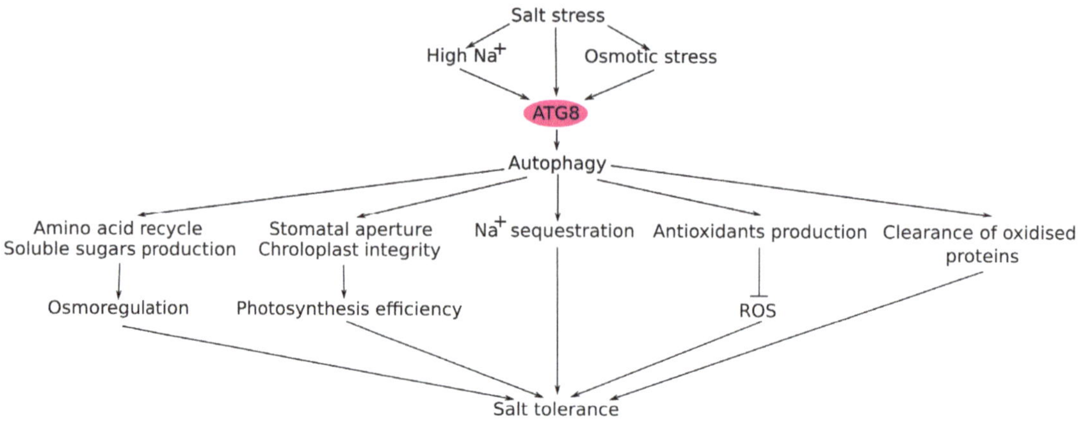

Figure 3. A proposed model on the regulatory roles of ATG8 and autophagy in salinity stress tolerance. There are five beneficial effects of autophagy: (1) soluble sugars' production and amino acids' recycling to improve plant osmotic status; (2) ATG8 contributes to the maintenance of optimal stomatal aperture and chloroplasts integrity, thus improving photosynthesis efficiency; (3) modulates expression/activity of various transporters, resulting in Na^+ sequestration into the vacuole; (4) increased production of antioxidants, which reduces ROS levels and decreases oxidative stress; (5) salt stress-activated autophagy increases the clearance of oxidised and damaged proteins. In total, these beneficial effects increase plant tolerance to salt stress. Arrow lines indicate activation; blunt line—negative regulation.

Moreover, we unveil the interconnectedness of autophagy with key phytohormone signalling pathways (such as abscisic and salicylic acids, brassinosteroids, and auxins). However, further exploration should focus on the molecular mechanisms underlying interplay with other hormones (such as cytokinins, and gibberellic and jasmonic acids), because their role in salt stress-stimulated autophagy is still unknown. Additionally, investigating the potential synergies between autophagy and other stress response pathways could unveil novel avenues for enhancing plant resilience to salt stress resistance.

Overall, the discussed results suggested that understanding the molecular mechanisms underlying the role of ubiquitin-based protein degradation systems and particularly autophagy in salt stress tolerance may offer valuable insights for the design of innovative strategies to improve crop productivity and ensure global food security in the face of the globally increasing severity of soil salinisation. By harnessing the potential of autophagy and other molecular mechanisms, we can pave the way for sustainable agriculture practices that mitigate the adverse effects of salt stress on crop yields and develop novel stress-tolerant crops.

Author Contributions: Conceptualisation, methodology, and formal analysis, S.A.D. and S.V.I.; writing—original draft preparation, S.A.D.; supervision, S.V.I.; writing—review and editing, S.A.D. and S.V.I. All authors have read and agreed to the published version of the manuscript.

Funding: This research received no external funding.

Institutional Review Board Statement: Not applicable.

Data Availability Statement: Not applicable.

Conflicts of Interest: The authors declare no conflicts of interest.

References

1. Shah, I.H.; Manzoor, M.A.; Jinhui, W.; Li, X.; Hameed, M.K.; Rehaman, A.; Li, P.; Zhang, Y.; Niu, Q.; Chang, L. Comprehensive Review: Effects of Climate Change and Greenhouse Gases Emission Relevance to Environmental Stress on Horticultural Crops and Management. *J. Environ. Manag.* **2024**, *351*, 119978. [CrossRef]
2. Benitez-Alfonso, Y.; Soanes, B.K.; Zimba, S.; Sinanaj, B.; German, L.; Sharma, V.; Bohra, A.; Kolesnikova, A.; Dunn, J.A.; Martin, A.C.; et al. Enhancing Climate Change Resilience in Agricultural Crops. *Curr. Biol.* **2023**, *33*, R1246–R1261. [CrossRef]
3. Varshney, V.; Majee, M. Emerging Roles of the Ubiquitin–Proteasome Pathway in Enhancing Crop Yield by Optimizing Seed Agronomic Traits. *Plant Cell Rep.* **2022**, *41*, 1805–1826. [CrossRef]
4. Doroodian, P.; Hua, Z. The Ubiquitin Switch in Plant Stress Response. *Plants* **2021**, *10*, 246. [CrossRef]
5. Raffeiner, M.; Zhu, S.; González-Fuente, M.; Üstün, S. Interplay between Autophagy and Proteasome during Protein Turnover. *Trends Plant Sci.* **2023**, *28*, 698–714. [CrossRef]
6. Mukhopadhyay, R.; Sarkar, B.; Jat, H.S.; Sharma, P.C.; Bolan, N.S. Soil Salinity under Climate Change: Challenges for Sustainable Agriculture and Food Security. *J. Environ. Manag.* **2021**, *280*, 111736. [CrossRef]
7. Xiao, F.; Zhou, H. Plant Salt Response: Perception, Signaling, and Tolerance. *Front. Plant Sci.* **2023**, *13*, 1053699. [CrossRef]
8. Zhao, S.; Zhang, Q.; Liu, M.; Zhou, H.; Ma, C.; Wang, P. Regulation of Plant Responses to Salt Stress. *Int. J. Mol. Sci.* **2021**, *22*, 4609. [CrossRef]
9. Mittler, R. ROS Are Good. *Trends Plant Sci.* **2017**, *22*, 11–19. [CrossRef]
10. Balasubramaniam, T.; Shen, G.; Esmaeili, N.; Zhang, H. Plants' Response Mechanisms to Salinity Stress. *Plants* **2023**, *12*, 2253. [CrossRef]
11. Signorelli, S.; Tarkowski, Ł.P.; Van den Ende, W.; Bassham, D.C. Linking Autophagy to Abiotic and Biotic Stress Responses. *Trends Plant Sci.* **2019**, *24*, 413–430. [CrossRef]
12. Xu, J.; Liu, H.; Zhou, C.; Wang, J.; Wang, J.; Han, Y.; Zheng, N.; Zhang, M.; Li, X. The Ubiquitin-Proteasome System in the Plant Response to Abiotic Stress: Potential Role in Crop Resilience Improvement. *Plant Sci.* **2024**, *342*, 112035. [CrossRef]
13. Mackinnon, E.; Stone, S.L. The Ubiquitin Proteasome System and Nutrient Stress Response. *Front. Plant Sci.* **2022**, *13*, 867419. [CrossRef]
14. Saxena, H.; Negi, H.; Sharma, B. Role of F-Box E3-Ubiquitin Ligases in Plant Development and Stress Responses. *Plant Cell Rep.* **2023**, *42*, 1133–1146. [CrossRef]
15. Ghimire, S.; Tang, X.; Liu, W.; Fu, X.; Zhang, H.; Zhang, N.; Si, H. SUMO Conjugating Enzyme: A Vital Player of SUMO Pathway in Plants. *Physiol. Mol. Biol. Plants* **2021**, *27*, 2421–2431. [CrossRef]
16. Jmii, S.; Cappadocia, L. Plant SUMO E3 Ligases: Function, Structural Organization, and Connection With DNA. *Front. Plant Sci.* **2021**, *12*, 652170. [CrossRef]
17. Srivastava, M.; Sadanandom, A.; Srivastava, A.K. Towards Understanding the Multifaceted Role of SUMOYLATION in Plant Growth and Development. *Physiol. Plant.* **2021**, *171*, 77–85. [CrossRef]
18. Morrell, R.; Sadanandom, A. Dealing With Stress: A Review of Plant SUMO Proteases. *Front. Plant Sci.* **2019**, *10*, 1122. [CrossRef]
19. Yau, R.; Rape, M. The Increasing Complexity of the Ubiquitin Code. *Nat. Cell Biol.* **2016**, *18*, 579–586. [CrossRef]
20. Sharma, S.; Prasad, A.; Sharma, N.; Prasad, M. Role of Ubiquitination Enzymes in Abiotic Environmental Interactions with Plants. *Int. J. Biol. Macromol.* **2021**, *181*, 494–507. [CrossRef]
21. Marshall, R.S.; Vierstra, R.D. Dynamic Regulation of the 26S Proteasome: From Synthesis to Degradation. *Front. Mol. Biosci.* **2019**, *6*, 40. [CrossRef] [PubMed]
22. Mao, Y. Structure, Dynamics and Function of the 26S Proteasome. In *Macromolecular Protein Complexes III: Structure and Function*; Subcellular Biochemistry; Harris, J.R., Marles-Wright, J., Eds.; Springer International Publishing: Cham, Switzerland, 2021; Volume 96, pp. 1–151. ISBN 978-3-030-58970-7.
23. Schwihla, M.; Korbei, B. The Beginning of the End: Initial Steps in the Degradation of Plasma Membrane Proteins. *Front. Plant Sci.* **2020**, *11*, 680. [CrossRef] [PubMed]
24. Martinez-Fonts, K.; Davis, C.; Tomita, T.; Elsasser, S.; Nager, A.R.; Shi, Y.; Finley, D.; Matouschek, A. The Proteasome 19S Cap and Its Ubiquitin Receptors Provide a Versatile Recognition Platform for Substrates. *Nat. Commun.* **2020**, *11*, 477. [CrossRef] [PubMed]
25. Han, G.; Qiao, Z.; Li, Y.; Yang, Z.; Wang, C.; Zhang, Y.; Liu, L.; Wang, B. RING Zinc Finger Proteins in Plant Abiotic Stress Tolerance. *Front. Plant Sci.* **2022**, *13*, 877011. [CrossRef] [PubMed]
26. Trenner, J.; Monaghan, J.; Saeed, B.; Quint, M.; Shabek, N.; Trujillo, M. Evolution and Functions of Plant U-Box Proteins: From Protein Quality Control to Signaling. *Annu. Rev. Plant Biol.* **2022**, *73*, 93–121. [CrossRef] [PubMed]
27. Lan, W.; Miao, Y. New Aspects of HECT-E3 Ligases in Cell Senescence and Cell Death of Plants. *Plants* **2019**, *8*, 483. [CrossRef] [PubMed]
28. Hwang, S.; Kim, J.J.; Lim, S.D.; Park, Y.C.; Moon, J.; Jang, C.S. Molecular Dissection of *Oryza Sativa Salt-induced* RING *Finger Protein 1* (OsSIRP1): Possible Involvement in the Sensitivity Response to Salinity Stress. *Physiol. Plant.* **2016**, *158*, 168–179. [CrossRef] [PubMed]
29. Park, Y.C.; Lim, S.D.; Moon, J.; Jang, C.S. A Rice Really Interesting New Gene H 2-type E 3 Ligase, OsSIRH2-14, Enhances Salinity Tolerance via Ubiquitin/26 S Proteasome-mediated Degradation of Salt-related Proteins. *Plant Cell Environ.* **2019**, *42*, 3061–3076. [CrossRef] [PubMed]

30. Lv, Q.; Zhang, L.; Zan, T.; Li, L.; Li, X. Wheat RING E3 Ubiquitin Ligase TaDIS1 Degrade TaSTP via the 26S Proteasome Pathway. *Plant Sci.* **2020**, *296*, 110494. [CrossRef]
31. Augustine, R.C.; Vierstra, R.D. SUMOylation: Re-Wiring the Plant Nucleus during Stress and Development. *Curr. Opin. Plant Biol.* **2018**, *45*, 143–154. [CrossRef]
32. Castro, P.H.; Santos, M.Â.; Freitas, S.; Cana-Quijada, P.; Lourenço, T.; Rodrigues, M.A.A.; Fonseca, F.; Ruiz-Albert, J.; Azevedo, J.E.; Tavares, R.M.; et al. Arabidopsis Thaliana SPF1 and SPF2 Are Nuclear-Located ULP2-like SUMO Proteases That Act Downstream of SIZ1 in Plant Development. *J. Exp. Bot.* **2018**, *69*, 4633–4649. [CrossRef]
33. Nukarinen, E.; Tomanov, K.; Ziba, I.; Weckwerth, W.; Bachmair, A. Protein Sumoylation and Phosphorylation Intersect in Arabidopsis Signaling. *Plant J.* **2017**, *91*, 505–517. [CrossRef]
34. Han, D.; Chen, C.; Xia, S.; Liu, J.; Shu, J.; Nguyen, V.; Lai, J.; Cui, Y.; Yang, C. Chromatin-Associated SUMOylation Controls the Transcriptional Switch between Plant Development and Heat Stress Responses. *Plant Commun.* **2021**, *2*, 100091. [CrossRef]
35. Benlloch, R.; Lois, L.M. Sumoylation in Plants: Mechanistic Insights and Its Role in Drought Stress. *J. Exp. Bot.* **2018**, *69*, 4539–4554. [CrossRef]
36. Clark, L.; Sue-Ob, K.; Mukkawar, V.; Jones, A.R.; Sadanandom, A. Understanding SUMO-Mediated Adaptive Responses in Plants to Improve Crop Productivity. *Essays Biochem.* **2022**, *66*, 155–168. [CrossRef]
37. Smalley, S.; Hellmann, H. Review: Exploring Possible Approaches Using Ubiquitylation and Sumoylation Pathways in Modifying Plant Stress Tolerance. *Plant Sci.* **2022**, *319*, 111275. [CrossRef] [PubMed]
38. Hua, Z.; Doroodian, P.; Vu, W. Contrasting Duplication Patterns Reflect Functional Diversities of *Ubiquitin* and *Ubiquitin-like Protein Modifiers* in Plants. *Plant J.* **2018**, *95*, 296–311. [CrossRef] [PubMed]
39. Singh, M.; Singh, A.; Yadav, N.; Yadav, D.K. Current Perspectives of Ubiquitination and SUMOylation in Abiotic Stress Tolerance in Plants. *Front. Plant Sci.* **2022**, *13*, 993194. [CrossRef]
40. Pan, W.; Lin, B.; Yang, X.; Liu, L.; Xia, R.; Li, J.; Wu, Y.; Xie, Q. The UBC27–AIRP3 Ubiquitination Complex Modulates ABA Signaling by Promoting the Degradation of ABI1 in Arabidopsis. *Proc. Natl. Acad. Sci. USA* **2020**, *117*, 27694–27702. [CrossRef] [PubMed]
41. Liu, Y.; Peng, L.; Gao, X.; Liu, Y.; Liu, Z.; Li, X.; Yang, Y.; Wang, J. AtPPRT3, a Novel E3 Ubiquitin Ligase, Plays a Positive Role in ABA Signaling. *Plant Cell Rep.* **2020**, *39*, 1467–1478. [CrossRef]
42. Srivastava, A.K.; Zhang, C.; Yates, G.; Bailey, M.; Brown, A.; Sadanandom, A. SUMO Is a Critical Regulator of Salt Stress Responses in Rice. *Plant Physiol.* **2016**, *170*, 2378–2391. [CrossRef] [PubMed]
43. Mishra, N.; Srivastava, A.P.; Esmaeili, N.; Hu, W.; Shen, G. Overexpression of the Rice Gene OsSIZ1 in Arabidopsis Improves Drought-, Heat-, and Salt-Tolerance Simultaneously. *PLoS ONE* **2018**, *13*, e0201716. [CrossRef] [PubMed]
44. Zhang, Y.; Lyu, S.; Hu, Z.; Yang, X.; Zhu, H.; Deng, S. Identification and Functional Characterization of the SUMO System in Sweet Potato under Salt and Drought Stress. *Plant Sci.* **2023**, *330*, 111645. [CrossRef] [PubMed]
45. Wang, H.; Wang, M.; Xia, Z. Overexpression of a Maize SUMO Conjugating Enzyme Gene (ZmSCE1e) Increases Sumoylation Levels and Enhances Salt and Drought Tolerance in Transgenic Tobacco. *Plant Sci.* **2019**, *281*, 113–121. [CrossRef] [PubMed]
46. Wang, S.; Hu, W.; Liu, F. Autophagy in the Lifetime of Plants: From Seed to Seed. *Int. J. Mol. Sci.* **2022**, *23*, 11410. [CrossRef] [PubMed]
47. Su, W.; Bao, Y.; Yu, X.; Xia, X.; Liu, C.; Yin, W. Autophagy and Its Regulators in Response to Stress in Plants. *Int. J. Mol. Sci.* **2020**, *21*, 8889. [CrossRef] [PubMed]
48. Qi, H.; Xia, F.; Xiao, S. Autophagy in Plants: Physiological Roles and Post-translational Regulation. *JIPB* **2021**, *63*, 161–179. [CrossRef] [PubMed]
49. Sieńko, K.; Poormassalehgoo, A.; Yamada, K.; Goto-Yamada, S. Microautophagy in Plants: Consideration of Its Molecular Mechanism. *Cells* **2020**, *9*, 887. [CrossRef] [PubMed]
50. Li, Y.; Xu, X.; Qi, G.; Cui, D.; Huang, C.; Sui, X.; Li, G.; Fan, Q. Mechanisms of Autophagy Function and Regulation in Plant Growth, Development, and Response to Abiotic Stress. *Crop J.* **2023**, *11*, 1611–1625. [CrossRef]
51. Shemi, A.; Ben-Dor, S.; Vardi, A. Elucidating the Composition and Conservation of the Autophagy Pathway in Photosynthetic Eukaryotes. *Autophagy* **2015**, *11*, 701–715. [CrossRef]
52. Li, F.; Vierstra, R.D. *Arabidopsis* ATG11, a Scaffold That Links the ATG1-ATG13 Kinase Complex to General Autophagy and Selective Mitophagy. *Autophagy* **2014**, *10*, 1466–1467. [CrossRef] [PubMed]
53. McCready, K.; Spencer, V.; Kim, M. The Importance of TOR Kinase in Plant Development. *Front. Plant Sci.* **2020**, *11*, 16. [CrossRef]
54. Soto-Burgos, J.; Bassham, D.C. SnRK1 Activates Autophagy via the TOR Signaling Pathway in Arabidopsis Thaliana. *PLoS ONE* **2017**, *12*, e0182591. [CrossRef]
55. Huang, X.; Zheng, C.; Liu, F.; Yang, C.; Zheng, P.; Lu, X.; Tian, J.; Chung, T.; Otegui, M.S.; Xiao, S.; et al. Genetic Analyses of the Arabidopsis ATG1 Kinase Complex Reveal Both Kinase-Dependent and Independent Autophagic Routes during Fixed-Carbon Starvation. *Plant Cell* **2019**, *31*, 2973–2995. [CrossRef]
56. Liu, F.; Hu, W.; Li, F.; Marshall, R.S.; Zarza, X.; Munnik, T.; Vierstra, R.D. AUTOPHAGY-RELATED14 and Its Associated Phosphatidylinositol 3-Kinase Complex Promote Autophagy in Arabidopsis. *Plant Cell* **2020**, *32*, 3939–3960. [CrossRef]
57. Zhuang, X.; Chung, K.P.; Cui, Y.; Lin, W.; Gao, C.; Kang, B.-H.; Jiang, L. ATG9 Regulates Autophagosome Progression from the Endoplasmic Reticulum in *Arabidopsis*. *Proc. Natl. Acad. Sci. USA* **2017**, *114*, E426–E435. [CrossRef]

58. Huang, L.; Wen, X.; Jin, L.; Han, H.; Guo, H. HOOKLESS1 Acetylates AUTOPHAGY-RELATED PROTEIN18a to Promote Autophagy during Nutrient Starvation in Arabidopsis. *Plant Cell* **2023**, *36*, 136–157. [CrossRef]
59. Seo, E.; Woo, J.; Park, E.; Bertolani, S.J.; Siegel, J.B.; Choi, D.; Dinesh-Kumar, S.P. Comparative Analyses of Ubiquitin-like *ATG8* and Cysteine Protease *ATG4* Autophagy Genes in the Plant Lineage and Cross-Kingdom Processing of ATG8 by ATG4. *Autophagy* **2016**, *12*, 2054–2068. [CrossRef]
60. Zhen, X.; Li, X.; Yu, J.; Xu, F. OsATG8c-Mediated Increased Autophagy Regulates the Yield and Nitrogen Use Efficiency in Rice. *Int. J. Mol. Sci.* **2019**, *20*, 4956. [CrossRef]
61. Le Bars, R.; Marion, J.; Satiat-Jeunemaitre, B.; Bianchi, M.W. Folding into an Autophagosome: ATG5 Sheds Light on How Plants Do It. *Autophagy* **2014**, *10*, 1861–1863. [CrossRef]
62. Luo, S.; Li, X.; Zhang, Y.; Fu, Y.; Fan, B.; Zhu, C.; Chen, Z. Cargo Recognition and Function of Selective Autophagy Receptors in Plants. *Int. J. Mol. Sci.* **2021**, *22*, 1013. [CrossRef] [PubMed]
63. Boycheva Woltering, S.; Isono, E. Knowing When to Self-Eat–Fine-Tuning Autophagy Through ATG8 Iso-Forms in Plants. *Front. Plant Sci.* **2020**, *11*, 579875. [CrossRef] [PubMed]
64. Shen, J.Z.; Wu, G.; Guo, S. Amino Acids in Autophagy: Regulation and Function. In *Amino Acids in Nutrition and Health*; Advances in Experimental Medicine and Biology; Wu, G., Ed.; Springer International Publishing: Cham, Switzerland, 2021; Volume 1332, pp. 51–66. ISBN 978-3-030-74179-2.
65. Fu, L.; Wang, P.; Xiong, Y. Target of Rapamycin Signaling in Plant Stress Responses. *Plant Physiol.* **2020**, *182*, 1613–1623. [CrossRef]
66. Naumann, C.; Müller, J.; Sakhonwasee, S.; Wieghaus, A.; Hause, G.; Heisters, M.; Bürstenbinder, K.; Abel, S. The Local Phosphate Deficiency Response Activates Endoplasmic Reticulum Stress-Dependent Autophagy. *Plant Physiol.* **2019**, *179*, 460–476. [CrossRef] [PubMed]
67. Cao, P.; Kim, S.-J.; Xing, A.; Schenck, C.A.; Liu, L.; Jiang, N.; Wang, J.; Last, R.L.; Brandizzi, F. Homeostasis of Branched-Chain Amino Acids Is Critical for the Activity of TOR Signaling in Arabidopsis. *Elife* **2019**, *8*, e50747. [CrossRef] [PubMed]
68. Shinozaki, D.; Merkulova, E.A.; Naya, L.; Horie, T.; Kanno, Y.; Seo, M.; Ohsumi, Y.; Masclaux-Daubresse, C.; Yoshimoto, K. Autophagy Increases Zinc Bioavailability to Avoid Light-Mediated Reactive Oxygen Species Production under Zinc Deficiency. *Plant Physiol.* **2020**, *182*, 1284–1296. [CrossRef]
69. Pottier, M.; Dumont, J.; Masclaux-Daubresse, C.; Thomine, S. Autophagy Is Essential for Optimal Translocation of Iron to Seeds in Arabidopsis. *J. Exp. Bot.* **2018**, *70*, 859–869. [CrossRef]
70. Schepetilnikov, M.; Makarian, J.; Srour, O.; Geldreich, A.; Yang, Z.; Chicher, J.; Hammann, P.; Ryabova, L.A. GTP Ase ROP 2 Binds and Promotes Activation of Target of Rapamycin, TOR, in Response to Auxin. *EMBO J.* **2017**, *36*, 886–903. [CrossRef]
71. Yin, R.; Liu, X.; Yu, J.; Ji, Y.; Liu, J.; Cheng, L.; Zhou, J. Up-Regulation of Autophagy by Low Concentration of Salicylic Acid Delays Methyl Jasmonate-Induced Leaf Senescence. *Sci. Rep.* **2020**, *10*, 11472. [CrossRef]
72. Rigault, M.; Citerne, S.; Masclaux-Daubresse, C.; Dellagi, A. Salicylic Acid Is a Key Player of Arabidopsis Autophagy Mutant Susceptibility to the Necrotrophic Bacterium Dickeya Dadantii. *Sci. Rep.* **2021**, *11*, 3624. [CrossRef]
73. Gong, W.; Li, B.; Zhang, B.; Chen, W. ATG4 Mediated Psm ES4326/AvrRpt2-Induced Autophagy Dependent on Salicylic Acid in Arabidopsis Thaliana. *Int. J. Mol. Sci.* **2020**, *21*, 5147. [CrossRef] [PubMed]
74. Wang, P.; Zhao, Y.; Li, Z.; Hsu, C.-C.; Liu, X.; Fu, L.; Hou, Y.-J.; Du, Y.; Xie, S.; Zhang, C.; et al. Reciprocal Regulation of the TOR Kinase and ABA Receptor Balances Plant Growth and Stress Response. *Mol. Cell* **2018**, *69*, 100–112. [CrossRef] [PubMed]
75. Liao, C.-Y.; Pu, Y.; Nolan, T.M.; Montes, C.; Guo, H.; Walley, J.W.; Yin, Y.; Bassham, D.C. Brassinosteroids Modulate Autophagy through Phosphorylation of RAPTOR1B by the GSK3-like Kinase BIN2 in Arabidopsis. *Autophagy* **2023**, *19*, 1293–1310. [CrossRef] [PubMed]
76. Wang, Y.; Cao, J.-J.; Wang, K.-X.; Xia, X.-J.; Shi, K.; Zhou, Y.-H.; Yu, J.-Q.; Zhou, J. BZR1 Mediates Brassinosteroid-Induced Autophagy and Nitrogen Starvation in Tomato. *Plant Physiol.* **2019**, *179*, 671–685. [CrossRef] [PubMed]
77. Kurusu, T.; Koyano, T.; Kitahata, N.; Kojima, M.; Hanamata, S.; Sakakibara, H.; Kuchitsu, K. Autophagy-Mediated Regulation of Phytohormone Metabolism during Rice Anther Development. *Plant Signal. Behav.* **2017**, *12*, e1365211. [CrossRef] [PubMed]
78. Marshall, R.S.; Vierstra, R.D. Autophagy: The Master of Bulk and Selective Recycling. *Annu. Rev. Plant Biol.* **2018**, *69*, 173–208. [CrossRef] [PubMed]
79. Sharma, I.; Kirti, P.B.; Pati, P.K. Autophagy: A Game Changer for Plant Development and Crop Improvement. *Planta* **2022**, *256*, 103. [CrossRef]
80. Wang, P.; Nolan, T.M.; Yin, Y.; Bassham, D.C. Identification of Transcription Factors That Regulate *ATG8* Expression and Autophagy in *Arabidopsis*. *Autophagy* **2020**, *16*, 123–139. [CrossRef]
81. Pérez-Pérez, M.E.; Lemaire, S.D.; Crespo, J.L. The ATG4 Protease Integrates Redox and Stress Signals to Regulate Autophagy. *J. Exp. Bot.* **2021**, *72*, 3340–3351. [CrossRef]
82. Michaeli, S.; Honig, A.; Levanony, H.; Peled-Zehavi, H.; Galili, G. *Arabidopsis* ATG8-INTERACTING PROTEIN1 Is Involved in Autophagy-Dependent Vesicular Trafficking of Plastid Proteins to the Vacuole. *Plant Cell* **2014**, *26*, 4084–4101. [CrossRef]
83. Yang, F.; Kimberlin, A.N.; Elowsky, C.G.; Liu, Y.; Gonzalez-Solis, A.; Cahoon, E.B.; Alfano, J.R. A Plant Immune Receptor Degraded by Selective Autophagy. *Mol. Plant* **2019**, *12*, 113–123. [CrossRef] [PubMed]
84. Hachez, C.; Veljanovski, V.; Reinhardt, H.; Guillaumot, D.; Vanhee, C.; Chaumont, F.; Batoko, H. The *Arabidopsis* Abiotic Stress-Induced TSPO-Related Protein Reduces Cell-Surface Expression of the Aquaporin PIP2;7 through Protein-Protein Interactions and Autophagic Degradation. *Plant Cell* **2014**, *26*, 4974–4990. [CrossRef] [PubMed]

85. Nolan, T.M.; Brennan, B.; Yang, M.; Chen, J.; Zhang, M.; Li, Z.; Wang, X.; Bassham, D.C.; Walley, J.; Yin, Y. Selective Autophagy of BES1 Mediated by DSK2 Balances Plant Growth and Survival. *Dev. Cell* **2017**, *41*, 33–46. [CrossRef]
86. Li, X.; Liu, Q.; Feng, H.; Deng, J.; Zhang, R.; Wen, J.; Dong, J.; Wang, T. Dehydrin MtCAS31 Promotes Autophagic Degradation under Drought Stress. *Autophagy* **2020**, *16*, 862–877. [CrossRef] [PubMed]
87. Zhan, N.; Wang, C.; Chen, L.; Yang, H.; Feng, J.; Gong, X.; Ren, B.; Wu, R.; Mu, J.; Li, Y.; et al. S-Nitrosylation Targets GSNO Reductase for Selective Autophagy during Hypoxia Responses in Plants. *Mol. Cell* **2018**, *71*, 142–154. [CrossRef]
88. Sun, J.; Wang, W.; Zheng, H. ROOT HAIR DEFECTIVE3 Is a Receptor for Selective Autophagy of the Endoplasmic Reticulum in Arabidopsis. *Front. Plant Sci.* **2022**, *13*, 817251. [CrossRef] [PubMed]
89. Li, B.; Niu, F.; Zeng, Y.; Tse, M.K.; Deng, C.; Hong, L.; Gao, S.; Lo, S.W.; Cao, W.; Huang, S.; et al. Ufmylation Reconciles Salt Stress-Induced Unfolded Protein Responses via ER-Phagy in *Arabidopsis*. *Proc. Natl. Acad. Sci. USA* **2023**, *120*, e2208351120. [CrossRef]
90. Chen, Q.; Soulay, F.; Saudemont, B.; Elmayan, T.; Marmagne, A.; Masclaux-Daubresse, C. Overexpression of *ATG8* in Arabidopsis Stimulates Autophagic Activity and Increases Nitrogen Remobilization Efficiency and Grain Filling. *Plant Cell Physiol.* **2019**, *60*, 343–352. [CrossRef]
91. Li, B.; Liu, G.; Wang, Y.; Wei, Y.; Shi, H. Overexpression of Banana ATG8f Modulates Drought Stress Resistance in Arabidopsis. *Biomolecules* **2019**, *9*, 814. [CrossRef]
92. Rasmussen, N.L.; Kournoutis, A.; Lamark, T.; Johansen, T. NBR1: The Archetypal Selective Autophagy Receptor. *J. Cell Biol.* **2022**, *221*, e202208092. [CrossRef]
93. Zhang, Y.; Chen, Z. Broad and Complex Roles of NBR1-Mediated Selective Autophagy in Plant Stress Responses. *Cells* **2020**, *9*, 2562. [CrossRef] [PubMed]
94. Ji, C.; Zhou, J.; Guo, R.; Lin, Y.; Kung, C.-H.; Hu, S.; Ng, W.Y.; Zhuang, X.; Jiang, L. AtNBR1 Is a Selective Autophagic Receptor for AtExo70E2 in Arabidopsis. *Plant Physiol.* **2020**, *184*, 777–791. [CrossRef] [PubMed]
95. Tarnowski, L.; Collados Rodriguez, M.; Brzywczy, J.; Cysewski, D.; Wawrzynska, A.; Sirko, A. Overexpression of the Selective Autophagy Cargo Receptor NBR1 Modifies Plant Response to Sulfur Deficit. *Cells* **2020**, *9*, 669. [CrossRef] [PubMed]
96. Jung, H.; Lee, H.N.; Marshall, R.S.; Lomax, A.W.; Yoon, M.J.; Kim, J.; Kim, J.H.; Vierstra, R.D.; Chung, T. Arabidopsis Cargo Receptor NBR1 Mediates Selective Autophagy of Defective Proteins. *J. Exp. Bot.* **2020**, *71*, 73–89. [CrossRef] [PubMed]
97. Zhou, J.; Zhang, Y.; Qi, J.; Chi, Y.; Fan, B.; Yu, J.-Q.; Chen, Z. E3 Ubiquitin Ligase CHIP and NBR1-Mediated Selective Autophagy Protect Additively against Proteotoxicity in Plant Stress Responses. *PLoS Genet.* **2014**, *10*, e1004116. [CrossRef] [PubMed]
98. Ma, A.; Nan, N.; Shi, Y.; Wang, J.; Guo, P.; Liu, W.; Zhou, G.; Yu, J.; Zhou, D.; Yun, D.-J.; et al. Autophagy Receptor OsNBR1 Modulates Salt Stress Tolerance in Rice. *Plant Cell Rep.* **2024**, *43*, 17. [CrossRef] [PubMed]
99. Chen, L.; Lv, Q.; Yang, W.; Yang, H.; Chen, Q.; Wang, B.; Lei, Y.; Xie, Y. TaNBR1, a Novel Wheat NBR1-like Domain Gene Negatively Regulates Drought Stress Tolerance in Transgenic Arabidopsis. *Int. J. Mol. Sci.* **2022**, *23*, 4519. [CrossRef]
100. Thirumalaikumar, V.P.; Gorka, M.; Schulz, K.; Masclaux-Daubresse, C.; Sampathkumar, A.; Skirycz, A.; Vierstra, R.D.; Balazadeh, S. Selective Autophagy Regulates Heat Stress Memory in Arabidopsis by NBR1-Mediated Targeting of HSP90.1 and ROF1. *Autophagy* **2021**, *17*, 2184–2199. [CrossRef] [PubMed]
101. Üstün, S.; Hofius, D. Anti- and pro-Microbial Roles of Autophagy in Plant-Bacteria Interactions. *Autophagy* **2018**, *14*, 1465–1466. [CrossRef]
102. Hafrén, A.; Macia, J.-L.; Love, A.J.; Milner, J.J.; Drucker, M.; Hofius, D. Selective Autophagy Limits Cauliflower Mosaic Virus Infection by NBR1-Mediated Targeting of Viral Capsid Protein and Particles. *Proc. Natl. Acad. Sci. USA* **2017**, *114*, E2026–E2035. [CrossRef]
103. Tarnowski, L.; Rodriguez, M.C.; Brzywczy, J.; Piecho-Kabacik, M.; Krčkova, Z.; Martinec, J.; Wawrzynska, A.; Sirko, A. A Selective Autophagy Cargo Receptor NBR1 Modulates Abscisic Acid Signalling in Arabidopsis Thaliana. *Sci. Rep.* **2020**, *10*, 7778. [CrossRef] [PubMed]
104. Su, W.; Bao, Y.; Lu, Y.; He, F.; Wang, S.; Wang, D.; Yu, X.; Yin, W.; Xia, X.; Liu, C. Poplar Autophagy Receptor NBR1 Enhances Salt Stress Tolerance by Regulating Selective Autophagy and Antioxidant System. *Front. Plant Sci.* **2021**, *11*, 568411. [CrossRef] [PubMed]
105. Chen, X.; Tian, D.; Kong, X.; Chen, Q.; Ef, A.; Hu, X.; Jia, A. The Role of Nitric Oxide Signalling in Response to Salt Stress in Chlamydomonas Reinhardtii. *Planta* **2016**, *244*, 651–669. [CrossRef] [PubMed]
106. Panahi, B.; Frahadian, M.; Dums, J.T.; Hejazi, M.A. Integration of Cross Species RNA-Seq Meta-Analysis and Machine-Learning Models Identifies the Most Important Salt Stress–Responsive Pathways in Microalga Dunaliella. *Front. Genet.* **2019**, *10*, 752. [CrossRef] [PubMed]
107. Li, W.; Chen, M.; Wang, E.; Hu, L.; Hawkesford, M.J.; Zhong, L.; Chen, Z.; Xu, Z.; Li, L.; Zhou, Y.; et al. Genome-Wide Analysis of Autophagy-Associated Genes in Foxtail Millet (*Setaria italica* L.) and Characterization of the Function of SiATG8a in Conferring Tolerance to Nitrogen Starvation in Rice. *BMC Genom.* **2016**, *17*, 797. [CrossRef] [PubMed]
108. Fang, Y.; Wang, S.; Wu, H.; Li, C.; Zhao, H.; Chen, H.; Wang, X.; Wu, Q. Genome-Wide Identification of ATG Gene Family Members in Fagopyrum Tataricum and Their Expression during Stress Responses. *Int. J. Mol. Sci.* **2022**, *23*, 14845. [CrossRef] [PubMed]

109. Zhai, Y.; Guo, M.; Wang, H.; Lu, J.; Liu, J.; Zhang, C.; Gong, Z.; Lu, M. Autophagy, a Conserved Mechanism for Protein Degradation, Responds to Heat, and Other Abiotic Stresses in *Capsicum annuum* L. *Front. Plant Sci.* **2016**, *7*, 175103. [CrossRef] [PubMed]
110. Sun, X.; Pan, B.; Xu, W.; Chen, Q.; Wang, Y.; Ban, Q.; Xing, C.; Zhang, S. Genome-Wide Identification and Expression Analysis of the Pear Autophagy-Related Gene PbrATG8 and Functional Verification of PbrATG8c in Pyrus Bretschneideri Rehd. *Planta* **2021**, *253*, 32. [CrossRef] [PubMed]
111. Fu, X.-Z.; Zhou, X.; Xu, Y.-Y.; Hui, Q.-L.; Chun, C.-P.; Ling, L.-L.; Peng, L.-Z. Comprehensive Analysis of Autophagy-Related Genes in Sweet Orange (*Citrus sinensis*) Highlights Their Roles in Response to Abiotic Stresses. *Int. J. Mol. Sci.* **2020**, *21*, 2699. [CrossRef]
112. Song, I.; Hong, S.; Huh, S.U. Identification and Expression Analysis of the Solanum Tuberosum StATG8 Family Associated with the WRKY Transcription Factor. *Plants* **2022**, *11*, 2858. [CrossRef]
113. Luo, L.; Zhang, P.; Zhu, R.; Fu, J.; Su, J.; Zheng, J.; Wang, Z.; Wang, D.; Gong, Q. Autophagy Is Rapidly Induced by Salt Stress and Is Required for Salt Tolerance in Arabidopsis. *Front. Plant Sci.* **2017**, *8*, 1459. [CrossRef] [PubMed]
114. Yue, W.; Zhang, H.; Sun, X.; Su, N.; Zhao, Q.; Yan, Z.; Weining, S.; Yue, H. The Landscape of Autophagy-Related (ATG) Genes and Functional Characterization of TaVAMP727 to Autophagy in Wheat. *Int. J. Mol. Sci.* **2022**, *23*, 891. [CrossRef] [PubMed]
115. Zhang, T.; Zhou, T.; Zhang, Y.; Chen, J.; Song, H.; Wu, P.; Yue, C.; Huang, J.; Zhang, Z.; Hua, Y. Genome-Wide Identification and Functional Characterization Reveals the Pivotal Roles of BnaA8.ATG8F in Salt Stress Tolerance and Nitrogen Limitation Adaptation in Allotetraploid Rapeseed. *Int. J. Mol. Sci.* **2022**, *23*, 11318. [CrossRef]
116. Yu, X.-Q.; Su, W.; Zhang, H.; Niu, M.; Liu, X.; Li, Z.; Liu, C.; Wang, H.-L.; Yin, W.; Xia, X. Genome-Wide Analysis of Autophagy-Related Gene Family and PagATG18a Enhances Salt Tolerance by Regulating ROS Homeostasis in Poplar. *Int. J. Biol. Macromol.* **2023**, *224*, 1524–1540. [CrossRef] [PubMed]

plants

MDPI

Article

Growth, Productivity, Yield Components and Seasonality of Different Genotypes of Forage Clover *Lotus corniculatus* L. under Varied Soil Moisture Contents

Sahara Xolocotzi-Acoltzi [1], Aurelio Pedroza-Sandoval [1,*], Gabino García-De los Santos [2], Perpetuo Álvarez-Vázquez [3] and Isaac Gramillo-Ávila [1]

1 Unidad Regional Universitaria de Zonas Áridas de la Universidad Autónoma Chapingo, Km. 40 Carretera Gómez Palacio—Ciudad Juárez, Bermejillo C.P. 35230, Durango, Mexico; xolocotzi18@gmail.com (S.X.-A.); igramillo4@gmail.com (I.G.-Á.)

2 Departamento de Recursos Naturales del Colegio de Posgraduados, Campus Montecillo, Km. 36.5 Carretera México-Texcoco, Montecillo, Texcoco C.P. 56230, Estado de México, Mexico; garciag@colpos.mx

3 Departamento de Recursos Naturales de la Universidad Autónoma Agraria Antonio Narro, Calzada Antonio Narro 1923, Buenavista, Saltillo C.P. 25315, Coahuila, Mexico; perpetuo.alvarezv@uaaan.edu.mx

* Correspondence: apedroza@chapingo.uruza.edu.mx

Abstract: This study aimed to evaluate the response to water deficit of different ecotypes and a variety of *Lotus corniculatus* on growth, productivity, and yield components, through seasonal times. A randomized block experimental design in a 2 × 5 factorial arrangement with three replicates was used. The first variation factor was soil moisture contents: field capacity (FC) was 26.5% ± 1.5, and water deficit (WD) was 85% of FC (22.5% ± 1.5); the second variation factor was the ecotypes identified with the codes 255301, 255305, 202700 and 226792 and of the variety Estanzuela Ganador. The best responses in plant cover and weight of accumulated fresh biomass were obtained in the ecotype 202700 under WD, with values of 1649.0 cm^2 and 583.7 g plant^{-1}, and 1661.2 cm^2 and 740.1 g plant^{-1} in ecotype 255305 under FC. The leaf clover was the main component of yield during the summer and autumn seasons. Ecotype 226792 was tolerant to low temperatures during the winter season with better leaf development. Ecotype 202700 is the best option for forage clover production when water is limited, and ecotype 255305 when water is not resource-limited, but these preliminary conclusions need to be confirmed in field studies.

Keywords: drought; stress physiology; agronomic indicators; plant stress

Citation: Xolocotzi-Acoltzi, S.; Pedroza-Sandoval, A.; García-De los Santos, G.; Álvarez-Vázquez, P.; Gramillo-Ávila, I. Growth, Productivity, Yield Components and Seasonality of Different Genotypes of Forage Clover *Lotus corniculatus* L. under Varied Soil Moisture Contents. *Plants* **2024**, *13*, 1407. https://doi.org/10.3390/plants13101407

Academic Editors: Geovani Soares de Lima, Lauriane Almeida dos Anjos Soares and Francisco Vanies Da Silva Sá

Received: 19 April 2024
Revised: 8 May 2024
Accepted: 15 May 2024
Published: 18 May 2024

1. Introduction

Climate change and environmental impact are some of the causes of the degradation of natural resources, in addition to anthropogenic action, which compromises the stability of agroecosystems and the consequent agri-food productivity given the increase in the world population [1]. Agriculture is the productive activity with the greatest demand for resources and inputs, especially water resources, with a consumption of 85% of the water available used for agri-food production [2].

Water scarcity is the most recurrent problem due to the lack of rain with increasingly intense and frequent droughts in the world [3]. Due to droughts, bodies of water are reduced, and the recharge of aquifers is poor due to the overexploitation of water resources in irrigation areas with deep well water, with negative effects of low quantity and quality of the water for various domestic and productive uses [4,5].

This imbalance between the volume of water extracted and the increasingly diminished recharge of aquifers already has serious economic, social, and environmental impacts in different forage-producing regions [6–8]. An alternative to the problem of forage monoculture with high water requirements in areas with low availability of quality water is the diversification of crops with the same forage type, with productive potential but with lower water demand [9–11]. Biodiversity through the introduction of alternative forages holds promise in mitigating these challenges due to their drought tolerance and adaptability to various environmental conditions [12–14]. In contexts such as those observed by this study, livestock producing milk and meat on a large scale are the most demanding of forage for feed under stable conditions [15]. Alfalfa (*Medicago sativa* L.) is a forage crop that is excellent at producing great quantities of fresh matter with high nutritional content to produce milk and meat; however, this species is a semi-perennial crop requiring irrigation throughout the year. Alfalfa is a plant with the greatest water demand for growing and development, which makes it a crop species with low efficiency in water use [7,16].

The diversification of forage crops with productive potential but with lower water demand is an alternative to the problem of monoculture with high water requirements in areas with low availability of water [10]. Some alternatives are native crops such as nopal [17,18] or introduced crops such as forage clover [19,20] which are viable alternatives for irrigated agricultural areas affected by the problem of water scarcity.

Recently, the performance of legume species has taken on special relevance, which have diverse adaptation properties to marginal conditions and contribute to sustainable agriculture and agri-food systems [11,21]. Different forage species of clover are being explored through research and technological development programs that can compete with crops such as alfalfa to maintain productive quality and quantity with more efficient use of water [22]; bird's foot clover *Lotus corniculatus* L. is the most outstanding of these. This forage species is a forage legume of Mediterranean European and African origin, which has a high capacity to adapt to different environments, such as water deficit, nutritional deficiencies, low temperatures, and tolerance to soils with acidic pH, as well as to high levels of aluminum and manganese [23]. This species has a productive performance comparable to alfalfa or forage clover *Trifolium repens* L. [24]. The growth of *L. corniculatus* is successful in countries with dry summers and marked seasonality, such as New Zealand and Uruguay, where it has shown good response behavior as a forage crop [25] and tolerance to water deficit by drought [26].

Bird's-foot trefoil, a prevalent grass-legume, thrives in poorly drained acid soils and demonstrates adaptability to sandy soils [27,28]. Unlike other forage legumes in eastern North America, it does not induce bloat in ruminants [29]. The species has three distinct growth habit types—prostrate, erect, and semi-erect—with each type catering to specific agricultural purposes. Seedling vigor is comparatively lower, and the species utilizes root reserves for early spring growth, unlike most legumes that replenish root reserves in late summer [27]. However, the widespread establishment of this crop has not been easy to solve since the complexity of its reproduction by seed and its relatively slow growth make it difficult to expand it in intensive production systems [30]. This study aimed to evaluate different genotypes of *L. corniculatus* in their capacity to respond to growth, production, and yield under water deficit through seasonal times.

2. Results

2.1. Climatic Conditions within the Shade Mesh

According to the temperature report during the study time, an average maximum of 42.3 °C and an average minimum of 7.7 °C were identified, with a maximum of 46.9 °C and a minimum of −4.6 °C. The average relative humidity recorded ranged between 44 and 73%, with a minimum between 5 and 10% in the months from May to July and a maximum of 100% in the rainy periods distributed across the months July, August, and September; both variables were measured during the period from April 2021 to May 2022 (Figure 1).

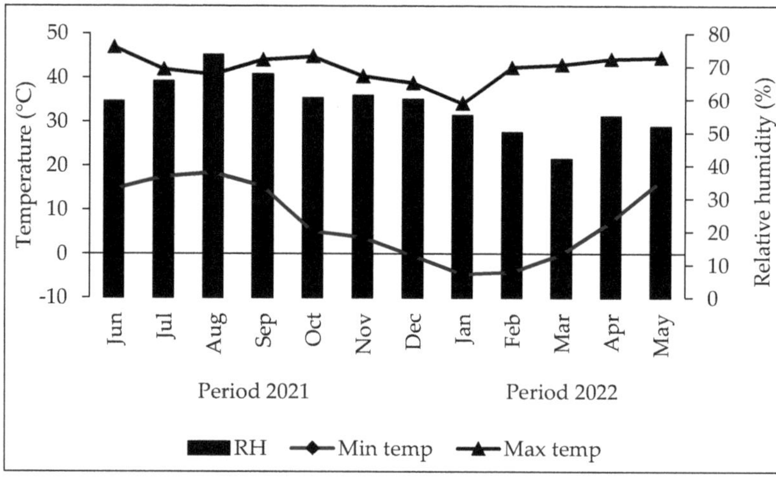

Figure 1. Variation in temperature (°C), and relative humidity (%) within the shadow mesh of the experimental area from June 2021 to May 2022.

2.2. Stem Length

The length of stems did not vary significantly ($p \leq 0.05$) in soil moisture content at FC in the different genetic materials of *L. corniculatus* that were evaluated, while in the WD treatments, a genotypic variation was recorded with better behavior of ecotype 202700, with a length of 23 cm, compared to ecotypes 255305, 226792 and the Estanzuela Ganador variety, which recorded the lowest growth with values of 19.5, 19.1 and 19.7 cm, respectively (Table 1).

Table 1. Growth, development, and forage productivity of different ecotypes and varieties of *L. corniculatus* under two soil moisture contents.

Ecotypes/Variety	Stem Length (cm)		Plant Coverage (cm^2)		PAFB (g plant^{-1})	
	FC	WD	FC	WD	FC	WD
255301	20.5 a ± 6.0	21.6 ab ± 7.2	1299.7 b ± 643.7	1252.8 ab ± 880.1	605.8 ab ± 54.5	410.6 b ± 118.7
255305	22.0 a ± 5.2	19.5 b ± 4.8	1661.2 a ± 875.5	1244.3 ab ± 681.4	740.1 a ± 58.9	497.6 ab ± 94.6
202700	21.8 a ± 7.0	23.0 a ± 4.3	1530.0 ab ± 920.2	1649.0 a ± 756.5	571.3 b ± 160.9	583.7 a ± 45.7
226792	20.3 a ± 5.8	19.1 b ± 5.3	1249.0 b ± 589.1	1121.5 b ± 597.7	554.7 b ± 322.2	497.7 ab ± 278.5
Estanzuela Ganador	21.2 a ± 4.6	19.7 b ± 4.4	1334.1 ab ± 671.8	1117.1 b ± 498.9	640.4 ab ± 116.9	486.5 ab ± 283.0

Tukey test ($p \leq 0.05$). Figures with the same letters within the same column are statistically equal. FC is field capacity (26.5% ± 1.5); WD is water deficit at 85% FC (22.5% ± 1.5); PAFB is the production of accumulated fresh biomass; values after ± are standard deviations.

2.3. Plant Coverage

Plant cover was significantly greater ($p \leq 0.05$) in ecotypes 255305 and 202700 under FC and WD, with values of 1661.2 and 1649 cm^2, respectively, compared to ecotypes 226792 and 255301, which showed the lowest values of 1249 and 1299 cm^2 in FC, respectively. The ecotype 226792 and the Estanzuela Ganador variety had the lowest values under WD, with values of 1121.5 and 1117.1 cm^2, respectively (Table 1).

2.4. Production of Accumulated Fresh Biomass (PAFB)

The PAFB of different Lotus genotypes was like that of plant coverage ($p \leq 0.05$) due to the relationship between both variables; ecotypes 255305 and 202700 were the ones with the best response, with values of 740.1 and 583.7 g plant^{-1} under FC and WD, respectively.

Ecotype 255301 was the one with the lowest response, with 410.6 g plant^{-1}. The rest of the genetic materials had an intermediate behavior across both response effects (Table 1).

2.5. Performance Components

The contribution of the components to the forage yield of the genetic materials evaluated showed variability depending on the season but was similar in both soil moisture contents FC and WD. The range of contribution of leaves was 50–80%, stems 15–40%, and dry stems 1–10%. Inflorescences showed percentages of 0.5–4%, but only in 202700 and Estanzuela Ganador in the periods from Spring to Summer (SP-SU) under FC, and SU under both FC and WD.

The winter was the period with the lowest contribution of leaves to the yield, with a range of 46–60%, except for ecotype 226792, which maintained a contribution of 60–70% under FC and WD. Ecotypes 255301 and 226792 registered the highest values of leaf contribution with 60–80%, stems 15 to 25%, and dry stems 0–5% on the day of cutting; ecotype 202700 recorded the lowest values in this variable, with 45–55%, stems 20–45%, dry stems 0–30%, and inflorescences 0–4% (Figure 2).

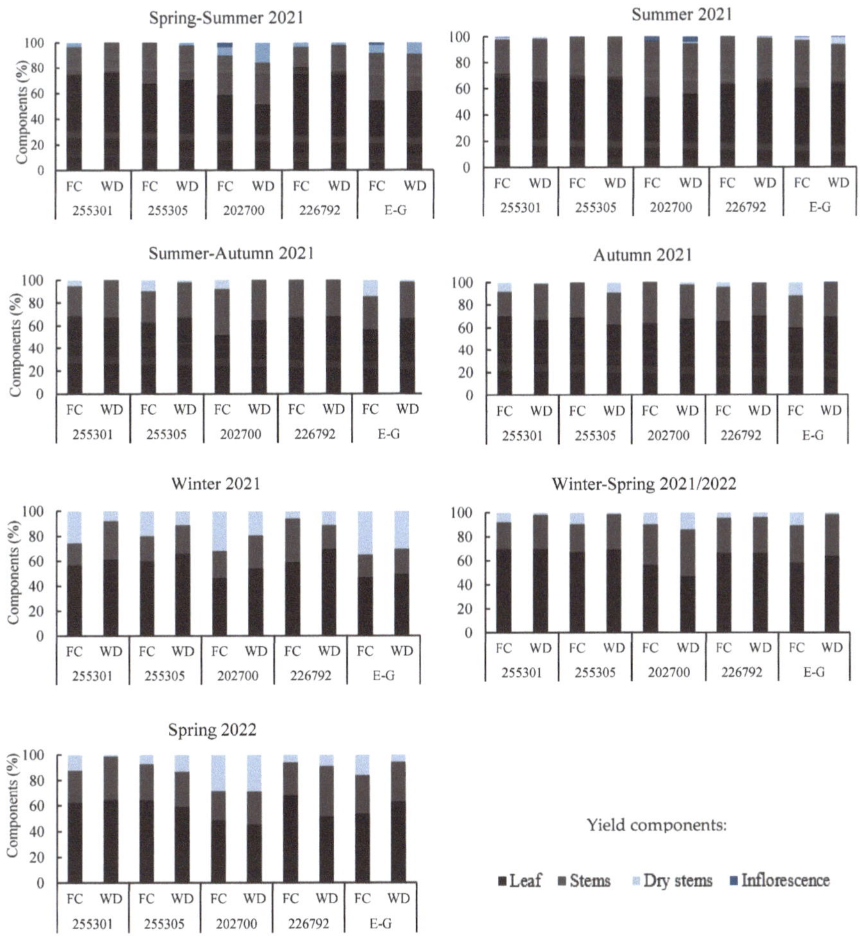

Figure 2. Performance components of different genetic materials of *L. corniculatus* according to season at field capacity (FC) and water deficit (WD: 85% FC).

2.6. Temporal Dynamics of Plant Growth

The variables stem length, plant cover, and accumulated fresh biomass did not have a significant difference ($p \leq 0.05$) between genotypes, but they did differ according to the harvest period. The highest values of stem length corresponded to the Summer–Winter (SU-W) periods, where lengths of 25.7 cm were observed under WD and 29.1 cm under FC. The vegetation cover on the day of cutting in the SU-W period observed very similar behavior in both moisture contents, with values of 2363.2 cm^2 and 2300.7 cm^2 under FC and WD, respectively (Figure 3).

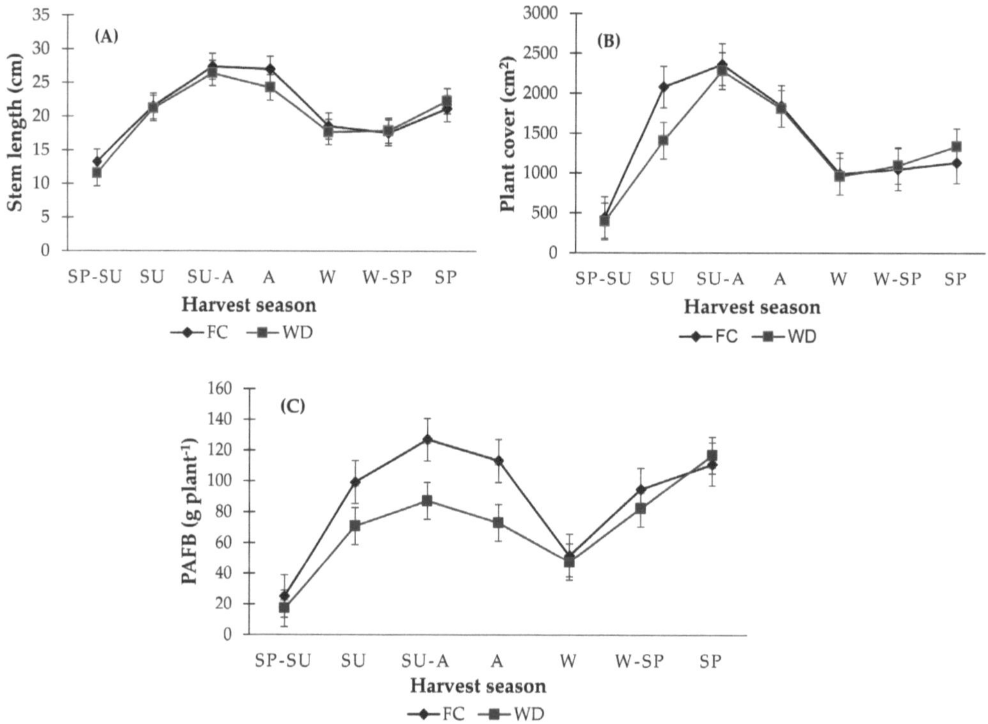

Figure 3. Temporal dynamics of stem length (**A**), plant cover, and (**B**) accumulated fresh biomass production (PAFB) (**C**) of *L. corniculatus* at different seasonal times of the year in optimal content (CF) and water deficit (WD: 85% FC) during the period from April 2021 to May 2022. Spring–Summer (S-SU), Spring (SP), Summer (SU), Summer–Autumn (SU-A), Autumn (A), Winter (W), and Winter–Spring (W-SP) correspond to periods of harvest seasons.

The lowest values of this variable occurred in the SP-SU periods, with 443.2 and 396.2 cm^2, and W, with 992.8 and 960.6 cm^2, in both soil moisture contents FC and WD, respectively. In terms of PAFB at 85% CC, the Summer–Autumn (SU-A) and SP times were the most productive, and the winter period had the lowest productivity; however, in W-SP and SP, there was a similar production in both soil moisture contents (Figure 3).

3. Discussion

According to Tyree et al. [31] the first stages of plant development are critical, as in the results obtained in our study, since the materials evaluated showed a pattern of slow growth and low productivity in the first months of evaluation, tolerating abiotic stress factors, mainly thermal with the presence of high temperatures (46.9 °C) and low temperatures (−4.6 °C). These results coincide with those reported by García-Bonilla et al. [32] when evaluating different genetic materials of *L. corniculatus*, but in temperate climates, where the highest values of plant height (23–24 cm) were recorded in periods with temperatures above 30 °C, and less growth (10 cm) in the winter season, with temperatures of 0–14 °C. Álvarez-Vázquez et al. [33] point out that the optimal growth temperature for Lotus is 22 °C, which corresponds to average temperatures that occur in most of the year in the region of this study.

The low response effect on *L. corniculatus* growth, measured in terms of stem length, indicates that, although the plants are different in growth habit, all ecotypes/varieties develop at similar growth rates under an optimal regime of soil moisture, but with variation between genotypes under deficient irrigation. That means that the genetic materials identified in this study differ in their tolerance to water deficit, which allows us the opportunity to select the best phytogeneric materials to tolerate water stress in plants in regions with dry climates [34]. Plant covering was favored by the optimal irrigation condition (FC) in the ecotype 255305, which stood out, while 202700 was the most tolerant to WD (85% FC).

The best effect in PAFB in ecotype 202700 under water-deficient conditions suggests the good performance of the plant's metabolism under these restrictive conditions of soil moisture content. Tolerance to water deficit depends largely on the characteristics of the species, related to the ability to optimize carbon assimilation processes and moderate levels of transpiration [35]. Reports from IICA [26] show that *L. corniculatus* registered the highest values of stomatal conductance and the highest water potentials under water stress, which suggests that this species increases efficiency in water use under water deficit.

The best response in terms of yield components is a high leaf, which is the main forage component of the evaluated plants, and the ecotypes 255301 and 226792 stood out the most in this regard. This effect was maintained in terms of plant covering, but not in the production of accumulated fresh biomass. The importance of leaf production in forages greater than 50% lies in the fact that it is the most active organ of metabolism in the plant, and consequently has the highest concentration of easily degradable metabolites [36], proteins, and other nitrogenous compounds [37].

Regarding PAFB, the ecotypes with the best response were 255305 and 202700 under FC and WD, respectively; however, 202700 under water deficit had a high percentage of stem (>35%) as a component of the forage yield, which is considered an undesirable characteristic since stems are relatively rigid and contain more lignified tissues that are less digestible [36].

The results obtained coincide with those reported by Álvarez-Vázquez et al. [38], whereby the productive evaluation of ectype 202700 showed that the leaf component was 53% of the annual yield, followed by stems with 32%, dry steams 8%, and weeds 7%. Similar values were reported by the same authors in another study, where leaves contributed 56%, stems 30.5%, dry steams 8.5%, and weeds 4.5% [33]. In the case of our study, the presence of dead material was greater in the periods W, W-SP, and SU, with low-temperature stress being the one that notably affected the productivity of clover *L. corniculatus*. However, ecotype 226792 was identified as the most tolerant to this type of stress, regardless of the soil moisture content established in this study.

The seasonal effects on forage growth and the variation of the components of its yield are influenced by environmental conditions such as temperature variation rather than the moisture content in the soil, which coincides with Festo et al. [39] who reported that plants increase leaf coverage in temperatures between 20 and 35.2 °C, but decrease when they exceed 35 °C.

In the results obtained, a significant contribution of dry stems of up to 10% was observed on the day of cutting in the winter and spring periods, which could affect the quality of the forage, since the age of the plant is a determining factor in the distribution of dry matter in the tissues; with increasing plant age, the proportion of stems and senescent material increases, and leaf formation decreases [40]. Therefore, the growing periods of

L. corniculatus are reduced to less than 45 days in summer and autumn and less than 90 days in winter.

4. Materials and Methods

4.1. Geographic Location of the Study Area

The experiment was carried out at the Unidad Regional Universitaria de Zonas Áridas of the Universidad Autónoma Chapingo in Bermejillo, Durango, Mexico, located at 25.8° LN and 103.6° LW at an altitude of 1130 m. The area corresponds to a desert climate, with rain in summer and cool winters, an average annual rainfall of 258 mm, an average annual potential evaporation of 2000 mm, and an average annual temperature of 21 °C, with a maximum of 33.7 °C and a minimum of 7.5 °C [41].

4.2. Preparing the Experiment

In the experiment, rigid plastic pots of 20 kg capacity with dimensions of 35 cm in diameter and 31.3 cm in height were used, each one containing 18 kg of a substrate mixture: 50% soil, 30% compost, and 20% sand (Figure 4). The physical and chemical characteristics were 22% silt, pH of 8.69, electric conductance (EC) of 10.76 dS m^{-1}, and apparent density of 1.46 g cm^{-3}.

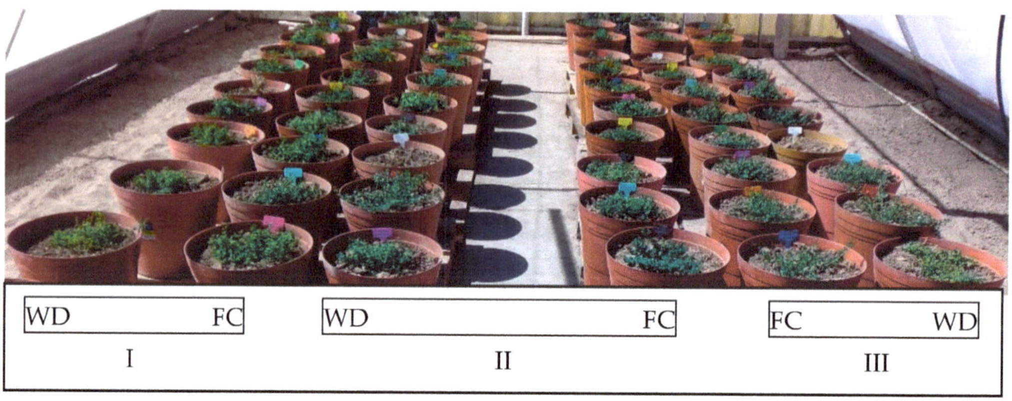

Figure 4. General view of the experiment of different ecotypes and a variety of *Lotus corniculatus* L. at different soil moisture contents under mesh-shade conditions. WD is water deficit; FC is field capacity; I, II, and III are the replicates. Each pot row is a large plot, and each pot is an ecotype/variety of *L. corniculatus* L.

Transplanting each ecotype/variety into a pot was completed using a seedling rhizome, which was previously grown in a black plastic bag of 1 kg capacity for 90 days; after this time, the transplant was carried out into the pot when the rhizome of the seedling had sufficient growth, development, and vigor. For 62 days after transplanting, irrigation was maintained at FC, and subsequently, a standardization foliage cut was made at a height of 6 cm from the base of the substrate.

For irrigation of the clover plants in each pot, the soil moisture content was monitored in real time with the use of the Model HB-2 digital tensiometer equipment (Ontario, Canada). Once the soil moisture content treatments were differentiated (FC, and 85% FC), the moisture content of each treatment decreased to a minimum level corresponding to 25% and 21% in FC and WD conditions, respectively, and water levels were manually raised to their maximum levels, corresponding to 28% (0.5% more than FC) and 24% in the WD condition, applying 0.54 L for 18 kg of soil per pot for each period of irrigation to raise the maximum level of each pot's soil moisture content. This volume of water was applied at each irrigation time, varying the interval time according to the evaporation rate of the season. A margin of 3.5% of the usable moisture range (17.5–21%) was left to prevent reaching PWP.

4.3. Experimental Design

A randomized block experimental design in split plot arrangement was used with three replicates, under shade mesh conditions. This experimental design was used for practical reasons, where each row of pots, named large plots, corresponded to a soil moisture content in an experimental field, and within each row (large plot), the five Lotus ecotypes were placed. Then, the large plots were assigned two soil moisture contents: field capacity (FC), maintaining an interval of soil moisture content of 26.5% ± 1.5 (25–28%), and water deficit (WD: 85% FC), 22.5% ± 1.5 (21–24%), according to the soil moisture properties obtained using the pot method membrane [42], where FC was 27.5% and the permanent wilting point (PWP) was 17.5% (Figure 5). For the optimal level of soil moisture, the field capacity was maintained between the range of 25 and 28%, where the upper limit is slightly higher than FC, only 0.5% more, due to ongoing evaporation. This is not a significant excess, and nor does it imply a water leak, since it soon dropped to its lower limit (25%), almost always remaining at an average level of 26.5%, which corresponds to the FC. Small plots were four ecotypes and one variety of *Lotus corniculatus*, with IDs 255301, 255305, 202700, 226792, and Estanzuela Ganador variety (Table 2).

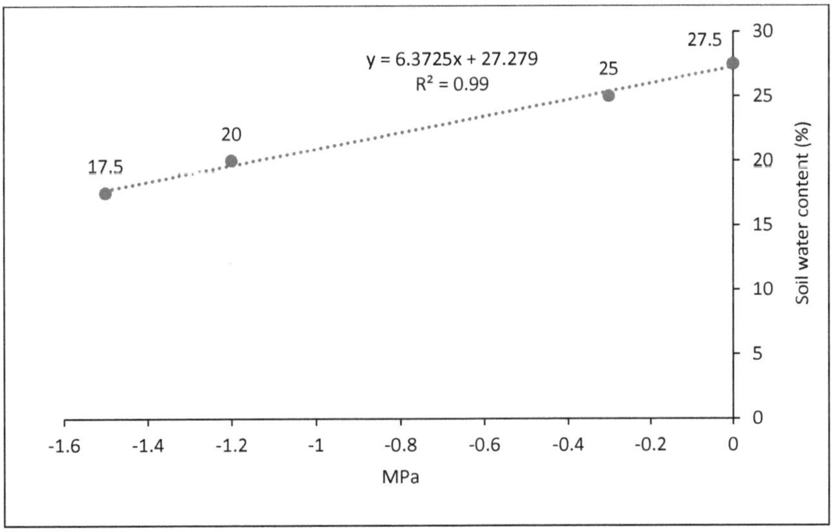

Figure 5. Soil moisture abatement curve [42] of the substrate used in the pots containing *Lotus corniculatus* L. under shade mesh conditions.

Table 2. The relationship, origin, and growth habit of *Lotus corniculatus* ecotypes were evaluated in the experiment.

Ecotypes/Variety	Country of Origin	Growth Habit
255301	Francia	Semi erecto
255305	Italia	Semi erecto
202700	Uruguay	Erecto
226792	Canadá	Semi erecto
Estanzuela Ganador	Uruguay	Erecto

An ORIA brand digital thermometer–hygrometer (Model OUS-WA62, China) was placed inside the shade mesh, with which the daily maximum and minimum temperatures and relative humidity were recorded during the study period from March 2021 to May 2022.

The experiment was covered with a plastic roof during the rainy period to avoid alterations in the soil moisture content. The measurements of the respective variables began in April 2021, and a total of seven harvest cuts of fresh foliage (forage) were made across the seasons of the year and the periods between them: spring–summer, summer, summer–autumn, autumn, winter, winter–spring, and spring. The first cut was carried out on 20 July 2021, and the last one on 12 May 2022. No fertilizer application was carried out in the experiment.

4.4. Measured Variables

4.4.1. Stem Length

Stem length (cm) was measured with a graduated ruler from the basal part of the plant for both the erect and prostrate ecotypes/variety.

4.4.2. Plant Coverage

Plant coverage (cm^2), which is an area obtained from the diameters of the foliage crown measured from the center of the bud with a tape measure, was obtained using the following equation [43]:

$$PC = \frac{\pi * d_1 * d_2}{4}$$

where PC = plant cover (cm^2); d_1 = diameter in the north–south direction (cm); and d_2 = diameter in the east–west direction.

4.4.3. Production Accumulated Fresh Biomass

The production of accumulated fresh biomass (PAFB) (g $plant^{-1}$) was quantified by making foliage cuts from a height of 6 cm above the ground on each sampling date, and the accumulated production was obtained by summing all the cuts made during the year.

4.4.4. Yield Components of Forage

Yield components of forage refer to the proportion of each morphological component of the plant in its contribution to biomass production, for which 10 stems were taken at random per treatment, separating the inflorescences, dry stems, leaves, and stem, each part used as subsamples of the total return component. The subsamples were dried in a HAFO® Brand forced air oven (Model 1600, SHEL-LAB, Cornelius, OR, USA) at a temperature of 60 °C for 24 h or until constant weight. After this time, the partial and total dry weight were recorded to determine the percentage contribution to yield, for which the following equation was used [33]:

$$PMC = \left[\frac{SC}{TDM}\right] \times 100$$

where PMC = percentage per morphological component (%), SC = subsample of the component (g of dry matter), and TDM = total dry matter (g dry matter).

4.5. Analysis of Data

The database was processed using Statistical Analysis System Software [44] version 9.0, with which an analysis of variance and multiple range test of Tukey means ($p \leq 0.05$) were performed to identify the treatment effect. Additionally, Excel program version 6.0 was used for regression analysis.

5. Conclusions

The best response in plant cover and weight of accumulated fresh biomass with values of 1649.0 cm^2 and 583.7 g $plant^{-1}$ was obtained in the 202700 ecotypes under water deficit and 1661.2 cm^2 and 740.1 g $plant^{-1}$ in ecotype 255305 under optimum soil moisture content.

The leaf clover was the main component of yield during the summer and autumn seasons. Ecotype 226792 was tolerant to low temperatures during the winter season, with better leaf development. Thus, ecotype 202700 is the best option for forage clover production when water is limited, and ecotype 255305 when water is not resource-limited, but these preliminary conclusions need to be confirmed in field studies

Author Contributions: Conceptualization, A.P.-S., S.X.-A. and G.G.-D.l.S.; methodology, A.P.-S. and S.X.-A.; software, P.Á.-V. and I.G.-Á.; validation, A.P.-S., S.X.-A. and G.G.-D.l.S.; formal analysis, A.P.-S. and S.X.-A.; investigation, A.P.-S., S.X.-A. and I.G.-Á.; resources, A.P.-S.; data curation, P.Á.-V. and I.G.-Á.; writing—original draft preparation, A.P.-S., S.X.-A. and G.G.-D.l.S.; writing—review and editing, A.P.-S. and S.X.-A.; visualization, P.Á.-V. and I.G.-Á.; supervision, A.P.-S.; project administration, A.P.-S.; funding acquisition, A.P.-S. All authors have read and agreed to the published version of the manuscript.

Funding: This research was supported by Dirección General de Investigación y Posgrado—Universidad Autónoma Chapingo through the project ID: 21017-EI.

Data Availability Statement: Data are available on request to the corresponding author's email with appropriate justification.

Acknowledgments: Thanks to the Postgraduate College for the donation of Lotus genetic materials.

Conflicts of Interest: The authors declare no conflicts of interest.

References

1. Martin, M.L.; Rivera, A.J.Y.; Castizo, R.R. Cambio Climático y Desarrollo Sustentable. Informe La Rábida, Huelva. 2018, p. 13. Available online: https://www.fundacioncarolina.es (accessed on 22 March 2023).
2. Neri, G.M.C.Y.; Ortega, M.A. Cultivos alternativos como un mecanismo para el desarrollo de zonas áridas: El caso de jojoba en el Altiplano Potosino. *Rev. Int. Estadística Geogr.* **2019**, *10*, 44–63.
3. Pedrozo, A.A. *Escasez de Agua, Disponibilidad y Agricultura*; Perspectivas IMTA, Medio Ambiente: 2022; p. 5. [CrossRef]
4. Corral-Bermudez, M.L.; Sánchez-Ortiz, E.; Álvarez-Bernal, D.; Gutiérrez-Montenegro, M.O.; Cassio-Madrazo, E. Scenarios of availability of water due to overexploitation of the aquifer in the basin of Laguna de Santiaguillo, Durango, Mexico. *PeerJ* **2019**, *7*, e6814. [CrossRef] [PubMed]
5. Azpilcueta-Pérez, M.E.; Pedroza-Sandoval, A.; Trejo-Calzada, R.; Sánchez-Cohen, I.Y.; Jacobo-Salcedo, M.; Del, R. Chemical residuality in maize (*Zea mays* L.) fields irrigated with deep well water. *Rev. Ecosistemas Recur. Agropecu.* **2018**, *5*, 111–117. [CrossRef]
6. Varijakshapanicker, P.; Mckune, S.; Miller, L.; Hendrickx, S.; Balehegn, M.; Dahl, G.E.; Adesogan, A.T. Sustainable livestock systems to improve human health, nutrition, and economic status. *Anim. Front.* **2019**, *9*, 39–50. [CrossRef] [PubMed]
7. Ramírez Barraza, B.A.; González Estrada, A.; Valdivia Alcalá, R.; Salas González, J.M.; García Salazar, J.A. Tarifas eficientes para el agua de uso agrícola en la Comarca Lagunera. *Rev. Mex. Cienc. Agrícolas* **2019**, *10*, 539–550. [CrossRef]
8. Çakmakçı, R.; Salık, M.A.; Çakmakçı, S. Assessment and Principles of Environmentally Sustainable Food and Agriculture Systems. *Agriculture* **2023**, *13*, 1073. [CrossRef]
9. Scott, R.C.; Murari, J.; Serkan, A.; Faik, Y.B.; Mohamed, K.O.; Loss, S. Potential legume alternatives to fallow and wheat monoculture for Mediterranean environments. *Crop Pasture Sci.* **2014**, *66*, 113–121.
10. Capstaff, N.; Miller, A. Improving the Yield and Nutritional Quality of Forage Crops. *Front. Plant Sci.* **2018**, *9*, 535. [CrossRef]
11. Hasanuzzaman, M. *Plant Ecophysiology and Adaptation under Climate Change: Mechanisms and Perspectives I*; Springer: Singapore, 2020.
12. Yimer, E.A.; De Trift, L.; Lobkowicz, I.; Villani, L.; Nossent, J.; van Griensven, A. The underexposed nature-based solutions: A critical state-of-art review on drought mitigation. *J. Environ. Manag.* **2024**, *352*, 119903. [CrossRef]
13. Sinclair, F.; Wezel, A.; Mbow, C.; Chomba, S.; Robiglio, V.; Harrison, R. *The Contribution of Agroecological Approaches To Realizing Climate-Resilient Agriculture*; Global Commission on Adaptation to Inform Its 2019 Flagship Report; GCA: Rotterdam, The Netherlands; Washington, DC, USA, 2019; Volume 46. Available online: https://gca.org/wp-content/uploads/2020/12/TheContributionsOfAgroecologicalApproaches.pdf (accessed on 5 April 2023).
14. Grant, W.F.; Small, E. The origin of the Lotus corniculatus (Fabaceae) complex: A synthesis of diverse evidence. *Can. J. Bot.* **1996**, *74*, 975–989. [CrossRef]
15. Qi, A.; Whatford, L.; Payne-Gifford, S.; Cooke, R.; Van Winden, S.; Häsler, B.; Barling, D. Can 100% Pasture-Based Livestock Farming Produce Enough Ruminant Meat to Meet the Current Consumption Demand in the UK? *Grasses* **2023**, *2*, 185–206. [CrossRef]
16. Sitterson, J.M.; Andales, A.A.; Mooney, D.F.; Capurro, M.C.; Brummer, J.E. Developing a Crop Water Production Function for Alfalfa under Deficit Irrigation: A Case Study in Eastern Colorado. *Agriculture* **2023**, *13*, 831. [CrossRef]

17. Flores-Hernández, A.; Araújo-Filho, J.T.; Gomes da Silva, F.; Ramírez-Ordoñez, S.; Murillo-Amador, B. Dietas a base de forraje tradicional y nopal (*Opuntia* spp.) enriquecido con proteínas para alimentar cabras. *Nova Sci.* **2017**, *9*, 149–166. [CrossRef]
18. Bacarrillo-López, R.; Pedroza-Sandoval, A.; Inzunza Ibarra, M.A.; Flores-Hernández, A.; Macías-Rodríguez, F.J. Productividad de forraje de variedades de nopal (*Opuntia* spp.) bajo diferentes regímenes de humedad del suelo. *Ecosistemas Recur. Agropecu.* **2021**, *8*, 1–10. [CrossRef]
19. Pedroza-Sandoval, A.; Xolocotzi-Acoltzi, S.; Trejo-Calzada, R.; García-De Los Santos, G.; Álvarez-Vázquez, P.Y.; Arreola-Ávila, J.G. Leaf area index and forage productivity indicators of *Lotus corniculatus* L. at different soil moisture contents and seasons of the year. *Rev. Mex. Cienc. Pecu.* **2024**, *15*, 17–31.
20. González-Espíndola, L.A.; Pedroza-Sandoval, A.; de los Santos, G.G.; Trejo-Calzada, R.; Álvarez-Vázquez, P.Y.; María del Rosario Jacobo-Salcedo, M. Secondary Metabolites and Their Antioxidant Activity Enhance the Tolerance to Water Deficit on Clover *Lotus corniculatus* L. through Different Seasonal Times. *Int. J. Plant Biol.* **2024**, *15*, 175–186. [CrossRef]
21. Voisin, A.S.; Guéguen, J.; Huyghe, C.; Jeuffroy, M.H.; Magrini, M.B.; Meynard, J.M.; Mougel, C.; Pellerin, S.; Pelzer, E. Legumes for feed, food, biomaterials and bioenergy in Europe: A review. *Agron. Sustain. Dev.* **2014**, *34*, 361–380. [CrossRef]
22. Ventura Ríos, J.; Hernández Moreno, E.; Santiago Ortega, M.A.; Wilson García, C.Y.; Maldonado Peralta, M.D.L.Á.; Rojas García, A.R. Rendimiento de trébol blanco asociado con pasto ovillo a diferentes frecuencias de pastoreo. *Rev. Mex. Cienc. Agrícolas* **2020**, *11*, 1–12. [CrossRef]
23. Sujkowska-Rybkowska, M.; Kasowska, D.Y.; Gediga, K. *Lotus corniculatus*-rhizobia symbiosis under Ni, Co and Cr stress on ultramafic soil. *Plant Soil* **2020**, *451*, 459–484. [CrossRef]
24. Marley, C.L.; Fychan, R.; Jones, R. Yield, persistency and chemical composition of Lotus species and varieties (birdsfoot trefoil and greater birdsfoot trefoil) when harvested for silage in the UK. *Grass Forage Sci.* **2006**, *61*, 134–145. [CrossRef]
25. Barry, T.N.; Kemp, P.D.; Ramírez-Restrepo, C.A.Y.; López-Villalobos, N. Sheep production and agronomic performance of *Lotus corniculatus* under dryland farming. *Res. Pract. Ser.* **2003**, *11*, 109–115. [CrossRef]
26. IICA. *LOTASSA: Un Puente Entre la Genómica y las Pasturas del Siglo XXI*; PROCISUR: Montevideo, Uruguay, 2010.
27. Casler, M.D.; Undersander, D.J. Identification of temperate pasture grasses and legumes. In *Horse Pasture Management*; Sharpe, P., Ed.; Academic Press: London, UK, 2018; pp. 11–36.
28. Clua, A.; Paez, M.; Orsini, H.; Beltrano, J. Incidence of drought stress and rewatering on Lotus tenuis. Effects on cell membrane stability. *Lotus Newsl.* **2009**, *29*, 21–27.
29. Christensen, R.G.; Eun, J.S.; Yang, S.Y.; Min, B.R.; MacAdam, J.W. In vitro effects of birdsfoot trefoil (*Lotus corniculatus* L.) pasture on ruminal fermentation, microbial population, and methane production. *Prof. Anim. Sci.* **2017**, *33*, 451–460. [CrossRef]
30. Ixtaina, V.Y.Y.; Mujica, M.M. Seedling vigor response of Lotus tenuis populations to contrasting variations of water and nutrient availability. *Agrociencia* **2010**, *44*, 31–41.
31. Tyree, M.T.; Engelbrecht, B.M.; Vargas, G.; Kursar, T.A. Desiccation tolerance of five tropical seedlings in Panama: Relationship to a field assessment of drought performance. *Plant Physiol.* **2003**, *132*, 1439–1447. [CrossRef] [PubMed]
32. García-Bonilla, D.V.; Guerrero-Rodríguez, J.D.; García-de los Santos, G.Y.; Lagunes-Rivera, S. Rendimiento y calidad de forraje de genotipos *Lotus corniculatus* L., en el Estado de México. *Nova Sci.* **2014**, *7*, 170–189. [CrossRef]
33. Álvarez-Vázquez, P.; García-de los Santos, G.; Guerrero-Rodríguez, J.; Mendoza-Pedroza, S.I.; Ortega-Cerilla, M.Y.; Hernández-Garay, A. Productive performance of *Lotus corniculatus* L. dependent on harvest strategy. *Agrociencia* **2018**, *52*, 1081–1093.
34. Corrêa, T.R.; Picoli, E.A.d.T.; Pereira, W.L.; Condé, S.A.; Resende, R.T.; de Resende, M.D.V.; da Costa, W.G.; Cruz, C.D.; Zauza, E.A.V. Very Early Biomarkers Screening for Water Deficit Tolerance in Commercial Eucalyptus Clones. *Agronomy* **2023**, *13*, 937. [CrossRef]
35. Medrano, H.; Bota, J.; Cifre, J.; Flexas, J.; Ribas-Carbó, M.Y.; Gulías, J. Eficiencia del uso del agua por las plantas. *Investig. Geográficas* **2007**, *43*, 63–84. [CrossRef]
36. Moore, K.J.; Lenssen, A.W.; Fales, S.L. Factors Affecting Forage Quality. In *Forages: The Science of Grassland Agriculture II*; Moore, K.J., Collins, M., Nelson, C.J., Redfearn, D., Eds.; Wiley: Hoboken, NJ, USA, 2020; pp. 701–717.
37. Andrzejewska, J.; Ignaczak, S.; Albrecht, K.A.; Surucu, M. Fractional harvest of perennial legumes can improve forage quality and their exploitation. *Grassl. Sci. Eur.* **2017**, *22*, 509–511.
38. Álvarez-Vázquez, P.; Hernández-Garay, A.; García-de los Santos, G.; Guerrero-Rodríguez, J.D.; Mendoza-Pedroza, S.I.; Ortega-Cerilla, M.E.; Rojas-García, A.R.Y.; Wilson-García, C.Y. Potencial Forrajero de *Lotus corniculatus* L. con diferentes estrategias de manejo. *Agroproductividad* **2018**, *11*, 24–28.
39. Festo, J.M.; Sabed, N.A.Y.; Jeremy, A.R. The impact of temperature on leaf appearance in bamba groundnut landraces. *Crop Sci.* **2003**, *43*, 1375–1379.
40. Tafolla-Arellano, J.C.; González-León, A.; Tiznado-Hernández, M.E.; Zacarías-García, L.Y. Báez-Sañudo, R. Composición, fisiología y biosíntesis de la cutícula en plantas. *Rev. Fitotec. Mex.* **2013**, *36*, 3–12.
41. Medina, G.; Díaz, G.; López, J.; Ruíz, J.A.Y.; Marín, M.S. *Estadísticas Climatológicas Básicas del Estado de Durango (Periodo 1961–2003)*; Libro Técnico No. 1; Campo Experimental Valle del Guadiana, CIRNOC-INIFAP: Durango, Mexico, 2005.
42. Richards, L.A. Porous plate apparatus for measuring moisture retention and transmission by soil. *Soil Sci.* **1948**, *66*, 105–110. [CrossRef]

43. Caleño, R.B.L. *Caracterización de Poblaciones y Producción Silvestre de Frutos de Agraz (Vaccinium meridionale Sw.) en Ráquira, Boyacá: Bases para su Protocolo de Manejo Sostenible en el Altiplano Cundiboyacense*; Instituto de Investigación de Recursos Biológicos Alexander von Humboldt: Bogotá, Colombia, 2022; p. 87.
44. SAS, Institute. *SAS/STAT®9.2. Use's Guide Release*; SAS Institute Inc.: Cary, NC, USA, 2009.

plants

MDPI

Article

Isolation and Characterization of Plant-Growth-Promoting, Drought-Tolerant Rhizobacteria for Improved Maize Productivity

Victor Funso Agunbiade, Ayomide Emmanuel Fadiji, Nadège Adoukè Agbodjato and Olubukola Oluranti Babalola *

Food Security and Safety Focus Area, Faculty of Natural and Agricultural Sciences, North-West University, Mmabatho 2735, South Africa

* Correspondence: olubukola.babalola@nwu.ac.za; Tel.: +27-18-389-2568

Abstract: Drought is one of the main abiotic factors affecting global agricultural productivity. However, the application of bioinocula containing plant-growth-promoting rhizobacteria (PGPR) has been seen as a potential environmentally friendly technology for increasing plants' resistance to water stress. In this study, rhizobacteria strains were isolated from maize (*Zea mays* L.) and subjected to drought tolerance tests at varying concentrations using polyethylene glycol (PEG)-8000 and screened for plant-growth-promoting activities. From this study, 11 bacterial isolates were characterized and identified molecularly, which include *Bacillus licheniformis* A5-1, *Aeromonas caviae* A1-2, *A. veronii* C7_8, *B. cereus* B8-3, *P. endophytica* A10-11, *B. halotolerans* A9-10, *B. licheniformis* B9-5, *B. simplex* B15-6, *Priestia flexa* B12-4, *Priestia flexa* C6-7, and *Priestia aryabhattai* C1-9. All isolates were positive for indole-3-acetic acid (IAA), siderophore, 1-aminocyclopropane-1-carboxylate (ACC) deaminase, ammonia production, nitrogen fixation, and phosphate solubilization, but negative for hydrogen cyanide production. *Aeromonas* strains A1-2 and C7_8, showing the highest drought tolerance of 0.71 and 0.77, respectively, were selected for bioinoculation, singularly and combined. An increase in the above- and below-ground biomass of the maize plants at 100, 50, and 25% water-holding capacity (WHC) was recorded. Bacterial inoculants, which showed an increase in the aerial biomass of plants subjected to moderate water deficiency by up to 89%, suggested that they can be suitable candidates to enhance drought tolerance and nutrient acquisition and mitigate the impacts of water stress on plants.

Keywords: plant-microbial interaction; sustainable agriculture; 16S rRNA gene; drought tolerance; biological control; stress condition

Citation: Agunbiade, V.F.; Fadiji, A.E.; Agbodjato, N.A.; Babalola, O.O. Isolation and Characterization of Plant-Growth-Promoting, Drought-Tolerant Rhizobacteria for Improved Maize Productivity. *Plants* **2024**, *13*, 1298. https://doi.org/10.3390/plants13101298

Academic Editors: Cristiana Sbrana, Geovani Soares de Lima, Lauriane Almeida dos Anjos Soares, Francisco Vanies Da Silva Sá and Davide Savy

Received: 10 November 2023
Revised: 5 February 2024
Accepted: 20 February 2024
Published: 8 May 2024

1. Introduction

The impact of abiotic-induced drought stress on plant growth and agricultural productivity has been a major global concern that requires urgent attention to avert future food scarcity [1]. Drought conditions occur due to the continuous increase in water scarcity required for plant use, most especially in the arid or semi-arid areas of the world [2]. Drought stress negatively affects the plant's biochemical, physiological, and morphological processes, causing a reduction in plant yield, root architecture, photosynthetic rate, and nutrient assimilation [3]. The reduction in the yield of maize has been linked to limited water supply due to climate change [4,5]. Promisingly, exploring drought-tolerant plant-growth-promoting bacteria (PGPB) can be a suitable strategy for boosting plant adaptation and survival in low-water-regime areas [6]. Many studies have validated the potential of PGPB associated with the rhizosphere of tomato, wheat, maize, soybean, etc., as bioinoculants to enhance plant growth and ameliorate drought-induced stress on plants [5,7–9].

Plant-growth-promoting rhizobacteria (PGPR) are a group of bacteria found in the rhizosphere, on surfaces of the root, and around plant roots, which can indirectly or directly enhance the quality or yield of plants [10]. Plants' reactions to abiotic or biotic stresses may be influenced by the close interface between rhizospheric microbes and the host plants [11,12]. The rhizosphere can be regarded as the narrow region in the soil environment, which can be directly influenced by root exudate secretions. Indirect enhancement in crop plant growth occurs when PGPR lessens or prevents adverse effects of phytopathogens, while the direct enhancement involves either inoculating plants with growth-promoting substances that are bacteria-synthesized or enabling specific plant nutrients' uptake from the soil environment [13]. Reports indicated that plants affect native populations of soil microbes, with each plant species believed to establish a community of a particular microbial biomass [14–16].

Equally, root exudates alter microbial community structures in response to environmental stimuli or stress that can additionally influence microbial diversity. Recently, approachable techniques have been employed especially for root microbiome isolation and characterization, which expedites their recruitment to promote plant development and growth in stressed environments [17]. The microbes also assist in cycling nutrients by aiding nitrogen and phosphorus uptake while abetting the modification of root architecture under drought conditions [18]. Also, 1-aminocyclopropane-1-carboxylate (ACC) deaminase production, which regulates plants' level of ethylene and inhibits leaf senescence under drought conditions, has been reported as one of the astonishing impacts of PGPR on crops [1,19]. Similarly, rhizobacteria can produce 1-aminocyclopropane-1-carboxylate (ACC) deaminase, lower the detrimental impact of ethylene, mitigate stress in plants, and enhance plant growth in drought stress conditions [20].

Drought is among the foremost stresses being confronted by a host of crop plants as a result of extreme scarcity of water and constitutes one of the key factors ascribed to substantial global crop production losses [21]. Many investigations in the past few decades have paid attention to transgenic approaches and molecular breeding for fortifying drought-resistant characteristics in varieties of crops [22,23]. As an alternative to the aforementioned approaches, the use of plant microbiomes has been dubbed as one of the finest approaches, considering its very low-cost input and ecofriendly nature, and is consequently amassing worldwide attention for drought stress management [24].

Frantic efforts are, therefore, being made to cultivate drought-tolerant plants via biotechnological approaches and not excluding breeding. Although these procedures may have high-cost implications, a few studies have explored using helpful microorganisms that already exist in the root microbiome to improve drought tolerance in a variety of crops [25–27]. However, the precise mechanisms supporting these are yet to be fully understood. Furthermore, it has been demonstrated that rhizobacteria-producing 1-aminocyclopropane-1-carboxylate (ACC) deaminase can enhance permeability by boosting soil aggregation and maintaining higher water potential around the roots, thus resulting in increased nutrient uptake, increased plant growth, and a defense against drought stress [15]. Authors hypothesized that the soils used in this study harbored notable PGPR capable of increasing drought tolerance in maize in water-scarcity-prone areas. Although studies on the PGPR-inducing drought stress in maize are evident in the literature, nevertheless, the use of *Aeromonas caviae* A1-2 and *Aeromonas veronii* C7_8 as bioinoculants to boost maize resilience to drought stress in South Africa is less studied. Hence, our findings provided new insights on the two potent rhizobacteria, *Aeromonas caviae* A1-2 and *A. veronii* C7_8, isolated from the maize rhizosphere, by assessing their plant-growth-promoting and drought-tolerant abilities with recommendations for future exploration in ameliorating effects of drought stress on plant development and productivity.

2. Materials and Methods

2.1. Study Site and Collection of Soil Samples

Soil samples were collected from the rhizosphere of two different locations where maize plants were cultivated within the Agricultural Farmland of North-West University, Molelwane, South Africa. The maize farm was located behind the Animal Health Centre, North-West University, South Africa, with sampling sites' geographical coordinates (25°47′25.24056″ S 25°37′8.17464″ E) and (25°47′30.14056″ S 25°37′9.27464″ E). Healthy maize plants with intact root systems were uprooted and the soil samples attached to the roots were taken into sterilized plastic bags, labeled, and transported in an ice-packed box to the Microbial Biotechnology Research Laboratory, North-West University, in South Africa. Samples were stored at −20 °C before use and for a further analysis.

2.2. Isolation of Culturable Rhizobacteria

Rhizosphere soil was collected from the maize root environment at a depth of 2–4 cm using a sterile soil auger. A Radwag weighing machine was used to weigh about 1 g of soil, which was then positioned in a test tube holding 9 mL of sterile distilled water. The soil samples functioned as the source material for rhizobacteria isolation. Approximately 0.1 mL from the dilutions 10^{-7} and 10^{-6} was plated on sterilized Luria agar, nutrient agar, and tryptic agar plates, respectively, prepared according to the manufacturer's instructions. Pure cultures were obtained from the mixed cultures by repeated streaking on nutrient agar following the method of Ahmad et al. [28]. The plates were incubated for 24 h at 30 °C. The obtained pure cultures were maintained on nutrient agar slants for further characterization.

2.3. Morphological, Microscopic, and Biochemical Characterization of Rhizobacteria

Morphological characteristics of the isolated rhizobacteria were assayed on nutrient agar by utilizing the pure culture of the bacterial isolates that was incubated for 24 h at 28 ± 2 °C. Colonies' morphology, including shape, size, pigmentation, and color, was properly documented after incubation, while Gram staining of the bacterial isolates was also carried out. Furthermore, the biochemical characterization tests of the bacterial isolates, such as Voges Proskauer, oxidase, nitrate reduction, catalase, indole, citrate, and starch hydrolysis, were carried out using the standardized procedure as described by Clarke and Cowan [29]. The carbohydrate utilization ability of the bacterial isolates was carried out using sugars such as galactose, maltose, xylose, lactose, fructose, glucose, and sucrose.

2.4. Drought Tolerance Screening of the Rhizobacteria

The evaluation of the bacterial isolates' drought tolerance was carried out in a nutrient-broth medium amended with varied PEG8000 concentrations (0, 5, 10, 15, 20, 25, and 30%). The PEG800 has a CAS number (25322683) and Catalog number (P0131). A single colony from a plate for each strain was inoculated into the broth medium and incubated for 48 h at 28 °C under continuous shaking at 200 rpm [2]. The estimation of bacterial growth was conducted by recording the cultures' optical density with the use of Merck's UV spectrophotometer at 600 nm. The best isolates were selected and subsequently examined for their capacity to promote plant growth under drought conditions.

2.5. Screening of Rhizobacteria for Plant-Growth-Promoting Traits

2.5.1. Ammonia Production

The methods described by Cappuccino and Sherman [30] were employed for the ammonia production screening of the bacterial isolates in peptone water. The pure cultures of each drought-tolerant bacterial strain were grown freshly and inoculated separately in test tubes containing 10 mL of peptone water and incubated at 28 ± 2 °C for 4 days on a rotary shaker (SI-600, LAB Companion (Republic of Korea (South Korea, Seoul))) at 120 rpm. Nessler's reagent (0.5 mL) was added to each test tube and allowed to stay for 5 min for color development. A positive test for the production of ammonia was confirmed with the development of a yellow-to-orange coloration. The uninoculated test tube served as a control. The experiment was performed in triplicate.

2.5.2. Phosphate Solubilization

Phosphate solubilization screening was carried out by a spot inoculation of each bacteria isolate on Pikovskaya's media [31]. The Pikovskaya agar was composed of 5 g of tricalcium phosphate ($Ca_3(PO_4)_2$), 10 g of glucose ($C_6H_{12}O_6$), 0.002 g of manganese sulfate ($MnSO_4{\cdot}H_2O$), 0.2 g of sodium chloride (NaCl), 0.2 g of potassium chloride (KCl), 0.1 g of magnesium sulfate ($MgSO_4$), 0.5 g of ammonium sulfate ($(NH_4)_2SO_4$), 0.5 g of a yeast extract, 15 g of agar, and 1000 mL of sterile distilled water at pH 7.0. The media were sterilized and allowed to cool before pour-plating. The rhizobacteria were spot-inoculated at the center of the Petri plates and incubated at 28 ± 1 °C for 4–5 days. The emergence of a clear zone around the bacteria colony implied a positive result.

2.5.3. Siderophore Production

The methods of Schwyn and Neilands [32] were employed to determine the siderophore production of each bacterial isolate in a chrome-azurol S (CAS) medium composed of 10 mL of an Fe(III) solution (27 mg of $FeCl_3{\cdot}6H_2O$ and 83.3 µL of concentrated HCl in 100 mL of ddH_2O) along with 72.9 mg of hexadecyltrimethyl ammonium bromide (HDTMA). Each fresh bacterial culture 24 h old on nutrient agar was aseptically picked with a sterilized inoculating loop and streaked on the sterilized chrome-azurol S (CAS) medium and incubated for 48–72 h at 37 °C. The appearance of yellow/orange coloration around the colonies implied siderophore production.

2.5.4. Nitrogen Fixation

The rhizobacteria nitrogen fixation characteristics were determined according to AlAli et al. [33] in Jensen's medium composed of sucrose at 20 g; K_2HPO_4 at 1.0 g; $MgSO_4{\cdot}7H_2O$ at 0.5 g; NaCl at 0.5 g; $FeSO_4{\cdot}7H_2O$ at 0.1 g; $CaCO_3$ at 2.0 g; agar at 15.0 g; Na_2MoO_4 at 0.005 g; and 1 liter of sterile distilled water. The medium was adjusted to pH 7.2 before sterilization at 121 °C for 15 min. Twenty-four-hour-old rhizobacterial cultures were inoculated into the Petri plates containing Jensen's medium and incubated at 28 ± 2 °C for 7 days. The presence of bacterial colonies on the Petri dishes during the incubation implied the ability of nitrogen fixation by the bacterial cultures.

2.5.5. Indole-Acetic Acid (IAA) Production

The qualitative IAA was determined according to the methods of Loper and Schroth [34] at different 5 mM (D)L-tryptophan concentrations of 0, 100, and 200 mg in LB broth. Pure bacterial isolates were inoculated in the broth and incubated in a shaking incubator at an ambient temperature (28 ± 2 °C) for 5 days. Bacterial growth suspensions were subjected to centrifugation at 3000 rev/min for 30 min. In total, 2 mL of the bacterial supernatant was mixed with 4 mL of a Salkowski reagent (prepared by mixing 49 mL of 35% perchloric acid in 1 mL of 0.5 M ferric chloride solution) and two drops of orthophosphoric acid. The development of a pink color indicated the production of IAA.

2.5.6. Hydrogen Cyanide (HCN) Determination

A slight modification of Bakker and Schippers' [35] procedure was employed for HCN production by each bacterial isolate. Glycine at 4.4 g was used to enrich the nutrient agar medium. Whatman No. 1 paper was immersed in 1% sodium carbonate, and 0.5% picric acid was placed on the lid of the Petri plates sealed with paraffin and incubated at 28 ± 1 °C for 5 days. The Petri plates were examined daily for color changes in the filter paper. A color change from yellow to brown indicated a positive result. The uninoculated tube served as a control. The experiment was carried out in triplicate.

2.5.7. 1-Aminocyclopropane-1-carboxylate (ACC) Deaminase Activity

Qualitative ACC deaminase activity for the bacterial isolates was determined according to the method of Duan et al. [36]. Screening for ACC deaminase activity of drought-tolerant rhizobacterial isolates was conducted based on their ability to use ACC as a sole nitrogen source. All the bacterial isolates were grown in 5 mL of the LB medium incubated at 28 °C at 120 rpm for 4 days. The cells were harvested by centrifugation at 3000× *g* for 5 min and washed twice with sterile 0.1 M Tris-HCl (pH 7.5) and resuspended in 1 mL of 0.1 M Tris-HCl (pH 7.5) and spot-inoculated on Petri plates containing modified DF (Dworkin and Foster) salts' minimal medium [37]. The medium composition includes glucose, 2.0 g; gluconic acid, 2.0 g; citric acid, 2.0 g; KH_2PO_4, 4.0 g; Na_2HPO_4, 6.0 g; $MgSO_4{\cdot}7H_2O$, 0.2 g; a micronutrient solution ($CaCl_2$, 200 mg; FeSO4$\cdot 7H_2O$, 200 mg; H_3BO_3, 15 mg; $ZnSO_4{\cdot}7H_2O$, 20 mg; Na_2MoO_4, 10 mg; KI, 10 mg; NaBr, 10 mg; $MnCl_2$, 10 mg; $COCl_2$, 5 mg; $CuCl_2$, 5 mg; $AlCl_3$, 2 mg; $NiSO_4$, 2 mg; distilled water, 1000 mL), 10 mL; and distilled water, 990 mL, supplemented with 3 mM ACC as the sole nitrogen source. Plates contained only DF salts' minimal medium without ACC as a negative control and with $(NH4)_2SO_4$ (0.2% *w/v*) as a positive control. The plates were incubated at 28 °C for 72 h. Growths of isolates on ACC-supplemented plates were compared to negative and positive controls and were selected based on growth by utilizing ACC as a nitrogen source.

2.6. *Tolerance of Bacterial Isolates to pH, Temperatures, Salt Concentrations, and Heavy Metals*

2.6.1. Bacterial Tolerance to Different pH

The nutrient broth, Luria–Bertani Broth, tryptic soy broth, and Reasoner's 2A broth were used for this test. The broth medium was prepared and sterilized at 121 °C for 15 min to test the tolerance of bacterial isolates to various pH levels. The pH of the medium was adjusted to acidic (pH 4) by adding diluted 1 M NaOH, neutral (pH 7), and alkaline (pH 10) by adding diluted 1 M HCl. Some 20 mL of each bacteria overnight culture was divided equally and then added to a sterile 10 mL LB broth and homogenized. The bacterial inoculation was performed in triplicate, and the inoculated Luria–Bertani broth was incubated at 30 °C for 24 h on a rotary shaker at 150 rpm. The optical density of the bacterial growth was taken at 600 nm using a UV spectrophotometer (Thermo Spectronic; Merck, Tokyo, Japan).

2.6.2. Tolerance of Bacterial Isolates to Different Temperatures

The nutrient broth, Luria–Bertani Broth, tryptic soy broth, and Reasoner's 2A broth were used for the test. The broth medium was prepared and sterilized at 121 °C for 15 min. Some 25 mL of Luria–Bertani broth was gently vortexed after each bacterial strain was inoculated. The inoculated broth from each media was incubated at 25, 30, 35, and 40 °C. A spectrophotometer (Thermo Spectronic; Meck, Pretoria, South Africa) was used to measure the bacterial growth's optical density at 600 nm after 24 h of incubation.

2.6.3. Bacterial Growth Response to Different Salt Concentrations

The nutrient broth, Luria–Bertani Broth, tryptic soy broth, and Reasoner's 2A broth were used for the test. The broth medium was prepared and sterilized at 121 °C for 15 min to test the effect of bacterial strain tolerance to various salt concentrations (NaCl). While the control included no salt (0%), each broth from the mentioned media was produced following the manufacturer's instructions and supplemented with salt concentrations at 1%, 3%, and 5%. The tubes containing 10 mL of each broth added with various salt concentrations were incubated at 30 °C on a shaking incubator for 72 h after inoculation with 100 µL of an overnight bacteria culture. A UV spectrophotometer (Thermo Spectronic; Merck) was used to detect the absorbance of the bacterial growth at 600 nm.

2.6.4. Bacterial Tolerance to Different Heavy Metals

The nutrient broth, Luria−Bertani broth, tryptic soy broth, and Reasoner's 2A broth were prepared and sterilized at 121 °C for 15 min. The media were supplemented with 100 mg/L of various heavy metals, Pb^{2+}, $Cr_2O_7{}^{2-}$, and Cd^{2+}, in test tubes to test for the tolerance of each bacterial isolate to various heavy metals. Test tubes containing precisely 10 mL of each broth medium containing a different heavy metal were filled, sterilized at 121 °C for 15 min, inoculated with 1 mL of each bacteria strain overnight culture, and then incubated for 72 h at 30 °C on a rotary shaker at 150 rpm. After incubation at 600 nm, the optical density of the bacterial growth was measured using a UV spectrophotometer (Thermo Spectronic; Merck).

2.7. Extraction of DNA and Genotypic Identification

Fresh bacterial cultures on nutrient agar plates were used for the DNA extraction. The DNA of bacteria was extracted with the Zymo DNA extraction kit (Zymo Research, Irvine, CA, USA; Cat. No. D6005) as specified in the manufacturer's instructions. A NanoDrop spectrophotometer was, thereafter, used to determine extracted DNA samples' concentrations at a 260 nm wavelength; agarose gel electrophoresis was used to determine the presence of DNA. Also, 10 μL of Hyper Ladder™ 1 kb and 4 μL of bacterial DNA samples were run in 1.0% agarose concocted through the heating of 1 g of agarose powder in 100 mL of 1 × TAE at 65 °C that lasted 4 min, followed by the addition of 10 μL of ethidium bromide. The bands of DNA were observed in a gel Doc machine.

2.7.1. Polymerase Chain Reaction (PCR) Amplification

Partial nucleotide sequences of 16S rRNA and PCR amplification were used to carry out genotypic identification. The bacteria 16S rRNA gene partial nucleotide sequences were obtained using the direct sequencing process of amplified PCR products. The bacteria 16S rRNA gene was amplified by PCR for bar-coded pyrosequencing. The full-length 16S rRNA gene sequences of all bacteria were amplified by utilizing universal forward, 27F (5′-AGAGTTTGATCCTGGCTCAG-3′) as well as the reverse, 1541R (5′-AAGGAGGTGATCCAGCCGCA-3′) primers [38]. A 25 μL reaction volume for an individual bacteria isolate was made up of 12.5 μL of OneTaq 2X MasterMix with the Standard Buffer, ~5 ng of genomic DNA, 1 μM for each primer, and 9.5 μL of nuclease-free water. A PCR thermal cycler machine (BIO-RAD Laboratories, Hercules, CA, USA, C1000 Touch) was used to carry out PCR amplification based on the corresponding program conditions: 95 °C for 2 min; 30 cycles at 95 °C for 30 s, 72 °C for 30 s, and 55 °C for 30 s, with a last extension of 72 °C for approximately 5 min. The standard procedure was used to perform quality control of PCR amplification and sequence preprocessing as well as raw data processing. Amplified DNA was checked by running DNA samples in 1% agarose gel as well as in a gel Doc machine (BIO-RAD Laboratories), as previously highlighted. DNA samples of each bacterial isolate of about 20 μL were forwarded for sequencing (INQABA, Pretoria, South Africa).

2.7.2. Alignment Sequence and Construction of the Phylogenetic Tree

To determine sequence identities and similarities of bacterial isolates, the BLAST program of the nucleotide sequence on the National Centre for Biotechnology Information (NCBI) database was employed. Accession numbers were given to the identified bacteria after deposition on the GenBank on the NCBI web server. The sequences isolated were additionally exposed to the alignment of multiple sequences via ClustalX through the use of a Bio-Edit program. The Maximum Parsimony Analysis of taxa with the Kimura 2-parameter model was used to plot the phylogenetic tree using the MEGA-X program (Version 11) [39]. The bootstrap method was used to test the phylogeny.

2.8. Experimental Set-up and Maize Growth Conditions

The experimental set-up for maize growth under a greenhouse was achieved using a three-factorial complete randomized design (CRD). The three factors were considered, which include *Aeromonas caviae* A1-2, *A. veronii* C7_8, and drought or water stress. With a shovel and plastic bags, soil samples were taken from topsoil (0–25 cm) near the North-West University garden on the Mahikeng campus. Between July 2021 and February 2022, soil samples were collected, homogenized, and sieved using a circular sieve with a 5.6 mm diameter. The sieved soil samples were sealed in autoclavable plastic bags and sterilized for 1 h at 121 °C (autoclave temperature). To ensure complete sterilization, this process was repeated three times.

In this study, three drought or water regimes—100, 50, and 25% field capacity (FC)—were employed. The FC calculation has been included in the Supplementary File. *A. caviae* A1-2 and *A. veronii* C7_8 were grown in broth cultures and harvested according to the method of Prakamhang et al. [40] with little modifications. In flasks containing 1000 mL of nutrient broth, fresh bacterial cultures were cultivated for six days at 180 rpm in a shaker incubator (FMH200 Instruments manufactured by FireChief, located in East Sussex, United Kingdom). After incubation, the broth culture was centrifuged at 1000× *g* for 10 min, followed by two-step washing with a sterile 0.85% saline solution. The initial optical density (OD) of the bacterial strains was adjusted to 1.3 OD, and the bacterial titer was 30×10^9 (colony-forming unit) mL^{-1}. The experiment was carried out in plastic pots (30 cm in diameter and 29 cm in height), each with 6 kg of sterilized soil. A total of 192 pots containing two maize plants were produced by setting up 24 treatments with four replicates per treatment.

The PAN 413 cultivar of maize seeds was cleaned in sterile distilled water, surface-sterilized in 75% ethanol, and then thoroughly rinsed with 1% sodium hypochlorite to eliminate all traces of the chemicals. Then, approximately 240 surface-sterilized maize seeds were dispensed in 20 mL broth cultures of both *A. caviae* A1-2 and *A. veronii* C7_8 adjusted to 1.3 OD of an equal volume for 1 h at 28 °C. An equal volume (10 mL) of each bacterial strain was combined for seed bacterialization. The flasks containing seeds were shaken for 24 h at 180 rpm at 28.2 °C to properly combine the seeds and inocula. In a sterilized laminar flow cabinet (Filta Matix Laminar Flow Cabinet), the liquid suspension was decanted and air-dried on sterile aluminum foil sheets before seeding. Approximately 6 kg of sterilized soil was placed in each container in the greenhouse. Ten maize seeds were planted per pot. The seeds and seedlings in the container were inoculated with 10 mL bacterial cultures separately at growth intervals over time. The seeds dispensed inside sterile distilled water without bacterial inoculation served as the control. The seed inoculation was carried out, but direct inoculation onto the seed inside the pots and seedling roots appeared after 7 days of growth. The seeds' viability was validated at the Agricultural Research Council, Pretoria, South Africa (supplier of the maize seeds).

Two weeks following emergence, the seedlings were reduced to one per container and harvested at 45 days of maturity. The above- and below-ground yield parameters were measured. For about three weeks after germination, all the water treatments (100, 50, and 25% FC) were fully watered to the field capacity at 72-hour intervals. Thereafter, drought or water stress was initiated in the 50 and 25% FC treatments, and they were watered every 48 h with 335 mL and 170 mL of water, respectively. However, the 100% FC treatments (the watering control) continued to be watered with 670 mL of water. The pots were watered every 72 hours until the experiment's end, increasing drought stress three weeks after it was first introduced [41].

Below-Ground–Above-Ground/Morphological Parameters

The chlorophyll content, plant height, leaf count, plant diameter, fresh total plant mass (FTPM), fresh root mass (FRM), fresh shoot mass (FSM), dry total mass (DTM), dry root mass (DRM), and dry shoot mass (DSM) were measured and determined after harvesting the mature maize plants. The roots of the maize plant were thoroughly cleaned, and then the plant's fresh weights were ascertained right away by weighing the entire plant on a scale (Wagi Elek'roniczne, city of Bieniądzice, located in Wieluń, Poland). The shoot was then cut away from the roots. Using a measuring tape, the shoot height and root length were determined; the number of completely grown branches, leaves, and lateral roots (excluding the root) were then counted to determine the branch, leaf, and lateral root number. After being separated, the shoots and roots were oven-dried at 70 °C for 48 h and then weighed to determine their dry weights. A chlorophyll content meter (CCM-200 plus) was used to measure the chlorophyll content of the youngest completely grown leaf at the distal end of the maize plant.

2.9. Parameters Measured

2.9.1. Leaf Relative Water Content

The leaf relative water content was determined according to the methods described by Ortiz et al. [42] with little modifications, and the "youngest fully developed leaves of each plant" were used. Fresh leaf samples were weighed (fresh weight—FW), placed in test tubes saturated with water, and kept at 4 °C for 48 h. Thereafter, the leaf samples were weighed again to obtain the turgid weight (TW) and oven-dried at 60 °C for 24 h to obtain the dry weight (DW). The leaf relative water content was calculated as follows:

$$\text{Leaf relative water content } (\%) = \frac{(\text{FW} - \text{DW})}{(\text{TW} - \text{DW})} \times 100$$

2.9.2. Electrolyte Leakage

The Ortiz et al. [42] approach was used to assess electrolyte leakage. From each of the four replicates for each treatment, a young leaf of nearly the same size from the distal end of the maize stalk was taken and thoroughly washed with deionized water to remove any electrolytes that could have adhered to the leaf surface. After being incubated for 24 h at 28 ± 2 °C on a rotational shaker, leaf samples were deposited in sealed test tubes. The electrical conductivity (Lt) of the sample solution was then calculated using a Taiwanese-made PL-700AL conductivity meter. Before determining the final electrical conductivity (L_0), samples were autoclaved at 120 °C for 20 min and then allowed to cool to 25 °C. The following formula was used to determine the electrolyte leakage:

$$\text{Electrolyte leakage } (\%) \; \frac{Lt}{L0} \times 100\%$$

2.9.3. Soluble Sugar Content in Maize Leaves

One hundred milligrams (100 mg) of maize leaves was homogenized in 4 mL of 80% ethanol and centrifuged for 10 min at 5000 rpm. In total, 1 mL of the supernatant was mixed with 5 mL of concentrated sulphuric acid (H_2SO_4) and 1 mL of 5% phenol. Samples were vigorously agitated and allowed to cool for 30 min, and absorbance was measured at 485 nm using a spectrophotometer (Thermo Spectronic, South Africa). The concentration of soluble sugar was calculated from a standard curve "established with a reference glucose solution", as described by Zarik et al. [43] with little modifications. Briefly, stock solutions of glucose were prepared by weighing 10, 20, 30, 40, 50, 60, and 70 mg of D(+)-glucose in 10 mL of sterile distilled water. Two millileters of each stock solution was pipetted into three cleaned test tubes. In total, 1 mL of 5% phenol was added, and 5 mL of concentrated H_2SO_4 was quickly added. The tubes were allowed to stand for 10 min, then shaken and placed in a water bath for 10 min at 51–57 °C. Tubes were allowed to cool for 30 min, and absorbance was measured at 485 nm using a spectrophotometer and a standard curve was plotted from the values obtained.

2.9.4. Proline Content in Maize Leaves

Proline in maize leaves was extracted using the Bates et al. [44] technique with only minor modifications. Briefly, 1.25 g of ninhydrin was dissolved in 20 mL of 6 M phosphoric acid and 30 mL of glacial acetic acid by heating on a hot plate with agitation. The solution was allowed to cool and kept at 4 °C and the solution became stable after 24 h. Approximately 1 g of a fresh maize leaf sample was ground in 10 mL of 3% aqueous sulfur-salicylic acid and centrifuged at 10,000× *g* for 10 min. In total, 2 mL of the supernatant was reacted with 2 mL of glacial acetic acid and 2 mL of an acid–ninhydrin solution in 45 mL falcon tubes at 100 °C in a water bath for 60 min, and the reaction was stopped in an ice box. Four millileters of toluene was added to extract the mixture and agitated vigorously for 15–20 s in a shaker incubator at 250 rpm. The mixture was kept in the dark for 30 min, the "chromophore containing toluene was aspirated from the aqueous phase, and the absorbance was read at 520 nm using toluene for a blank. The concentration of proline was estimated from a standard curve" established with a reference proline solution. Briefly, a 1 mg/mL stock solution of proline was prepared by weighing 10 mg of proline (DL-Proline, Hefei, China) in 10 mL of sterile water. A stock solution of 0, 50, 100, 150, 200, 250, and 300 μL was pipetted into seven tubes containing 300, 250, 200, 150, 100, 50, and 0 μL of sterile water, respectively. The mixtures were then reacted with 2 mL of glacial acetic acid and 2 mL of the acid–ninhydrin solution in 45 mL falcon tubes at 100 °C in a water bath for 60 min, and the reaction was stopped in an ice box. The mixture was vigorously agitated using a vortex (Vortex Genie, New York, NY, USA) after adding 4 mL of toluene. The mixture was kept in the dark for 30 min, and the absorbance of the proline-containing upper layer was read at 520 nm using toluene for a blank, and the proline standard curve was plotted from the absorbance values.

2.10. Statistical Analyses

All the study's data were collected in three replicates at various time intervals, and they were all processed using SPSS version 26 and Microsoft Excel software version 26. To evaluate the effects of each treatment at various drought levels, the data were subjected to a general linear model analysis of variance (ANOVA). Differences between means (post hoc analysis) were determined by Duncan's multiple range test (DMRT) [45], and significant differences were set at $p \leq 0.05$.

3. Data Availability Statement

The sequence data of the bacteria strains A5-1, A1-2, B8-3, B12-4, B9-5, B15-6, C6-7, C7_8, C1-9, A9-10, and A10-11 were deposited in the NCBI GenBank with accession numbers ON745408, ON745409, ON745410, ON745411, ON745412, ON745413, ON745414, ON745415, ON745416, ON745417, and ON745418, respectively.

4. Results

4.1. Characterization of the Rhizobacteria Strains

The microscopic appearance and biochemical characteristics of the rhizobacterial isolates are presented (Figure S1 and Table S1). The microscopic observations of the bacterial isolates revealed different structural appearances under the light microscope (Figure S1). Also, the biochemical tests showed different results (Table S1). All the bacterial strains showed 100% similarity compared with the NCBI database, except for strain A1-2, which showed 99.83% similarity.

The biochemical characteristics and Gram reaction test showed that A5-1 was Gram-positive short-rod motile, while B8-3, B12-4, B9-5, B15-6, and C1-9 were Gram-positive rod motile; similarly, bacterial strain C6-7 was Gram-positive motile and Bacilli, respectively. Furthermore, bacteria strains A9-10 and A10-11 were Gram-positive nonmotile cocci and rods, respectively. All the bacterial isolates were positive for catalase, in addition to being oxidase-positive for the entire strains except bacterial B8-3 and B15-6, which were negative for catalase, and bacteria strains B12-4, C6-7, C1-9, and A9-10, which were negative for oxidase. All the isolates tested positive for maltose, glucose, and sucrose. Isolate B12-4 was not determined for nitrate reduction, while isolate A9-10 was not determined for xylose. The phylogeny of the *Aeromonas* strains A1-2 and C7_8 with high drought tolerance is presented in Figures 1 and 2.

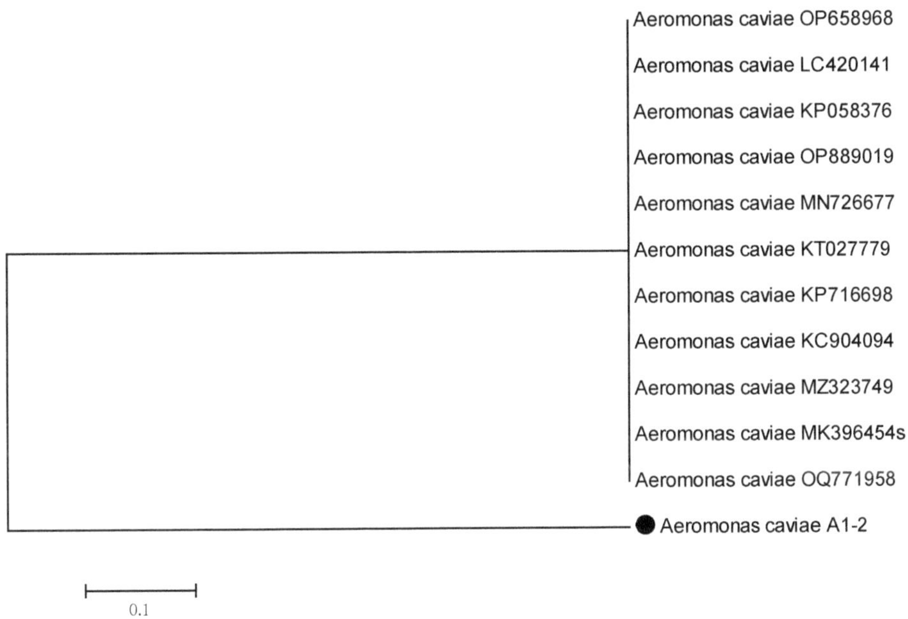

Figure 1. Evolutionary relationships of the taxa tree based on partial 16S rRNA sequences using maximum likelihood. The tree shows correlations between the rhizospheric *A. caviae* A1-2 and their closely associated strains from the NCBI GenBank. The boxes represent the sequences generated in this study.

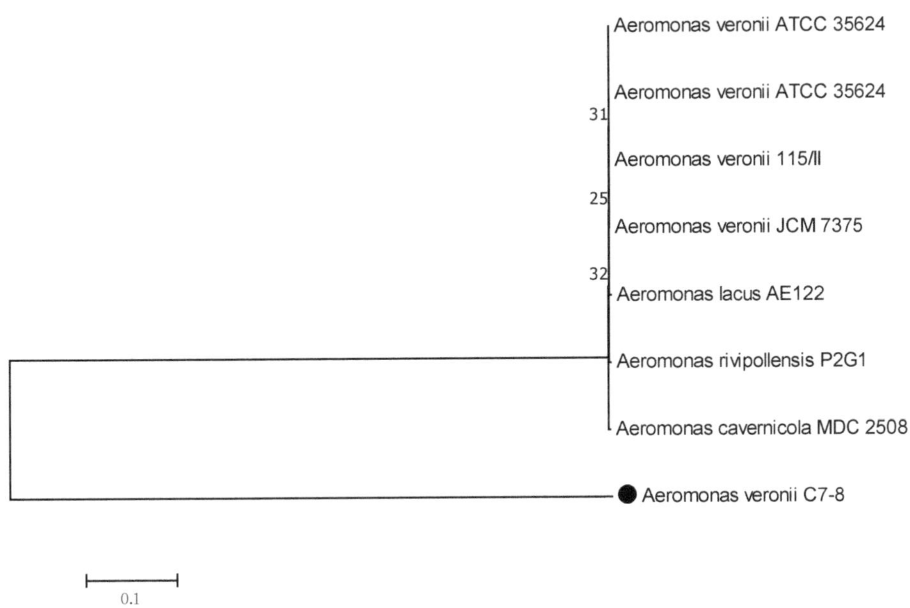

Figure 2. Evolutionary relationships of the taxa tree based on partial 16S rRNA sequences utilizing maximum likelihood that shows correlations between the rhizospheric *A. veronii* C7_8 and their closely associated strains from the NCBI GenBank. The boxes represent the sequences generated in this study.

4.2. Selection of Drought-Tolerant Rhizobacteria Strains

The identifiable rhizobacteria strains with distinctive drought tolerance levels in a broth medium amended with PEG-8000 are presented in Table 1. The rhizobacterial strains C7_8 and A1-2 were selected based on their high response drought tolerance values of 0.77 and 0.71 in a medium supplemented with 30% PEG-8000 compared with the control that showed low drought tolerance. Similarly, the rhizobacterial strain A5-1 showed a negligible value of 0.12 at 30% of PEG amendments.

Table 1. Response of rhizobacterial strains to drought stress amended with various concentrations of PEG-8000.

Treatment	Control	5%	10%	15%	20%	25%	30%
A5-1	0.35 ± 0.17fg	0.18 ± 0.01c	0.16 ± 0.01d	0.16 ± 0.00f	0.16 ± 0.00d	0.13 ± 0.00f	0.12 ± 0.00c
A1-2	1.01 ± 0.07b	0.67 ± 0.23a	0.80 ± 0.15a	0.89 ± 0.03a	0.87 ± 0.03a	0.81 ± 0.01a	0.71 ± 0.01a
B8-3	0.73 ± 0.03bc	0.38 ± 0.11ab	0.41 ± 0.02bc	0.39 ± 0.00c	0.37 ± 0.02b	0.30 ± 0.00c	0.26 ± 0.03bc
B12-4	0.65 ± 0.08cde	0.40 ± 0.11abc	0.41 ± 0.01bc	0.41 ± 0.01c	0.41 ± 0.00b	0.35 ± 0.01b	0.30 ± 0.00bc
B9-5	0.21 ± 0.03g	0.24 ± 0.05c	0.21 ± 0.02cd	0.20 ± 0.00ef	0.20 ± 0.00d	0.18 ± 0.01e	0.15 ± 0.01c
B15-6	0.43 ± 0.08egf	0.27 ± 0.05c	0.31 ± 0.02cbd	0.31 ± 0.00d	0.28 ± 0.00c	0.22 ± 0.00d	0.20 ± 0.00c
C6-7	0.84 ± 0.06bc	0.40 ± 0.11abc	0.45 ± 0.04b	0.41 ± 0.00c	0.40 ± 0.0.00b	0.32 ± 0.01bc	0.29 ± 0.00bc
C7_8	1.31 ± 0.13a	0.53 ± 0.20ab	0.73 ± 0.15a	0.84 ± 0.03b	0.84 ± 0.03a	0.81 ± 0.04a	0.77 ± 0.03a
C1-9	0.35 ± 0.01fg	0.33 ± 0.08bc	0.31 ± 0.02bcd	0.21 ± 0.00e	0.20 ± 0.00d	0.19 ± 0.00de	0.14 ± 0.01c
A9-10	0.23 ± 0.03g	0.24 ± 0.04c	0.17 ± 0.00d	0.17 ± 0.01ef	0.16 ± 0.00d	0.17 ± 0.01e	0.16 ± 0.01c
A10-11	0.55 ± 0.05edf	0.25 ± 0.04c	0.20 ± 0.01cd	0.18 ± 0.01ef	0.17 ± 0.01d	0.17 ± 0.01e	0.15 ± 0.01c
LSD	0.22	0.23	0.20	0.04	0.04	0.04	0.24

The values are represented as the mean ± standard error with the lowercase letters indicating significant differences down the column.

4.3. Plant-Growth-Promoting Abilities of Rhizobacterial Strains and Phylogeny

Table 2 shows the plant-growth-promoting traits of the screened rhizobacteria. It was observed that all the rhizobacterial strains exhibited positive reactions to nitrogen fixation, siderophore, ammonia, IAA, phosphate solubilization, and ACC deaminase activity. All the bacterial strains tested negative for hydrogen cyanide production. Similarly, seven out of eleven bacterial isolates exhibited a relatively high level of ammonia production. Meanwhile, the bacteria strain *B. cereus* B8-3 exhibited moderate ammonia production. Based on the additional evaluation, the rhizospheric bacteria that belong to *Bacillus licheniformis*, *Aeromonas caviae*, *Priestia flexa*, and *Aeromonas veronii* exhibited high positive reactions to siderophore production compared to other strains. *Aeromonas caviae* A1-2, *Priestia flexa* C6-7, and *Aeromonas veronii* C7_8 showed a high level of indole-acetic acid (IAA) compared to other strains.

Table 2. Plant-growth-promoting characteristics of rhizobacterial strains.

Rhizobacterial Isolates	Plant-Growth-Promoting Traits						
	Nitrogen Fixation	HCN Production	Siderophore Production	Ammonia Production	IAA Production	Phosphate Solubilization	ACC Deaminase
B. licheniformis A5-1	+	−	++	+++	+	+	+
A. caviae A1-2	+	−	++	++++	++	+	+
B. cereus B8-3	+	−	+	+++	+	+	+
P. flexa B12-4	+	−	+	++	+	+	+
B. licheniformis B9-5	+	−	+	++	+	+	+
B. simplex B15-6	+	−	+	++	+	+	+
P. flexa C6-7	+	−	++	+++	++	+	+
A. veronii C7_8	+	−	++	++++	++	+	+
P. aryabhattai C1-9	+	−	+	++	+	+	+
B. halotolerans A9-10	+	−	+	+++	+	+	+
P.endophytica A10-11	+	−	+	+++	+	+	+

Key: + = positive; − = negative.

4.4. Rhizobacteria Tolerance to Drought Stress under Different Growth Conditions

The rhizobacteria tolerance to drought under different growth conditions is presented in Tables 3–6. The rhizobacteria *Aeromonas caviae* (A1-2) and *Aeromonas veronii* (C7_8) were subjected to salt stress at 0% (control), 1%, 3%, and 5% concentrations in different nutrient media, nutrient broth (NB), Luria−Bertani broth (LBB), tryptic soy broth (TSB), and Reasoner's 2A broth (R2AB). Both rhizobacteria strain A1-2 and C7_8 displayed various levels of salt tolerance at the 5% salt concentration compared with the control (Table 3).

Table 3. Growth of isolated rhizobacterial strains exposed to salt-stimulated stress.

Rhizobacteria	Media	0% Salt	1% Salt	3% Salt	5% Salt
Aeromonas caviae (A1-2)	NB	1.64 ± 0.18^{a}	1.42 ± 0.56^{a}	1.37 ± 0.60^{a}	0.55 ± 0.02^{b}
	LB	1.62 ± 0.18^{a}	1.34 ± 0.56^{ab}	0.96 ± 0.56^{b}	0.50 ± 0.02^{c}
	TSB	1.47 ± 0.12^{a}	1.28 ± 0.56^{ab}	0.90 ± 0.56^{b}	0.45 ± 0.02^{c}
	R2AB	1.37 ± 0.12^{a}	1.22 ± 0.51^{a}	1.18 ± 0.56^{a}	0.40 ± 0.02^{b}
Aeromonas veronii (C7_8)	NB	1.61 ± 0.18^{a}	1.42 ± 0.54^{a}	0.70 ± 0.543^{bc}	0.45 ± 0.12^{c}
	LB	1.60 ± 0.13^{a}	1.35 ± 0.61^{ab}	1.00 ± 0.61^{abc}	0.41 ± 0.12^{c}
	TSB	1.52 ± 0.25^{a}	1.28 ± 0.55^{ab}	0.96 ± 0.56^{abc}	0.37 ± 0.12^{c}
	R2AB	1.4224 ± 0.3154^{a}	1.22 ± 0.60^{ab}	0.93 ± 0.57^{abc}	0.38 ± 0.18^{c}

The mean value ± SD with the same superscript letter across each row was not significantly different according to the Duncan multiple range test (DMRT) at $p \leq 0.05$. Number of replicates (n) = 4. NB = nutrient broth, LB = Luria−Bertani broth, TSB = tryptic soy broth, R2AB = Reasoner's 2A broth.

Table 4. Growth of isolated rhizobacterial strains exposed to pH-stimulated stress.

Rhizobacteria	Media	pH4	pH7	pH10
Aeromonas caviae (A1-2)	NB	0.97 ± 0.05^{ab}	1.11 ± 0.02^{a}	0.56 ± 0.32^{c}
	LB	0.88 ± 0.05^{b}	1.03 ± 0.27^{a}	0.50 ± 0.29^{c}
	TSB	0.79 ± 0.04^{b}	0.96 ± 0.30^{a}	0.45 ± 0.26^{c}
	R2AB	0.71 ± 0.04^{b}	1.03 ± 0.56^{a}	0.27 ± 0.01^{c}
Aeromonas veronii (C7_8)	NB	0.91 ± 0.09^{ab}	1.53 ± 0.60^{a}	0.40 ± 0.01^{c}
	LB	0.82 ± 0.07^{ab}	1.11 ± 0.56^{a}	0.36 ± 0.01^{c}
	TSB	0.74 ± 0.07^{ab}	1.36 ± 0.59^{a}	0.32 ± 0.01^{c}
	R2AB	0.66 ± 0.06^{ab}	1.29 ± 0.60^{a}	0.30 ± 0.02^{c}

The mean value ±SD with the same superscript letter across each row is not significantly different according to the Duncan multiple range test (DMRT) at $p \leq 0.05$. Number of replicates (n) = 4. NB = nutrient broth, LB = Luria−Bertani Broth, TSB = tryptic soy broth, R2AB = Reasoner's 2A broth.

Table 5. Growth of isolated rhizobacterial strains exposed to temperature-stimulated stress.

Rhizobacteria	Media	25 °C	30 °C	35 °C	40 °C
Aeromonas caviae (A1-2)	NB	0.93 ± 0.03^{a}	1.20 ± 0.01^{a}	1.72 ± 0.01^{a}	0.83 ± 0.03^{a}
	LB	0.87 ± 0.07^{a}	1.08 ± 0.01^{a}	1.59 ± 0.17^{a}	0.75 ± 0.032^{b}
	TSB	0.99 ± 0.42^{a}	0.97 ± 0.01^{a}	1.21 ± 0.48^{a}	0.67 ± 0.02^{b}
	R2AB	0.88 ± 0.01^{b}	1.01 ± 0.59^{a}	1.32 ± 0.58^{a}	0.61 ± 0.02^{b}
Aeromonas veronii (C7_8)	NB	0.97 ± 0.01^{a}	1.09 ± 0.01^{a}	1.64 ± 0.05^{a}	0.82 ± 0.04^{a}
	LB	0.87 ± 0.01^{b}	0.98 ± 0.01^{b}	1.55 ± 0.01^{a}	0.74 ± 0.04^{c}
	TSB	0.88 ± 0.01^{b}	1.02 ± 0.40^{a}	1.40 ± 0.01^{a}	0.66 ± 0.03^{c}
	R2AB	1.04 ± 0.57^{a}	1.26 ± 0.01^{a}	0.79 ± 0.01^{b}	0.60 ± 0.03^{b}

The mean value ± SD with the same superscript letter across each row was not significantly different according to the Duncan multiple range test (DMRT) at $p \leq 0.05$. Number of replicates (n) = 4. NB = nutrient broth, LB = Luria−Bertani broth, TSB = tryptic soy broth, R2AB = Reasoner's 2A broth.

Table 6. Growth of isolated rhizobacterial strains exposed to heavy-metal-stimulated stress.

Rhizobacteria	Media	0 (Control)	Pb^{2+}	$Cr_2O_7^{2-}$	Cd^{2+}
Aeromonas caviae (A1-2)	NB	1.28 ± 0.54 ^{a}	0.76 ± 0.08 ^{ab}	0.63 ± 0.10 ^{b}	0.50 ± 0.09 ^{b}
	LB	1.56 ± 0.22 ^{a}	0.68 ± 0.08 ^{b}	0.56 ± 0.09 ^{bc}	0.44 ± 0.08 ^{c}
	TSB	1.51 ± 0.13 ^{a}	0.61 ± 0.07 ^{b}	0.51 ± 0.08 ^{c}	0.40 ± 0.07 ^{c}
	R2AB	1.08 ± 0.36 ^{a}	0.55 ± 0.06 ^{b}	0.46 ± 0.07 ^{b}	0.36 ± 0.06 ^{c}
Aeromonas veronii (C7_8)	NB	1.26 ± 0.53 ^{a}	0.74 ± 0.08 ^{b}	0.61 ± 0.10 ^{b}	0.49 ± 0.09 ^{c}
	LB	1.26 ± 0.35 ^{a}	0.67 ± 0.01 ^{b}	0.55 ± 0.09 ^{b}	0.44 ± 0.08 ^{c}
	TSB	1.15 ± 0.34 ^{a}	0.60 ± 0.07 ^{b}	0.50 ± 0.08 ^{b}	0.40 ± 0.07 ^{c}
	R2AB	1.0787 ± 0.3217 ^{a}	0.54 ± 0.06 ^{b}	0.45 ± 0.07 ^{c}	0.36 ± 0.06 ^{c}

The mean value ± SD with the same superscript letter across each row was not significantly different according to the Duncan multiple range test (DMRT) at $p \leq 0.05$. Number of replicates (n) = 4. NB = nutrient broth, LB = Luria−Bertani broth, TSB = tryptic soy broth, R2AB = Reasoner's 2A broth.

The growth of isolated rhizobacterial strains A1-2 and C7_8 exposed to pH-stimulated stress is presented in Table 4. The bacteria showed varied tolerance to drought at different pH measurements of 4, 7, and 10 in the growth medium. Furthermore, the rhizobacterial isolates showed tolerance by varying the temperatures of the medium at 25, 30, 35, and 40 °C (Table 5). High bacterial tolerance to drought was recorded at 35 °C compared with the control. This study, in addition, affirms the tolerance of isolated rhizobacteria *Aeromonas caviae* (A1-2) and *Aeromonas veronii* (C7_8) to heavy metals, Pb^{2+}, $Cr_2O_7^{2-}$, and Cd^{2+} (Table 6). Similarly, rhizobacterium A1-2 exhibited high activity of 0.76, 0.63, and 0.50 under Pb^{2+}, $Cr_2O_7^{2-}$, and Cd^{2+} stress, while the rhizobacterium C7_8 metal tolerance level was 0.74, 0.61, and 0.49 under Pb^{2+}, $Cr_2O_7^{2-}$, and Cd^{2+} stress, respectively. The use of nutrient broth shows considerable and potential support for the growth of the bacteria under different growth conditions. Therefore, NB is considered the best medium for the selected strains under different conditions.

4.5. Rhizobacteria Improve the Growth of Maize Plants Exposed to Varying Levels of Drought Stress in a Greenhouse

The untreated and treated maize with rhizobacteria, *A. caviae* (A1-2), and *A. veronii* (C7_8) against drought stress in the greenhouse is presented in Table 7. The response of the maize plants against drought stress was measured concerning the plant height (cm), number of leaves, leaf area (cm^2), stem girth (cm), and chlorophyll content. It was observed that a combination of both bacteria offered the best stress tolerance level for the maize against drought stress, as maize treated with both bacteria showed better performance at 100, 50, and 25 field capacity (FC). The combined treatment of bacteria strains (A1-2+C7_8) displayed a high plant height of 9.33 cm, 8.77 cm, and 8.38 cm compared to the control with the low plant height of 7.27 cm, 6.08 cm, and 6.10 cm at 100, 50, and 25 FC, respectively (Table 7). Similar observations were recorded for stem girth, number of leaves, leaf area, and chlorophyll content. The combined treatment of strains A1-2+C7_8 had the highest leaf numbers of 7.08, 6.50, and 6.50; stem girth of 4.58, 4.53, and 4.18 cm; and chlorophyll content of 14.13, 14.64, and 13.64 at 100, 50, and 25 FC, respectively, compared with the control. In terms of leaf area, the highest values of 85.13, 72.37, and 73.80 cm^2 at 100, 50, and 25 FC were recorded compared with the control.

The response of the plants to varied drought stress was shown every week (Figure 3). The combined treatment of selected rhizobacteria strains, A1-2+C7_8, showed a better drought tolerance response compared to the control. All the treatments showed similar pattern increments in terms of plant height at 100 FC (Figure 3). However, the combined treatments showed an upward considerable drought tolerance response at day 25 concerning plant length (Figure 3), stem girth (Figure 3), leaf number (Figure 3), chlorophyll content (Figure 3), and leaf area (Figure 3) at 100, 50, and 25 FC. Hence, the response of the maize plants to drought stress can be arranged in decreasing order as A1-2+C7_8 > A1-2 > C7_8 > control, corresponding to extremely effective > moderately effective > slightly effective, respectively.

Table 7. Effects of rhizobacterial inoculation on the growth of maize plants exposed to varying periods of drought stress.

Water Regime	Growth Parameters	Control	A1-2	C7_8	A1-2+C7_8
100 WC	Plant height (cm)	7.27 ± 0.95 c	8.43 ± 0.43 ab	7.95 ± 0.50 b	9.33 ± 0.34 a
	Number of leaves	5.67 ± 0.29 d	6.67 ± 0.14 b	6.50 ± 0.29 c	7.08 ± 0.29 a
	Stem girth (cm)	3.08 ± 0.21 c	4.19 ± 0.31 b	3.73 ± 0.18 bc	4.58 ± 0.27 a
	Chlorophyll content	9.12 ± 0.71 bcd	11.78 ± 2.22 b	10.13 ± 0.92 bc	14.13 ± 2.74 a
	Leaf area (cm^2)	52.60 ± 5.35 c	79.67 ± 4.30 ab	63.81 ± 2.89 abc	85.13 ± 14.13 a
50 WC	Plant height (cm)	6.08 ± 0.45 d	8.11 ± 0.63 ab	6.83 ± 0.27 c	8.71 ± 0.23 a
	Number of leaves	5.92 ± 0.14 a	6.58 ± 0.43 a	6.08 ± 0.14 a	6.50 ± 0.29 a
	Stem girth (cm)	3.06 ± 0.23 b	3.75 ± 0.41 ab	3.45 ± 0.20 ab	4.53 ± 0.32 a
	Chlorophyll content	8.44 ± 0.52 c	10.69 ± 1.49 b	9.57 ± 0.89 b	14.64 ± 3.56 a
	Leaf area (cm^2)	46.88 ± 1.76 c	66.57 ± 8.72 ab	44.44 ± 0.52 abc	72.37 ± 0.73 a
25 WC	Plant height (cm)	6.10 ± 0.30 d	7.68 ± 0.75 ab	6.62 ± 0.29 c	8.38 ± 0.53 a
	Number of leaves	6.00 ± 0.29 a	6.42 ± 0.39 a	6.08 ± 0.14 a	6.50 ± 0.00 a
	Stem girth (cm)	3.17 ± 0.11 b	3.73 ± 0.36 ab	3.43 ± 0.13 b	4.18 ± 0.34 a
	Chlorophyll content	9.09 ± 0.92 c	12.93 ± 1.98 ab	10.92 ± 1.57 abc	13.64 ± 1.50 a
	Leaf area (cm^2)	42.67 ± 11.19 c	55.94 ± 2.03 a	47.78 ± 1.29 bc	73.80 ± 0.82 a

Values are means of three replicates ± SD; values across each row having the same superscript letters are not significant according to the Duncan multiple range test (DMRT) at $p \leq 0.05$. TO = uninoculated plants (Control), A1-2 = plants whose roots were inoculated with rhizobacterium *Aeromonas caviae* (Treatment 1), C7_8 = plants whose roots were inoculated with rhizobacterium *Aeromonas veronii* (Treatment 2), A1-2+C7_8 = plants whose roots were co-inoculated with rhizobacteria *Aeromonas caviae* and *Aeromonas veronii* (Treatment 3).

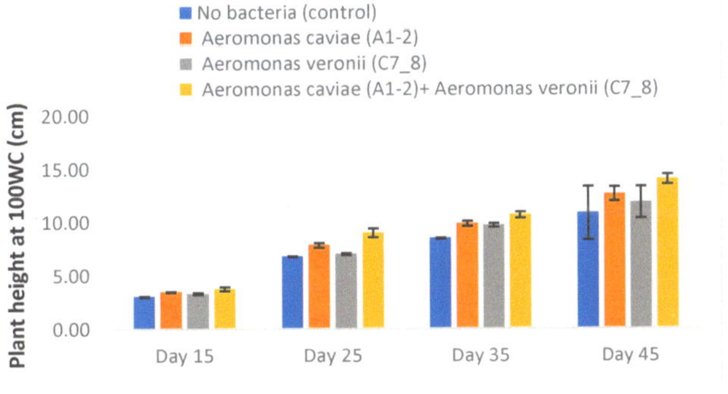

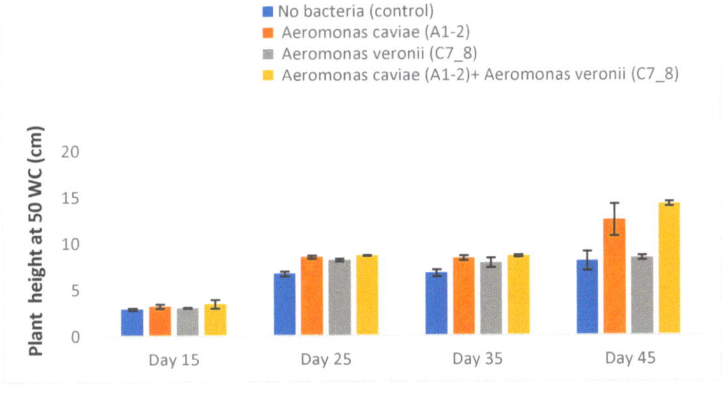

Figure 3. *Cont.*

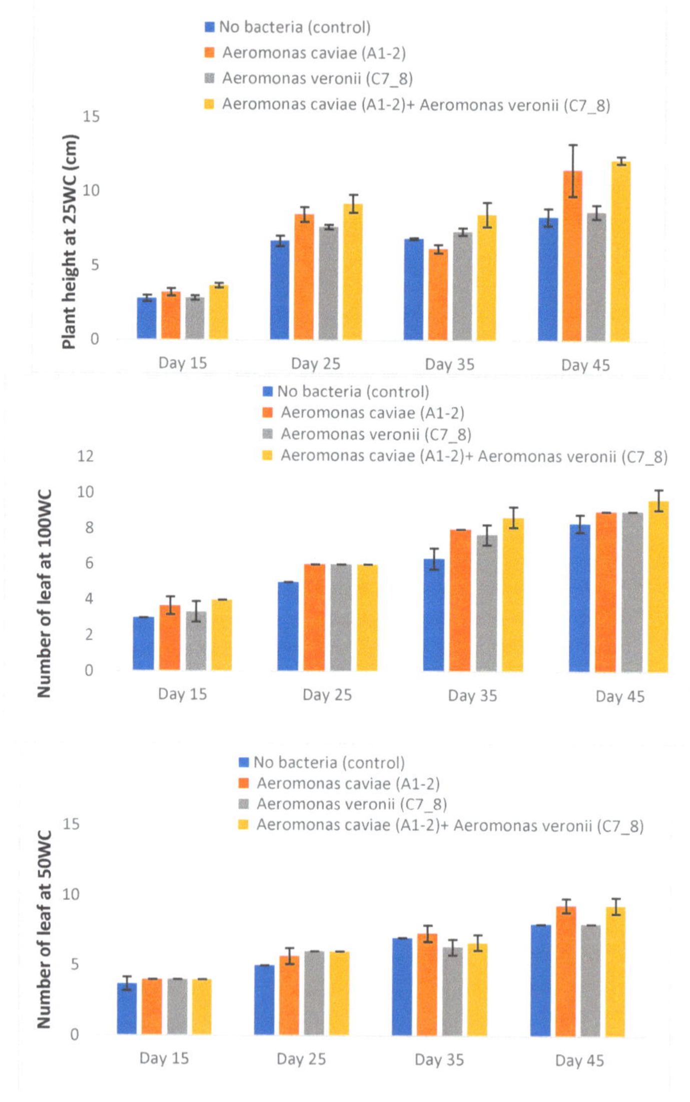

Figure 3. *Cont.*

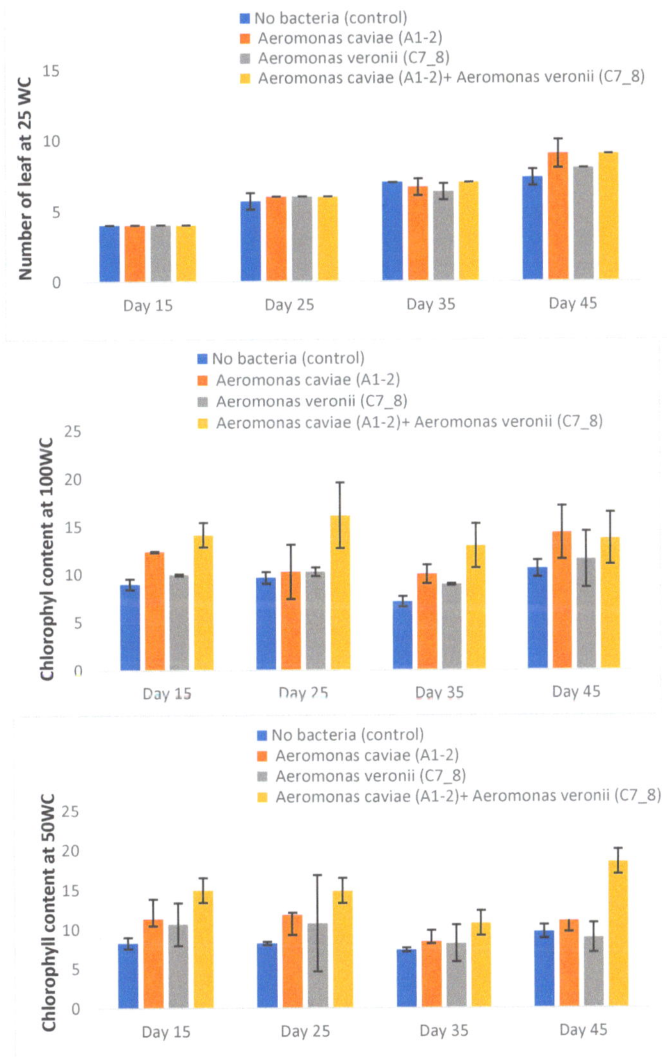

Figure 3. *Cont.*

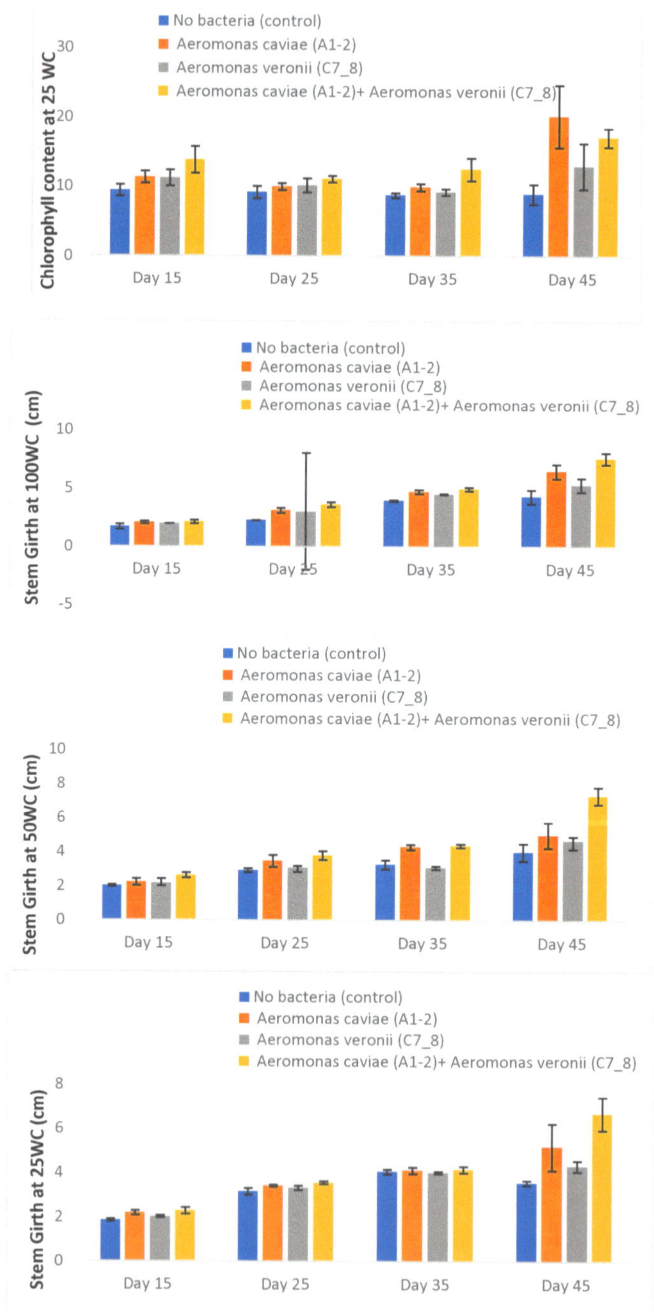

Figure 3. The growth response of maize plants exposed to rhizobacterial treatments at varying periods of drought stress. No bacteria = Uninoculated plants (control), Treatment 1 = Maize plants with rhizobacterium *Aeromonas caviae* A1_2, Treatment 2 = Maize plants with rhizobacterium *Aeromonas veronii* C7_8, Treatment 1 and 2 = Maize plants with rhizobacteria *Aeromonas caviae* A1_2 and C7_8.

4.6. Rhizobacteria Improving the Biomass of Maize Exposed to Drought Stress in the Greenhouse

The response of maize plants to drought stress enhanced by rhizobacteria treatment was evaluated in the greenhouse. The plants treated with 20 mL each of the rhizobacteria, *A. caviae* (A1-2), and *A. veronii* (C7_8) showed improved biomass compared to the control (Table 8). In detail, the combined treatments of A1-2+C7_8 had high fresh total plant mass of 19.41 g, 16.68 g, and 10.29 g, while maize treated with *A. caviae* (A1-2) had fresh total plant biomass of 18.70 g, 13.45 g, and 10.52 g at WC of 100, 50, and 25, respectively, compared with the control. A fresh root mass of 8.42 g, 4.35 g, and 3.91 g of treated maize with strains A1-2+C7_8 at water capacities of 100, 50, and 25 was recorded. The maize treated with *A. caviae* (A1-2) showed higher performance than those treated with *A. veronii* (C7_8) compared with the control. Based on the biomass, the response of the maize to drought stress can be arranged in a decreasing order as follows: A1-2+C7_8> A1-2 > C7_8 > control, corresponding to extremely effective > moderately effective > slightly effective, respectively.

Table 8. Effects of rhizobacterial inoculation on the biomass of maize exposed to drought stress.

Water Capacity	Parameters (g)	Control	A1-2	C7_8	A1-2+C7_8
100	Fresh Total Plant Mass	10.32 ± 0.22^{c}	18.70 ± 4.36^{ab}	11.85 ± 1.01^{b}	19.41 ± 4.01^{a}
	Fresh Root Mass	3.59 ± 1.69^{d}	6.28 ± 1.75^{ab}	3.52 ± 0.45^{c}	8.42 ± 2.31^{a}
	Fresh Shoot Mass	6.49 ± 1.58^{d}	10.06 ± 2.14^{b}	8.28 ± 0.98^{c}	13.19 ± 2.53^{a}
	Dry Total Plant Mass	1.45 ± 0.03^{d}	3.22 ± 0.68^{ab}	1.81 ± 0.23^{cd}	3.45 ± 0.62^{a}
	Dry Root Mass	0.53 ± 0.17^{cd}	1.24 ± 0.35^{ab}	0.58 ± 0.06^{cd}	1.35 ± 0.34^{a}
	Dry Shoot Mass	1.03 ± 0.09^{bcd}	1.87 ± 0.34^{ab}	1.21 ± 0.19^{abc}	2.12 ± 0.28^{a}
50	Fresh Total Plant Mass	5.75 ± 2.12^{cd}	13.45 ± 3.58^{b}	7.53 ± 1.31^{bc}	16.68 ± 0.90^{a}
	Fresh Root Mass	1.70 ± 0.85^{c}	4.59 ± 0.98^{a}	3.93 ± 0.58^{b}	4.35 ± 0.48^{a}
	Fresh Shoot Mass	4.03 ± 1.27^{cd}	8.72 ± 2.52^{ab}	3.58 ± 1.86^{d}	12.31 ± 1.36^{a}
	Dry Total Plant Mass	1.07 ± 0.06^{bcd}	2.15 ± 0.71^{ab}	1.33 ± 0.14^{bc}	2.80 ± 0.24^{a}
	Dry Root Mass	0.37 ± 0.05^{d}	0.91 ± 0.36^{ab}	0.47 ± 0.08^{bc}	0.96 ± 0.03^{a}
	Dry Shoot Mass	0.64 ± 0.03^{d}	1.43 ± 0.49^{ab}	0.78 ± 0.08^{abc}	1.77 ± 0.22^{a}
25	Fresh Total Plant Mass	5.12 ± 0.44^{cd}	10.52 ± 3.86^{ab}	6.69 ± 1.20^{bc}	10.29 ± 2.85^{a}
	Fresh Root Mass	1.49 ± 0.21^{c}	2.03 ± 0.33^{ab}	1.75 ± 1.08^{abc}	3.91 ± 0.70^{a}
	Fresh Shoot Mass	2.94 ± 0.07^{bcd}	6.61 ± 3.04^{b}	5.14 ± 1.36^{bc}	8.54 ± 1.78^{a}
	Dry Total Plant Mass	0.91 ± 0.07^{d}	1.72 ± 0.49^{ab}	1.21 ± 0.13^{c}	1.86 ± 0.56^{a}
	Dry Root Mass	0.39 ± 0.05^{bcd}	0.50 ± 0.13^{b}	0.43 ± 0.03^{ac}	0.74 ± 0.21^{a}
	Dry Shoot Mass	0.51 ± 0.01^{cd}	0.98 ± 0.27^{ab}	0.78 ± 0.13^{c}	1.27 ± 0.34^{a}

Values are means of three replicates ± SD; values across each row having the same superscript letters are not significant according to the Duncan multiple range test (DMRT) at $p \leq 0.05$. Uninoculated plants (Control), A1-2 = plants whose roots were inoculated with rhizobacterium *A. caviae* (Treatment 1), C7_8 = plants whose roots were inoculated with rhizobacterium *A. veronii* (Treatment 2), A1-2+C7_8 = plants whose roots were co-inoculated with *A. caviae* and *A. veronii* (Treatment 3).

4.7. Rhizobacteria Improve the Morphophysiological Attributes of Maize Plants Exposed to Drought Stress in the Greenhouse

The rhizobacterial treatment of maize plants to withstand varying levels of drought stress was also established on some morphophysiological characteristics of the plant concerning the % leaf relative water content (Figure 4), % electrolyte leakages (Figure 5), sugar content (mg/mL) (Figure 6), and proline content (μmol/g of dry weight) (Figure 7). It was observed that the co-inoculation of *A. caviae* A1-2 and *A. veronii* C7_8 enhanced the relative water content (RWC) of the plant leaves both under drought stress at 50 and 25% water capacity (WC) and under unstressed conditions at 100 WC (Figure 4). The singular inoculation of the bacterial strains exerted a slight enhancement in the RWC of the plants at 100, 50, and 25 WC compared with the control, which showed a decrease with an increase in drought stress. The leaves of the uninoculated maize plants significantly increased the electrolyte leakages from 50 to 25 WC, while the singular and combined treatment showed a reduction in the electrolyte leakages (Figure 5). The soluble sugar content in the maize plants both under stress and under unstressed conditions is presented in Figure 6. It was observed that the combined rhizobacterial strains' treatment enhanced the sugar content of the leaves compared with the control ($p \leq 0.05$). The combined maize treatment with A1-2+C7_8 had higher glucose content than that treated with either *A. caviae* (A1-2) or *A. veronii* (C7_8) singularly. In Figure 7, the rhizobacterial treatment significantly enhanced the proline content in maize both under drought-stressed (50 and 25 WC) and unstressed conditions (100 WC) compared to the control.

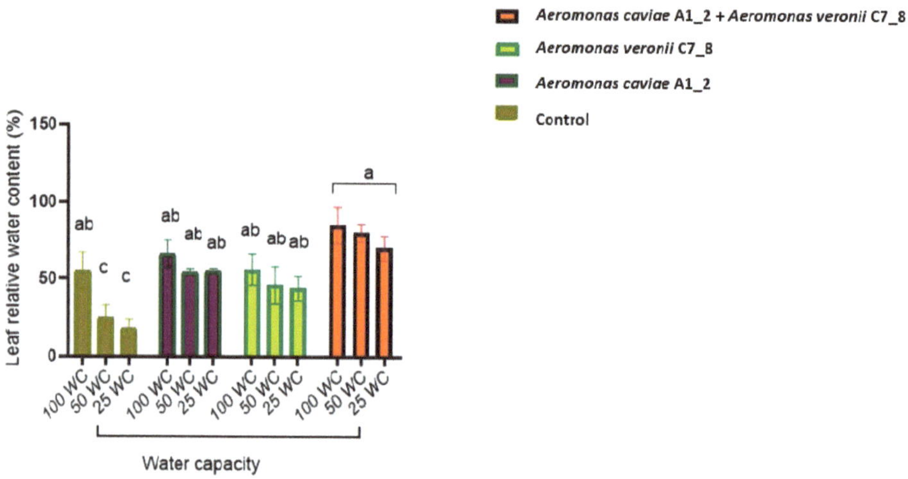

Figure 4. Relative water content of maize leaves inoculated and noninoculated with rhizobacteria, *A. caviae* (A1-2), and *A. veronii* (C7_8) under varying levels of drought stress. Bars are means of three replicates ± SD. Control—noninoculated treatment, and combined strains A1-2+C7_8.

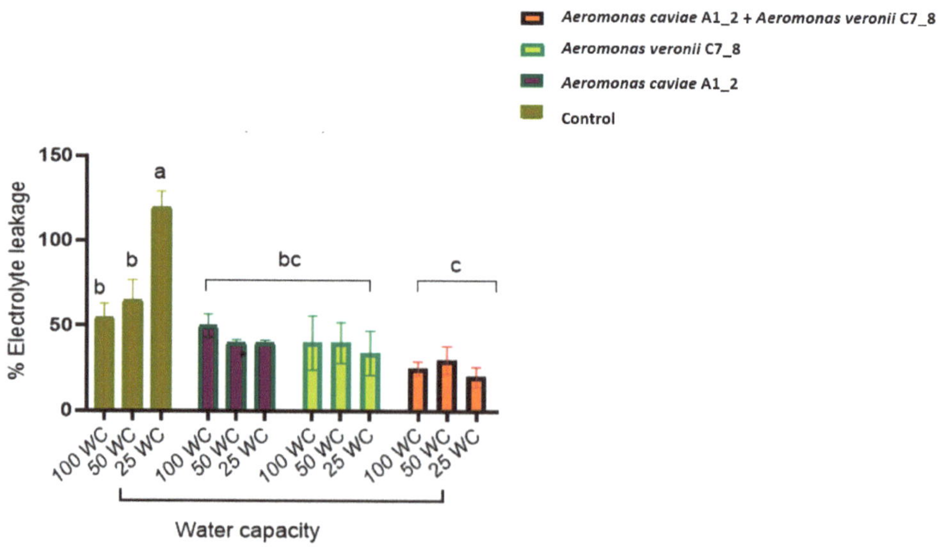

Figure 5. Electrolyte leakage of maize leaves inoculated and noninoculated with rhizobacteria *A. caviae* (A1-2) and *A. veronii* (C7_8) under varying levels of drought stress. Bars are means of three replicates ± SD. Control—noninoculated treatment, and combined strains A1-2+C7_8.

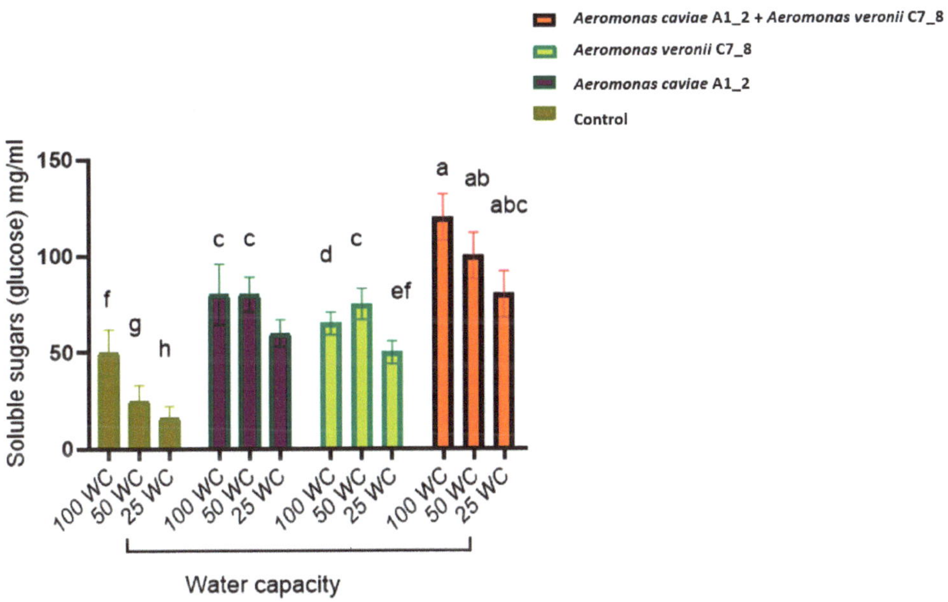

Figure 6. Soluble sugar accumulation in maize of maize leaves inoculated and noninoculated with rhizobacteria, *A. caviae* (A1-2), and *A. veronii* (C7_8) under varying levels of drought stress. Bars are means of three replicates ± SD. Control—noninoculated treatment.

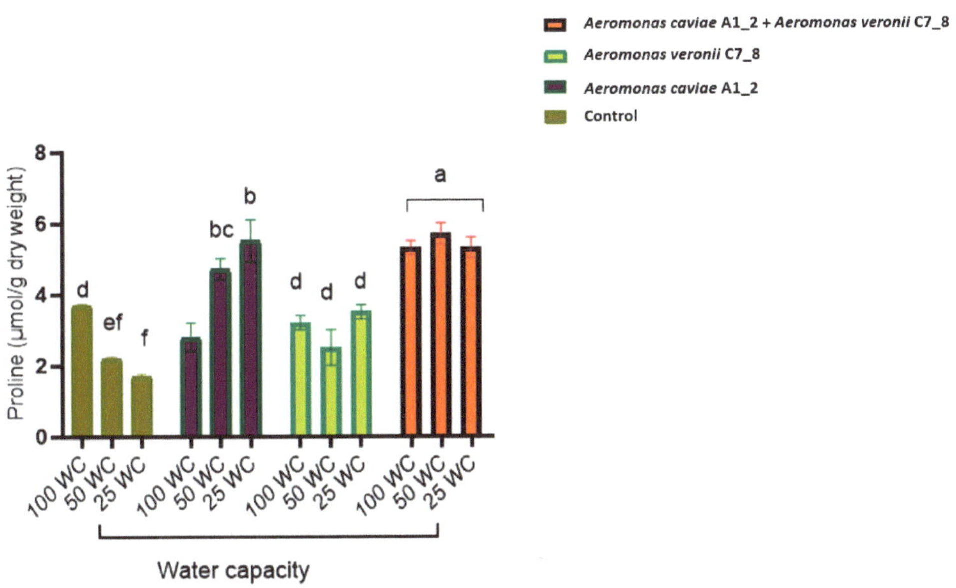

Figure 7. Proline accumulation in maize of maize leaves inoculated and noninoculated with rhizobacteria, *A. caviae* (A1-2), and *A. veronii* (C7_8) under varying levels of drought stress. Bars are means of three replicates ± SD. Control—noninoculated treatment.

5. Discussion

The rhizosphere represents discreet regions of plant—soil—microbe interactions with the host-plant—root interface [46]. As such, the rhizosphere is home to many plant-growth-promoting rhizobacteria (PGPR), with literature evidence related to their beneficial effects on plants and how they can be better exploited for sustainable agriculture and crop production [16,47,48]. The plant-growth-enhancing efficiency of rhizobacteria is considerably dependent on their capability to subsist and create an effective colonization of the plant root [49,50]. In addition, efficient plant root colonization by PGPR is crucial in promoting growth regardless of their mode of action [51]. Therefore, screening of drought tolerance PGPR should be carried out to evaluate their potential and drought stress reduction in plants. Limited studies have identified PGPR *Aeromonas* species from some plants in arid and semi-arid regions [52,53]. Therefore, there is a need to study further the potential impact of the mono- and co-inoculation of the rhizobacteria strains on maize in the greenhouse and field experimental conditions. Many identifiable PGPR belonging to the genera *Bacillus*, *Pantoea*, *Gluconacetobacer*, *Pseudomonas*, and *Paenibacillus* with plant-growth-promoting attributes have been reported [53–56]. Hence, this study affirms the biostimulatory potentials of rhizobacteria for maize improvement under drought stress in a greenhouse.

The isolation and characterization of drought-tolerant PGPR associated with maize were carried out in this study. Two of the identifiable bacterial isolates exhibited high drought tolerance, coupled with the PGP properties. The rhizobacterial strains, *Aeromonas caviae* A1-2 and *A. veronii* C7_8, at 30% PEG amendment induced high drought tolerance, thus alleviating the effect of drought stress on the inoculated maize and enhancing their growth. The phylogeny revealed that the two bacterial strains belong to the genera *Aeromonas*, which are most prevalent in agrarian soils. A greater number of *Aeromonas* species have been extensively investigated for plant improvement [57,58]. According to Arshad et al. [59], PGPR that possess ACC deaminase activity aid crop plants to withstand drought stress by lowering the ethylene (stress hormone) production level through the hydrolysis of ACC into ammonia and α-ketobutyrate. These ACC-deaminase-producing bacteria have already been documented to enhance plant growth, particularly in stress conditions, such as heavy metals, flooding, high salt, and drought [21].

In a study carried out by Nawaz et al. [52], the authors characterized PGPR, *Aeromonas* strains SAL-17 and SAL-21, isolated from two wheat genotypes and soybean to produce acyl-homoserine lactone (AHL), which assisted plants to withstand stress conditions. Sarkar et al. [60] reported ACC deaminase activity of halotolerant *Enterobacter* sp. to enhance plant growth. Furthermore, the high ACC-deaminase-producing potential of *Aeromonas* strains A1-2 and C7_8 corroborates the findings of Cruz et al. [61], who reported *Aeromonas* sp. in other crops by confirming their potentialities in ameliorating drought stress for plant survival. The ACC deaminase activity of the *Aeromonas* strains A1-2 and C7_8 may be associated with their viability/survival in the PEG-amended medium and genetic makeup to assist crops' survival under stress [62]. Drought tolerance genes aid bacteria in withstanding adverse conditions and colonizing endorhizosphere regions through the linkage of fibrillar material, which fosters bacterial attachment to the surfaces while protecting plants [63].

The PGPR are well known to support plant growth via symbiotic association. Typically, rhizobacteria inoculation can cause an improvement in plant growth through diverse mechanisms, such as the production of siderophores, IAA, ACC deaminase activity, nitrogen fixation, and other growth-promoting features [64]. The production of phytohormones can alter plant root architecture to improve water and nutrient uptake and retention. The ability of PGPR to ameliorate abiotic stressors in the environment is just one of the advantages underlining various metabolic and physiological inductions in the host plants. In addition to the in vitro growth assessment, greenhouse studies have shown that growth traits exhibited by the rhizobacteria strains A1-2 and C7_8 contributed significantly to the seed germination index compared to the control.

The present studies focus on the enhancement in the symbiotic plant–soil–microbe mechanisms for the improvement in maize production. We employed rhizobacteria strains A1-2 and C7_8 as PGPR and demonstrated how their inoculation into maize roots supports the growth and development of maize under the greenhouse experiment. This mechanism relies on the exchange of materials between the plant and the microbes encompassing a better decomposition of organic, nutrient release, and cycling, as well as the induction of systemic resistance against biotic and abiotic stresses [65]. Niu et al. [62] and Danish et al. [6] had demonstrated earlier the induction of drought tolerance in plants using rhizosphere microbes. The observed increase in the maize seed germination rate was consistent with earlier reports, which support the use of microbial agents, singularly or combined with other agents, to mediate physiological changes and improve crop growth [66,67].

Consequently, plant inoculation with drought-tolerant rhizobacteria with manifold growth-promoting tendencies can enhance the effectiveness of the bacteria inoculants, especially in semi-arid or arid regions [62]. The rhizobacteria strains, A1-2 and C7_8, exhibited varied drought tolerance at varied salinity, pH, temperature, and heavy metal stress, which stimulated plant growth when they co-existed to ameliorate effects of the stressors and corroborated the findings of Abedinzadeh et al. [68], who affirmed the potential effects of abiotic stressors on plant growth, through enzyme and protein synthesis. The results from this study are similar to the findings of Fasusi et al. [69], who affirmed the plant-growth-promoting tendencies of PGPR isolated from food plants in South Africa.

It is well known that cereals and legumes like soybean and maize are popular food crops of economic importance and are of immense value for human diets and feed consumption. Their cultivation, however, can easily be affected by different factors ranging from soil infertility, pest disease, and abiotic stressors. Hence, the need to improve their production to meet the demand of the increasing world population remains paramount [70]. The use of many synthetic agrochemicals in modern crop production is on the rise, with concurrent increases in cost and residual harmful effects on the environment and resultant health risks to the consumers along the food value chain [71,72]. Many PGPR have been reported to stimulate plant growth and survival against different environmental stresses [73–75].

Among several recent studies on rhizobacteria as biostimulants, Kunwar et al. [76] reported the use of rhizobacteria *Pseudomonas* species to stimulate maize plants through nutrient bioavailability and acquisition. Albareda et al. [77] suggested the use of peat as a carrier of Rhizobia. Kunwar et al. [76] reported on *P. syringe* + peat and *P. putida* + clay; they demonstrated that the solid biostimulants improved the plant yield in terms of height, stem diameter, and leaf number and area. In another study, Siddiq et al. [78] showed that newly isolated *B. thuringiensis* from Pakistan significantly increased bean growth and yield. Scott et al. [79] established the enhancement in corn growth, nutrients, and chlorophyll content using rhizobacterium *Azospirillum brasilense* HM053.

Our study established the importance of rhizobacteria strains A1-2 and C7_8 in the improvement in maize plants against drought stress, which corroborates previous findings of Vacheron et al. [80], who reported the effect of biostimulatory agents on maize yield. Diarrassouba et al. [81] employed *P. fluorescens* as a bioinoculant to improve the growth of two varieties of *Zea mays* L. in soils under low water capacity. Likewise, Noumavo et al. [82] also reported the induction of metabolic and fungicidal potentials of maize using some locally isolated rhizobacteria strains. In a more recent study, Amogou et al. [83] established the application of some locally isolated rhizobacteria from Benin to enhance maize growth. Notably, Mejri et al. [84] formulated a deleterious rhizobacterium, *Pseudomonas trivialis* X33d, as a biocontrol agent against the Durum wheat pathogen *Bromus diandrus*. Additionally, Agbodjato et al. [85] reported synergistic interactions between rhizobacteria and chitosan for the in vitro acclimatization of maize under germination and nutrient uptake under greenhouse conditions. The improvement in the maize growth and yield subjected to drought stress under the greenhouse experiment agrees with the findings of Janardan et al. [86] and Kumar et al. [87], who demonstrated the effect of some rhizobacteria in seed germination and growth enhancement in chickpea (*Cicer arietinum* L.) and wheat (*Triticum aestivum* L.) under greenhouse conditions. Therefore, biostimulants using PGPR can be considered over synthetic chemicals, which can pose residual effects and health hazards.

Under harsh conditions, the host plant tends to respond to drought-induced stress by producing osmolytes to help the plant sustain osmotic balancing in econiches [88]. The ability of plants to produce osmolytes, such as glycine betaine, total soluble sugars, proline, and total protein under drought stress, can help mitigate the effects of oxidative stress on plants. Osmolyte production in plants can be enhanced by bioinoculation with the copious agriculturally important rhizobacteria with diverse growth-promoting characteristics [89]. The maize treated with the combined *A. caviae* A1-2 + *A. veronii* C7_8 enhanced the sugar contents of the leaves compared with the control. The combined maize treatment with A1-2+C7_8 also exhibited a higher glucose content than the maize treated with either *A. caviae* A1-2 or *A. veronii* C7_8 singularly under drought stress. The massive production of osmolytes in maize inoculated with *A. caviae* A1-2 + *A. veronii* C7_8 can signally help in sustaining cell membrane integrity and water status and the prevention of protein degradation under drought stress. The results obtained in this study established the bacterial inoculant co-existing with maize and production of enhanced soluble sugar under drought stress, which indicated the drought stress amelioration potential of *A. caviae* A1-2 and *A. veronii* C7_8 on the maize plants grown in water-deficit areas.

The seed inoculation enhanced the crop yield, and this can be linked to the ability of the bacterial strains to adapt to and survive harsh conditions. Our current studies align with the literature stating crops within naturally drought-stressed conditions are expedient cradles of drought-tolerant bacteria with plant-growth-enhancing prospects. The inferences from this study support the hypothesis that plant PGPR can exhibit important input in the drought locale plant adaptation, such as maize. Hence, we propose that drought-tolerant rhizobacteria, *A. caviae* A1-2, and *A. veronii* C7_8, can further be explored as bioinoculants in managing abiotic drought stress in maize and other staple food crops.

6. Conclusions

This study affirmed that drought tolerance and notable PGPR strains A1-2 and *A. veronii* C7_8 can alleviate drought stress in maize plants upon inoculation, either singularly or combined. Notably, the results of this study established that the inoculation of identifiable rhizobacteria *Aeromonas* strains A1-2 and C7_8 had significant effects on maize growth under drought stress. The bioinoculation supported the physiological and biochemical functions of the maize plants. It also culminated in improving both the below-ground–above-ground and/or morphological components of maize plants compared to the noninoculated plants. The combined effects of *A. caviae* A1-2 and *A. veronii* C7_8 were observed to significantly enhance maize yields, such as plant height, leaf number, stem girth, leaf area and chlorophyll content, and plant shoot and maize biomass, such as fresh total plant mass, fresh root mass, fresh shoot mass, dry total plant mass, dry root mass, and dry shoot mass under drought stress. The results obtained can provide more insights into the impacts of the rhizobacteria on maize growth and provide theoretical background for the real-life application of these strains in field experimental conditions. This biorational strategy can help resolve or abate food insecurity problems in most developing countries in the semi-arid and arid regions of the world. Nevertheless, an additional investigation of these bacterial strains can help unravel their effectiveness as suitable candidates in the formulation of biofertilizers. Furthermore, omics studies can promisingly help to unravel the identity and actual mechanisms of *Aeromonas* species in a natural plant environment for future exploration in enhancing crop yield, promoting plant growth under drought stress, and ensuring food security.

Supplementary Materials: The following supporting information can be downloaded at https://www.mdpi.com/article/10.3390/plants13101298/s1. Figure S1. Microscopic View of Rhizobacterial Strains Isolated from Maize Plants; Table S1. Morpho-biochemical Characteristics of The Isolated Rhizobacterial Strains.

Author Contributions: O.O.B. conceptualized the research. V.F.A. and O.O.B. designed the study. V.F.A. managed the literature searches, conducted laboratory work, interpreted the results, wrote the first draft, and revised and formatted the manuscript. N.A.A. and A.E.F. assisted in the result analysis and draft review. O.O.B. provided academic input, thoroughly critiqued the manuscript drafts, proofread the draft, secured research funds, and supervised all co-authors. All authors approved the article for publication. All authors have read and agreed to the published version of the manuscript.

Funding: This research was funded by the National Research Foundation of South Africa for the grants (UID: 132595).

Institutional Review Board Statement: This article does not contain any studies with human participants or animals performed by any of the authors.

Data Availability Statement: The original contributions presented in the study are included in the article/Supplementary Material, further inquiries can be directed to the corresponding author/s.

Acknowledgments: VFA would like to thank North-West University for the postgraduate doctoral bursary. OOB recognizes the National Research Foundation of South Africa for the grants (UID: 132595) that support work in her research group.

Conflicts of Interest: Authors declare that there are no known conflicts of interest for this manuscript.

References

1. Adeleke, B.S.; Babalola, O.O. Meta-omics of endophytic microbes in agricultural biotechnology. *Biocatal. Agric. Biotechnol.* **2022**, *42*, 102332. [CrossRef]
2. Sandhya, V.; SK, A.; Grover, M.; Reddy, G.; Venkateswarlu, B. Alleviation of drought stress effects in sunflower seedlings by the exopolysaccharides producing Pseudomonas putida strain GAP-P45. *Biol. Fertil. Soils* **2009**, *46*, 17–26. [CrossRef]
3. Robert, G.A.; Rajasekar, M.; Manivannan, P. Triazole-induced drought stress amelioration on growth, yield, and pigments composition of *Helianthus annuus* L.(sunflower). *Int. Multidiscip. Res. J.* **2015**, *5*, 6–15.
4. Aroca, R.; Irigoyen, J.J.; Sánchez-Díaz, M. Drought enhances maize chilling tolerance. II. Photosynthetic traits and protective mechanisms against oxidative stress. *Physiol. Plant.* **2003**, *117*, 540–549. [CrossRef] [PubMed]
5. Naveed, M.; Mitter, B.; Reichenauer, T.G.; Wieczorek, K.; Sessitsch, A. Increased drought stress resilience of maize through endophytic colonization by *Burkholderia phytofirmans* PsJN and *Enterobacter* sp. FD17. *Environ. Exp. Bot.* **2014**, *97*, 30–39. [CrossRef]
6. Danish, S.; Zafar-ul-Hye, M.; Mohsin, F.; Hussain, M. ACC-deaminase producing plant growth promoting rhizobacteria and biochar mitigate adverse effects of drought stress on maize growth. *PLoS ONE* **2020**, *15*, e0230615. [CrossRef] [PubMed]
7. Eke, P.; Kumar, A.; Sahu, K.P.; Wakam, L.N.; Sheoran, N.; Ashajyothi, M.; Patel, A.; Fekam, F.B. Endophytic bacteria of desert cactus (*Euphorbia trigonas* Mill) confer drought tolerance and induce growth promotion in tomato (*Solanum lycopersicum* L.). *Microbiol. Res.* **2019**, *228*, 126302. [CrossRef] [PubMed]
8. Naveed, M.; Hussain, M.B.; Zahir, Z.A.; Mitter, B.; Sessitsch, A. Drought stress amelioration in wheat through inoculation with *Burkholderia phytofirmans* strain PsJN. *Plant Growth Reg.* **2014**, *73*, 121–131. [CrossRef]
9. Silva, E.R.; Zoz, J.; Oliveira, C.E.S.; Zuffo, A.M.; Steiner, F.; Zoz, T.; Vendruscolo, E.P. Can co-inoculation of *Bradyrhizobium* and *Azospirillum* alleviate adverse effects of drought stress on soybean (*Glycine max* L. Merrill.)? *Arch. Microbiol.* **2019**, *201*, 325–335. [CrossRef]
10. Drogue, B.; Doré, H.; Borland, S.; Wisniewski-Dyé, F.; Prigent-Combaret, C. Which specificity in cooperation between phytostimulating rhizobacteria and plants? *Res. Microbiol.* **2012**, *163*, 500–510. [CrossRef]
11. Gepstein, S.; Glick, B.R. Strategies to ameliorate abiotic stress-induced plant senescence. *Plant Mol. Biol.* **2013**, *82*, 623–633. [CrossRef]
12. Bhattacharyya, A.; Pablo, C.H.; Mavrodi, O.V.; Weller, D.M.; Thomashow, L.S.; Mavrodi, D.V. Rhizosphere plant-microbe interactions under water stress. In *Advances in Applied Microbiology*; Elsevier: Amsterdam, The Netherlands, 2021; Volume 115, pp. 65–113.
13. Kushwaha, P.; Kashyap, P.L.; Bhardwaj, A.K.; Kuppusamy, P.; Srivastava, A.K.; Tiwari, R.K. Bacterial endophyte mediated plant tolerance to salinity: Growth responses and mechanisms of action. *World J. Microbiol. Biotechnol.* **2020**, *36*, 26. [CrossRef] [PubMed]
14. Ngoma, L.; Mogatlanyane, K.; Babalola, O.O. Screening of endophytic bacteria towards the development of cottage industry: An in vitro study. *J. Hum. Ecol.* **2014**, *47*, 45–63. [CrossRef]
15. Babalola, O.O.; Sanni, A.I.; Odhiambo, G.D.; Torto, B. Plant growth-promoting rhizobacteria do not pose any deleterious effect on cowpea and detectable amounts of ethylene are produced. *World J. Microbiol. Biotechnol.* **2007**, *23*, 747–752. [CrossRef]
16. Gupta, A.; Gopal, M.; Thomas, G.V.; Manikandan, V.; Gajewski, J.; Thomas, G.; Seshagiri, S.; Schuster, S.C.; Rajesh, P.; Gupta, R. Whole genome sequencing and analysis of plant growth promoting bacteria isolated from the rhizosphere of plantation crops coconut, cocoa and arecanut. *PLoS ONE* **2014**, *9*, e104259. [CrossRef] [PubMed]
17. Mathur, P.; Roy, S. Insights into the plant responses to drought and decoding the potential of root associated microbiome for inducing drought tolerance. *Physiol. Plant.* **2021**, *172*, 1016–1029. [CrossRef] [PubMed]

18. Huang, X.-F.; Chaparro, J.M.; Reardon, K.F.; Zhang, R.; Shen, Q.; Vivanco, J.M. Rhizosphere interactions: Root exudates, microbes, and microbial communities. *Botany* **2014**, *92*, 267–275. [CrossRef]
19. Naing, A.H.; Maung, T.T.; Kim, C.K. The ACC deaminase-producing plant growth-promoting bacteria: Influences of bacterial strains and ACC deaminase activities in plant tolerance to abiotic stress. *Physiol. Plant.* **2021**, *173*, 1992–2012. [CrossRef]
20. Prischl, M.; Hackl, E.; Pastar, M.; Pfeiffer, S.; Sessitsch, A. Genetically modified Bt maize lines containing cry3Bb1, cry1A105 or cry1Ab2 do not affect the structure and functioning of root-associated endophyte communities. *Appl. Soil Ecol.* **2012**, *54*, 39–48. [CrossRef]
21. Ali, S.Z.; Sandhya, V.; Venkateswar Rao, L. Isolation and characterization of drought-tolerant ACC deaminase and exopolysaccharide-producing fluorescent *Pseudomonas* sp. *Ann. Microbiol.* **2014**, *64*, 493–502. [CrossRef]
22. Oladosu, Y.; Rafii, M.Y.; Samuel, C.; Fatai, A.; Magaji, U.; Kareem, I.; Kamarudin, Z.S.; Muhammad, I.i.; Kolapo, K. Drought resistance in rice from conventional to molecular breeding: A review. *Int. J. Mol. Sci.* **2019**, *20*, 3519. [CrossRef] [PubMed]
23. Qaim, M. Role of new plant breeding technologies for food security and sustainable agricultural development. *Appl. Econ. Perspect. Policy* **2020**, *42*, 129–150. [CrossRef]
24. Igiehon, N.O.; Babalola, O.O.; Aremu, B.R. Genomic insights into plant growth promoting rhizobia capable of enhancing soybean germination under drought stress. *BMC Microbiol.* **2019**, *19*, 159. [CrossRef] [PubMed]
25. Zhang, L.; Zhang, W.; Li, Q.; Cui, R.; Wang, Z.; Wang, Y.; Zhang, Y.-Z.; Ding, W.; Shen, X. Deciphering the root endosphere microbiome of the desert plant *Alhagi sparsifolia* for drought resistance-promoting bacteria. *Appl. Environ. Microbiol.* **2020**, *86*, 02819–02863. [CrossRef] [PubMed]
26. Johnson, J.M.; Alex, T.; Oelmüller, R. *Piriformospora indica*: The versatile and multifunctional root endophytic fungus for enhanced yield and tolerance to biotic and abiotic stress in crop plants. *J. Tropical Agric.* **2014**, *52*, 103–122.
27. Viscardi, S.; Ventorino, V.; Duran, P.; Maggio, A.; De Pascale, S.; Mora, M.; Pepe, O. Assessment of plant growth promoting activities and abiotic stress tolerance of *Azotobacter chroococcum* strains for a potential use in sustainable agriculture. *J. Soil Sci. Plant Nutr.* **2016**, *16*, 848–863. [CrossRef]
28. Ahmad, F.; Ahmad, I.; Khan, M. Screening of free-living rhizospheric bacteria for their multiple plant growth promoting activities. *Microbiol. Res.* **2008**, *163*, 173–181. [CrossRef]
29. Clarke, P.H.; Cowan, S. Biochemical methods for bacteriology. *Microbiology* **1952**, *6*, 187–197. [CrossRef] [PubMed]
30. Cappuccino, J.; Sherman, N. *Microbiology: A Laboratory Manual*; Pearson: New York, NY, USA, 1992; pp. 125–179.
31. Pikovskaya, R. Mobilization of phosphorus in soil in connection with vital activity of some microbial species. *Microbiologiya* **1948**, *17*, 362–370.
32. Schwyn, B.; Neilands, J. Universal chemical assay for the detection and determination of siderophores. *Anal. Biochem.* **1987**, *160*, 47–56. [CrossRef]
33. AlAli, H.A.; Khalifa, A.; Al-Malki, M. Plant growth-promoting rhizobacteria from *Ocimum basilicum* improve growth of *Phaseolus vulgaris* and *Abelmoschus esculentus*. *S. Afr. J. Bot.* **2021**, *139*, 200–209. [CrossRef]
34. Loper, J.; Schroth, M. Influence of bacterial sources of indole-3-acetic acid on root elongation of sugar beet. *Phytopathology* **1986**, *76*, 386–389. [CrossRef]
35. Bakker, A.W.; Schippers, B. Microbial cyanide production in the rhizosphere in relation to potato yield reduction and *Pseudomonas* spp-mediated plant growth-stimulation. *Soil Biol. Biochem.* **1987**, *19*, 451–457. [CrossRef]
36. Duan, J.; Müller, K.M.; Charles, T.C.; Vesely, S.; Glick, B.R. 1-aminocyclopropane-1-carboxylate (ACC) deaminase genes in rhizobia from southern Saskatchewan. *Microb. Ecol.* **2009**, *57*, 423–436. [CrossRef]
37. Dworkin, M.; Foster, J. Experiments with some microorganisms which utilize ethane and hydrogen. *J. Bacteriol.* **1958**, *75*, 592–603. [CrossRef]
38. Devereux, R.; Willis, S.G. Amplification of ribosomal RNA sequences. In *Molecular Microbial Ecology Manual*; Springer: Berlin/Heidelberg, Germany, 1995; pp. 277–287.
39. Kumar, S.; Stecher, G.; Li, M.; Knyaz, C.; Tamura, K. MEGA X: Molecular evolutionary genetics analysis across computing platforms. *Mol. Biol. Evol.* **2018**, *35*, 1547. [CrossRef]
40. Prakamhang, J.; Tittabutr, P.; Boonkerd, N.; Teamtisong, K.; Uchiumi, T.; Abe, M.; Teaumroong, N. Proposed some interactions at molecular level of PGPR coinoculated with *Bradyrhizobium diazoefficiens* USDA110 and *B. japonicum* THA6 on soybean symbiosis and its potential of field application. *Appl. Soil Ecol.* **2015**, *85*, 38–49. [CrossRef]
41. Yallou, C.; Aïhou, K.; Adjanohoun, A.; Toukourou, M.; Sanni, O.; Ali, D. Itinéraires techniques de production de maïs au Bénin. *Fiche Technique. Dépôt Légal* **2010**, *4992*, 18.
42. Ortiz, N.; Armada, E.; Duque, E.; Roldán, A.; Azcón, R. Contribution of *arbuscular mycorrhizal* fungi and/or bacteria to enhancing plant drought tolerance under natural soil conditions: Effectiveness of autochthonous or allochthonous strains. *J. Plant Physiol.* **2015**, *174*, 87–96. [CrossRef]
43. Zarik, L.; Meddich, A.; Hijri, M.; Hafidi, M.; Ouhammou, A.; Ouahmane, L.; Duponnois, R.; Boumezzough, A. Use of *Arbuscular mycorrhizal* fungi to improve the drought tolerance of *Cupressus atlantica* G. *Comptes Rendus Biol.* **2016**, *339*, 185–196. [CrossRef]
44. Bates, L.; Waldren, R.A.; Teare, I. Rapid determination of free proline for water-stress studies. *Plant Soil* **1973**, *39*, 205–207. [CrossRef]
45. Duncan, D.B. Multiple range and multiple F tests. *Biometrics* **1955**, *11*, 1–42. [CrossRef]

46. Ding, L.-J.; Cui, H.-L.; Nie, S.-A.; Long, X.-E.; Duan, G.-L.; Zhu, Y.-G. Microbiomes inhabiting rice roots and rhizosphere. *FEMS Microbiol. Ecol.* **2019**, *95*, fiz040. [CrossRef]
47. Mommer, L.; Kirkegaard, J.; van Ruijven, J. Root–root interactions: Towards a rhizosphere framework. *Trends Plant Sci.* **2016**, *21*, 209–217. [CrossRef]
48. Perez-Perez, R.; Oudot, M.; Hernandez, I.; Napoles, M.C.; Perez-Martinez, S.; Sosa-Del Castillo, D.; Postal, G.; San José de las Lajas, M. Isolation and characterization of *Stenotrophomonas* asociated to maize (*Zea mays* L.) rhizosphere. *Cultiv. Trop.* **2020**, *41*, e03.
49. Bulgarelli, D.; Schlaeppi, K.; Spaepen, S.; Van Themaat, E.V.L.; Schulze-Lefert, P. Structure and functions of the bacterial microbiota of plants. *Ann. Rev. Plant Biol.* **2013**, *64*, 807–838. [CrossRef]
50. Lugtenberg, B.; Kamilova, F. Plant-growth-promoting rhizobacteria. *Annu. Rev. Microbiol.* **2009**, *63*, 541–556. [CrossRef]
51. Davey, M.E.; O'toole, G.A. Microbial biofilms: From ecology to molecular genetics. *Microbiol. Mol. Biol. Rev.* **2000**, *64*, 847–867. [CrossRef]
52. Nawaz, M.S.; Arshad, A.; Rajput, L.; Fatima, K.; Ullah, S.; Ahmad, M.; Imran, A. Growth-stimulatory effect of quorum sensing signal molecule N-acyl-homoserine lactone-producing multi-trait *Aeromonas* spp. on wheat genotypes under salt stress. *Front. Microbiol.* **2020**, *11*, 2164. [CrossRef]
53. Noori, F.; Etesami, H.; Noori, S.; Forouzan, E.; Jouzani, G.S.; Malboobi, M.A. Whole genome sequence of *Pantoea agglomerans* ANP8, a salinity and drought stress–resistant bacterium isolated from alfalfa (*Medicago sativa* L.) root nodules. *Biotechnol. Rep.* **2021**, *29*, e00600. [CrossRef]
54. Filgueiras, L.; Silva, R.; Almeida, I.; Vidal, M.; Baldani, J.I.; Meneses, C.H.S.G. *Gluconacetobacter diazotrophicus* mitigates drought stress in *Oryza sativa* L. *Plant Soil* **2020**, *451*, 57–73. [CrossRef]
55. Annapurna, K.; Govindasamy, V.; Sharma, M.; Ghosh, A.; Chikara, S.K. Whole genome shotgun sequence of *Bacillus paralicheniformis* strain KMS 80, a rhizobacterial endophyte isolated from rice (*Oryza sativa* L.). *3 Biotech* **2018**, *8*, 223. [CrossRef]
56. Rani, R.; Kumar, V.; Usmani, Z.; Gupta, P.; Chandra, A. Influence of plant growth promoting rhizobacterial strains *Paenibacillus* sp. IITISM08, *Bacillus* sp. PRB77 and *Bacillus* sp. PRB101 using *Helianthus annuus* on degradation of endosulfan from contaminated soil. *Chemosphere* **2019**, *225*, 479–489. [CrossRef] [PubMed]
57. Hol, W.G.; Bezemer, T.M.; Biere, A. Getting the ecology into interactions between plants and the plant growth-promoting bacterium *Pseudomonas fluorescens*. *Front. Plant Sci.* **2013**, *4*, 81. [CrossRef]
58. Tiwari, S.; Lata, C.; Chauhan, P.S.; Nautiyal, C.S. *Pseudomonas putida* attunes morphophysiological, biochemical and molecular responses in *Cicer arietinum* L. during drought stress and recovery. *Plant Physiol. Biochem.* **2016**, *99*, 108–117. [CrossRef] [PubMed]
59. Arshad, M.; Saleem, M.; Hussain, S. Perspectives of bacterial ACC deaminase in phytoremediation. *Trends Biotechnol.* **2007**, *25*, 356–362. [CrossRef] [PubMed]
60. Sarkar, A.; Ghosh, P.K.; Pramanik, K.; Mitra, S.; Soren, T.; Pandey, S.; Mondal, M.H.; Maiti, T.K. A halotolerant Enterobacter sp. displaying ACC deaminase activity promotes rice seedling growth under salt stress. *Res. Microbiol.* **2018**, *169*, 20–32. [CrossRef]
61. Cruz, J.A.; Tubana, B.S.; Fultz, L.M.; Dalen, M.S.; Ham, J.H. Identification and profiling of silicate-solubilizing bacteria for plant growth-promoting traits and rhizosphere competence. *Rhizosphere* **2022**, *23*, 100566. [CrossRef]
62. Niu, X.; Song, L.; Xiao, Y.; Ge, W. Drought-tolerant plant growth-promoting rhizobacteria associated with foxtail millet in a semi-arid agroecosystem and their potential in alleviating drought stress. *Front. Microbiol.* **2018**, *8*, 2580. [CrossRef]
63. Bashan, Y.; Holguin, G. *Azospirillum*–plant relationships: Environmental and physiological advances (1990–1996). *Canad. J. Microbiol.* **1997**, *43*, 103–121. [CrossRef]
64. Zaveri, S.H.; Dasgupta, S. The Interaction between Plant and Endophytes for Conferring Tolerance to Biotic and Abiotic Stress in Presence of ACC (1-Aminocyclopropane-1-Carboxylate) Deaminase: A Stress Weapon. *Acta Sci. Biotechnol.* **2023**, *4*, 9–14.
65. Liu, Y.; Gao, J.; Bai, Z.; Wu, S.; Li, X.; Wang, N.; Du, X.; Fan, H.; Zhuang, G.; Bohu, T. Unraveling mechanisms and impact of microbial recruitment on oilseed rape (*Brassica napus* L.) and the rhizosphere mediated by plant growth-promoting rhizobacteria. *Microorganisms* **2021**, *9*, 161. [CrossRef] [PubMed]
66. Khanna, A.; Raj, K.; Kumar, P.; Wati, L. Antagonistic and growth-promoting potential of multifarious bacterial endophytes against Fusarium wilt of chickpea. *Egypt. J. Biol. Pest Contr.* **2022**, *32*, 17. [CrossRef]
67. Chaudhary, P.; Xu, M.; Ahamad, L.; Chaudhary, A.; Kumar, G.; Adeleke, B.S.; Verma, K.K.; Hu, D.-M.; Širić, I.; Kumar, P. Application of synthetic consortia for improvement of soil fertility, pollution remediation, and agricultural productivity: A review. *Agronomy* **2023**, *13*, 643. [CrossRef]
68. Abedinzadeh, M.; Etesami, H.; Alikhani, H.A. Characterization of rhizosphere and endophytic bacteria from roots of maize (*Zea mays* L.) plant irrigated with wastewater with biotechnological potential in agriculture. *Biotechnol. Rep.* **2019**, *21*, e00305. [CrossRef] [PubMed]
69. Fasusi, O.A.; Amoo, A.E.; Babalola, O.O. Characterization of plant growth-promoting rhizobacterial isolates associated with food plants in South Africa. *Antonie Van Leeuwenhoek* **2021**, *114*, 1683–1708. [CrossRef] [PubMed]
70. Raklami, A.; Bechtaoui, N.; Tahiri, A.-I.; Anli, M.; Meddich, A.; Oufdou, K. Use of rhizobacteria and mycorrhizae consortium in the open field as a strategy for improving crop nutrition, productivity and soil fertility. *Front. Microbiol.* **2019**, *10*, 1106. [CrossRef] [PubMed]
71. Nehra, M.; Dilbaghi, N.; Marrazza, G.; Kaushik, A.; Sonne, C.; Kim, K.-H.; Kumar, S. Emerging nanobiotechnology in agriculture for the management of pesticide residues. *J. Hazard. Mater.* **2021**, *401*, 123369. [CrossRef] [PubMed]

72. Yadav, A.N.; Kour, D.; Kaur, T.; Devi, R.; Yadav, A.; Dikilitas, M.; Abdel-Azeem, A.M.; Ahluwalia, A.S.; Saxena, A.K. Biodiversity, and biotechnological contribution of beneficial soil microbiomes for nutrient cycling, plant growth improvement and nutrient uptake. *Biocatal. Agric. Biotechnol.* **2021**, *33*, 102009. [CrossRef]
73. Asari, S.; Tarkowská, D.; Rolčík, J.; Novák, O.; Palmero, D.V.; Bejai, S.; Meijer, J. Analysis of plant growth-promoting properties of Bacillus amyloliquefaciens UCMB5113 using Arabidopsis thaliana as host plant. *Planta* **2017**, *245*, 15–30. [CrossRef]
74. Goswami, D.; Thakker, J.N.; Dhandhukia, P.C. Portraying mechanics of plant growth promoting rhizobacteria (PGPR): A review. *Cogent Food Agric.* **2016**, *2*, 1127500. [CrossRef]
75. Majeed, A.; Abbasi, M.K.; Hameed, S.; Imran, A.; Rahim, N. Isolation and characterization of plant growth-promoting rhizobacteria from wheat rhizosphere and their effect on plant growth promotion. *Front. Microbiol.* **2015**, *6*, 198. [CrossRef]
76. Kunwar, V.S.; Lamichhane, J.; Gauchan, D.P. Strategies to improve phosphorus availability in a sustainable agricultural system. *Int. J. Innov. Sci. Res. Technol.* **2018**, *3*, 323–331.
77. Albareda, M.; Rodríguez-Navarro, D.N.; Camacho, M.; Temprano, F.J. Alternatives to peat as a carrier for rhizobia inoculants: Solid and liquid formulations. *Soil Biol. Biochem.* **2008**, *40*, 2771–2779. [CrossRef]
78. Siddiq, S.; Saleem, U.; Ahmad, K.; Anayat, A.; Affan, Q.; Anwar, M.; Nazir, H.; Asghar, N. Comparison of conventional and non-conventional carriers for bacterial survival and plant growth. *Int. J. Agric. Innovations Res.* **2018**, *6*, 126–129.
79. Scott, S.; Housh, A.; Powell, G.; Anstaett, A.; Gerheart, A.; Benoit, M.; Wilder, S.; Schueller, M.; Ferrieri, R. Crop yield, ferritin and Fe (II) boosted by *Azospirillum brasilense* (HM053) in corn. *Agronomy* **2020**, *10*, 394. [CrossRef]
80. Vacheron, J.; Renoud, S.; Muller, D.; Babalola, O.O.; Prigent-Combaret, C. Alleviation of abiotic and biotic stresses in plants by Azospirillum. In *Handbook for Azospirillum: Technical Issues Protocols*; Springer: Berlin/Heidelberg, Germany, 2015; pp. 333–365.
81. Diarrassouba, N.; Dago, D.N.; Soro, S.; Fofana, I.J.; Silué, S.; Coulibaly, A. Multi-variant statistical analysis evaluating the impact of rhizobacteria (*Pseudomonas fluorescens*) on growth and yield parameters of two varieties of maize (*Zea mays.* L). *Int. J. Contemp. Appl. Sci.* **2015**, *2*, 206–224.
82. Noumavo, P.A.; Agbodjato, N.A.; Gachomo, E.W.; Salami, H.A.; Baba-Moussa, F.; Adjanohoun, A.; Kotchoni, S.O.; Baba-Moussa, L. Metabolic and biofungicidal properties of maize rhizobacteria for growth promotion and plant disease resistance. *Afr. J. Biotechnol.* **2015**, *14*, 811–819.
83. Amogou, O.; Dagbénonbakin, G.; Agbodjato, N.A.; Noumavo, P.A.; Salako, K.V.; Adoko, M.Y.; Kakaï, R.G.; Adjanohoun, A.; Baba-Moussa, L. Applying rhizobacteria on maize cultivation in Northern Benin: Effect on growth and yield. *Agric. Sci.* **2019**, *10*, 763–782. [CrossRef]
84. Mejri, D.; Gamalero, E.; Souissi, T. Formulation development of the deleterious rhizobacterium *Pseudomonas trivialis* X33d for biocontrol of brome (*Bromus diandrus*) in durum wheat. *J. Appl. Microbiol.* **2013**, *114*, 219–228. [CrossRef]
85. Agbodjato, N.A.; Noumavo, P.A.; Adjanohoun, A.; Agbessi, L.; Baba-Moussa, L. Synergistic effects of plant growth promoting rhizobacteria and chitosan on in vitro seeds germination, greenhouse growth, and nutrient uptake of maize (*Zea mays* L.). *Biotechnol. Res. Int.* **2016**, *2016*, 7830182. [CrossRef] [PubMed]
86. Janardan, Y.; Verma, J.P.; Tiwari, K.N. Effect of plant growth promoting rhizobacteria on seed germination and plant growth chickpea (*Cicer arietinum* L.) under in vitro conditions. *Biol. Forum* **2010**, *2*, 15–18.
87. Kumar, A.; Maurya, B.; Raghuwanshi, R. Isolation and characterization of PGPR and their effect on growth, yield and nutrient content in wheat (*Triticum aestivum* L.). *Biocatal. Agric. Biotechnol.* **2014**, *3*, 121–128. [CrossRef]
88. Hashem, A.; Tabassum, B.; Fathi Abd_Allah, E. *Bacillus subtilis*: A plant-growth promoting rhizobacterium that also impacts biotic stress. *Saudi J. Biol. Sci.* **2019**, *26*, 1291–1297. [CrossRef]
89. Vurukonda, S.S.K.P.; Vardharajula, S.; Shrivastava, M.; SkZ, A. Enhancement of drought stress tolerance in crops by plant growth promoting rhizobacteria. *Microbiol. Res.* **2016**, *184*, 13–24. [CrossRef]

plants

Article

Effect of Secondary Paper Sludge on Physiological Traits of *Lactuca sativa* L. under Heavy-Metal Stress

Marija Yurkevich, Arkadiy Kurbatov and Elena Ikkonen *

Institute of Biology of the Karelian Research Centre, Russian Academy of Sciences, 185910 Petrozavodsk, Russia; svirinka@mail.ru (M.Y.); arkadiy1416@gmail.com (A.K.)
* Correspondence: likkonen@gmail.com

Abstract: To eliminate the negative effect of soil contamination with heavy metals on plant growth and crop yield, different methods and techniques are the subject of discussion and study. In this study, we aimed to evaluate the effect of secondary pulp and paper-mill sludge application to soil on the response of the main physiological processes such as the growth, photosynthesis, and respiration of lettuce (*Lactuca sativa* L.) plants to soil contamination with Pb. For the pot experiment, Pb was added to sandy loam soil at concentrations of 0, 50, and 250 mg $Pb(NO_3)_2$ per kg of the soil, and secondary sludge was added to a 0, 20, or 40% sludge solution during each plant watering. The Pb-mediated change in plant biomass allocation, decrease in the photosynthetic rate, increase in leaf respiration rate, and the degree of light inhibition of respiration were closely associated with increases in both root and shoot Pb content. For the Pb-free soil condition, secondary sludge application contributed to the allocation of plant biomass towards a greater accumulation in the shoots than in the roots. Although stomatal opening was not affected by either Pb or sludge, sludge application increased photosynthetic CO_2 assimilation regardless of soil Pb content, which was associated with an increase in the electron-transport rate and carboxylase activity of Rubisco. Soil contamination with Pb significantly increased the ratio of respiration to photosynthesis, reflecting a shift in the carbon balance toward carbon losses in the leaves, but sludge application modified the coupling between the processes with a decrease in the proportion of respiratory carbon losses. The sludge-mediated recovery of the physiological processes of *L. sativa* reflected an increase in plant tolerance to soil contamination with heavy metals, the formation of which is associated with plant and soil adjustments initiated by secondary sludge application.

Keywords: Pb; pulp and paper sludges; soil properties; lettuce; photosynthesis; respiration

Citation: Yurkevich, M.; Kurbatov, A.; Ikkonen, E. Effect of Secondary Paper Sludge on Physiological Traits of *Lactuca sativa* L. under Heavy-Metal Stress. *Plants* **2024**, *13*, 1098. https://doi.org/10.3390/plants13081098

Academic Editor: Valeria Spagnuolo

Received: 14 March 2024
Revised: 7 April 2024
Accepted: 11 April 2024
Published: 14 April 2024

1. Introduction

Pulp and paper-mill sludges are generated from the wastewater treatments of the pulp and paper industry. For different plant species, soil types and climatic zones, the positive effect of pulp and paper sludge on soil properties and crop yield has been established [1,2]. Recent studies have shown that pulp and paper sludges cause an increase in the soil organic matter and nutrient content, soil microbial complex and its activity, and improve soil physical properties [3,4]. Plant growth and productivity are largely dependent on nutrients and water ability in soils, as well as on soil aeration, so improving the soil with pulp and paper sludges can have a positive effect on plants. In line with this assumption, the agricultural use of pulp and paper sludges has been found to have a positive effect on crop growth and yield [5,6]. However, the effect largely depends on the pulp and paper sludges' application rate and sludge type [5–7].

Pulp and paper-mill sludges are mainly classified as primary and secondary sludge with clear distinctions between the types in terms of chemical and physical properties [2]. The biological treatment of primary sludge with microorganisms in order to reduce the charge in dissolved organic substrates results in the formation of secondary sludge [8].

Secondary sludge has a higher organic content and a higher nitrogen and phosphorus content due to the addition of nutrients used to stimulate microbial activity during secondary treatment [9]. The beneficial uses and effects of pulp and paper-sludge application in agriculture on plant productivity have been studied mainly for primary and mixed primary and secondary sludges [2,7], but much less information is available about the use of secondary sludge in this regard. Moreover, the effect of paper sludges, including secondary, has been studied mainly on crop growth and yield [5,10], which are one of the main indicators in agricultural processes, without focusing on the plant physiological traits including fundamental processes, such as photosynthesis, respiration, and their coupling.

Unlike most domestic and industrial wastewater, only a low level of pollutants, including heavy metals, was found in the pulp and paper sludges [11]. Moreover, Lister and Line [12] showed that paper sludges can be effective biosorbents for Pb and Zn ions. Heavy metals can adsorb to certain natural substances and become immobile as a result, and soil organic matter is considered an important component controlling heavy-metal adsorption in soil [13,14]. Pulp and paper-mill sludges are rich in organic matter [9], with the organic-matter content ranges from 40% for primary sludge to 50% for secondary sludge [2]. So, it can be expected that paper sludges' application can increase the adsorption properties of soil, which in turn can positively affect plant growth under soil contamination with heavy metals. However, there has been no published evidence supporting this state.

High levels of heavy-metal accumulation in agricultural soils negatively affect plant development and crop yields contributing to threats to global food security [15]. It is also important that soil contamination with heavy metals is associated with potential risks to the environment and human health [16]. Lead (Pb), one of the most common soil pollutants, cannot be naturally removed from soil [17] and can therefore reach high levels in soils. In addition to natural sources such as igneous and sedimentary rocks, the long-term use of pesticides and fertilizers is believed to be the main source of Pb contamination in agricultural soils [18]. The irrigation of agricultural land with domestic and industrial wastewater also contributes to Pb accumulation in soils [19]. Increased soil Pb content leads to the transformation of the many physiological processes of plants with a decrease in seed germination rate, biomass accumulation and allocation, a disruption of gene expression, enzymatic activity, nutrient uptake and water balance, and an increase in the production of reactive oxygen species [15,20–22]. The Pb-mediated degradation of chlorophyll, as well as the decrease in the energy transformation efficiency of photosystem II, reduced electron transport, and the low activity of Calvin cycle enzymes, can be responsible for a decline in the photosynthetic CO_2 assimilation rate under the condition of soil contamination with Pb [23,24]. Plants may also respond to soil Pb by increasing the respiration rate [25] due to an increased demand of intermediates and energy under stress conditions [26]. The stress-mediated suppression of photosynthesis, accompanied by increased respiration, reflects a shift in the balance of plant carbon towards carbon losses, negatively affecting plant growth.

As was mentioned above, improving the chemical and physical properties of soils using pulp and paper sludges can stimulate plant growth and productivity. In addition, paper-sludge application to soil may reduce heavy metals' load on plants. To test this hypothesis, our study aimed to evaluate secondary-sludge effects on plant physiological responses using *Lactuca sativa* L. plants as the study object. This crop species is important to the human diet and is widely cultivated all over the world. As photosynthesis is the main driver for crop production, and a balance between photosynthesis and respiration is responsible for plant biomass accumulation, our study focused on these processes and associated traits of lettuce plants. We also assessed the effect of secondary sludge on soil chemical properties, which are closely related to soil fertility and plant productivity.

2. Results

2.1. Soil Properties

Increased soil Pb concentration was accompanied by a decrease in soil pH and an increase in the content of total N in the soil (Table 1). While secondary-sludge application significantly increased carbon content only at the highest soil Pb level used in this study, for P content, this tendency was recorded at all Pb rates examined. In accordance with the increase in the sludge-application rate, soil Pb concentrations decreased especially in highly contaminated soil. The sludge application decreased soil pH, but this effect was statistically significant only for Pb-free soil. No significant effects of both Pb and sludge on the content of available Ca and Mg were found in this study ($p > 0.05$).

Table 1. Chemical and physical properties of the soil substrate containing 0, 50, or 250 mg $Pb(NO_3)_2$ kg^{-1} and watered with the 0, 20, or 40% of secondary pulp and paper-mill sludge.

Parameter	0 mg $Pb(NO_3)_2$ kg^{-1}			50 mg $Pb(NO_3)_2$ kg^{-1}			250 mg $Pb(NO_3)_2$ kg^{-1}		
	0%	20%	40%	0%	20%	40%	0%	20%	40%
total C, %	4.3 ± 0.1 abc	4.4 ± 0.1 abc	4.7 ± 0.1 a	4.2 ± 0.2 b	4.5 ± 0.2 ab	4.4 ± 0.1 abc	4.0 ± 0.2 c	4.3 ± 0.1 abc	4.7 ± 0.2 a
total N, %	0.27 ± 0.00 c	0.29 ± 0.00 b	0.28 ± 0.00 b	0.29 ± 0.01 b	0.28 ± 0.00 b	0.28 ± 0.01 b	0.40 ± 0.00 a	0.41 ± 0.00 a	0.40 ± 0.00 a
available P, g kg^{-1}	0.91 ± 0.04 c	1.16 ± 0.08 a	1.26 ± 0.02 a	0.91 ± 0.06 c	1.00 ± 0.09 bc	1.14 ± 0.06 ab	0.88 ± 0.03 c	0.85 ± 0.02 c	0.95 ± 0.04 c
available Mg, mg kg^{-1}	196 ± 2 abc	196 ± 9 abc	198 ± 6 abc	212 ± 8 a	189 ± 10 c	190 ± 1 bc	193 ± 10 abc	210 ± 4 ab	202 ± 6 abc
available Ca, g kg^{-1}	1.95 ± 0.05 bc	2.07 ± 0.06 bc	2.22 ± 0.06 ab	1.93 ± 0.24 bc	1.86 ± 0.10 c	1.99 ± 0.06 bc	2.15 ± 0.08 abc	2.34 ± 0.03 a	2.39 ± 0.05 a
available Pb, mg kg^{-1}	0.1 ± 0.0 c	0.1 ± 0.0 c	0.1 ± 0.0 c	4.0 ± 0.5 c	2.8 ± 0.1 c	3.0 ± 0.5 c	41.6 ± 3.6 a	45.4 ± 2.3 a	36.6 ± 1.5 b
pH	5.40 ± 0.03 a	5.28 ± 0.04 b	5.26 ± 0.03 bc	5.25 ± 0.02 bc	5.19 ± 0.02 c	5.19 ± 0.01 c	5.29 ± 0.02 b	5.25 ± 0.01 bc	5.28 ± 0.03 b

Different letters indicate significant differences between the means.

2.2. Plant-Growth Parameters

According to the two-way ANOVA, the effect of soil contamination with Pb was statistically significant for the leaf number per plant and the root–shoot ratio, but not for biomass accumulation in roots or shoots. Secondary sludges, on the contrary, affected plant biomass and did not affect the number of leaves and biomass allocation ($p > 0.05$). However, sludges only increased the leaf number of the plants grown on Pb-free soil, with no effect on plants grown on Pb-rich soil (Figure 1). Sludge application tended to increase biomass accumulation in both shoots and roots regardless of the Pb content in the soil. The contamination of the soil with the heavy metal resulted in a reduced root–shoot ratio, indicating a more strong inhibitory effect of soil Pb on roots than shoots. Sludges decreased the value of the root–shoot ratio in the plants grown on the Pb-free soil, but no significant differences in these values were found between sludge-treated and untreated plants grown on the Pb-rich soil.

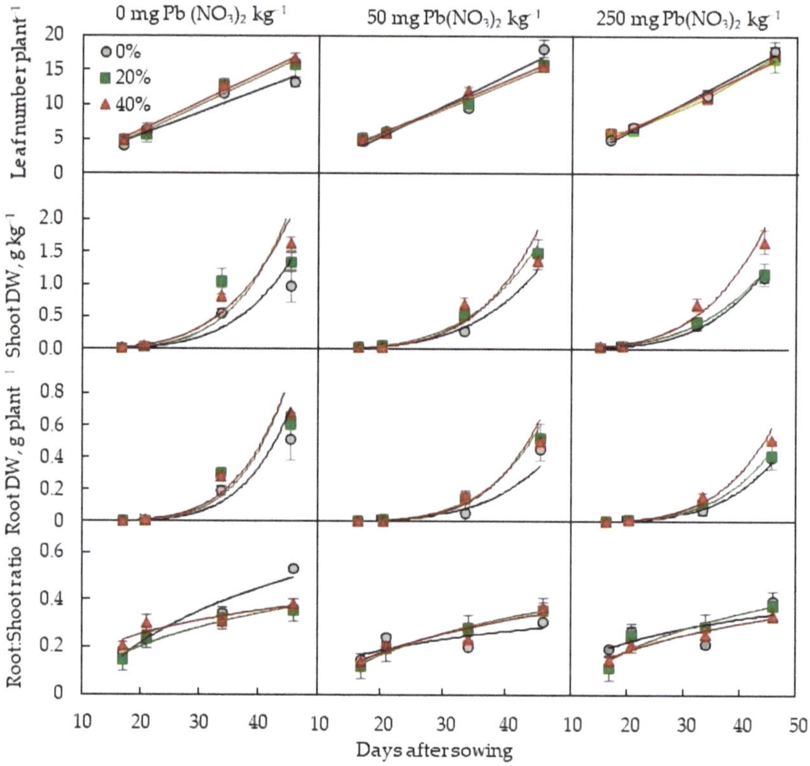

Figure 1. Leaf number per plant, shoot dry weight (DW), root dry weight, and root–shoot ratio of *L. sativa* grown on the soil containing 0, 50, or 250 mg $Pb(NO_3)_2$ kg^{-1} and watered with 0, 20, or 40% secondary pulp and paper-mill sludge.

2.3. Plant Physiological Parameters

2.3.1. Photosynthesis

Both soil Pb and secondary sludges significantly affected the gross photosynthetic CO_2 assimilation (A_g) rate of lettuce plants under study. In accordance with the increase in soil Pb content, the A_g rate and α values decreased, LCP increased, and C_i, J, v_c, v_o, F_0, F_v:F_m, and SPAD values were not affected by Pb regardless of its content in the soil (Table 2). Secondary-sludge application led to an increase in the A_g, J, v_c, and α, but the increases were statistically significant mainly for plants grown in the soil with 250 mg $Pb(NO_3)_2$ kg^{-1}. The highest sludge-modulated increase in the net CO_2 assimilation rate was observed for the plants grown under the condition of the highest Pb content in the soil (Figure 2). Among the studied Pb content, no statistical differences in C_i:C_a, v_o, F_o, F_v:F_m, or SPAD index were found between the sludge treatments.

Table 2. Physiological traits of *L. sativa* grown on the soil containing 0, 50, or 250 mg $Pb(NO_3)_2$ kg^{-1} and watered with 0, 20, or 40% secondary pulp and paper-mill sludge.

Parameter	0 mg $Pb(NO_3)^2$ kg^{-1}			50 mg $Pb(NO_3)^2$ kg^{-1}			250 mg $Pb(NO_3)^2$ kg^{-1}		
	0%	20%	40%	0%	20%	40%	0%	20%	40%
A_g, µmol m^{-2} s^{-1}	7.6 ± 0.5 abc	7.9 ± 0.6 ab	8.9 ± 0.6 a	6.6 ± 0.5 bc	7.6 ± 0.5 abc	7.5 ± 0.7 abc	5.9 ± 0.4 c	7.7 ± 0.5 abc	8.7 ± 0.4 a
Ci:Ca	0.59 ± 0.04 ab	0.55 ± 0.05 ab	0.59 ± 0.03 ab	0.61 ± 0.02 ab	0.56 ± 0.02 ab	0.53 ± 0.04 b	0.62 ± 0.07 ab	0.63 ± 0.03 a	0.60 ± 0.04 ab
J, µmol m^{-2} s^{-1}	48 ± 4 b	53 ± 3 ab	58 ± 3 a	46 ± 4 ab	53 ± 4 ab	48 ± 5 ab	43 ± 6 b	46 ± 3 b	55 ± 4 a
ν_c, µmol m^{-2} s^{-1}	8.9 ± 0.6 b	9.7 ± 0.7 ab	10.8 ± 0.6 a	8.7 ± 0.7 ab	9.7 ± 0.6 ab	9.1 ± 0.9 ab	8.1 ± 1.0 b	8.7 ± 0.8 ab	10.4 ± 0.7 a
ν_o, µmol m^{-2} s^{-1}	1.9 ± 0.3 a	1.7 ± 0.4 a	2.2 ± 0.4 a	2.1 ± 0.1 a	1.7 ± 0.1 a	2.1 ± 0.3 a	1.9 ± 0.3 a	2.2 ± 0.2 a	2.4 ± 0.2 a
α, µmol $µmol^{-1}$	0.018 ± 0.001 a	0.020 ± 0.002 a	0.018 ± 0.001 a	0.017 ± 0.001 ab	0.017 ± 0.001 a	0.018 ± 0.001 a	0.014 ± 0.002 b	0.017 ± 0.001 ab	0.018 ± 0.001 a
F_o	449 ± 8 a	427 ± 8 a	417 ± 20 a	449 ± 35 a	413 ± 9 a	427 ± 15 a	396 ± 22 a	419 ± 14 a	437 ± 23 a
F_v:F_m	0.844 ± 0.004 a	0.846 ± 0.004 a	0.834 ± 0.012 a	0.827 ± 0.014 a	0.839 ± 0.007 a	0.843 ± 0.009 a	0.842 ± 0.007 a	0.833 ± 0.009 a	0.826 ± 0.016 a
SPAD index	15.3 ± 0.6 ab	16.1 ± 0.9 a	14.2 ± 0.5 ab	14.1 ± 0.6 ab	14.5 ± 0.6 ab	15.8 ± 1.6 a	14.0 ± 1.1 ab	13.0 ± 0.5 b	13.0 ± 0.8 b
LCP, µmol m^{-2} s^{-1}	33 ± 5 cd	24 ± 6 d	42 ± 7 bcd	54 ± 6 b	43 ± 5 bcd	37 ± 4 bcd	78 ± 15 a	59 ± 6 b	52 ± 9 c
R_d, µmol m^{-2} s^{-1}	0.71 ± 0.07 bc	0.61 ± 0.13 c	0.87 ± 0.11 ab	0.99 ± 0.08 ab	0.90 ± 0.07 bc	0.74 ± 0.06 b	1.14 ± 0.18 a	1.07 ± 0.09 a	0.97 ± 0.14 ab
R_l, µmol m^{-2} s^{-1}	0.58 ± 0.08 bc	0.49 ± 0.15 c	0.74 ± 0.11 abc	0.86 ± 0.08 ab	0.73 ± 0.09 abc	0.64 ± 0.07 bc	1.00 ± 0.17 a	0.99 ± 0.09 a	0.93 ± 0.15 ab
R_l:R_d	0.81 ± 0.04 b	0.76 ± 0.09 b	0.83 ± 0.04 ab	0.87 ± 0.01 ab	0.87 ± 0.03 ab	0.86 ± 0.04 ab	0.87 ± 0.01 ab	0.93 ± 0.03 a	0.95 ± 0.05 a
1 − R_l:R_d, %	19 ± 4 a	24 ± 9 a	17 ± 4 ab	13 ± 1 ab	13 ± 1 ab	16 ± 3 ab	13 ± 1 ab	7 ± 3 a	5 ± 2 a
Tr, mmol m^{-2} s^{-1}	1.1 ± 0.1 a	1.0 ± 0.2 a	1.2 ± 0.1 a	0.9 ± 0.1 a	1.1 ± 0.1 a	0.9 ± 0.1 a	0.9 ± 0.1 a	1.1 ± 0.1 a	1.2 ± 0.1 a
g_s, mmol m^{-2} s^{-1}	81 ± 9 a	73 ± 11 a	94 ± 9 a	69 ± 7 a	78 ± 8 a	69 ± 12 a	70 ± 10 a	82 ± 6 a	89 ± 7 a
RWC, %	71 ± 2 a	69 ± 3 a	70 ± 1 a	69 ± 2 a	73 ± 2 a	72 ± 1 a	76 ± 1 a	70 ± 2 a	70 ± 2 a
PWUE, µmol $mmol^{-1}$	6.8 ± 0.7 a	7.3 ± 0.7 a	6.8 ± 0.2 a	6.4 ± 0.6 a	6.6 ± 0.3 a	7.2 ± 0.7 a	6.4 ± 0.9 a	6.4 ± 0.7 a	6.6 ± 0.7 a

Different letters indicate significant differences between the means.

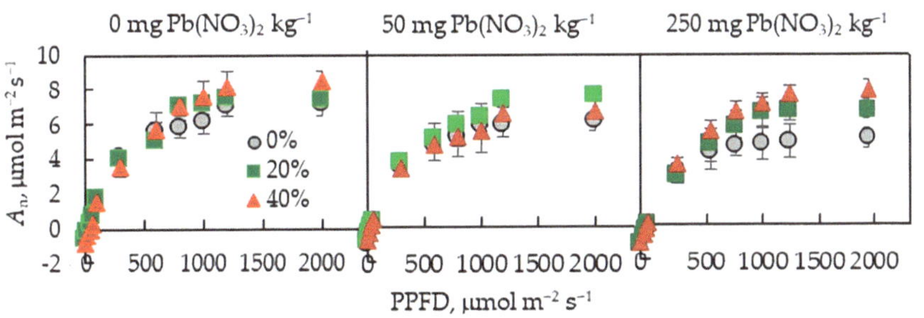

Figure 2. Light curve of net CO_2 assimilation rate of *L. sativa* grown in the soil containing 0, 50, or 250 mg $Pb(NO_3)_2$ kg^{-1} and watered with 0 (grey), 20 (green), or 40 (red)% secondary pulp and paper-mill sludge.

2.3.2. Leaf Respiration

According to the two-way ANOVA test, leaf respiration was significantly affected by Pb but not sludge application, regardless of whether respiration occurred in light or darkness. The R_d and R_l rates were significantly higher in plants grown on Pb-containing soil than on Pb-free soil (Table 2). For all treatments, the degree the of light inhibition of leaf respiration (1 − R_l:R_d) varied between 5 and 24%, with an average value of 14%.

2.3.3. Respiration-to-Photosynthesis Ratio

The two-way ANOVA revealed a significant effect of both Pb content and sludge dose on the ratio of respiration to photosynthesis regardless of whether respiration occurred in light or darkness ($p > 0.05$). The R_d:A_g and R_l:A_g ratios increased dramatically in accordance with the increase in Pb content in the soil (Figure 3), reflecting the shift of the plant carbon balance toward carbon losses. For the plants grown on Pb-free soil, and for both R_d and R_l, no significant differences in the ratio were found between the sludge treatments. However, sludge application resulted in a decrease in the ratios of plants grown on the soil with both Pb levels tested, with a recovery of carbon balances to the levels close to plants grown on the Pb-free soil.

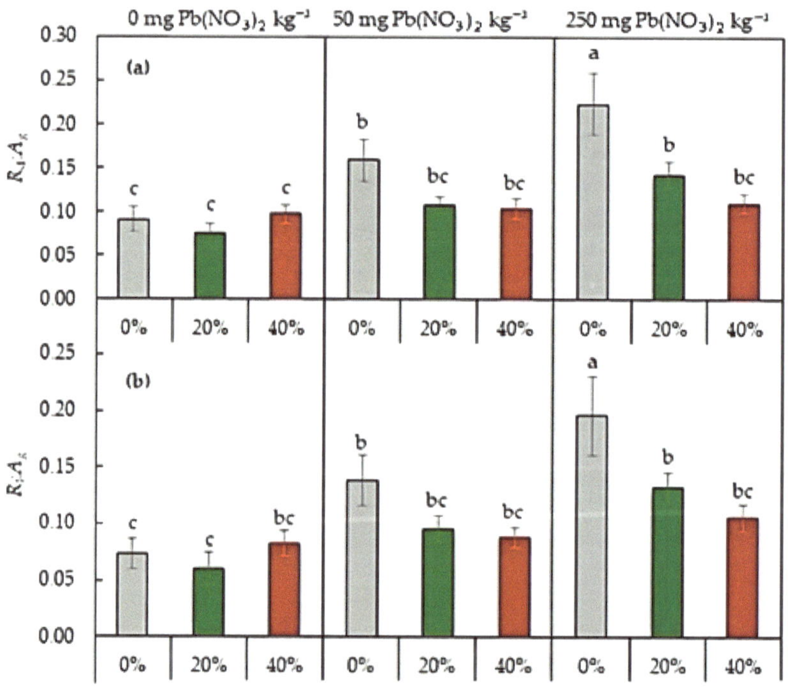

Figure 3. Ratios of leaf respiration in darkness (R_d, **a**) or in the light (R_l, **b**) to gross CO_2 assimilation rate (A_g) of *L. sativa* grown in the soil containing 0, 50, or 250 mg $Pb(NO_3)_2$ kg^{-1} and watered with 0 (grey), 20 (green), or 40 (red)% secondary pulp and paper-mill sludge. Different letters indicate significant differences between the means.

2.3.4. Leaf H_2O Exchange

No statistically significant effect of either Pb or sludge on Tr and g_s, or for RWC and PWUE values, was found, regardless of soil Pb content or sludge rate. Although the Tr and g_s values tended to decrease with an increase in the soil Pb level, and tended to increase with increasing sludge dose, these changes were not supported by statistical tests (Table 2).

2.4. Pb Content in Shoots and Roots

Soil contamination significantly affected the concentration of Pb in both the roots and shoots of lettuce plants. As expected, the Pb content in plant organs increased in accordance with the increase in soil Pb among all sludge treatments (Figure 4). No significant differences in both Pb contents were found between the sludge treatments under the conditions of Pb-free soil. However, secondary-sludge application led to a decrease in Pb content in the roots under the highest soil Pb level used in this study, and in the shoots under the rate of 50 mg $Pb(NO_3)_2$ per kg of the soil.

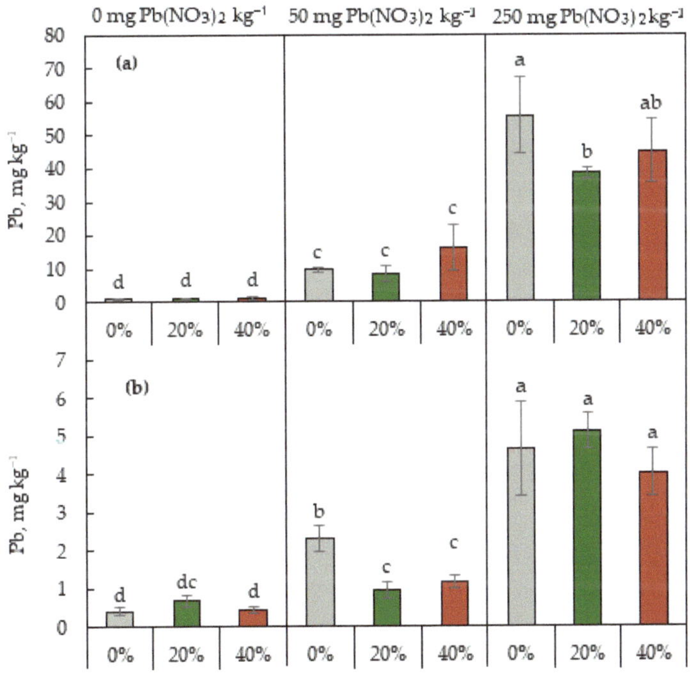

Figure 4. Pb content in roots (**a**) and shoots (**b**) of *L. sativa* plants grown on the soil containing 0, 50, or 250 mg $Pb(NO_3)_2$ kg^{-1} and watered with 0 (grey), 20 (green), or 40 (red)% secondary pulp and paper-mill sludge. Different letters indicate significant differences between the means.

2.5. Leaf Mass per Area and Area-Based Leaf Nutrient Content

The ANOVA test revealed no significant effect of soil Pb on the LMA, LNC, or LPC values, but the effect of sludge application was statistically significant for LMA and LNC. Under Pb-free soil and the soil containing 250 mg $Pb(NO_3)_2$ kg^{-1}, LMA increased with increasing sludge dosage (Figure 5). While both Pb and sludge had no effect on LPC, sludge application increased LNP in the soil without Pb and in the soil with 250 mg $Pb(NO_3)_2$ kg^{-1}. For A_n, a positive significant correlation was found with LNC, but not with LPC (Figure 6).

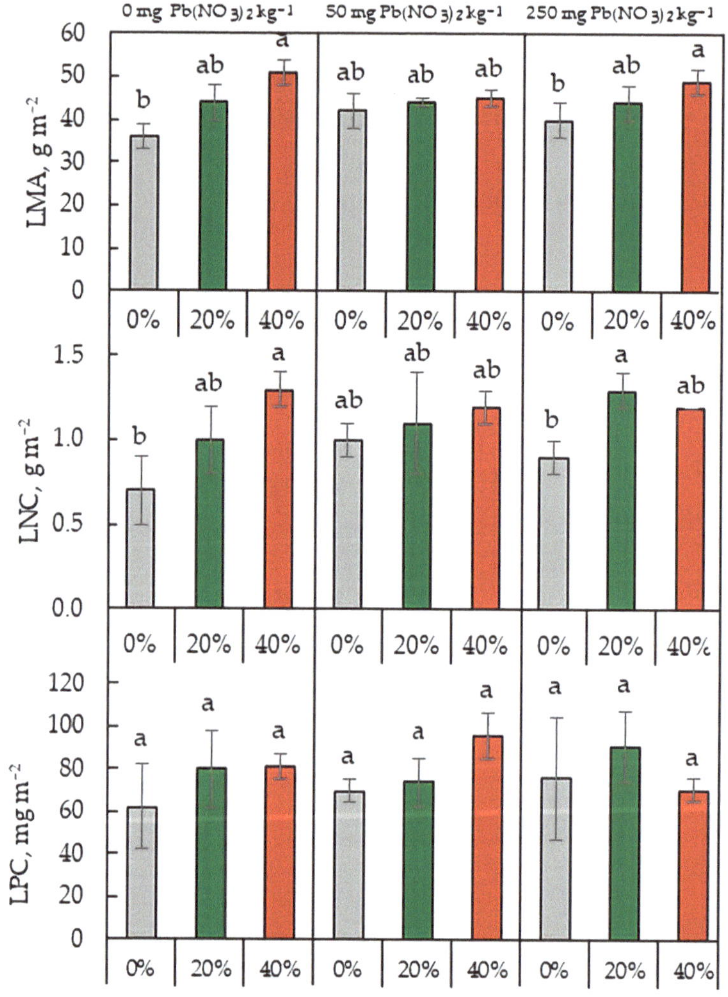

Figure 5. Leaf mass per area (LMA), area-based leaf N content (LNC), and area-based leaf P content (LPC) in *L. sativa* plants grown on the soil containing 0, 50, or 250 mg $Pb(NO_3)_2$ kg^{-1} and watered with 0 (grey), 20 (green), or 40 (red)% secondary pulp and paper-mill sludge. Different letters indicate significant differences between the means.

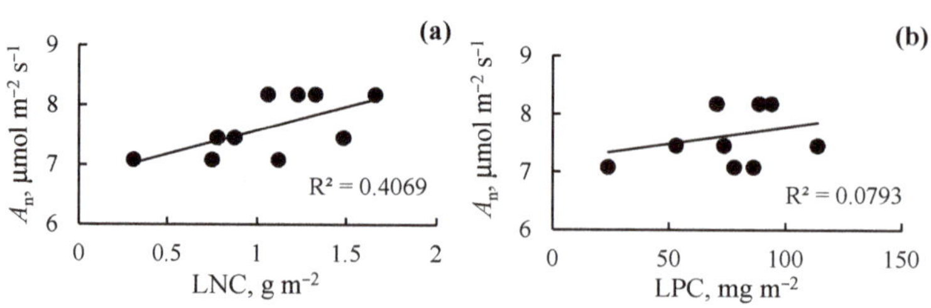

Figure 6. Relationship between net CO_2 assimilation rate (A_n), area-based leaf N content (LNC, **a**), and area-based leaf P content (LPC, **b**) in *L. sativa*.

2.6. Photosynthetic Nutrient-Use Efficiency

Among all treatments, the highest PNUE and PPUE values were found in plants grown on Pb-free soil without sludge applications (Figure 7). Soil contamination with Pb caused a statistically significant decrease in PNUE value. For plants grown on Pb-rich soil, secondary-sludge application had no effect on either PNUE or PPUE.

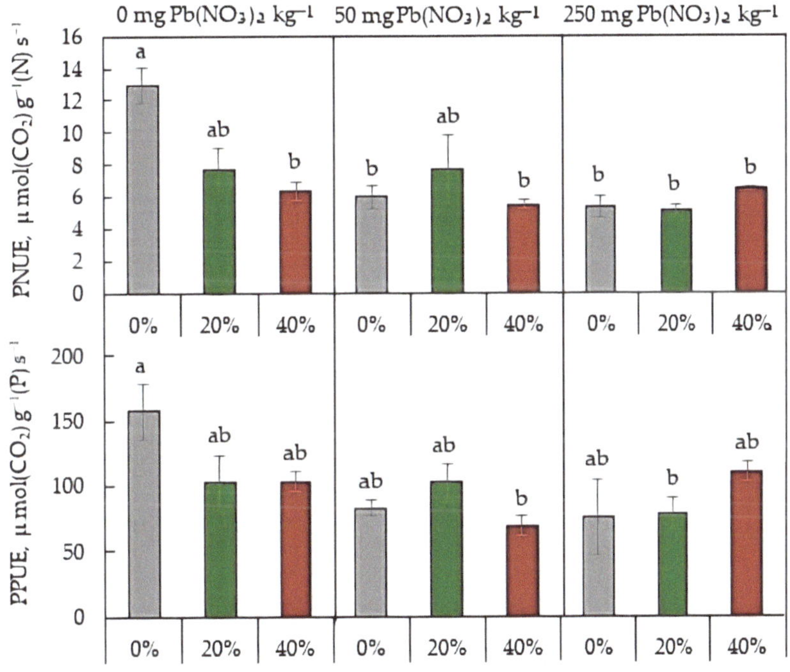

Figure 7. Photosynthetic nitrogen-use efficiency (PNUE) and photosynthetic phosphorus-use efficiency (PPUE) in *L. sativa* grown in soil containing 0, 50, or 250 mg $Pb(NO_3)_2$ kg^{-1} and watered with 0 (grey), 20 (green), or 40 (red)% secondary pulp and paper-mill sludge. Different letters indicate significant differences between the means.

3. Discussion

Soil contamination with Pb can affect the availability of soil macronutrients to plants, as found in this study (Table 1). High concentrations of soil Pb were accompanied by an increase in the content of total N and available K in the soil, as well as a decrease in the level of available P. Phosphorus is involved in primary metabolic processes such as photosynthesis and respiration and plays a key role in complex energy transformations as the main structural component of ATP (adenosine triphosphate) [27]. A decreased availability of soil P may be one of the reasons for the inhibition of photosynthesis in the studied plants when exposed to high concentrations of Pb in the soil. The negative effect of Pb on the photosynthetic apparatus is well studied [28–30], and our study confirms the conclusions. Pb accumulation in the lettuce organs, along with a decrease in the photosynthetic rate, caused an increase in respiratory costs (Figures 2 and 4, Table 3), which is consistent with early studies [24] and appears to be associated with an increase in energy and intermediate demand under the stress. Both Pb-mediated inhibition in the CO_2 assimilation rate and increased respiration rate caused a shift in the plant carbon balance towards carbon losses (Figure 3), resulting in a change in plant biomass allocation between roots and shoots (Figure 1).

Table 3. Chemical and physical properties of secondary sludge under the study.

C, %	pH	N, mg L^{-1}	P, mg L^{-1}	K, mg L^{-1}	Na, mg L^{-1}	Ca, mg L^{-1}	Mg, mg L^{-1}
56.7	7.16	238	64	9	47	233	22

Secondary paper-sludge application affected macronutrients content in the soil with a more pronounced impact on C and P concentration (Tables 1 and 2). The sludge-mediated increase in soil P may be due to the introduction of phosphorus into the biological treatment processes of paper sludges, which results in an increase in the phosphorus content of the secondary sludge [2]. A positive effect of secondary pulp and paper sludge on the content of some macronutrients, including phosphorus, was also shown by Nunes et al. [5] for Mediterranean agricultural soils. Increased soil P may be partly responsible for the recovering of photosynthetic activity and stabilization of the respiration-to-photosynthesis ratio of lettuce plants grown in the Pb-rich soil. Multiple regulatory mechanisms determine plant photosynthesis and its tolerance to stress. For the lettuce plants exposed to high Pb concentrations in the soil, our study identified electron-transport and carboxylation rates as modifying traits that improve the rate of CO_2 assimilation under the secondary-sludge application. The sludge-mediated increase in A_n was accompanied by an increase in LMA, indicating structural changes in leaf tissue that may be responsible for enhanced photosynthesis. The A_n rate was also positively associated with area-based leaf N content (Figure 6), supporting earlier studies [31,32]. The positive correlation occurs because most of the N is integrated into photosynthetic proteins, especially Rubisco [30]. The net photosynthetic rate per unit N in a leaf, termed photosynthetic nitrogen-use efficiency, has been considered an important functional trait in characterizing leaf physiology [33]. The PNUE values were drastically declined by both soil Pb and sludge application (Figure 7). Numerous studies have shown a negative correlation between PNUE and LMA [30,33], and our results are in agreement with this finding (Figures 5 and 7). The higher LMA in plants grown under sludge application can relate to a higher fraction of leaf dry mass and cell walls. Greater N partitioning into cell-wall proteins against lesser N partitioning into Rubisco proteins can be responsible for the negative relation between LMA and PNUE [32]. Given the negative relationship between A_n and PNUE and the positive relationship between A_n and LMA, the sludge-mediated increase in the net CO_2 assimilation rate may be due to structural changes in leaf tissue as well as carboxylation capacity, but not to an increase in photosynthetic N-use efficiency.

The coupling between the photosynthetic and respiration processes is an important indicator of the carbon balance of plants and is useful in assessing the ability of plants to adapt to stress conditions [34]. The stabilized ratio of respiration to photosynthesis under the sludge application, whether respiration occurs in light or darkness, may be responsible for improving lettuce growth on the Pb-contaminated soil. The leaf respiration in the light is considered a more stable process than respiration in darkness [35]. Our results revealed high variability in both R_l and R_d rates under zero and low soil Pb content, but under a high Pb level, the effect of sludge-amendment rate was greater for R_d than R_l. This caused a change in the $1 - R_l{:}R_d$ value, which reflects the extent of the light inhibition of respiration, with a decrease in inhibition following an increase in the sludge-amendment rate.

Not only the chemical but also the physical properties of soils can be transformed when introducing secondary paper sludge into soils [4]. Soil texture and structure play an important role in soil water availability for plants and nutrient uptake [36] and therefore successful plant growth and productivity. Secondary-sludge application when watering plants can lead to the formation of stable organo-mineral particles with an increase in the proportion of large drainage pores and saturated water content, providing increased water supply to plants [4,37]. The secondary sludge decreased the effective thermal conductivity of the soil in an air-dried state but increased the heat transfer in wet soil, which may be associated with the sludge-mediated increase in the soil water content [4]. Improving the physical properties of the soil using secondary sludge, as well as the chemical properties, could have a positive effect on the physiological traits of lettuce plants under the study.

The lettuce plants grown in the Pb-rich soil displayed elevated Pb contents in both the roots and leaves (Figure 4) with, as expected, a higher metal level in leaves than in roots. Secondary-sludge application provided a reduction in Pb levels in roots under the highest soil Pb content for this study. Experimental evidence supports the role of soil organic matter in reducing the mobility and phytoavailability of heavy metals including Pb, by the formation of organo-mineral complexes [38,39]. Moreover, Pb mobility can be reduced by phosphate application into soil [39,40]. Since the secondary sludge contains organic matter and phosphorus, its application to the soil could contribute to Pb immobilization reducing its uptake by plants, thereby helping to reduce heavy-metal stress on the plants.

4. Materials and Methods

4.1. Plant Materials Preparation

The sandy loam soil used for this pot experiment was collected randomly from the topsoil layer in the northwest region of Russia (61.826573, 33.179712). The collected soil is characterized by a low natural fertility, low water holding capacity, and low pH (5.46) and humus content (0.5–2.5%), and had 0.39% of total N and 1.6 g kg^{-1} of available P and K [41]. The available Pb content of collected soil was 0.13 ± 0.01 mg kg^{-1}. After air drying, the soil was sieved with a 2 mm sieve, then fertilized with 150 mg kg^{-1} of N, P, and K before being mixed with $Pb(NO_3)_2$ at a rate of 0, 50, or 250 mg kg^{-1}. The soil substrates were well watered and incubated under 70–80% of the maximum soil water-holding capacity and air temperature of 21 ± 1 °C for two weeks, and then placed in 0.8 L pots (D = 15 cm) with a soil bulk density of about 1.4 g cm^{-3}. Each Pb treatment included 24 pots that were divided into three parts and served as units for secondary-sludge treatments.

Six seeds of lettuce (*Lactuca sativa* L. var. Medvezhje ushko) were sown per pot. All pots were subjected to controlled conditions of 23/20 °C day/night temperature, a 16 h photoperiod, and 250 µmol m^{-2} s^{-1} of photosynthetic photon flux density (PPFD). The pots were randomly repositioned daily to avoid plant-location effects within the location area. The pots were watered through the soil surface every two days following a design for secondary-sludge treatment. The first part of the pots of each Pb treatment was watered with distilled water (0% treatment), and the second and third parts were watered with 20% and 40% solutions of the secondary pulp and paper-mill sludges (20% and 40% treatment, accordingly). The sludge concentrations were chosen in a preliminary experiment. In this way, the completely randomized experimental design included three levels of Pb content in the soil (0, 50, or 250 mg $Pb(NO_3)_2$ kg^{-1}) and three sludge treatments for each Pb treatment. Each unit treatment included eight pots. The design and execution of the experiment followed the strategies recently reviewed [42]. During the 46-day experiment, 0, 139, and 278 mL of the secondary sludge was added to 1 L of soil, accordingly, for 0%, 20%, and 40% treatments. The sludge used in this study was generated after the secondary aeration of wastewater from a pulp-and-paper mill located in the northwestern region of Russia. The chemical and physical characteristics of secondary sludge used in the study are presented in Table 3. To achieve a sludge content of 20 and 40%, it was mixed with distilled water. Two weeks after sowing, the seedlings were thinned to three per pot.

4.2. Plant-Growth Parameters

For each treatment, four 17-, 21-, 34-, and 46-day-old lettuce plants were harvested and separated into shoots and roots. The roots were carefully separated from the soil and washed. The total leaf number per plant was counted. The collected shoots and roots were dried under a temperature of 70 °C to a constant weight and weighed. The ratio of root dry biomass to shoot dry biomass (root–shoot) was calculated. The total leaf area per plant was determined by leaf scanning and using the program "AreaS". The leaf mass per area (LMA) was calculated as dry leaf mass divided by leaf area.

4.3. Gas-Exchange Parameters

The leaf area-based CO_2 and H_2O gas-exchange parameters were measured on mature *L. sativa* leaves of 43–46-day plants using a portable photosynthesis system (HCM-1000, Walz, Effeltrich, Germany). The leaves were randomly selected from plants from different pots of each treatment. During each measurement, the temperature and air humidity in the leaf chamber were maintained at 20 °C and 60–70%, respectively. The net CO_2 assimilation rate (A_n), stomatal conductance (g_s), transpiration water losses (Tr), and intercellular (C_i) and ambient (C_a) CO_2 concentration were determined at 1200 μmol m^{-2} s^{-1} of PPFD. In addition, to determine the light-response curves of photosynthesis, the A_n rate was measured at 300, 2000, 1200, 1000, 800, 300, 60, 40, 20, and 0 μmol m^{-2} s^{-1} of PPFD. Readings were taken after a steady-state rate of A_n was reached. The apparent quantum yield of photosynthesis (α) was calculated as the slope of A_n versus irradiances of 20, 40 and 60 μmol PPFD m^{-2} s^{-1}, according to Garmash and Golovko [43]. The photosynthetic electron-transport rate (J), carboxylase activity of Rubisco (v_c), and oxygenase activity of Rubisco (v_o) were calculated according to Farquhar and von Caemmerer [44]. The leaf respiration rate in darkness (R_d) was taken after 30 min of zero irradiance. The rate of leaf respiration in the light (R_l) was calculated according the Kok method [45]. The gross photosynthesis rate ($A_g = A_n + R_l$) with A_n at 1200 μmol m^{-2} s^{-1} of PPFD, and the ratios R_d to A_g (R_d:A_g), R_l to A_g (R_l:A_g), and R_l to R_d (R_l:R_d), were calculated. The light-compensation point (LCP), which is the light level at which the rate of photosynthesis is equal to the rate of respiration, was determined using the light-response curve of photosynthesis. The photosynthetic water-use efficiency (PWUE) value at the leaf level was defined as the A_n-to-Tr ratio.

4.4. Chlorophyll Fluorescence and Content; Relative Water Content of Leaves

Chlorophyll-fluorescence parameters and chlorophyll content were measured on the same leaves as the gas-exchange parameters. The minimum and maximum fluorescence (F_o and F_m, respectively) and the maximum photochemical quantum efficiency of PSII (F_v:F_m) of leaves were measured using MINI-PAM (Walz, Effeltrich, Germany). The leaves were previously dark-adapted for 30 min. The F_v:F_m values were calculated as ($F_m - F_o$):F_m. The chlorophyll content was measured using a SPAD-502 chlorophyll meter (Minolta, Tokyo, Japan).

For measuring relative water content (RWC), leaf discs were collected and the fresh weight (FW) of the disks was determined. After that, the disks were placed in distilled water at 4 °C for 24 h for rehydration. The leaf turgid weight (TW) was determined after the incubation period, and leaf disks were dried at 70 °C before the constant weight. The dry weight (DW) of leaf disks was measured, and the RWC value was calculated as (FW − DW)/(TW − DW) × 100%, according to González and González-Vilar [46].

4.5. Chemical Analyses

Samples of soil and plant organs (shoots and roots) from each treatment were collected after the completion of the measurements of plant physiological parameters. The soils from five pots of each treatment were collected and mixed, and then three subsamples were used for chemical analyses. All plants of each treatment were collected from all pots, and plant organs were mixed and divided into three repetitions. The analyses were carried out in the Core Facility "Analytical laboratory" of the Forest institute of KRC of RAS. For C and N analysis, the soil samples were sieved with a less-than-0.05 mm sieve. The pH of each soil sample was measured in 1 M L^{-1} KCl solution (pHKCl) at a ratio of 1:2.5 (w:w). Total soil carbon was analyzed with a total organic C analyzer (TOC-L CPN, Shimadzu, Japan). The leaf samples of 0.2–0.3 g were preliminarily homogenized and digested with HNO_3 and HCl (Vekton, Saint Petersburg, Russia). The total N content in the samples was determined by the Kjeldahl method [47]. The available P in soil samples was extracted with 0.2 M HCl. Soil Ca and Mg were extracted with 1MNH_4Ac buffered to pH 7. The mass-based concentrations of Pb in the plant tissue and soil were determined by spectrophotometric atomic absorption (Shimadzu AA-7000, Kyoto, Japan). The N content in plant samples was determined by the nesslerization of the ammonia, and the P content was determined by the ammonium molybdate method. Spectrophotometric and flame photometric methods were used to measure mass-based N and P, accordingly (SF 2000 OKB Spectrum, Saint Petersburg, Russia). Area-based leaf N concentration (LNC) and P concentration (LPC) were calculated using the values of LMA and the mass-based element contents. Photosynthetic nitrogen-use efficiency (PNUE) and photosynthetic phosphorus-use efficiency (PPUE) were calculated by dividing A_n by LNC and LPC, accordingly.

4.6. Statistical Analysis

For each treatment, the means ± SE were determined with at least five replicates for the physiological parameters of plants and three replicates for chemical analyses. To evaluate the significant differences between the means, the LSD test at a $p < 0.05$ level was used (Statistica software, v. 8.0.550.0, StatSoft, Inc., Tulsa, OK, USA).

5. Conclusions

In summary, this study demonstrated the benefits of using secondary paper sludge for *L. sativa* in optimizing plant carbon balance, especially under soil contamination with Pb. The sludge application contributed to the recovery of the photosynthetic activity of plants grown in Pb-rich soil to the level of their counterparts grown in Pb-free soil with the stabilization of coupling between photosynthesis and respiration. The sludge-mediated recovery of photosynthetic capacity was mainly attributed to structural adjustments and enhanced carboxylation capacity, but not to an increase in photosynthetic nutrient-use efficiency. The positive effect of secondary sludge on plants could presumably be achieved both through improving the chemical and physical properties of the soil as well as through the immobilization of Pb in the soil.

Author Contributions: Conceptualization, E.I. and M.Y.; methodology, E.I. and A.K.; validation, E.I.; formal analysis, A.K.; investigation, E.I.; resources, M.Y.; writing—original draft preparation, E.I.; writing—review and editing, M.Y.; project administration, M.Y.; funding acquisition, M.Y. All authors have read and agreed to the published version of the manuscript.

Funding: This research was funded by the Ministry of Science and Higher Education of the Russian Federation, grant number FMEN-2022-0012.

Data Availability Statement: The original contributions presented in the study are included in the article, further inquiries can be directed to the corresponding author.

Acknowledgments: Experimental facilities for this study were offered by the Core Facility of the Karelian Research Centre of the Russian Academy of Sciences.

Conflicts of Interest: The authors declare no conflicts of interest. The funders had no role in the design of the study; in the collection, analyses, or interpretation of data; in the writing of the manuscript; or in the decision to publish the results.

References

1. Kinnula, S.; Toivonen, M.; Soinne, H.; Joona, J.; Kivelä, J. Effects of mixed pulp mill sludges on crop yields and quality. *Agric. Food Sci.* **2020**, *29*, 276–286. [CrossRef]
2. Turner, T.; Wheeler, R.; Oliver, I.W. Evaluating land application of pulp and paper mill sludge: A review. *J. Environ. Manag.* **2022**, *317*, 115439. [CrossRef] [PubMed]
3. Camberato, J.J.; Gagnon, B.; Angers, D.A.; Chantigny, M.H.; Pan, W.L. Pulp and paper mill by-products as soil amendments and plant nutrient sources. *Can. J. Soil Sci.* **2006**, *86*, 641–653. [CrossRef]
4. Butylkina, M.; Ikkonen, E. Physical Properties of Retisol under Secondary Pulp and Paper Sludge Application. *Land* **2023**, *12*, 2022. [CrossRef]
5. Nunes, J.R.; Cabral, F.; López-Piñeiro, A. Short-term effects on soil properties and wheat production from secondary paper sludge application on two Mediterranean agricultural soils. *Bioresour. Technol.* **2008**, *99*, 4935–4942. [CrossRef] [PubMed]

6. Fahim, S.; Nisar, N.; Ahmad, Z.; Asghar, Z.; Said, A.; Atif, S.; Ghani, N.; Qureshi, N.; Soomro, G.A.; Iqbal, M.; et al. Managing Paper and Pulp Industry By-Product Waste Utilizing Sludge as a Bio-Fertilizer. *Pol. J. Environ. Stud.* **2019**, *28*, 83–90. [CrossRef] [PubMed]
7. Faubert, P.; Barnabé, S.; Bouchard, S.; Côté, R.; Villeneuve, C. Pulp and paper mill sludge management practices: What are the challenges to assess the impacts on greenhouse gas emissions? *Resour. Conserv. Recycl.* **2016**, *108*, 107–133. [CrossRef]
8. Mahmood, T.; Elliott, A. A review of secondary sludge reduction technologies for the pulp and paper industry. *Water Res.* **2006**, *40*, 2093–2112. [CrossRef]
9. Bajpai, P. *Management of Pulp and Paper Mill Waste*, 1st ed.; Springer: Cham, Switzerland, 2015; p. 197.
10. Curnoe, W.E.; Irving, D.C.; Dow, C.B.; Velema, G.; Unc, A. Effect of spring application of a paper mill soil conditioner on corn yield. *Agron. J.* **2006**, *98*, 423–429. [CrossRef]
11. Kuokkanen, T.; Nurmesniemi, H.; Pöykiö, R.; Kujala, K.; Kaakinen, Y.; Kuokkanen, M. Chemical and leaching properties of paper mill sludge. *Chem. Spec. Bioavailab.* **2008**, *20*, 111–122. [CrossRef]
12. Lister, S.K.; Line, M.A. Potential utilization of sewage sludge and paper mill waste for biosorption of metals from polluted waterways. *Bioresour. Technol.* **2021**, *79*, 35–39. [CrossRef] [PubMed]
13. Kwiatkowska-Malina, J. Functions of organic matter in polluted soils: The effect of organic amendments on phytoavailability of heavy metals. *Appl. Soil Ecol.* **2018**, *123*, 542–545. [CrossRef]
14. Campillo-Cora, C.; Conde-Cid, M.; Arias-Estévez, M.; Fernández-Calviño, D.; Alonso-Vega, F. Specific Adsorption of Heavy Metals in Soils: Individual and Competitive Experiments. *Agronomy* **2020**, *10*, 1113. [CrossRef]
15. Angulo-Bejarano, P.I.; Puente-Rivera, J.; Cruz-Ortega, R. Metal and Metalloid Toxicity in Plants: An Overview on Molecular Aspects. *Plants* **2021**, *10*, 635. [CrossRef] [PubMed]
16. Kumar, B.; Smita, K.; Flores, L.C. Plant mediated detoxification of mercury and lead. *Arab. J. Chem.* **2017**, *10*, 2335–2342. [CrossRef]
17. Xu, Y.; Deng, G.; Guo, H.; Yang, M.; Yang, Q. Accumulation and sub cellular distribution of lead (Pb) in industrial hemp grown in Pb contaminated soil. *Ind. Crops Prod.* **2021**, *161*, 113220. [CrossRef]
18. Alengebawy, A.; Abdelkhalek, S.T.; Qureshi, S.R.; Wang, M.-Q. Heavy Metals and Pesticides Toxicity in Agricultural Soil and Plants: Ecological Risks and Human Health Implications. *Toxics* **2021**, *9*, 42. [CrossRef]
19. Alloway, B.J. Sources of Heavy Metals and Metalloids in Soils. In *Heavy Metals in Soils. Trace Metals and Metalloids in Soils and their Bioavailability*; Alloway, B.J., Ed.; Springer: Dordrecht, The Netherlands, 2013; pp. 11–50.
20. Sharma, P.; Dubey, R. Lead toxicity in plants. *Braz. J. Plant Physiol.* **2005**, *17*, 35–52. [CrossRef]
21. Kaur, G.; Kaur, S.; Singh, H.P.; Batish, D.R.; Kohli, R.K.; Rishi, V. Biochemical Adaptations in *Zea mays* Roots to Short-Term Pb2+ Exposure: ROS Generation and Metabolism. *Bull. Environ. Contam. Toxicol.* **2015**, *95*, 246–253. [CrossRef]
22. Kohli, S.K.; Handa, N.; Bali, S.; Khanna, K.; Arora, S.; Sharma, A.; Bhardwaj, R. Current Scenario of Pb Toxicity in Plants: Unraveling Plethora of Physiological Responses. *Residue Rev.* **2019**, *249*, 153–197.
23. Madhu, P.M.; Sadagopan, R.S. Effect of Heavy Metals on Growth and Development of Cultivated Plants with Reference to Cadmium, Chromium and Lead—A Review. *J. Stress Physiol. Biochem.* **2020**, *16*, 84–102.
24. Shu, X.; Yin, L.Y.; Zhang, Q.F.; Wang, W.B. Effect of Pb toxicity on leaf growth, antioxidant enzyme activities, and photosynthesis in cuttings and seedlings of *Jatropha curcas* L. *Environ. Sci. Pollut. Res.* **2012**, *19*, 893–902. [CrossRef] [PubMed]
25. Romanowska, E.; Igamberdiev, A.U.; Parys, E.; Gardeström, P. Stimulation of respiration by Pb^{2+} in detached leaves and mitochondria of C_3 and C_4 plants. *Physiol. Plant.* **2002**, *116*, 148–154. [CrossRef]
26. Semikhatova, O.A. Evaluation of plant adaptation potential by assessing dark respiration. *Russ. J. Plant Physiol.* **1998**, *45*, 122–128.
27. Marschner, H. *Marschner's Mineral Nutrition of Higher Plants*, 3rd ed.; Academic Press: Amsterdam, The Netherlands, 2012.
28. Kumar, A.; Prasad, M.N.V. Plant-lead interactions: Transport, toxicity, tolerance, and detoxification mechanisms. *Ecotoxicol. Environ. Saf.* **2018**, *166*, 401–418. [CrossRef]
29. Collin, S.; Baskar, A.; Geevarghese, D.M.; Ali, M.N.V.S.; Bahubali, P.; Choudhary, R.; Lvov, V.; Tovar, G.I.; Senatov, F.; Koppala, S.; et al. Bioaccumulation of lead (Pb) and its effects in plants: A review. *J. Hazard. Mater.* **2022**, *3*, 100064. [CrossRef]
30. Ikkonen, E.; Kaznina, N. Physiological Responses of Lettuce (*Lactuca sativa* L.) to Soil Contamination with Pb. *Horticulturae* **2022**, *8*, 951. [CrossRef]
31. Evans, J.R. Photosynthesis and nitrogen relationship in leaves of C_3 plants. *Oecologia* **1989**, *78*, 9–19. [CrossRef]
32. Poorter, H.; Evans, J.R. Photosynthetic nitrogen-use efficiency of species that differ inherently in specific leaf area. *Oecologia* **1998**, *116*, 26–37. [CrossRef]
33. Hikosaka, K. Interspecific difference in the photosynthesis-nitrogen relationship: Patterns, physiological causes, and ecological importance. *J. Plant Res.* **2004**, *117*, 481–494. [CrossRef]
34. Yamori, W.; Noguchi, K.; Hikosaka, K.; Terashima, I. Cold tolerant crop species have greater temperature homeostasis of leaf respiration and photosynthesis than cold-sensitive species. *Plant Cell Physiol.* **2009**, *50*, 203–215. [CrossRef] [PubMed]
35. Atkin, O.K.; Scheurwater, I.; Pons, T.L. High thermal acclimation potential of both photosynthesis and respiration in two lowland *Plantago* species in contrast to an alpine congeneric. *Glob. Chang. Biol.* **2006**, *12*, 500–515. [CrossRef]
36. Passioura, J.B. Soil structure and plant growth. *Soil Res.* **1991**, *29*, 717–728. [CrossRef]
37. Zibilske, L.; Clapham, W.; Rourke, R. Multiple applications of paper mill sludge in an agricultural system: Soil effects. *J. Environ. Qual.* **2000**, *29*, 1975–1981. [CrossRef]

38. Hamid, Y.; Tang, L.; Yaseen, M.; Hussain, B.; Zehra, A.; Aziz, M.Z.; He, Z.; Yang, X. Comparative efficacy of organic and inorganic amendments for cadmium and lead immobilization in contaminated soil under rice-wheat cropping system. *Chemosphere* **2019**, *214*, 259–268. [CrossRef] [PubMed]
39. Popescu, S.M.; Zheljazkov, V.D.; Astatkie, T.; Burducea, M.; Termeer, W.C. Immobilization of Pb in Contaminated Soils with the Combination Use of Diammonium Phosphate with Organic and Inorganic Amendments. *Horticulturae* **2023**, *9*, 278. [CrossRef]
40. Cao, X.D.; Ammar, L.W.; Bing, M.; Yongliang, L.Y. Immobilization of Zn, Cu, and Pb in contaminated soils using phosphate rock and phosphoric acid. *J. Hazard. Mater.* **2009**, *164*, 555–564. [CrossRef] [PubMed]
41. Ikkonen, E.; Chazhengina, S.; Jurkevich, M. Photosynthetic Nutrient and Water Use Efficiency of *Cucumis sativus* under Contrasting Soil Nutrient and Lignosulfonate Levels. *Plants* **2021**, *10*, 340. [CrossRef] [PubMed]
42. Brien, C.J.; Sermarini, R.A.; Borges Demétrio, C. Exposing the confounding in experimental designs to understand and evaluate them, and formulating linear mixed models for analyzing the data from a designed experiment. *Biom. J.* **2023**, *65*, 2200284. [CrossRef]
43. Garmash, E.V.; Golovko, T.K. CO_2 gas exchange and growth in *Rhaponticum carthamoides* under the condition of middle taiga subzone of Northeastern Europe: Dependence of photosynthesis and respiration on environmental factors. *Russ. J. Plant Physiol.* **1997**, *44*, 737–745.
44. Farquhar, G.D.; von Caemmerer, S. Modelling of photosynthetic response to environmental conditions. In *Encyclopedia of Plant Physiology. V. 12B. Physiological Plant Ecology II. Water Relations and Carbon Assimilation*; Lange, O.L., Nobel, P.S., Osmond, C.B., Ziegler, H., Eds.; Springer: Berlin/Heidelberg, Germany, 1982; Volume 12, pp. 549–587.
45. Kok, B. A critical consideration of the quantum yield of Chlorella-photosynthesis. *Enzymologia* **1948**, *13*, 1–5.
46. González, L.; González-Vilar, M. Determination of relative water content. In *Handbook of Plant Ecophysiology Techniques*; Reigosa Roger, M.J., Ed.; Kluwer Academic Publishers: Amsterdam, The Netherlands, 2001; pp. 207–212.
47. Sáez-Plaza, P.; Michałowski, T.; Navas, M.J.; Asuero, A.G.; Wybraniec, S. An overview of the Kjeldahl method of nitrogen determination. Part I. Early history, chemistry of the procedure, and titrimetric finish. *Crit. Rev. Anal. Chem.* **2013**, *43*, 178–223. [CrossRef]

plants

MDPI

Article

The Physiological Response of Apricot Flowers to Low-Temperature Stress

Jingjing Gao †, Wenbo Guo †, Qingwei Liu, Meige Liu, Chen Shang, Yuqin Song, Ruijie Hao, Liulin Li * and Xinxin Feng *

College of Horticulture, Shanxi Agricultural University, Taigu, Jinzhong 030801, China; gaojj@sxau.edu.cn (J.G.); gwbo0903@163.com (W.G.); 15582894301@163.com (Q.L.); liumeige265@126.com (M.L.); triciashang@163.com (C.S.); songyuqin_yyjc@163.com (Y.S.); hrj000@126.com (R.H.)

* Correspondence: sxaulll@sxau.edu.cn (L.L.); fengxx@sxau.edu.cn (X.F.)

† These authors contributed equally to this work.

Abstract: The growth and development of apricot flower organs are severely impacted by spring frosts. To better understand this process, apricot flowers were exposed to temperatures ranging from 0 °C to −8 °C, including a control at 18 °C, in artificial incubators to mimic diverse low-temperature environments. We aimed to examine their physiological reactions to cold stress, with an emphasis on changes in phenotype, membrane stability, osmotic substance levels, and antioxidant enzyme performance. Results reveal that cold stress induces significant browning and cellular damage, with a sharp increase in browning rate and membrane permeability below −5 °C. Soluble sugars and proteins initially rise as osmoprotectants, but their content decreases at lower temperatures. Proline content consistently increases, suggesting a protective role. Antioxidant enzyme activities, including catalase (CAT), peroxidase (POD), superoxide dismutase (SOD), and ascorbate peroxidase (APX), exhibit a complex pattern, with initial increases followed by declines at more severe cold conditions. Correlation and principal component analyses highlight the interplay between these responses, indicating a multifaceted adaptation strategy. The findings contribute to the understanding of apricot cold tolerance and inform breeding efforts for improved crop resilience.

Keywords: apricot flowers; low-temperature stress; antioxidant enzymes; osmotic adjustment; cellular damage

Citation: Gao, J.; Guo, W.; Liu, Q.; Liu, M.; Shang, C.; Song, Y.; Hao, R.; Li, L.; Feng, X. The Physiological Response of Apricot Flowers to Low-Temperature Stress. *Plants* **2024**, *13*, 1002. https://doi.org/10.3390/plants13071002

Academic Editor: Daniela Trono

Received: 5 March 2024
Revised: 28 March 2024
Accepted: 28 March 2024
Published: 31 March 2024

1. Introduction

The apricot flowers, a symbol of spring in China, are particularly vulnerable to low-temperature frost damage, especially in regions like Shanxi [1]. In recent years, sudden temperature drops during the blooming season have frequently occurred, hindering pollination and causing significant production losses [2]. To mitigate this threat, farmers have implemented strategies including improved orchard management and advanced weather tracking [3]. However, due to the uncertainty of climate change, late frost damage remains a significant issue in apricot production. Therefore, further research is needed to enhance the cold tolerance of apricot trees and ensure sustainable production.

Many studies have been dedicated to elucidating the mechanisms by which apricot flowers endure and respond to cold stress, as well as the genetic diversity in frost tolerance among various cultivars. In a study, Pakkish and Tabatabaienia [4] explored the application and mechanism of nitric oxide (NO) in preventing frost damage to apricot flowers. They demonstrated that NO treatment significantly reduced freezing injury, lipid peroxidation, and ion leakage, suggesting a role for NO in enhancing the antioxidant capacity of apricot flowers and altering solute content to improve cold tolerance. Kaya and Kose [5] utilized differential thermal analysis (DTA) to determine the cell death points (CDPs) in different apricot flower organs at subzero temperatures. Their findings revealed distinct sensitivities among various organs, with the receptacle being the most sensitive and the pistil the most tolerant, providing valuable insights into the structural vulnerabilities of apricot flowers to frost. Extending the research to a broader temporal scale, Kaya et al. [6] investigated the exothermic processes associated with late spring frost injury in apricot flower buds. Their work highlighted the changes in cold tolerance as the buds developed, indicating a dynamic response to environmental conditions. In a long-term study spanning 15 years, Szalay et al. [7] assessed the cold hardiness of apricot flower buds and flowers during the blooming period. They found that both genotype and year were significant factors affecting frost tolerance, with the later stages of flowering showing a greater influence of genotype. This research underscores the importance of considering both genetic and environmental factors in breeding programs aimed at improving frost resistance in apricot.

While previous studies have delved into the physiological reactions of apricot flowers to cold stress [8,9], a systematic and comprehensive understanding of the plant's adaptive strategies remains elusive. There remains a gap in knowledge regarding the integrated response of apricot flowers to the full spectrum of low-temperature conditions, from mild to severe frost. A more thorough investigation is required to elucidate the complex interplay between various physiological responses and to identify the key factors that determine the success or failure of cold tolerance in apricot flowers.

This study aims to address this knowledge gap by conducting a comprehensive analysis of the physiological responses of apricot flowers to a gradient of low temperatures. The research will focus on several critical aspects, including changes in the phenotypic appearance and anatomical structure of the flowers, alterations in membrane permeability and lipid peroxidation, fluctuations in the content of osmoregulatory substances, and the enzymatic activities of the antioxidant system. By examining these elements in a systematic manner, we seek to uncover the underlying mechanisms that enable apricot flowers to withstand cold stress and to identify potential targets for genetic improvement or management practices aimed at enhancing cold tolerance.

2. Results

2.1. Changes in the Phenotype and Anatomical Structure of Apricot Flowers

The impact of low-temperature stress on the browning process of apricot flowers is depicted in Figure 1A. When subjected to temperatures ranging from 18 °C to −3 °C, the flowers exhibit minimal browning. However, as the temperature further descends to −4 °C to −6 °C, a noticeable increase in browning is observed. At temperatures between −7 °C and −8 °C, the flowers are almost entirely browned. Following exposure to frost, the apricot flowers initially take on a yellow-brown or black hue, with the scales becoming limp and loose. After a period of recovery at room temperature, they dry out and fall off. The petal tissue hardens upon initial freezing, and as the temperature warms, the petals gradually shift from white to yellow-brown before withering and dropping off. The pistils and stamens also undergo browning and drying, eventually curling up. At lower temperatures, the ovary may gradually wilt and shrivel due to freezing, leading to eventual detachment.

The influence of low-temperature stress on the anatomic structure of apricot flowers is illustrated in Figure 1B. Paraffin section observations of apricot flowers that have been subjected to low-temperature stress show that under mild low-temperature conditions (0 °C to −4 °C), there are no significant differences in the anatomical structure of the ovary compared to the control at 18 °C. However, at a treatment temperature of −5 °C, minor fissures or cavities start to form in the ovules and the ovary wall. As the temperature drops further (−7 °C to −8 °C), these fissures and cavities gradually increase, indicating a progressive worsening of damage to the apricot flowers.

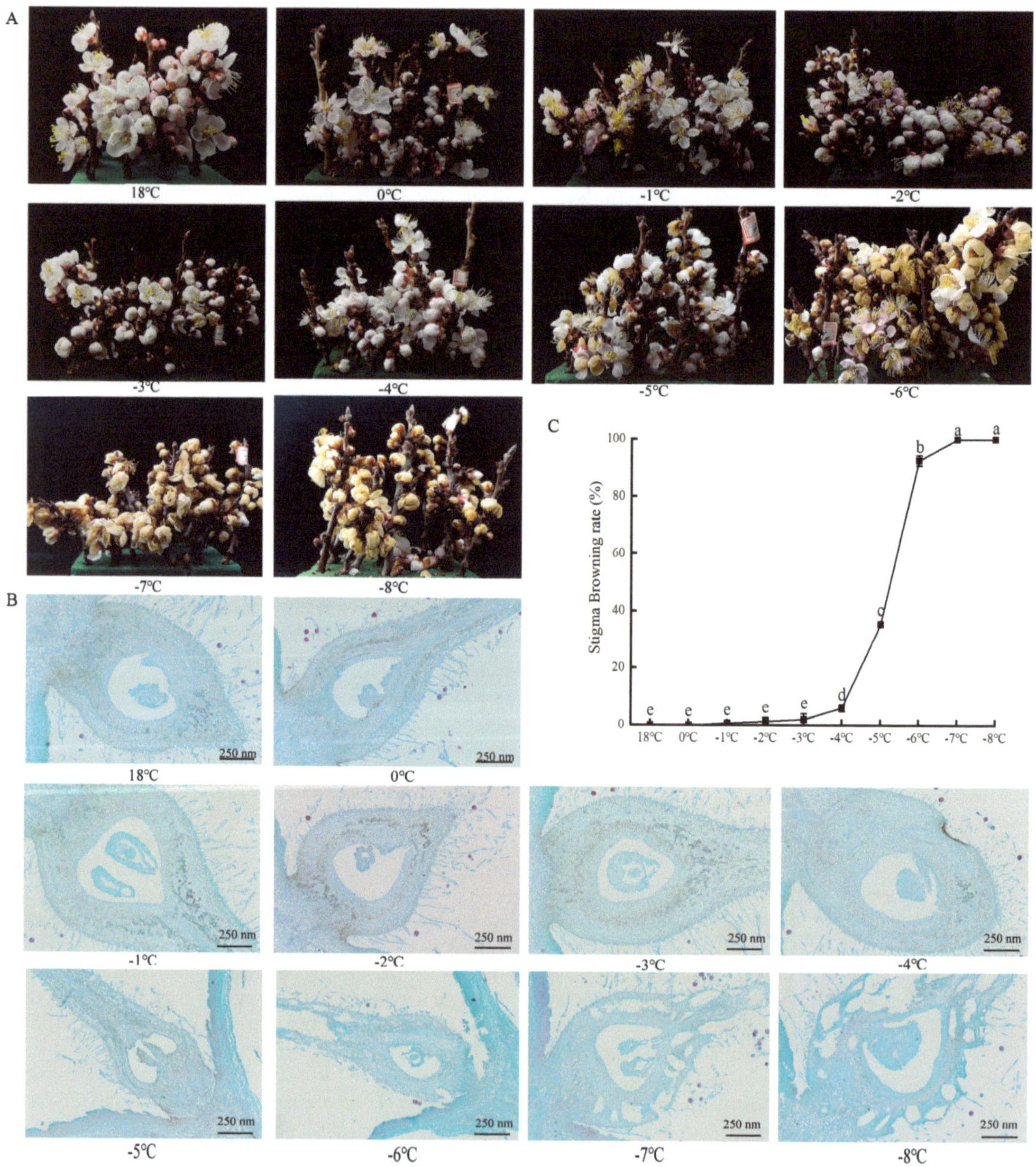

Figure 1. Browning phenomena in apricot flowers (**A**), anatomic structure of apricot ovaries (**B**), and browning rates analysis of apricot flowers (**C**) under low-temperature stress. Each data type has three sets of replicates, and superscripts of data in the different temperature treatments with the same lowercase letters indicate nonsignificant differences ($p > 0.05$).

Simultaneously, we assessed the browning rate of apricot flowers across a range of temperature conditions (Figure 1C). The rate remained relatively stable from 18 °C down to −3 °C, with no significant alterations observed. Yet, a marked escalation occurred as the temperature further descended; between −4 °C and −6 °C, the browning rate climbed significantly at a rate of 44.7%. This trend culminated in a complete browning of the apricot blossoms when subjected to temperatures between −7 °C and −8 °C, reaching a 100% browning rate.

2.2. Changes in Membrane Permeability and Membrane Lipid Peroxidation

As shown in Figure 2, the relative electrical conductivity (REC) of apricot blossoms follows an "S"-shaped trajectory in response to the progressive drop in temperature. Within the range of 18 °C to −3 °C, there is a negligible rise in the relative conductivity. However, a marked increase is observed, with a 69.8% surge, as the temperature plummets from −4 °C to −6 °C, after which it plateaus at approximately 90%. In parallel, the accumulation of the metabolic byproduct hydrogen peroxide (H_2O_2) commences in earnest when the blossoms encounter low-temperature stress at 0 °C, escalating at a rate of 66%. Simultaneously, the levels of malondialdehyde (MDA), a marker of membrane lipid peroxidation, rise by 26.3% from 0 °C to −6 °C. This escalation is further compounded between −7 °C and −8 °C, with an accelerated rate of increase, reaching 71.2%. Consequently, the notable uptick in relative conductivity, H_2O_2, and MDA content with decreasing temperatures underscores the escalating vulnerability of apricot blossom cells to cold-induced damage.

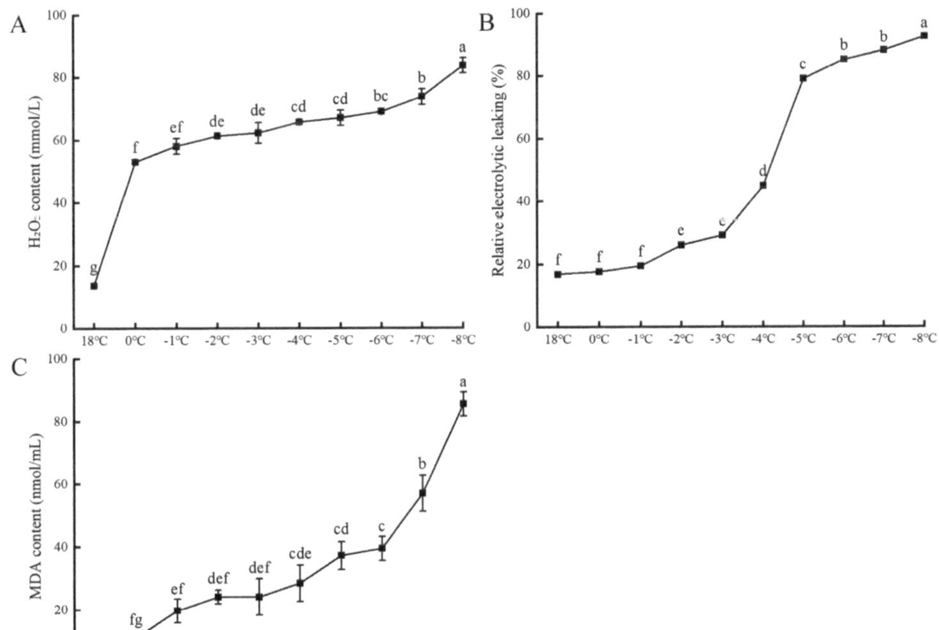

Figure 2. The effects of low-temperature stress on the hydrogen peroxide content (**A**), relative electrical conductivity (**B**), and malondialdehyde content (**C**) of apricot flowers. Each data type has three sets of replicates, and superscripts of data in the different temperature treatments with the same lowercase letters indicate nonsignificant differences ($p > 0.05$).

The semi-lethal temperature (LT_{50}), a critical threshold for plant survival, is determined by fitting a logistic regression model to REC data, which serves as a predictive indicator of a plant's cold tolerance. In the context of this research, the semi-lethal low temperature for apricot blossoms in full bloom has been pinpointed at −5.2 °C, delineating the boundary of their frost resistance.

2.3. Changes in the Content of Osmoregulatory Substances

Figure 3 illustrates that as the temperature progressively declines, the levels of soluble sugars (SSs) and soluble proteins (SPs) in apricot flowers exhibit an initial ascent followed by a descent. These components increase at rates of 34.3% and 96%, respectively, peaking at −5 °C with concentrations of 132.4 g/L and 56.9 mg/g. Beyond −5 °C, the continued temperature drop may impair the cells' normal metabolic synthesis, resulting in a slight but nondramatic reduction in these contents. Conversely, the proline (Pro) content maintains an upward trajectory, escalating at a rate of 7.1% from 18 °C to −6 °C. This growth rate intensifies between −7 °C and −8 °C, with a 24.5% increase, potentially linked to disruptions in protein synthesis. Consequently, under low-temperature stress, apricot blossoms initially boost their SS and SP levels, which then peak and decline as temperatures fall further. In contrast, Pro content steadily ascends throughout the temperature range.

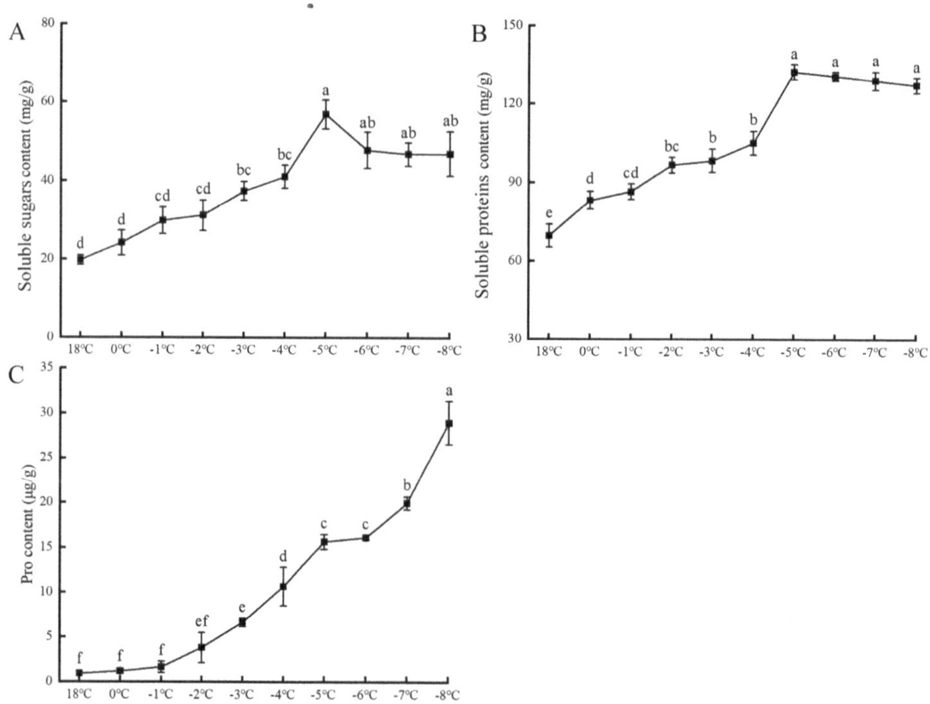

Figure 3. The effects of low-temperature stress on the soluble sugar content (**A**), soluble protein content (**B**), and proline content (**C**) of apricot flowers. Each data type has three sets of replicates, and superscripts of data in the different temperature treatments with the same lowercase letters indicate nonsignificant differences ($p > 0.05$).

2.4. Changes in the Enzymatic Activity of the Antioxidant System

Figure 4 illustrates the impact of low-temperature stress on the activities of the antioxidant enzymes catalase (CAT), peroxidase (POD), superoxide dismutase (SOD), and ascorbate peroxidase (APX) in apricot flowers. As the temperature progressively drops, these enzymes' activities undergo a dual response, initially spiking and then subsequently declining. CAT and POD activities climb at rates of 0.5% and 24%, respectively, from 18 °C down to −4 °C, where they reach peak values of 0.82 U·mgprot^{-1} and 71.9 U·g^{-1} FW, respectively. Beyond this point, their activities taper off at rates of 0.6% and 60.3%, respectively. In a similar vein, SOD and APX activities surge at rates of 250.6% and 0.5%, respectively, from 18 °C to −5 °C, achieving their maximum activities of 329 U·g^{-1} FW and 0.82 U·g^{-1} FW, respectively, at −5 °C. Following this peak, their activities diminish at rates of 284.7% and 0.5%, respectively. This suggests that CAT and POD may have distinct thresholds for low-temperature tolerance, peaking at −4 °C, while SOD and APX reach their optimal activity at −5 °C. Thus, within a specific low-temperature window, the antioxidant enzyme activities in apricot flowers initially ascend before exhibiting a decline.

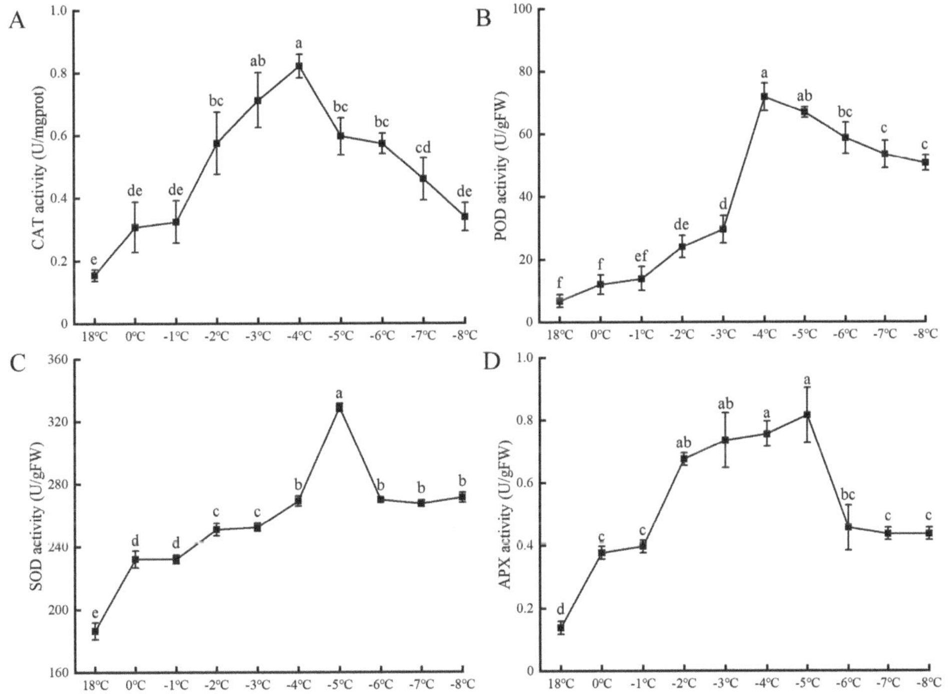

Figure 4. The effects of low-temperature stress on the catalase activity (**A**), peroxidase activity (**B**), superoxide dismutase activity (**C**), and ascorbate peroxidase activity (**D**) of apricot flowers. Each data type has three sets of replicates, and superscripts of data in the different temperature treatments with the same lowercase letters indicate nonsignificant differences ($p > 0.05$).

2.5. Correlation Analysis

Under low-temperature stress, the physiological indices of apricot flowers exhibit significant correlations that vary across different temperature ranges. As depicted in Figure 5A, within the temperature range of 18 °C to −4 °C, a strong positive correlation ($p < 0.01$) is observed between the osmotic adjustment substances in apricot flowers and the activity of the antioxidant enzyme system. The membrane permeability indicator REC is significantly positively correlated ($p < 0.05$) with the metabolic product H_2O_2 and shows an extremely strong positive correlation ($p < 0.01$) with other indicators, including the antioxidant enzyme POD and the membrane lipid peroxidation product MDA. Moreover, H_2O_2 is significantly positively correlated ($p < 0.05$) with the osmotic adjustment substance Pro, while MDA is extremely significantly positively correlated ($p < 0.01$) with all other indicators except Pro content.

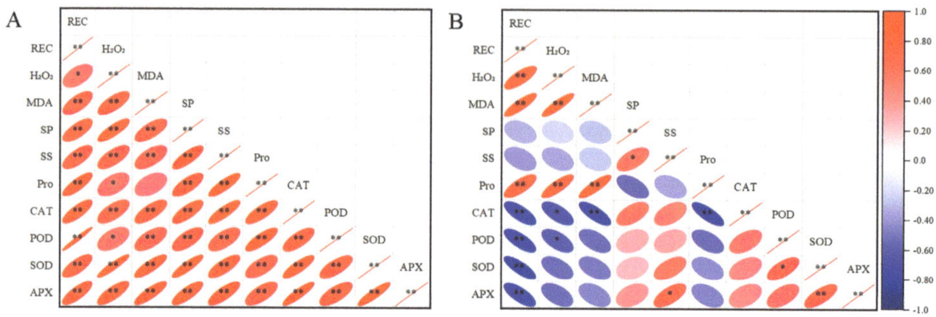

Figure 5. Correlation analysis of physiological indices of cold resistance in apricot flowers under 18 °C~−4 °C treatment (**A**) and −5 °C to −8 °C treatment (**B**). There are three replicates of each dataset and * in the graph indicates a significant correlation ($p < 0.05$) and ** indicates a highly significant correlation ($p < 0.01$).

In the more severe low-temperature stress of −5 °C to −8 °C (Figure 5B), REC of apricot flowers maintains an extremely significant positive correlation ($p < 0.01$) with H_2O_2, MDA content, and Pro content. However, it shows an extremely significant negative correlation ($p < 0.01$) with the activities of the antioxidant enzyme system, including CAT, POD, SOD, and APX. The positive correlation between H_2O_2 and both MDA and Pro content is more pronounced ($p < 0.01$), and there is a significant negative correlation with CAT and POD activities ($p < 0.05$). The positive correlation between MDA content and Pro content remains extremely significant ($p < 0.01$), while the negative correlation with CAT activity is equally pronounced ($p < 0.01$). There is a significant positive correlation ($p < 0.05$) between the contents of osmotic adjustment substances SPs and SSs, as well as between SS content and APX activity. Pro content is extremely significantly negatively correlated ($p < 0.01$) with CAT activity, while POD activity is significantly positively correlated ($p < 0.05$) with SOD activity. Additionally, a strong positive correlation ($p < 0.01$) is observed between SOD and APX activities. These findings shed light on the physiological adjustments that apricot flowers make to respond to low-temperature stress across varying cold conditions.

2.6. Principal Component Analysis (PCA)

PCA was performed on the physiological indicators of cold resistance in apricot flowers, yielding two principal components, each with an eigenvalue exceeding 1 (Table 1). These two principal components cumulatively account for 85.583% (>80%) of the variance in the physiological indicators of cold resistance, suggesting that they encapsulate 85.583% of the original data's information without variable loss. The first principal component is defined by SP, SS, REC, H_2O_2, MDA, POD, and SOD, which together explain 67.962% of the original data's information. The second principal component, determined by CAT and APX, represents 17.621% of the original data's information. This analysis underscores the key factors contributing to the cold resistance of apricot flowers and their interrelationships.

Table 1. Principal component analysis results of apricot flower physiological indicators under various low-temperature stress conditions.

	PC1	PC2	PC3
Eigenvalue	8.36	2.17	0.56
Proportion %	67.96	17.62	4.57
Cumulative %	67.96	85.58	90.15

Based on the principal component analysis, a principal component double-labeled plot was made based on the first and second component values extracted from the information on the physiological indicators of apricot flowers. If the angle between different physiological indicators was less than 90°, it indicated a positive correlation, and if it was greater than 90°, it indicated a negative correlation. The results showed that there was a significant positive correlation between the physiological indicators of apricot flowers under low-temperature stress from 18 °C~−8 °C. The degree of frost damage in apricot flowers increased with decreasing temperature. By connecting the two treatments furthest from the origin in the same direction with a straight line, the final graph will form the closed polygon that will frame all the treatments in it. Then, making plumb lines from the origin to each side of the polygon, these plumb lines divide the whole double-labeled diagram into sectors. The four sectors show that at 18 °C~−4 °C, the cold resistance of the indicators is not outstanding; at −4 °C~−5 °C, the antioxidant enzyme systems CAT, POD, SOD, APX, and the osmotically conditioned substance SS is outstanding; at −6 °C~−8 °C, as the temperature continues to fall, the cell membrane is disrupted, membrane lipid peroxidation is aggravated, and the antioxidant enzymes are inactivated, resulting in the outstanding performance of SP, REC, MDA, and Pro indicators (Figure 6).

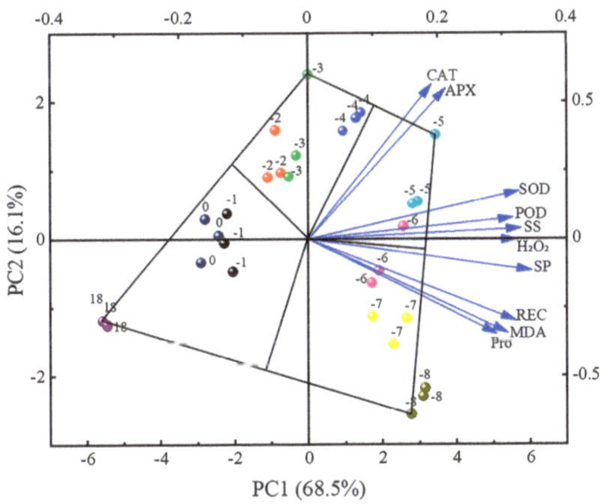

Figure 6. Interrelationship of physiological indicators in apricot flowers under various low-temperature stress conditions.

3. Discussion

The experimental results presented in this study offer a comprehensive understanding of the physiological responses of apricot flowers to low-temperature stress, revealing the intricate interplay between cellular damage, osmoregulation, and antioxidant defense mechanisms. The findings provide valuable insights into the adaptive strategies of apricot flowers and the critical thresholds for cold tolerance.

Numerous studies have demonstrated that the impact of low temperatures can lead to significant physiological and morphological changes in plants [8,10]. In this study, the phenotypic and anatomical changes observed in apricot flowers under low-temperature gradients are indicative of the severity of cold stress. The progressive browning and structural deterioration, particularly at temperatures below −5 °C, suggest a loss of cellular integrity and function. These changes are consistent with the increased browning rate and REC, which are markers of cellular damage. The accumulation of H_2O_2 and the rise in MDA content further confirm the oxidative stress and membrane lipid peroxidation experienced by the flowers, leading to cell death.

The increase in osmoregulatory substances such as soluble sugars and proteins up to −5 °C indicates an initial adaptive response to cold stress. These substances likely serve as osmoprotectants, helping to maintain cell turgor and stabilize cellular structures [11,12]. However, the subsequent decrease in content at lower temperatures may be due to the inhibition of normal metabolic processes, suggesting that the cold tolerance mechanisms of apricot flowers have limitations.

Proline (Pro), also a versatile osmoprotectant [13,14], is pivotal in enhancing plant resistance. It contributes to cellular osmotic equilibrium and structural integrity, particularly under stress [15,16]. Pro's accumulation is instrumental in stabilizing proteins and membranes under various stress conditions, reducing oxidative stress [17,18]. Its antioxidant capabilities are crucial for neutralizing reactive oxygen species (ROS), thus protecting cells from oxidative harm [19,20]. These attributes render Pro an essential factor in plants' adaptation to harsh conditions. In this study, the consistent increase in Pro content across all temperatures is particularly noteworthy. Its accumulation may be a key factor in the cold acclimation process of apricot flowers, although further research is needed to elucidate its exact role.

The enzymatic activities within the antioxidant system are instrumental in bolstering plant resilience [21,22]. Enzymes, including CAT, POD, SOD, and APX, synergistically combat the harmful effects of ROS, safeguarding plant cells from oxidative stress [23,24]. Their catalytic actions in the reduction and decomposition of ROS help to uphold the cellular redox equilibrium, which is essential for maintaining cellular integrity and enhancing the plant's adaptive capacity to environmental challenges [25,26]. In the present study, the enzymatic activities of the antioxidant system showed a complex response pattern to low temperatures. The initial increase in activity up to -4 °C suggests an active defense against ROS generated during cold stress. However, the subsequent decline in activity at temperatures below -4 °C may indicate a saturation point or the inactivation of these enzymes due to extreme cold. The differential response of CAT and POD versus SOD and APX could be attributed to their distinct roles and sensitivities to temperature changes.

Correlation analysis revealed significant relationships between various physiological indicators, highlighting the interconnected nature of cold stress responses. The highly significant positive correlations between osmoregulatory substances and antioxidant enzyme activities, as well as between H_2O_2 and membrane permeability indicators, suggest a coordinated response to mitigate cold damage. The significant negative correlation between H_2O_2 and antioxidant enzyme activities, particularly CAT and POD, may indicate a regulatory mechanism where ROS levels are tightly controlled to prevent excessive oxidative stress.

PCA further supported the findings by identifying the key determinants of cold resistance in apricot flowers. The first principal component, which included SP, SS, H_2O_2, and MDA, reflects the overall cellular damage and osmotic stress. The second principal component, determined by CAT and APX, underscores the importance of antioxidant enzymes in the cold stress response. The PCA results indicate that while osmotic adjustment and antioxidant defense are crucial for cold tolerance, the effectiveness of these mechanisms is temperature-dependent.

This study focuses on the flowering stage because there is a significant difference in the tree's sensitivity to cold stress between the preflowering period, when buds are dormant, and the peak blooming period. Typically, dormant buds, due to their reduced metabolic rate, possess enhanced cold resistance, allowing them to more effectively endure cold temperatures [6–8]. Furthermore, the capacity for cold resistance during the flowering period varies among different *Prunus* species. Certain species, such as the *Prunus sibirica*, demonstrate superior cold hardiness, attributed to their adaptation to chillier climates; in contrast, others like the *Prunus mume* show greater sensitivity to cold, likely due to their natural habitat in more temperate zones [27]. This diversity in cold resistance is presumably influenced by a combination of genetic predispositions and the distinct environmental challenges that have molded the evolutionary journey of each species.

In conclusion, the study provides a detailed analysis of the physiological responses of apricot flowers to low-temperature stress, highlighting the importance of temperature in determining the success of these responses. The findings suggest that while apricot flowers have certain adaptive mechanisms to cope with cold stress, there are critical thresholds beyond which these mechanisms are overwhelmed. This knowledge can be instrumental in developing strategies to improve the cold tolerance of apricot cultivars, potentially through breeding or genetic engineering approaches. Future research should focus on the molecular mechanisms underlying these responses and the identification of genetic markers for cold tolerance, which could lead to the development of more resilient apricot varieties for cultivation in colder climates.

4. Materials and Methods

4.1. Plant Materials

The apricots (*Prunus armeniaca* L. cv. Golden Sun) were sourced from the orchards of Taigu County, situated in Jinzhong City, Shanxi Province. The samples were harvested during early flowering stage when a flower's buds have grown significantly and are on the verge of opening. To ensure methodological rigor, a selection criterion was established, mandating uniformity across orchard cultivation practices, tree vitality, environmental settings, and agricultural management standards, leading to the random selection of a cohort of 10 trees. From each of these trees, a set of 20–25 fruit-bearing branches, each measuring between 15 to 20 cm, were excised from the cardinal points—east, west, south, and north—at a standardized height of 1.5 m from the ground. After harvest, the branches were immersed in a 3% sucrose solution for an acclimatization treatment, with 50–60% of the apricot blossoms in full bloom. Subsequently, they were carefully transported to the laboratory for further experimental procedures.

4.2. Low-Temperature Treatment

Ten treatment temperatures were established at 18 °C, 0 °C, −1 °C, −2 °C, −3 °C, −4 °C, −5 °C, −6 °C, −7 °C, and −8 °C. Apricot short fruiting branches during the flowering stage were inserted into flower foam that had been soaked in water for 24 h (20 branches per block of foam, with each branch serving as one replicate). These were then placed in a PERCIVAL low-temperature incubator to simulate a low-temperature environment. The low-temperature incubator starts by cooling down first. Then, it slowly drops 1 °C every hour until it hits the target temperature, staying there for 1 h. After that, it warms up by 10 °C every 30 min until it reaches room temperature. Observations were made after a 12 h recovery period at room temperature. During the low-temperature stress, a small number of pistils were collected and fixed with FAA fixative. After a 3 h recovery at room temperature, a portion of the apricot flowers was used for the statistical analysis and measurement of browning rate and electrical conductivity (with the calyx removed). The other portion of the apricot flowers was preserved in liquid nitrogen and stored in a −80 °C ultra-low-temperature freezer for the determination of physiological indicators.

4.3. Measurement Methods

4.3.1. Observation of Flower Phenotype and Browning Rate

After the low-temperature stress treatment, 100 flowers were randomly selected and observed for the degree of browning after a 3 h recovery period at room temperature. The browning rates of petals and stigmas were meticulously observed and documented. Floral organs that had suffered from frost damage took on shades of brown or black, in contrast to those that remained unscathed, which retained their fresh and vibrant green color. The results were statistically analyzed and the process was repeated three times.

4.3.2. Observation of the Anatomical Structure of the Apricot Flower

The pistil materials, fixed with FAA (formaldehyde 38%, glacial acetic acid, and 70% alcohol in a ratio of 1:1:18), were dehydrated sequentially using ethanol of varying concentrations [28]. They were then cleared in a mixture of ethanol and n-butanol in different ratios, followed by infiltration with a solution of n-butanol and paraffin wax in varying proportions. Finally, the samples were embedded in embedding boxes, cooled, and subjected to continuous sectioning at a thickness of 8 μm. The sections were then stained with safranin and fast green and mounted.

4.3.3. Relative Conductivity Determination

We weighed out 0.5 g of the treated fresh flowers, cut them into small pieces, and placde them into a conical flask containing 10 mL of distilled water. After allowing the mixture to stand for 24 h, it was shaken for 20 min to measure the initial electrical conductivity. Then, we immersed the flask in boiling water for 30 min. After cooling to room temperature, we measured the final electrical conductivity [29]. Measurement of electrical conductivity was performed by laboratory conductivity meter Cond 720 (WTW Inolab, Weilheim, Germany).

The calculation of the lethal temperature (LT_{50}) was performed by measuring the relative electrical conductivity of apricot flowers after low-temperature stress [30]. The data were fitted to a logistic regression equation: $Y = K/(1 + ae^{-bx})$. In this equation, Y represents the relative electrical conductivity, K is the asymptotic relative electrical conductivity, and a and b are parameters of the equation, with x denoting the temperature. By calculating the second derivative of the logistic regression equation and setting it to zero, the inflection point temperature (LT_{50}) can be determined. The LT_{50} is then used to calculate the lethal temperature using the formula $LT_{50} = (\ln a)/b$.

4.3.4. Membrane Lipid Peroxisome Determination

The determination of H_2O_2 and MDA in apricot flowers was conducted using assay kits from Nanjing Jiancheng, following the manufacturer's instructions. Samples were ground in a mortar with liquid nitrogen, and 0.15 g was immediately weighed out after thorough grinding. This was then added to 10 mL of 0.1 mol/L phosphate buffer solution at pH = 7.8. After centrifugation at 4 °C, the supernatant was placed in an ice bath for the measurement of various physiological parameters. The reaction of H_2O_2 with molybdic acid produced a complex, and its absorbance was measured at 405 nm; MDA reacted with thiobarbituric acid to form a red product, and its absorbance was measured at 532 nm.

4.3.5. Determination of the Content of Osmoregulatory Substances

Plant soluble sugars, soluble proteins, and proline were determined using the Plant Soluble Sugars, Soluble Proteins, and Proline Assay Kit (Nanjing Jiancheng, Nanjing, China), following the instructions. Samples were ground in a mortar with liquid nitrogen, and immediately after thorough grinding, 0.15 g was weighed out and mixed with 2 mL of 0.1 mol/L phosphate buffer solution at pH 7.8. After centrifugation at 4 °C, the supernatant was placed in an ice bath for the subsequent measurement of various physiological parameters. The $-NH^{3+}$ groups of proteins were stained with Coomassie Brilliant Blue G-250, resulting in a purple color, and the absorbance was measured at 595 nm. Soluble sugars, under the action of concentrated sulfuric acid, reacted with anthrone to form blue–green sugar aldehyde derivatives, and their absorbance was measured at 630 nm. Proline reacted with acidic ninhydrin to produce a red product, and its absorbance was measured at 520 nm.

4.3.6. Determination of the Activity of the Antioxidant Enzyme System

The plant CAT, POD, SOD, and APX activity assay kits (Nanjing Jiancheng, Nanjing, China) were used and the method was carried out according to the instructions. Samples were finely ground in a mortar with liquid nitrogen, and immediately after adequate grinding, 0.15 g was weighed out and combined with 2 mL of 0.1 mol/L phosphate buffer solution at pH 7.8. Following centrifugation at 4 °C, the supernatant was placed in an ice bath for the measurement of various physiological parameters. The reaction of CAT decomposing H_2O_2 was rapidly halted by the addition of ammonium molybdate, and the residual H_2O_2 reacted with the molybdate to form a pale-yellow complex, with absorbance measured at 405 nm. The activity of POD was determined using the guaiacol method, which is based on the principle of POD catalyzing the reaction of H_2O_2, with absorbance measured at 420 nm. SOD activity was assayed by the colorimetric method, where the superoxide anion O^{2-} is oxidized by the dye to form nitrite, resulting in a purple–red complex, and absorbance was measured at 550 nm. The activity of APX was measured by its catalysis of the reaction between ascorbic acid and H_2O_2, converting ascorbic acid to monodehydroascorbic acid, with absorbance measured at 290 nm.

4.4. Processing of Data

The test data were collated and calculated by WPS 2021, and one-way ANOVA, LSD, and Duncan multiple analysis comparisons and PCA were performed using SPSS 22 statistical software, with graphical and correlational analytical analyses completed with the aid of WPS 2021 and Origin 2021.

5. Conclusions

This research provided a detailed analysis of the physiological reactions of apricot flowers to low temperatures, highlighting the critical role of temperature in inducing cellular damage and the activation of defense mechanisms. The study observed significant browning, structural deterioration, and increased oxidative stress markers, such as H_2O_2 and MDA, indicating severe cold stress. The initial increase in osmoregulatory substances and antioxidant enzyme activities suggests an adaptive response, though this was limited by extreme temperatures. Correlation and principal component analyses revealed strong interconnections between these physiological indicators, pointing to a complex acclimation process. The findings are valuable for guiding future breeding efforts to develop cold-resistant apricot cultivars and for deepening our understanding of plant cold tolerance, which is crucial for agricultural sustainability in the face of climate change.

Author Contributions: Conceptualization, W.G., X.F. and L.L.; data curation, J.G., W.G. and R.H.; formal analysis, J.G., Q.L., M.L. and Y.S.; writing—original draft preparation, J.G., W.G. and C.S.; writing—review and editing, X.F. and L.L. All authors have read and agreed to the published version of the manuscript.

Funding: This research was funded by the Collaborative Research Project between Datong City and Shanxi Agricultural University, and the Science and Technology Innovation Fund Project of Shanxi Agricultural University (Grant No. 2017YJ45).

Data Availability Statement: Data are contained within the article.

Conflicts of Interest: The authors declare no conflicts of interest.

References

1. Yang, J.Y.; Zhang, L.; Huo, Z.G.; Wang, P.J.; Wu, D.R.; Ma, Y.P. Disaster process–based spatiotemporal characteristics of apricot frost in the warm temperate zone (WTZ), China. *Int. J. Biometeorol.* **2023**, *67*, 1733–1744. [CrossRef]
2. Millard, J.; Outhwaite, C.L.; Ceaușu, S.; Carvalheiro, L.G.; Silva e Silva, F.D.; Dicks, L.V.; Ollerton, J.; Newbold, T. Key tropical crops at risk from pollinator loss due to climate change and land use. *Sci. Adv.* **2023**, *9*, eadh0756. [CrossRef]
3. Malhi, G.S.; Kaur, M.; Kaushik, P. Impact of climate change on agriculture and its mitigation strategies: A review. *Sustainability* **2021**, *13*, 1318. [CrossRef]
4. Pakkish, Z.; Tabatabaienia, M.S. The use and mechanism of NO to prevent frost damage to flower of apricot. *Sci. Hortic.* **2016**, *198*, 318–325. [CrossRef]
5. Kaya, O.; Kose, C. Cell death point in flower organs of some apricot (*Prunus armeniaca* L.) cultivars at subzero temperatures. *Sci. Hortic.* **2019**, *249*, 299–305. [CrossRef]
6. Kaya, O.; Kose, C.; Gecim, T. An exothermic process involved in the late spring frost injury to flower buds of some apricot cultivars (*Prunus armeniaca* L.). *Sci. Hortic.* **2018**, *241*, 322–328. [CrossRef]
7. Szalay, L.; Bakos, J.; Teweldemedhin Keleta, B.; Froemel-Hajnal, V.; Karsaib, I. A 15-year long assessment of cold hardiness of apricot flower buds and flowers during the blooming period. *Sci. Hortic.* **2021**, *290*, 110520. [CrossRef]
8. Viti, R.; Bartolini, S.; Andreini, L. Apricot flower bud dormancy: Main morphological, anatomical and physiological features related to winter climate influence. *Adv. Hort. Sci.* **2013**, *27*, 5–17.
9. Rodrigo, J.; Herrero, M. Effects of pre-blossom temperatures on flower development and fruit set in apricot. *Sci. Hortic.* **2002**, *93*, 1–13. [CrossRef]
10. Khodorova, N.V.; Boitel-Conti, M. The role of temperature in the growth and flowering of geophytes. *Plants* **2013**, *2*, 699–711. [CrossRef]

11. Zulfiqar, F.; Akram, N.A.; Ashraf, M. Osmoprotection in plants under abiotic stresses: New insights into a classical phenomenon. *Planta* **2020**, *251*, 3. [CrossRef] [PubMed]
12. Ghosh, U.K.; Islam, M.N.; Siddiqui, M.N.; Khan, M.A.R. Understanding the roles of osmolytes for acclimatizing plants to changing environment: A review of potential mechanism. *Plant Signal. Behav.* **2021**, *16*, e1913306. [CrossRef]
13. Ghosh, U.K.; Islam, M.N.; Siddiqui, M.N.; Cao, X.; Khan, M.A.R. Proline, a multifaceted signalling molecule in plant responses to abiotic stress: Understanding the physiological mechanisms. *Plant Biol.* **2022**, *24*, 227–239. [CrossRef] [PubMed]
14. Kaur, G.; Asthir, B. Proline: A key player in plant abiotic stress tolerance. *Biol. Plant.* **2015**, *59*, 609–619. [CrossRef]
15. Zouari, M.; Ben Ahmed, C.; Elloumi, N.; Bellassoued, K.; Delmail, D.; Labrousse, P.; Ben Abdallah, F.; Ben Rouina, B. Impact of proline application on cadmium accumulation, mineral nutrition and enzymatic antioxidant defense system of *Olea europaea* L. cv Chemlali exposed to cadmium stress. *Ecotox. Environ. Saf.* **2016**, *128*, 195–205. [CrossRef] [PubMed]
16. Vives-Peris, V.; Gómez-Cadenas, A.; Pérez-Clemente, R.M. Citrus plants exude proline and phytohormones under abiotic stress conditions. *Plant Cell Rep.* **2017**, *36*, 1971–1984. [CrossRef]
17. Slabbert, M.M.; Krüger, G.H.J. Antioxidant enzyme activity, proline accumulation, leaf area and cell membrane stability in water stressed *Amaranthus* leaves. *S. Afr. J. Bot.* **2014**, *95*, 123–128. [CrossRef]
18. Liang, X.W.; Zhang, L.; Natarajan, S.K.; Becker, D.F. Proline mechanisms of stress survival. *Antioxid. Redox Signal.* **2013**, *19*, 998–1011. [CrossRef] [PubMed]
19. Viehweger, K. How plants cope with heavy metals. *Bot. Stud.* **2014**, *55*, 35. [CrossRef]
20. Yan, Z.M.; Guo, S.R.; Shu, S.; Sun, J.; Tezuka, T. Effects of proline on photosynthesis, root reactive oxygen species (ROS) metabolism in two melon cultivars (*Cucumis melo* L.) under NaCl stress. *Afr. J. Biotechnol.* **2011**, *10*, 18381–18390.
21. Rajput, V.D.; Harish, H.; Singh, R.K.; Verma, K.K.; Sharma, L.; Quiroz-Figueroa, F.R.; Meena, M.; Gour, V.S.; Minkina, T.; Sushkova, S.; et al. Recent developments in enzymatic antioxidant defence mechanism in plants with special reference to abiotic stress. *Biology* **2021**, *10*, 267. [CrossRef] [PubMed]
22. Barreca, D. Mechanisms of plant antioxidants action. *Plants* **2021**, *10*, 35. [CrossRef] [PubMed]
23. Irato, P.; Santovito, G. Enzymatic and non-enzymatic molecules with antioxidant function. *Antioxidants* **2021**, *10*, 579. [CrossRef]
24. Kapoor, D.; Singh, S.; Kumar, V.; Romero, R.; Prasad, R.; Singh, J. Antioxidant enzymes regulation in plants in reference to reactive oxygen species (ROS) and reactive nitrogen species (RNS). *Plant Gene* **2019**, *19*, 100182. [CrossRef]
25. Mehla, N.; Sindhi, V.; Josula, D.; Bisht, P.; Wani, S.H. An introduction to antioxidants and their roles in plant stress tolerance. In *Reactive Oxygen Species and Antioxidant Systems in Plants: Role and Regulation under Abiotic Stress*; Khan, M., Khan, N., Eds.; Springer: Singapore, 2017; pp. 1–23.
26. Miller, G.; Suzuki, N.; Ciftci-Yilmaz, S.; Mittler, R. Reactive oxygen species homeostasis and signalling during drought and salinity. *Plant Cell Environ.* **2010**, *33*, 453–467. [CrossRef] [PubMed]
27. Ding, A.Q.; Bao, F.; Zhang, T.X.; Yang, W.R.; Wang, J.; Cheng, T.R.; Zhang, Q.X. Screening of optimal reference genes for qRT-PCR and preliminary exploration of cold resistance mechanisms in *Prunus mume* and *Prunus sibirica* varieties. *Mol. Biol. Rep.* **2020**, *47*, 4945–4957. [CrossRef]
28. Huang, B.; Yeung, E. Chemical and physical fixation of cells and tissues: An overview. In *Plant Microtechniques and Protocols*; Yeung, E., Stasolla, C., Sumner, M., Huang, B., Eds.; Springer: Cham, Switzerland, 2015; pp. 23–43.
29. Zhao, S.J.; Shi, G.A.; Dong, X.C. *Techniques of Plant Physiological Experiment*; China Agriculture Science and Technology Press: Beijing, China, 2002; pp. 84–85. (In Chinese)
30. Wang, Y.X.; Hu, Y.; Chen, B.H.; Zhu, Y.F.; Dawuda, M.M.; Sofkova, S. Physiological mechanisms of resistance to cold stress associated with 10 elite apple rootstocks. *J. Integr. Agric.* **2018**, *17*, 857–866. [CrossRef]

plants

MDPI

Article

Ability of Different Growth Indicators to Detect Salt Tolerance of Advanced Spring Wheat Lines Grown in Real Field Conditions

Muhammad Mubushar, Salah El-Hendawy *, Yaser Hassan Dewir and Nasser Al-Suhaibani

Department of Plant Production, College of Food and Agriculture Sciences, King Saud University, Riyadh 11451, Saudi Arabia

* Correspondence: mosalah@ksu.edu.sa

Abstract: Plant growth indicators (GIs) are important for evaluating how different genotypes respond to normal and stress conditions separately. They consider both the morphological and physiological components of plants between two successive growth stages. Despite their significance, GIs are not commonly used as screening criteria for detecting salt tolerance of genotypes. In this study, 36 recombinant inbred lines (RILs) along with four genotypes differing in their salt tolerance were grown under normal and 150 mM NaCl in a two-year field trial. The performance and salt tolerance of these germplasms were assessed through various GIs. The analysis of variance showed highly significant variation between salinity levels, genotypes, and their interaction for all GIs and other traits in each year and combined data for two years, with a few exceptions. All traits and GIs were significantly reduced by salinity stress, except for relative growth rate (RGR), net assimilation rate (NAR), and specific leaf weight (SLW), which increased under salinity conditions. Traits and GIs were more correlated with each other under salinity than under normal conditions. Principal component analysis organized traits and GIs into three main groups under both conditions, with RGR, NAR, and specific leaf area (SLA) closely associated with grain yield (GY) and harvest index, while leaf area duration (LAD) was closely associated with green leaf area (GLA), plant dry weight (PDW), and leaf area index (LAI). A hierarchical clustering heatmap based on GIs and traits organized germplasms into three and four groups under normal and salinity conditions, respectively. Based on the values of traits and GIs for each group, the germplasms varied from high- to low-performing groups under normal conditions and from salt-tolerant to salt-sensitive groups under salinity conditions. RGR, NAR, and LAD were important factors determining genotypic variation in GY of high- and low-performing groups, while all GIs, except leaf area duration (LAR), were major factors describing genotypic variation in GY of salt-tolerant and salt-sensitive groups. In conclusion, different GIs that reveal the relationship between the morphological and physiological components of genotypes could serve as valuable selection criteria for evaluating the performance of genotypes under normal conditions and their salt tolerance under salinity stress conditions.

Keywords: grain yield; green leaf area; leaf area duration; net assimilation rate; plant dry weight; relative growth rate; specific leaf weight

Citation: Mubushar, M.; El-Hendawy, S.; Dewir, Y.H.; Al-Suhaibani, N. Ability of Different Growth Indicators to Detect Salt Tolerance of Advanced Spring Wheat Lines Grown in Real Field Conditions. *Plants* **2024**, *13*, 882. https://doi.org/10.3390/plants13060882

Academic Editors: Geovani Soares de Lima, Lauriane Almeida dos Anjos Soares and Francisco Vanies Da Silva Sá

Received: 18 January 2024
Revised: 10 March 2024
Accepted: 15 March 2024
Published: 19 March 2024

1. Introduction

The world is currently facing major salinity issues, which are causing a significant decrease in the growth and production of essential food crops, especially in arid and semiarid regions. Several studies have forecasted that around 50% of productive lands will become unproductive by 2050 due to salinity, raising serious concerns about global food security in the near future [1]. Salinity problems affect more than 7% of the world's total land area, 20% of arable land, and 33% of irrigated agricultural lands. Additionally, salinity issues in the agricultural sector result in an annual loss of more than USD 12 billion due to a 30–50% decrease in the production of essential food crops in salinity-affected areas [2–4].

Addressing the salinity issue in agriculture is crucial in achieving a 50–70% increase in grain yields for key food crops such as wheat, rice, and maize. This increase is essential to meet the needs of the projected 10 billion people by 2050 [5].

Wheat is the dominant crop on 2.1 million km^2 of arable land, but its productivity is hindered by several environmental stresses, especially salinity stress, which is common in arid and semiarid regions [6,7]. Bread wheat is a moderately salt-tolerant crop, but high salinity levels can reduce its yield by more than 50% [3,8]. Therefore, cultivating wheat in saline conditions is challenging without implementing practices to mitigate the negative effects of salinity stress on its growth and yield. Agronomic practices such as applying mineral gypsum, organic amendments, and effective drainage schemes can help alleviate these negative impacts [9,10]. However, these practices can be costly and require expertise, particularly when implemented on a large scale. Previous studies have concurred that providing farmers with salt-tolerant genotypes is a feasible and effective strategy for addressing salinity in agriculture and achieving reasonable yields [8,11–13]. Despite the research community's great efforts to enhance the salinity tolerance of wheat genotypes, the number of salt-tolerant genotypes worldwide remains very limited. Based on published studies, the reasons behind this include (1) the salt tolerance of genotypes is usually assessed under tightly controlled conditions with few experiments carried out under real environmental conditions [3,8,14]; (2) the salt tolerance of genotypes is usually assessed based on one or two developmental growth stages, particularly the early ones (germination and seedling), despite the fact that the ranking of genotypes for their salt tolerance often varies across different growth stages [15]; (3) evaluating the salt tolerance of genotypes often requires finding suitable screening criteria and accurate evaluation methods that have the potential to make the assessment of salt tolerance among different genotypes more precise and efficient, particularly across multiple traits [16–18]; (4) the lack of specific indicators that accurately reflect both the physiological and agronomic behavior of genotypes [3,19]; and (5) the use of different growth indicators (GIs) as screening criteria is not common. These GIs can provide insight into the strong correlation between physiological tolerance and agronomic traits, as well as the close relationship between source and sink organs. They also shed light on how biomass is allocated between different plant organs and the duration of active photosynthesis [20–22].

Salinity stress significantly impacts plant growth and production through osmotic stress, ion toxicity, and essential nutrient deficiencies [11,23]. This leads to the excessive buildup of toxic ions in the leaf blade, restricted water and essential nutrient uptake, and significant alterations in physiological and biochemical processes at various plant levels. These changes include inhibiting cell division and elongation, accelerating cell death and leaf senescence, increasing leaf degradation, reducing leaf growth, inhibiting photosystem II (PSII) activity, decreasing stomatal conductance, destroying photosynthetic pigments, reducing carbohydrate supply to young leaves and grains, limiting photosynthetic rate, biomass accumulation, and source-sink activity, and altering plant water status [3,11,24–27]. In the end, all of the aforementioned negative effects of salinity stress have a detrimental impact on the various agro-morphological traits of wheat plants [15,24,28,29]. Thus, different agro-morphological traits related to salt tolerance can be used as screening criteria to improve the adaptation of wheat genotypes to salinity stress. These traits can be assessed at different growth stages, and therefore, understanding their contribution to salt tolerance mechanisms at specific developmental stages is essential in improving breeding techniques and developing salt-tolerant genotypes. This will make the assessment of salt tolerance among genotypes more effective. Generally, to enhance the precision and efficiency of evaluating salt tolerance among genotypes, it is essential to simultaneously assess multiple traits at various growth stages [30,31]. However, this approach can be costly and time-consuming, particularly when dealing with a large number of genotypes.

In general, changes in plant performance between two successive growth stages are closely related to changes in several morphological and physiological components under normal or stress conditions. Previous studies have reported that various GIs, including relative growth rate (RGR), net assimilation rate (NAR), leaf area duration (LAD), leaf area ratio (LAR), specific leaf area (SLA), leaf weight ratio (LWR), and specific leaf weight (SLW), can be used to comprehensively understand how different genotypes respond to various environmental conditions at the morphological and/or physiological levels. Therefore, they could provide a suitable approach to crop improvement programs in both normal and stress conditions [21,32–34]. For instance, RGR, the most commonly used growth analysis component, is used to assess how dry matter is accumulated and distributed between different parts of the plant, and to determine if differences in growth performance among different genotypes are attributed to changes in morphological and/or physiological components. This is because RGR is a combination of morphological components (LAR) and physiological components (NAR) of growth [35]. NAR is a growth metric that quantifies the increase in dry biomass per unit leaf area, and is primarily influenced by the balance between photosynthesis and respiration rates of the entire plant. On the other hand, LAR represents the ratio between the photosynthetic organs (leaf area) and the photosynthetic product (total dry mass), making it a product of SLA (the ratio of surface leaf area to leaf biomass) and LWR (the ratio of leaf biomass to total plant biomass) [35]. Cramer et al. [36] found that salinity stress at the early growth stage significantly affected RGR, LAR, SLA, and LWR, and these factors were correlated with differences in salt tolerance between maize hybrids. However, NAR was not significantly affected by salinity stress and was not associated with differences in salt tolerance between hybrids. This suggests that the salt tolerance of hybrids is linked to reduced growth (dry matter accumulation) and leaf expansion, rather than the photosynthetic rate. In contrast, other studies have shown that differences in salt tolerance between wheat genotypes were associated with both RGR and NAR, indicating that both leaf expansion and photosynthetic rate play important roles in the salt tolerance of wheat genotypes [23]. Hence, different GIs, which can be calculated using simple mathematical equations, can help us understand the variations in salt tolerance among different genotypes. This makes them useful screening criteria for evaluating the salt tolerance of wheat genotypes.

Despite the importance of various GIs in understanding the complex relationships between morphological and physiological components and their response to various environmental growth conditions, there is a lack of research on using GIs as screening criteria to evaluate the growth and production performance of genotypes under normal conditions and their salt tolerance under salinity conditions. Additionally, there is also a lack of studies on using GIs to assess whether genotypic variation in production under normal conditions and salt tolerance under salinity conditions is linked to differences in morphological and/or physiological characteristics. Therefore, this study aimed to assess the effectiveness of different GIs in identifying genotypes with good performance under normal conditions and salt tolerance under salinity stress. Additionally, the study aimed to determine if genotypic variation in salt tolerance is associated with differences in morphological and/or physiological characteristics.

2. Results

2.1. *Effects of Different Treatments and Their Interaction on Agro-morphological Traits and Growth Indicators*

The F-values from the analysis of variance (ANOVA) show that both salinity levels (SL) and genotypes (G) had significant effects on all agro-morphological traits (traits) and growth indicators (GIs) in each year (Y) and in the combined analysis of data for both years, except for harvest index (HI) and LAR in the first year, which were not affected by SL and G, respectively (Table 1). The SL × G interaction had significant effects on all traits and GIs, except for LAR in the first year and green leaf area measured at 90 days after sowing (GLA-90), LAD, and SLA in the second year. The Y had significant effects on all traits and GIs except for plant dry weight measured at 90 days after sowing (PDW-90) and HI. Three out of eight traits (GLA-90, PDW-75, and LAI-90) and GIs related to morphological components (LAD, SLA, and SLW) were significantly affected by the

SL × Y interaction. The G × Y interaction had significant effects on all traits except for GLA-90 and PDW-70. Additionally, two GIs related to physiological components (RGR, NAR) and one related to morphological components (LAD) were also significantly affected by this interaction. The G × SL × Y interaction had significant effects on all traits except for PDW-75, GY, and HI. Additionally, RGR, NAR, and LAR, which are GIs related to physiological and morphological components, were also significantly affected by this three-way interaction (Table 1).

2.2. Variation of Agro-Morphological Traits and Growth Indicators among Genotypes under Normal and Salinity Conditions

Figure 1 (box plot) illustrates the variability of traits and GIs among genotypes under both normal and salinity conditions. Meanwhile, Table 2 provides a summary of the descriptive statistics for traits and GIs in response to both conditions across 36 genotypes. According to Figure 1, most of the traits and GIs showed high variability among genotypes under both conditions. According to Table 2, the maximum values for the traits and GIs were approximately one to two times greater than the minimum values, regardless of whether under normal or salinity conditions. The descriptive statistics values, including range, 1st quartile, and 3rd quartile, also indicated a wide range of variation in traits and GIs among genotypes in both conditions. Additionally, the values of traits and GIs were notably reduced under high salinity levels (150 mM NaCl) compared to the normal treatment, with the exception of GIs related to physiological components (RGR and NAR) and SLW, which exhibited a significant increase under salinity stress. For instance, when compared to the control treatment, salinity stress resulted in a decrease of several traits by 7.1% to 49.3% and GIs related to morphological components (LAD, LAR, and SLA) by 14.0% to 46.5% (Table 2).

2.3. Correlational Studies under Normal and Salinity Conditions

The degree of association between all variables (traits and GIs) under both conditions was assessed using a Pearson correlation matrix, as shown in Table 3. Overall, the salinity stress conditions showed a stronger correlation between variables compared to the normal conditions. Additionally, all variables displayed positive correlations with each other, except for LAR and SLW, which had negative correlations with the other variables (Table 3). Morphological traits (GLA, PDW, and LAI) showed a strong positive correlation with each other (r ranged from 0.78 to 0.95) and a moderate to strong correlation with GY (r ranged from 0.44 to 0.70) under salinity conditions. Meanwhile, under normal conditions, these traits exhibited a moderate to strong correlation with each other (r ranged from 0.31 to 0.86) and a weak to moderate correlation with GY (r ranged from 0.15 to 0.49; Table 3). Three GIs related to morphological components (LAD, SLA, and SLW) exhibited a strong correlation with morphological traits and GY under salinity conditions (r ranged from 0.63 to 0.97), while they showed a weak to strong correlation under normal conditions (r ranged from 0.25 to 0.81). Under salinity conditions, GIs related to physiological components (RGR and NAR) exhibited a moderate correlation with PDW-90, LAI-75, and LAI-90 (r = 0.34–0.45), and a strong correlation with GY (r = 0.84–0.89). Under normal conditions, they showed a moderate correlation with only PDW-90 (r = 0.41–0.53) and still a strong correlation with GY (r = 0.79–0.82). LAR exhibited a moderate and negative correlation with only PDW-75 and PDW-90 under both conditions (r = 0.38–0.54). HI showed a moderate correlation with only GY, SLA, and SLW under both conditions, as well as with RGR under salinity conditions (Table 3).

Table 1. F-values of analysis of variance (ANOVA) for the effects of salinity level (SL), genotype (G), and their interaction on different agro-morphological traits measured at 75 and 90 days after sowing and growth indicators for each year (Y) and combined two years. The units of traits and GIs are mentioned in Figure 1.

Source	First Year			Second Year			Combined Two Years						
	SL	G	G × SL	SL	G	G × SL	Y	SL	SL × Y	G	G × Y	G × SL	G × SL × Y
df	1	35	35	1	35	35	1	1	1	35	35	35	35
Green leaf area (GLA-75)	391.35 **	10.83 ***	4.69 ***	399.72 **	4.24 ***	2.49 ***	769.70 **	788.19 ***	3.99 ns	11.17 ***	2.29 ***	4.57 ***	2.07 ***
Green leaf area (GLA-90)	3784.02 ***	7.40 ***	3.74 ***	884.86 **	3.07 ***	1.31 ns	28.81 *	3090.96 ***	11.47 *	7.45 ***	1.45 ns	2.23 ***	1.94 **
Plant dry weight (PDW-75)	2217.97 ***	6.13 ***	2.80 ***	762.07 **	4.35 ***	2.02 **	136.76 **	2423.42 ***	12.00 *	9.13 ***	1.20 ns	3.89 ***	0.88 ns
Plant dry weight (PDW-90)	12,883.85 ***	8.99 ***	5.60 ***	573.37 **	7.69 ***	4.32 ***	1.19 ns	2351.10 ***	3.19 ns	14.66 ***	2.00 **	8.93 ***	1.49 *
Leaf area index (LAI-75)	1183.05 ***	7.37 ***	2.72 ***	8918.05 ***	2.78 ***	1.68 *	4851.17 ***	4228.80 ***	0.26 ns	6.52 ***	2.90 ***	2.77 ***	1.47 *
Leaf area index (LAI-90)	980.37 **	6.85 ***	3.16 ***	13,123.10 ***	3.87 ***	2.09 **	130.68 **	3109.88 ***	28.29 **	7.56 ***	2.92 ***	3.40 ***	1.77 **
Grain yield (GY)	10,364.12 ***	9.71 ***	5.25 ***	2767.47 ***	8.30 ***	3.36 ***	21.48 *	8794.04 ***	0.28 ns	16.23 ***	1.65 *	7.63 ***	0.81 ns
Harvest index (HI)	2.11 ns	2.56 ***	1.70 *	27.83 *	5.46 ***	1.67 *	2.36 ns	14.47 *	1.58 ns	5.62 ***	1.48 *	2.81 ***	0.58 ns
Relative growth rate (RGR)	104.73 **	7.60 ***	3.90 ***	375.69 **	4.53 ***	2.07 **	234.51 **	291.81 ***	2.91 ns	9.84 ***	2.11 ***	4.25 ***	1.61 *
Net assimilation rate (NAR)	203.95 **	6.66 ***	3.82 ***	1136.09 ***	4.74 ***	3.45 ***	22.07 *	642.28 ***	1.82 ns	9.46 ***	1.95 **	5.30 ***	1.98 **
Leaf area duration (LAD)	2149.09 ***	8.57 ***	2.93 ***	237,457.21 ***	2.96 ***	1.47 ns	685.05 **	7846.93 ***	14.30 *	8.23 ***	2.73 ***	2.95 ***	1.30 ns
Leaf area ratio (LAR)	534.74 **	1.33 ns	1.41 ns	533.14 **	1.61 *	2.40 ***	1009.90 ***	1066.07 ***	1.68 ns	2.23 ***	0.72 ns	2.16 ***	1.68 *
Specific leaf area (SLA)	1753.83 ***	2.13 **	1.89 **	866.78 **	1.74 *	1.45 ns	87.53 *	2497.53 ***	41.95 **	3.23 ***	0.66 ns	2.68 ***	0.67 ns
Specific leaf weight (SLW)	1072.18 ***	3.25 ***	2.77 ***	584.00 **	2.06 **	1.81 **	75.43 *	1633.05 ***	53.51 **	4.52 ***	1.03 ns	3.82 ***	0.95 ns

* $p \leq 0.05$, ** $p \leq 0.01$, *** $p \leq 0.001$, ns—not significant.

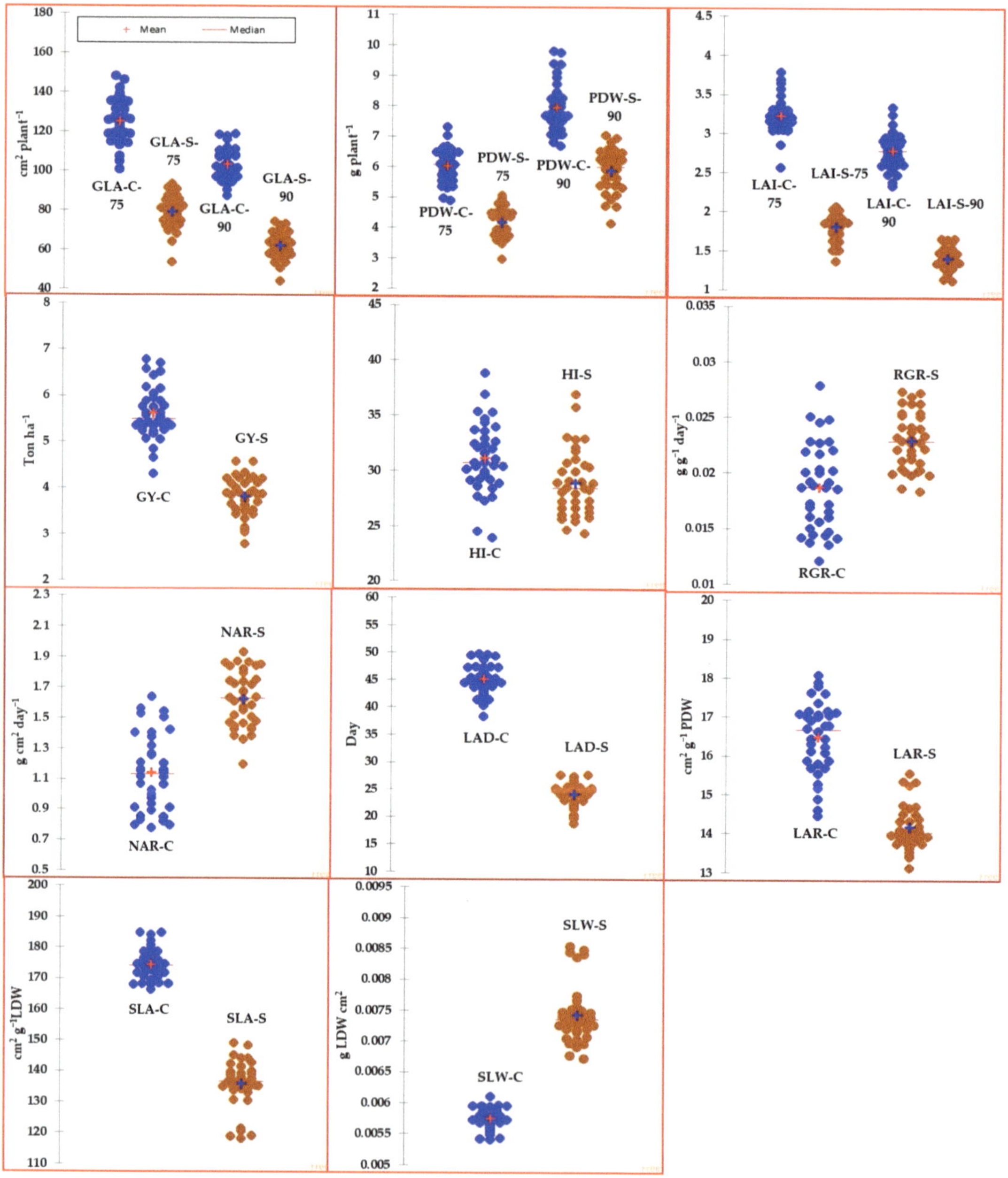

Figure 1. Box plots representing the variability of different agro-morphological traits measured at 75 and 90 days after sowing along with growth indicators for 36 genotypes grown under normal (C) and salinity (S) conditions. Data are the mean of pool data for two years. The full names of traits and GIs are mentioned in Table 1.

Table 2. Descriptive statistics of different agro-morphological traits measured at 75 and 90 days after sowing along with growth indicators (GIs) among 36 genotypes under normal and salinity conditions. Data are the mean of pool data for two years. The full names of traits and GIs are mentioned in Table 1. The units of traits and GIs are mentioned in Figure 1.

Statistic	GLA-75	GLA-90	PDW-75	PDW-90	LAI-75	LAI-90	GY	HI	RGR	NAR	LAD	LAR	SLA	SLW
						Normal Conditions								
Minimum	100.79	87.36	4.90	6.70	2.56	2.33	4.30	23.93	0.012	0.779	38.39	14.46	166.34	0.0054
Maximum	148.32	119.10	7.32	9.82	3.79	3.34	6.78	38.82	0.028	1.638	49.80	18.08	184.83	0.0061
Range	47.53	31.74	2.43	3.12	1.22	1.01	2.48	14.89	0.016	0.859	11.41	3.62	18.49	0.0007
1st Quartile	117.72	97.43	5.59	7.45	3.08	2.61	5.31	29.19	0.015	0.913	43.55	15.86	170.58	0.0057
3rd Quartile	134.55	107.94	6.48	8.31	3.30	2.92	5.90	33.08	0.022	1.331	47.26	17.10	177.23	0.0059
Mean	125.21	103.32	6.03	7.97	3.23	2.78	5.61	31.13	0.019	1.139	45.06	16.48	174.41	0.0058
Standard error	1.88	1.32	0.09	0.14	0.04	0.04	0.09	0.53	0.001	0.042	0.46	0.15	0.83	0.0000
Variation	**1.5**	**1.4**	**1.5**	**1.5**	**1.5**	**1.4**	**1.6**	**1.6**	**2.3**	**2.1**	**1.3**	**1.3**	**1.1**	**1.1**
						Salinity conditions								
Minimum	53.81	44.19	2.98	4.16	1.37	1.11	2.79	24.29	0.018	1.198	18.76	13.13	118.10	0.0067
Maximum	93.53	74.51	5.06	7.04	2.07	1.66	4.58	36.93	0.027	1.936	27.68	15.56	149.03	0.0085
Range	39.71	30.32	2.08	2.88	0.70	0.55	1.79	12.63	0.009	0.737	8.92	2.42	30.93	0.0018
1st Quartile	74.90	58.04	3.76	5.42	1.72	1.34	3.55	26.63	0.021	1.468	22.96	13.84	134.21	0.0072
3rd Quartile	85.12	65.96	4.49	6.37	1.90	1.49	4.14	30.59	0.025	1.764	25.31	14.40	140.31	0.0075
Mean	79.28	62.10	4.17	5.88	1.80	1.41	3.82	28.90	0.023	1.623	24.10	14.17	135.83	0.0074
Standard error	1.36	1.09	0.08	0.11	0.03	0.02	0.07	0.50	0.000	0.030	0.37	0.09	1.31	0.0001
Variation	**1.7**	**1.7**	**1.7**	**1.7**	**1.5**	**1.5**	**1.6**	**1.5**	**1.5**	**1.6**	**1.5**	**1.2**	**1.3**	**1.3**
Reduction (%)	**36.7**	**39.9**	**30.8**	**26.2**	**44.1**	**49.3**	**31.9**	**7.1**	**−22.6**	**−42.5**	**46.5**	**14.0**	**22.1**	**−29.1**

Reduction (%) refers to the percentage decrease in the values of traits or GIs under salinity conditions compared to control conditions.

Table 3. Correlation matrix between all variables under normal (upper right) and salinity stress (lower left) conditions across two years. The full names of traits and GIs are mentioned in Table 1. The units of traits and GIs are mentioned in Figure 1.

Variables	GLA-75	GLA-90	PDW-75	PDW-90	LAI-75	LAI-90	GY	HI	RGR	NAR	LAD	LAR	SLA	SLW
GLA-75		0.68 ***	0.58 ***	0.62 ***	0.63 ***	0.43 **	0.42 **	0.08	0.16	0.09	0.65 ***	−0.25	0.48 ***	−0.49 ***
GLA-90	0.91 ***		0.86 ***	0.78 ***	0.61 ***	0.65 ***	0.24	0.00	0.02	0.03	0.78 ***	−0.18	0.43 ***	−0.44 ***
PDW-75	0.92 ***	0.88 ***		0.83 ***	0.66 ***	0.46 ***	0.09	0.09	0.17	0.02	0.69 ***	−0.51 ***	0.34 *	−0.36 *
PDW-90	0.93 ***	0.92 ***	0.95 ***		0.54 ***	0.57 ***	0.55 ***	0.06	0.41 **	0.53 ***	0.69 ***	−0.54 ***	0.56 ***	−0.56 ***
LAI-75	0.90 ***	0.87 ***	0.82 ***	0.90 ***		0.31	0.15	0.27	0.13	0.10	0.81 ***	−0.09	0.25	−0.28
LAI-90	0.78 ***	0.90 ***	0.77 ***	0.85 ***	0.87 ***		0.49 ***	0.18	0.25	0.27	0.81 ***	−0.12	0.40 *	−0.40 *
GY	0.53 ***	0.57 ***	0.44 ***	0.68 ***	0.70 ***	0.68 ***		0.42 **	0.82 ***	0.79 ***	0.39 **	−0.03	0.60 ***	−0.58 ***
HI	0.15	0.16	0.02	0.13	0.21	0.25	0.33 *		0.25	0.21	0.05	−0.12	0.47 ***	−0.47 ***
RGR	0.21	0.29	0.02	0.34 *	0.41 **	0.41 **	0.84 ***	0.35 *		0.96 ***	0.07	−0.10	0.40 *	−0.37 *
NAR	0.24	0.29	0.16	0.45 ***	0.44 **	0.44 **	0.89 ***	0.28	0.94 ***		0.10	−0.26	0.43 **	−0.39 **
LAD	0.87 ***	0.91 ***	0.82 ***	0.91 ***	0.97 ***	0.96 ***	0.71 ***	0.24	0.42 **	0.45 ***		−0.13	0.40 *	−0.42 **
LAR	−0.13	−0.05	−0.42**	−0.38 *	−0.15	−0.13	−0.24	0.20	0.08	−0.25	−0.14		−0.08	0.06
SLA	0.63 ***	0.67 ***	0.63 ***	0.72 ***	0.72 ***	0.69 ***	0.84 ***	0.34 *	0.71 ***	0.72 ***	0.73 ***	−0.11		−0.99 ***
SLW	−0.66 ***	−0.69 ***	−0.56 ***	−0.74 ***	−0.75 ***	−0.70 ***	−0.83 ***	−0.33 *	−0.69 ***	−0.71 ***	−0.75 ***	0.11	−1.00***	

* $p \leq 0.05$, ** $p \leq 0.01$, *** $p \leq 0.001$.

2.4. Principal Component and Hierarchical Clustering Analyses

We conducted principal component analysis (PCA) to investigate the relationship between variables and genotypes, as well as the interconnections among the variables under normal and salinity conditions (Figure 2). The results showed that the first five principal components (PCs) accounted for 92.1% and 96.9% of the overall variation in variables and genotypes under normal and salinity conditions, respectively (Figure 2). The first PC accounted for the highest percentage of the total variability among variables, explaining 44.4% and 62.8% of the total variability under normal and salinity conditions, respectively. It was significantly associated with three morphological traits (GLA, PDW, and LAI) measured at 75 and 90 DAS, as well as GY and three GIs related to morphological components (LAD, SLA, and SLW) under both conditions. Additionally, it was associated with GIs related to physiological components (RGR and NAR) but only under salinity conditions. The second PC accounted for 22.1% and 17.2% of the total variability under normal and salinity conditions, respectively. It was primarily associated with PDW measured at

75 DAS and GIs related to physiological components (RGR, and NAR) under both conditions, as well as with LAI measured at 75 DAS, GY, and HI under normal conditions. The third, fourth, and fifth PCs explained only 11.8%, 8.1%, and 5.7% of the total variability under normal conditions and 9.1%, 5.0%, and 2.8% under salinity conditions, respectively. The third PC was strongly associated with LAR under both conditions, the fourth PC with HI under salinity conditions, and the fifth PC with LAI measured at 90 DAS under normal conditions (Figure 2).

THE PCA-biplot, based on the first two PCs, organized the variables (traits and GIs) into three main groups, with each group including the same variables under both conditions (Figure 2). The first group consisted of GY, HI, and GIs related to physiological components (RGR and NAR), and one GI related to morphological components (SLA). The second group included all morphological traits measured at 75 and 90 DAS and one GI related to morphological components (LAD). The third group included two GIs related to morphological components (SLW and LAR). The variable vectors within each group were closely aligned and formed acute angles (less than 90°), which indicated a close relationship between them. The variable vectors of the third group formed an angle of 180° with the variable vectors of the first and second groups, indicating a negative relationship between the variables of the third group and those of the first and second groups (Figure 2). Additionally, the PCA-biplot successfully separated the genotypes and distributed them across the four quadrants of the biplot, which reflected a significant variation among genotypes under both conditions (Figure 2). Interestingly, there was a clear difference between the salt-tolerant genotypes Sakha 93 and Kharchia and the salt-sensitive genotype Sakha 61. The two salt-tolerant genotypes were always located in the opposite direction to the salt-sensitive genotype under both conditions. The salt-tolerant genotypes were located in the quadrant with the highest first and second PCs, while the salt-sensitive genotype was found in the quadrant with the lowest first and second PCs under salinity conditions. Under normal conditions, the two salt-tolerant genotypes were located in the quadrant with the highest PC1 and lowest PC2, while the salt-sensitive genotype was found in the quadrant with the highest first and second PCs (Figure 2).

The hierarchical clustering heatmap (HCH) confirmed the results of the PCA analysis and effectively grouped the variables into three distinct categories in both conditions. Under both conditions, the two GIs related to physiological components (RGR and NAR), one GI related to morphological components (SLA), as well as GY and HI, formed a separate group. One GI related to morphological components (LAD) was grouped with all morphological traits measured at 75 and 90 DAS to form a separate group. Two GIs related to morphological components (SLW and SLA) were separated from other variables and formed their own group (Figure 3). Additionally, the HCH effectively separated the genotypes into three and four groups under normal and salinity conditions, respectively (Figure 3). It is noteworthy that under both conditions, the salt-tolerant genotypes Sakha 93 and Kharchia were grouped together, while the salt-sensitive genotype Sakha 61 was placed in a separate group. Furthermore, the majority of genotypes in each group did not match under both conditions. For example, the first group, which exhibited the highest values for all variables except LAR and SLW, included Sakha 61 and 6 RILs from the first crossing (2, 3, 4, 5, 7, and 9) under normal conditions. However, it included three RILs from the first crossing (3, 4, and 9) and three RILs from the second crossing (2, 5, and 15) under salinity conditions. The two salt-tolerant genotypes were located in the third group under normal conditions and in the second group under salinity conditions. Genotypes in the third group had the lowest values for all morphological traits and LAD and moderate values for other variables, while the genotypes in the second group showed moderate values for all variables (Figure 3 and Table 4). The genotypes in the fourth group under salinity conditions exhibited the lowest values for all variables except SLW. This group consisted of Sakha 61, three RILs from the first crossing (5, 11, and 13), and one RIL from the second crossing (14). However, under normal conditions, Sakha 61 and RIL-5 exhibited the highest values, RIL-11 and RIL-14 exhibited the lowest values, and RIL-13 exhibited moderate values for most variables (Figure 3 and Table 4). Therefore, the genotypes and RILs in groups 1, 2, and 3 under normal conditions could be categorized as high-, intermediate-, and low-performing genotypes, respectively. Meanwhile, the genotypes and RILs in groups 1, 2, 3, and 4 under saline conditions could be classified as salt-tolerant, moderately salt-tolerant, moderately salt-sensitive, and salt-sensitive genotypes, respectively (Table 4).

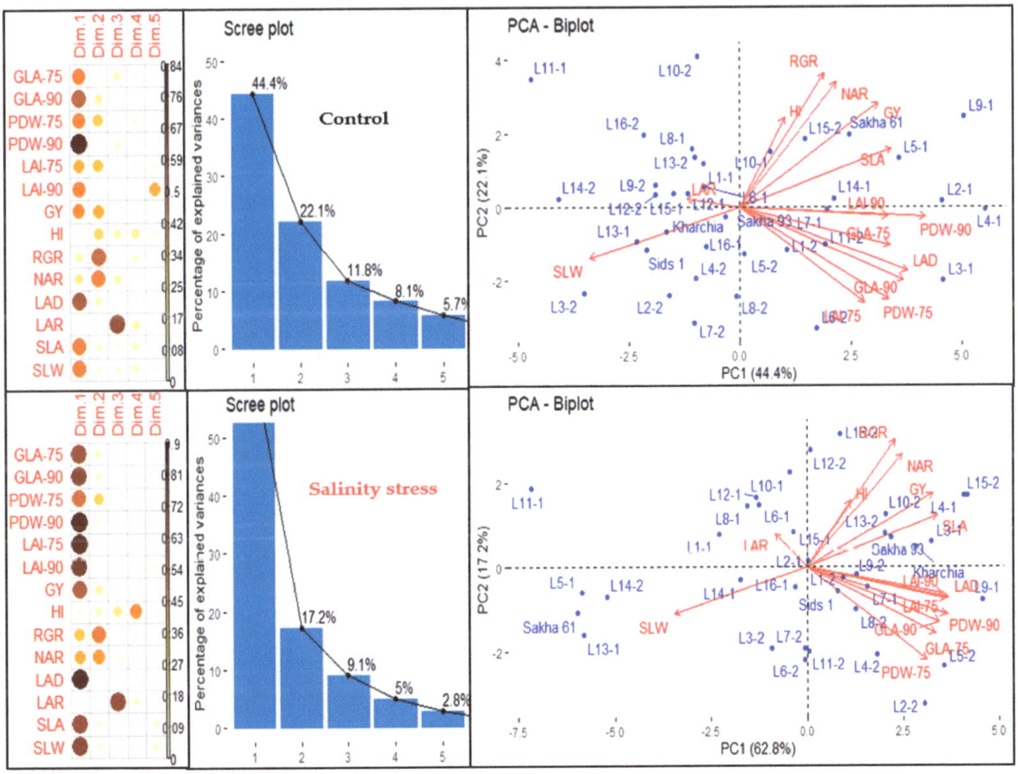

Figure 2. Principal component analysis (PCA) of different traits and growth indicators (GIs) in 36 genotypes under normal and salinity conditions. PCA biplot illustrates the grouping of traits and GIs in the first two principal components (PCs). Scree plot shows the percentage of explained variances of each PC, as well as the contribution of traits and GIs on extracted PCs based on square cosine and squared coordinates. The full names of the traits and GIs are mentioned in Table 1.

Table 4. Mean values of different traits measured at 75 and 90 days after sowing along with growth indicators (GIs) for different groups of genotypes under normal and salinity conditions. The full names of traits and GIs are mentioned in Table 1.

	GLA-75	GLA-90	PDW-75	PDW-90	LAI-75	LAI-90	GY	HI	RGR	NAR	LAD	LAR	SLA	SLW
	Normal Conditions													
	First Year													
Group-1 (7)	125.82	108.86	6.72	9.17	3.29	2.95	6.17	32.88	0.021	1.399	46.84	14.89	180.78	0.0055
Group-2 (14)	123.42	105.60	6.43	7.86	3.15	2.78	5.01	29.42	0.013	0.833	44.52	16.08	171.29	0.0059
Group-3 (15)	113.31	96.14	5.78	7.59	2.89	2.57	5.54	31.46	0.018	1.162	40.98	15.83	172.16	0.0058
	Second year													
Group-1 (7)	139.40	115.05	6.62	9.45	3.55	3.01	6.34	30.55	0.024	1.498	49.23	16.04	180.73	0.0055
Group-2 (14)	127.35	106.30	5.99	7.75	3.38	2.84	5.35	30.36	0.017	1.015	46.65	17.08	174.11	0.0058
Group-3 (15)	129.86	97.54	5.35	7.42	3.31	2.74	5.88	32.56	0.022	1.225	45.38	17.89	173.93	0.0058

Table 4. *Cont.*

Groups	GLA-75	GLA-90	PDW-75	PDW-90	LAI-75	LAI-90	GY	HI	RGR	NAR	LAD	LAR	SLA	SLW
	Salinity conditions													
	First year													
Group-1 (6)	87.09	71.27	4.83	6.86	1.86	1.46	3.99	30.13	0.024	1.725	24.89	13.66	137.27	0.0073
Group-2 (15)	82.41	61.07	4.57	6.25	1.75	1.28	3.70	28.58	0.021	1.583	22.73	13.30	132.93	0.0076
Group-3 (10)	72.20	54.75	3.74	5.27	1.61	1.16	3.75	31.37	0.023	1.630	20.76	14.14	130.54	0.0077
Group-4 (5)	58.34	43.38	3.46	4.50	1.22	0.95	2.93	27.44	0.017	1.371	16.29	12.78	110.82	0.0091
	Second year													
Group-1 (6)	89.59	71.89	4.51	6.64	2.11	1.79	4.51	29.35	0.026	1.776	29.27	14.60	145.70	0.0069
Group-2 (15)	83.62	67.09	4.29	6.14	1.98	1.61	4.03	28.06	0.024	1.659	26.94	14.56	143.89	0.0070
Group-3 (10)	78.11	62.63	3.90	5.67	1.89	1.54	3.94	29.32	0.025	1.697	25.75	14.86	142.84	0.0070
Group-4 (5)	72.53	59.81	3.56	4.94	1.77	1.40	3.28	26.09	0.022	1.417	23.80	15.70	128.38	0.0078

2.5. Relationship between Growth Indicators and Grain Yield for Each Group under Normal and Salinity Conditions

Figure 4 represents the relationship between different GIs and GY under normal and salinity conditions. Under normal conditions, the two GIs related to physiological components (RGR and NAR) and one GI related to morphological components (LAD) showed a moderate and quadratic relationship (R^2 = 0.42–0.59) with the GY of both high- and low-performing genotype groups. RGR exhibited a moderate and linear relationship with the GY of the intermediate-performing genotype group (R^2 = 0.44), while NAR and LAD did not show a significant relationship with the GY of this group (Figure 4). The other three GIs related to morphological components (LAR, SLA, and SLW) did not show a significant relationship with the GY of both high- and low-performing genotype groups. However, they did show a weak to moderate relationship with the GY of the intermediate-performing genotype group (R^2 = 0.31–0.54) and this relationship was linear for SLA and SLW, and quadratic for LAR (Figure 4).

Under salinity conditions, the two GIs related to physiological components (RGR and NAR) and two GIs related to morphological components (SLA and SLW) showed a strong and linear relationship (R^2 = 0.76–0.98) with the GY of salt-tolerant and salt-sensitive genotype groups. RGR and NAR still had a moderate to strong and linear relationship (R^2 = 0.60–0.87) with the GY of intermediate salt-tolerant and intermediate salt-sensitive genotype groups. However, SLA and SLW had a moderate and quadratic relationship (R^2 = 0.50) with the GY of the intermediate salt-tolerant genotype group, while they did not show a significant relationship with the GY of the intermediate salt-sensitive genotype group. The other GIs related to morphological components (LAD and LAR) showed a strong (R^2 = 0.86) and moderate (R^2 = 0.57) relationship, respectively, with only the GY of the salt-sensitive genotype group. However, they did not show a significant relationship with the other three groups (Figure 4).

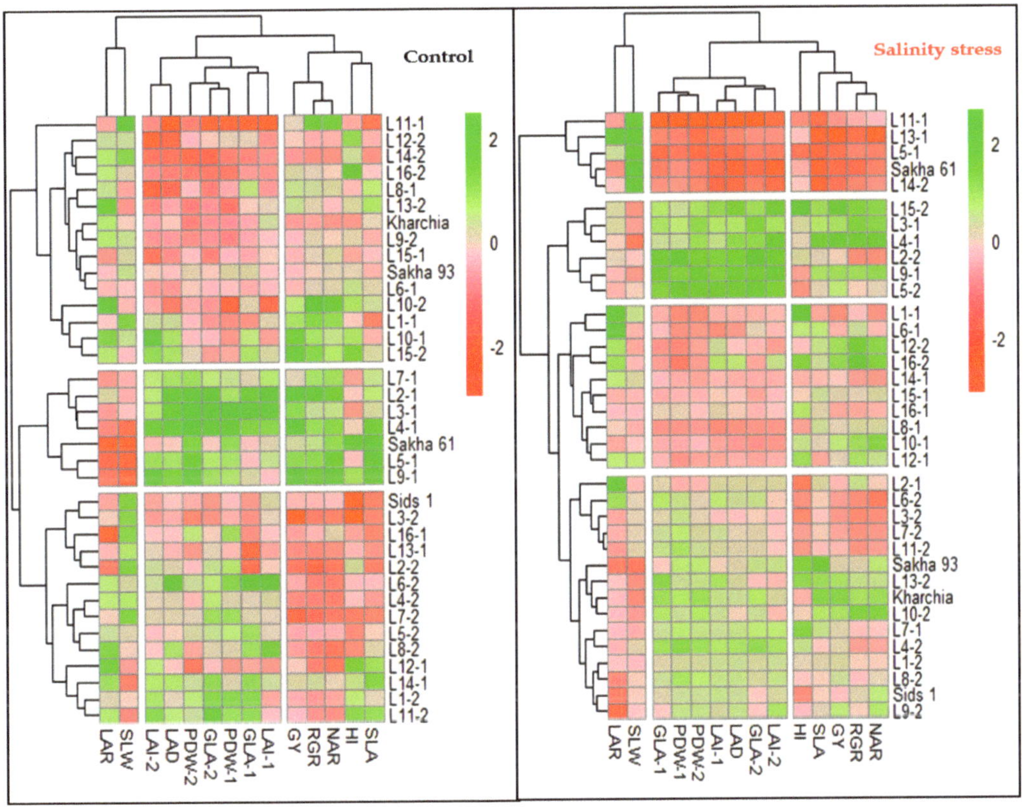

Figure 3. Hierarchical clustering heatmap displaying the relationships between the 36 genotypes and all variables under both normal and salinity conditions. Columns represent the variables and rows represent the genotypes. The different colors and their intensities represent the degree associations between variables and genotypes. Darker green and darker red indicate higher and lower values of the mean valuables, respectively.

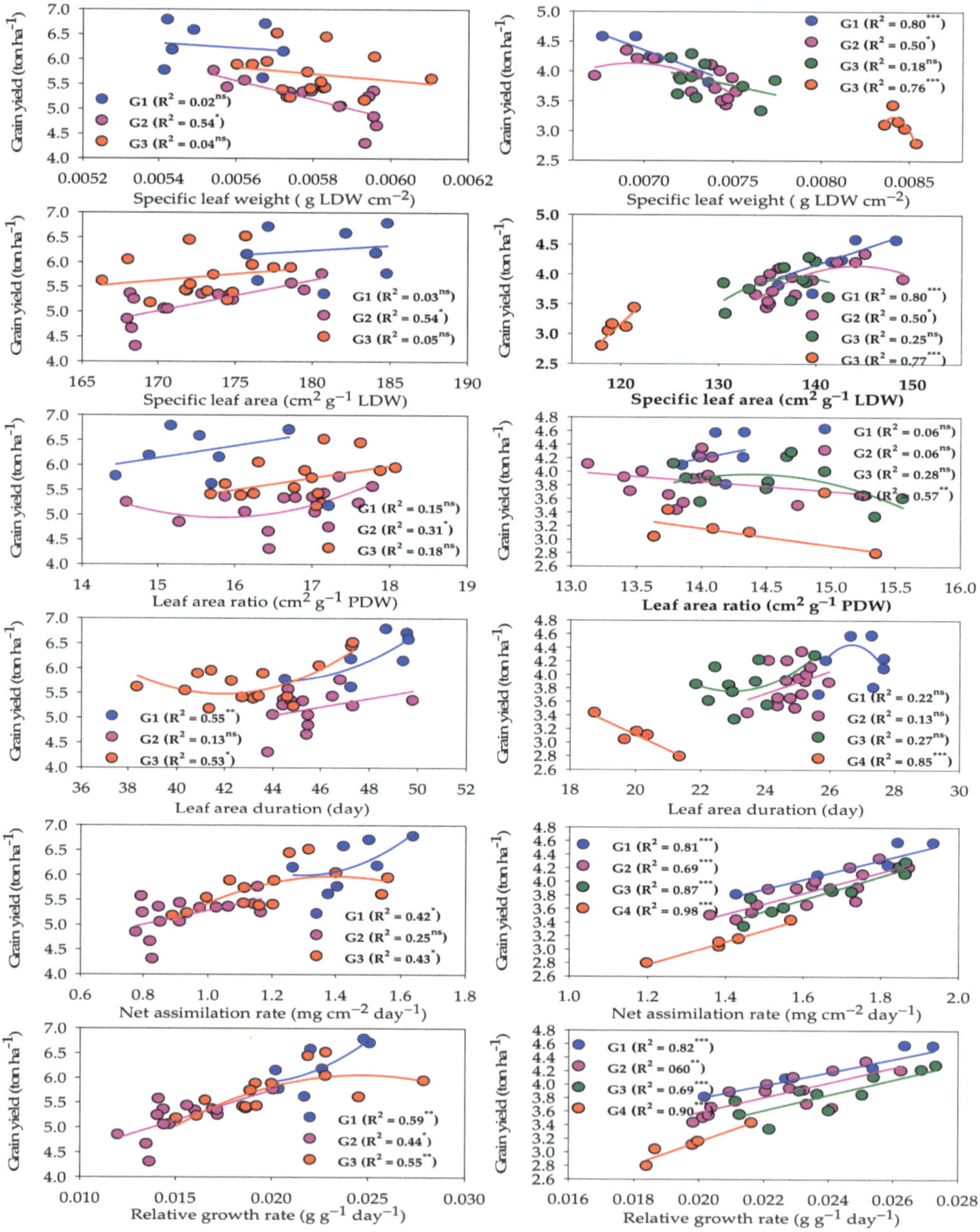

Figure 4. Fit regression models of the relationship between different growth indicators and grain yield for each genotype group (G) clustered under normal (left side) and salinity (right side) conditions. *** Significant at $p < 0.0001$; ** significant at $p < 0.001$; * significant at $p < 0.05$; ns: not significant.

3. Discussion

Developing salt-tolerant genotypes is one of the most effective options for mitigating the negative impacts of salinity stress on crop growth and production. Nevertheless, creating salt-tolerant genotypes is a challenging task. Exploring the genetic diversity among different wheat genotypes is an important step in breeding programs that aim to identify potential candidates with salt tolerance. Achieving this step requires two important things: assessing the salt tolerance of wheat genotypes in real field conditions, as well as identifying traits that can be used as reliable screening criteria for distinguishing salt tolerance among wheat genotypes. In this study, we investigated the salt-tolerance of 36 wheat germplasms, including 4 cultivars differing in their salt tolerance and 32 RILs. The salt-tolerance of these germplasms was evaluated in real field conditions using different traits and GIs as screening criteria. To our knowledge, the use of GIs as screening criteria for assessing the salt tolerance of wheat genotypes is not widely used in breeding programs. The findings of this study showed a highly genetic variability for all traits and GIs between SL and G (Table 1). Furthermore, the traits and GIs varied by approximately 1.5-fold between different genotypes under both normal and salinity stress conditions (Figure 1 and Table 2). These findings indicate that both traits and GIs could effectively classify genotypes into distinct groups under normal and salinity stress conditions in real field conditions.

Generally, genotypes can enhance their performance (growth and production) by adjusting their morphological and/or physiological components in response to environmental conditions. These adjustments can lead to improved growth and production under normal conditions, as well as better adaptation and tolerance to environmental stress. Because GIs are indices that reflect the interaction of physiological, morphological, and phenological aspects, as well as being largely affected by the balance between source and sink organs and biomass allocation [21,32,34,36–40], we believe that these GIs can be used as screening criteria to understand genotypic differences in growth and yield under either normal or stress conditions. For instance, RGR, which is a compound rate of change, reflects the increase of PDW per unit of biomass and per unit of time. Furthermore, this index is a product of NAR (a physiological component) and LAR (a morphological component). NAR is an index that shows the balance between the rates of photosynthesis and respiration in the whole plant. This indicates the rate at which a plant converts energy from photosynthesis into biomass. LAR is an index that measures the efficiency of plants in capturing light and converting it into biomass. A high LAR indicates that the plants have a large leaf area relative to their biomass, while a low LAR indicates the opposite. SLA is an index that measures the ratio of leaf area per unit leaf mass. A high SLA indicates the plants tend to have thinner leaves and a larger leaf area relative to their mass, while a low SLA indicates the opposite. SLW is another important index that refers to the weight of a leaf per unit leaf area. It is often related to chlorophyll content, photosynthetic function, and the impact of environmental factors on leaf structure and function. Therefore, it is often used as an indicator of plant health and stress [41]. The aforementioned facts about different GIs indicate that, by analyzing these indicators, it would be possible to explain whether genotypic variation in growth and production under normal and salinity stress conditions, as well as salt tolerance under salinity stress conditions, can be attributed to morphological and/or physiological components. In this study, we observed a strong positive correlation of GY with the two GIs related to physiological components (RGR and NAR) under both conditions. Furthermore, one GI related to morphological components (LAD) showed a strong positive correlation with the three morphological traits (GLA, PDW, and LAI) measured at 75 and 90 DAS under both conditions. However, the correlation of the three morphological traits with the other GIs related to morphological components (SLA and SLW) was tighter under salinity than under normal conditions. Under both conditions, the correlation of one GI related to morphological components (LAR) with all traits was weak and non-significant, except for that of PDW (Table 3). All of these findings indicate that (1) the genotypic variation in GY under normal and salinity conditions was strongly attributed to the variation in physiological components (RGR and NAR), but not with LAR; (2) one GI related to morphological components (LAD) played a vital role in enhancing growth and production of genotypes under both conditions; and (3) SLA and SLW, which are morphological components associated with leaf structure and function, were significantly affected by salinity stress, with SLA decreasing significantly due to salinity stress compared to under normal conditions, while the opposite was found with SLW (Figure 1 and Table 2). These findings are consistent with the results reported by Cramer et al. [36], Murillo-Amador et al. [42], and Chen et al. [43]. They found that SLA values were higher under salinity stress compared to the control, and vice versa for SLW. The low SLA under salinity stress indicates that plants develop thicker leaves and a smaller leaf area relative to their mass. This suggests that leaf expansion may play an important role in the salt tolerance of genotypes. This may be due to the different components of salinity stress (osmotic stress, ionic toxicity, and essential nutrient deficit) interacting together to limit cell division and elongation. In addition, increased leaf thickness due to decreased SLA could create a longer path for carbon dioxide diffusion from stomata to chloroplasts, leading to a significant reduction in the photosynthetic capacity of plants. The increase in SLW under salinity conditions indicates that salinity stress affected the dry weight of leaves to a greater extent than their leaf area. Meanwhile, salinity stress resulted in the allocation of less biomass and photosynthetic products in leaves due to the accumulation of toxic ions and their negative impacts on the photosynthesis rate. Therefore, the sensitivity of GIs related to morphological components, particularly LAD, SLA, and SLW, to salinity stress suggests that they could be used as screening criteria to evaluate the salt tolerance of wheat genotypes.

To understand the relationship between traits and genotypes, as well as classify traits and genotypes into distinct groups, both PCA and HCH were used to achieve this goal (Figures 2 and 3). In PCA, the GIs related to physiological components (RGR and NAR) and SLA were grouped with GY under both conditions. The vectors of these GIs and GY formed acute angles between each other. This indicates that improving RGR, NAR, and SLA in wheat breeding could be recognized as an effective strategy to enhance production under normal and salinity stress conditions. Our results also indicate that LAD is an important index associated with wheat growth under both conditions. This index is closely related to the three morphological traits (GLA, PDW, and LAI) measured at 75 and 90 DAS, and their vectors also form acute angles with the vectors of morphological traits (Figure 2). The high LAD refers to the higher photosynthetic capacity of leaves and, therefore, the amount of biomass produced by the plant [44]. Generally, salinity stress leads to modifications in plant morphological features such as LAD [3], which results in a reduction in the rate and duration of photosynthetic efficiency and consequently the amount of dry matter produced by photosynthesis. This may explain why the three morphological traits are closely associated with LAD under both conditions. This result also supports that LAD could be used as an effective screening criterion for evaluating the salt tolerance of wheat genotypes at early growth stages. The results also found that the vectors of SLA formed a straight angle with GY under both conditions. This confirms that improving SLW plays a vital role in enhancing crop production, whether crops are grown under normal or stressful conditions. Chen et al. [43] reported that an increase in SLW under salinity stress reflects the ability of plants to allocate more organic matter to RuBP carboxylase and photosynthesis-related thylakoids. This ultimately leads to a higher capacity of plants for photosynthetic rate and the accumulation of photosynthetic products in leaves. Therefore, we confirm that SLW could be related to enhanced leaf photosynthesis and yield potential of the wheat crop, which suggests that this indicator could be used as a screening criterion for evaluating genotypes under either normal or salinity stress conditions. Several studies have reported that SLW is highly correlated with photosynthetic capacity and dry matter accumulation under both normal and stress conditions, as well as under different agronomic practices. This ultimately contributes favorably to final GY [34,42,45–47].

One important finding of this study is that the different GIs did not classify the genotypes into similar groups under both normal and salinity stress conditions. For instance, the group that included the salt-sensitive genotype Sakha 61 had the lowest values for LAR and SLW and the highest values for other GIs under normal conditions. However, the group that included this genotype under salinity conditions showed opposite results compared to under the control conditions. The same situation was observed also with the salt-tolerant genotypes Sakha 93 and Kharchia (Figure 3 and Table 4). This finding suggests that when using GIs as screening criteria for grouping genotypes, it is necessary to group genotypes separately for each growth condition. This is because the growth behaviors of genotypes may vary under different environmental conditions. Therefore, we recommend using different GIs to explain the variation in growth and yield potential among genotypes under normal conditions, as well as to explain the variation in salt tolerance among genotypes under salinity stress. This is because the different GIs are efficient in detecting changes that occur simultaneously in morphological and physiological components at a whole-plant level. Therefore, this helps to determine whether variations among genotypes are due to morphological and/or physiological changes, which mainly depend on the behavior of genotypes under specific growth conditions. In this study, when the GIs were related to GY for different genotype groups under each condition, we found that, under normal conditions, GIs related to physiological components (RGR and NAR) and LAD showed a significant relationship with the GY of both high- and low-performing genotype groups. Meanwhile, GIs related to morphological components (SLA and SLW) did not show a significant relationship with the GY of both groups. Under salinity conditions, all GIs related to both morphological and physiological components showed a strong relationship with the GY of salt-tolerant and salt-sensitive genotype groups (Figure 4). These findings indicate that physiological components are the main factor explaining variations in yield performance among genotypes under normal conditions, while both physiological and morphological components play a major role in explaining salt tolerance variations among genotypes under salinity stress conditions. The studies of Praxedes et al. [21], Bayuelo-Jimenez et al. [22], and Simane et al. [48] also found that the differences in stress tolerance (salinity or drought) among genotypes were due to both physiological and morphological components of GIs. An explanation for this finding is that the different components of salinity stress (osmotic stress, ion toxicity, and nutrient imbalance) not only inhibit the physiological processes of plants but also lead to substantial changes in leaf traits such as leaf area, leaf expansion, necrosis, browning, and senescence.

4. Materials and Methods

4.1. Plant Materials and Experimental Conditions

A total of 36 wheat germplasms was used in this study and evaluated under normal and salinity stress conditions in real field conditions. These germplasms included 36 F8 recombinant inbred lines (RILs), developed by crossing the salt-sensitive genotype Sakha-61 with the salt-tolerant genotype Sakha-93 (16 RILs), and the moderately salt-tolerant genotype Sids 1 with Sakha-93 (16 RILs), as well as the salt-tolerant genotype Kharchia. The three parents and Kharchia were chosen based on previous studies that examined their salt tolerance in real field and greenhouse conditions using various agro-physiological traits [12,30,49,50]. All genotypes were evaluated under normal and high salinity concentration (150 mM NaCl) at the Experimental Farm Station of the College of Food and Agriculture Sciences at King Saud University in Riyadh, Saudi Arabia (24°25′N, 46°34′E, 400 m a.s.l.) during the winter seasons of 2019–2020 and 2020–2021. The soil of the experimental site was characterized as a sandy loam (14.9% clay, 28.4% silt, and 56.7% sand) with a bulk density of 1.48 g cm^{-3}, organic matter of 0.45%, pH range of 7.80–7.90, and electrical conductivity (EC) of 1.12 dS m^{-1}. The winter season at the experimental site is often sunny, with precipitation, relative humidity, and mean temperature of 8.0–25.0 mm, 17.7–47.5%, and 12.9−32.2 °C, respectively.

4.2. Experimental Design, Agronomic Practices, and Salinity Treatments

The field trial was carried out in a split-plot design with three replications, with the salinity treatments arranged in the main plot and the genotypes in the subplots. The size of the subplot was 1.5 m^2 (five 1.5 m long rows spaced 20 cm apart). The seeds of the genotypes were sown by hand on 25 November in the first growing season, and 17 November in the second growing season at a seeding rate of 15 g m^{-2}. The plants of wheat were fertilized with 150 kg of N, 100 kg of P_2O_5, and 90 kg of K_2O per hectare. The nitrogen (N), phosphorus (P), and potassium (K) fertilizers were applied in the form of ammonium nitrate (33.5% N), calcium superphosphate (18.5% P_2O_5), and potassium chloride (50% K_2O), respectively. The entire amount of P was applied before sowing. The amounts of N and K were divided into three and two equal doses, respectively. The first dose of each was applied before sowing; the second dose of N was applied at the late tillering growth stage; and the last dose of each was applied at the booting growth stage.

All genotypes in the normal treatment were irrigated with fresh water throughout the growth cycle, whereas in the salinity treatment, fresh water was used for the first 21 days after sowing in order to obtain a high germination percentage and good seedling establishment. Afterward, saline water containing 8.8 g NaCl L^{-1} (150 mM NaCl) was used for irrigation until the last irrigation. The fresh and saline water were applied through a low-pressure surface irrigation system, which consisted of a mainline (76 mm in diameter) connected to a five-cubic-meter plastic water tank and branched off at each subplot to the sub-main hoses. To control the amount of water delivered to each subplot, each sub-main hose was equipped with a manual control valve. The irrigation rate and frequency were determined according to wheat plant phenology and climatic conditions at the experimental site. In the salinity treatment, the build-up of salt in the root zone was monitored throughout the growth cycle by collecting soil samples from various locations at depths of 0–60 cm, and recording their EC. Based on the analysis of the different soil samples, the mean EC value did not exceed 16.3 dS m^{-1}.

4.3. Measurements of Various Traits

At 75 and 90 days after sowing, which are the booting and anthesis growth stages in wheat, respectively, 10 plants from each subplot in both the normal and salinity treatments were manually pulled, packed in plastic bags, and brought to the laboratory to measure various growth traits. The green leaves of the 10 plants were separated and their area was measured using an area meter (LI 3100; LI-COR Inc., Lincoln, NE, USA) to determine green leaf area (GLA). The leaves and shoots of the 10 plants were oven-dried at 80 °C until their weight became constant and then weighed to record the dry weight of the green leaves (LDW) and shoots (SDW). The LDW and SDW were summed to record the final plant dry weight (PDW). GLA and ground area per plant were used to calculate leaf area index (LAI) according to Yue et al. [51]

Upon plants reaching the phase of full maturity (at approximately 150 days from sowing), two inner rows of 1.25 m long (totaling 0.75 m^2 in area) were harvested by hand from each subplot. The plants were then air-dried for 7 days and weighed to record the biological yield (BY). The grains were then collected, cleaned, adjusted to moisture content of 14.0%, and weighed to record the grain yield (GY). Both BY and GY were finally expressed as tons per hectare and used to calculate the harvest index (HI), which is the ratio of GY to BY.

4.4. Calculation of Different Growth Indicators

Based on the values of PDW (W1 and W2), GLA (GLA1 and GLA2), and GLDW (GLDW1 and GLDW2) measured at 75 (T1) and 90 (T2) days after sowing, various growth indicators were calculated according to Hunt [52]. The abbreviation, full name, and formula of each growth indicator are presented in Table 5.

Table 5. The abbreviation (Abb.), full name, and formula of various growth indicators used in this study.

Abb.	Index Full Name	Index Formula
RGR	Relative growth rate (g g^{-1} day^{-1})	$\frac{\text{LnW}_2-\text{LnW}_1}{\text{T}_2-\text{T}_2}$
NAR	Net assimilation rate (g cm^2 day^{-1})	$\frac{(W_2-W_1)(\text{Ln}GLA_2-\text{Ln}GLA_1)}{(GLA_2-GLA_1)(T_2-T_1)}$

Table 5. *Cont.*

Abb.	Index Full Name	Index Formula
LAD	Leaf area duration (day)	$\frac{(GLA_1+GLA_2)(T_2-T_1)}{2}$
LAR	Leaf area ratio (cm^2 g^{-1} PDW)	$\frac{(GLA_2-GLA_1)(\text{Ln}W_2-\text{Ln}W_1)}{(\text{Ln}GLA_2-\text{Ln}GLA_1)(W_2-W_1)}$
SLA	Specific leaf area (cm^2 g^{-1} LDW)	$\frac{(GLA_2-GLA_1)(\text{Ln}GLDW_2-\text{Ln}GLDW_1)}{(\text{Ln}GLA_2-\text{Ln}GLA_1)(GLDW_2-GLDW_1)}$
SLW	Specific leaf weight (g LDW cm^2)	$\frac{(W_2-W_1)(\text{Ln}GLA_2-\text{Ln}GLA_1)}{(GLA_2-GLA_1)(\text{Ln}W_2-\text{Ln}W_1)}$

PDW and LDW indicate plant dry weight and leaf dry weight, respectively.

4.5. Data Analysis

Before statistical analysis, the normality distribution and homogeneity of variances of all traits were tested using Shapiro–Wilk and Bartlett's chi-square tests, respectively. Sub-sequently, the data of various traits were subjected to analysis of variance (ANOVA) ap-propriate to a split-plot design to detect the effects of the main factors (salinity and genotypes) and their interaction on these traits: ANOVA was carried out for each year separately and combined over two years. The analysis of the level of correlation among traits and GIs of each other was performed using the non-parametric Spearman test. This analysis was performed using RStudio 1.3.959 (RStudio Team 2020). All traits and genotypes were subjected to PCA to detect the interrelationships among various variables and reduce the dimensionality of the dataset without losing important information. Eigenvalues, cumulative variability, and a PCA-biplot constructed between the first two components (PC1 and PC2) were extracted using the XLSTAT package version 2021. Hierarchical clustering heatmap analysis was performed for different GIs to group genotypes into distinct groups under normal and salinity conditions. Relationship analyses between different GIs and GY of each group under normal and salinity conditions were performed using GIs as independent variables, while GY was the dependent variable. Figures of regression were derived using the Sigma Plot 14.0 software program (Systat software, Inc., Chicago, IL, USA).

5. Conclusions

From the results of the different analyses applied in this study, we concluded that significant differences in various traits and GIs were observed between SL, G, and their interaction in each growing season and the combined data of the two seasons. The correlation between traits and GIs was higher under salinity stress than under normal conditions. PCA and HCA organized the traits and GIs into three main groups, with each group including the same variables under both conditions. Traits and GIs successfully categorized the genotypes into different groups, but did not group the genotypes similarly under both normal and salinity stress conditions. Based on the relationship between the GY of different groups and GIs, the physiological components (RGR and NAR) were the main factors explaining variations in yield performance among genotypes under normal conditions, while both physiological and morphological components (all GIs) were the major factors explaining variations in salt tolerance among genotypes under salinity stress conditions. Collectively, our findings in this study indicate that different GIs are effective as screening criteria for understanding and selecting genotypes with good performance under normal conditions and their salt tolerance under salinity stress conditions.

Author Contributions: Conceptualization, S.E.-H., M.M. and N.A.-S.; methodology, S.E.-H., N.A.-S., and M.M.; software, S.E.-H., Y.H.D., and M.M.; validation, S.E.-H., N.A.-S., and M.M.; formal analysis, S.E.-H. and M.M.; investigation, S.E.-H.; resources, S.E.-H. and Y.H.D.; data curation, S.E.-H., Y.H.D., N.A.-S., and M.M.; writing—original draft preparation, S.E.-H.; writing—review and editing, S.E.-H.; visualization, S.E.-H., N.A.-S., and M.M.; supervision, S.E.-H. and N.A.-S.; project administration, S.E.-H.; funding acquisition, S.E.-H., Y.H.D., and N.A.-S. All authors have read and agreed to the published version of the manuscript.

Funding: This research was funded by the Researchers Supporting Project number (RSPD2024R730), King Saud University, Riyadh, Saudi Arabia.

Data Availability Statement: All data are presented in article.

Acknowledgments: The authors acknowledge the Researchers Supporting Project number (RSPD2024R730), King Saud University, Riyadh, Saudi Arabia.

Conflicts of Interest: The authors declare no conflicts of interest.

References

1. Wang, Z.; Fan, B.; Guo, L. Soil salinization after long-term mulched drip irrigation poses a potential risk to agricultural sustainability. *Eur. J. Soil Sci.* **2019**, *70*, 20–24. [CrossRef]
2. Qadir, M.; Quillérou, E.; Nangia, V.; Murtaza, G.; Singh, M.; Thomas, R.J.; Drechsel, P.; Noble, A.D. Economics of salt-induced land degradation and restoration. *Nat. Resour. Forum* **2014**, *38*, 282–295. [CrossRef]
3. El-Hendawy, S.E.; Hassan, W.M.; Al-Suhaibani, N.A.; Refay, Y.; Abdella, K.A. Comparative performance of multivariable agro-physiological parameters for detecting salt tolerance of wheat cultivars under simulated saline field growing conditions. *Front. Plant Sci.* **2017**, *8*, 435. [CrossRef] [PubMed]
4. Ivushkin, K.; Bartholomeus, H.; Bregt, A.K.; Pulatov, A.; Kempen, B.; de Sousa, L. Global mapping of soil salinity change. *Remote Sens. Environ.* **2019**, *231*, 111260. [CrossRef]
5. FAO. How to Feed the World—2050: High-Level Expert Forum, Rome 12–13 October 2009—Investment—World | ReliefWeb. Available online: https://reliefweb.int/report/world/how-feed-world-2050-high-level-expert-forum-rome-12-13-oct-2009-investment (accessed on 1 January 2020).
6. Grote, U.; Fasse, A.; Nguyen, T.T.; Erenstein, O. Food security and the dynamics of wheat and maize value chains in Africa and Asia. *Front. Sustain. Food Syst.* **2021**, *4*, 617009. [CrossRef]
7. Shiferaw, B.; Smale, M.; Braun, H.-J.; Duveiller, E.; Reynolds, M.; Muricho, G. Crops that feed the world 10. past successes and future challenges to the role played by wheat in global food security. *Food Sec.* **2013**, *5*, 291–317. [CrossRef]
8. Tao, R.; Ding, J.; Li, C.; Zhu, X.; Guo, W.; Zhu, M. Evaluating and screening of agro-physiological indices for salinity stress tolerance in wheat at the seedling stage. *Front. Plant Sci.* **2021**, *12*, 646175. [CrossRef]
9. Sheoran, P.; Kumar, A.; Kumar, A.; Raju, R.; Sharma, R.; Parjapat, K.; Barman, A.; Singh, R.K.; Kumar, S.; Sharma, P.C.; et al. Impact of pressmud application in reclamation of high RSC irrigation water induced soil sodification and sustaining rice (*Oryza sativa*)—Wheat (*Triticum aestivum*) production in indo-gangetic plains. *Indian J. Agric. Sci.* **2020**, *90*, 206–211. [CrossRef]
10. Minhas, P.S.; Bali, A.; Bhardwaj, A.K.; Singh, A.; Yadav, R.K. Structural stability and hydraulic characteristics of soils irrigated for two decades with waters having residual alkalinity and its neutralization with gypsum and sulfuric acid. *Agric. Water Manag.* **2021**, *244*, 106609. [CrossRef]
11. Munns, R.; Tester, M. Mechanisms of salinity tolerance. *Annu. Rev. Plant Biol.* **2008**, *59*, 651–681. [CrossRef]
12. Mansour, E.; Moustafa, E.S.A.; Desoky, E.-S.M.; Ali, M.M.A.; Yasin, M.A.T.; Attia, A.; Alsuhaibani, N.; Tahir, M.U.; El-Hendawy, S. Multidimensional evaluation for detecting salt tolerance of bread wheat genotypes under actual saline field growing conditions. *Plants* **2020**, *9*, 1324. [CrossRef]
13. Sheoran, P.; Kumar, A.; Sharma, R.; Prajapat, K.; Kumar, A.; Barman, A.; Raju, R.; Kumar, S.; Dar, Y.J.; Singh, R.K.; et al. Quantitative dissection of salt tolerance for sustainable wheat production in sodic agro-ecosystems through farmers' participatory approach: An Indian experience. *Sustainability* **2021**, *13*, 3378. [CrossRef]
14. Tavakkoli, E.; Rengasamy, P.; McDonald, G.K. The response of barley to salinity stress differs between hydroponic and soil systems. *Funct. Plant Biol.* **2009**, *37*, 621–633.
15. Matković Stojšin, M.; Petrović, S.; Banjac, B.; Zečević, V.; Roljević Nikolić, S.; Majstorović, H.; Đorđević, R.; Knežević, D. Assessment of genotype stress tolerance as an effective way to sustain wheat production under salinity stress conditions. *Sustainability* **2022**, *14*, 6973. [CrossRef]
16. Mubushar, M.; El-Hendawy, S.; Tahir, M.U.; Alotaibi, M.; Mohammed, N.; Refay, Y.; Tola, E. Assessing the suitability of multivariate analysis for stress tolerance indices, biomass, and grain yield for detecting salt tolerance in advanced spring wheat lines irrigated with saline water under field conditions. *Agronomy* **2022**, *12*, 3084. [CrossRef]
17. Negarestani, M.; Tohidi-Nejad, E.; Khajoei-Nejad, G.; Nakhoda, B.; Mohammadi-Nejad, G. Comparison of different multivariate statistical methods for screening the drought tolerant genotypes of pearl millet (*Pennisetum americanum* L.) and sorghum (*Sorghum bicolor* L.). *Agronomy* **2019**, *9*, 645. [CrossRef]
18. Pongprayoon, W.; Tisarum, R.; Theerawittaya, C.; Cha-Um, S. Evaluation and clustering on salt-tolerant ability in rice genotypes (*Oryza sativa* L. subsp. *indica*) using multivariate physiological indices. *Physiol. Mol. Biol. Plants* **2019**, *25*, 473–483. [CrossRef]
19. Miransari, M.; Smith, D. Sustainable wheat (*Triticum aestivum* L.) Production in saline fields: A review. *Crit. Rev. Biotechnol.* **2019**, *39*, 999–1014. [CrossRef]
20. Konings, H. Physiological and morphological differences between plants with a high NAR or a high LAR as related to environmental conditions. In *Causes and Consequences of Variation in Growth Rate and Productivity of Higher Plants*; Lambers, H., Cambridge, M.L., Konings, H., Pons, T.L., Eds.; SPB Academic Publishing: The Hague, The Netherlands, 1989; pp. 101–123.

21. Praxedes, S.C.; De Lacerda, C.F.; DaMatta, F.M.; Prisco, J.T.; Gomes-Filho, E. Salt tolerance is associated with differences in ion accumulation, biomass allocation and photosynthesis in cowpea cultivars. *J. Agron. Crop Sci.* **2010**, *196*, 193–204. [CrossRef]
22. Bayuelo-Jiménez, J.S.; Jasso-Plata, N.; Ochoa, I. Growth and physiological responses of *Phaseolus* species to salinity stress. *Int. J. Agron.* **2012**, *2012*, 527673. [CrossRef]
23. El-Hendawy, S.E.; Hu, Y.; Schmidhalter, U. Growth, ion content, gas exchange, and water relations of wheat genotypes differing in salt tolerances. *Aust. J. Agric. Res.* **2005**, *56*, 123–134. [CrossRef]
24. El-Hendawy, S.E.; Ruan, Y.; Hu, Y.; Schmidhalter, U. A comparison of screening criteria for salt tolerance in wheat under field and controlled environmental conditions. *J. Agron. Crop Sci.* **2009**, *195*, 356–367. [CrossRef]
25. Basu, S.; Kumar, A.; Benazir, I.; Kumar, G. Reassessing the role of ion homeostasis for improving salinity tolerance in crop plants. *Physiol. Plant.* **2021**, *171*, 502–519. [CrossRef]
26. Kiani-Pouya, A.; Rasouli, F.; Rabbi, B.; Falakboland, Z.; Yong, M.; Chen, Z.-H.; Zhou, M.; Shabala, S. Stomatal traits as a determinant of superior salinity tolerance in wild barley. *J. Plant Physiol.* **2020**, *245*, 153108. [CrossRef] [PubMed]
27. EL Sabagh, A.; Islam, M.S.; Skalicky, M.; Ali Raza, M.; Singh, K.; Anwar Hossain, M.; Hossain, A.; Mahboob, W.; Iqbal, M.A.; Ratnasekera, D.; et al. Salinity stress in wheat (*Triticum aestivum* L.) in the changing climate: Adaptation and management strategies. *Front. Agron.* **2021**, *3*, 661932. [CrossRef]
28. Dadshani, S.; Sharma, R.C.; Baum, M.; Ogbonnaya, F.C.; Léon, J.; Ballvora, A. Multi-dimensional evaluation of response to salt stress in wheat. *PLoS ONE* **2019**, *14*, e0222659. [CrossRef]
29. Sheoran, P.; Basak, N.; Kumar, A.; Yadav, R.K.; Singh, R.; Sharma, R.; Kumar, S.; Singh, R.K.; Sharma, P.C. Ameliorants and salt tolerant varieties improve rice-wheat production in soils undergoing sodification with alkali water irrigation in indo–gangetic plains of India. *Agric. Water Manag.* **2021**, *243*, 106492. [CrossRef]
30. El-Hendawy, S.E.; Hu, Y.; Yakout, G.M.; Awad, A.M.; Hafiz, S.E.; Schmidhalter, U. Evaluating salt tolerance of wheat genotypes using multiple parameters. *Eur. J. Agron.* **2005**, *22*, 243–253. [CrossRef]
31. Zeng, L.; Shannon, M.C.; Grieve, C.M. Evaluation of salt tolerance in rice genotypes by multiple agronomic parameters. *Euphytica* **2002**, *127*, 235–245. [CrossRef]
32. Villar, R.; Marañón, T.; Quero, J.L.; Panadero, P.; Arenas, F.; Lambers, H. Variation in relative growth rate of 20 *Aegilops* species (*Poaceae*) in the field: The importance of net assimilation rate or specific leaf area depends on the time scale. *Plant Soil* **2005**, *272*, 11–27. [CrossRef]
33. Shipley, B. Net assimilation rate, specific leaf area and leaf mass ratio: Which is most closely correlated with relative growth rate? A meta-analysis. *Funct. Ecol.* **2006**, *20*, 565–574. [CrossRef]
34. Zhu, G.; Ren, Z.; Liu, Y.; Lu, F.; Gu, L.; Liu, J.; Zhou, G.; Ahmed Nimir, N.E.; Mohapatra, P.K. Optimization of leaf properties and plant phenotype through yield-based genetic improvement of rice over a period of seventy years in the Yangtze river basin of China. *Food Energy Secur.* **2020**, *9*, e223. [CrossRef]
35. Hunt, R. *Encyclopedia of Applied Plant Sciences*; Academic Press: Cambridge, MA, USA, 2016; Volume 2, ISBN 978-0-12-394808-3.
36. Cramer, G.; Alberico, G.; Schmidt, C. Leaf expansion limits dry matter accumulation of salt-stressed maize. *Funct. Plant Biol.* **1994**, *21*, 663. [CrossRef]
37. Shipley, B. Trade-offs between net assimilation rate and specific leaf area in determining relative growth rate: Relationship with daily irradiance. *Funct. Ecol.* **2002**, *16*, 682–689. [CrossRef]
38. Li, X.; Schmid, B.; Wang, F.; Paine, C.E.T. Net assimilation rate determines the growth rates of 14 species of subtropical forest trees. *PLoS ONE* **2016**, *11*, e0150644. [CrossRef] [PubMed]
39. Karki, H.; Bargali, K.; Bargali, S. Effect of sowing time on germination and early seedling growth of *Quercus floribunda* Lindl. *J. For. Environ. Sci.* **2018**, *34*, 199–208. [CrossRef]
40. Assefa, M. Performance and growth analysis of three mung bean (*Vigna radiate*) genotypes at Hawassa, Ethiopia. *J. Agric. Sci. Food Res.* **2021**, *12*, 277.
41. Niinemets, Ü. Research review. Components of leaf dry mass per area—Thickness and density—Alter leaf photosynthetic capacity in reverse directions in woody plants. *New Phytol.* **1999**, *144*, 35–47. [CrossRef]
42. Murillo-Amador, B.; Nieto-Garibay, A.; Troyo-Diéguez, E.; García-Hernández, J.L.; Hernández-Montiel, L.; Valdez-Cepeda, R.D. Moderate salt stress on the physiological and morphometric traits of *Aloe vera* L. *Bot. Sci.* **2015**, *93*, 639–648. [CrossRef]
43. Chen, W.; Mou, X.; Meng, P.; Chen, J.; Tang, X.; Meng, G.; Xin, K.; Zhang, Y.; Wang, C. Anatomical responses of leaf and stem of *Catalpa bungei* C.A.Mey. to nitrogen application and rhizophagus intraradices inoculation. *Environ. Exp. Bot.* **2023**, *209*, 105308. [CrossRef]
44. Corinzia, S.A.; Crapio, E.; Testa, G.; Cosentino, S.L.; Patanè, C.; Scordia, D. Leaf area duration and crop radiation use efficiency determine biomass yield of lignocellulosic perennial grasses under different soil water content. *Agronomy* **2023**, *13*, 2270. [CrossRef]
45. Amanullah specific leaf area and specific leaf weight in small grain crops wheat, rye, barley, and oats differ at various growth stages and NPK source. *J. Plant Nutr.* **2015**, *38*, 1694–1708. [CrossRef]
46. Zhu, G.; Peng, S.; Huang, J.; Cui, K.; Nie, L.; Wang, F. Genetic improvements in rice yield and concomitant increases in radiation- and nitrogen-use efficiency in middle reaches of Yangtze river. *Sci. Rep.* **2016**, *6*, 21049. [CrossRef]
47. Chen, B.; Bian, X.; Tu, M.; Yu, T.; Jiang, L.; Lu, Y.; Chen, X. Moderate Salinity Stress Increases the Seedling Biomass in Oilseed Rape (*Brassica napus* L.). *Plants* **2023**, *12*, 1650. [CrossRef]
48. Simane, B.; Peacock, J.M.; Struik, P.C. Differences in developmental plasticity and growth rate among drought-resistant and susceptible cultivars of durum wheat (*Triticum turgidum* L. Var. durum). *Plant Soil* **1993**, *157*, 155–166. [CrossRef]

49. Ashraf, M. Exploitation of genetic variation for improvement of salt tolerance in spring wheat. In *Prospects for Saline Agriculture*; Ahmad, R., Malik, K.A., Eds.; Tasks for Vegetation Science; Springer Netherlands: Dordrecht, The Netherlands, 2002; Volume 37, pp. 113–121. ISBN 978-90-481-6019-8.
50. El-Hendawy, S.E.; Hu, Y.; Schmidhalter, U. Assessing the suitability of various physiological traits to screen wheat genotypes for salt tolerance. *J. Integr. Plant Biol.* **2007**, *49*, 1352–1360. [CrossRef]
51. Yue, J.; Feng, H.; Yang, G.; Li, Z. A Comparison of regression techniques for estimation of above-ground winter wheat biomass using near-surface spectroscopy. *Remote Sens.* **2018**, *10*, 66. [CrossRef]
52. Hunt, R. Relative growth rates. In *Basic Growth Analysis: Plant Growth Analysis for Beginners*; Springer Dordrecht: Dordrecht, The Netherlands, 1990; pp. 25–34.

plants

Article

Transcriptome Profiling of a Salt Excluder Hybrid Grapevine Rootstock 'Ruggeri' throughout Salinity

Pranavkumar Gajjar [1], Ahmed Ismail [1,2,3], Tabibul Islam [4], Md Moniruzzaman [1], Ahmed G. Darwish [1,5], Ahmed S. Dawood [6], Ahmed G. Mohamed [1], Amr M. Haikal [3], Abdelkareem M. El-Saady [7], Ashraf El-Kereamy [2], Sherif M. Sherif [8], Michael D. Abazinge [9], Devaiah Kambiranda [10] and Islam El-Sharkawy [1,*]

1 Center for Viticulture and Small Fruit Research, College of Agriculture and Food Sciences, Florida A&M University, Tallahassee, FL 32308, USA; pranavkumar1.gajjar@famu.edu (P.G.); ahmed.ismail@ucr.edu (A.I.); md.moniruzzaman@famu.edu (M.M.); ahmed.darwish@famu.edu (A.G.D.); ahmed.mohamed2@famu.edu (A.G.M.)

2 Department of Botany and Plant Sciences, University of California Riverside, Riverside, CA 92521, USA; ashrafe@ucr.edu

3 Department of Horticulture, Faculty of Agriculture, Damanhour University, Damanhour 22516, Egypt; amr.haikal@agr.dmu.edu.eg

4 Plant Sciences Department, University of Tennessee, Knoxville, TN 37996, USA; islamt@utk.edu

5 Department of Biochemistry, Faculty of Agriculture, Minia University, Minia 61519, Egypt

6 Horticulture Department, Faculty of Agriculture, Al-Azhar University, Cairo 11884, Egypt; ahmed8dawood@azhar.edu.eg

7 Fertilization Technology Department, National Research Center (NRC), Cairo 12622, Egypt; elsaadyam@nrc.sci.eg

8 Alson H. Smith Jr. Agricultural Research and Extension Center, School of Plant and Environmental Sciences, Virginia Tech, Winchester, VA 22602, USA; ssherif@vt.edu

9 School of the Environment, Florida A&M University, Tallahassee, FL 32307, USA; michael.abazinge@famu.edu

10 Department of Plant and Soil Sciences, Southern University Agricultural Research and Extension Center, Baton Rouge, LA 70813, USA; devaiah_kambiranda@subr.edu

* Correspondence: islam.elsharkawy@famu.edu; Tel.: +1-850-599-8685

Abstract: Salinity is one of the substantial threats to plant productivity and could be escorted by other stresses such as heat and drought. It impairs critical biological processes, such as photosynthesis, energy, and water/nutrient acquisition, ultimately leading to cell death when stress intensity becomes uncured. Therefore, plants deploy several proper processes to overcome such hostile circumstances. Grapevine is one of the most important crops worldwide that is relatively salt-tolerant and preferentially cultivated in hot and semi-arid areas. One of the most applicable strategies for sustainable viticulture is using salt-tolerant rootstock such as Ruggeri (RUG). The rootstock showed efficient capacity of photosynthesis, ROS detoxification, and carbohydrate accumulation under salinity. The current study utilized the transcriptome profiling approach to identify the molecular events of RUG throughout a regime of salt stress followed by a recovery procedure. The data showed progressive changes in the transcriptome profiling throughout salinity, underpinning the involvement of a large number of genes in transcriptional reprogramming during stress. Our results established a considerable enrichment of the biological process GO-terms related to salinity adaptation, such as signaling, hormones, photosynthesis, carbohydrates, and ROS homeostasis. Among the battery of molecular/cellular responses launched upon salinity, ROS homeostasis plays the central role of salt adaptation.

Keywords: grapevine rootstock; Ruggeri; photosynthesis; carbohydrates; ROS detoxification; salinity; transcriptome profiling

Citation: Gajjar, P.; Ismail, A.; Islam, T.; Moniruzzaman, M.; Darwish, A.G.; Dawood, A.S.; Mohamed, A.G.; Haikal, A.M.; El-Saady, A.M.; El-Kereamy, A.; et al. Transcriptome Profiling of a Salt Excluder Hybrid Grapevine Rootstock 'Ruggeri' throughout Salinity. *Plants* **2024**, *13*, 837. https://doi.org/10.3390/plants13060837

Academic Editors: Geovani Soares de Lima, Lauriane Almeida dos Anjos Soares and Francisco Vanies Da Silva Sá

Received: 3 February 2024
Revised: 5 March 2024
Accepted: 12 March 2024
Published: 14 March 2024

1. Introduction

Salt stress negatively contributes to agriculture worldwide and, hence, represents one of the major threats to plant productivity. Salt accumulation in soil solution results in ionic anxiety (sodium–Na^+ and chloride–Cl^-), osmotic stress (physiological drought), and the interaction between both pressure types [1–3]. Further, salinity could be escorted by other stresses such as heat and/or drought. In addition, similar to other biotic/abiotic stresses, salinity leads to oxidative stress as the generation of excessive reactive oxygen species (ROS) [4,5]. Approximately 65–87% of the global crop production is annually lost due to plant stresses through which salinity only affects more than 833 million hectares worldwide, most notably in (semi)-arid regions [6,7]. It hinders plant growth and development by impairing critical biological processes such as photosynthesis, energy, and water/nutrient acquisition that ultimately culminate in cell death when stress intensity becomes uncured [8,9].

To survive under salinity, plants have to respond effectively by recruiting a set of proper actions [3,8]. The rapid access of NaCl into the root system triggers a battery of signal cascades that act in harmony with cellular secondary messengers such as Ca^{2+}, H^+, ROS, and plant hormones. Moreover, the rise of NaCl at the cellular level can act to a certain extent as a cheap osmolyte to counteract NaCl-imposed osmotic stress, hence turning a foe into a friend [2,10,11]. However, such a rapid influx of NaCl must be restricted to prevent its cellular accumulation, along with the fast decline in leaf stomatal conductance to restrict water loss [12]. Furthermore, NaCl must be effluxed out of the cell, compartmentalized into the vacuole, and damped inside old leaves [8]. Likewise, plants have evolved other efficient machinery to maintain redox homeostasis and osmotic balance. For instance, ROS detoxification is achieved by engaging antioxidant enzymes (e.g., superoxide dismutase-SOD, catalase-CAT, glutathione reductase-GR, glutathione peroxidase, and ascorbate peroxidase-APX) and non-enzymatic antioxidants, including stilbenes, flavonoids, carotenoids, and sugar alcohols [13,14]. Similarly, osmotic adjustment can be executed by accumulating osmolytes (osmoprotectants) such as proline, proline betaine, glycine betaine, alanine betaine, pipecolate betaine, hydroxyproline betaine, trehalose, and polyols [15].

Grapevine (*Vitis vinifera*) is one of the most important crops worldwide. The United States was the world's third-largest grape cultivator in 2023, with roughly 6.9 million tons (www.worldpopulationreview.com; accessed on 5 March 2024). Grapes are relatively salt-tolerant, known for their higher sensitivity to Cl^- toxicity than Na^+ [16]. The hot and semi-arid areas are the grapes' preferentially cultivated regions, where soil salinization is predicted to be exaggerated due to global climate change [17]. Unfortunately, some 25–73% of current Mediterranean grapevine-growing regions are at risk of becoming uncultivable due to soil desertification, a climate-change-related process in which different types of water-related stresses (e.g., salinity and drought) can coexist [18]. Hence, their production is challenged by several types of plant stresses, most notably water-related stresses [19]. Several practical measures are taken to alleviate the negative impacts of stresses such as salt- and drought-related damages. For instance, economically important salt-sensitive cultivars are grafted on rootstocks derived from salt-tolerant sand grape (*V. rupestris*) and winter grape (*V. berlandieri*) [16,20]. In our previous study, we evaluated the physiological impacts of salinity on two grape hybrid rootstocks, 140 Ruggeri (*V. berlandieri* × *V. rupestris*) and Millardet et de Grasset 420A (*V. berlandieri* × *V. riparia*) [21]. They were considered highly competent salt excluder rootstocks that reduce salt accumulation in the leaves and berries of their grafted scions [22,23]. We showed that RUG maintains higher photosynthetic capacity under salinity and, accordingly, accumulates more sugars [21]. In the current study, we extend our efforts to understand the molecular events that occurred throughout a regime of salt stress followed by re-watering. Therefore, transcriptomic data was generated to address in time-resolved detail the response of grapevine rootstock RUG to salinity and recovery treatments. The pairwise comparisons between transcriptome profiles of salt-stressed on non-stressed RUG showed substantial changes in transcriptomic profiles, particularly after 24 h and 48 h. The employment of the WGCNA system biology approach linked the transcriptome data with the previously published results [21] (i.e., carotenoids, Fru, TSS, GPX, SOD, and Pro) demonstrated a considerable enrichment of Biological process GO terms (BP GOs) related to salinity adaptation in the positively correlated modules M1 and M11. For instance, the BP GOs of hormones (e.g., "hormone-mediated signaling pathway" and "ethylene-activated signaling pathway"), polysaccharide (e.g., "cell wall polysaccharide metabolic process"), photosynthesis (e.g., "chlorophyll metabolic process", "chlorophyll biosynthetic process", "carotenoid metabolic process", "pigment metabolic process", and "pigment biosynthetic process"), and ROS detoxification (e.g., "superoxide metabolic process", "response to superoxide", "removal of superoxide radicals", "cellular response to superoxide", "cellular response to oxygen radical", and "response to reactive oxygen species") were highly represented in both modules. The considerable enrichment of BP GO terms for signaling, hormones, photosynthesis, carbohydrates, ROS homeostasis, and other stress-related terms in M1 and M11 modules and their associated K-means clusters pointed to a battery of molecular/cellular responses established during salt stress. However, the ROS homeostasis activity stemmed from a wide range of gene-encoding proteins of different pathways, highlighting the central roles of ROS detoxification and homeostasis mechanisms in salt adaptation. This work would help in promoting the generation of new salt-adapted rootstocks to better cope with global climate changes.

2. Results

2.1. Global Changes in RUG Transcriptome during Salt Stress

In the current study, we investigated the changes in the transcriptome profile of RUG rootstock throughout salinity and recovery procedures to acquire an overview of the associated molecular events underlying its physiological responses. All collected samples were used for the RNA-seq experiment, excluding 8 h salt stress and 8 days recovery. Unless otherwise specified, the sampling order R1–R8 will refer to the sampling time; control, 0 h, 0.5 h, 2 h, 24 h, 48 h, 4 d, and 12 d, respectively.

The sequenced, pair-ended 24 libraries (3 biological replicates × 8-time points) resulted in 620.1 Gb of high-quality clean data ranging from 21.1 Gb to 35.3 Gb per replicate. The clean reads were aligned to the V. vinifera reference genome sequence and projected to its transcriptome using STAR, which resulted in an 84.2% to 90.8% mapping rate (Table S1) [24,25]. Subsequently, the output transcriptomic coordinates files were quantified by Salmon in alignment mode [26]. Samples hierarchical clustering showed progressive changes in their transcript abundance during salt stress (Figure S1). Likewise, the principal component analysis (PCA) exhibited high consistency among transcript profiles that were separated along two main components. The first component (PC1) was accountable for 66% of the variance and was associated with the course of salinity (Figure 1A). The second component (PC2) was responsible for 20% of the variance and was essentially linked to the effect of recovery and sampling time. Interestingly, when the two recovery samples (R7 and R8) were excluded, the PC1 increased by 14% and, therefore, represented 80% of the variance (Figure 1B). Contrary, the PC2 diminished by half, signifying 10% of the variance. The data confirmed that salinity causes the primary source of variance with more than 65%, while the recovery treatment counteracts its impact by ~10–14%. However, other factors, such as sampling time and/or the circadian clock, impact the variance by 10% only.

Seven pairwise transcriptome comparisons between each consecutive time point of salinity and recovery (R2–R1, R3–R2, R4–R3, R5–R4, R6–R5, R7–R6, or R8–R7) resulted in 13,214 and 6736 redundant and non-redundant differentially expressed genes (DEGs) with PFDR < 0.05 and log2fold change > ±1.5, respectively (Table S2). The total non-redundant DEGs (6736) were approximately equal to the number of redundant up- (6505) and down-regulated (6709) transcripts. Interestingly, ~52% of the non-redundant DEGs (3492 out of 6736) showed up- or down-regulation at one of the studying points (Figure 1C). In addition, the total and exclusive number of DEGs, and the up- and down-regulated transcripts in each comparison differed significantly throughout stress (Figures 1C,D, S2 and S3). For instance, the R5–R4 comparison showed the highest number of up- (2627) and down-regulated (2019) transcripts, which roughly represented one-third of the total number of redundant DEGs (Figure 1D). Similarly, the highest number of unique DEGs (1385), and the up- (1574) and down-regulated (1661) genes, respectively, was also detected in the R5-R4 comparison (Figures 1C, S2 and S3). Contrary, the R7–R6 comparison displayed the lowest total number of redundant DEGs (353), while R4–R3 exhibited the smallest number of non-redundant DEGs (35) (Figure 1C,D). The data showed that there were a large number of genes involved in transcriptional reprogramming during salinity, most notably at 24 h–48 h of stress, confirming the previous physiological results where RUG accumulated higher amounts of the photosynthetic pigments (total chlorophyll–Chl-T and chlorophyll a–Chl-a) and the soluble sugar (e.g., total soluble sugar (TSS), glucose (Glu), fructose (Fru), and sucrose (Suc)), but ROS-related compounds at 48 h only (mainly proline and SOD) [21].

2.2. Cluster-Based Analysis Showed Dynamic Transcript Abundance during Salinity

To investigate the temporal pattern of significantly expressed genes throughout stress and recovery, the unsupervised clustering approach "K-means clustering" was used to partition transcripts into specific accumulation profiles. The analysis classified the 12,760 non-redundant expressed genes into 15 clusters, which could be manually fitted into six groups (Figures 2A and S4 and Table S3). In all cases; however, the kinetic patterns of the transcript abundance within each group were not completely identical. The first group (I) includes clusters K5 and K10, in which transcript abundance peaked at time-0 but then declined sharply (K5) or gradually (K10). The second group (II) consists of three clusters, where transcripts continued to increase within 0.5 h–2 h of salt stress, then decreased along with stress progression. These transcripts were increased during the recovery procedure (K3) or increased 4 days post-recovery and declined after that (K7 and K15). The third group (III), composed of clusters K2, K6, and K9, exhibited an initial drop at time-0 compared to control, followed by an increase in the transcript abundance, then dropped again at 24 h (K2), or dropped earlier (K6 and K9). By contrast, clusters that showed induction between 2 h–48 h of salt stress with inconsistent patterns were placed in the fourth grope (IV), including K4, K8, and K12. The fifth group showed a sharp induction at a particular time point (24 h) as in K1 and K13; however, K1 exhibited a second increase at 12 days of recovery. Finally, clusters K11 and K14 with inconsistent patterns were placed in the sixth group (IV).

To provide a broad overview of the types of transcripts in each group, the Gene Ontology (GO; i.e., Molecular Function MF, Cellular Component CC, and Biological Process BP) terms and Kyoto Encyclopedia of Genes and Genomes (KEGG) enrichment strategies were used to analyze each cluster [27]. Several stress- and salinity adaptation-related GO-term categories displayed high enrichment with a wide range of cluster differences. For instance, cluster K10 was enriched in ROS homeostasis MF subcategories ("oxidoreductase activity, acting on the CH-CH group of donors" and "oxidoreductase activity, acting on the CH-NH group of donors and quinone or similar compound as acceptor") (Figures 2B and S5 and Table S4). Moreover, K10 with K8 clusters showed considerable enrichment of BP GO terms for signaling, hormones, photosynthesis, carbohydrates, and other stress-related terms (e.g., cell communication, intracellular signal transduction, signal transduction, abscisic acid-activated signaling pathway, polyamine biosynthetic process, fructose 6-phosphate metabolic process, sucrose biosynthetic process, oligosaccharide biosynthetic process, fructose 1,6-bisphosphate metabolic process, fructose metabolic process, carbohydrate derivative metabolic process, disaccharide biosynthetic process, carbohydrate metabolic process, oligosaccharide metabolic process, monosaccharide metabolic process, disaccharide metabolic process, protein dephosphorylation, sucrose metabolic process, gluconeogenesis, oxidative photosynthetic carbon pathway, and photosynthesis) (Figures 2C,D, S6 and S7 and Table S4).

The previous data analysis showed the global significant changes in transcriptome profile and, subsequently, the GO term enrichments that eventually differentiated in response to salt stress. Most notably, the photosynthesis, carbohydrates, and ROS homeostasis-related GO terms were overrepresented with the progression of salinity.

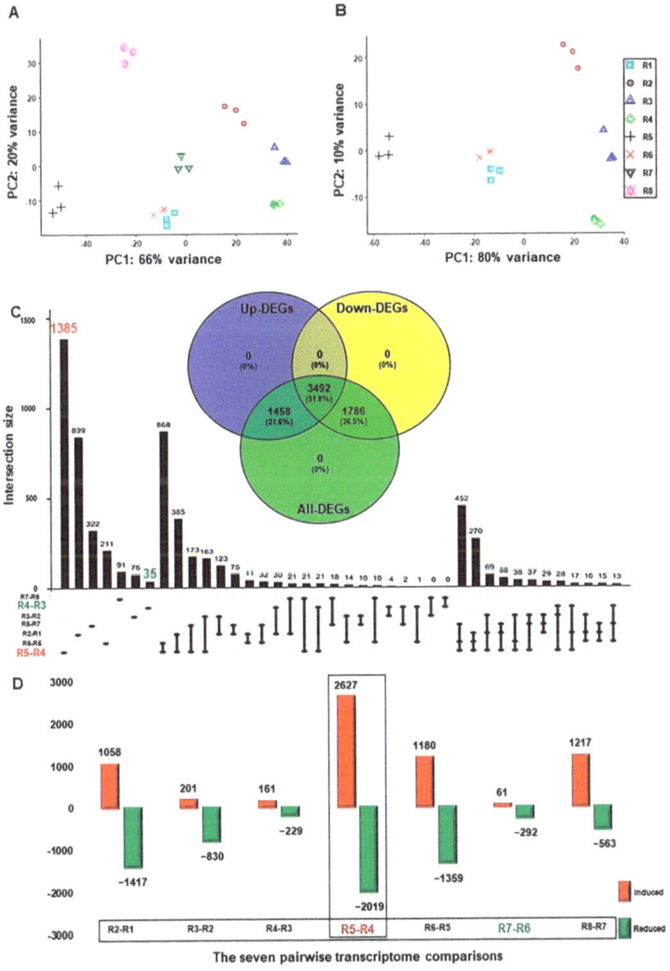

Figure 1. Temporal dynamics of RUG rootstock transcriptome during salt and recovery treatments. (**A**) Principal Component Analysis (PCA) showing the global similarity among RNA-seq samples of RUG rootstock, where salt stress was applied gradually, followed by a recovery procedure. (**B**) The PCA of the transcriptome data after excluding the recovery samples. (**C**) The Upset plot of the total number of the differentially expressed genes (DEGs) of each comparison. By using the DESeq2 pipeline, each two consecutive time points during the salinity and recovery procedure were compared, resulting in 13,214 (6736 non-redundant), with $|\log2FC| > \pm1.5$ and p-adjusted < 0.05. The included Venn diagram showed the non-redundant DEGs among the total, up-, and down-regulated transcripts. (**D**) Bar plots of DEGs that demonstrated the temporal expression patterns of genes in RUG during salinity and recovery procedures. The sampling order R1–R8 refers to the sampling time as follows: control, 0 h, 0.5 h, 2 h, 24 h, 48 h, 4 days, and 12 days, respectively. Each time point during the salinity and recovery treatments was given a distinctive color and symbol.

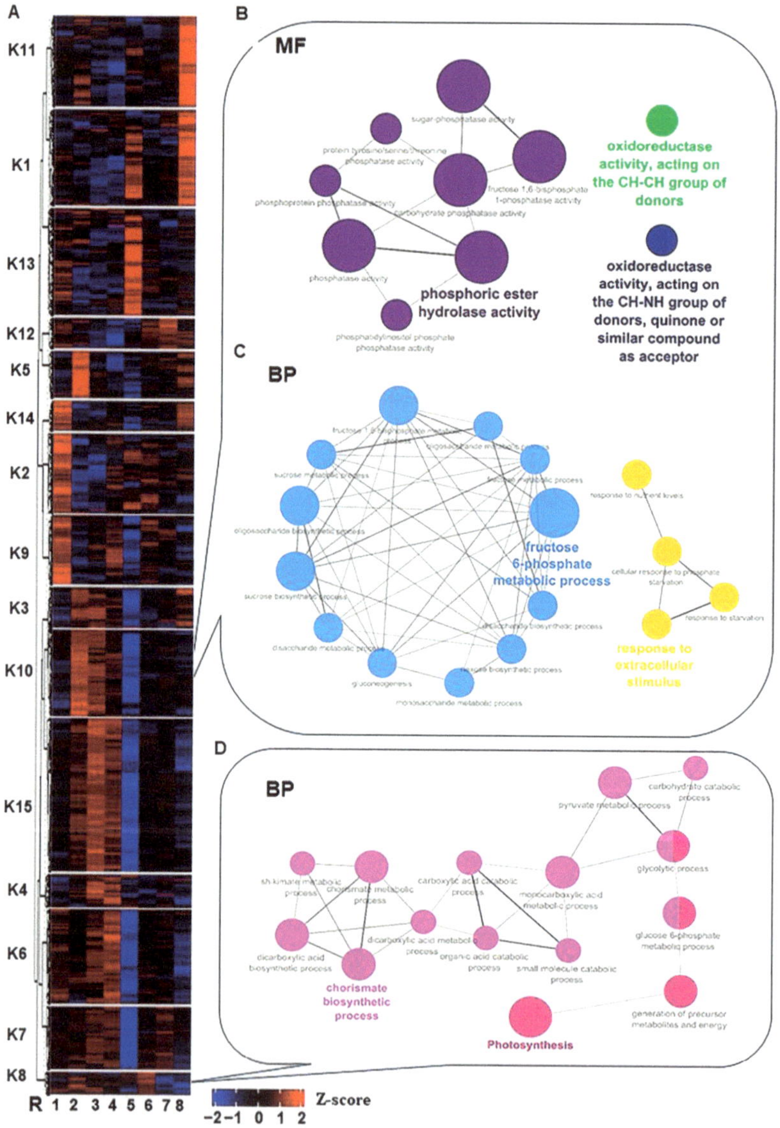

Figure 2. Heatmap of the K-means clusters of the 12,760 non-redundant significantly expressed genes throughout salinity stress in RUG. (**A**) By using transcript per million (TPM) of the 12,760 non-redundant significantly expressed genes during salinity stress, the K-means algorithm generated 15 clusters (K1 to K15) that were hierarchically clustered utilizing Pearson pairwise correlation. The sampling order R1–R8 refers to the sampling time, as described in Figure 1. (**B–D**) Network views for specific predefined Biological processes GO terms pathways for significantly expressed genes within K8 and K10, and the Molecular Function (MF) in K10 only. All MF and BP GO terms for each cluster (p-adjusted < 0.05) were extracted by the g:Profiler website with the Benjamini-Hochberg FDR multiple testing correction method. The default ClueGO settings were applied, and the terms are functionally grouped based on shared genes (kappa score), shown with different colors. The size of the nodes indicates the number of mapped genes ranged from 0–5, 5–10, 10–20, 20–30, and ≥30 genes. The most significant term defines the name of the group.

2.3. Analysis of Expression Patterns and Identification of WGCNA Modules Associated with Salinity Adaptation

In our previous study, we examined the physiological responses of RUG rootstock to salinity, including photosynthetic pigments (chlorophyll "Chl-T, Chl-a, Chl-b", and carotenoids "Caro") and sugar contents (total soluble sugar "TSS", sucrose "Suc", glucose "Glu", and fructose "Fru") as a manifestation of the efficacy of photosynthesis and sugar/energy metabolism. In addition, the enzymatic antioxidant activities of superoxide dismutase "SOD", catalase "CAT", and glutathione peroxidase "GPX" along with the determination of proline "Pro" content as an indication of redox homeostasis and two oxidative stress markers (malondialdehyde "MDA", hydrogen peroxide "H_2O_2") were assessed in the same samples used in this study [21]. The physiological results and the transcriptomic data suggested that RUG maintains efficient photosynthesis and antioxidative machinery under salinity by accumulating high amounts of sugars and suppressing the induction of ROS, respectively. The WGCNA system biology approach was exploited to investigate the gene modules and physiological data to get a better understanding of the molecular events under salt stress. However, as the recovery treatment alleviates the impact of salinity, its data was omitted from this analysis.

The pairwise correlations among the 18 samples (control, 0 h, 0.5 h, 2 h, 24 h, and 48 h) identified 46 modules, which were labeled in distinct colors shown in a hierarchical clustering dendrogram and network heatmap (Figures S8 and S9). The analysis of module-trait correlations among the identified modules and the physiological data enabled us to identify 6 and 2 modules with positive and negative correlations, respectively (Figure 3). Among the positively correlated modules, the M1 module was highly correlated with Fru and TSS modules ($r^2 \geq 0.81$ and 0.72, respectively), holding 327 significant genes. The M3 and M32 (10 and 18 significant genes, respectively) displayed a substantial positive correlation with GPX ($r^2 \geq 0.72$). Finally, the M11, M28, and M29 modules involving 1664, 17, and 29 significant genes were positively correlated with Caro, SOD, and Pro ($r^2 \geq 0.71$), respectively. On the other hand, M16 module showed a negative correlation with SOD and GPX ($r^2 \geq -0.73$), while M45 was negatively associated with Fru ($r^2 \geq -0.77$).

Consequently, we focused on the most two prominent modules (M1 and M11), exhibiting positive correlation and holding 2065 significant genes (Figure 3). The GO terms and KEGG enrichment analysis of M1 and M11 modules exposed high enrichment of GO terms associated with stress adaptation (Table S5). For instance, M1 displayed considerable enrichment of BP GO terms for "cellular response to hormone stimulus", "hormone-mediated signaling pathway", "ethylene-activated signaling pathway", "regulation of biosynthetic process", "regulation of cellular metabolic process", "regulation of primary metabolic process", glycosyltransferase activity, and "cell wall polysaccharide metabolic process" (Figure S10 and Table S5). However, M11 was highly enriched for BP GO terms for "plastid organization", "chloroplast organization", "pigment metabolic process", "pigment biosynthetic process", "cellular response to oxidative stress", "chlorophyll metabolic process", "superoxide metabolic process", "chlorophyll biosynthetic process", "protein import into chloroplast stroma", "thylakoid membrane organization", "plastid membrane organization", "response to superoxide", "removal of superoxide radicals", "cellular response to superoxide", "cellular response to oxygen radical", "response to oxygen radical", "regulation of chlorophyll metabolic process", "flavonoid metabolic process", "regulation of cellular biosynthetic process", "carbohydrate metabolic process", "response to reactive oxygen species", "regulation of biosynthetic process", and "carotenoid metabolic process" (Figures 4A and S11 and Table S5). Altogether, the WGCNA approach and the GO term enrichments analysis of positively correlated modules (M1 and M11) confirmed the high correlation of photosynthesis, carbohydrates, and ROS homeostasis with salinity adaptation of RUG rootstock.

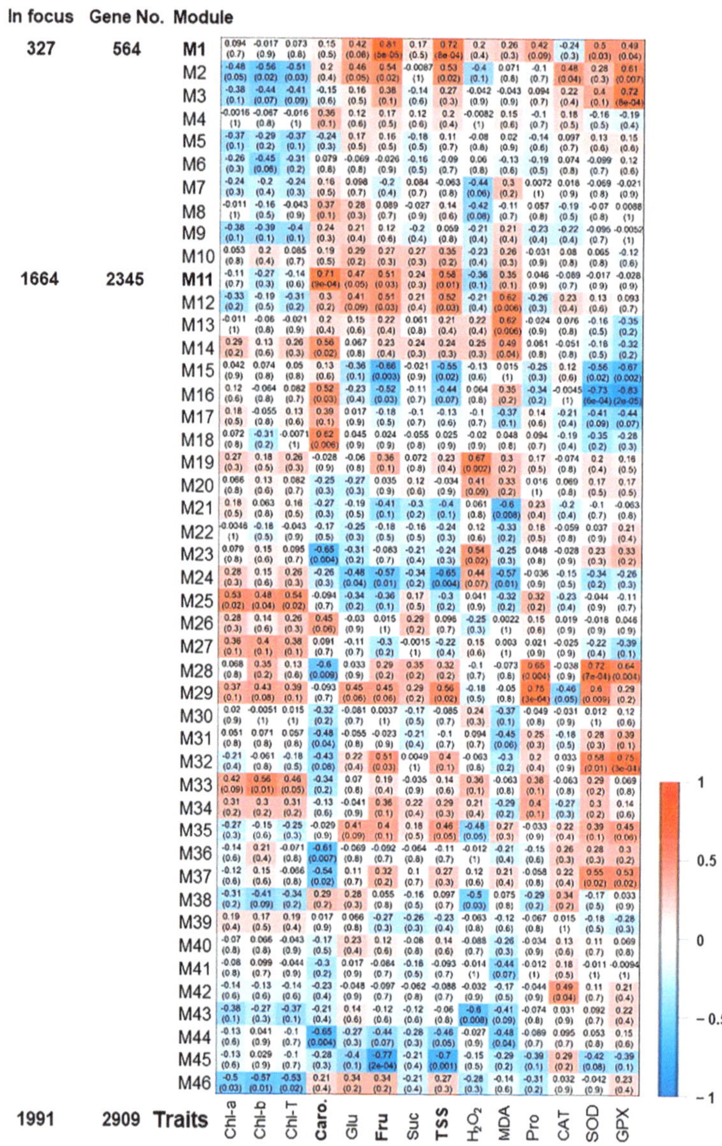

Figure 3. Expression heatmap of co-expression modules. Module-trait associations between RNA-seq data and evaluated physiological traits of photosynthetic pigments (chlorophyll "Chl-T, Chl-a, Chl-b", and carotenoids "Caro"), sugar contents (total soluble sugar "TSS", sucrose "Suc", glucose "Glu", and fructose "Fru"), antioxidant enzymatic activities of superoxide dismutase "SOD", catalase "CAT", and glutathione peroxidase "GPX", and antioxidant non-enzymatic activity represented by proline "Pro" content [21]. The color of the cell indicates the correlation coefficient between a given module and the applicable trait at the row-column intersection. Each row corresponds to a module (M1–M46). Modules of interest (M1 and M11) were shown in bold-line and were selected for further analysis. The left panel displayed the assigned number of genes to two selected modules, whether the number of total input genes or significantly expressed genes from seven pairwise transcriptome comparisons between each two consecutive time points during salinity (R2–R1, R3–R2, R4–R3, R5–R4, R6–R5, R7–R6 or R8–R7).

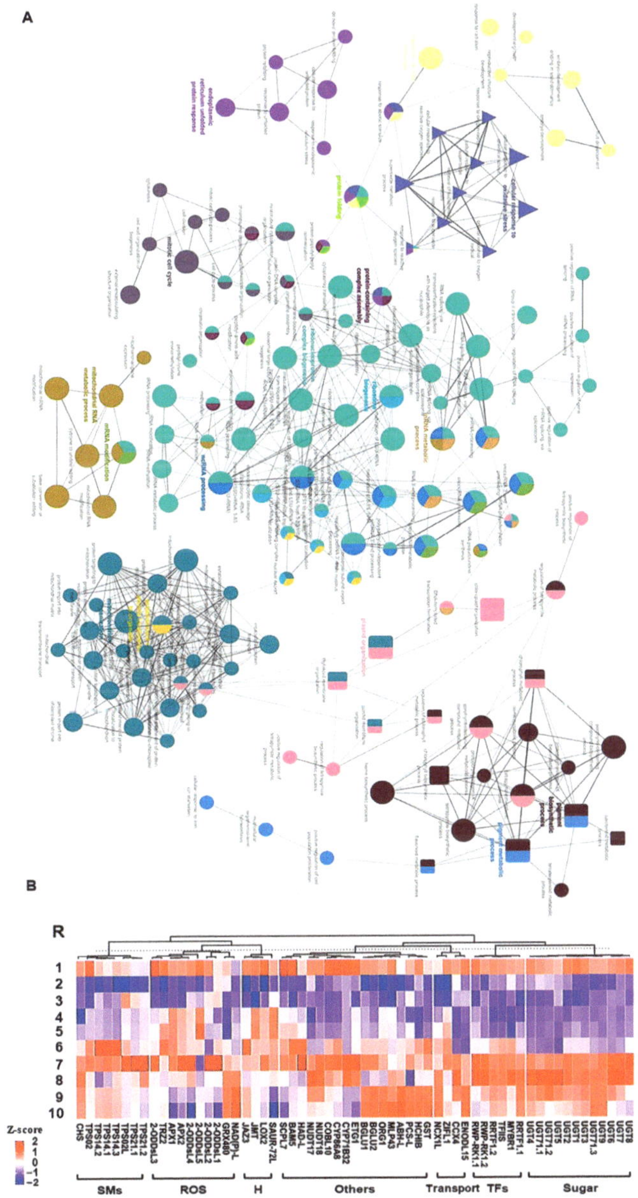

Figure 4. (**A**) Network views for specific predefined Biological processes GO terms pathways for significantly expressed genes within WGCNA module M11 with *p*-adjusted < 0.05. The GO terms were extracted by g:Profiler website with the Benjamini-Hochberg FDR multiple testing correction method. The default ClueGO settings were applied, and the terms are functionally grouped based on shared genes (kappa score), shown with different colors. The size of the nodes indicates the number of mapped genes ranged from 0–5, 5–10, 10–20, 20–30, and ≥30 genes. The most significant term defines the name of the group. (**B**) Heatmap of the 61 hub genes significantly expressed throughout salinity and recovery in RUG using qPCR gene expression data. Genes were hierarchically clustered using Pearson pairwise correlation. The sampling order R1-R10 refers to the sampling time as follows: control, 0 h, 0.5 h, 2 h, 8 h, 24 h, 48 h, 4 d, 8 d, and 12 d, respectively.

2.4. Identification of Hub Genes Regulating Salinity Adaptation

To identify essential genes involved in salinity adaptation, annotation information of significantly expressed genes within M1 and M11 modules was extracted from the *V. vinifera* PN40024 grapevine reference genome [28,29]. A total of 61 genes were identified and classified into 7 groups based on their predicted functions as follows: ROS detoxification, sugars metabolism, hormones, transcription factors (TFs), secondary metabolites, transport, and other functions (Figure 4B). Despite that they were exclusively located in 2 WGCNA modules, they exhibited diverse expression profiles (K1, K2, K8, K9, K11, K12, 13, K14, or no cluster) that was validated by quantitative real-time PCR (qPCR) assay. In addition to the biological samples used for transcriptome analysis, two extra points were included in the qPCR assay at 8 h of salinity and 8 days of recovery. The qPCR results of hub genes exhibited a significant correlation with mRNA-seq data ($r^2 \geq 0.90$), except for one gene that showed $r^2 \geq 0.73$, validating the transcriptome profiles (Figure 4B and Table S6).

For ROS detoxification-related genes, 9 out of 10 transcripts started to accumulate early after 0.5 h of salinity, including genes encoding ascorbate peroxidase (*APX1* and *2*) and seven oxidoreductase superfamily protein (*2-ODDsL* (5 isoforms), *NAD(P)-L*, *TRZ*). Only *GRX480*, which encodes thioredoxin superfamily protein, was considerably induced after 24 h of stress. With few exceptions, the expression levels of these ROS detoxification-related genes reached their maximal levels after 48 h of salinity that exceeds their basal levels during typical conditions; however, many of them tended to decline during the recovery procedure. For genes related to sugar metabolism, 12 transcripts encoding UDP-glycosyltransferases (*UGTs*) superfamily proteins were identified. Generally, their strong induction took place after 48 h of salt stress with a tendency to accumulate during the recovery process, establishing a peak at 4 days. Regarding the hormone-related genes, 3 out of 4 genes were jasmonic acid (JA)-related genes, while the fourth gene belongs to auxin. Except for the JA carboxyl methyltransferase (*JMT*), they showed fast accumulation during early salt stress (0.5 h). Of particular interest is the jasmonate-zim-domain protein 3 (*JAZ3*), which negatively affects the JA signal, playing a crucial role in fine-tuning the JA pathway [30,31]. Similarly, the gene encoding the SAUR-like auxin-responsive protein family (*SAUR72*) was strongly expressed between 2–24 h of stress but reached around null expression at 8–12 days of recovery.

For TF-related genes, the 6 hub genes located in module M1 showed their strong expression within 48 h of stress and 4 days of recovery compared to control, including the redox responsive transcription factor 1 (*RRTF1*, two isoforms), myb domain protein r1 (*MYBR1*), plant regulator RWP-RK family protein (*RWP-RK1*, two isoforms), E2F target gene 1 (*ETG1*), and transcription elongation factor (*TFIIS*) family protein. Genes related to secondary metabolites accumulated mainly during late stress (24–48 h). In addition to one gene for chalcone and stilbene synthase family protein (*CHS*), this set of genes mainly includes terpene synthases (*TPSs*, 7 isoforms), which are liable for the structural diversity of the superfamily of terpenoid products (also referred to as terpenoids or isoprenoids) with diverse expression profiles that decline during the recovery. For transport-related genes, four hub genes displayed distinct accumulation patterns, two of which are sodium (cation)/calcium exchanger (*NCX1L* and *CCX4*), one early nodulin-like protein 15 (*EN-ODL15*), and one for the zinc-induced facilitator-like 1 (*ZIFL1*). Finally, for other genes involved in the salinity tolerance, 17 genes were assessed located mainly in M11 with different expression profiles, including cytochromes P450 (*CYP71B32* and *CYP86A8*), nudix hydrolases (*NUDT17* and *18*), major latex protein-like protein 43 (*MLP43*), basic chitinase (*HCHIB*), alpha/beta hydrolase (ABH) superfamily (*ABH-L*), serine carboxypeptidase-like 7 (*SCPL7*), beta-amylase (*BGLU1*, *BGLU2*, and *BAM5*), halo acid dehalogenase (HAD) superfamily (*HAD-L*), glutathione S-transferase (*GST*), COBRA-like protein 10 precursor (*COBL10*), Eukaryotic aspartyl protease family protein (*PCS-L*), and the OBP3-responsive gene 1 (*ORG1*).

3. Discussion

In this study, leaves of RUG grapevine rootstock showed transcriptome reprogramming profiles over salinity and recovery conditions. Additional levels of complication arose from the experimental procedure with different time points of both salinity and recovery. Moreover, the re-watering application counteracted salinity, alleviating its adverse effects on transcriptome, as shown by PCA results. Monitoring the growth status of grapevine plants indicated that this influence is stress-dependent, and significantly impacted by its gravity and duration [32]. In addition, the genetic background of plants determines the effect of salinity and, accordingly, the transcriptome remodeling as shown by the coastal salt-tolerant wild grapevine AS1B compared to salt-sensitive rootstock Richter 110 [33]. The pairwise comparisons between transcriptome profiles of consecutive time points of RUG during salinity and recovery showed substantial changes in its transcriptomic profiles. Notably, when the 24 h of stress was compared against the 2 h of stress (R5–R4), it exhibited the highest number of up- and down-regulated transcripts. Likewise, transcriptomic analysis of salt-tolerant wild grapevine Tebaba roots throughout salinity depicted similar results commensurate with the accumulation of polyphenol content and enhancement of antioxidant enzymatic activities [34]. The longer the time under salinity (in the range of 24–48 h), the higher the changes at the molecular, metabolic, and physiological levels.

It is tempted to indicate that RUG is a salt excluder rootstock with salt tolerance capacity accumulating higher levels of photosynthetic pigments and sugars along with enhanced ROS detoxification activity under salinity compared to the less tolerant Millardet et de Grasset 420A [21,23]. Therefore, we were guided by our previous results, where physiological parameters, including photosynthetic pigments, and sugar contents, along with enzymatic and non-enzymatic antioxidant activities were estimated in the same RUG samples used for the transcriptome analysis [21]. From a physiological perspective, salt-stressed plants exhibited a considerable reduction in vital biological processes such as photosynthesis, energy, and water/nutrient acquisition, but induction of ROS might lead to cell death based on stress severity and plant tolerance capacity [8,9]. The WGCNA strategy was able to identify 8 modules, exhibiting significant correlation with the recently published physiological data [21] in which 6 modules were positively correlated with carotenoids, Fru, TSS, GPX, SOD, and Pro. The most prominent two modules of M1 and M11 encompass 327 and 1664 significant genes that were highly enriched of BP GO terms related to salinity adaptation. While M1 was enriched with GO terms linked to hormones and polysaccharides, M11 was highly enriched with GO terms for photosynthesis, ROS detoxification, and carbohydrate metabolic processes. However, the K-means clustering showed that each module could include different clusters with distinct kinetic behaviors and, thus, a wide range of GO terms and KEGG enrichment analysis for each cluster.

The considerable enrichment of BP GO terms for signaling, hormones, photosynthesis, carbohydrates, ROS homeostasis, and other stress-related terms in M1 and M11 modules suggested a battery of molecular/cellular responses established during salinity. Certainly, salt-challenged plants accumulate high inorganic ions, especially Na^+ and Cl^-, resulting in cellular toxicity, nutritional and energetic imbalances, and lipid peroxidation. The stress ultimately leads to the production of ROS and metabolic dysfunction, which impair photosynthesis and nutrient acquisition, causing cell death based on stress severity [2–4]. Therefore, the identified 61 hub genes were selected to cover a wide range of salt-adaptation processes, including ROS detoxification, sugar metabolism, hormones, transcription factors (TFs), secondary metabolites, transport, and other functions. Salinity is generally accompanied by overproduction of ROS and alterations in secondary metabolite accumulation [35]. Hence, plants have to swiftly re-establish their ROS homeostasis. Our results showed that 9 out of 10 ROS detoxification-related genes started accumulating earlier after 0.5 h of salinity, such as *APX1/2* and seven genes encoding oxidoreductase superfamily protein. The role of the thioredoxin superfamily and *APX* in redox homeostasis is not negotiable [36]. For example, APX is an antioxidant enzyme that functions in synthesizing ascorbate (AsA) and scavenging the H_2O_2. Ectopic expression of Celery (*Apium graveolens*) *APX1* into *Arabidopsis* resulted in the induction of AsA content and antioxidant capacity but less reduction in photosynthetic rate and chlorophyll content that culminated in drought resistance [37].

Interestingly, other gene-encoded proteins from different pathways also have ROS detoxification activity besides their primary function. For instance, the UDP-glycosyltransferases (*UGTs*) and terpene synthases (*TPSs*) have a crucial role in detoxification and homeostasis besides their prominent role of catalyzing a sugar conjugation with small lipophilic compounds and generating the structural diversity of terpenoids superfamily, respectively [38,39]. Further, terpenes and terpenoids have a wide range of biological activities, acting as anticancer, antimicrobial, anti-inflammatory, and anti-allergic [39]. Similarly, TFs such as *RRTF1* and *MYBR1* were up-regulated by stress and H_2O_2, resulting in less ROS accumulation due to efficient antioxidant machinery [40,41]. On the contrary, the TFs of *RWP-RK* play essential roles in nitrate starvation responses and root nodulation [42], while the activity of TFIIS is substantial for heat stress adaptation in plants, while its absence negatively affects the expression and the alternative splicing pattern of hundreds of heat-regulated transcripts [43].

ROS can highly coordinate and regulate multiple signaling pathways such as antioxidants, kinases, defense-related genes, Ca^{2+} influx, and hormones (e.g., JA, ethylene, and salicylic acid) [2,44]. Our data showed that transport-related genes encoding cation/Ca^{2+} exchanger were strongly induced during late stress and recovery. Similarly, the *ZIFL1* is critical in regulating stomatal closure during stress in *Arabidopsis* [45]. Moreover, the *ENODL* genes, such as *GhENODL6*, were shown to exhibit SA-induction under biotic stress and regulate the ROS production in cotton [46]. Furthermore, 3 out of 4 hormone-related genes are JA-related, especially the *JAZ3* that showed strong induction at 2–48 h compared to 0 h. It reported that salinity adaptation in grapevine is highly correlated with tight control of JA and its isoleucine conjugate (JA-Ile) production and, hence, fine-tuning their signals [3]. In contrast, salt sensitivity is more likely attributed to the unconstrained jasmonate (JA) signaling accompanied by ROS burst. Finally, our results revealed a list of genes with different functions that are up-regulated under salinity with involved stress tolerance and antioxidant activities during stress such as cytochromes P450, nudix hydrolases, major latex protein-like protein, serine carboxypeptidase-like, and glutathione S-transferase [47–51].

In conclusion, this investigation addressed the transcriptional remodeling of grapevine rootstock RUG throughout salinity and recovery, complementing our previous physiological study [21] by improving our understanding of the molecular dynamic coordinating grapevine's salt tolerance mechanism. The data demonstrated the significant impact of establishing different sets of processes to cope with salinity stress, most notably ROS detoxification and homeostasis. Future research would allow further in-depth validation and functional characterization of pivotal stress biomarkers to select suitable traits for generating innovative salt-tolerant rootstocks.

4. Materials and Methods

4.1. Plant Materials, Salinity Application, and Sampling

Leave samples were collected from 3-year-old *Vitis* hybrid 140 Ruggeri (*V. berlandieri* × *V. rupestris*) grown under greenhouse conditions during the summer season of 2020 at the Florida A&M University (Tallahassee, FL, USA). The hybrid rootstock was selected according to its salinity adaptation [21,22]. A total of 50 grapevine rootstock plants were used for the experiment through which each plant was grown in a one-gallon pot supported with a based plate and peat moss mixed with perlite (3:1). Salt stress was applied gradually by irrigating plants with 2.5 L of 30 mM NaCl daily for 5 days to reach 150 mM NaCl as a final concentration. Once the fifth addition of NaCl was received, samples were collected from mature and healthy leaves, mainly the third and/or fourth ones. At each time point, fifteen leaves were collected and distributed into 3 biological replicates (5 leaves/replicate). For the salinity time course, leave samples were collected at time points 0, 0.5, 2, 8, 24, and 48 h. Control plants were irrigated using distilled water. Similarly, treated plants were re-watered with distilled water for the recovery experiment, and leaf samples were collected at 4, 8, and 12 days. All samples were immediately flash-frozen in liquid nitrogen and stored at −80 °C for further analysis.

4.2. Nucleic Acid Extraction and RNA-Seq Library Construction

Total RNA was extracted from leaf samples as described previously [52]. All RNA samples were treated with the RNase-Free DNase Set (Qiagen, Valencia, CA, USA), then cleaned up with the RNeasy Mini Kit (Qiagen). A total of 24 RNA-seq libraries (three biological replicates at 8 salt-stressed time points, R1–R8) were constructed using NEBNext Ultra II RNA Library Prep Kit for Illumina (New England Biolabs, Ipswich, MA, USA). The libraries were multiplexed equally for paired-end 150-base sequencing in two lanes of NovaSeq 6000 (Illumina, San Diego, CA, USA) at the Novogene Co., Ltd. (Sacramento, CA, USA).

4.3. RNA-Seq Data Preprocessing and Identification of Differentially Expressed Genes

Illumina sequencing of the multiplexed RNA-seq libraries yielded 24 FASTQ files of sequences (GenBank accession number: PRJNA1036264). As previously described [53], reads quality was examined by FastQC (https://www.bioinformatics.babraham.ac.uk/projects/fastqc/, accessed on 10 February 2023) twice before and after trimming using Trimmomatic v0.39 [54]. Trimmed reads were aligned to the *Vitis* genome (*V. vinifera*_457_Genoscope.12X) and projected to the transcriptome using STAR (Table S1) [24,25]. The later transcriptomic coordinates files (*.BAM) were then quantified by Salmon in alignment mode (Table S1) [26].

Differentially expressed genes (DEGs) during salinity and recovery times were identified between consecutive time points (R2–R1, R3–R2, R4–R3, R5–R4, R6–R5, R7–R6 or R8–R7) using DESeq2 package setting on default parameters [55]. DEGs of each comparison (DEGs, $P_{FDR} < 0.05$, log2fold change > ±1.5) were considered to be expressed (Table S2). The open-source R package UpSetR was used to generate a scalable matrix-based visualization for all intersections among comparisons [56]. Finally, the web-based tool Venny was used to construct the consensus result (https://bioinfogp.cnb.csic.es/tools/venny/index.html, accessed on 12 February 2023).

4.4. Weighted Gene Co-Expression Network Analysis

Co-expression network modules were constructed using the variance stabilizing transformation values and the R package WGCNA (v. 1.72-1) [57]. Lowly expressed genes among all samples were removed by DESeq2, and the remaining 20,717 genes were used in module construction. The co-expression modules were obtained using the default settings, except that the soft threshold power was 4, TOMType was signed, minModuleSize was 30, mergeCutHeight was 0.25, and scale-free topology fit index 0.8 ($R^2 = 0.8$). A module eigengene (ME) value, which summarizes the expression profile of a given module as the first principal component, was calculated and used to evaluate the association of modules with leaf physiological parameters, including photosynthetic pigments (chlorophyll "Chl-T, Chl-a, Chl-b", and carotenoids "Caro"), sugar contents (total soluble sugar "TSS", sucrose "Suc", glucose "Glu", and fructose "Fru"), enzymatic antioxidant activities (superoxide dismutase "SOD", catalase "CAT", and glutathione peroxidase "GPX"), non-enzymatic antioxidant activities (proline content "Pro"), and two stress markers malondialdehyde "MDA" and hydrogen peroxide "H_2O_2" [21]. As a result, the final matrix for WGCNA was assigned to 46 modules (M1-M46) with 20,717 genes. The module membership (MM) and gene significance (GS) values were calculated, and the intra-modular hub genes were identified (GS > 0.2, MM > 0.8, and p-value < 0.05).

4.5. GO Enrichment and KEGG Pathway Analyses

GO and KEGG enrichment analyses were assigned using the g:Profiler website by applying the Benjamini-Hochberg multiple testing correction method with $P_{FDR} < 0.05$ [27]. The Cytoscape plug-in ClueGO was used to visualize the non-redundant GO terms and KEGG pathways [58].

4.6. Validation of DEG Subsets by qPCR

DNase-treated RNA (4 μg) was reverse transcribed in a reaction of 50 μL using the High Capacity cDNA Reverse Transcription Kit (Applied Biosystems, Foster City, CA, USA). Gene-specific primers were designed using Primer Express (v3.0, Applied Biosystems) (Table S6). The qPCR assays were performed using 20 ng of cDNA and 300 nM of each primer in a 10 μL reaction volume with SsoAdvanced Universal SYBR Green Supermix (Bio-Rad Laboratories, Hercules, CA, USA). Three biological and three technical replicates for each reaction were analyzed on a CFX384 Touch Real-Time PCR Detection System instrument (Bio-Rad Laboratories) with the first step of 95 °C for 5 min followed by 40 cycles of 95 °C for 10 s, 60 °C for 10 s, and 72 °C for 20 s. Melting curves were generated using the program: 95 °C for 15 s, 60 °C for 15 s, and 95 °C for 15 s. Transcript abundance was quantified using standard curves for the target and reference genes generated from serial dilutions of PCR products from corresponding cDNAs. Transcript abundance was normalized to the reference genes *MrActin* and *MrEF1*, which showed high stability across the different muscadine genotypes and tissues. The geometric mean of the selected housekeeping genes was validated as an accurate normalization factor.

Supplementary Materials: The following supporting information can be downloaded at: https://www.mdpi.com/article/10.3390/plants13060837/s1, Figure S1: Heatmap of sample-to-sample distances of RNA-seq data from Ruggeri leaves during the salt stress and recovery procedure, Figure S2: The Upset plot of the total number of the upregulated differentially expressed genes (DEGs), Figure S3: The Upset plot of the total number of the down-regulated differentially expressed genes (DEGs), Figure S4: K-means clustering of non-redundant significantly expressed genes during salt stress and recovery procedure, Figure S5: A network view for the pre-defined molecular function (MF) GO terms pathways for the K10 cluster in-group (I), Figure S6: A network view for the pre-defined Biological process (BP) GO terms and KEEG pathways for the K10 cluster in the group (I), Figure S7: A network view for the pre-defined Biological process (PB) GO terms and KEEG pathways for K8 cluster in the group (IV), Figure S8: Sample dendrogram and trait heatmap, Figure S9: The hierarchical cluster dendrogram of genes showing co-expressed modules identified by WGCNA, Figure S10: A network view for the pre-defined Biological process (BP) GO terms pathways for the WGCNA M1 module, Figure S11: A network view for the predefined Biological process (BP) GO terms pathways for the WGCNA M11 module; Table S1: Summary of read mapping rate of RNA-Seq library of Ruggeri leaves at different time points of salt stress or recovery experiment, Table S2: Significantly expressed genes over the time course of salt stress and recovery in Ruggeri rootstock, Table S3: The gene list of each K-mean cluster including their transcript per million (TPM) values and annotation information, Table S4: The GO-term information of the gene list of each K-means cluster, Table S5: The GO-term information of the gene list of each significantly correlated WGCNA module, and Table S6: The oligonucleotide primers of assessed genes by qPCR and the correlation coefficient between RNA-seq and qPCR data.

Author Contributions: Conceptualization, A.I., A.E.-K., D.K., A.M.E.-S., I.E.-S., M.D.A. and S.M.S.; Methodology, A.G.D., A.I., A.G.M., A.S.D., A.M.H., I.E.-S., T.I., M.M. and P.G.; Software, A.I. and I.E.-S.; Validation, A.I., A.S.D., A.M.H., I.E.-S. and P.G.; Formal Analysis, A.I., A.G.M., A.S.D., I.E.-S., T.I. and P.G.; Investigation, A.I., I.E.-S. and P.G.; Resources, D.K., I.E.-S. and M.D.A.; Data Curation, I.E.-S.; Writing—Original Draft Preparation, A.I., T.I. and P.G.; writing—review and editing, A.I., A.E.-K., D.K., A.M.E.-S., I.E.-S., M.D.A. and S.M.S.; Visualization, A.I., I.E.-S. and P.G.; Supervision, I.E.-S. and M.D.A.; Project Administration, I.E.-S.; Funding Acquisition, I.E.-S. All authors have read and agreed to the published version of the manuscript.

Funding: I.E.-S. received support from the 1890 Institution Teaching, Research, and Extension Capacity Building Grants (CBG) Program (grant no. 2023-38821-39592; project accession no. 1030371) from the USDA National Institute of Food and Agriculture.

Data Availability Statement: The data presented in this study are available upon request.

Conflicts of Interest: The authors declare no conflict of interest.

References

1. Ismail, A.A.A.M. Grapes for the Desert: Salt Stress Signaling in *Vitis*. Ph.D. Thesis, Karlsruhe Institute of Technology, Karlsruhe, Germany, 2013.
2. Ismail, A.; Takeda, S.; Nick, P. Life and death under salt stress: Same players, different timing? *J. Exp. Bot.* **2014**, *65*, 2963–2979. [CrossRef]
3. Ismail, A.; Seo, M.; Takebayashi, Y.; Kamiya, Y.; Eiche, E.; Nick, P. Salt adaptation requires efficient fine-tuning of jasmonate signalling. *Protoplasma* **2014**, *251*, 881–898. [CrossRef]
4. Ismail, A.; Riemann, M.; Nick, P. The jasmonate pathway mediates salt tolerance in grapevines. *J. Exp. Bot.* **2012**, *63*, 2127–2139. [CrossRef]
5. Isayenkov, S.V.; Maathuis, F.J.M. Plant salinity stress: Many unanswered questions remain. *Front. Plant Sci.* **2019**, *10*, 80. [CrossRef]
6. Shinozaki, K.; Uemura, M.; Bailey-Serres, J.; Bray, E.A.; Weretilnyk, E. "Plant environment and agriculture: Responses to abiotic stress". In *Biochemistry & Molecular Biology of Plants*, 2nd ed.; Buchanan, B.B., Gruissem, W., Jones, R.L., Eds.; John Wiley & Sons, Ltd.: Hoboken, NJ, USA, 2015; pp. 1051–1100.
7. FAO. *The Global Map of Salt-Affected Soils (GSASmap)*; FAO: Rome, Italy, 2021.
8. Ismail, A.; El-Sharkawy, I.; Sherif, S. Salt stress signals on demand: Cellular events in the right context. *Int. J. Mol. Sci.* **2020**, *21*, 3918. [CrossRef]
9. Yang, Z.; Li, J.L.; Liu, L.N.; Xie, Q.; Sui, N. Photosynthetic regulation under salt stress and salt-tolerance mechanism of sweet sorghum. *Front. Plant Sci.* **2020**, *10*, 1722. [CrossRef]
10. Ismail, A.; Seo, M.; Takebayashi, Y.; Kamiya, Y.; Nick, P. A balanced JA/ABA status may correlate with adaptation to osmotic stress in *Vitis* cells. *J. Plant Physiol.* **2015**, *185*, 57–64. [CrossRef]
11. Bazihizina, N.; Colmer, T.D.; Cuin, T.A.; Mancuso, S.; Shabala, S. Friend or Foe? Chloride Patterning in Halophytes. *Trends Plant Sci.* **2019**, *24*, 142–151. [CrossRef]
12. Karimi, S.M.; Freund, M.; Wager, B.M.; Knoblauch, M.; Fromm, J.; Mueller, H.M.; Ache, P.; Krischke, M.; Mueller, M.J.; Müller, T.; et al. Under salt stress guard cells rewire ion transport and abscisic acid signaling. *New Phytol.* **2021**, *231*, 1040–1055. [CrossRef]
13. Wani, S.H.; Singh, N.B.; Haribhushan, A.; Mir, J.I. Compatible solute engineering in plants for abiotic stress tolerance-role of glycine betaine. *Curr. Genom.* **2013**, *14*, 157–165. [CrossRef]
14. Kesawat, M.S.; Satheesh, N.; Kherawat, B.S.; Kumar, A.; Kim, H.U.; Chung, S.M.; Kumar, M. Regulation of reactive oxygen species during salt stress in plants and their crosstalk with other signaling molecules-current perspectives and future directions. *Plants* **2023**, *12*, 864. [CrossRef]
15. Singh, M.; Kumar, J.; Singh, S.; Singh, V.P.; Prasad, S.M. Roles of osmoprotectants in improving salinity and drought tolerance in plants: A review. *Rev. Environ. Sci. Biotechnol.* **2015**, *14*, 407–426. [CrossRef]
16. Downton, W.J.S. Influence of rootstocks on the accumulation of chloride, sodium and potassium in grapevines. *Aust. J. Agr. Res.* **1977**, *28*, 879–889. [CrossRef]
17. Phogat, V.; Pitt, T.; Stevens, R.M.; Cox, J.W.; Šimůnek, J.; Petrie, P.R. Assessing the role of rainfall redirection techniques for arresting the land degradation under drip irrigated grapevines. *J. Hydrol.* **2020**, *587*, 125000. [CrossRef]
18. Hannah, L.; Roehrdanz, P.R.; Ikegami, M.; Shepard, A.V.; Shaw, M.R.; Tabor, G.; Zhi, L.; Marquet, P.A.; Hijmans, R.J. Climate change, wine, and conservation. *Proc. Natl. Acad. Sci. USA* **2013**, *110*, 6907–6912. [CrossRef]
19. Walker, R.R.; Blackmore, D.H.; Clingeleffer, P.R.; Holt, H.; Pearson, W.; Francis, I.L. Effect of rootstock on yield, grape composition and wine sensory attributes of Shiraz grown in a moderately saline environment. *Aust. J. Grape Wine Res.* **2019**, *25*, 414–429. [CrossRef]
20. Sauer, M.R. Effects of vine rootstocks on chloride concentration in Sultana scions. *Vitis* **1968**, *7*, 223–226.
21. Gajjar, P.; Ismail, A.; Islam, T.; Darwish, A.G.; Moniruzzaman, M.; Abuslima, E.; Dawood, A.S.; El-Saady, A.M.; Tsolova, V.; El-Kereamy, A.; et al. Physiological comparison of two salt-excluder hybrid grapevine rootstocks under salinity reveals different adaptation qualities. *Plants* **2023**, *12*, 3247. [CrossRef]
22. Hillin, D.; Helwi, P.; Scheiner, J.J. Comparison of salt exclusion in muscadine and interspecific hybrid grapes using a greenhouse screening procedure. *Horttechnology* **2021**, *31*, 771–779. [CrossRef]
23. Zhou-Tsang, A.; Wu, Y.; Henderson, S.W.; Walker, A.R.; Borneman, A.R.; Walker, R.R.; Gilliham, M. Grapevine salt tolerance. *Aust. J. Grape Wine Res.* **2021**, *27*, 149–168. [CrossRef]
24. Dobin, A.; Davis, C.A.; Schlesinger, F.; Drenkow, J.; Zaleski, C.; Jha, S.; Batut, P.; Chaisson, M.; Gingeras, T.R. STAR: Ultrafast universal RNA-seq aligner. *Bioinformatics* **2013**, *29*, 15–21. [CrossRef]
25. Srivastava, A.; Malik, L.; Sarkar, H.; Zakeri, M.; Almodaresi, F.; Soneson, C.; Love, M.I.; Kingsford, C.; Patro, R. Alignment and mapping methodology influence transcript abundance estimation. *Genome Biol.* **2020**, *21*, 239. [CrossRef]
26. Patro, R.; Duggal, G.; Love, M.I.; Irizarry, R.A.; Kingsford, C. Salmon: Fast and bias-aware quantification of transcript expression using dual-phase inference. *Nat. Methods* **2017**, *14*, 417–419. [CrossRef]
27. Raudvere, U.; Kolberg, L.; Kuzmin, I.; Arak, T.; Adler, P.; Peterson, H.; Vilo, J.G. Profiler: A web server for functional enrichment analysis and conversions of gene lists (2019 update). *Nucleic Acids Res.* **2019**, *47*, W191–W198. [CrossRef]
28. Jaillon, O.; Aury, J.M.; Noel, B.; Policriti, A.; Clepet, C.; Casagrande, A.; Choisne, N.; Aubourg, S.; Vitulo, N.; Jubin, C.; et al. The grapevine genome sequence suggests ancestral hexaploidization in major angiosperm phyla. *Nature* **2007**, *449*, 463–467.

29. Canaguier, A.; Grimplet, J.; Di Gaspero, G.; Scalabrin, S.; Duchêne, E.; Choisne, N.; Mohellibi, N.; Guichard, C.; Rombauts, S.; Le Clainche, I.; et al. A new version of the grapevine reference genome assembly (12X.v2) and of its annotation (VCost.v3). *Genom. Data* **2017**, *14*, 56–62. [CrossRef]
30. Chini, A.; Fonseca, S.; Fernàndez, G.; Adie, B.; Chico, J.M.; Lorenzo, O.; García-Casado, G.; López-Vidriero, I.; Lozano, F.M.; Ponce, M.R.; et al. The JAZ family of repressors is the missing link in jasmonate signalling. *Nature* **2007**, *448*, 666–671. [CrossRef]
31. Thines, B.; Katsir, L.; Melotto, M.; Niu, Y.; Mandaokar, A.; Liu, G.; Nomura, K.; He, S.Y.; Howe, G.A.; Browse, J. JAZ repressor proteins are targets of the SCF^{COI1} complex during jasmonate signalling. *Nature* **2007**, *448*, 661–665. [CrossRef]
32. Guan, L.; Haider, M.S.; Khan, N.; Nasim, M.; Jiu, S.; Fiaz, M.; Zhu, X.; Zhang, K.; Fang, J. Transcriptome sequence analysis elaborates a complex defensive mechanism of grapevine (*Vitis vinifera* L.) in response to salt stress. *Int. J. Mol. Sci.* **2018**, *19*, 4019. [CrossRef]
33. Carrasco, D.; Zhou-Tsang, A.; Rodriguez-Izquierdo, A.; Ocete, R.; Revilla, M.A.; Arroyo-García, R. Coastal wild grapevine accession (*Vitis vinifera* L. ssp. sylvestris) shows distinct late and early transcriptome changes under salt stress in comparison to commercial rootstock Richter 110. *Plants* **2022**, *11*, 2688. [CrossRef]
34. Daldoul, S.; Hanzouli, F.; Hamdi, Z.; Chenenaoui, S.; Wetzel, T.; Nick, P.; Mliki, A.; Gargouri, M. The root transcriptome dynamics reveals new valuable insights in the salt-resilience mechanism of wild grapevine (*Vitis vinifera* subsp. sylvestris). *Front Plant Sci.* **2022**, *13*, 1077710. [CrossRef]
35. Daldoul, S.; Gargouri, M.; Weinert, C.; Jarrar, A.; Egert, B.; Mliki, A.; Nick, P. A Tunisian wild grape leads to metabolic fingerprints of salt tolerance. *Plant Physiol.* **2023**, *193*, 371–388. [CrossRef]
36. Mata-Pérez, C.; Spoel, S.H. Thioredoxin-mediated redox signalling in plant immunity. *Plant Sci.* **2019**, *279*, 27–33. [CrossRef]
37. Liu, J.X.; Feng, K.; Duan, A.Q.; Li, H.; Yang, Q.Q.; Xu, Z.H.; Xiong, A.S. Isolation, purification, and characterization of an ascorbate peroxidase from celery and overexpression of the *AgAPX1* gene enhanced ascorbate content and drought tolerance in *Arabidopsis*. *BMC Plant Biol.* **2019**, *19*, 488. [CrossRef]
38. Bock, K.W. The UDP-glycosyltransferase (UGT) superfamily expressed in humans, insects and plants: Animal-plant arms-race and co-evolution. *Biochem. Pharmacol.* **2016**, *99*, 11–17. [CrossRef]
39. Rudolf, J.D.; Chang, C.Y. Terpene synthases in disguise: Enzymology, structure, and opportunities of non-canonical terpene synthases. *Nat. Prod. Rep.* **2020**, *37*, 425–463. [CrossRef]
40. Matsuo, M.; Johnson, J.M.; Hieno, A.; Tokizawa, M.; Nomoto, M.; Tada, Y.; Godfrey, R.; Obokata, J.; Sherameti, I.; Yamamoto, Y.Y.; et al. High Redox Responsive Transcription FACTOR1 levels result in accumulation of reactive oxygen species in *Arabidopsis thaliana* shoots and roots. *Mol Plant.* **2015**, *8*, 1253–1273. [CrossRef]
41. Tiwari, P.; Indoliya, Y.; Chauhan, A.S.; Pande, V.; Chakrabarty, D. Over-expression of rice R1-type *MYB* transcription factor confers different abiotic stress tolerance in transgenic *Arabidopsis*. *Ecotoxicol. Environ. Saf.* **2020**, *206*, 111361. [CrossRef]
42. Chardin, C.; Girin, T.; Roudier, F.; Meyer, C.; Krapp, A. The plant *RWP-RK* transcription factors: Key regulators of nitrogen responses and of gametophyte development. *J. Exp. Bot.* **2014**, *65*, 5577–5587. [CrossRef]
43. Szádeczky-Kardoss, I.; Szaker, H.M.; Verma, R.; Darkó, É.; Pettkó-Szandtner, A.; Silhavy, D.; Csorba, T. Elongation factor *TFIIS* is essential for heat stress adaptation in plants. *Nucleic Acids Res.* **2022**, *50*, 1927–1950. [CrossRef]
44. Vinay, K.; Tushar, K.; Mansi, S.; Shabir, H.W. ROS-induced signaling and gene expression in crops under salinity stress. Role and regulation under abiotic stress. In *Reactive Oxygen Species and Antioxidant Systems in Plants*; Iqbal, M., Khan, R., Khan, N.A., Eds.; Springer: Singapore, 2017; pp. 159–184.
45. Remy, E.; Cabrito, T.R.; Baster, P.; Batista, R.A.; Teixeira, M.C.; Friml, J.; Sá-Correia, I.; Duque, P. A major facilitator superfamily transporter plays a dual role in polar auxin transport and drought stress tolerance in *Arabidopsis*. *Plant Cell* **2013**, *25*, 901–926. [CrossRef]
46. Zhang, M.; Wang, X.; Yang, J.; Wang, Z.; Chen, B.; Zhang, X.; Zhang, D.; Sun, Z.; Wu, J.; Ke, H.; et al. *GhENODL6* isoforms from the phytocyanin gene family regulated verticillium wilt resistance in cotton. *Int. J. Mol. Sci.* **2022**, *23*, 2913. [CrossRef]
47. Gullner, G.; Komives, T.; Király, L.; Schröder, P. Glutathione S-transferase enzymes in plant-pathogen interactions. *Front. Plant Sci.* **2018**, *9*, 1836. [CrossRef]
48. Liu, H.; Ma, X.; Liu, S.; Du, B.; Cheng, N.; Wang, Y.; Zhang, Y. The *Nicotiana tabacum* L. major latex protein-like protein 423 (*NtMLP423*) positively regulates drought tolerance by ABA-dependent pathway. *BMC Plant Biol.* **2020**, *20*, 475. [CrossRef]
49. Xu, X.; Zhang, L.; Zhao, W.; Fu, L.; Han, Y.; Wang, K.; Yan, L.; Li, Y.; Zhang, X.-H.; Min, D.-H. Genome-wide analysis of the serine carboxypeptidase-like protein family in *Triticum aestivum* reveals *TaSCPL184-6D* is involved in abiotic stress response. *BMC Genom.* **2021**, *22*, 350. [CrossRef]
50. Kondo, Y.; Rikiishi, K.; Sugimoto, M. Rice nudix hydrolase *OsNUDX2* sanitizes oxidized nucleotides. *Antioxidants* **2022**, *11*, 1805. [CrossRef]
51. Minerdi, D.; Savoi, S.; Sabbatini, P. Role of Cytochrome P450 enzyme in plant microorganisms' communication: A focus on grapevine. *Int. J. Mol. Sci.* **2023**, *24*, 4695. [CrossRef]
52. Gambino, G.; Perrone, I.; Gribaudo, I. A rapid and effective method for RNA extraction from different tissues of grapevine and other woody plants. *Phytochem. Anal.* **2008**, *19*, 520–525. [CrossRef]
53. Ismail, A.; Darwish, A.G.; Park, M.; Gajjar, P.; Tsolova, V.; Soliman, K.F.A.; El-Sharkawy, I. Transcriptome profiling during muscadine berry development reveals the dynamic of polyphenols metabolism. *Front. Plant Sci.* **2022**, *12*, 818071. [CrossRef]

54. Bolger, A.M.; Lohse, M.; Usadel, B. Trimmomatic: A flexible trimmer for Illumina sequence data. *Bioinformatics* **2014**, *30*, 2114–2120. [CrossRef]
55. Love, M.I.; Huber, W.; Anders, S. Moderated estimation of fold change and dispersion for RNA-seq data with DESeq2. *Genome Biol.* **2014**, *15*, 550. [CrossRef] [PubMed]
56. Conway, J.R.; Lex, A.; Gehlenborg, N. UpSetR: An R package for the visualization of intersecting sets and their properties. *Bioinformatics* **2017**, *33*, 2938–2940. [CrossRef] [PubMed]
57. Langfelder, P.; Horvath, S. WGCNA: An R package for weighted correlation network analysis. *BMC Bioinform.* **2008**, *9*, 559. [CrossRef]
58. Bindea, G.; Mlecnik, B.; Hackl, H.; Charoentong, P.; Tosolini, M.; Kirilovsky, A.; Fridman, W.H.; Pagès, F.; Trajanoski, Z.; Galon, J. ClueGO: A Cytoscape plug-in to decipher functionally grouped gene ontology and pathway annotation networks. *Bioinformatics* **2009**, *25*, 1091–1093. [CrossRef] [PubMed]

plants

MDPI

Article

Interactive Effects of Temperature, Water Regime, and [CO_2] on Wheats with Different Heat Susceptibilities

Rong Zhou [1,2,*], Benita Hyldgaard [2], Lamis Abdelhakim [2], Thayna Mendanha [2], Steven Driever [3], Davide Cammarano [4], Eva Rosenqvist [5] and Carl-Otto Ottosen [2]

1. College of Horticulture, Nanjing Agricultural University, Nanjing 210095, China
2. Department of Food Science, Aarhus University, DK-8200 Aarhus, Denmark; behy@seges.dk (B.H.); lamisabdelhakim@gmail.com (L.A.); tm@food.au.dk (T.M.); coo@food.au.dk (C.-O.O.)
3. Centre for Crop Systems Analysis, Wageningen University, Bornsesteeg 48, 6708 PE Wageningen, The Netherlands; steven.driever@wur.nl
4. Department of Agroecology, iClimate, CBIO, Aarhus University, 8830 Tjele, Denmark; davide.cammarano@agro.au.dk
5. Department of Plant and Environmental Sciences, University of Copenhagen, DK-2630 Taastrup, Denmark; ero@plen.ku.dk

* Correspondence: zhour@njau.edu.cn

Abstract: Plants' response to single environmental changes can be highly distinct from the response to multiple changes. The effects of a single environmental factor on wheat growth have been well documented. However, the interactive influences of multiple factors on different wheat genotypes need further investigation. Here, treatments of three important growth factors, namely water regime, temperature, and CO_2 concentration ([CO_2]), were applied to compare the response of two wheat genotypes with different heat sensitivities. The temperature response curves showed that both genotypes showed more variations at elevated [CO_2] (e[CO_2]) than ambient [CO_2] (a[CO_2]) when the plants were treated under different water regimes and temperatures. This corresponded to the results of water use efficiency at the leaf level. At e[CO_2], heat-tolerant 'Gladius' showed a higher net photosynthetic rate (P_n), while heat-susceptible 'Paragon' had a lower P_n at reduced water, as compared with full water availability. The temperature optimum for photosynthesis in wheat was increased when the growth temperature was high, while the leaf carbon/nitrogen was increased via a reduced water regime. Generally, water regime, temperature and [CO_2] have significant interactive effects on both wheat genotypes. Two wheat genotypes showed different physiological responses to different combinations of environmental factors. Our investigation concerning the interactions of multi-environmental factors on wheat will benefit the future wheat climate-response study.

Keywords: wheat; physiological response; elevated CO_2 concentration; reduced watering; increased temperature

check for updates

Citation: Zhou, R.; Hyldgaard, B.; Abdelhakim, L.; Mendanha, T.; Driever, S.; Cammarano, D.; Rosenqvist, E.; Ottosen, C.-O. Interactive Effects of Temperature, Water Regime, and [CO_2] on Wheats with Different Heat Susceptibilities. *Plants* **2024**, *13*, 830. https://doi.org/10.3390/plants13060830

Academic Editors: Zhaoshi Xu, Geovani Soares de Lima, Lauriane Almeida dos Anjos Soares and Francisco Vanies Da Silva Sá

Received: 28 November 2023
Revised: 11 February 2024
Accepted: 8 March 2024
Published: 13 March 2024

1. Introduction

The effect of adverse climate on terrestrial ecosystems has been an ongoing source of concern over the past decades. Anthropogenic activities have led to an increase in the atmospheric CO_2 concentration ([CO_2]), ascribed to be the driving force of climate change [1]. Future agriculture production is expected to face not only rising [CO_2] levels but also a combination of other factors, such as warmer global temperatures, drought stress, and an increase in extreme climatic events. Crop yields have already been impacted by climate change as indicated by a global annual yield decrease in rice and wheat (0.3% and 0.9%, respectively) due to climate change reported by Ray et al. (2019) [2]. In Europe, yields of all non-tropical crops have decreased (6.3–21.2%) due to climate change [2]. Despite the continuous genetic progress, climate change may be the primary cause behind the decline in the yield growth trend for cereals in Europe [3–5].

Even though the response of plants to individual stresses, such as high temperature, drought, and elevated [CO_2] (e[CO_2]), are extensively discussed in the literature, relatively little is known about the physiological response of plants to the interaction of those factors. Multifactor stresses are complex, and a positive plastic response to a single stress may play an antagonistic role when other stresses are added [6]. For instance, stomatal regulation contrasts upon its response to drought and heat stresses: while drought closes stomata to prevent water loss, heat stress increases stomatal conductance to cool the leaves by transpiration.

Photosynthesis is one of the highly thermo-sensitive processes in plants, which can be affected by temperature changes in many ways [7,8]. Meanwhile, the water deficit can affect the photosynthesis of wheat in various degrees depending on genotypes [9]. Overlapping changes in temperature and water might intensify the injuries experienced by the photosynthetic apparatus [10]. Still, plants can acclimate to stressful conditions. For example, wheat and rice are able to acclimate their photosynthesis to temperature and elevated [CO_2] during the growing season [11]. The acclimation of photosynthesis in a valuable crop such as wheat is crucial to study, since the inhibition of photosynthesis is directly associated with a decreased yield and optimizing photosynthesis serves to maximize carbon gain [12].

Water use efficiency (WUE; g kg^{-1}) is a key indicator of drought tolerance. Crop management increased the yield, which also resulted in high WUE [13], indicating the positive relation between crop yield and WUE. Numerous authors have demonstrated substantial differences between the WUE values of cereal species [14–16] but have also emphasized the fact that changes in WUE are especially important if the water supply to the plants is limited [17]. A synergetic response of WUE is usually reported to the combination of drought and e[CO_2] on plants [6,18]. Similarly, while higher temperatures increase respiration more than photosynthesis, e[CO_2] alters the ratio of [CO_2]/[O_2] towards [CO_2], favoring Rubisco carboxylation over oxygenation, hence alleviating carbon loss due to photorespiration under heat stress conditions [19]. Moreover, carbon and nitrogen metabolism of plants under drought are strongly interrelated [20]. Additional information is required about actual crop evapotranspiration and water use levels when plants suffer different levels of stress during different growth stages [21].

Plants experience oxidative damage due to the overaccumulation of reactive oxygen species (ROS) and shortage in the production of enzymatic and non-enzymatic antioxidants under drought and heat stress. For instance, carotenoids play a crucial role in antioxidant defense systems, as they can serve as antioxidants in crops in adverse environments. Plants grown in e[CO_2] have been reported to up-regulate the antioxidant defense by the excessive carbohydrate production and the decrease in photorespiration, therefore attenuating oxidative stress [22,23]. In addition, under heat, drought, and their combination, wheat yield reduction is reported to be alleviated by e[CO_2] [24], due to an increase in photosynthesis [25]. Yet, multigenerational exposure to e[CO_2] aggravated grain quality reduction by reducing N, K, Ca, protein, and total amino-acid concentration in wheat grains [24]. Clearly, the effects of e[CO_2] on wheat being subjected to temperature changes and water regimes are complicated and require further exploration.

In this study, two wheat cultivars with different levels of susceptibility to heat were exposed to two temperature levels during the earlier vegetative phase with full and reduced watering under ambient [CO_2] (a[CO_2]) and e[CO_2]. Photosynthetic temperature response curves, WUE, optimum temperature, integrated carbon gain, leaf area, and the C/N ratio were investigated. We aimed to clarify the interactive effects of three factors, namely temperature, water, and [CO_2], on wheat physiology. We hypothesized that (1) the detrimental effect of reduced water, temperature changes, and their combination on plant photosynthesis could be alleviated by e[CO_2], as it allowed photosynthetic rates to be maintained under non-optimum conditions; and (2) the physiological response of the two wheat genotypes to temperature changes, water regime, and [CO_2] could differ, as heat-tolerant wheat showed high resilience. Studies on multifactor stresses in cereals will contribute to an increase in the understanding of plant responses under realistic field situations and can improve current stress management and CO_2 fertilization strategies. The inclusion of different genotypes in such studies can add to the understanding of crucial adaptation traits to environmental changes and the pursuit of engineering a better stress-tolerant germplasm.

2. Materials and Methods

2.1. Plant Material and Experimental Treatments

Two commercial spring wheat cultivars (*Triticum aestivum* L.) were studied: heat-tolerant 'Gladius' from Australia and heat-susceptible 'Paragon' from the UK [26]. One seed was sown in 0.6 L plastic pots. The pots were filled with commercial peat substrate (Pindstrup Færdigblanding 2, Pindstrup Mosebrug A/S, Ryomgaard, Denmark). The seedlings were cultivated in a greenhouse for 14 days during the summer of the year 2015, at the Department of Food Science, Aarhus University, Aarslev, DK (55.30 N, 10.44 E). Seedlings were grown at ambient CO_2 (400 ppm), with an average air temperature of 24 °C. Supplementary light (high-pressure sodium lamps, SON-T Agro, 600 W, Philips, Eindhoven, The Netherlands) was provided whenever the natural photosynthetic photon flux density (PPFD) reached below 150 μmol m^{-2} s^{-1}. Once the seeds sprouted, pots were fertigated daily by flooding the greenhouse table with a nutrient solution (N:P:K of 190:35:275 ppm, pH 5.8, electrical conductivity of 1.9 mS m^{-1}).

Vigorous uniform plants at growth stage Zadoks 12–13 [27] were moved into two controlled climate chambers in two batches separated in time due to space limitations. During the first batch, plants were grown at 400 ppm CO_2 concentration (a[CO_2]), and during the second round, at 800 ppm CO_2 concentration (e[CO_2]). There were four environmental settings in the climate chambers for each CO_2 concentration: (1) 18/14°C air temperature and full watering (FW) in which pots were kept at 100% field capacity as control; (2) 28/24 °C air temperature and FW in which pots were kept at 100% field capacity; (3) 18/14 °C air temperature and reduced watering (RW) by keeping pots at 50% field capacity; and (4) 28/24 °C air temperature and RW by keeping pots at 50% field capacity. In total, eight treatments were established, as shown in Figure 1. Values of air relative humidity (RH) were adjusted to 45/47% at 18/14 °C and 73/67% at 28/24 °C to ensure a vapor deficit temperature (VPD) of approximately 1 kPa. For all treatments, the light level was set at 280 μmol m^{-2} s^{-1} (Fionia Lighting LED FL300 sunlight, Søndersø, Denmark) at plant canopy level supplied in a 14/10 h day/night photoperiod regime corresponding to a daily light integral (DLI) of 14.11 mol m^{-2} day^{-1}. The water regime was imposed using drought spotter units (Phenospex, Heerlen, The Netherlands) installed at each chamber. Pots were assigned to the scales in a completely randomized design (n = 6). The treatments with 18/14 °C and 28/24 °C lasted 21 days and 14 days, respectively, in order to obtain harvest plants at the same developmental stage.

2.2. Leaf Area and Carbon/Nitrogen (C/N)

Plants grown in a control chamber (18/14 °C) were harvested seven days later (21 days) than plants grown in a warm chamber (14 days) in an attempt to end the experiment with all treatments at a similar developmental stage. Leaf area (LA) was determined using a LI-3100C area meter (Li-Cor, Lincoln, NE, USA). A 3–4 cm piece of primary stem (3 cm above soil surface) was also harvested for C/N determination. C/N determination was performed at Aberystwyth University (Aberystwyth, Wales, UK) via an elemental analysis (Elementar Vario Max Q CN analyser) [28]. There were six replicates.

2.3. Leaf Photosynthetic Pigments

The leaf used for photosynthesis measurements was sampled after 14 days of the treatments, weighted, and immediately stored at −80 °C for pigment analysis. Leaf samples were freeze-dried, weighed, and milled in a mixer mill (MM200, Retsch Inc., Haan, Germany), using a steel ball. Photosynthetic pigments were extracted with 96% Ethanol overnight at room temperature. Contents of pigments, including chlorophyll a (Chl a), chlorophyll b (Chl b), and carotenoids (Car), were determined spectrophotometrically

(Shimadzu UV-1700, Kyoto, Japan), according to the method of Lichtenthaler (1987) [29]. The ratios of chlorophylls a:b (Chl a:b) and total chlorophylls to carotenoids (Chl:Car) were calculated. There were six replicates.

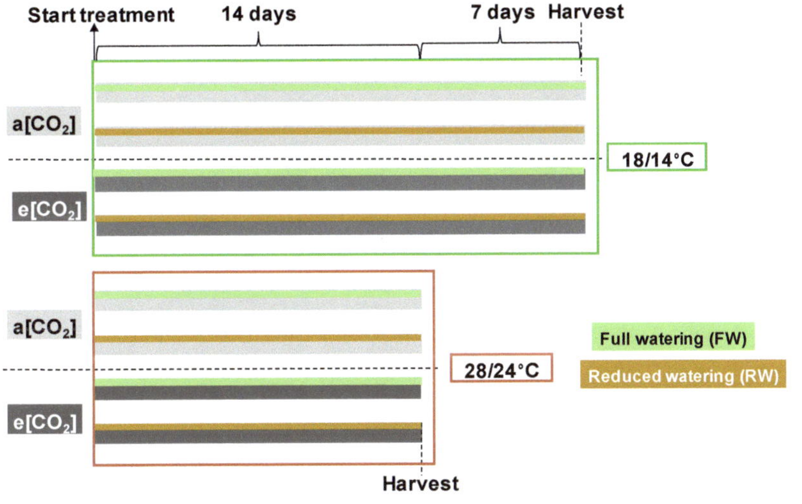

Figure 1. Illustration of the eight established treatments. To ensure plant harvest at the same developmental stage, the plants grown at 18/14 °C were harvested one week after those grown at 28/24 °C.

2.4. Photosynthetic Temperature Response Curves

Photosynthetic temperature response curves (A/T) were performed after 14 days of treatment, using a portable gas exchange fluorescence system (CIRAS-2, PP systems, Amesbury, MA, USA). The last young fully developed leaf was affixed in a 1.7 cm^2 leaf cuvette with a 195 mL min^{-1} flow rate and 1500 µmol photons m^{-2} s^{-1} PPFD controlled by an LED light unit. The CO_2 concentration of the cuvette was set to 400 and 800 ppm for the plants grown under a[CO_2] and e[CO_2], respectively. The CIRAS cuvette and the plant were placed in a controlled climate cabinet during the measurements, and the temperature in the cabinet was adjusted according to the following set points: 15 °C, 18 °C, 25 °C, 30 °C, 35 °C, and 40 °C. Once the steady leaf temperature was reached, the gas exchange rates were recorded. A moist cloth was placed around the grid of the water vapor equilibrator of the gas analyzer to increase relative humidity when needed and maintain VPD around 1 kPa. There were three replicates.

2.5. Leaf Water Use Efficiency

The water use efficiency at the leaf level (WUE_{leaf}) was calculated by P_n/E (net photosynthetic rate and transpiration). The optimum temperature (T_{opt}) was calculated as the vertex of a hyperbola fitted for each A/T response curve. The integrated carbon gain ($\int P_n$) was calculated by the sum of the trapezoid area of each temperature point. There were three replicates.

$$\int P_n = (P_{n1} + P_{n2}) \times \frac{(T_2 - T_1)}{2} \quad (1)$$

where T_1 and T_2 indicate two temperatures, while P_{n1} and P_{n2} indicate the net photosynthetic rate at the two temperatures.

2.6. Data Analysis

Statistical analyses were performed using R open-source statistical computing software (Version 3.4.3, The R Foundation, Vienna, Austria). Data were checked for variance homogeneity and normal distribution before statistical analysis. Means were compared using a pairwise comparison procedure adjusted with Bonferroni correction, and the significant difference was indicated by different small letters. The data were analyzed for each cultivar separately. A three-way ANOVA was performed to indicate the effect of [CO_2], water regime, temperature, and their interaction (CO_2 × water × temperature), and the level of significance of each factor is indicated as * $p < 0.05$, ** $p < 0.01$, and *** $p < 0.0001$.

3. Results

3.1. Leaf Area and C/N

The e[CO_2] decreased the leaf area in 'Paragon' at 18 °C + FW and 28 °C + RW and in 'Gladius' at 18 °C + FW and 18 °C + RW (Figure 2). Reduced water decreased the leaf area in both cultivars at 18 °C regardless of [CO_2] (Figure 2). Three individual factors had significant effects on leaf area in both cultivars with interaction (Table 1).

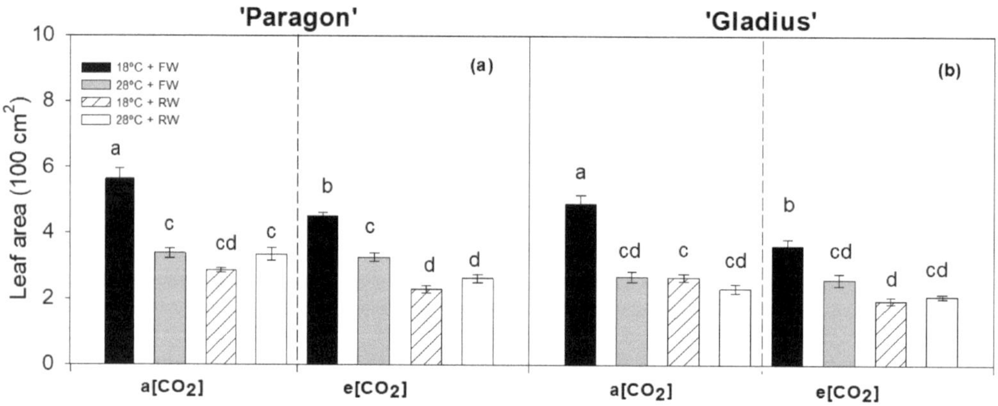

Figure 2. Leaf area of two wheat cultivars, 'Paragon' (**a**) and 'Gladius' (**b**), for each treatment under a[CO_2] and e[CO_2]. Different lowercase letters show significant difference (post hoc adjusted Bonferroni test, $p < 0.05$) between treatments within each panel. 'Paragon' data of leaf area were squared root-transformed to pass normality and homogeneity test. Data represent mean values +/− S.E.M. (n = 6).

Table 1. Output of the three-way ANOVA for net photosynthetic rate (P_n), water use efficiency (WUE), optimum temperature (T_{opt}), integrated carbon gain ($\int P_n$), leaf area, carbon/nitrogen ratio in leaves (C:N), chlorophyll a (Chl a), chlorophyll b (Chl b), carotenoids content (Car), and ratios of chlorophylls a:b (Chl a:b) and total chlorophylls to carotenoids (Chl:Car) at each temperature point of two wheat cultivars, 'Gladius' and 'Paragon'.

Source of Variation	P_{n15}	P_{n18}	P_{n25}	P_{n28}	P_{n35}	P_{n40}	WUE_{15}	WUE_{18}	WUE_{25}	WUE_{28}	WUE_{35}	WUE_{40}	T_{opt}	$\int P_n$	Leaf Area	C:N	Chl a	Chl b	Car	Chl a:b	Chl: Car
'Paragon'																					
CO_2 × water × temp.	ns	ns	ns	ns	ns	ns	*	ns	ns	*	ns	ns	ns	ns	*	-	*	*	ns	*	*
Water × temp.	*	*	*	ns	ns	ns	ns	ns	ns	ns	ns	ns	ns	ns	***	-	*	*	ns	ns	ns
Temp. × CO_2	ns	ns	ns	ns	ns	*	ns	*	ns	*	*	ns	ns	ns	ns	-	**	**	**	ns	**
Water × CO_2	*	*	*	ns	ns	ns	**	ns	ns	ns	ns	ns	**	ns	ns	-	***	ns	ns	***	ns
Water	***	***	**	*	ns	ns	*	ns	ns	ns	ns	ns	**	*	***	-	*	***	*	***	***
Temp.	ns	ns	ns	ns	**	***	*	***	***	***	**	ns	***	*	**	-	ns	ns	*	**	*
CO_2	*	**	***	***	***	***	***	***	***	***	***	***	***	***	**	-	*	ns	ns	***	ns
'Gladius'																					
CO_2 × water × temp.	-	*	ns	ns	-	ns	*	ns	**	**	**	*	ns	-	ns	ns	**	***	***	***	***
Water × temp.	-	ns	ns	ns	-	ns	ns	ns	ns	ns	ns	ns	ns	-	***	**	***	***	***	***	***
Temp. × CO_2	-	**	*	*	-	*	ns	ns	ns	ns	ns	*	ns	-	**	ns	*	ns	ns	ns	ns
Water × CO_2	-	ns	ns	ns	-	ns	ns	ns	ns	*	ns	***	***	-	ns	ns	***	**	**	ns	*
Water	-	ns	*	ns	-	***	*	ns	*	*	ns	*	ns	-	***	***	***	**	***	ns	**
Temp.	-	ns	ns	ns	-	**	**	***	**	ns	ns	ns	***	-	***	ns	ns	ns	*	***	*
CO_2	-	*	***	***	-	***	*	***	***	***	***	***	***	-	**	ns	ns	ns	ns	***	ns

Note: Level of significance of each factor is indicated as * $p < 0.05$, ** $p < 0.01$, *** $p < 0.0001$ within each cultivar; 'ns' indicates no statistically significant difference; '-' indicates parameters that did not fulfil the assumption of the parametric method. 'Temp.' is the abbreviation for temperature.

In 'Paragon', the data for C:N in the leaves did not fulfil the assumption of the parametric method; therefore, the data were compared using a non-parametric Kruskal–Wallis test. The treatments followed a similar trend observed for 'Gladius' (Figure 3), as the C:N ratio of leaves was unaffected by either the different [CO_2] or temperature regimes (Table 1). The RW treatment, however, significantly increased the C:N values under both [CO_2] for 'Paragon' at 18 °C and for 'Gladius' at both temperatures (Figure 3).

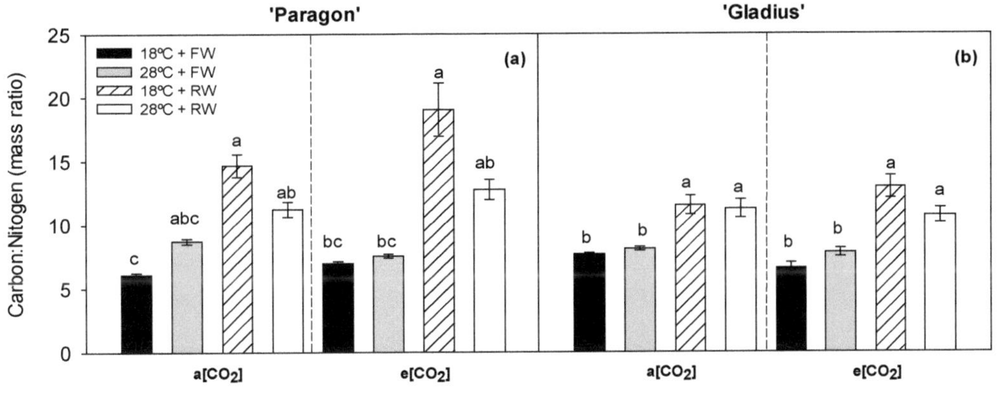

Figure 3. The carbon/nitrogen ratio (C:N ratio) in leaves of two wheat cultivars, 'Paragon' (**a**) and 'Gladius' (**b**). Different lowercase letters show significant difference (post hoc adjusted Bonferroni test, $p < 0.05$) between treatments within each panel. 'Gladius' data were log-transformed to pass the normality and homogeneity test. 'Paragon' data did not fulfil the assumption of the parametric method; therefore, they were compared using non-parametric Kruskal–Wallis test. Data represent mean values +/− S.E.M. (n = 6).

3.2. Leaf Pigment Content

For 'Paragon', Chl a was decreased by e[CO_2] in plants at RW (Table 2). Reduced water decreased Chl a, Chl b, and Chl:Car in leaves of 'Paragon' at e[CO_2] (Table 2). Increased temperature decreased Chl a content and Chl:Car in leaves of 'Paragon' at a[CO_2] + FW (Table 2). e[CO_2] increased the Chl a, Chl b, Car content, and Chl:Car, but it decreased Chl a:b in the leaves of 'Gladius' at 28 °C + FW (Table 2). Reduced water increased Chl a, Chl b, Car content, and Chl:Car, but it decreased Chl a:b in the leaves of 'Gladius' at a[CO_2] + 28 °C (Table 2). Increased temperature decreased Chl a, Chl b, Car content, and Chl:Car in the leaves of 'Gladius' at a[CO_2] + FW, but it increased Chl b and Car content at a[CO_2] + RW (Table 2). For both cultivars, all pigments and pigment ratios showed a three-factorial interaction, except for the Car content of 'Paragon' (Table 2).

3.3. Photosynthetic Temperature Response

When each point of the temperature curve was compared separately, an interaction between the water regime and [CO_2] was observed for 'Paragon' at temperatures $\leq$ 25 °C (Table 1). e[CO_2] alleviated the detrimental effect of reduced water at P_n set points 15 °C, 18 °C, and 25 °C when compared to a[CO_2] (Figure 4a,b and Table 1). For 'Gladius', the P_n at individual set points of the curves was influenced by the interaction between temperature and [CO_2] (Table 1). e[CO_2] alleviated the reduction in P_n caused by elevated temperature (28 °C) when compared to a[CO_2] at all set points, when compared by the parametric method (Figure 4c,d and Table 1).

Table 2. Pigment content of wheat leaves.

	Chl *a* (mg/g DW)	Chl *b* (mg/g DW)	Car (mg/g DW)	Chl *a:b*	Chl:Car
			'Paragon'		
a[CO_2]					
18 °C + FW	12.49 ± 0.63 a	3.34 ± 0.20 ab	2.64 ± 0.14 a	3.74 ± 0.04 c	6.00 ± 0.10 a
18 °C + RW	10.57 ± 0.51 abc	2.23 ± 0.13 cd	2.25 ± 0.12 ab	4.76 ± 0.07 a	5.69 ± 0.07 a
28 °C + FW	9.26 ± 0.83 bcd	2.58 ± 0.28 bcd	2.25 ± 0.15 ab	3.64 ± 0.10 c	5.19 ± 0.19 b
28 °C + RW	11.94 ± 0.84 ab	2.79 ± 0.26 abc	2.49 ± 0.19 a	4.30 ± 0.08 b	5.91 ± 0.04 a
e[CO_2]					
18 °C + FW	10.21 ± 0.58 abc	2.78 ± 0.17 abc	2.15 ± 0.13 ab	3.67 ± 0.04 c	6.05 ± 0.06 a
18 °C + RW	6.52 ± 0.78 d	1.82 ± 0.24 d	1.70 ± 0.19 b	3.61 ± 0.07 cd	4.84 ± 0.07 b
28 °C + FW	12.06 ± 0.41 ab	3.59 ± 0.12 a	2.75 ± 0.07 a	3.35 ± 0.01 d	5.68 ± 0.09 a
28 °C + RW	8.76 ± 0.62 cd	2.60 ± 0.17 bcd	2.27 ± 0.01 ab	3.35 ± 0.02 d	4.99 ± 0.04 b
			'Gladius'		
a[CO_2]					
18 °C + FW	10.75 ± 0.41 b	2.32 ± 0.14 bc	2.34 ± 0.07 bc	4.63 ± 0.05 a	5.57 ± 0.10 a
18 °C + RW	9.98 ± 0.73 ab	2.65 ± 0.17 c	2.35 ± 0.14 bc	3.73 ± 0.07 b	5.31 ± 0.11 bc
28 °C + FW	5.55 ± 0.55 c	1.44 ± 0.17 d	1.71 ± 0.09 d	3.89 ± 0.07 b	4.01 ± 0.20 d
28 °C + RW	13.29 ± 0.83 a	3.94 ± 0.24 a	3.27 ± 0.19 a	3.37 ± 0.04 c	5.26 ± 0.04 c
e[CO_2]					
18 °C + FW	9.63 ± 0.80 b	2.57 ± 0.20 c	2.25 ± 0.11 c	3.73 ± 0.05 b	5.37 ± 0.18 ab
18 °C + RW	9.52 ± 0.69 b	2.58 ± 0.19 c	2.33 ± 0.12 bc	3.69 ± 0.04 b	5.15 ± 0.12 c
28 °C + FW	10.06 ± 0.28 b	2.94 ± 0.13 bc	2.57 ± 0.04 bc	3.43 ± 0.07 c	5.05 ± 0.08 c
28 °C + RW	11.42 ± 0.54 ab	3.37 ± 0.15 ab	2.81 ± 0.12 ab	3.28 ± 0.02 c	5.25 ± 0.02 c

Note: chlorophyll a (Chl a), chlorophyll b (Chl b), carotenoids content (Car), and ratios of chlorophylls a:b (Chl a:b) and total chlorophylls to carotenoids (Chl:Car). Different lowercase letters show significant difference (post hoc adjusted Bonferroni test, $p < 0.05$) between treatments within each cultivar. Data represent mean values ± S.E.M. (n = 6).

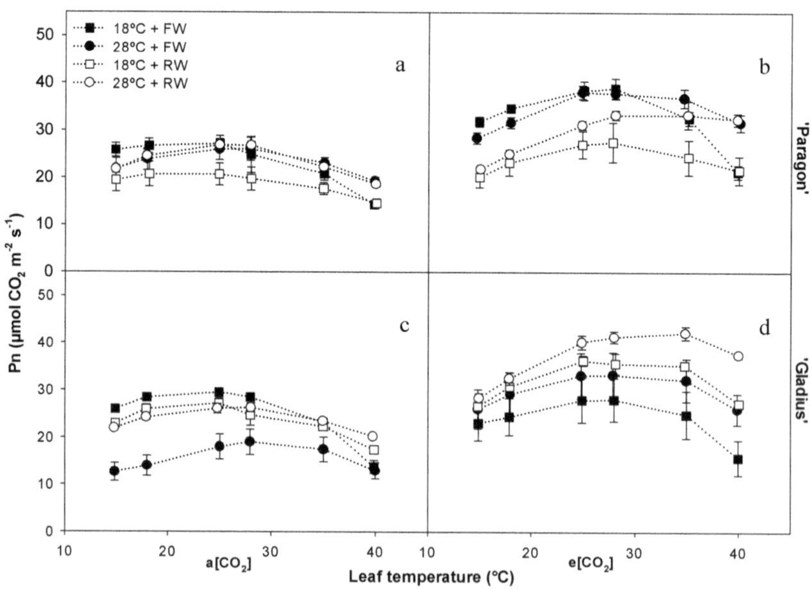

Figure 4. The effects of different treatments in plants grown and measured under a[CO_2] and e[CO_2], respectively, on temperature response of net photosynthesis of the wheat cultivars 'Paragon' (**a**,**b**) and 'Gladius' (**c**,**d**). Data represent mean values +/− S.E.M. (n = 3).

3.4. WUE and Integrated Carbon Gain

When comparing the WUE of the treatments, [CO_2] and temperature were the factors that consistently influenced most points of the response curve (Table 1). For both cultivars, e[CO_2] increased WUE as compared with a[CO_2], while the elevated temperature of 28 °C lowered the parameter in most temperature points as compared with 18 °C (Figure 5).

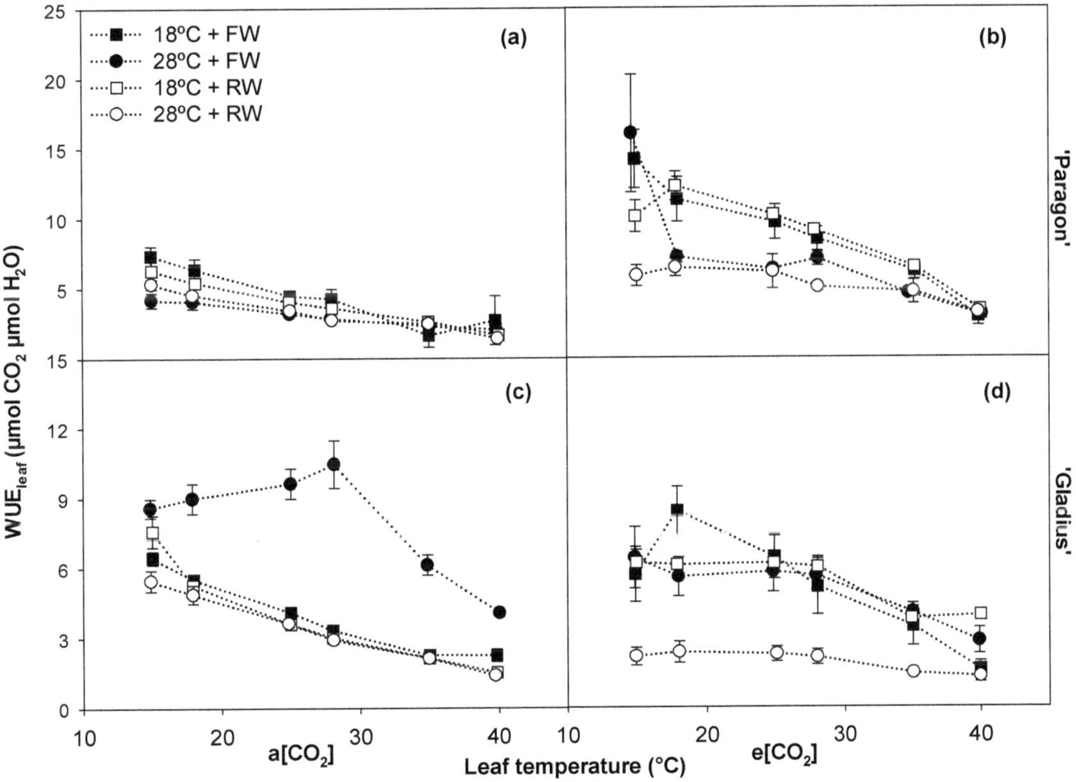

Figure 5. The effects of different treatments under a[CO_2] and e[CO_2] on water use efficiency at the leaf level (WUE_{leaf}) of two wheat cultivars, 'Paragon' (**a**,**b**) and 'Gladius' (**c**,**d**). Data represent mean values +/− S.E.M. (n = 3).

The e[CO_2] increased the T_{opt} for 'Paragon' grown at 18 °C + FW, 18 °C + RW, and 28 °C + RW and for 'Gladius' grown at 18 °C + RW and 28 °C + RW (Figure 6a,b). Reduced water increased the T_{opt} in both cultivars at e[CO_2], except for 'Paragon' grown at 18 °C (Figure 6a,b). A higher temperature increased the T_{opt}, except for 'Paragon' grown at RW with a[CO_2] and 'Gladius' at FW with e[CO_2] (Figure 6a,b). Regarding the T_{opt}, the only significant interaction noticed was between [CO_2] and the water regime for both cultivars, even though the individual factor significantly affected T_{opt} (Table 1). The e[CO_2] increased $\int P_n$ only in 'Paragon' grown at FW as compared with the a[CO_2] (Figure 6c,d). Accordingly, three individual factors had significant effects on $\int P_n$ only in 'Paragon' without interaction (Table 1).

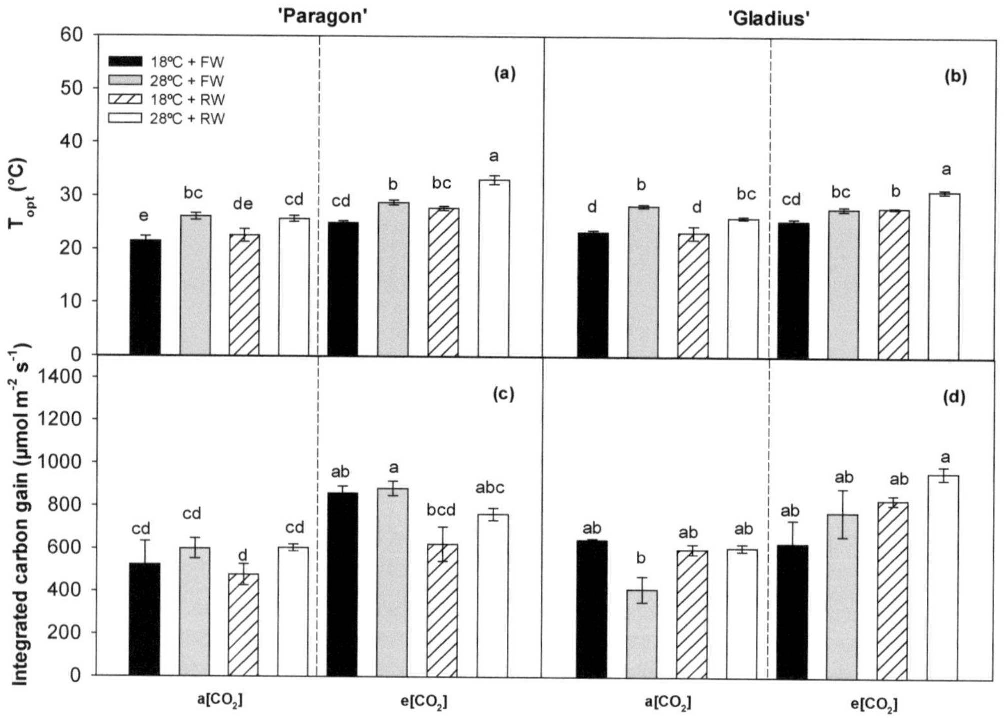

Figure 6. The effects of different treatments under a[CO_2] and e[CO_2] on the optimum temperature and the integrated carbon gain of photosynthesis of two wheat cultivars, 'Paragon' (**a**,**c**) and 'Gladius' (**b**,**d**). Different lowercase letters show significant difference (post hoc adjusted Bonferroni test, $p < 0.05$) between treatments within each panel. 'Gladius' data of integrated carbon gain did not fulfil the assumption of the parametric method; therefore, the data were compared using non-parametric Kruskal–Wallis test. Data represent mean values +/− S.E.M. (n = 3).

4. Discussions

4.1. Physiological Responses of Wheat to Changes on Growth Temperature, Water Regime, and [CO_2]

Most plants, including wheat, can acclimate their photosynthetic characteristics with continuing increases in growth temperature [7,30]. Even though the general temperature response of photosynthesis is well studied, wheat genotypic variation in photosynthetic thermal acclimation to high temperature needs further investigation [8]. In this study, not only two wheat genotypes with different heat susceptibilities were included but also the interaction between temperature, water regime, and [CO_2] on wheats were studied (Figure 1). The photosynthetic temperature response curves indicated the effects of temperature changes on photosynthetic CO_2 fixation. It was found that the P_n and WUE_{leaf} curves at different measured temperatures were altered by [CO_2], with more variation pronounced in two wheat genotypes grown at e[CO_2] (Figures 4 and 5). This indicated that wheat grown with increased [CO_2] was more sensitive to the changes in the temperature and water regime. The pigment content analysis presented a three-factorial interaction to chlorophylls *a* and *b* and the ratios between Chl *a:b* and Chl:Car, while the values of the carotenoids content showed two-way interactions (Tables 1 and 2). Changes in chlorophyll content could partially explain the response of P_n in wheat cultivars at different treatments. Moreover, at e[CO_2], a higher carotenoids content in 'Paragon' was induced at 28 °C + FW than that at 18 °C + RW; while that in 'Gladius' was higher at 28 °C + RW than that at 18 °C + FW (Table 2). The significant increase in carotenoids content showed that the non-enzyme antioxidant system in wheat actively responded to the environmental changes. In accordance with Berry and Björkman (1980) [7] and Posch et al. (2019) [8], P_n increases when the leaf temperature rises, peaking at an T_{opt} and then dropping. However, the T_{opt} can be shifted by the growth environments of the plants. We found that higher growth temperature (28 °C) generally increased T_{opt} as compared with 18 °C (Figure 6). The changes in T_{opt} can be the result of thermal acclimation of wheat, a process by which plants adjust Rubisco activity, the enzyme being involved in CO_2 fixation [31]. A significantly increased C/N ratio in rice and wheat plants at e[CO_2] were observed as compared with ambient [CO_2] [32]. Even though there was no increase in leaf C/N ratio induced by e[CO_2], reduced water availability increased the C/N ratio in 'Paragon' at 18 °C regardless of [CO_2] and in 'Gladius' regardless of temperature and [CO_2] (Figure 3). A significant increased leaf C/N under reduced water as compared with full water indicated the increased availability of carbon and reduced availability of nitrogen in wheat with less irrigation. The reduced water could decrease the content of N being absorbed by roots, resulting in less N being transported to leaves and higher leaf C/N ratio. The increased C/N ratio in 'Gladius' by reduced water could also be explained by the higher or non-significantly changed P_n in plants at RW as compared with FW (Figure 4).

4.2. Genotype-Dependent Interactive Effects on Wheats with Different Heat Susceptibilities

Water, temperature, and [CO_2] have significant individual and interactive effects on both wheat genotypes (Tables 1 and 2). The interactive effects of multiple stresses could be antagonistic, synergistic, and additive, which corresponds to less, more, and equal effects as compared with the sum of effects of single stressors, respectively [33]. In the current study, the interactive effects of water regime, temperature, and [CO_2] on wheat were complex and genotype dependent; however they cannot be simply concluded to be individual antagonistic, synergistic, and additive effects. For instance, the P_n in 'Paragon' at 18 °C + RW was low at both a[CO_2] and e[CO_2] as compared with the other three treatments, while that in 'Gladius' was not. Moreover, at e[CO_2], the water regime induced an opposite effect on the photosynthetic temperature response curve of the two cultivars.

Wheat genotypic variation could benefit our understanding of genotype by environment interaction, as genotypes with high physiological and phenotypic plasticity might have greater resilience against climatic changes [34]. A previous study showed that the acclimation strategies of 'Paragon' and 'Gladius' to climate factors varied significantly, as indicated by the developmental rate, physiological response pattern, and parameters responsible for most of the variation [34]. In accordance, the current study shows that the physiological response of wheat to temperature changes, water regime, and [CO_2] is also genotype dependent, as shown by not only temperature response curves but also the WUE_{leaf}, T_{opt}, leaf C/N ratio, and pigment content. More significant differences in carotenoids between the eight treatments caused by environmental changes were observed in heat-tolerant 'Gladius' than heat-susceptible 'Paragon' (Table 2). Moreover, not only the plant genotypes will alter the effects of multiple stresses; the type, intensity, start time, and duration of each stress will also have a significant influence when two or more stresses happen simultaneously [33,35,36]. Thereby, different studies might obtain distinct conclusions on the influences of multiple stresses due to experimental design.

The two wheat genotypes showed similar physiological responses, but with some differences. Single factors made some significant effects on wheat physiology, while there were interactions of two or three factors (Figure 7). Together with our results that both two genotypes were clustered firstly based on water regimes from Eller et al. (2020) [34], reduced water played a predominant role among the three environmental factors in the current study (Figure 7). The interactive effects of three key environmental factors on wheat depending on genotype were complicated, which cannot be deducted from the effects of a single factor. Overall, photosynthesis was generally increased by e[CO_2] being accompanied by more fluctuations when the plants were treated by different water regimes and temperatures. The T_{opt} was up-regulated when the growth temperature was high, and the leaf C/N ratio was increased by reduced water. The growth of both wheat genotypes was linked to the climatic effects on photosynthesis being affected by the temperature, water regime, and [CO_2].

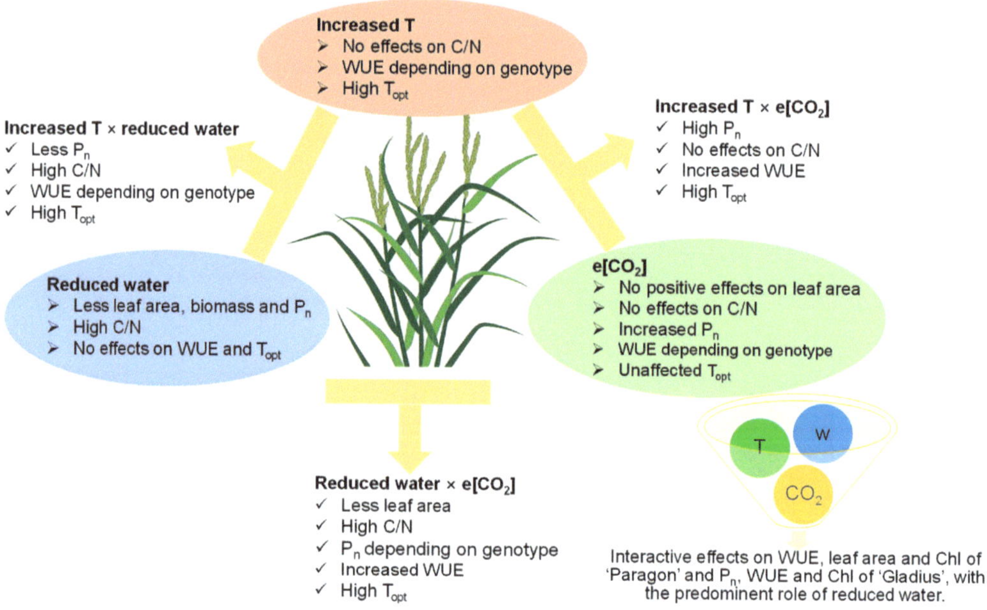

Figure 7. The single and interactive effects of increased temperature, reduced water, and elevated [CO_2] on wheat physiology.

Author Contributions: Conceptualization, R.Z. and B.H.; investigation and validation, B.H. and T.M.; formal analysis, L.A. and T.M.; writing—original draft preparation, R.Z., B.H., L.A. and T.M.; writing—review and editing, S.D., D.C., E.R. and C.-O.O. All authors have read and agreed to the published version of the manuscript.

Funding: The work was supported by funding from Aarhus University Research Foundation (AUFF Grant No. 30379).

Data Availability Statement: Data was contained within the article.

Acknowledgments: We acknowledged the kind help from Kaj Ole Dideriksen for plant management.

Conflicts of Interest: The authors declare no conflict of interest.

References

1. Arias, P.; Bellouin, N.; Coppola, E.; Jones, R.; Krinner, G.; Marotzke, J.; Joeri, R. IPCC Climate Change 2021: The Physical Science Basis. Contribution of Working Group I to the Sixth Assessment Report of the Intergovernmental Panel on Climate Change. 2021. Available online: https://www.ipcc.ch/report/ar6/wg1/ (accessed on 4 May 2021).
2. Ray, D.K.; West, P.C.; Clark, M.; Gerber, J.S.; Prishchepov, A.V.; Chatterjee, S. Climate change has likely already affected global food production. *PLoS ONE* **2019**, *14*, e0217148. [CrossRef]
3. Brisson, N.; Gate, P.; Gouache, D.; Charmet, G.; Oury, F.-X.; Huard, F. Why are wheat yields stagnating in Europe? A comprehensive data analysis for France. *Field Crops Res.* **2010**, *119*, 201–212. [CrossRef]
4. Lobell, D.B.; Gourdji, S.M. The influence of climate change on global crop productivity. *Plant Physiol.* **2012**, *160*, 1686–1697. [CrossRef]
5. Hou, P.; Liu, Y.; Liu, W.; Yang, H.; Xie, R.; Wang, K.; Ming, B.; Liu, G.; Xue, J.; Wang, Y.; et al. Quantifying maize grain yield losses caused by climate change based on extensive field data across China. *Resour. Conserv. Recycl.* **2021**, *174*, 105811. [CrossRef]

6. Xu, Z.; Shimizu, H.; Yagasaki, Y.; Ito, S.; Zheng, Y.; Zhou, G. Interactive effects of elevated CO_2, drought, and warming on plants. *J. Plant Growth Regul.* **2013**, *32*, 692–707. [CrossRef]
7. Berry, J.A.; Björkman, O. Photosynthetic response and adaptation to temperature in higher plants. *Annu. Rev. Plant Biol.* **1980**, *31*, 491–543. [CrossRef]
8. Posch, B.C.; Kariyawasam, B.C.; Bramley, H.; Coast, O.; Richards, R.A.; Reynolds, M.P.; Trethowan, R.; Atkin, O.K. Exploring high temperature responses of photosynthesis and respiration to improve heat tolerance in wheat. *J. Exp. Bot.* **2019**, *70*, 5051–5069. [CrossRef] [PubMed]
9. Inoue, T.; Inanaga, S.; Sugimoto, Y.; An, P.; Eneji, A.E. Effect of drought on ear and flag leaf photosynthesis of two wheat cultivars differing in drought resistance. *Photosynthetica* **2004**, *42*, 559–565. [CrossRef]
10. Kadam, N.N.; Xiao, G.; Melgar, R.J.; Bahuguna, R.N.; Quiñones, C.; Tamilselvan, A.; Prasad, P.V.V.; Jagadish, K.S.V. Agronomic and physiological responses to high temperature, drought, and elevated CO_2 interactions in cereals. *Adv. Agron.* **2014**, *127*, 111–156.
11. Cai, C.; Yin, X.; He, S.; Jiang, W.; Si, C.; Struik, P.C.; Luo, W.; Li, G.; Xie, Y.; Xiong, Y.; et al. Responses of wheat and rice to factorial combinations of ambient and elevated CO_2 and temperature in FACE experiments. *Glob. Change Biol.* **2016**, *22*, 856–874. [CrossRef]
12. Scafaro, A.P.; Atkin, O.K. The impact of heat stress on the proteome of crop species. In *Agricultural Proteomics Volume*; Springer: Berlin/Heidelberg, Germany, 2016; pp. 155–175.
13. Ritchie, J.T.; Basso, B. Water use efficiency is not constant when crop water supply is adequate or fixed: The role of agronomic management. *Eur. J. Agron.* **2008**, *28*, 273–281. [CrossRef]
14. Morell, F.J.; Lampurlanes, J.; Alvaro-Fuentes, J.; Cantero-Martinez, C. Yield and water use efficiency of barley in a semiarid Mediterranean agroecosystem: Long-term effects of tillage and N fertilization. *Soil. Tillage Res.* **2011**, *117*, 76–84. [CrossRef]
15. Rizza, F.; Ghashghaie, J.; Meyer, S.; Matteu, L.; Mastrangelo, A.M.; Badeck, F.-W. Constitutive differences in water use efficiency between two durum wheat cultivars. *Field Crops Res.* **2012**, *125*, 49–60. [CrossRef]
16. Zhang, S.; Sadras, V.; Chen, X.; Zhang, F. Water use efficiency of dryland wheat in the Loess Plateau in response to soil and crop management. *Field Crops Res.* **2013**, *151*, 9–18. [CrossRef]
17. Varga, B.; Varga-Laszlo, E.; Bencze, S.; Balla, K.; Veisz, O. Water use of winter cereals under well-watered and drought-stressed conditions. *Plant Soil. Environ.* **2013**, *59*, 150–155. [CrossRef]
18. Xu, Z.; Jiang, Y.; Zhou, G. Response and adaptation of photosynthesis, respiration, and antioxidant systems to elevated CO_2 with environmental stress in plants. *Front. Plant Sci.* **2015**, *6*, 159764. [CrossRef] [PubMed]
19. Ainsworth, E.A.; Rogers, A. The response of photosynthesis and stomatal conductance to rising CO_2: Mechanisms and environmental interactions. *Plant Cell Environ.* **2007**, *30*, 258–270. [CrossRef] [PubMed]
20. Nicolas, M.E.; Simpson, R.J.; Lambers, H.; Dalling, M.J. Effects of drought on partitioning of nitrogen in two wheat varieties differing in drought-tolerance. *Ann. Bot.* **1985**, *55*, 743–754. [CrossRef]
21. Evans, R.G.; Sadler, E.J. Methods and technologies to improve efficiency of water use. *Water Resour. Res.* **2008**, *44*, 1–15. [CrossRef]
22. Foyer, C.H.; Noctor, G. Redox Regulation in Photosynthetic Organisms: Signaling, Acclimation, and Practical Implications. *Antioxid. Redox Signal.* **2009**, *11*, 861–905. [CrossRef]
23. Zinta, G.; AbdElgawad, H.; Domagalska, M.A.; Vergauwen, L.; Knapen, D.; Nijs, I.; Janssens, I.A.; Beemster, G.T.S.; Asard, H. Physiological, biochemical, and genome-wide transcriptional analysis reveals that elevated CO_2 mitigates the impact of combined heat wave and drought stress in Arabidopsis thaliana at multiple organizational levels. *Glob. Change Biol.* **2014**, *20*, 3670–3685. [CrossRef] [PubMed]
24. Li, X.; Kristiansen, K.; Rosenqvist, E.; Liu, F. Elevated CO_2 modulates the effects of drought and heat stress on plant water relations and grain yield in wheat. *J. Agron. Crop Sci.* **2019**, *205*, 362–371. [CrossRef]
25. Wang, L.; Feng, Z.; Schjoerring, J.K. Effects of elevated atmospheric CO_2 on physiology and yield of wheat (*Triticum aestivum* L.): A meta-analytic test of current hypotheses. *Agric. Ecosyst. Environ.* **2013**, *178*, 57–63. [CrossRef]
26. Mendanha, T.; Rosenqvist, E.; Hyldgaard, B.; Ottosen, C.-O. Heat priming effects on anthesis heat stress in wheat cultivars (*Triticum aestivum* L.) with contrasting tolerance to heat stress. *Plant Physiol. Biochem.* **2018**, *132*, 213–221. [CrossRef] [PubMed]
27. Zadoks, J.C.; Chang, T.T.; Konzak, C.F. A decimal code for the growth stages of cereals. *Weed Res.* **1974**, *14*, 415–421. [CrossRef]
28. Jacques, R.G. Vermicomposting manure: Ecology and horticultural use. Master's Thesis, Aberystwyth University, Aberystwyth, UK, 2019.
29. Lichtenthaler, H.K. Chlorophylls and carotenoids: Pigments of photosynthetic biomembranes. In *Methods in Enzymology*; Academic Press: Cambridge, MA, USA, 1987; Volume 148, pp. 350–382.
30. Yamori, W.; Hikosaka, K.; Way, D.A. Temperature response of photosynthesis in C_3, C_4, and CAM plants: Temperature acclimation and temperature adaptation. *Photosynth. Res.* **2014**, *119*, 101–117. [CrossRef]
31. Yamasaki, T.; Yamakawa, T.; Yamane, Y.; Koike, H.; Satoh, K.; Katoh, S. Temperature acclimation of photosynthesis and related changes in photosystem II electron transport in winter wheat. *Plant Physiol.* **2002**, *128*, 1087–1097. [CrossRef]
32. Ma, H.; Zhu, J.; Xie, Z.; Liu, G.; Zeng, Q.; Han, Y. Responses of rice and winter wheat to free-air CO_2 enrichment (China FACE) at rice/wheat rotation system. *Plant Soil.* **2007**, *294*, 137–146. [CrossRef]
33. Zandalinas, S.I.; Mittler, R. Plant responses to multifactorial stress combination. *New Phytol.* **2022**, *234*, 1161–1167. [CrossRef]
34. Eller, F.; Hyldgaard, B.; Driever, S.M.; Ottosen, C.-O. Inherent trait differences explain wheat cultivar responses to climate factor interactions: New insights for more robust crop modelling. *Glob. Change Biol.* **2020**, *26*, 5965–5978. [CrossRef] [PubMed]

35. Zhang, H.; Sonnewald, U. Differences and commonalities of plant responses to single and combined stresses. *Plant J.* **2017**, *90*, 839–855. [CrossRef] [PubMed]
36. Zandalnas, S.I.; Fritschi, F.B.; Mittler, R. Signal transduction networks during stress combination. *J. Exp. Bot.* **2020**, *71*, 1734–1741. [CrossRef] [PubMed]

plants

MDPI

Article

Comparison of Growth and Physiological Effects of Soil Moisture Regime on *Plantago maritima* Plants from Geographically Isolated Sites on the Eastern Coast of the Baltic Sea

Katrīna Anna Ozoliņa, Astra Jēkabsone, Una Andersone-Ozola and Gederts Ievinsh *

Department of Plant Physiology, Faculty of Biology, University of Latvia, 1 Jelgavas Str., LV-1004 Riga, Latvia; katrina_anna.ozolina@lu.lv (K.A.O.); astra.jekabsone@lu.lv (A.J.); una.andersone-ozola@lu.lv (U.A.-O.)

* Correspondence: gederts.ievins@lu.lv

Abstract: The aim of the present study was to evaluate the morphological and physiological responses of *P. maritima* plants from five geographically isolated sites growing in habitats with different conditions to different substrate moisture levels in controlled conditions. Plants were produced from seed and cultivated in a greenhouse at four relatively constant soil moisture regimes: at 25, 50, and 75% soil water content and in soil flooded 3 cm above the surface (80% F). The two morphological traits that varied most strikingly among *P. maritima* accessions were the number of flower stalks and the number of leaves. Only plants from two accessions uniformly produced generative structures, and allocation to flowering was suppressed by both low moisture and flooding. Optimum shoot biomass accumulation for all accessions was at 50 and 75% soil moisture. The Performance Index Total was the most sensitive among the measured photosynthesis-related parameters, and it tended to decrease with an increase in soil water content for all *P. maritima* accessions. The initial hypothesis—that plants from relatively dry habitats will have a higher tolerance against low soil water levels, but plants from relatively wet habitats will have a higher tolerance against waterlogged or flooded soil—was not proven. The existence of three ecotypes of *P. maritima* within the five accessions from geographically isolated subpopulations on the eastern coast of the Baltic Sea at the level of morphological responses to soil water content can be proposed. *P. maritima* plants can be characterized as extremely tolerant to soil waterlogging and highly tolerant to soil flooding and low soil water content.

Keywords: chlorophyll; chlorophyll *a* fluorescence; coastal habitats; ion accumulation; osmotic adjustment; *Plantago maritima*; soil flooding; soil water content; waterlogging

Citation: Ozoliņa, K.A.; Jēkabsone, A.; Andersone-Ozola, U.; Ievinsh, G. Comparison of Growth and Physiological Effects of Soil Moisture Regime on *Plantago maritima* Plants from Geographically Isolated Sites on the Eastern Coast of the Baltic Sea. *Plants* **2024**, *13*, 633. https://doi.org/10.3390/plants13050633

Academic Editors: Geovani Soares de Lima, Lauriane Almeida dos Anjos Soares and Francisco Vanies Da Silva Sá

Received: 22 January 2024
Revised: 23 February 2024
Accepted: 23 February 2024
Published: 25 February 2024

1. Introduction

The physiological and molecular mechanisms of the environmental resilience of wild plants have gained considerable interest for a prolonged period of time [1]. In addition to solving purely natural conservation problems, the assessment of plant features associated with existence in highly heterogeneous conditions, such as ones in coastal habitats, can provide an understanding of plant adaptation mechanisms, which can be useful in various practical developments in the creation of new crop varieties resistant to environmental extremes [2–4].

Several environmental factors are important for shaping characteristic vegetation in coastal habitats [5–7]. In addition to heterogeneous salinity, soil moisture is another factor that strongly affects the growth and performance of plants native to coastal habitats. The sea coast is a very contrasting place with respect to moisture availability, which is reflected in the different plant adaptations required. Plants in dune habitats usually have distinctly xeromorphic features, allowing them to withstand prolonged periods of water shortage. On the other hand, plants in coastal wetlands are frequently subjected to inundation, and they have properties to tolerate a lack of rootzone oxygen. Also, plants on different types of beaches can exhibit episodes of both drought and flooding. In the non-tidal conditions of the seashore meadows and beaches of the Baltic Sea, flooding can be caused either by seawater as a result of wave activity and changes in sea level or by freshwater due to prolonged rain [8].

Some predominantly coastal-specific plant species occupy a relatively wide range of habitats with different sets of environmental conditions, possibly exhibiting a high degree of phenotypic plasticity. *Plantago maritima* L. is one such species, being found in various salt-affected coastal habitats. *P. maritima* is a self-incompatible, wind-pollinated perennial hemicryptophyte, flowering multiple times during its life span [9]. It is not listed as being a clonal species [10]; however, incidents of clonal expansion have been found both in the field [11] and in controlled conditions [12].

With respect to the EUNIS habitat classifications system, *P. maritima* is a constant species of three coastal habitats, i.e., "Baltic coastal meadow" (MA232), "Atlantic and Baltic rocky sea cliff and shore" (N31), and "Atlantic and Baltic soft sea cliff" (N34) [13]. In general, the species can also be found in other coastal habitats, including salt marshes, mud flats, sea dikes, etc. [14]. According to the ecological indicator values for Sweden, *P. maritima* has been found on mesic–moist soils (indicator value 3 out of 12) and is "competitive only under moderate–high salinity" (indicator value 4 out of 5) [15]. However, according to the indicator list for Great Britain, *P. maritima* is characterized as a "dampness indicator, mainly on constantly moist or damp, but not on wet soils" (indicator value 7 out of 12) and the "species most common in coastal sites but regularly present in freshwater or on non-saline soils inland" (indicator value 3 out of 9) [16].

The salinity responses of *P. maritima* have been widely studied, and it was shown that salt accumulation in shoots [17], the synthesis of organic osmolytes [18], and the induction of the enzymatic antioxidative system [19] are important constituents of the relatively high salinity tolerance of the species. However, given the absence of any growth stimulation due to increased salinity, *P. maritima* is designated as a facultative halophyte [20].

So far, differences in the life characteristics of *P. maritima* populations growing in different conditions have only been seldom assessed. Thus, two salt marsh populations in the Netherlands were compared in a study, and it appeared that the one from the lower zones subjected to tidal flooding and grazing by cattle was long-lived and propagated mainly vegetatively, but the one from the upper zones with low flooding intensity was short-lived and propagated via seeds [11]. These differences appear to be caused by phenotypic plasticity as the result of differences in biotic and abiotic factors. It was concluded that the vegetative spread seen as the formation of daughter rosettes was due to trampling by cattle, and this was confirmed in further greenhouse experiments. Another early study of *P. maritima* in the natural conditions of a seashore meadow along a distributional gradient confirmed that the species, indeed, has high phenotypic plasticity due to differences in salinity, flooding, and light [21]. So far, no conclusive evidence has been obtained on the existence of specific ecotypes of *P. maritima* as a result of local genetic adaptation.

Only a limited amount of information can be found on the responses of *P. maritima* plants to increased soil water content and/or flooding with water. Thus, the waterlogging of *P. maritima* plants from a tidal salt marsh cultivated in salt marsh soil does not result in significant changes in growth irrespective of salinity regime, but the number of dead leaves increases only because of waterlogging in non-saline conditions [22]. Similarly, the growth of *P. maritima* plants is not negatively affected by soil waterlogging or submergence, but the number of dead and severely affected leaves increases by 35 and 50%, respectively [23]. In addition, submergence has a clearly negative effect on the flowering and generative reproduction of *P. maritima* plants [24]. Thus, results from ecological and ecophysiological studies confirm that *P. maritima* is well adapted to high soil water content and even partial submergence, which is in contrast with the reported values of ecological indicators, especially in the Baltic region [15,16]. In addition, the optimum soil moisture regime of *P. maritima* plants in controlled conditions has not been established.

The osmotic protection of the cellular environment is an important feature of coastal plants both in conditions of increased soil salinity and in water shortage situations [5]. It has been argued that halophytic species mostly accumulate inorganic ions (Na^+ and Cl^-) to maintain osmotic balance in salinity conditions [25]. However, the accumulation of K^+ is associated with tolerance to xeric conditions [26]. In particular, the vacuolar pool of K^+ is an important constituent of osmotic regulation for plants in non-saline conditions [27]. In addition, osmotically active organic compounds (compatible solutes) are important constituents of the water maintenance system in plant tissues exposed to low-water-potential media [28].

In conditions of suboptimal soil water supplies, plant growth is diminished, but the functional reasons for that can differ in the case of low soil moisture (drought) and excessive soil moisture (soil waterlogging). During soil water shortage, the interruption of water flow and decreased turgor pressure in tissues result in negative consequences for cell division and elongation [29–31]. Soil waterlogging results in decreased oxygen content, further leading to rootzone hypoxia, the suppression of root respiration, decreased mineral nutrition and photosynthesis, and eventual plant dieback [32]. In addition, soil flooding conditions leading to partial or full plant submergence, the hypoxia of aboveground parts, and post-hypoxia reoxygenation result in the death of plant parts or the whole plant [33]. Endogenous oxidative stress is an inevitable component of functional deterioration in all situations of critically inadequate water provision [31,34,35].

Besides changes in growth, the physiological status of plants is an important indication of plant adaptive success in particular conditions [36]. Usually, photosynthesis-related traits are used as indicators for plant functional status, as photosynthesis is relatively more sensitive to changes in environmental conditions in comparison with growth-related indices [37]. In particular, chlorophyll *a* fluorescence parameters have been used to assess plant tolerance to both drought [38] and flooding [39].

The aim of the present study was to evaluate the morphological and physiological responses of *P. maritima* plants from several geographically isolated sites growing in habitats with different conditions to different substrate moisture levels in controlled conditions. It was hypothesized that plants from relatively dry habitats (such as dry dune grasslands) would have a higher tolerance against low soil water levels, but plants from relatively wet habitats (such as flood-prone grasslands or beaches) would have a higher tolerance to waterlogged or flooded soil. Special emphasis was placed on the possible existence of ecotypes of *P. maritima* in isolated subpopulations in contrast to the phenotypic plasticity of the species.

2. Materials and Methods

2.1. Plant Material and Seedling Establishment

Seed material from different geographically isolated subpopulations (accessions) of *P. maritima* was collected in the territory of Estonia in 2022 (Table 1, Figure 1). Sites with *P. maritima* were selected with the idea to include different habitats in terms of environmental conditions (Figure S1). Two sites were located on the island of Kihnu and had contrasting soil moisture conditions (dry sandy dune grassland vs. wet salt-affected meadow). Distance by the coastline between these two sites was about 5 km. In addition to subpopulations of *P. maritima* from the islands of Saaremaa and Hiiumaa, one more accession was obtained from the mainland coastal part of Estonia. Seeds were dried in laboratory conditions for four weeks and then stored at 4 °C until use.

Table 1. Sites and habitat conditions of *Plantago maritima* accessions in Estonia used in the present study.

Accession	Site	Location	Habitat	Coordinates
PM1	Linaküla	Island of Kihnu	Dry dune grassland	58°07′51.9″ N 23°57′35.8″ E
PM2	Sääre	Island of Kihnu	Wet saline grassland with high plant diversity	58°09′16.4″ N 24°00′13.0″ E
PM3	Kassari	Island of Hiiumaa	Wet saline grazed grassland with low plant diversity	58°48′36.6″ N 22°52′00.2″ E
PM4	Ohesaare	Island of Saaremaa	Dominant *Plantago maritima* stand on pebbly beach	58°00′06.4″ N 22°01′26.3″ E
PM5	Matsi-Sõmeri	Pärnu County	Flood-prone stony beach	58°21′02.9″ N 23°44′30.3″ E

Figure 1. Map of the eastern Baltic Sea region indicating sites for collection of different accessions of *Plantago maritima* in Estonia. Kihnu, Hiiumaa, and Saaremaa are Estonian islands.

The experiment was performed in the winter season. Seeds were germinated in 1 L tissue culture containers filled with autoclaved commercial garden soil (Biolan, Eura, Finland) [40] mixed with deionized sterile water and placed in a plant growth cabinet, MLR-352H (Sanyo Electric, Osaka, Japan), with a photoperiod of 16 h (photon flux density of photosynthetically active radiation, 40 µmol m^{-2} s^{-1}) and day/night temperatures of 20/15 °C. After the formation of the third true leaf, the seedlings were individually transplanted into 220 mL plastic containers containing a mixture of heat-treated (60 °C, 24 h) garden soil (Biolan, Eura, Finland) and quartz sand (Saulkalne S, Saulkalne, Latvia), 4:1 (*v*/*v*). Seedlings were placed in 48 L lid-closed plastic boxes, laid out in a greenhouse, and gradually adapted to greenhouse conditions. An automated experimental greenhouse (HortiMaX, Maasdijk, the Netherlands) with supplemented light provided by Master SON-TPIA Green Power CG T 400W (Philips, Amsterdam, The Netherlands) and Powerstar HQI-BT 400 W/D PRO (Osram, Munich, Germany) lamps (photon flux density of photosynthetically active radiation, 380 µmol m^{-2} s^{-1} at the plant level) with a 16 h photoperiod was used for plant cultivation. The day/night temperature was 23/16 °C, and the relative air humidity was maintained at 60 to 70%.

After seven weeks, plants were individually transplanted into 1.3 L plastic containers with a mixture of garden soil (Biolan, Eura, Finland) and quartz sand (Saulkalne S, Saulkalne, Latvia), 4:1 (*v*/*v*). Plants were watered with deionized water when necessary to maintain substrate water content of 50–60% and measured with an HH2 moisture meter equipped with a WET-2 sensor (Delta-T Devices, Burwell, UK). Individual containers were randomly placed on a greenhouse bench and repositioned once a week. One week after additional acclimatization in the greenhouse, the treatments were started.

2.2. Treatments

For each of the five *P. maritima* accessions, four water content regimes were maintained, with five individual plants as replicates. For 25 and 50% treatments, soil water content was measured daily in all containers with an HH2 moisture meter at the four sides of the container. The substrate was supplemented with the necessary amount of deionized water according to a previously performed calibration so that the water content was within 20–30% or 45–55% for the 25 and 50% treatments, respectively. For the 75% treatment, which corresponded to the substrate waterlogging, containers with plants were placed inside 1.5 L plastic containers where 0.5 L of deionized water was maintained. This resulted in a relatively constant substrate water content of 75%. Flooding treatment was performed by placing containers with *P. maritima* plants inside a 4 L plastic container and maintaining a water level at least 3 cm above the soil surface. This resulted in a relatively constant substrate water content of 80%. All treatments lasted for three full weeks, but after that, the flooding treatment was terminated, and the respective plants were maintained in a waterlogged state. Immediately after that, the plants were fertilized with Yara Tera Kristalon Red and Yara Tera Calcinit fertilizers (Yara International, Oslo, Norway). A stock solution was prepared for each fertilizer (100 g L^{-1}), and the working solution contained 25 mL of each per 10 L of deionized water, used at a rate of 100 mL per container. The experiment continued for another three weeks.

2.3. Plant Harvest and Measurements

One week prior to termination of the experiment, leaf chlorophyll concentration and chlorophyll a fluorescence were measured as described previously using a chlorophyll meter CCM300 (Opti-Sciences, Hudson, NH, USA) and a Handy PEA fluorometer (Hansatech Instruments, King's Lynn, UK), respectively [40]. For each individual plant, two independent measurements were performed, with 10 measurements per accession per treatment. A fluorescence data analysis was performed using the PEA Plus software (v. 3.11, Hansatech Instruments, King's Lynn, UK). A number of parameters derived from the fast fluorescence induction curve were used [41,42]. F_v/F_m, calculated as $(F_m - F_0)/F_m$, represents the maximum quantum efficiency of photosystem II, indicating the probability that a trapped photon will perform a further photochemical energy transfer, and is used as a stress indicator. F_v/F_0, calculated as $(F_m - F_0)/F_0$, reflects the instant photochemical activity on the donor side of photosystem II. Performance Index Total is used as a relative indication of plant vitality and includes information on the status of both photosystem II and photosystem I, in addition to characterizing the electron flow between the two systems, which is on an absorption basis.

At termination, plants were individually separated into different parts according to morphological traits (roots, old leaves, middle leaves, new leaves, flower stalks, and flowers (both flowers and fruits)). Inflorescences were counted, and the height of flower stalks was measured. All plant material was weighed separately before and after drying in an oven at 60 °C for 72 h. Tissue water content was calculated as g of H_2O per g of dry mass.

Electrical conductivity, concentrations of Na^+ and K^+, and osmotic values in water extracts of dry plant material were measured as described previously [43]. Electrical conductivity was measured using a LAQUAtwin B-771 conductivity meter, Na^+ concentration was measured using a LAQUAtwin B-722 compact meter, and K^+ concentration was measured using a LAQUAtwin B-731 compact meter (Horiba, Kyoto, Japan). Osmotic activity in tissue extracts was measured by a freezing point osmometer Osmomat 3000 Basic (Gonotec Meß- und Regeltechnik, Berlin, Germany). Using a standard curve for different concentrations of NaCl and KCl, the osmotic value caused by the total concentration of Na^+ and K^+ was calculated according to the actual Na^+ and K^+ concentrations of each sample extract. For each sample, the difference between the total osmotic value and the osmotic value due to Na^+, K^+, and Cl^- ions was calculated and designated as the "non-ionic osmotic value", which showed the osmotic effect of other osmotically active ions (besides Na^+, K^+, and

Cl^-) or non-ionic compounds. Six tissue samples were independently measured for each accession/soil water content combination.

2.4. Data Analysis

All data were entered into MS Excel 365 (v. 16.81, Microsoft Corporation, Redmond, WA, USA). Data were analyzed using JASP (v. 0.17.2.1, University of Amsterdam, Amsterdam, the Netherlands). ANOVA analysis with a Tukey HSD test was performed at $p < 0.05$. Graphs were generated using RStudio (v. 4.2.2, R Foundation for Statistical Computing, Vienna, Austria). Principal component analysis and heat map generation were performed using ClustVis (http://biit.cs.ut.ee/clustvis/, accessed on 5 January 2024), a freely available web program [44]. Hierarchical clusters were generated using the average linkage method with correlation distance.

3. Results

3.1. Morphological Parameters

At the end of the experiment, *P. maritima* plants showed morphological differences between various accessions, as well as because of variations in soil moisture (Figures S2–S6). The two morphological traits that varied most strikingly among *P. maritima* accessions were the number of flower stalks and the number of leaves. Individuals of accessions PM1 and PM2 showed a low and variable ability to form generative organs (Figure 2) and had a relatively high number of leaves (Figure 3). Plants of accessions PM3 and PM4 formed many flower stalks and had a lower number of leaves in comparison with PM1 and PM2. However, PM5 plants had both a very low capacity for generative reproduction, as well as a low number of leaves. Given the clear genotype-related differences between *P. maritima* accessions, the effect of soil moisture on these two parameters also varied. For PM3 and PM4, plants in flooded conditions had a significantly decreased number of flower stalks, and the same was evident for PM2 (Figure 2). The highest number of leaves developed for PM1, PM2, PM4, and PM5 plants cultivated at 50% soil moisture, but for PM3 plants, this occurred at 75% soil moisture (Figure 3). However, because of the relatively large individual variability in leaf number, these effects for PM3, PM4, and PM5 plants were not statistically significant.

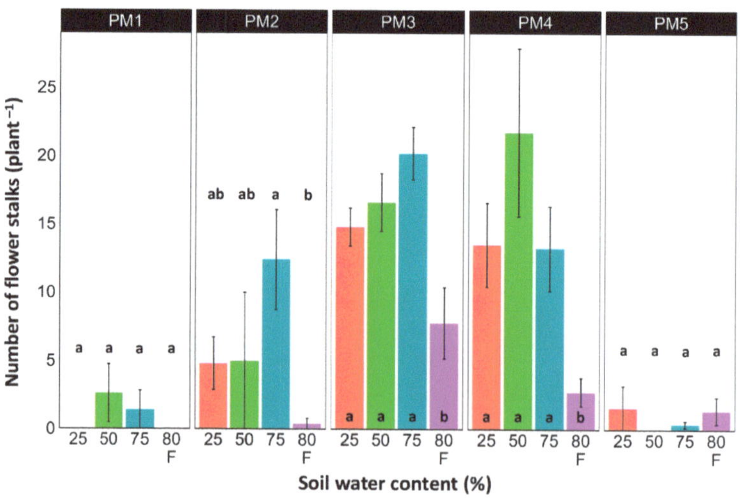

Figure 2. Effects of soil moisture and flooding (F) on the number of flower stalks of *Plantago maritima* plants from different accessions. Data are means ± SE from five replicates. Different letters indicate statistically significant differences according to the Tukey HSD test ($p < 0.05$) within the particular accession.

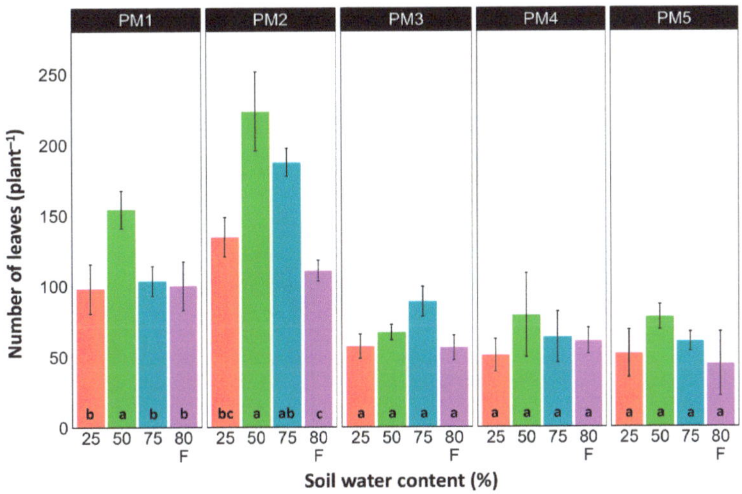

Figure 3. Effects of soil moisture and flooding (F) on the number of leaves of *Plantago maritima* plants from different accessions. Data are means ± SE from five replicates. Different letters indicate statistically significant differences according to the Tukey HSD test ($p < 0.05$) within the particular accession.

Other parameters of PM3 and PM4 related to generative reproduction were more sensitive to differences in soil moisture. Thus, biomass allocated to flower stalks was low at 25% soil moisture and increased for plants at 50 and 75% moisture, but the effect was statistically significant only for PM3 plants (Table 2). In the case of PM4, a significant decrease in flower stalk biomass was observed in flooded plants. Similarly, the biomass of flowers and fruits was low at 25% moisture, increased at 50 and 75% moisture, and decreased again for flooded plants (Table 2). However, for PM4 plants, this effect was significant only for flooded plants.

Table 2. Effects of soil moisture and flooding (F) on average height of flower stalks, dry mass of flower stalks, and dry mass of flowers and fruits of *Plantago maritima* plants of accessions PM3 and PM4.

Soil Moisture	Average Height of Flower Stalk (mm)		Dry Mass of Flower Stalks (g plant^{-1})		Dry Mass of Flowers and Fruits (g plant^{-1})	
	PM3	PM4	PM3	PM4	PM3	PM4
25%	144 ± 16 c	199 ± 18 bc	0.55 ± 0.05 cd	0.71 ± 0.16 bcd	1.15 ± 0.07 c	0.91 ± 0.19 c
50%	242 ± 7 ab	226 ± 17 ab	1.71 ± 0.13 a	1.42 ± 0.28 ab	2.43 ± 0.29 a	1.62 ± 0.28 abc
75%	189 ± 22 bc	221 ± 22 ab	1.33 ± 0.16 ab	1.33 ± 0.07 ab	2.13 ± 0.15 ab	1.34 ± 0.20 bc
80% F	273 ± 10 a	215 ± 5 ab	1.08 ± 0.21 abc	0.22 ± 0.03 d	1.14 ± 0.24 c	0.18 ± 0.01 d

Data are means ± SE from five replicates. Different letters indicate statistically significant differences according to the Tukey HSD test ($p < 0.05$) for the particular parameter.

In spite of different biomass investment strategies between plants from the PM1, PM2 (toward leaf development and growth), PM3, and PM4 (toward generative reproduction) accessions, the total shoot biomass at the optimum soil moisture (50 and 75%) of these plants was similar (Figure 4). However, PM5 plants had lower biomass values. For roots, the highest biomass for all accessions was in 50% soil moisture conditions, but these differences were statistically significant in all other moisture regimes only for PM2 plants (Figure 5). In addition, for all accessions, root biomass was significantly lower for plants cultivated at 25% soil moisture in comparison with that at 50% moisture. Moreover, a significant decrease in root biomass was also evident for flooded PM5 plants.

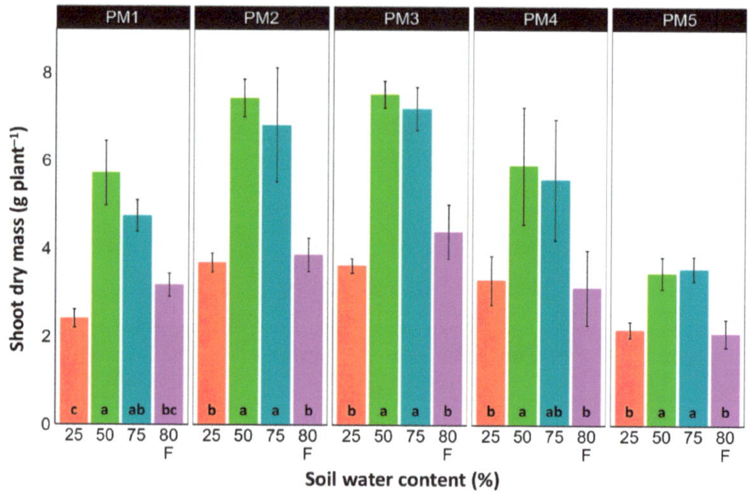

Figure 4. Effects of soil moisture and flooding (F) on the dry mass of shoots of *Plantago maritima* plants from different accessions. Data are means ± SE from five replicates. Different letters indicate statistically significant differences according to the Tukey HSD test ($p < 0.05$) within the particular accession.

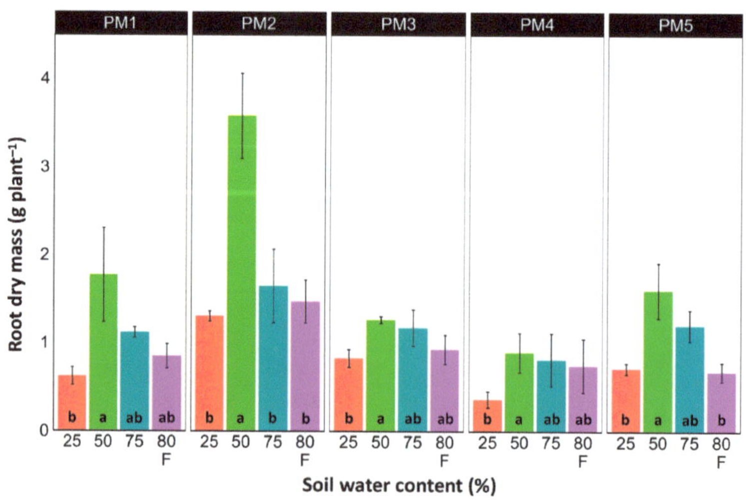

Figure 5. Effects of soil moisture and flooding (F) on the dry mass of roots of *Plantago maritima* plants from different accessions. Data are means ± SE from five replicates. Different letters indicate statistically significant differences according to the Tukey HSD test ($p < 0.05$) within the particular accession.

Water content was estimated in the middle (Figure 6) and old leaves (Figure 7) of *P. maritima* plants cultivated at different soil moisture regimes in order to monitor any preliminary signs of leaf senescence. While plants at 25, 50, and 75% soil moisture showed no significant changes in leaf water content, flooding resulted in a significant decrease in this parameter for plants of all accessions, except PM3 in the case of middle leaves, but for old leaves, this effect was significant only for PM1 and PM4 plants.

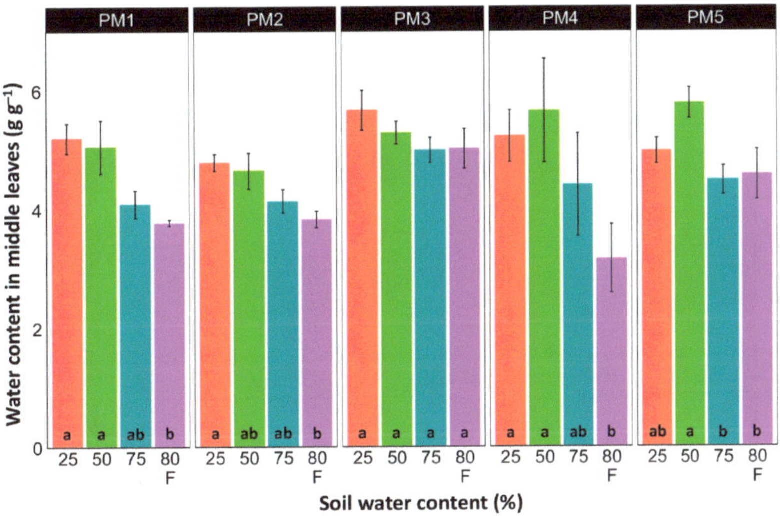

Figure 6. Effects of soil moisture and flooding (F) on water content in middle leaves of *Plantago maritima* plants from different accessions. Data are means ± SE from five replicates. Different letters indicate statistically significant differences according to the Tukey HSD test ($p < 0.05$) within the particular accession.

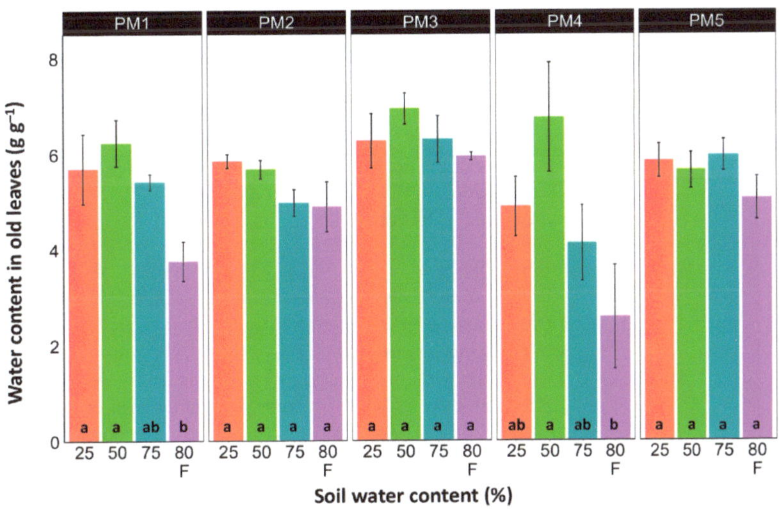

Figure 7. Effects of soil moisture and flooding (F) on water content in old leaves of *Plantago maritima* plants from different accessions. Data are means ± SE from five replicates. Different letters indicate statistically significant differences according to the Tukey HSD test ($p < 0.05$) within the particular accession.

To characterize the similarity between genotypes in relation to morphological parameters and their changes under the influence of changing soil moisture, multivariate analysis was performed (Figure 8). Accessions PM1 and PM2 were the most similar, and another pair with a somewhat lower degree of similarity was formed by PM3 and PM4. However, PM5 formed a separate cluster. It is also interesting to note that, among morphological traits, those related to leaf number and biomass, as well as root biomass, clustered together.

Another more structured cluster was formed by traits such as water content and the number of flower stalks.

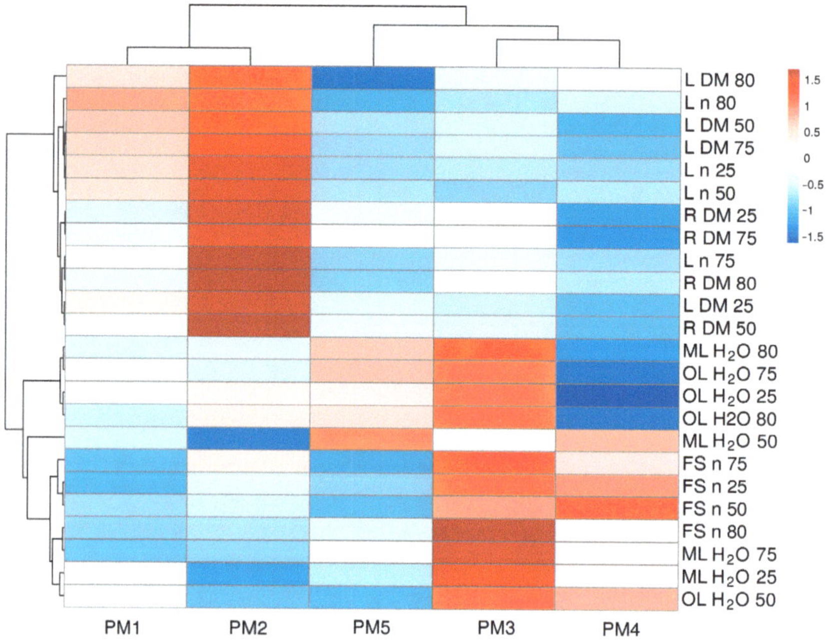

Figure 8. Generated heat map and cluster analysis of the effect of soil moisture and flooding on morphological parameters in *Plantago maritima* plants from different accessions. Hierarchical clusters were generated via the average linkage method with correlation distance; color scale shows relative intensity of normalized parameter values. DM, dry mass; FS, flower stalks; H_2O, water content; L, leaves; ML, middle leaves; n, number; OL, old leaves; R, roots. The numbers 25, 50, 75, and 80 indicate soil water content (%).

3.2. Accumulation of Ions and Osmotic Activity

For the characterization of total soluble ion content, electrical conductivity in leaf extracts was measured in the middle leaves of *P. maritima* plants cultivated at different soil moisture levels (Table 3). In general, the highest level of electrical conductivity was in plants cultivated at 25% soil moisture, which decreased with increasing soil moisture, with the lowest level in flooded plants. This effect was statistically significant already at 50% soil moisture for PM1 and PM3 plants but only for flooded plants of accessions PM2, PM3, and PM4. The concentration of Na^+ also showed a tendency to decrease with increasing soil water content, but statistically significant differences were evident for PM1, PM2, and PM4 plants only in flooded conditions and for PM3 plants also at 75% soil moisture (Table 3). The concentration of K^+ significantly decreased for PM2 in leaves of plants cultivated at 75% and flooded plants, as well as for PM3 at 50% and flooded PM5 plants, in comparison with the values at 25% (Table 3). The total osmotic activity was relatively stable, especially for PM3 and PM5 plants and other accessions at low or moderate soil moisture (PM1), or decreased significantly only for flooded plants (PM2, PM4). The activity of non-ionic osmolytes did not change with differences in soil water content in PM1 and PM5, but it increased at 50 and 75% (PM2) or only 50% (PM3), while a significant decrease was evident for flooded PM4 plants.

Table 3. Effects of soil moisture and flooding (F) on electrical conductivity, Na^+ concentration, K^+ concentration, and osmotic activity in middle leaves of *Plantago maritima* plants from different accessions.

Soil Moisture	PM1	PM2	PM3	PM4	PM5
			Electrical conductivity (mS m^{-1} kg^{-1} DM)		
25%	108.6 ± 3.4 a	107.9 ± 3.3 a	106.1 ± 3.3 ab	95.9 ± 6.9 abcd	88.2 ± 1.7 cde
50%	90.5 ± 5.2 bcde	98.1 ± 4.1 abcd	88.3 ± 0.9 cde	92.7 ± 5.1 abcd	100.6 ± 1.8 abc
75%	85.2 ± 4.0 cde	93.4 ± 2.2 abcd	85.2 ± 3.2 cde	91.3 ± 5.5 abcde	81.2 ± 3.8 def
80% F	73.9 ± 1.9 efg	66.2 ± 2.2 fg	73.7 ± 4.0 efg	62.3 ± 3.4 g	63.4 ± 2.6 g
			Na^+ concentration (g kg^{-1} DM)		
25%	5.22 ± 0.21 a	4.03 ± 0.20 abcd	5.23 ± 0.23 a	5.18 ± 0.19 a	3.10 ± 0.33 cde
50%	4.24 ± 0.43 abcd	4.05 ± 0.33 abcd	5.08 ± 0.38 a	4.94 ± 0.75 ab	4.49 ± 0.13 abc
75%	3.90 ± −0.26 abcd	3.16 ± 0.14 cde	3.44 ± 0.28 bcd	4.44 ± 0.36 abc	2.92 ± 0.23 de
80% F	3.07 ± 0.13 cde	2.37 ± 0.07 e	3.28 ± 0.07 e	2.90 ± 0.14 de	2.44 ± 0.09 e
			K^+ concentration (g kg^{-1} DM)		
25%	23.2 ± 1.4 ab	22.1 ± 1.4 ab	18.6 ± 1.4 abcd	22.1 ± 4/5 ab	21.7 ± 0.9 abc
50%	23.3 ± 1.2 a	16.5 ± 0.9 bcdef	9.8 ± 0.8 f	20.1 ± 0.8 abc	24.1 ± 0.8 a
75%	21.3 ± 1.1 abc	15.1 ± 0.6 cdef	12.9 ± 1.8 def	22.8 ± 2.5 ab	20.0 ± 1.1 abc
80% F	18.5 ± 0.6 abcd	11.6 ± 0.7 ef	12.0 ± 0.6 def	17.4 ± 0.6 def	12.5 ± 0.6 def
			Osmotic activity (osmol kg^{-1} DM)		
25%	3.25 ± 0.12 a	2.85 ± 0.04 abcdef	3.17 ± 0.09 abc	3.09 ± 0.05 abc	2.84 ± 0.08 bcdef
50%	3.09 ± 0.11 abc	3.04 ± 0.08 abcd	3.07 ± 0.07 abc	3.13 ± 0.11 abc	3.18 ± 0.03 aab
75%	2.82 ± 0.10 bcdef	2.78 ± 0.03 cdef	2.97 ± 0.09 abcdef	3.12 ± 0.05 abc	2.81 ± 0.10 bcdef
80% F	2.65 ± 0.08 defg	2.27 ± 0.05 g	2.99 ± 0.06 abcde	2.61 ± 0.06 efg	2.57 ± 0.07 fg
			Non-ionic osmotic activity (osmol kg^{-1} DM)		
25%	1.42 ± 0.06 defg	1.18 ± 0.06 g	1.57 ± 0.09 bcde	1.61 ± 0.06 abcde	1.26 ± 0.08 efg
50%	1.33 ± 0.05 defg	1.63 ± 0.04 abcd	1.90 ± 0.08 a	1.44 ± −0.07	1.38 ± 0.03 defg
75%	1.19 ± 0.05 g	1.52 ± 0.05 cdef	1.78 ± −0.08 abc	1.41 ± 0.08 defg	1.32 ± 0.05 defg
80% F	1.22 ± 0.07 fg	1.24 ± 0.06 fg	1.86 ± 0.04 ab	1.24 ± 0.03 fg	1.49 ± 0.09 cdefg

Data are means ± SE from five replicates. Different letters indicate statistically significant differences according to the Tukey HSD test ($p < 0.05$) for the particular parameter. DM, dry mass.

The variation in these parameters in roots between various accessions under the effect of differences in soil moisture was less pronounced (Table S1). For electrical conductivity, a statistically significant difference was observed between plants at 50 and 70% soil moisture only for PM3. For Na^+ concentration, PM2 plants at 75% and PM4 plants at 50% had values significantly higher than those of the rest of the moisture levels, and Na^+ at 25% was higher than at 50 and 75% for PM5. The only statistically significant differences for root K^+ concentration were between 50 and 75% for PM2 and 25, 50%, 75, and 80% for PM3. More differences were evident for total osmotic activity in roots, where there was a tendency for plants at low or moderate soil water levels (25 and 50%) to have lower values in comparison with those of flooded plants at the high water level. There were no statistically significant differences for root non-ionic osmotic activity except between plants at 50 and 75% soil moisture for PM3.

Because of the rather variable responses in the ion accumulation and osmotic activity of different *P. maritima* accessions cultivated at different soil moisture percentages, the cluster analysis revealed that each accession had a rather unique response distribution picture, but the highest similarity was between accessions PM4 and PM5 (Figure S7).

3.3. Photosynthesis-Related Parameters

Leaf chlorophyll concentration in *P. maritima* accessions PM3 and PM5 showed a tendency to decrease with increasing soil water content, but a statistically significant effect was evident only at 75% and with flooding (PM3) or flooding alone (PM5) (Figure 9). In addition, accessions PM1 and PM2 also had lower chlorophyll concentrations during flooding in comparison with those at 25% soil moisture. The chlorophyll *a* fluorescence parameter F_v/F_m was relatively insensitive to changes in soil water content, but significantly lower values of the parameter were seen in all accessions except PM4 for flooded plants (Figure 10). The character of changes in another fluorescence parameter, F_v/F_0, was similar to that of F_v/F_m, but to a greater extent (Figure 11). Significantly lower values of F_v/F_0 for flooded plants were evident for accessions PM2, PM3, and PM4. The Performance Index Total was the most sensitive among the measured photosynthesis-related parameters, and it tended to decrease with increases in soil water content for all *P. maritima* accessions, but the differences were not always statistically significant (Figure 12).

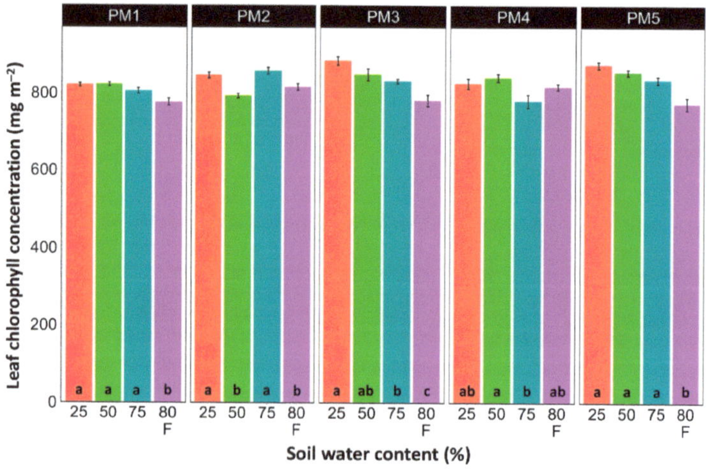

Figure 9. Effects of soil moisture and flooding (F) on chlorophyll concentration in middle leaves of *Plantago maritima* plants from different accessions. Data are means ± SE from five replicates, with two measurements each. Different letters indicate statistically significant differences according to the Tukey HSD test ($p < 0.05$) within the particular accession.

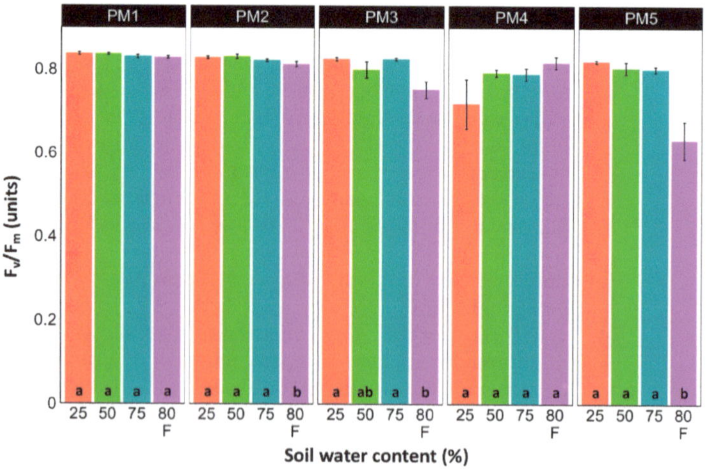

Figure 10. Effects of soil moisture and flooding (F) on chlorophyll *a* fluorescence parameter F_v/F_m in middle leaves of *Plantago maritima* plants from different accessions. Data are means ± SE from five replicates, with two measurements each. Different letters indicate statistically significant differences according to the Tukey HSD test ($p < 0.05$) within the particular accession.

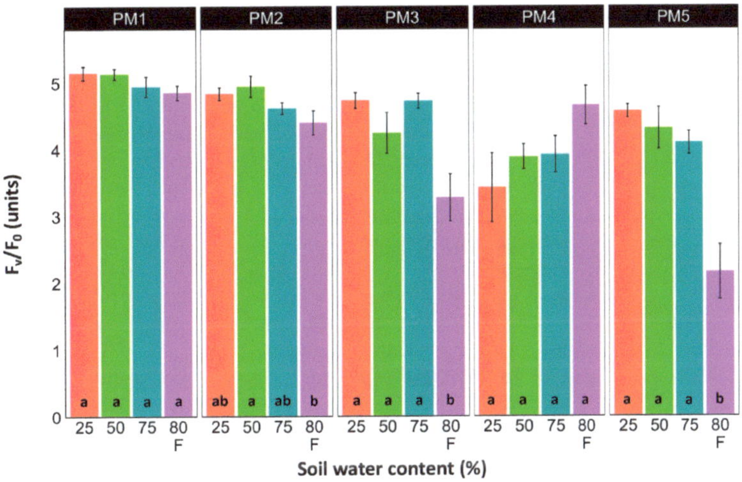

Figure 11. Effects of soil moisture and flooding (F) on chlorophyll *a* fluorescence parameter F_v/F_0 in middle leaves of *Plantago maritima* plants from different accessions. Data are means ± SE from five replicates, with two measurements each. Different letters indicate statistically significant differences according to the Tukey HSD test ($p < 0.05$) within the particular accession.

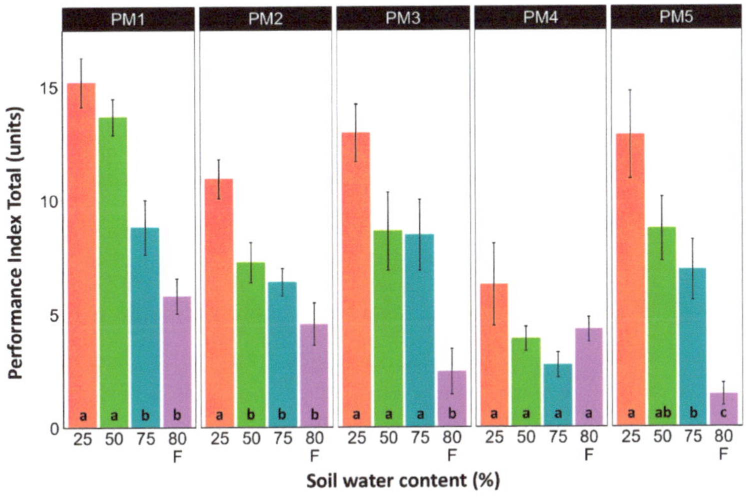

Figure 12. Effects of soil moisture and flooding (F) on chlorophyll *a* fluorescence parameter Performance Index Total in middle leaves of *Plantago maritima* plants from different accessions. Data are means ± SE from five replicates, with two measurements each. Different letters indicate statistically significant differences according to the Tukey HSD test ($p < 0.05$) within the particular accession.

Multivariate analysis revealed a relatively close similarity between PM3 and PM5 at the level of photosynthesis-related parameter responses (Figure S8). At the association level of the measured parameters, leaf chlorophyll concentration showed a close relationship in plants at all soil moisture levels except flooding. On the other hand, all parameters for flooded plants are also clustered together.

4. Discussion

Usually, the effects of either drought episodes or high soil moisture and flooding on plants are studied separately in the context of plant moisture regimes [45]. Very often, changes in soil moisture are not precisely monitored, and because of variations in other factors, the results obtained in such experiments are difficult to generalize. The present results show that differences in plant growth and the physiological state of plants due to stable differences in soil moisture from moderate insufficiency to waterlogging can be viewed as a continuous phenomenon. However, flooding-dependent plant submergence results in qualitatively distinct reactions, both at the signal perception and morphophysiological response levels [35,46]. It needs to be noted, however, that soil moisture in coastal habitats is extremely heterogeneous both in space and time [5]. Therefore, plants can be exposed to drought/excess moisture and flooding episodes of different durations and intensities, which can have different effects.

In general, *P. maritima* plants were extremely tolerant to soil waterlogging and highly tolerant to both soil flooding and low water content. The growth and development of plants were significantly reduced both in low moisture and flooded conditions, but no visual signs of disturbance were evident. Another characteristic species of salinity- and flooding-affected coastal habitats, *Trifolium fragiferum*, showed signs of metabolic disturbance even after several weeks of flooding, such as leaf chlorosis, accompanied by a severe reduction in plant growth [47]. *Saussurea esthonica* plants, typical for calcareous fens, had maximum growth and development in waterlogged conditions but showed symptoms of metabolic disorder in the form of purple regions between main leaf veins (after four weeks of waterlogging) and chlorotic lesions (after six weeks of waterlogging) [48].

There was no relationship between potential differences in daily water regime at the growth sites and morphological responses to experimentally induced water changes in the different *P. maritima* accessions. Thus, the initial hypothesis was not confirmed. All accessions had the highest shoot biomass at 50 and 75% soil moisture, showing exceptional tolerance to soil waterlogging (Figure 4). The relative decrease in biomass in low soil moisture (25%) or flooded conditions (80% F) was similar between both regimes and for all accessions. However, significantly lower water content in the older leaves of previously flooded plants of PM1 and PM4 indicated that, over a longer period of time, the effects of the treatment could manifest as the dieback of the older leaves (Figure 7). This was in part confirmed by the significant decrease in leaf chlorophyll concentrations in flooded plants (Figure 9). In addition, chlorophyll *a* fluorescence parameter F_v/F_m, an indicator of long-term unfavorable environmental influence [49], significantly decreased only in plants in flooded conditions for PM2, PM 3, and PM 5 (Figure 10), and for PM5, it was below 0.7, possibly showing the existence of negative metabolic changes. The reduced water content of the middle leaves in flooded plants (Figure 6) is more difficult to explain but could also be related to adverse metabolic changes. In comparative experiments with several plant species, it has been proven that chlorophyll fluorescence measurements can be used to distinguish flooding-sensitive plants from flooding-tolerant plants [39], indicating that the ability to sustain photochemical activity during flooding is an important adaptive mechanism. However, even native wetland species with a high tolerance to submergence show a decrease in photosynthesis-related parameters in parallel to growth inhibition with an increase in flooding depth [50].

Initial experiments looked for *P. maritima* ecotypes within the same geographic population among plants that grow in different areas of salt marshes [11] or along a distributional gradient in seashore meadows [21] and show distinguishing morphological characteristics. However, when cultivated in identical environmental conditions, *P. maritima* plants from different microenvironments show uniform growth and morphology, confirming the high phenotypic plasticity of the species. In the present study, however, seed material from *P. maritima* plants growing in geographically isolated sites was used. Only PM1 and PM2 plants were from the same island, but they were found to persist in different conditions (dry dune grasslands vs. wet saline grasslands).

One of the most striking morphological differences between the accessions of *P. maritima* was at the level of flower induction (Figure 2). The photoperiodic requirements for flowering in *P. maritima* are not known. However, a 16 h photoperiod has been used to induce flowering in *Plantago lanceolata* [51]. Similarly, an increase in the light period from 12 to 16 h increases the number of flowering *Plantago major* individuals [52]. The present experiment was performed in the winter season, meaning that the length of natural daylight had no effect on the plants, and a 16 h photoperiod with 8 h of darkness was not an appropriate induction factor for flowering for several accessions of *P. maritima*, which probably require even shorter nights. Low light intensity due to dense vegetation prevents flowering in *P. maritima* [21]. However, in natural conditions, flowering is a factor that increases individual mortality, especially during low salinity and decreased flooding frequency and duration [21].

P. maritima can be characterized as a salt-accumulating species that controls the total electrolyte content in its leaves in native coastal habitat conditions via changes in Na^+ concentration against the background of stable and low K^+ concentrations [53]. Soil waterlogging did not affect Na^+ and K^+ concentrations in the shoots of *P. maritima* plants [22]. In the present study, soil moisture affected the accumulation of soluble ions in the leaves of *P. maritima*, but the effect was clearly genotype-specific (Table 3). In general, there was a decrease in the electrical conductivity of tissue extracts with increasing soil moisture, which was associated in part with a decrease in Na^+ and K^+ concentrations. No positive relationship between growth at low soil moisture and K^+ accumulation was found.

Osmotic protection in *P. maritima* in saline conditions has been attributed to the accumulation of both inorganic ions (Na^+, Cl^-) [20] and sorbitol [18]. Total osmotic activity in the leaves of *P. maritima* was stable between 25 and 50% soil moisture but decreased significantly for flooded PM1, PM2, and PM4 plants (Table 3). However, the activity of non-ionic osmolytes showed differential responses to soil moisture, as it was higher for PM2 plants at 50 and 75% and PM3 plants at 50% while lower for flooded PM4 plants. Thus, in general, the osmotic adjustment system was not involved in adaptation to various moisture regimes, at least on the dry soil side.

Among the photosynthesis-related indicators of plant physiological status, the Performance Index Total was the most sensitive parameter to changes in soil moisture (Figure 12). However, as the optimum growth of *P. maritima* plants was at 50–75% soil moisture and the highest values of the Performance Index Total were at 25% soil moisture, this indicates that this parameter cannot be used as the sole indicator of the physiological status of plants. A similar discrepancy between fluorescence indicators and biomass production under the influence of various factors, including soil moisture, has also been shown in other studies [48]. From a physiological point of view, it has been argued that photosynthesis and the expansive growth of plants under different water availability situations are independently controlled [30]. As a result, these two processes can be uncoupled even for prolonged periods of time and show different responses to changes in soil water content.

The existence of three morphological ecotypes can be proposed between the studied accessions of *P. maritima*. First, two accessions, PM1 and PM2, from the island of Kihnu showed close association at the level of responses to variation in soil moisture and flooding, irrespective of their native location in habitats with contrasting soil moisture regimes. Thus, the initial hypothesis—that the plants from relatively dry habitats will have higher tolerance against low soil water levels, but the plants from relatively wet habitats will have higher tolerance against waterlogged or flooded soil—was not approved. Second, only plants from accessions PM3 and PM4 showed the full potential of generative reproduction in the conditions of the experiment. Third, PM5 plants stood out with a low number of leaves with low total biomass (similar to those of PM3 and PM4), but the plants did not flower in the conditions of the experiment. Consequently, there is reason to believe that geographically isolated but relatively close subpopulations contain genetically distinct material, possibly resulting from local genetic adaptation or, more likely, from different seed sources during post-glacial colonization. There is no indication of the water-assisted seed dispersal of *P.*

maritima or any other *Plantago* species [15], but the close association of *P. maritima* with coastal habitats in the Baltic Sea region, as well as the complete recovery of seed germination after prior exposure to saline water [54,55], means that we cannot discount this possibility. Molecular genetic studies of *P. maritima* accessions from a wider geographical region of the Baltic Sea are necessary to find out the degree of genetic relatedness between isolated subpopulations in order to determine the biological diversity of the species.

Supplementary Materials: The following supporting information can be downloaded at https://www.mdpi.com/article/10.3390/plants13050633/s1: Figure S1: Sites with *Plantago maritima* accessions in Estonia. Figure S2: Typical *Plantago maritima* PM1 plants six weeks after the start of the treatment (two weeks after the flooding treatment was terminated). Figure S3: Typical *Plantago maritima* PM2 plants six weeks after the start of the treatment (two weeks after the flooding treatment was terminated). Figure S4: Typical *Plantago maritima* PM3 plants six weeks after the start of the treatment (two weeks after the flooding treatment was terminated). Figure S5: Typical *Plantago maritima* PM4 plants six weeks after the start of the treatment (two weeks after the flooding treatment was terminated). Figure S6: Typical *Plantago maritima* PM5 plants six weeks after the start of the treatment (two weeks after the flooding treatment was terminated). Figure S7: Generated heat map and cluster analysis of the effect of soil moisture and flooding on ion accumulation and osmotic activity in *Plantago maritima* plants from different accessions. Figure S8: Generated heat map and cluster analysis of the effect of soil moisture and flooding on photosynthesis-related parameters in middle leaves of *Plantago maritima* plants from different accessions. Table S1: Effects of soil moisture and flooding (F) on electrical conductivity, Na^+ concentration, K^+ concentration, and osmotic activity in roots of *Plantago maritima* plants from different accessions.

Author Contributions: G.I. proposed the research. K.A.O., A.J., U.A.-O. and G.I. performed the experiments and analyzed the data. G.I. and K.A.O. drafted the manuscript. A.J. revised the manuscript. All authors have read and agreed to the published version of the manuscript.

Funding: This study was supported by the University of Latvia's "Functional diversity of ecosystems and their contribution to ecosystem services II" project. The funding source had no involvement in any phase of the study.

Data Availability Statement: All data reported here are available from the authors upon request.

Conflicts of Interest: The authors declare no conflicts of interest.

References

1. Anderson, J.T.; Willis, J.H.; Mitchell-Olds, T. Evolutionary genetics of plant adaptation. *Trends Genet.* **2011**, *27*, 258–266. [CrossRef]
2. Henry, R.J.; Nevo, E. Exploring natural selection to guide breeding for agriculture. *Plant Biotechnol. J.* **2014**, *12*, 655–662. [CrossRef]
3. Razzaq, A.; Wani, S.H.; Saleem, F.; Wu, M.; Zhou, M.; Shabala, S. Rewilding crops for climate resilience: Economic analysis and de novo domestication strategies. *J. Exp. Bot.* **2021**, *72*, 6123–6139. [CrossRef]
4. Kouhen, M.; García-Caparrós, P.; Twyman, R.M.; Abdelly, C.; Mahmoudi, H.; Schillnerg, S. Improving environmental stress resilience in crops by genome editing: Insights from extremophile plants. *Crit. Rev. Biotechnol.* **2023**, *43*, 559–574. [CrossRef]
5. Ievinsh, G. Biological basis of biological diversity: Physiological adaptations of plants to heterogeneous habitats along a sea coast. *Acta Univ. Latv.* **2006**, *710*, 53–79.
6. Fenu, G.; Carboni, M.; Acosta, A.T.R.; Baccheta, G. Environmental factors influencing coastal vegetation pattern: New insights from the Mediterranean Basin. *Folia Geobot.* **2013**, *48*, 493–508. [CrossRef]
7. Burden, A.; Smeaton, C.; Angus, S.; Garbutt, A.; Jones, L.; Lewis, H.D.; Rees, S.M. Impacts of climate change on coastal habitats relevant to coastal and marine environment around the UK. *MCCIP Sci. Rev.* **2020**, *2020*, 228–235.
8. Druva-Lūsīte, I.; Karlsons, A.; Andersone-Ozola, U.; Ieviņa, B.; Nečajeva, J.; Samsone, I.; Ievinsh, G. Physiological performance of a coastal marsh plant *Hydrocotyle vulgaris* in natural conditions in relation to mineral nutrition and mycorrhizal symbiosis. *Proc. Latv. Acad. Sci. B* **2020**, *74*, 252–262. [CrossRef]
9. Soekarjo, R. General morphology. In *Plantago: A Multidisciplinary Study*; Kuiper, D., Bos, M., Eds.; Springer: Berlin/Heidelberg, Germany, 1992; pp. 6–11.
10. Herben, T.; Klimešová, J. Evolution of clonal growth forms in angiosperms. *New Phytol.* **2020**, *225*, 999–1010. [CrossRef] [PubMed]
11. Blom, C.W.P.M. Plasticity of life characteristics in two different populations of *Plantago maritima* L. *Acta Oecol. Oecol. Plant.* **1983**, *4*, 377–394.
12. Ievinsh, G. Halophytic clonal plant species: Important functional aspects for existence in heterogeneous saline habitats. *Plants* **2023**, *12*, 1728. [CrossRef]

13. Tichý, L.; Axmanová, I.; Dengler, J.; Guarino, R.; Jansen, F.; Midolo, G.; Nobis, M.P.; Van Meerbeek, K.; Aćić, S.; Attorre, F.; et al. Ellenberg-type indicator values for European vascular plant species. *J. Veget. Sci.* **2023**, *34*, e13168. [CrossRef]
14. Blom, C.W.P.M. Phenotypic plasticity in *Plantago maritima*. In *Plantago: A Multidisciplinary Study*; Kuiper, D., Bos, M., Eds.; Springer: Berlin/Heidelberg, Germany, 1992; pp. 240–249.
15. Tyler, T.; Herbertsson, L.; Olofsson, J.; Olsson, P.A. Ecological indicator and traits values for Swedish vascular plants. *Ecol. Indic.* **2021**, *120*, 106923. [CrossRef]
16. Hill, M.O.; Ellenberg, H.H. *Ellenberg's Indicator Values for British Plants*; ECOFACT Research Report. Technical Annex; Institute of Terrestrial Ecology: Wallingford, UK, 1999; Volume 2.
17. Erdei, L.; Kuiper, P.J. The effect of salinity on growth, cation content, Na^+-uptake and translocation in salt-sensitive and salt-tolerant *Plantago* species. *Physiol. Plant.* **1979**, *47*, 95–99. [CrossRef]
18. Ahmad, I.; Larher, F.; Stewart, G.R. Sorbitol, a compatible osmotic solute in *Plantago maritima*. *New Phytol.* **1979**, *82*, 671–678. [CrossRef]
19. Sekmen, A.H.; Türkan, İ.; Takio, S. Differential responses of antioxidative enzymes and lipid peroxidation to salt stress in salt-tolerant *Plantago maritima* and salt-sensitive *Plantago media*. *Physiol. Plant.* **2007**, *131*, 399–411. [CrossRef]
20. Sleimi, N.; Guerfali, S.; Bankaji, I. Biochemical indicators of salt stress in *Plantago maritima*: Implications for environmental stress assessment. *Ecol. Indic.* **2015**, *48*, 570–577. [CrossRef]
21. Jerling, L. Population dynamics of *Plantago maritima* along a distributional gradient on a Baltic seashore meadow. *Vegetatio* **1985**, *61*, 155–161. [CrossRef]
22. Cooper, A. The effects of salinity and waterlogging on the growth and cation uptake of salt marsh plants. *New Phytol.* **1982**, *90*, 263–275. [CrossRef]
23. Groenendijk, A.M. Ecological consequences of tidal management for the salt-marsh vegetation. *Vegetatio* **1985**, *62*, 415–424. [CrossRef]
24. Groenendijk, A.M. Tidal management: Consequences for the salt marsh vegetation. *Water Sci. Technol.* **1984**, *16*, 79–86. [CrossRef]
25. Shabala, S.; Shabala, L. Ion transport and osmotic adjustment in plants and bacteria. *BioMol. Concepts* **2011**, *2*, 407–419. [CrossRef] [PubMed]
26. Glenn, E.; Pfister, R.; Brown, J.J.; Thompson, T.L.; O'Leary, J. Na and K accumulation and salt tolerance of *Atriplex canescens* (Chenopodiaceae) genotypes. *Am. J. Bot.* **1996**, *83*, 997–1005. [CrossRef]
27. Leigh, R.A.; Wyn Jones, R.G. A hypothesis relating critical potassium concentrations for growth to the distribution and function of this ion in the plant cell. *New Phytol.* **1984**, *97*, 1–13. [CrossRef]
28. Ashraf, M.; Akram, N.A.; Al-Qurainy, F.; Foolad, M.R. Drought tolerance: Roles of organic osmolytes, growth regulators, and mineral nutrients. *Adv. Agron.* **2011**, *111*, 249–296.
29. Nonami, H. Plant water relations and control of cell elongation at low water potentials. *J. Plant Res.* **1998**, *111*, 373–382. [CrossRef]
30. Tardieu, F.; Parent, B.; Caldeira, C.F.; Welcker, C. Genetic and physiological controls of growth under water deficit. *Plant Physiol.* **2014**, *164*, 1628–1635. [CrossRef]
31. Ilyas, M.; Nisar, M.; Khan, N.; Hazrat, A.; Khan, A.H.; Hayat, K.; Fahad, S.; Khan, A.; Ullah, A. Drought tolerance strategies in plants: A mechanistic approach. *J. Plant Growth Reg.* **2021**, *40*, 926–944. [CrossRef]
32. Pan, J.; Sharif, R.; Xu, X.; Chen, H. Mechanisms of waterlogging tolerance in plants: Research progress and prospects. *Front. Plant Sci.* **2021**, *11*, 627331. [CrossRef] [PubMed]
33. Yuan, L.-B.; Chen, M.-X.; Wang, L.-N.; Sasidharan, R.; Voesenek, L.A.C.J. Multi-stress resilience in plants recovering from submergence. *Plant Biotechnol. J.* **2023**, *21*, 466–481. [CrossRef] [PubMed]
34. de Carvalho, M.H.C. Drought stress and reactive oxygen species. *Plant Signal. Behav.* **2008**, *3*, 156–165. [CrossRef]
35. Fukao, T.; Barrera-Figueroa, B.E.; Juntawong, P.; Peña-Castro, J.M. Submergence and waterlogging stress in plants: A review highlighting research opportunities and understudied aspects. *Front. Plant Sci.* **2019**, *10*, 340. [CrossRef]
36. Morales, F.; Ancin, M.; Fakhet, D.; González-Torralba, J.; Gmez, A.L.; Seminario, A.; Soba, D.; Ben Mariem, S.; Garriga, M.; Aranjuelo, I. Photosynthetic metabolism under stressful growth conditions as a bases for crop breeding and yield improvement. *Plants* **2020**, *9*, 88. [CrossRef]
37. Lichtenthaler, H.K. Vegetation stress: An introduction to the stress concept in plants. *J. Plant Physiol.* **1996**, *148*, 4–14. [CrossRef]
38. Woo, N.S.; Badger, M.R.; Pogson, B.J. A rapid, non-invasive procedure for quantitative assessment of drought survival using chlorophyll fluorescence. *Plant Methods* **2008**, *4*, 27. [CrossRef]
39. Caudle, K.L.; Maricle, B.R. Effects of flooding on photosynthesis, chlorophyll fluorescence, and oxygen stress in plants of varying flooding tolerance. *Transact. Kansas Acad. Sci.* **2012**, *115*, 5–18.
40. Ose, A.; Andersone-Ozola, U.; Ievinsh, G. Substrate-dependent effect of vermicompost on yield and physiological indices of container-grown *Dracocephalum moldavica* plants. *Agriculture* **2021**, *11*, 1231. [CrossRef]
41. Banks, J.M. Continuous excitation chlorophyll fluorescence parameters: A review for practitioners. *Tree Physiol.* **2017**, *37*, 1128–1136. [CrossRef] [PubMed]
42. Tsimilli-Michael, M. Revisiting JIP-test: An educative review on concepts, assumptions, approximations, definitions and terminology. *Photosynthetica* **2020**, *58*, 275–292. [CrossRef]

43. Purmale, L.; Jēkabsone, A.; Andersone-Ozola, U.; Ievinsh, G. Salinity tolerance, ion accumulation potential and osmotic adjustment in vitro and in planta of different *Armeria maritima* accessions from a dry coastal meadow. *Plants* **2022**, *11*, 2570. [CrossRef]
44. Metsalu, T.; Vilo, J. ClustVis: A web tool for visualizing clustering of multivariate data using Principal Component Analysis and heatmap. *Nucleic Acids Res.* **2015**, *43*, W566–W570. [CrossRef]
45. Moreno, S.; Bedada, G.; Rahimi, Y.; Ingvarsson, P.K.; Westerbergh, A.; Lundquist, P.-O. Response to waterlogging stress in wild and domesticated accessions of timothy (*Phleum pratense*) and its relatives *P. alpinum and P. nodosum*. *Plants* **2023**, *12*, 4033. [CrossRef] [PubMed]
46. Voesenek, L.A.C.J.; Colmer, T.D.; Pierik, R.; Millenaar, F.F.; Peeters, A.J.M. How plants cope with complete submergence. *New Phytol.* **2006**, *170*, 213–226. [CrossRef] [PubMed]
47. Andersone-Ozola, U.; Jēkabsone, A.; Purmale, L.; Romanovs, M.; Ievinsh, G. Abiotic stress tolerance of coastal accessions of a promising forage legume species, *Trifolium fragiferum*. *Plants* **2021**, *10*, 1552. [CrossRef] [PubMed]
48. Gailite, A.; Andersone-Ozola, U.; Samsone, I.; Karlsons, A.; Ievinsh, G. Ecophysiology of endangered plant species *Saussurea esthonica*: Effect of mineral nutrient availability and soil moisture. *Plants* **2023**, *12*, 888. [CrossRef] [PubMed]
49. Malnoë, A. Photoinhibition or photoprotection of photosynthesis? Update on the (newly termed) sustained quenching component qH. *Environ. Exp. Bot.* **2018**, *154*, 123–133. [CrossRef]
50. Chen, H.; Zamorano, M.F.; Ivanoff, D. Effect of flooding depth on growth, biomass, photosynthesis, and chlorophyll fluorescence of *Typha domingensis*. *Wetlands* **2010**, *30*, 957–965. [CrossRef]
51. Lacey, E.P.; Herr, D. Parental effects in *Plantago lanceolata* L. III. Measuring parental temperature effects in the field. *Evolution* **2000**, *54*, 1207–1217.
52. Lotz, L.A.P. Variation in Life-History Characteristics between and within Populations of *Plantago major* L. Ph.D. Thesis, Katholieke Universiteit Nijmegen, Nijmegen, The Netherlands, 1989.
53. Ievinsh, G.; Ieviņa, S.; Andersone-Ozola, U.; Samsone, I. Leaf sodium, potassium and electrolyte accumulation capacity of plant species from salt-affected coastal habitats of the Baltic Sea: Towards a definition of Na hyperaccumulation. *Flora* **2021**, *274*, 151748. [CrossRef]
54. Binet, P. Action de la température et de la salinité sur la germination des graines de *Plantago maritima* (L.). *Bull. Soc. Bot. France* **1964**, *111*, 407–411. [CrossRef]
55. Hanslin, H.M.; Eggen, T. Salinity tolerance during germination of seashore halophytes and salt-tolerant grass cultivars. *Seed Sci. Res.* **2005**, *15*, 43–50. [CrossRef]

plants

Article

Synergistic Influence of *Arbuscular mycorrhizal* Fungi Inoculation with Nanoparticle Foliar Application Enhances Chili (*Capsicum annuum* L.) Antioxidant Enzymes, Anatomical Characteristics, and Productivity under Cold-Stress Conditions

Eman G. Sayed [1,*], S. F. Desoukey [2], Abeer F. Desouky [3], Mervat F. Farag [4], Ragab I. EL-kholy [5] and Samah N. Azoz [2]

1 Vegetable Crops Department, Faculty of Agriculture, Cairo University, Giza 12613, Egypt
2 Agricultural Botany Department, Faculty of Agriculture, Cairo University, Giza 12613, Egypt; sally.desouky@agr.cu.edu.eg (S.F.D.); samah.ali@agr.cu.edu.eg (S.N.A.)
3 Plant Biotechnology Department, Biotechnology Institute, National Research Center, Dokki, Cairo 12622, Egypt; abeerbiotech@yahoo.com
4 Horticulture Department, Faculty of Agriculture, Beni Suef University, Beni-Suef 62511, Egypt; mervat.farrag@agr.bsu.edu.eg
5 Agricultural Botany Department, Genetics Division Faculty of Agriculture, Al-Azhar University, Cairo 11884, Egypt; ragab_elkholy@azhar.edu.eg
* Correspondence: eman.ali@agr.cu.edu.eg; Tel.: +20-100-636-2604

Abstract: In this study, we aimed to evaluate the effects of *Arbuscular mycorrhiza* fungus (AMF) inoculation, foliar application of zinc oxide and selenium nanoparticles (ZnO-NPs and Se-NPs), and their combined interactions on the growth and productivity of chili pepper under cold-stress conditions. Two field experiments were successfully conducted during the winter seasons of 2021 and 2022 in an experimental field at the Faculty of Agriculture, Cairo University, Giza, Egypt. The results showed that, under cold stress, the combination of AMF inoculation and ZnO-NPs + Se-NPs as a foliar spray increased the average fruit weight by 92.4% and 98.7%, and the number of fruits by 34.6% and 54.8 compared to control treatment in the 2021 and 2022 seasons, respectively. Additionally, the combination of AMF and a mixture of nanoparticles (ZnO-NPs + Se-NPs) significantly increased the total marketable yield by 95.8% and 94.7% compared to the control, which recorded values of 2.4 and 1.9 kg m^{-2} in the 2021 and 2022 seasons, respectively. Furthermore, the combination of AMF and a mixture of nanoparticles (ZnO-NPs + Se-NPs) showed the highest total content of ascorbic acid and capsaicin in chili fruits compared to the other treatments. The combination of AMF and a mixture of nanoparticles (ZnO-NPs + Se-NPs) stimulated the accumulation of peroxidase (POD) and nitrogen glutamate dehydrogenase (GDH) while decreasing hydrogen peroxide (H_2O_2) and lipid peroxidation (MDA) contents. SDS analysis revealed that the application of ZnO-NPs, Se-NPs, AMF + ZnO-NPs, and AMF + ZnO-NPs + Se-NPs induced the emergence of new protein bands and reconstitution of those damaged by cold stress. Regarding histological structure, the combination of AMF inoculation and ZnO-NPs + Se-NPs as a foliar spray showed an enhancement in the thickness of grana thylakoids and increased the number of chloroplasts. Intriguingly, the findings showed that AMF and a mixture of nanoparticles (ZnO-NPs + Se-NPs) could offer guidance for increasing plant development and productivity under cold-stress conditions.

Keywords: *Capsicum annuum* L.; low temperature; *Glomus mosseae*; nano-selenium; nano-zinc oxide

Citation: Sayed, E.G.; Desoukey, S.F.; Desouky, A.F.; Farag, M.F.; EL-kholy, R.I.; Azoz, S.N. Synergistic Influence of *Arbuscular mycorrhizal* Fungi Inoculation with Nanoparticle Foliar Application Enhances Chili (*Capsicum annuum* L.) Antioxidant Enzymes, Anatomical Characteristics, and Productivity under Cold-Stress Conditions. *Plants* **2024**, *13*, 517. https://doi.org/10.3390/plants13040517

Academic Editors: Geovani Soares de Lima, Lauriane Almeida dos Anjos Soares and Francisco Vanies Da Silva Sá

Received: 1 January 2024
Revised: 5 February 2024
Accepted: 11 February 2024
Published: 14 February 2024

1. Introduction

Chili (*C. annuum* L.) is commercially grown and consumed for its great nutritional value and spicy flavor [1]. Low temperatures (below 15 °C) have a negative impact on plant growth, flowering, and productivity of chili plants [2]. Additionally, the primary repercussions of cold stress include damage to the chlorophyll content and photosynthesis.

Consequently, ROS production damages cell membranes. Plants have resistance systems called osmotic adjustment and antioxidant systems, which protect them from the adverse impacts of cold stress [3]. This enhanced osmolyte and antioxidant formation helps plants resist very low temperatures [4,5]. Furthermore, correlation studies of physiological and biochemical characteristics are very important for managing cold-stress resistance [6]. Hence, plants can adapt to changes in climate by changing their morphological and physiological acclimation [7].

Arbuscular mycorrhizal fungus (AMF) inoculation enhances soil structure and regulates nutrient uptake to promote the growth of the host plant [8]. Hence, AMF has been shown to improve plant growth under cold stress [9]. Furthermore, AMF colonization alters the morphological and physiological responses of host plants to stress, as well as increases plant growth and yield [10]. Moreover, AMF inoculation improves root architecture by increasing water and nutrient availability [11]. Additionally, AMF inoculation enhances the rate of photosynthesis by increasing leaf area and phosphorus content [12] and enhancing soil conditions for host plants [13]. The growth of *G. uralensis* plants was improved by inoculation with *Glomus mosseae* and *G. versiforme*, either separately or in combination. Additionally, mycorrhizal growth increases the amount of glycyrrhizic acid in roots [14]. Also, *Glomus. mosseae* and *Glomus. intraradices* can promote glycyrrhizic acid formation in plants of different species. Therefore, AMF inoculation has great potential for enhancing crop yields under field conditions [15].

Nanoparticles (NPs) reduce the use of chemical fertilizers and improve the growth of various plant species [16]. Selenium nanoparticles (Se-NPs) enhance root architecture and plant growth [17]. Additionally, Se-NPs alleviate the effects of stress conditions and improve plant growth and hydration [18]. Moreover, the application of selenium (Se) enhances vegetative growth and plant productivity by enhancing the antioxidant defense system under abiotic stress [19]. Foliar spray treatment, as opposed to soil supplementation, is an efficient alternative method for boosting Se levels in crops. Foliar nourishment with Se reduces the effects of soil chemistry, allowing for efficient absorption even at trace levels [20]. Thus, strawberry seedlings sprayed with Se have a higher photosynthetic rate under cold stress [20]. Zinc oxide nanoparticles (ZnO-NPs) are used worldwide because of their positive environmental impacts [21]. ZnO-NPs have exceptional photooxidation and photocatalytic abilities against chemical and biological species [22]. Furthermore, ZnO-NPs are crucial for both physiological and anatomical processes [23]. Hence, ZnO-NPs can improve the productivity of several vegetable crops [24]. Moreover, SDS proteins are useful in offering reactions for genetic conservation, where there is a need to assess genetic variability in plant species [25].

Based on the results of this study, we aimed to determine whether the combined effects of AMF inoculation and foliar application of Se-NPs and ZnO-NPs are more effective than their individual effects in improving chili plant growth and productivity under cold-stress conditions.

2. Results

2.1. Plant Growth and Flowering Traits of Chili Plants

As shown in Table 1, all treatments exhibited significantly ($p \leq 0.05$) increased plant growth characteristics compared to the control in both seasons. The greatest plant height, number of leaves, plant fresh weight, and plant dry weight were recorded for AMF inoculation combined with a mixture of nanoparticles (ZnO-NPs + Se-NPs) as a foliar spray in both seasons. Moreover, the leaf area increased by 54.8% and 50.5% following AMF inoculation combined with a nanoparticle mixture (ZnO-NPs + Se-NPs) compared to the control plants in 2021 and 2022, respectively. The flowering time and number of flowers were significantly influenced by all tested treatments in both seasons (Figure 1). Chili plants inoculated with AMF combined with the nanoparticle mixture (ZnO-NPs + Se-NPs) had the highest number of flowers compared to the control plants in both seasons. Furthermore, chili plants inoculated with AMF combined with a nanoparticle mixture (ZnO-NPs + -NPs) flowered earlier than control plants by between 23 and 22 days in the 2021 and 2022 seasons, respectively.

Table 1. Effect of inoculation with *Arbuscular mycorrhiza* fungus (AMF) and foliar applications of nanoparticles on the plant growth characteristics of chili plants during the 2021 and 2022 seasons.

Treatment	Plant Height (cm)	Number of Leaves	Plant Fresh Weight (g)	Plant Dry Weight (g)	Leaf Area (cm^2)
2021 Season					
Control	50.3 ^{h}	40.3 ^{h}	101.3 ^{f}	18.6 ^{f}	51.7 ^{h}
Se-NPs	59.0 ^{g}	52.0 ^{g}	156.0 ^{e}	29.7 ^{e}	55.3 ^{g}
ZnO-NPs	67.6 ^{f}	63.0 ^{f}	207.3 ^{d}	38.3 ^{d}	58.7 ^{f}
ZnO-NPs + Se-NPs	77.0 ^{e}	75.0 ^{e}	255.7 ^{c}	48.0 ^{c}	61.0 ^{e}
AMF	84.3 ^{d}	88.7 ^{d}	303.3 ^{b}	52.3 ^{bc}	65.0 ^{d}
AMF + Se-NPs	92.3 ^{c}	102.3 ^{c}	318.7 ^{b}	58.7 ^{b}	67.7 ^{c}
AMF + ZnO-NPs	102.3 ^{b}	117.7 ^{b}	312.0 ^{b}	58.0 ^{b}	71.0 ^{b}
AMF + ZnO-NPs + Se-NPs	115.7 ^{a}	135.7 ^{a}	388.3 ^{a}	74.7 ^{a}	80.0 ^{a}
LSD value at 0.05:	3.1	2.4	22.59	7.4	1.6
2022 Season					
Control	49.03 ^{g}	35.0 ^{g}	93.3 ^{h}	14.7 ^{g}	54.7 ^{h}
Se-NPs	58.1 ^{f}	55.3 ^{f}	142.7 ^{g}	25.0 ^{f}	58.3 ^{g}
ZnO-NPs	66.4 ^{e}	61.0 ^{f}	166.4 ^{f}	31.9 ^{e}	61.7 ^{f}
ZnO-NPs + Se-NPs	76.0 ^{d}	73.0 ^{e}	203.1 ^{e}	40.7 ^{d}	63.4 ^{e}
AMF	88.1 ^{c}	86.6 ^{d}	244.1 ^{d}	50.2 ^{c}	67.3 ^{d}
AMF + Se-NPs	90.8 ^{c}	99.7 ^{c}	282.5 ^{c}	61.1 ^{b}	69.9 ^{c}
AMF + ZnO-NPs	100.2 ^{b}	110.3 ^{b}	299.5 ^{b}	62.0 ^{b}	73.3 ^{b}
AMF + ZnO-NPs + Se-NPs	113.6 ^{a}	129.3 ^{a}	336.2 ^{a}	72.7 ^{a}	82.3 ^{a}
LSD value at 0.05:	6.3	8.09	7.1	3.3	1.7

Values followed by different letters are significant according to the LSD test ($p \leq 0.05\%$).

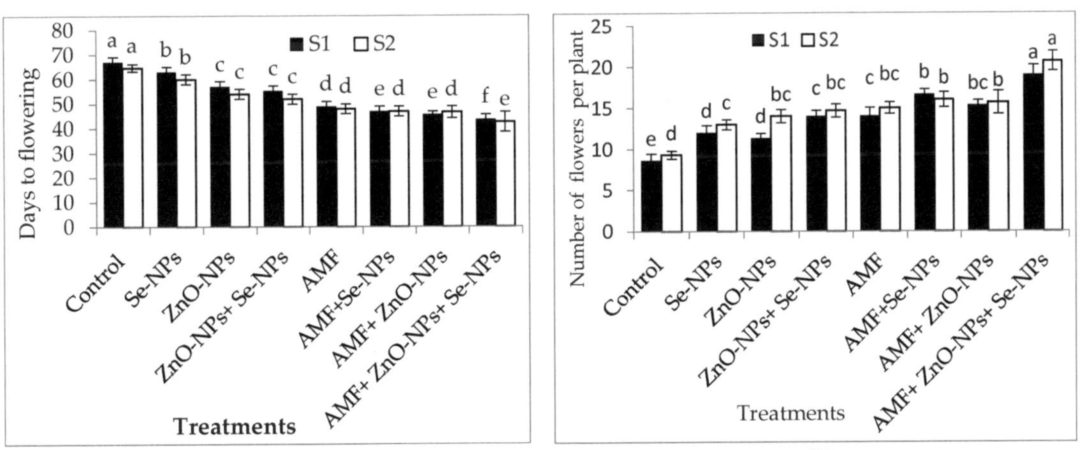

(A) (B)

Figure 1. Effect of inoculation with *Arbuscular mycorrhiza* fungus (AMF), foliar applications of ZnO-NPs, foliar applications of Se-NPs, ZnO-NPs + Se-NPs, AMF + ZnO-NPs, AMF + Se-NPs, AMF + ZnO-NPs + Se-NPs, and untreated plants (control) on (**A**) days to flowering, and (**B**) number of flowers of chili plants during (S1: 2021 and S2: 2022) seasons. Vertical bars indicate the standard error of the mean. The LSD test demonstrates that differences among values in each bar that are marked by different letters (a–f) are significant at $p \leq 0.05$.

2.2. Photosynthetic Pigments of Chili Plants

The data in Figure 2A–D clearly show that AMF inoculation and the foliar application of ZnO-NPs with or without + Se-NPs significantly ($p \leq 0.05$) affected the relative chlorophyll content (SPAD) and photosynthetic measurements compared to the control in both seasons. Chili plants inoculated with AMF combined with foliar application and a mixture of nanoparticles (ZnO-NPs + Se-NPs) exhibited better transpiration rate, stomatal conductance, and photosynthetic rate than the control plants in the 2021 and 2022 seasons. Moreover, the relative chlorophyll content significantly increased by 55.13% and 60.7% following AMF inoculation in combination with the nanoparticle mixture (ZnO-NPs + Se-NPs) as a foliar spray compared to the control chili plants during the 2021 and 2022 seasons, respectively. Likewise, the combination of inoculation with AMF and the mixture of nanoparticles (ZnO-NPs + Se-NPs) significantly increased the photosynthesis rate by 54.5% and 63.6% compared with the control plants in the 2021 and 2022 seasons, respectively. Furthermore, there was a significant increase in intercellular CO_2 concentrations in response to treatment with either Se-NPs or ZnO-NPs foliar application in the absence or presence of AMF on chili plants, without any significant difference between them, compared to the control in both seasons. In contrast, the control chili plants had the lowest intercellular CO_2 concentrations during both seasons.

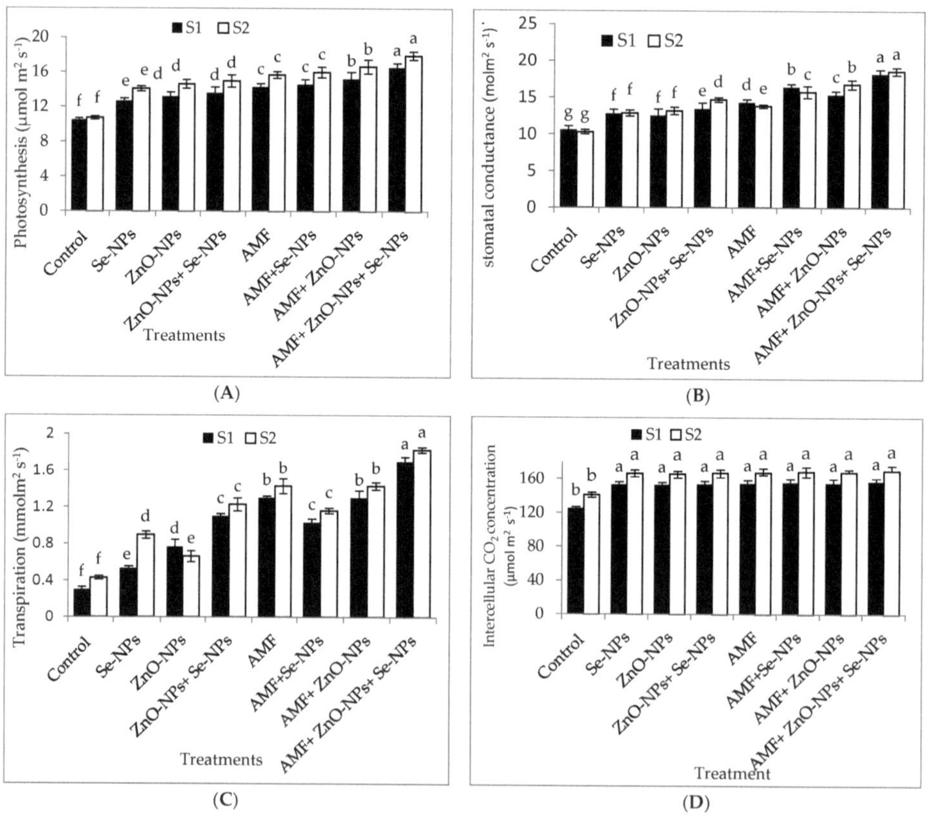

Figure 2. Effect of inoculation with *Arbuscular mycorrhiza* fungus (AMF), foliar applications of ZnO-NPs, foliar applications of Se-NPs, ZnO-NPs + Se-NPs, AMF + ZnO-NPs, AMF + Se-NPs, AMF + ZnO-NPs + Se-NPs, and untreated plants (control) on (**A**) photosynthesis rate, (**B**) stomatal conductance, (**C**) transpiration rate, and (**D**) intercellular CO_2 concentration of chili plants during (S1: 2021 and S2: 2022) seasons. Vertical bars indicate the standard error of the mean. The LSD test demonstrates that differences among values in each bar that are marked by different letters (a–f) are significant at $p \leq 0.05$.

2.3. Marketable Yield and Its Components

The results in Figure 3A–D clearly illustrate that chili pepper plants inoculated with AMF significantly ($p \leq 0.05$) influenced the number of fruits per plant, average fruit weight, total fruit yield per plant, and total marketable fruit yield per m^2 compared to the control plants in both seasons. Likewise, foliar spraying of ZnO-NPs with or without Se-NPs drastically ($p \leq 0.05$) enhanced the marketable yield compared to the control chili plants in both seasons. Chili plants inoculated with AMF and sprayed with (ZnO-NPs + Se-NPs) significantly increased average fruit weight per plant by (92.4% and 98.7%), number of fruits per plant by (34.6% and 54.8%), and total yield per plant by (93.8% and 90.4%) compared to the control in 2021 and 2022 seasons, respectively. Hence, the marketable yields of pepper plants recorded the highest values (4.7 and 3.8 kg fruits m^{-2}) when plants inoculated with AMF in combination with nanoparticles (ZnO-NPs + Se-NPs), which significantly increased by 95.8% and 94.7% in 2021 and 2022, respectively, compared to the control plants. Nevertheless, the control plants recorded the lowest number of fruits per plant, average fruit weight, total fruit yield per plant, and marketable yield during both seasons.

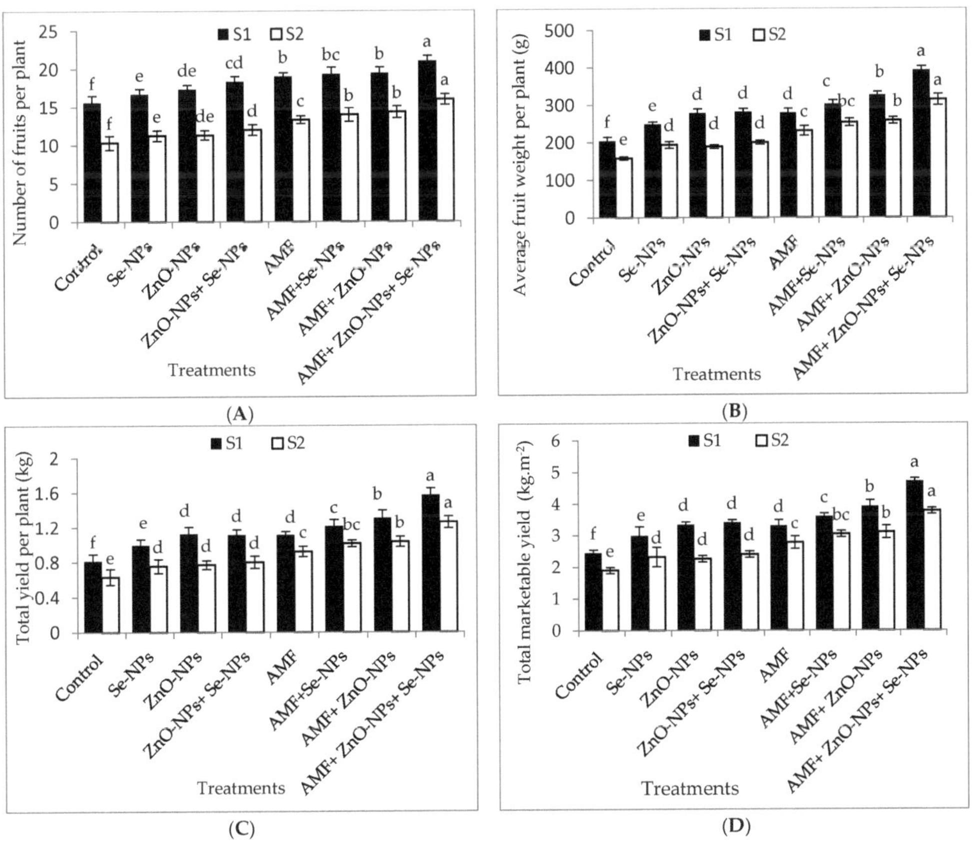

Figure 3. Effect of inoculation with *Arbuscular mycorrhiza* fungus (AMF), foliar applications of ZnO-NPs, foliar applications of Se-NPs, ZnO-NPs + Se-NPs, AMF + ZnO-NPs, AMF + Se-NPs, AMF + ZnO-NPs + Se-NPs, and untreated plants (control) on (**A**) number of fruits per plant, (**B**) average fruit weight per plant, (**C**) total yield per plant, and (**D**) total marketable yield per (m^2) of chili plants during (S1: 2021 and S2: 2022) seasons. Vertical bars indicate the standard error of the mean. The LSD test demonstrates that differences among values in each bar that are marked by different letters (a–f) are significant at $p \leq 0.05$.

2.4. Quality Parameters of Chili Fruits

Chili fruit quality traits were significantly ($p \leq 0.05$) affected by AMF inoculation, nanoparticles foliar application, and their combined interactions (Table 2). Fruit quality parameters, such as fruit length, single-fruit weight, fruit diameter, and capsaicin content, displayed significant variation between the tested treatments and control plants in both seasons (Table 2). The highest weight and length of fruits were recorded following the inoculation of AMF combined with the nanoparticle mixture ZnO-NPs and Se-NPs as a foliar application in the 2021 and 2022 seasons, respectively. Additionally, the inoculation of AMF with foliar spray of ZnO-NPs and Se-NPs drastically increased the average fruit weight by 43.6% and 28.3% compared to the control in the 2021 and 2022 seasons, respectively. Furthermore, AMF combined with ZnO-NPs with or without Se-NPs recorded the highest fruit diameter compared to the control in both years of the experiment. Additionally, harvested chili fruits from all treatments, except for the plants sprayed with ZnO-NPs, showed higher levels of ascorbic acid than the control in both seasons (Figure 3B). Moreover, the highest ascorbic acid content (125 and 129 mg/100 g FW) was recorded in chili fruits treated with AMF combined with nanoparticles mixture (ZnO-NPs + Se-NPs) foliar application, which significantly increased by 46.1% and 46.6%in the 2021 and 2022 seasons, respectively. Additionally, the highest content of capsaicin (142.9 and 143.3 mg/100 g FW) was recorded in chili fruits treated with AMF combined with nanoparticles mixture (ZnO-NPs + Se-NPs) foliar application, which significantly increased by 7.2 and 6.9% in 2021 and 2022 seasons, respectively, compared with untreated plants.

Table 2. Effect of inoculation with *Arbuscular mycorrhiza* fungus (AMF) and foliar applications of nanoparticles on the quality parameters of chili fruits during the 2021 and 2022 seasons.

Treatment	Single-Fruit Weight (g)	Fruit Length (cm)	Fruit Diameter (cm)	Capsaicin (µg/g)	Ascorbic Acid (mg/100 g)
2021 Season					
Control	13.0 ^{e}	8.0 ^{g}	0.67 ^{e}	133.3 ^{h}	85.5 ^{f}
Se-NPs	15.3 ^{c}	11.7 ^{f}	1.4 ^{d}	136.7 ^{f}	89.2 ^{de}
ZnO-NPs	14.3 ^{d}	10.5 ^{f}	1.03 ^{e}	134.6 ^{g}	88.4 ^{ef}
ZnO-NPs + Se-NPs	16.7 ^{b}	13.1 ^{e}	1.0 ^{d}	137.7 ^{e}	92.0 ^{d}
AMF	14.33 ^{d}	14.6 ^{d}	2.9 ^{b}	139.2 ^{d}	100.2 ^{c}
AMF + Se-NPs	15.6 ^{c}	17.8 ^{b}	3.9 ^{a}	140.5 ^{b}	115.4 ^{b}
AMF + ZnO-NPs	17.0 ^{b}	16.2 ^{c}	2.5 ^{c}	139.6 ^{c}	113.1 ^{b}
AMF + ZnO-NPs + Se-NPs	18.6 ^{a}	21.70 ^{a}	3.5 ^{a}	142.9 ^{a}	125.0 ^{a}
LSD value at 0.05:	0.93	1.38	0.39	0.34	3.3
2022 Season					
Control	15.3 ^{d}	8.7 ^{f}	0.63 ^{e}	134.0 ^{h}	88.0 ^{f}
Se-NPs	17.0 ^{bc}	14.4 ^{d}	1.30 ^{d}	137.5 ^{f}	91.7 ^{de}
ZnO-NPs	16.7 ^{c}	13.0 ^{e}	1.20 ^{d}	135.3 ^{g}	90.9 ^{de}
ZnO-NPs + Se-NPs	16.7 ^{c}	15.0 ^{d}	1.7 ^{c}	138.4 ^{e}	94.5 ^{d}
AMF	17.3 ^{bc}	15.5 ^{cd}	1.60 ^{c}	139.7 ^{d}	103.7 ^{c}
AMF + Se-NPs	17.6 ^{bc}	17.9 ^{b}	2.6 ^{a}	141.0 ^{b}	119.4 ^{b}
AMF + ZnO-NPs	18.0 ^{b}	16.2 ^{c}	2.1 ^{b}	140.1 ^{c}	117.1 ^{b}
AMF + ZnO-NPs + Se-NPs	19.6 ^{a}	21.0 ^{a}	2.7 ^{a}	143.3 ^{a}	129.0 ^{a}
LSD value at 0.05:	1.26	1.1	0.26	0.27	3.41

Values followed by different letters are significant according to the LSD test ($p \leq 0.05\%$).

2.5. Mineral Content in Chili Leaves

The results in Table 3 show that all tested treatments had a significant ($p \leq 0.05$) improvement in macronutrient concentrations, such as nitrogen (N), phosphorus (P), potassium (K), and calcium (Ca), in the leaves of chili plants compared with the control. The combination of AMF with a mixture of (ZnO-NPs + Se-NPs) increased the concentration of N in leaves by 56% and 82.6% compared to the control during the 2021 and 2022 seasons, respectively. In addition, the combination of AMF with a mixture of ZnO-NPs and Se-NPs resulted in maximum concentrations of P, K, and Ca compared to the control treatment.

In contrast, control plants showed the lowest concentrations of N, P, K, and Ca during the 2021 and 2022 seasons. Concerning the effect of treatments on the concentration of micronutrients in chili leaves (Table 4), selenium (Se), zinc (Zn), iron (Fe), copper (Cu), and manganese (Mn) were considerably influenced by AMF inoculation, foliar spraying of nanoparticles, and their combined interactions in both seasons. Moreover, the combination of AMF with a mixture of ZnO-NPs and Se-NPs exhibited the highest concentrations of Se, followed by AMF + Se-NPs in the 2021 and 2022 seasons, respectively, compared to the control treatment. Furthermore, the combination of AMF with a mixture of ZnO-NPs and Se-NPs showed the highest concentrations of Zn, followed by AMF + ZnO-NPs and ZnO-NPs + Se-NPs in the 2021 and 2022 seasons, respectively, compared to the control treatment. Additionally, the highest concentrations of Mn, Cu, and Fe were found in chili plants treated with AMF and combined nanoparticles (ZnO-NPs + Se-NPs) compared to the control in 2021 and 2022, respectively. On the other hand, the control plants recorded the lowest concentrations of Se, Zn, Mn, Cu, and Fe during the 2021 and 2022 seasons.

Table 3. Effect of inoculation with *Arbuscular mycorrhiza* fungus (AMF) and foliar applications of nanoparticles on the macronutrient content of chili leaves during the 2021 and 2022 seasons.

Treatment	N%	P%	K%	Ca%
2021 Season				
Control	2.5 f	0.08 d	2.1 d	0.6 d
Se-NPs	3.0 e	0.26 c	3.2 c	1.1 c
ZnO-NPs	2.9 e	0.27 bc	3.1 c	1.2 bc
ZnO-NPs + Se-NPs	3.2 d	0.31 bc	3.4 bc	1.3 bc
AMF	3.3 c	0.35 bc	3.6 bc	1.5 bc
AMF + Se-NPs	3.4 c	0.38 b	3.7 b	1.4 bc
AMF + ZnO-NPs	3.7 b	0.37 b	3.8 b	1.6 b
AMF + ZnO-NPs + Se-NPs	3.9 a	0.70 a	4.4 a	1.9 a
LSD value at 0.05:	0.14	0.11	0.46	0.35
2022 Season				
Control	2.3 f	0.12 h	1.9	0.5 h
Se-NPs	3.2 e	0.35 f	2.9 d	0.9 g
ZnO-NPs	3.3 e	0.28 g	2.9 cd	1.2 f
ZnO-NPs + Se-NPs	3.4 d	0.43 e	3.5 bc	1.53 e
AMF	3.5 cd	0.50 d	3.8 b	1.8 d
AMF + Se-NPs	3.6 c	0.64 b	3.9 b	2.15 c
AMF + ZnO-NPs	3.9 b	0.58 c	4.0 b	2.4 b
AMF + ZnO-NPs + Se-NPs	4.2 a	0.84 a	4.7 a	2.7 a
LSD value at 0.05:	0.44	0.055	0.61	0.11

Values followed by different letters are significant according to the LSD test ($p \leq 0.05\%$).

Table 4. Effect of inoculation with *Arbuscular mycorrhiza* fungus (AMF) and foliar applications of nanoparticles on the micronutrient content of chili leaves during the 2021 and 2022 seasons.

Treatment	Fe (ppm)	Mn (ppm)	Cu (ppm)	Zn (ppm)	Se (ppm)
2021 Season					
Control	56.2 h	12.7 h	2.1 g	32.9 f	0.5 g
Se-NPs	60.1 g	20.9 f	2.5 f	30.1 g	1.4 cd
ZnO-NPs	62.9 f	18.1 g	2.5 f	37.6 d	0.9 f
ZnO-NPs + Se-NPs	65.1 e	23.1 e	2.7 e	40.5 c	1.6 bc
AMF	67.6 d	25.6 d	2.9 d	23.7 h	1.3 de
AMF + Se-NPs	70.5 c	30.8 b	3.04 c	35.1 e	1.7 b
AMF + ZnO-NPs	72.8 b	28.5 c	3.3 b	42.8 b	1.1 ef
AMF + ZnO-NPs + Se-NPs	78.2 a	35.9 a	3.5 a	48.2 a	2.2 a
LSD value at 0.05:	1.14	1.41	0.078	0.99	0.22

Table 4. *Cont.*

Treatment	Fe (ppm)	Mn (ppm)	Cu (ppm)	Zn (ppm)	Se (ppm)
2022 Season					
Control	55.0 h	14.3 h	2.13 h	22.1 g	0.75 f
Se-NPs	60.50 g	23.3 f	2.8 f	30.5 f	1.18 cd
ZnO-NPs	63.33 f	20.2 g	2.5 g	35.3 de	0.94 e
ZnO-NPs + Se-NPs	66.0 e	25.5 e	3.16 e	40.7 c	1.3 bc
AMF	68.0 d	27.9 d	3.5 b	37.7 d	1.02 de
AMF + Se-NPs	71.0 c	33.2 b	3.4 c	33.3 e	1.4 b
AMF + ZnO-NPs	73.3 b	30.8 c	3.3 d	43.8 b	1.11 c–e
AMF + ZnO-NPs + Se-NPs	78.9 a	39.8 a	3.9 a	46.9 a	1.7 a
LSD value at 0.05:	1.113	1.66	0.055	2.5	0.17

Values followed by different letters are significant according to the LSD test ($p \leq 0.05\%$).

2.6. Plant Hormones, Nitrogen Metabolism Enzyme, Antioxidant Enzymes, Hydrogen Peroxide, and Lipid Peroxidation Content in Chili Leaves

The results in Figure 4A–E demonstrate the effects of AMF inoculation, nanoparticles foliar application, and their combined interactions on the content of the most crucial hormones, abscisic acid (ABA) and peroxidase (POD) antioxidant enzyme, and malondialdehyde (MDA), hydrogen peroxide (H_2O_2), and glutamate dehydrogenase (GDH) nitrogen metabolism enzyme. The combination of AMF with nanoparticles mixture (ZnO-NPs + Se-NPs) exhibited better POD activity than control in the 2021 and 2022 seasons. In contrast, the control plants registered the lowest POD content in both seasons. Furthermore, the combination of AMF with nanoparticles mixture (ZnO-NPs + Se-NPs) decreased the content of ABA by (86.45% and 83.3%) in chili leaves compared to the control in the 2021 and 2022 seasons, respectively. In contrast, the control chili plants had the highest ABA concentrations during both seasons. Correspondingly, AMF combined with a nanoparticle mixture (ZnO-NPs + Se-NPs) mitigated the adverse impact of cold stress on the activity of the nitrogen metabolism enzyme, with the activity of GDH increasing by 72.8% and 94.7%, respectively, compared with control plants in the 2021 and 2022 seasons, respectively (Figure 4E). By contrast, AMF combined with a nanoparticle mixture (ZnO-NPs + Se-NPs) decreased the content of MDA by (65.7% and 78.4%) and the content of H_2O_2 by (70.4% and 78.6%) in chili leaves compared to the control in the 2021 and 2022 seasons, respectively. However, untreated plants had the highest MDA and H_2O_2 contents in the leaves in both seasons.

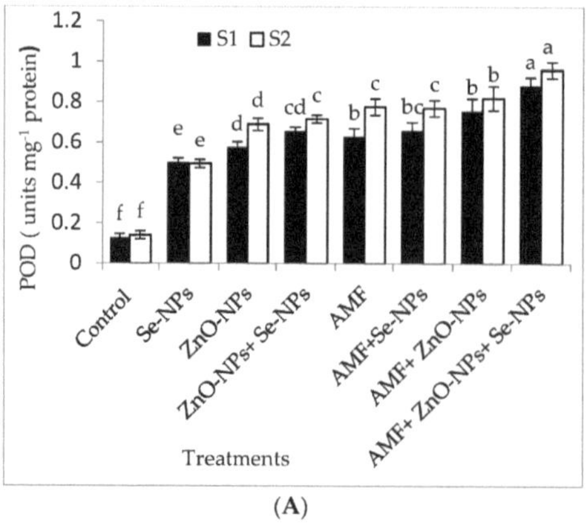

(A)

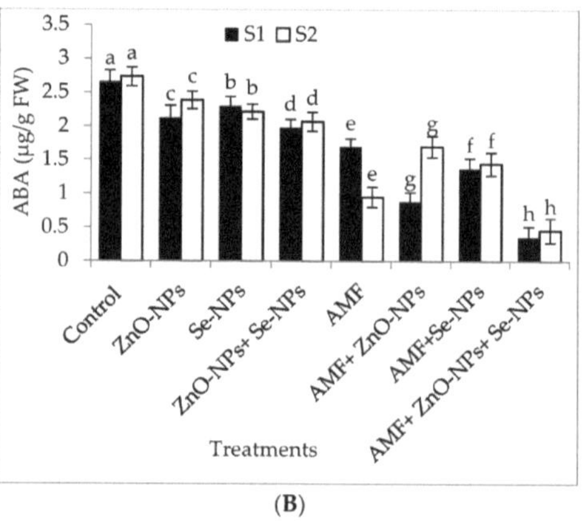

(B)

Figure 4. *Cont.*

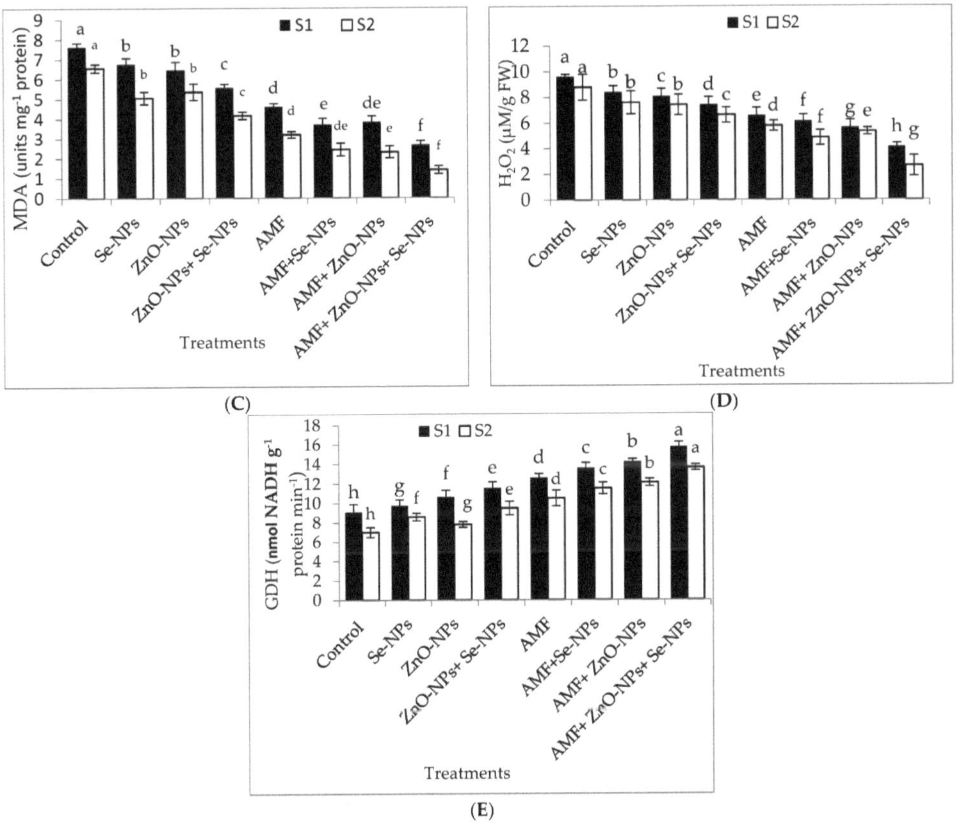

Figure 4. Effect of inoculation with *Arbuscular mycorrhiza* fungus (AMF), foliar application of ZnO-NPs, foliar application of Se-NPs, ZnO-NPs + Se-NPs, AMF + ZnO-NPs, AMF + Se-NPs, AMF + ZnO-NPs + Se-NPs, and untreated plants (control) on (**A**) peroxidase (POD), (**B**) abscisic acid (ABA), (**C**) malondialdehyde (MDA), (**D**) hydrogen peroxide (H_2O_2), and (**E**) glutamate dehydrogenase (GDH) of chili leaves during (S1: 2021 and S2: 2022) seasons. Vertical bars indicate the standard error of the mean. The LSD test demonstrates that differences among values in each bar that are marked by different letters (a–h) are significant at $p \leq 0.05$.

2.7. Protein Electrophoresis SDS-PAGE

AMF inoculation, nanoparticles foliar application, and their combined interactions stimulated the appearance or disappearance of different protein bands compared to control plants (Table 5 and Figure 5). There were seven protein bands ranging from 20.74 to 320.23 kDa in the chili pepper protein patterns. The control plants showed four bands with molecular weights of 133.56, 61.96, 38.82, and 20.74 kDa. Furthermore, ZnO-NPs, Se-NPs, AMF + ZnO-NPs + Se-NPs, and AMF + ZnO-NPs induced the synthesis of new bands with molecular weights of 320.23, 228, and 38.2 kDa. The treatment of AMF + Se-NPs led to the disappearance of 38.2 compared to the control plants. The treatment with AMF led to the disappearance of the 133.56 kDa compared to the control. Thus, we concluded that cold stress affected expression in the control, as the disappearance of three bands at various molecular weights appeared in different treatments, as ZnO-NPs, Se-NPs, AMF + ZnO-NPs + Se-NPs, and AMF + ZnO-NPs provided plants with a defense mechanism against cold stress compared with the control.

Table 5. The effect of AMF inoculation, nanoparticle foliar application, and their combined interactions on protein banding patterns of SDS-PAGE of pepper plants subjected to cold stress. T1: control, T2: Se-NPs, T3: ZnO-NPs, T4: AMF + ZnO-NPs + Se-NPs, T5: AMF + Se-NPs, T6: AMF + ZnO-NPs, T7: AMF, T8: ZnO-NPs + Se-NPs.

Mwt (kD)	T1	T2	T3	T4	T5	T6	T7	T8
320.23	-	+	+	+	+	+	+	-
228.02	-	+	+	+	+	+	+	+
133.56	+	+	+	+	+	+	-	+
61.96	+	+	+	+	+	+	+	+
38.82	+	+	+	+	-	+	+	+
28.09	-	+	+	+	+	+	+	+
20.74	+	+	+	+	+	+	+	+
Number of bands	7	7	4	7	6	7	6	7

Mwt: Molecular weight, +: Presence of band, -: absence of the band.

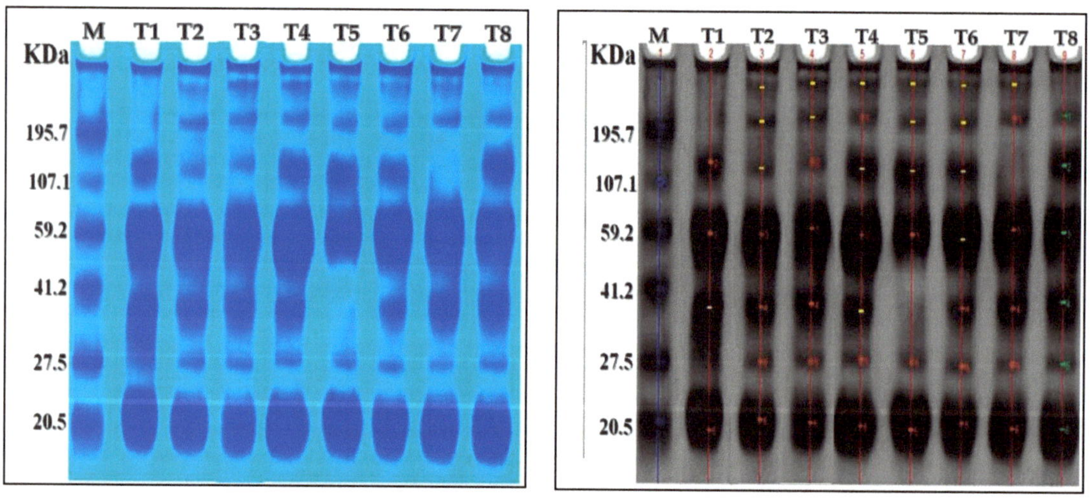

Figure 5. The change in protein bands (marked by arrowheads) in response to tested treatments on pepper plants under cold stress. Electrograph of soluble protein pattern by one-dimensional SDS-PAGE. As shown, lane one is a molecular weight marker with six bands of known size ranging from 20.74 to 320.23 KDA. T1: lane is control treatment, T2: is Se-NPs, T3: ZnO-NPs, T4: AMF + ZnO-NPs + Se-NPs, T5: AMF + Se-NPs, T6: AMF + ZnO-NPs, T7: AMF, T8: ZnO-NPs + Se-NPs.

2.8. *Leaf Anatomy*

Regarding anatomical maintenance, chili leaves have adaxial and abaxial epidermis with stomata on both sides. The epidermal cells were covered with a slim cuticle layer on the surface of the outer walls. The mesophyll was composed of 3–5 layers of spongy parenchyma cells with intercellular gaps and one layer of palisade parenchyma cells. In the midrib, the vascular bundle was bicollateral, with the phloem facing both sides of the leaf and surrounded by collenchyma cells Figures 6–8. As shown in Table 6 and Figures 6–8, the upper epidermis, thickness of the midvein, and lower epidermis of the chili plants inoculated with AMF combined with (ZnO-NPs + Se-NPs) increased by 20.41%, 34.15%, and 13.99%, respectively. Correspondingly, the thickness of the vascular bundle of the midvein was enhanced owing to the increase in the dimensions of the main midvein bundle and the mean diameter of the vessels by 67.72% and 12.84%, respectively, when compared with the control (Figures 6–8, T1 and T8). A promotive effect on leaf lamina thickness was observed in plants sprayed with ZnO-NPs + Se-NPs by 65.85%, owing to an enhancement in the thickness of palisade and spongy tissues by 62.0% and 68.15%, respectively, when compared with the control (Figure 8, T1 and T4). A minimal improvement was observed at Se-NPs in the thickness of the midvein, palisade, lamina, and spongy tissues by 18.16%, 4.47%, 10.84%, and 15.22%, respectively, compared to the control. Additionally, the upper and lower epidermis thickness increased by 0.44% and 3.43%, respectively, compared to the control (Figures 6–8, T1 and T2).

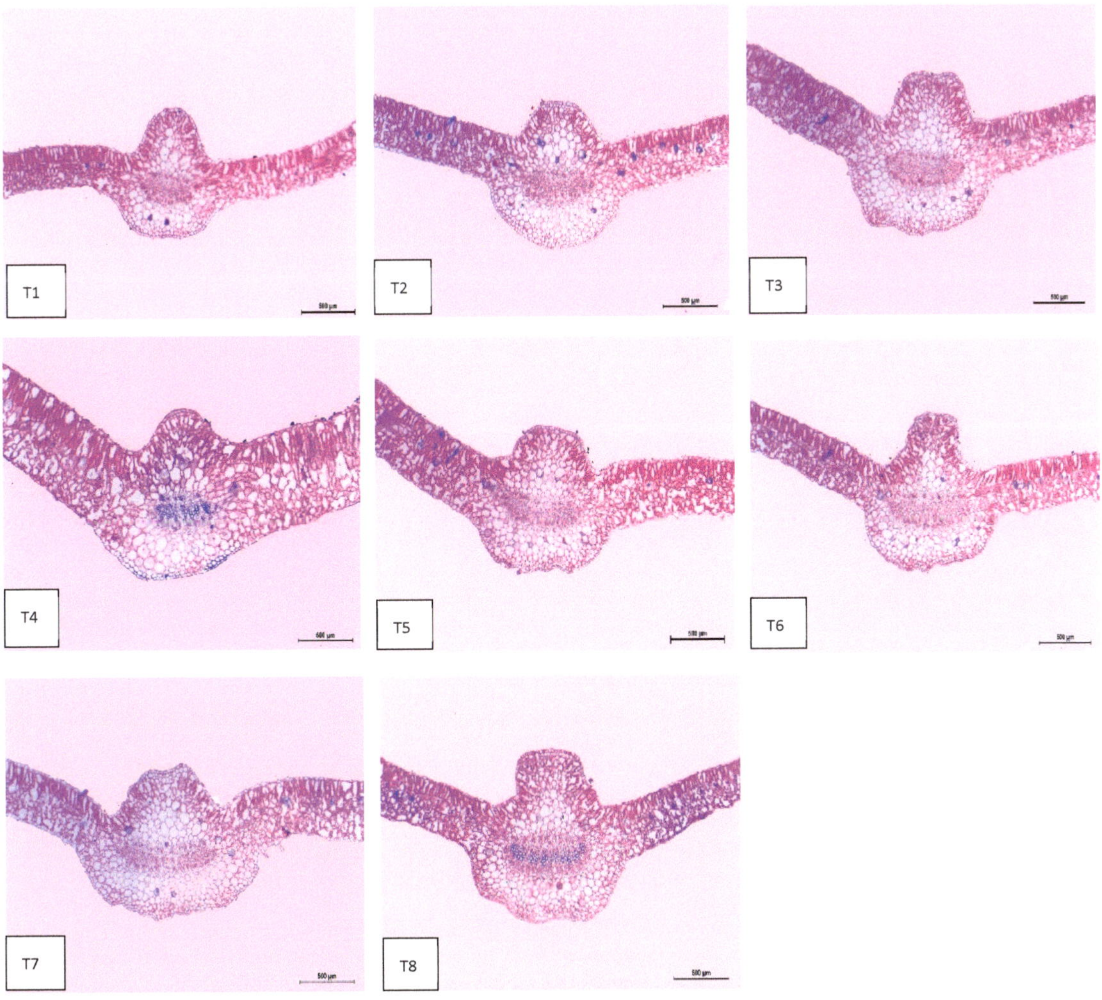

Figure 6. Microphotographs of cross sections through the chili pepper leaf, treated with *Arbuscular mycorrhiza* fungus (AMF) and foliar application of nanoparticles after 45 days from transplanting. Scale bars = 500 μm. (**T1**) control (untreated plants); (**T2**) Se-NPs; (**T3**) ZnO-NPs; (**T4**) ZnO-NPs + Se-NPs; (**T5**) AMF inoculation; (**T6**) AMF + Se-NPs; (**T7**) AMF + ZnO-NPs and (**T8**) AMF + (ZnO-NPs + Se-NPs).

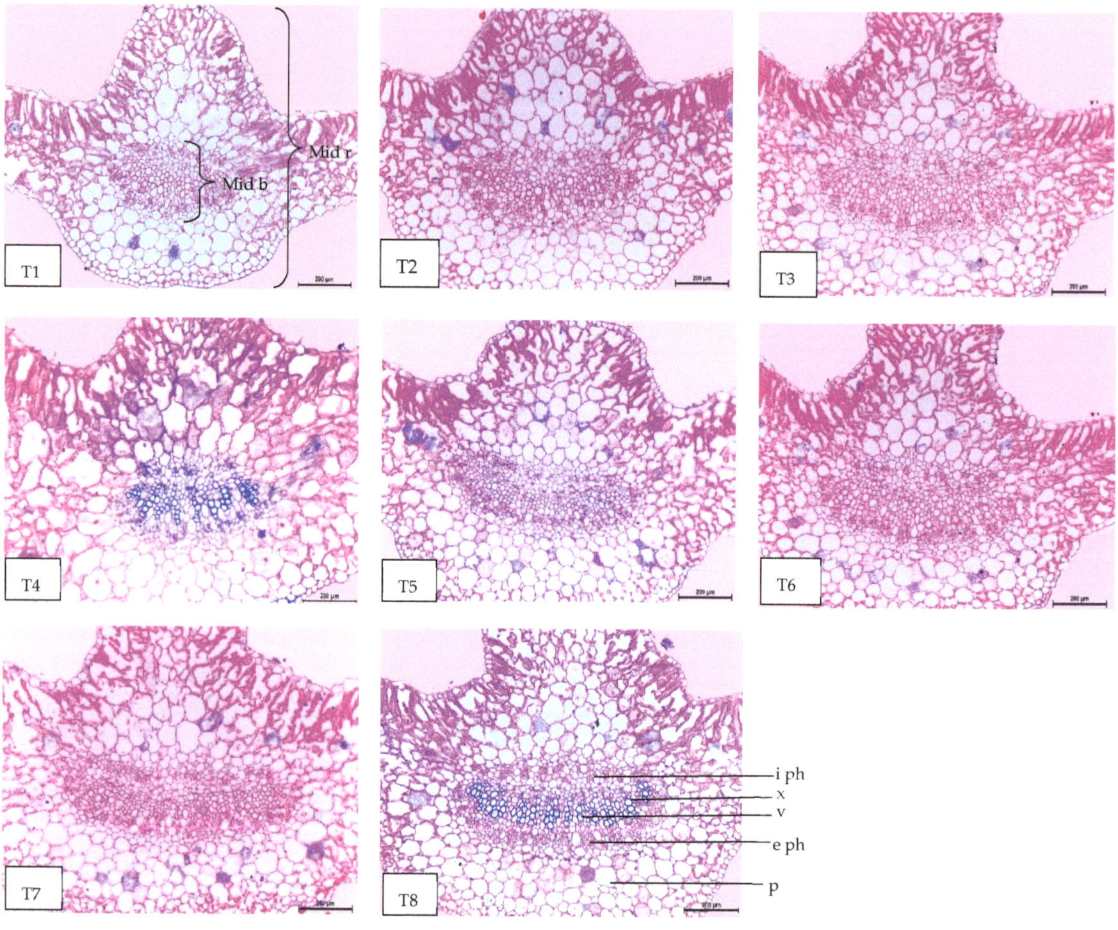

Figure 7. Magnified portions of the leaf midvein bundle of chili pepper plants after 45 days from transplanting. Scale bars = 200 µm. (**T1**) control (untreated plants); (**T2**) Se-NPs; (**T3**) ZnO-NPs; (**T4**) ZnO-NPs + Se-NPs; (**T5**) AMF inoculation; (**T6**) AMF + Se-NPs; (**T7**) AMF + ZnO-NPs and (**T8**) AMF + (ZnO-NPs + Se-NPs). Details: mid b, midvein bundle; mid r, midvein region; e ph, external phloem; i ph, internal phloem; v, vessel; x, xylem and p, parenchyma.

2.9. Mesophyll Parenchyma and Chloroplast Ultrastructure of Chili Pepper Leaves

The mesophyll parenchyma of the chili pepper leaves displayed significant differences between the tested treatments. The mesophyll cells of the chili plants treated with AMF with a mixture of nanoparticles (ZnO-NPs + Se-NPs) were larger than those of the control plants (Figure 9A,D). The chloroplasts in the mesophyll cells of the chili plants inoculated with AMF and sprayed with a mixture of nanoparticles (ZnO-NPs + Se-NPs) were more numerous and larger, in addition to being richer in starch grains than in the control cells (Figure 9A,B,D,E). Moreover, large plastoglobules were observed in the control plants (Figure 9B). The chloroplasts in the plants that were treated with AMF and sprayed with a mixture of nanoparticles (ZnO-NPs + Se-NPs) were enlarged and had thicker grana that contained multiple layers of thylakoids (stacked grana thylakoids), as well as non-stacked stroma thylakoids when compared to the control plants (Figure 10A,B).

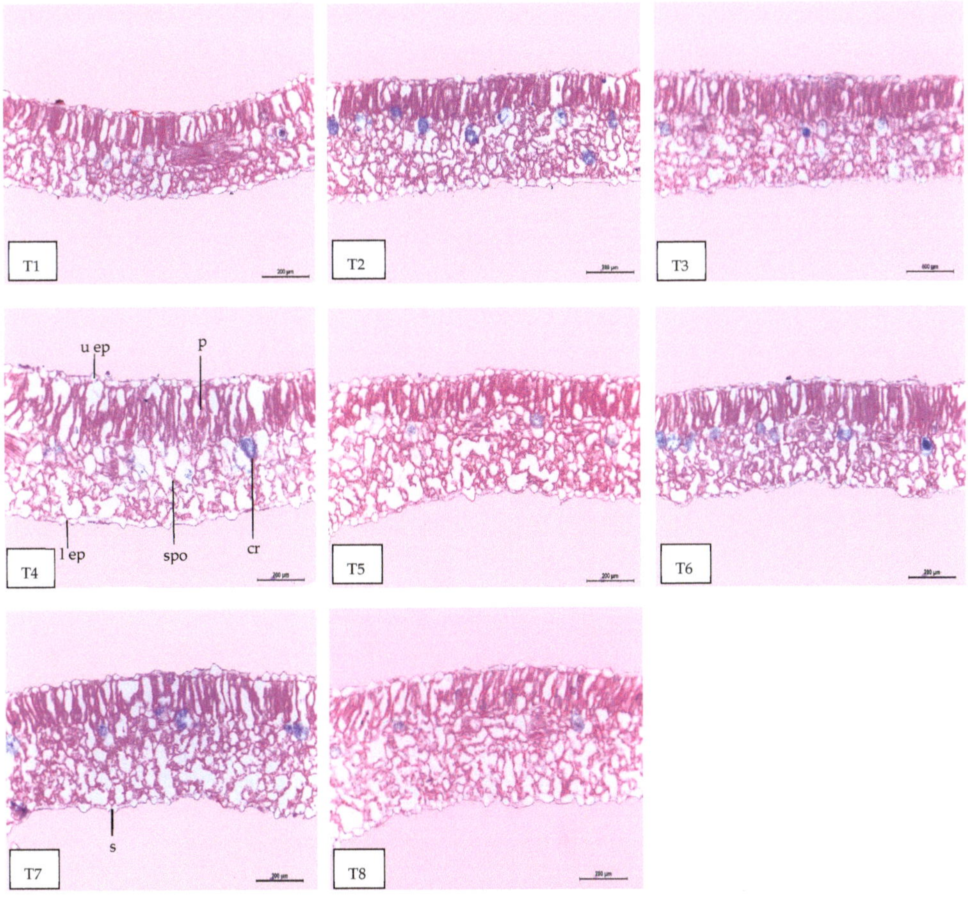

Figure 8. Magnified portions of the leaf blade lamina of chili pepper plants after 45 days from transplanting. Scale bars = 200 μm. (**T1**) control (untreated plants); (**T2**) Se-NPs; (**T3**) ZnO-NPs; (**T4**) ZnO-NPs + Se-NPs; (**T5**) AMF inoculation; (**T6**) AMF + Se-NPs; (**T7**) AMF + ZnO-NPs and (**T8**) AMF + (ZnO-NPs + Se-NPs). Details: l ep, lower epidermis; u ep, upper epidermis; cr, cells with dark content containing calcium oxalate crystals; s, stomata; spo, spongy tissue, and p, palisade tissue.

Table 6. Counts and measurements in micro-meters (μm) of certain histological characters in the transverse sections of the chili pepper leaf, treated with *Arbuscular mycorrhiza* fungus (AMF) and foliar applications of nanoparticles, 45 days from transplanting (Means of three sections from three specimens).

Treatments	Histological Aspects							
	Thickness of Midvein	Thickness of Upper Epiderm	Thickness of Lower Epiderm	Thickness of Lamina	Thickness of Pal-isade Tissue	Thickness of Spongy Tissue	Dimensions of Main Midvein Bundle	The Mean Diameter of the Vessel
Control	939.651	14.854	19.954	279.527	119.885	160.205	280.090	19.285
Se-NPs	1110.333	14.920	20.639	309.849	125.253	184.596	350.562	20.861
ZnO-NPs	1234.779	15.560	20.354	334.560	126.276	208.284	394.060	20.901
ZnO-NPs + Se-NPs	1255.086	17.208	21.758	463.622	194.225	269.397	286.452	20.333
AMF	1200.086	16.450	21.070	341.252	126.057	215.195	382.586	20.900
AMF + Se-NPs	1230.381	15.845	21.465	320.655	120.221	200.434	422.038	21.347
AMF + ZnO-NPs	1245.915	17.107	22.164	397.821	140.370	257.451	466.477	21.445
AMF + ZnO-NPs + Se-NPs	1260.588	17.886	22.747	419.633	144.979	275.654	469.774	21.762

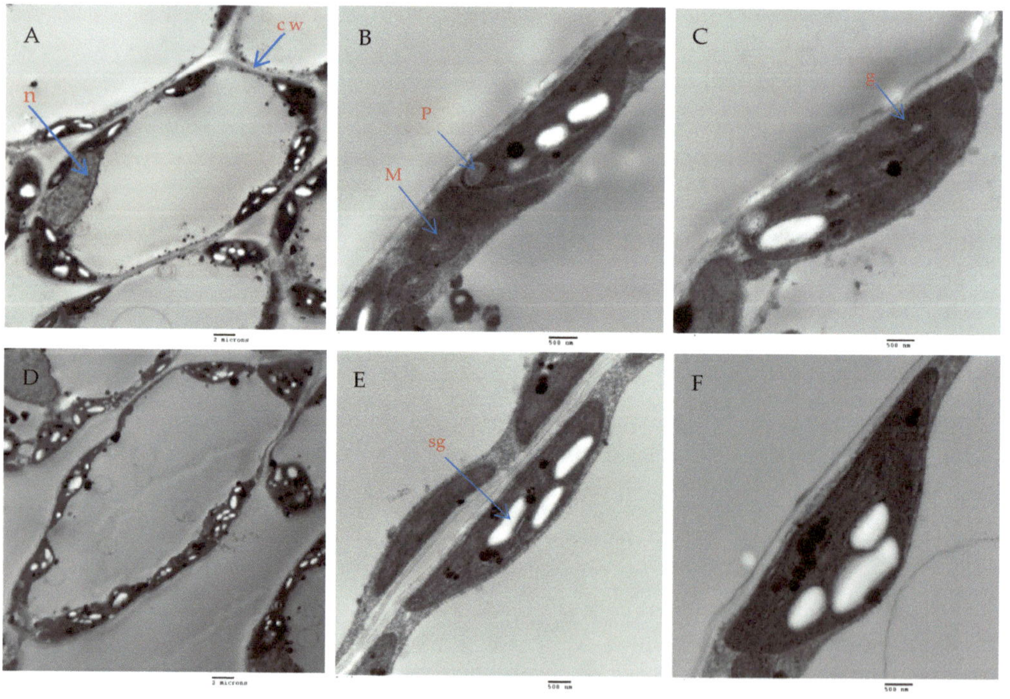

Figure 9. The effect of AMF + (ZnO-NPs + Se-NPs) on mesophyll parenchyma and chloroplast ultrastructure in pepper leaves: (**A–C**) Control plants, (**D–F**) Plants treated with AMF + (ZnO-NPs + Se-NPs). Details: cw, cell wall; n, nucleus; sg, starch granule; p, plastoglobule; m, mitochondria and g, grana. (**A**,**D**) scale bars = 2 m; (**B**,**E**) scale bars 500 nm; (**C**,**F**): scale bars = 500 nm.

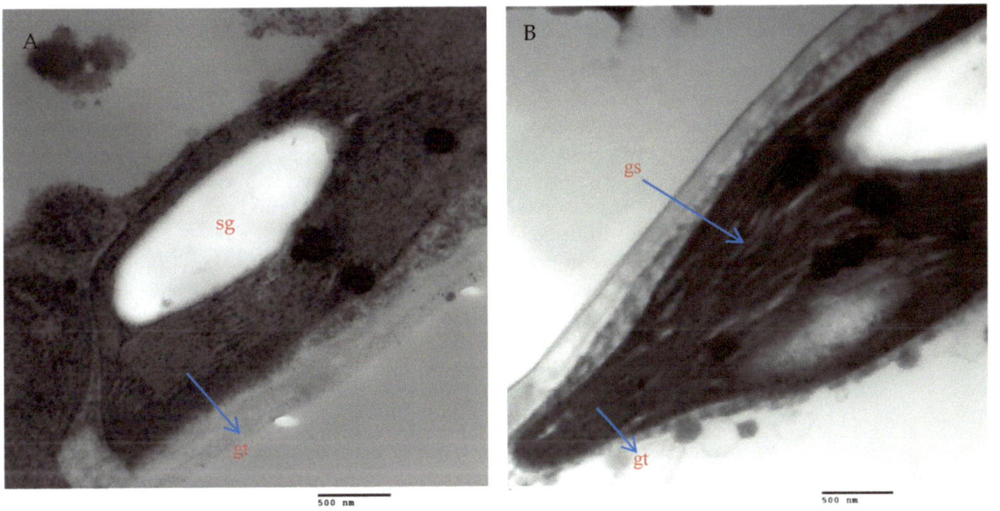

Figure 10. Magnified portions of the chloroplast ultrastructure in chili leaves. Scale bars = 500 nm. (**A**)—Control plants; (**B**)—Plants treated with AMF + (ZnO-NPs + Se-NPs). Details: sg, starch granule; gt, stacked grana thylakoids; gs, non-stacked stroma thylakoids.

3. Discussion

Cold winter temperatures can negatively affect the reproductive stage, productivity, and quality of chili plants [1,2]. Based on our observations, AMF colonization and spraying with ZnO-NPs or Se-NPs have the potential to enhance cold-stress tolerance in crop plants. Plant growth characteristics are strong indicators of abiotic stress. In our study, we observed that the growth characteristics of chili plants significantly decreased under cold stress (Table 1). These findings are consistent with those reported by the authors of [26]. However, AMF inoculation significantly increased the growth characteristics of chili plants in both seasons. This demonstrates the role of AMF colonization in chili plants, even when they are exposed to low optimum temperatures [27]. These increases in plant growth parameters have been linked to more efficient micro and macronutrient uptake [27]. These results are consistent with those obtained in [28]. Under cold stress, the spraying of nanoparticles was associated with significantly higher plant growth parameters in comparison with the control treatment, indicating that ZnO-NPs and Se-NPs enhanced photosynthetic activity and antioxidant enzyme levels [29,30]. In addition, plants colonized with AMF and sprayed with (ZnO-NPs + Se-NPs) flowered earlier than control plants in both seasons (Figure 1). This is a highly desirable trait because it can directly influence market product prices. Our findings are strongly supported by those in [31].

The findings also illustrated that suboptimal temperatures decreased the marketable yield and its components (Figure 3A–D). Cold stress reduces productivity by slowing CO_2 fixation and fruit photosynthetic partitioning [31]. These results are consistent with those previously reported [26,32]. In particular, AMF colonization combined with ZnO-NPs and Se-NPs spraying yielded drastically higher fruit weight, fruit number, and total marketable fruit yield in both seasons. Our observations are strongly supported by those of [33], who observed that AMF enhanced the productivity of snap bean crops. The positive impact of AMF on productivity could be owing to the production of glomalin (also known as glomalin-related soil protein (GRSP) or glycoprotein glomalin) released in the soil as an external secretion of AMF symbiosis, which has been proven to improve soil aggregate stability, soil water potential, and productivity. Moreover, the results demonstrated that AMF colonization combined with (ZnO-NPs + Se-NPs) spraying increased the total yield per plant by 95.8% and 94.7% in the 2021 and 2022 seasons, respectively. These increases could be attributed to the combination of nanoparticles and AMF, which have a synergistic effect against cold stress. Zinc oxide nanoparticles (ZnO-NPs) increased the productivity of wheat plants [34]. Zinc plays an important role in hormone metabolism regulation by altering auxin levels via tryptophan production and promoting cell division and expansion [35]. Furthermore, Se-NPs improved the total yield of groundnut cultivars [30].

The highest values of chlorophyll content, photosynthesis, transpiration rate, and stomatal conductance were recorded for AMF combined with nanoparticles mixture (ZnO-NPs + Se-NPs) in both seasons (Figure 2A–D). The observed impact of AMF on physiological features may be attributable to its contribution to decreasing N_2O emissions by increasing N uptake, therefore improving photosynthesis and potassium uptake [34]. Zn is required for the activities of enzymes [23,24]. These findings are consistent with those reported by the authors of [34]. Generally, metallic nanoparticles have active effects on photosynthetic efficiency, which increases light absorption by chlorophyll [36].

According to the results, the fruit quality traits changed due to cold stress. Similar results were obtained for tomatoes by the authors of [32]. Chili fruits harvested from AMF with nanoparticles mixture (ZnO-NPs + Se-NPs) recorded high values of single-fruit weight, fruit diameter, fruit length, ascorbic acid, and capsaicin content in both seasons compared to the control (Table 2). ZnO-NP application enhanced the quality and nutrient content of harvested bell pepper fruits [37].

Under cold stress, the control plants displayed the lowest content of nutrients, including N, K, P, Ca, Fe, Mn, Zn, and Se, in both seasons (Tables 3 and 4). In contrast, the results obtained from AMF combined with a nanoparticle mixture (ZnO-NPs + Se-NPs) showed better absorption of minerals under suboptimal temperatures in both seasons. AMF may exert its effect on the mineral content in plants by strigolactone production, revealed to be hyphal branching and host detection signals for AMF [38], which play an essential role in responses to abiotic stress conditions by improving growth and nutrient absorption. Se-NPs and ZnO-NPs exhibited remarkable nutrient content in the leaves of cowpea [39].

Plants have enzymatic defense systems to purify free radicals under cold stress. Changes in antioxidant enzyme levels have been associated with cold-stress reactions [32]. AMF combined with ZnO-NPs + Se-NPs improved the POD enzyme content in chili leaves in both seasons. These results are consistent with those reported by the authors [34]. AMF protects plants from cold stress and increases antioxidant enzyme levels [40]. ZnO-NPs cause changes in secondary chemicals relevant to plant defense against abiotic stress factors [41]. Se-NP treatment increased auxin production, therefore enhancing root architectural modifications in tomatoes [42]. GDH may have a special physiological role in the release of large quantities of ammonium in plants under cold stress [43]. These data indicated that AMF combined with ZnO-NPs + Se-NPs significantly improved the activity of the nitrogen metabolism enzyme (GDH) under cold stress. AMF improved N uptake and photosynthetic activity in chili leaves but led to a reduction in N_2O emissions in plant leaves [44]. Zn increased the synthesis of proteins as cofactors in specific enzymes (DNA and RNA), chlorophyll, carbohydrates, lipids, and hormones [45]. Spraying of ZnO-NPs + Se-NPs stimulated plant cells and boosted cell metabolic activity [39,41], therefore improving cold tolerance in chili plants.

ABA reactions control stomatal conductance and other physiological functions [45]. The results indicated that the lowest ABA content was registered in the chili plants colonized with AMF and sprayed with a nanoparticle mixture (ZnO-NPs + Se-NPs) in both seasons (Figure 4B). AMF inoculation affected stomatal function via abscisic acid modulate auxin concentration via tryptophan biosynthesis [45]. AMF combined with a nanoparticle mixture (ZnO-NPs + Se-NPs) decreased the MDA content and the H_2O_2 content in chili leaves compared to the control in the 2021 and 2022 seasons, respectively. Similar outcomes for wheat plants were reported in [34,46]. AMF increases antioxidant enzyme (POD), which eliminates ROS, therefore protecting plants from oxidative damage or membrane stabilization [43,45]. Zinc is a cofactor for numerous enzymes, allowing the antioxidant defense system under cold stress [45]. Selenium dismutation of O^{2-} and OH^- into H_2O_2 is followed by H_2O_2 scavenging by CAT antioxidant enzyme [47].

SDS-PAGE analysis demonstrated that ZnO-NPs, Se-NPs, AMF + ZnO-NPs + Se-NPs, and AMF + ZnO-NPs provided plants with a defense mechanism against cold stress compared to the control. Similar findings in wheat plants were reported by [34], who observed that AMF + ZnO-NPs enhanced the synthesis of new and distinctive bands in antioxidant enzymes as a stress-resistance strategy for wheat plants. Hence, the synthesis of unique bands is caused by several structural modifications of DNA that initiate variations in amino acids, resulting in protein formation [48]. In this regard, the authors of [49] hypothesized that an enzyme called Rubisco activase would be associated with a band with a molecular weight of 51 kDa. Furthermore, according to [50,51], under salinity stress, flax and sunflower plants express a protein with a molecular weight of 26 kDa to help their survival.

The combination of AMF inoculation and ZnO-NPs + Se-NPs as a foliar spray achieved an enhancement in most anatomical characteristics. The optimistic effect of AMF on the anatomical properties may be due to the vital role of AMF in mitigating crop abiotic stress by adjusting good root morphology using the extra-radical mycelium and improving the absorption of nutrients and water in the soil [52]. These findings are consistent with those reported by the authors of [53], who found that Zn and Se are crucial for promoting the chlorophyll scale and relieving all harmful outcomes caused by abiotic stress. In addition, Se-NPs improved the vascular bundle dimensions and thickness of mesophyll, xylem, and phloem, as well as the thickness of the lower and upper epidermis of Date Palm plants under salinity stress [54]. However, all anatomical characteristics decreased in the control plants. These results are in agreement with those reported by [55], who found that the cells of the upper and lower epidermis and mesophyll of wheat leaf plants that were not subjected to low temperature were closely coordinated, whereas the mesophyll cells of the plants exposed to low temperatures (0 °C at the elongation stage and 10 °C at tillering) decreased with increasing intercellular temperature.

Furthermore, AMF and a mixture of nanoparticles (ZnO-NPs + Se-NPs) increased the number of chloroplasts and improved the thickness of grana thylakoids. The stacked grana thylakoids in chili plants inoculated with AMF and sprayed with (ZnO-NPs + Se-NPs) were accompanied by the accumulation of starch granules, which indicates an increase in photosynthesis activity, as starch is the energy storage of the cells and is a product of photosynthesis. This agrees with the results obtained in [52] for peanuts and [56] for *Cucumis melo* L. In contrast, the number and size of plastoglobules within chloroplasts increased in the control plants. This agrees with the findings of [57], who observed that plastoglobules increased under cold stress because of their role in the active channeling of lipid molecules and lipid breakdown products.

4. Materials and Methods

4.1. The Experimental Location and Plant Materials

Two field trials were successfully performed throughout two winter seasons (December to June) in 2021 and 2022 at the Experimental Field, Vegetable Crops Department, Faculty of Agriculture, Cairo University, Giza, Egypt (30°1′10″ North latitude and 30°11′5″ East longitude). The climatic conditions at the growing location are shown in Figure 11. The chili (*C. annuum* L. Sina F1) cultivar was used as the experimental model in this study. The seeds of the chili pepper (Sina F1) cultivar were purchased from NEW Star Four Co., El Weili, Egypt. *Arbuscular mycorrhiza* fungus was produced by the Soil Department, Agricultural Research Centre, Ministry of Agriculture, Giza, Egypt. The AM fungal (*Glomus mosseae*) inoculum, containing hyphae and culture substrates, including spores (nearly 70 g^{-1}), was dried in air.

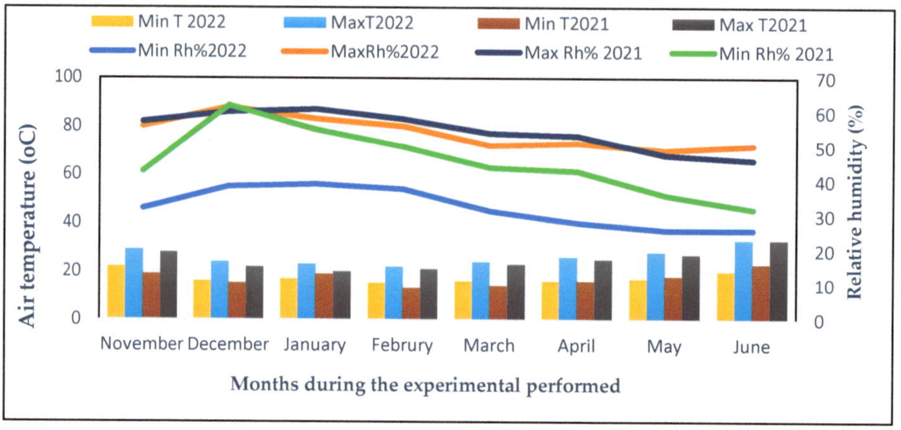

Figure 11. Climatic data for every month during the 2021 and 2022 seasons in The Experimental Field at Cairo University in Giza, Egypt. Minimum relative humidity (Min Rh%), minimum air temperature (Min T (°C)), and maximum air temperature (Max T, (°C)), maximum relative humidity (Max Rh%).

4.2. Mycorrhizal Inoculums and Growth Conditions

Before sowing, the seeds of chili (cv. Sina F1) were sterilized in sodium hypochlorite at 5% for 15 min and then rinsed with distilled water. On 1 November 2021 and 2nd 2022, seeds of chili plants were germinated in seedling trays containing 10 cm^3 cells filled with a mix of peat moss and vermiculite (1:1 v:v). Furthermore, half of the seedling trays received the AM fungi *(G. mosseae)* at a rate of 10 g inoculum per sowing cell, whereas the other half received non-AMF inoculum. The mycorrhizal inoculums were placed at 3–5 cm depth in compartments of trays directly before the growth of the chili seedlings to promote fungal inoculation of plant roots. The experimental trays were set up in a nursery with a relative humidity of 50–60%, an average temperature of 18–20 °C (day/night), with a 13/11 h (light/night) photoperiod, the photon flux density was in the range of 1100 μmol m^{-2} s^{-2}.

4.3. Field Experiments

Uniform chili pepper seedlings (30 days old) were carefully planted in an open field on 1 December 2021 and 2nd 2022 seasons at the Eastern Field of Vegetable Crops Department, Faculty of Agriculture, Cairo University, Giza, Egypt. The experimental plot area was 16 m^2 (8 m long and 2 m wide). The space between rows was 50 cm, and each transplant had a space of 30 cm. There were 48 plants in each replicate (six plants per treatment). Control chili seedlings (untreated seedlings) were grown under the same conditions. The application of Se-NPs or ZnO-NPs (10 mg/L) was performed three times. The first application was 10 days after transplanting (11 December 2021 and 13 December 2022), followed by two foliar sprays at intervals of 10 days. The experimental design included eight treatments: (T1) control (untreated plants), (T2) Se-NPs, (T3) ZnO-NPs, (T4) (ZnO-NPs + Se-NPs), (T5) AMF, (T6) AMF + Se-NPs, (T7) AMF + ZnO-NPs, (T8) AMF + (ZnO-NPs + Se-NPs). Six replicates were used for the treatments in a randomized complete block design (RCBD). Agricultural practices and soil preparation were implemented by the recommendations of the Egyptian Ministry of Agriculture. The physical characteristics of the experimental soils are presented in Table 7.

Table 7. Physical and chemical properties of the experimental soil location during the two cultivating seasons of 2021 and 2022.

Soil Parameters	2021	2022
	Physical properties	
pH	7.45 ± 0.5	7.35 ± 0.3
Clay (%)	42.2 ± 1.3	40.5 ± 1.2
Silt (%)	27.2 ± 5.1	28.7 ± 2.2
Sand (%)	20.5 ± 3.1	22.8 ± 5.1
Soil texture	Clay loam	Clay loam
	Chemical properties	
Soluble Cations (meq L^{-1})		
Mg^{2+}	2.45 ± 0.4	1.98 ± 0.2
K^{+}	0.30 ± 0.05	0.30 ± 0.03
Ca^{2+}	3.44 ± 0.4	3.50 ± 0.5
Na^{+}	1.61 ± 0.5	1.75 ± 0.4
Available phosphorus (mg L^{-1})	0.12 ± 0.03	0.1 ± 0.01
Total nitrogen (TN) (%)	0.19 ± 0.03	0.17 ± 0.01
Soluble Anions (meqL^{-1})		
HCO_{3-}	0.70 ± 0.2	0.78 ± 0.1
Cl^{-}	1.20 ± 0.3	1.50 ± 0.2
$SO_4{}^{2-}$	1.80 ± 0.5	1.35 ± 0.6

4.4. Preparation of Nanoparticles

Zinc oxide nanoparticles (ZnO-NPs) were formed from zinc chloride (General Purpose Reagent, minimum 98.5%, Sigma-Aldrich, Munich, Germany). Zink oxide nanoparticles (ZnO-NPs) (Figure 12A) were prepared using the method described in [58]. Zinc Oxide nanoparticles were synthesized from an aqueous solution containing zinc chloride. NaOH was gradually introduced at a molar ratio of 1:2 over 8 h. The obtained precipitate was thoroughly filtered and washed with ionized water in a mixture of water and toluene using a high-speed stirrer, followed by additional washing with only ionized water for 3 h. The precipitate was dried in an oven at 100 °C. Selenium nanoparticles (Se-NPs) (Figure 12B) were formed according to a previously described protocol [59]. Se-NPs were created by the chemical reduction of sodium selenite by glutathione. In particular, 9 mL of double-distilled water in a sterile cabinet was mixed with 3 mL of 25 mM Na_2SeO_3. A total of 1 M NaOH was added to the reactant solution after mixing to increase the pH to that of the alkaline media. Se-NPs were generated as soon as NaOH was added, as shown by the color change in the reactant solution from clear white to clear red. The structure and sizes of (ZnO-NPs) and (Se-NPs) were studied using a transmission electron microscope Jeol (JEM—1400 TEM). In addition, TEM imaging was performed in the TEM laboratory at the Research Park (CURP), Faculty of Agriculture, Cairo University, Cairo, Egypt.

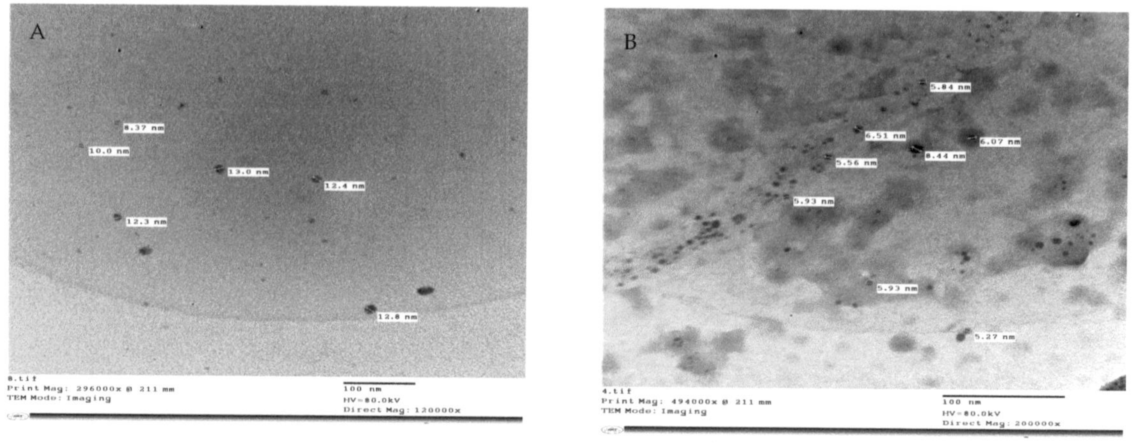

Figure 12. Transmission Electron Microscope (TEM) of (**A**) Zinc Oxide nanoparticles size from 8.37 to 12.8 nm, (**B**) selenium nanoparticles size from 5.27 to 6.51 nm. Scale bars = 100 nm.

4.5. Data Recorded

4.5.1. Plant Growth, Flowering, and Yield

Ten chili pepper plants were chosen from each treatment 60 days after transplanting to measure the plant height (cm) and the average number of leaves. In addition, a laser area meter (CI-202, Cedarhurst, NY, USA) was used to determine the leaf area (cm^2). To determine plant biomass, each fresh sample was weighed and dried at 70 °C in an oven until it reached a constant weight. Days to flowering following plant transplanting were estimated from second internode flowers, and the number of flowers was counted when 50% of the chili plants flowered; during harvest, chili fruits were harvested when they were marketable in size (fruit with no obvious flaws and brilliant green color as described by [28]. At each harvest, the fresh fruit weight per plant (g) and fruit number per plant were estimated, and the total yield per plant (kg) and total yield per square meter (m^2) were estimated after the fourth harvest.

4.5.2. Photosynthetic Parameters and Chlorophyll Content of Chili Plants

Photosynthetic parameters (transpiration rate, photosynthesis rate, stomatal conductance, and intercellular CO_2) were essayed using an infrared gas analyzer (LICOR 6400 Portable Photosynthesis System; IRGA, Licor Inc., Lincoln, NE, USA). To assess photosynthetic measurements, the fifth leaf of the chili plant was selected from each treatment 60 days after transplantation. Photosynthetic traits were evaluated between 10:00 a.m. and 1:30 p.m. with a light intensity of 1200 mol m^{-2} s^{-1} and 80% RH. Chlorophyll content was assessed using a SPAD meter (SPAD-502, Konica Minolta Sensing, Inc., Osaka, Japan) using the method described by [60], using the fifth leaf from each treatment.

4.5.3. Fruit Quality Parameters

At each harvest, four marketable chili fruits from each treatment were chosen to evaluate fruit diameter and length. The fruit diameter (cm) was assessed as the widest point in both perpendicular directions, and the two measurements were averaged to determine the fruit diameter. The fruit length (cm) was estimated from the tip point to the calyx. The total capsaicin content in the chili fruit was measured using high-performance liquid chromatography (HPLC) according to [61]. The ascorbic acid concentration was determined using a 2, 4 dinitrophenylhydrazine substance and a spectrophotometer at 540 nm, according to [62].

4.5.4. Mineral Content in Chili Leaves

Nutrient elements were determined by drying chili leaf samples at 70 °C for 24 h in an electric oven using the method outlined in [63]. Concentrated sulfuric acid (5 mL) was added to the samples, and the mixture was heated for 10 min to wet digest 0.2 g of the plant material. Perchloric acid (0.5 mL) was then added, and the mixture was heated until a clear solution was produced [63]. The total nitrogen content of the dried chili leaves was evaluated using the modified micro-Kjeldahl method, according to the authors of [63]. The phosphorus content of the dried chili leaves was estimated using the chlororstannousmolybdo phosphoric blue color method in sulfuric acid, as previously described [64]. Potassium content was estimated using a flame photometer apparatus (CORNING M 410, Germany), the calcium content in chili leaves was determined according to [65], and micronutrient concentrations (Zn, Se, Mn, Cu, and Fe) in chili leaves were estimated using an atomic absorption spectrophotometer with air-acetylene and fuel (PyeUnicam, model SP-1900, USA) [65,66].

4.5.5. Antioxidants Enzyme, Hormone, Lipid Peroxidation (MDA), Hydrogen Peroxide, and Nitrogen Metabolism Enzyme

The chili leaf samples were prepared using the method described in [67] to estimate peroxidase (POD)activity. Leaf tissue (0.5 g) was frozen in liquid nitrogen and used to extract peroxidase enzymes. The chili leaf samples were centrifuged at 3930× rpm for 20 min after being ground in 10 mL of extraction buffer (50 mM phosphate buffer, pH 7), which also contained 0.5 mM EDTA and 2% PVPP (*w*/*v*). Using a spectrophotometric method, the reaction mixture that included 450 mL of 25 mM guaiacol, 450 mL of 225 mM H_2O_2, and 100 mL of crude enzymes was used to produce the supernatant, which was then used to assess the peroxidase activity. The method described in [68] was employed to estimate the extraction of Abscisic acid (ABA) from chili leaves [68]. The homogenized freeze-dried materials were incorporated into a mixture that contained butylated hydroxytoluene and 80% *v*/*v* methanol. Malondialdehyde (MDA) levels in chili leaves were estimated using the thiobarbituric acid (TBA) method [69]. Hydrogen peroxide (H_2O_2) in chili leaves was evaluated using the trichloroacetic acid (TCA) method [70]. A total of 1 mL of 0.1% TCA was used to ground 0.1 g of the leaf. The homogenate was then centrifuged at 12,000× *g* rpm. The phosphate buffer (pH 7.0) and supernatant were combined in equal parts, and 1 mL of 1M potassium iodide was added. The absorbance of the reaction mixture was measured at 390 nm. The activity of glutamate dehydrogenase (GDH) was measured as described previously [71].

4.5.6. Protein Electrophoresis SDS-PAGE

Quick freeze-dried leaf samples (10 g) were extracted with 1 mL of protein buffer, maintained overnight in a freezer, vortexed for 5 min, and then centrifuged at 10,000× *g* rpm at 4 °C for 15 min. Electrophoretic analysis of proteins was performed using sodium dodecyl sulfate-polyacrylamide gel electrophoresis (SDS-PAGE). SDS-PAGE was performed as previously described [72]. Standard protein markers (20.7–320.23 kDa; Sigma-Aldrich) were used to measure the molecular weight of the isolated proteins. The protein bands were marked with Coomassie Brilliant Blue G-250 (Sigma–Aldrich). The molecular weight (Mwt) of each protein was related to standard markers using polyacrylamide gel plates, then photographed, scanned, and analyzed using Quantity One software (Version 4.6.2).

4.5.7. Anatomical of Chili Leaves

The tested specimens, including the upper leaf developed on the third branch from the base of the main stem of chili plants, were selected during the second growing season of 2022, 10 days after the third foliar application. A sample of approximately 1 cm of the specimens was taken from each treatment, killed, and fixed in FAA solution (5 mL of glacial acetic, 10 mL of formalin, 35 mL of water, and 50 mL of ethyl alcohol 70%), for 48 h at least. The chosen sample was washed in ethyl alcohol at 30%, dehydrated in a normal ethanol and butyl alcohol chain, buried in paraffin wax with a melting point of 56 °C, sliced to a thickness of 15 μm, stained with crystal violet and erythrosin, clarified in xylene, and mounted in Canada balsam. Transverse sections were obtained using a Leica Microtome RM 2125, micrographed, and measured using a Leica Light Image Analysis System DM 750 at the Research Park (CURP), Faculty of Agriculture, Cairo University. The following parameters were recorded: the thickness of the midvein (μm), lamina (μm), palisade tissue (μm), spongy tissue (μm), upper and lower epidermis (μm), bundles dimension (μm), and mean vessels diameter (μm).

4.5.8. Ultrastructure of Chili Leaves

Chili pepper leaf pieces at the age of 45 days from transplanting were processed for TEM by pre-fixing with 5% glutaraldehyde in 0.05 M phosphate buffer (pH 7.2) for 3 h, and post-fixed with 2% osmium tetroxide for 4 h, then dehydrated in ethanol alcohol, and mounted in epoxy resin. Semi-thin sections (500 nm) were obtained using a Leica Ultracut UCT Ultramicrotome, stained with toluidine blue. Ultra-thin sections were prepared at 100 nm thickness, stained with uranyl acetate and lead citrate in accordance with [73], and then examined using a transmission electron microscope Jeol (JEM—1400 TEM) at the candidate magnification at the TEM laboratory, Research Park (CURP), Faculty of Agriculture, Cairo University. Images were captured by CCD camera model AMT, an optronics camera with 1632 × 1632-pixel format as a side-mount configuration. This camera uses a 1394 firewire board for acquisition.

4.5.9. Statistical Analysis

MSTATC software was utilized to analyze the results of two successive growing seasons [74]. The LSD test was also performed to determine whether a statistical difference ($p \leq 0.05$) existed in the average values [75].

5. Conclusions

This study proves that cold stress can adversely affect the growth and productivity of chili plants. Nevertheless, according to SDS-PAGE analysis, the application of ZnO-NPs, Se-NPs, AMF + ZnO-NPs, and AMF + ZnO-NPs + Se-NPs provided plants with a defense mechanism against cold stress compared to the control. Chili plants inoculated with AMF combined with a mixture of nanoparticles (ZnO-NPs + Se-NPs) exhibited better transpiration rate, chlorophyll (SPAD), stomatal conductance, and photosynthetic rate, which subsequently led to higher vegetative growth, flowering, productivity, mineral components, antioxidant enzymes, and nitrogen metabolism enzymes. Conversely, the combination of AMF with a nanoparticle mixture (ZnO-NPs + Se-NPs) recorded the lowest content of ABA, MDA, and H_2O_2 in the first and second seasons. AMF combined with a nanoparticle mixture (ZnO-NPs + Se-NPs) is the best treatment, resulting in the highest number of chloroplasts and stacked grana thylakoids in chili plants. Finally, we recommend this scientific approach to farmers as a suitable strategy for reducing the harmful effects of cold stress on chili plants.

Author Contributions: Conceptualization, E.G.S., S.N.A. and S.F.D.; methodology, E.G.S., S.N.A., S.F.D. and A.F.D.; software E.G.S. and S.N.A.; validation E.G.S., A.F.D., R.I.E.-k. and A.F.D.; formal analysis E.G.S. and S.N.A.; investigation E.G.S., S.N.A. and A.F.D.; resources E.G.S., S.N.A. and S.F.D.; data curation, E.G.S., S.N.A. and S.F.D.; writing E.G.S. and S.N.A.—original draft preparation, E.G.S., S.N.A. and A.F.D.; writing—review and editing E.G.S., S.N.A. and S.F.D.; visualization E.G.S. and M.F.F.; supervision E.G.S.; project administration E.G.S. and S.F.D.; funding acquisition E.G.S., S.N.A., R.I.E.-k., A.F.D. and M.F.F. All authors have read and agreed to the published version of the manuscript.

Funding: This research was funded by Cairo University, the Faculty of Agriculture, Giza, Egypt.

Institutional Review Board Statement: Not applicable.

Informed Consent Statement: Not applicable.

Data Availability Statement: Data are contained within the article.

Acknowledgments: The authors are thankful to the Department of Vegetable Crops, Botany Department Faculty of Agriculture, Cairo University, for providing the equipment required to accomplish this research.

Conflicts of Interest: The authors declare no conflict of interest.

References

1. Mercado, A.J.; Mar, T.M.; Reid, M.S.; Valpuesta, V.; Quesada, M.A. Effects of low temperature on pepper pollen morphology and fertility: Evidence of cold induced exine alterations. *J. Hortic. Sci.* **1997**, *72*, 317–326. [CrossRef]
2. Ou, L.J.; Wei, G.; Zhang, Z.Q.; Dai, X.Z.; Zou, X.X. Effects of low temperature and low irradiance on the physiological characteristics and related gene expression of different pepper species. *Photosynthetica* **2015**, *53*, 85–94. [CrossRef]
3. Airaki, M.; Leterrier, M.; Mateos, R.M.; Valderrama, R.; Chaki, M.; Barroso, J.B.; del Río, L.A.; Palma, J.M.; Corpas, F.J. Metabolism of reactive oxygen species and reactive nitrogen species in pepper (*Capsicum annuum* L.) plants under low-temperature stress. *Plant Cell Environ.* **2012**, *35*, 281–295. [CrossRef]
4. Wang, X.; Mao, Z.; Zhang, J.; Hemat, M.; Huang, M.; Cai, J.; Zhou, Q.; Dai, T.; Jiang, D. Osmolyte accumulation plays important roles in the drought priming induced tolerance to post-anthesis drought stress in winter wheat (*Triticum aestivum* L.). *Environ. Exp. Bot.* **2019**, *166*, 103804. [CrossRef]
5. Jan, N.; Majeed, U.; Andrabi, K.I.; John, R. Cold stress modulates osmolytes and antioxidant system in Calendula officinalis. *Acta Physiol. Plant.* **2018**, *40*, 73. [CrossRef]
6. Megha, S.; Basu, U.N.K. Regulation of low-temperature stress in plants by microRNAs. *Plant Cell Environ.* **2017**, *41*, 12956. [CrossRef] [PubMed]
7. Karami, M.S.; Maali, A.R.; Kazemi, S.S. Effect of cold stress on oxidative damage and mitochondrial respiratory properties in chickpea. *Plant Physiol. Biochem.* **2018**, *122*, 31–39. [CrossRef] [PubMed]
8. Kumar, A.; Sharma, S.; Mishra, S.; Dames, J.F. Arbuscular mycorrhizal inoculation improves growth and antioxidative response of *Jatropha curcas* (L.) under Na_2SO_4 salt stress. *Plant Biosyst.* **2013**, *49*, 260–269. [CrossRef]
9. Kumar, A.; Sharma, S.; Mishra, S. Evaluating effect of arbuscular mycorrhizal fungal consortia and Azotobacter chroococcum in improving biomass yield of *Jatropha curcas*. *Plant Biosyst.* **2016**, *150*, 1056–1064. [CrossRef]
10. Ahanger, M.A.; Hashem, A.; AbdAllah, E.F.; Ahmad, P. Arbuscular mycorrhiza in crop improvement under environmental stress. In *Emerging Technologies and Management of Crop Stress Tolerance*; Academic Press: Cambridge, MA, USA, 2014; pp. 69–95. [CrossRef]
11. Hameed, A.; Wu, Q.S.; AbdAllah, E.F.; Hashem, A.; Kumar, A.; Lone, H.A.; Ahmad, P. Role of AM fungi in alleviating drought stress in plants. In *Use of Microbes for the Alleviation of Soil Stresses*; Miransari, M., Ed.; Springer Science Business Media: New York, NY, USA, 2014. [CrossRef]
12. Wahid, F.; Sharif, M.; Steinkellner, S.; Khan, M.A.; Marwat, K.B.; Khan, S.A. Inoculation of arbuscular mycorrhizal fungi and phosphate solubilizing bacteria in the presence of rock phosphate improves phosphorus uptake and growth of maize. *Pak. J. Bot.* **2016**, *48*, 739–747.
13. Piotrowski, J.S.; Denich, T.; Klironomos, J.N.; Graham, J.M.; Rillig, M.C. The effects of arbuscular mycorrhizas on soil aggregation depend on the interaction between plant and fungal species. *New Phytol.* **2004**, *164*, 365–373. [CrossRef] [PubMed]
14. Liu, J.; Wu, L.; Wei, S.; Xiao, X.; Su, C.; Jiang, P. Effects of arbuscular mycorrhizal fungi on the growth, nutrient uptake and glycyrrhizic acid production of licorice (*Glycyrrhiza uralensis* Fisch.). *Plant Growth Regul.* **2007**, *52*, 29–39. [CrossRef]
15. Ortas, I. The effect of mycorrhizal fungal inoculation on plant yield, nutrient uptake and inoculation effectiveness under long-term field conditions. *Field Crop. Res.* **2012**, *125*, 35–48. [CrossRef]
16. Acharya, P.; Jayaprakasha, G.K.; Crosby, K.M.; Jifon, J.L.; Patil, B.S. Greensynthesized nanoparticles enhanced seedling growth, yield, and quality of onion (*Allium cepa* L.). *ACS Sustain. Chem. Eng.* **2019**, *7*, 14580–14590. [CrossRef]
17. Ikram, M.; Raja, N.I.; Javed, B.; Mashwani, Z.R.; Hussain, M.; Hussain, M.; Ehsan, M.; Rafique, N.; Malik, K.; Sultana, T.; et al. Foliar applications of biofabricated selenium nanoparticles to improve the growth of wheat plants under drought stress. *Green. Process. Synth.* **2020**, *9*, 706–714. [CrossRef]
18. Hasanuzzaman, M. Silicon and selenium: Two vital trace elements that confer abiotic stress tolerance to plants. In *Emerging Technologies and Management of Crop Stress Tolerance*; Academic Press: San Diego, CA, USA, 2014; pp. 377–422.
19. Singh, R.; Upadhyay, A.K.; Singh, D.P. Regulation of oxidative stress and mineral nutrient status by selenium in arsenic treated crop plant *Oryza sativa*. *Ecotoxicol. Environ. Saf.* **2017**, *148*, 105–113. [CrossRef] [PubMed]
20. Huang, C.; Qin, N.; Sun, L.; Yu, M.; Hu, W.; Qi, Z. Selenium Improves Physiological Parameters and Alleviates Oxidative Stress in Strawberry Seedlings under Low-Temperature Stress. *Int. J. Mol. Sci.* **2018**, *19*, 1913. [CrossRef]
21. Feng, J.N.; Guo, X.P.; Chen, Y.R.; Lu, D.P.; Niu, Z.S.; Tou, F.Y.; Hou, L.J.; Xu, J.; Liu, M.; Yang, Y. Time-dependent effects of ZnO nanoparticles on bacteria in an estuarine aquatic environment. *Sci. Total Environ.* **2020**, *698*, 134298–134309. [CrossRef]
22. Rai-Kalal, P.; Jajoo, A. Priming with zinc oxide nanoparticles improve germination and photosynthetic performance in wheat. *Plant Physiol. Biochem.* **2021**, *160*, 341–351. [CrossRef]

23. Alharby, H.F.; Metwali, E.M.R.; Fuller, M.P.; Aldhebiani, A.Y. Impact of application of zinc oxide nanoparticles on callus induction, plant regeneration, element content and antioxidant enzyme activity in tomato (*Solanum lycopersicum* Mill.) under salt stress. *Arch. Biol. Sci.* **2016**, *68*, 723–735. [CrossRef]
24. Agarwal, H.; Kumar, S.V.; Rajeshkumar, S. A review on green synthesis of zinc oxide nanoparticles-An eco-friendly approach. *Resour. Effic. Technol.* **2017**, *3*, 406–413. [CrossRef]
25. Singh, A.; Singh, N.B.; Hussain, I.; Singh, H.; Yadav, V.; Singh, S.C. Green synthesis of nano zinc oxide and evaluation of its impact on germination and metabolic activity of *Solanum lycopersicum*. *J. Biotechnol.* **2016**, *233*, 84–94. [CrossRef] [PubMed]
26. Sharaf-Eldin, M.A.; Alshallash, K.S.; Alharbi, K.R.; Alqahtani, M.M.; Etman, A.A.; Yassin, A.M.; Azab, E.S.; El-Okkiah, S.A.F. Influence of Seed Soaking and Foliar Application Using Ozonated Water on Two Sweet Pepper Hybrids under Cold Stress. *Sustainability* **2022**, *14*, 13453. [CrossRef]
27. Hafez, O.M.; Saleh, M.A.; El-Lethy, S.R. Response of some seedlings olive cultivars to foliar spray of yeast and garlic extracts with or without vascular arbuscular mycorrhizal fungi. *J. World Appl. Sci.* **2013**, *24*, 1119–1129.
28. Latef, A.A.H.A.; Chaoxing, H. Effect of arbuscular mycorrhizal fungi on growth, mineral nutrition, antioxidant enzymes activity and fruit yield of tomatoes grown under salinity stress. *Sci. Hortic.* **2011**, *127*, 228–233. [CrossRef]
29. El-Badri, A.M.; Batool, M.; Wang, C.; Hashem, A.M.; Tabl, K.M.; Nishawy, E.; Kuai, J.; Zhou, G.; Wang, B. Selenium and zinc oxide nanoparticles modulate the molecular and morpho-physiological processes during seed germination of *Brassica napus* under salt stress. *Ecotoxicol. Environ. Saf.* **2021**, *225*, 112695. [CrossRef] [PubMed]
30. Hussein, H.A.A.; Darwesh, O.M.; Mekki, B.B. Environmentally friendly nano-selenium to improve the antioxidant system and growth of groundnut cultivars under sandy soil conditions. *Biocatal. Agric. Biotechnol.* **2019**, *18*, 101080. [CrossRef]
31. Nandagopal, J.G.T.; Harinarayanan, U.N.D.; Raghavan, S.; Girija, S. Foliar selenium application mitigates low-temperature stress in chilli (*Capsicum annuum* L.) seedlings. *Energy Nexus* **2022**, *6*, 100079. [CrossRef]
32. Sayed, E.G.; Mahmoud, A.W.M.; Abdel-Wahab, A.; El-bahbohy, R.M.; Azoz, S.N. Rootstock Priming with Shikimic Acid and *Streptomyces griseus* for Growth, Productivity, Physio-Biochemical, and Anatomical Characterisation of Tomato Grown under Cold Stress. *Plants* **2022**, *11*, 2822. [CrossRef]
33. Garcia, C.L.; Dattamudi, S.; Chanda, S.; Jayachandran, K. Effect of Salinity Stress and Microbial Inoculations on Glomalin Production and Plant Growth Parameters of Snap Bean (*Phaseolus vulgaris*). *Agronomy* **2019**, *9*, 545. [CrossRef]
34. El-Bassiouny, H.M.S.; Bakry, A.B.; Taha, M.H.; Younis, A.S.M. Alleviation of salt stress and improve crop productivity by using arbuscular mycorrhiza and ZnO-Nano or bulk particles in wheat. *Plant Arch.* **2019**, *19*, 205–214.
35. Ali, E.A.; Mahmoud, A.M. Effect of foliar spray by different salicylic acid and zinc concentrations on seed yield and yield components of mungbean in sandy soil. *Asian J. Crop Sci.* **2012**, *5*, 33–40. [CrossRef]
36. Smirnoff, N. Antioxidant systems and plant response to environment. In *Environmentand Plant Metabolism: Flexibility and Acclimation*; Smirnoff, N., Ed.; BIOS Scientific Publishers: Oxford, UK, 1995; pp. 217–243.
37. Uresti-Porras, J.G.; Cabrera, D.; Fuente, M.; Benavidez, M.A.; Olivares, S.E.; Cabrera, R.I.; Juárez, M.A. Effect of Graft and Nano ZnO on Nutraceutical and Mineral Content in Bell Pepper. *Plants* **2021**, *10*, 2793. [CrossRef]
38. Besserer, A.; Puech-Pagès, V.; Kiefer, P.; Gomez, R.V.; Jauneau, A.; Roy, S. Strigolactones stimulate arbuscular mycorrhizal fungi by activating mitochondria. *PLoS Biol.* **2006**, *4*, e226. [CrossRef] [PubMed]
39. El Lateef Gharib, F.A.; Zeid, I.M.; Ghazi, S.M.; Ahmed, E.Z. The response of cowpea (*Vigna unguiculata* L.) plants to foliar application of sodium selenate and selenium nanoparticles (SeNPs). *J. Nanomater. Mol. Nanotechnol.* **2019**, *8*, 4.
40. Zhu, X.; Song, F.; Liu, F. Arbuscular Mycorrhizal Fungi and Tolerance of Temperature Stress in Plants. In *Arbuscular Mycorrhizas and Stress Tolerance of Plants*; Springer Science and Business Media LLC: Berlin, Germany, 2017; Volume 33, pp. 163–194.
41. Vankova, R.; Landa, P.; Podlipna, R.; Dobrev, P.I.; Prerostova, S.; Langhansova, L.; Gaudinova, A.; Motkova, K.; Knirsch, V.; Vanek, T. ZnO nanoparticle effects on hormonal pools in *Arabidopsis thaliana*. *Sci. Total Environ.* **2017**, *593–594*, 535–542. [CrossRef]
42. Feng, R.; Wang, L.; Yang, J.; Zhao, P.; Zhu, Y.; Li, Y.; Yu, Y.; Liu, H.; Rensing, C.; Wu, Z. Underlying mechanisms responsible for restriction of uptake and translocation of heavy metals (metalloids) by selenium via root application in plants. *J. Hazard. Mater.* **2021**, *402*, 123570. [CrossRef] [PubMed]
43. Diab, H.; Limami, A.M. Metabolism upon Hypoxia Stress and Recovery: Roles of Alanine Aminotransferase (AlaAT) and Glutamate Dehydrogenase (GDH) Alternative route for nitrogen assimilation in higher plants. *Plants* **2016**, *5*, 25. [CrossRef] [PubMed]
44. Hajiboland, R.; Joudmand, A.; Aliasgharzad, N.; Tolrá, R.; Poschenrieder, C. Arbuscular mycorrhizal fungialleviate low-temperature stress and increase freezing resistance as a substitute for acclimation treatment in barley. *Crop. Pasture Sci.* **2019**, *70*, 218–233. [CrossRef]
45. Ebrahimian, E.; Bybordi, A. Exogenous silicium and zinc increase antioxidant enzyme activity and alleviate salt stress in leaves of sunflower. *J. Food Agric. Environ.* **2011**, *9*, 422–427.
46. Ouledali, S.; Ennajeh, M.; Ferrandino, A.; Khemira, H.; Schubert, A.; Secchi, F. Influence of arbuscular mycorrhizal fungi inoculation on the control of stomata functioning by abscisic acid (ABA) in drought-stressed olive plants. *S. Afr. J. Bot.* **2019**, *121*, 152–158. [CrossRef]
47. Xue, T.L.; Hou, S.F.; Tan, J.A.; Liu, G.L. The antioxidative function of selenium in higher plants: II. Non-enzymatic mechanisms. *Chin. Sci. Bull.* **1993**, *38*, 356–358.

48. Korkmazm, A.; Deger, O.; Szafranska, K.; Koklü, S.; Karaca, A.; Yakupoglu, G.; Kocacinar, F. Melatonin effects in enhancing chilling stress tolerance of pepper. *Sci. Hortic.* **2022**, *289*, 110434. [CrossRef]
49. Khater, M.; El-Awadi, M.; Elashtokhy, M.; Abdel-Baky, Y.; Shalaby, M. Physiological and Molecular Changes in Fenugreek (*Trigonella foenum-graecum* L.) as a Response to Gamma Rays. *Int. J. Pharm. Tech. Res.* **2016**, *9*, 306–316.
50. Abedi, T.; Alemzadeh, A.; Kazemeini, S.A. Wheat Yield and Grain Protein Response to Nitrogen Amount and Timing. *Aust. J. Crop Sci.* **2011**, *5*, 330–336.
51. Bassuony, F.; Hashem, H.A.; Hassanein, R.; Baraka, D.; Khalil, R.R. The Impact of Stigmasterol on Growth, Productivity and Biochemical Response of *Vicia faba* L. Plants Grown under Salt Stress. *Egypt. J. Bot.* **2014**, *54*, 203–218.
52. Liu, Y.; Lu, J.; Cui, L.; Tang, L.; Ci, D.; Zou, X.; Zhang, X.; Yu, X.; Wang, Y.; Si, Y. The multifaceted roles of Arbuscular Mycorrhizal Fungi in peanut responses to salt, drought, and cold stress. *BMC Plant Biol.* **2023**, *23*, 36. [CrossRef] [PubMed]
53. Ganguly, R.; Sarkar, A.; Dasgupta, D.; Acharya, K.; Keswani, C.; Popova, V.; Minkina, T.; Maksimov, A.; Chakraborty, N. Unravelling the Efficient Applications of Zinc and Selenium for Mitigation of Abiotic Stresses in Plants. *Agriculture* **2022**, *12*, 1551. [CrossRef]
54. Mahdi, A.; Abd, A.; Awad, K. Effect of Foliar Application of Nano-selenium on the Anatomical Characteristics of Date Palm *Phoenix dactylifera* L. Barhi Cultivar under salt stress. *Basrah J. Agric. Sci.* **2022**, *35*, 313–325. [CrossRef]
55. Zhang, W.; Huang, Z.; Wang, Q.; Guan, Y. Effects of Low Temperature on Leaf Anatomy and Photosynthetic Performance in Different Genotypes of Wheat Following a Rice Crop. *Int. J. Agric. Biol.* **2015**, *17*, 1165–1171. [CrossRef]
56. Kang, P.; Wu, Y.; Zhang, J.; An, Q.; Zhou, C.; Li, D.; Pan, C. Nano-selenium enhances the antioxidant capacity, organic acids and cucurbitacin B in melon (*Cucumis melo* L.) plants. *Ecotoxicol. Environ. Saf.* **2022**, *241*, 113777. [CrossRef]
57. Venzhik, Y.; Shchyogolev, S.; Dykman, L. Ultrastructural Reorganization of Chloroplasts during Plant Adaptation to Abiotic Stress Factors. *Russ. J. Plant Physiol.* **2019**, *66*, 850–863. [CrossRef]
58. Ejsmont, A.; Goscianska, J. Hydrothermal Synthesis of ZnO Superstructures with Controlled Morphology via Temperature and pH Optimization. *Materials* **2023**, *16*, 1641. [CrossRef] [PubMed]
59. Shahabadi, N.; Zendehcheshm, S.; Khademi, F. Selenium nanoparticles: Synthesis, in-vitro cytotoxicity, antioxidant activity and interaction studies with ct-DNA and HSA, HHb and Cyt c serum proteins. *Biotechnol. Rep.* **2021**, *30*, e00615. [CrossRef]
60. Boyhan, G.E.; McGregor, C.; O'Connell, S.; Biang, J.; Berle, D.A. Comparison of 13 sweet pepper varieties under an organic farming system. *HortTechnology* **2020**, *30*, 135–143. [CrossRef]
61. Thaib, N.; Katja, D.G.; Aritonang, H.F. Isolasi capsaicin dari oleoresin cabairawit (*Capsicum frutescens* L.). *Chem. Prog.* **2015**, *8*, 71–76.
62. Kumar, O.A.; Tata, S.S. Ascorbic Acid Contents in Chili Peppers (*Capsicum* L.). *Not. Sci. Biol.* **2009**, *1*, 50–52. [CrossRef]
63. Helrich, K. *Official Methods of Analysis*, 15th ed.; Association of Official Agricultural Chemist: Arlington, VA, USA, 1990; Volume 1, p. 673.
64. Singleton, V.L.; Rossi, J.A. Colorimetry of total phenolics with phosphomolybdicphosphotungstic acid reagents. *Am. J. Enol. Vitic.* **1965**, *16*, 144–158. [CrossRef]
65. Alfosea, S.M.; Simón, G.S.; Zavala, G.E.A.; Cámara, Z.J.M.; Simón, I.; Martínez, N.J.J.; Lidón, V.; García, S.F. Physiological, nutritional and metabolomic responses of tomato plants after the foliar application of amino acids Aspartic acid, Glutamic Acid and Alanine. *Front. Plant Sci.* **2020**, *11*, 581234. [CrossRef] [PubMed]
66. Jackson, M.L. *Soil Chemical Analysis*; Text Book; Printice-Hall of India Privat Limited: New Delhi, India, 1973; Volume 144–197, p. 381.
67. Polle, A.; Otter, T.; Mehne-Jakobs, B. Effect of magnesium deficiency on antioxidative systems in needles of Norway spruce [*Picea abies* (L.) Karst.] grown with different ratios of nitrate and ammonium as nitrogen souices. *New Phytol.* **1994**, *128*, 621–628. [CrossRef]
68. Fales, T.M.; Jaouni, J.F.; Babashak, I. Simple device for preparing ethereal diazomethane without resorting to distillation. *Ann. Chem.* **1973**, *45*, 2302–2303. [CrossRef]
69. Velikova, V.; Yordanov, I.; Edreva, A. Oxidative stress and some antioxidant systems in acid rain-treated bean plants protective role of exogenous polyamines. *Plant Sci.* **2000**, *151*, 59–66. [CrossRef]
70. Ashraf, M.A.; Rasheed, R.; Hussain, I.; Iqbal, M.; Haider, M.Z.; Parveen, S.; Sajid, M.A. Hydrogen peroxide modulates antioxidant system and nutrient relation in maize (*Zea mays* L.) under water-deficit conditions. *Arch. Agron. Soil Sci.* **2015**, *61*, 507–523. [CrossRef]
71. Majerowicz, N.; Kerbauy, G.; Nievola, C.; Suzuki, R. Growth and nitrogen metabolism of *Catasetum fimbriatum* (*Orchidaceae*) grown with different nitrogen sources. *Environ. Exp. Bot.* **2000**, *44*, 195–206. [CrossRef]
72. Laemmli, U.K. Cleavage of Structural Proteins during the Assembly of the Head of Bacteriophage T4. *Nature* **1970**, *227*, 680–685. [CrossRef] [PubMed]
73. Hunter, E. *Paractical Electron Microscopy: A Beginners Illustrated Guide*, 2nd ed.; Cambridge University Press: Cambridge, NY, USA, 1993.
74. Bricker, B. *MSTATC: A Micro Computer Program from the Design Management and Analysis of Agronomic Research Experiments*; Michigan State University: East Lansing, MI, USA, 1991.
75. Snedecor, G.W.; Cochran, W.G. *Statistical Methods*, 7th ed.; Iowa State University Press: Ames, IA, USA, 1980; p. 507.

plants

MDPI

Article

Beneficial Effect of Exogenously Applied Calcium Pyruvate in Alleviating Water Deficit in Sugarcane as Assessed by Chlorophyll a Fluorescence Technique

Mirandy dos Santos Dias [1,*], Francisco de Assis da Silva [1], Pedro Dantas Fernandes [1], Carlos Henrique de Azevedo Farias [1], Robson Felipe de Lima [1], Maria de Fátima Caetano da Silva [1], Vitória Régia do Nascimento Lima [1], Andrezza Maia de Lima [1], Cassiano Nogueira de Lacerda [1], Lígia Sampaio Reis [2], Weslley Bruno Belo de Souza [1], André Alisson Rodrigues da Silva [1] and Thiago Filipe de Lima Arruda [1]

1 Unidade Acadêmica de Engenharia Agrícola—UAEA, Centro de Tecnologia e Recursos Naturais—CTRN, Universidade Federal de Campina Grande–UFCG, Campus Campina Grande, Campina Grande 58428-830, PB, Brazil; agrofdsilva@gmail.com (F.d.A.d.S.); pedrodantasfernandes@gmail.com (P.D.F.); cahique.proojet@gmail.com (C.H.d.A.F.); robson.felipe@estudante.ufcg.edu.br (R.F.d.L.); maria.caetano@estudante.ufcg.edu.br (M.d.F.C.d.S.); regia7665@gmail.com (V.R.d.N.L.); andrezzamaia2010@hotmail.com (A.M.d.L.); cassianonogueiraagro@gmail.com (C.N.d.L.); weslley.bruno@estudante.ufcg.edu.br (W.B.B.d.S.); andre.alisson@estudante.ufcg.edu.br (A.A.R.d.S.); thiago.filipe@estudante.ufcg.edu.br (T.F.d.L.A.)

2 Campus de Engenharias e Ciências Agrárias—CECA, Universidade Federal de Alagoas—UFAL, Rio Largo 57100-000, AL, Brazil; lavenere_reis@hotmail.com

* Correspondence: mirandy.santos@estudante.ufcg.edu.br; Tel.: +55-82-99655-0778

Citation: Dias, M.d.S.; da Silva, F.d.A.; Fernandes, P.D.; Farias, C.H.d.A.; de Lima, R.F.; da Silva, M.d.F.C.; Lima, V.R.d.N.; de Lima, A.M.; de Lacerda, C.N.; Reis, L.S.; et al. Beneficial Effect of Exogenously Applied Calcium Pyruvate in Alleviating Water Deficit in Sugarcane as Assessed by Chlorophyll a Fluorescence Technique. *Plants* **2024**, *13*, 434. https://doi.org/10.3390/plants13030434

Academic Editor: Anja Schneider

Received: 23 December 2023
Revised: 26 January 2024
Accepted: 29 January 2024
Published: 1 February 2024

Abstract: The growing demand for food production has led to an increase in agricultural areas, including many with low and irregular rainfall, stressing the importance of studies aimed at mitigating the harmful effects of water stress. From this perspective, the objective of this study was to evaluate calcium pyruvate as an attenuator of water deficit on chlorophyll *a* fluorescence of five sugarcane genotypes. The experiment was conducted in a plant nursery where three management strategies (E1—full irrigation, E2—water deficit with the application of 30 mM calcium pyruvate, and E3—water deficit without the application of calcium pyruvate) and five sugarcane genotypes (RB863129, RB92579, RB962962, RB021754, and RB041443) were tested, distributed in randomized blocks, in a 3 × 5 factorial design with three replications. There is dissimilarity in the fluorescence parameters and photosynthetic pigments of the RB863129 genotype in relation to those of the RB041443, RB96262, RB021754, and RB92579 genotypes. Foliar application of calcium pyruvate alleviates the effects of water deficit on the fluorescence parameters of chlorophyll *a* and photosynthetic pigments in sugarcane, without interaction with the genotypes. However, subsequent validation tests will be necessary to test and validate the adoption of this technology under field conditions.

Keywords: *Saccharum officinarum* L.; water scarcity; water deficit mitigation

1. Introduction

The growing demand for food production has led to an increase in the number of agricultural areas, including the occupation of spaces that have never been farmed before. Sugarcane (*Saccharum officinarum* L.) is a crop that has been widely cultivated in these areas for many years, showing global importance and being used to produce sugar, biofuels, and energy [1]. However, due to its long growth cycle, sugarcane requires an abundant rainfall distribution (1850 to 2500 mm year^{-1}), showing greater sensitivity to water deficit during tillering and stem elongation [2,3].

For all crops, global warming and climate change increase the frequency and intensity of abiotic stress factors on a worldwide level, especially water stress [4,5]. In such conditions, the structure of the plant chloroplast is affected by reductions in the chlorophyll content caused by the photooxidation of photosynthetic pigments [6]. As a consequence, a smaller proportion of the incident energy is used to produce adenosine triphosphate (ATP) and nicotinamide adenine dinucleotide phosphate (NADPH), resulting in photoinhibition [7,8], negatively affecting the photosynthetic capacity of plants.

The photosynthetic efficiency of plants is directly related to several factors, including the chlorophyll and carotenoid content, whose leaf values indicate the damage that certain stresses can cause to the photosynthetic apparatus [9]. Furthermore, the evaluation of chlorophyll *a* fluorescence is essential in analyzing photochemical efficiency and damage to the photosynthetic system, providing important information about the inhibition or reduction in electron transfer between photosystems [10,11].

Although the negative effects of water stress on fluorescence emission and the photosynthetic pigment content of sugarcane plants are well reported in the literature [12–15], there are reports that organic substances can mitigate the harmful effects of drought by increasing the ability of plants to withstand such deleterious conditions, e.g., studies of pyruvate application carried out with peanuts [16] and cotton [17,18]. However, research focusing on the photosynthetic apparatus of sugarcane plants has not yet been carried out.

In the metabolism of C4 plants, pyruvate is responsible for the regeneration of phosphoenolpyruvate, an HCO^{3-} acceptor in mesophyll cells [19]. Furthermore, in the Krebs Cycle metabolism, this substance plays a vital role in converting glucose into energy, in a process in which glucose is broken down into two pyruvic acid molecules that, in the Krebs Cycle, are transformed into energy (ATP) [19]. In this context, it becomes relevant to test technologies that aim to alleviate the harmful effects of water deficit on the photosynthetic apparatus of sugarcane. From this perspective, this study aimed to evaluate calcium pyruvate as an attenuator of water deficit on chlorophyll *a* fluorescence in five sugarcane genotypes.

2. Results and Discussion

According to the F-test (Table 1), the initial fluorescence (Fo), the maximum fluorescence (Fm), the variable fluorescence (Fv), the maximum quantum efficiency of PSII (Fv/Fm), the photochemical efficiency (Fv/Fo), the basal quantum efficiency of the non-photochemical process (Fo/Fm), the initial fluorescence before the saturation pulse (F′), the maximum fluorescence after adaptation to saturating light (Fm′), the electron transport rate (ETR), the quantum efficiency of PSII (Y), the Stern–Volmer non-photochemical quenching (NPQ), the complete non-photochemical quenching of chlorophyll fluorescence (QCN), the quantum yield of non-regulated photochemical quenching (YNO), the quantum yield of regulated photochemical quenching (YNPQ), chlorophyll *a* (Chl *a*), chlorophyll *b* (Chl *b*), and carotenoids (Car) were significantly influenced by management strategies (E) ($p < 0.01$). For the genotypes, there was a significant effect on most of the fluorescence parameters studied, with the exception of Fv/Fm, Fv/Fo, Fo/Fm, Y, and YNPQ. The interaction between the management strategies (E) and genotypes (G) was not significant (Table 1).

An increase in Fo was observed (Figure 1A), reaching a mean value of 296.7 electrons quantum^{-1} in irrigated plants; this is significantly lower than the value recorded in E2 (415.8 electrons quantum^{-1}) and E3 (473.4 electrons quantum^{-1}), corresponding to an increase of 40.14% and 59.56%, respectively. As a result, there was a reduction of 178.4 quantum^{-1} electrons in the maximum fluorescence (−13.47%) and 355.1 electrons quantum^{-1} in the variable fluorescence (−34.55%), when related to the same treatments (Figure 1B,C). The increase in Fo and the reduction in Fv and Fm significantly affected the Fv/Fm and Fv/Fo ratios, with reductions of 15.38% and 24.60% (Fv/Fm) and 44.03% and 59.01% (Fv/Fo), respectively, for E2 and E3 in relation to E1, indicating that the photosynthetic apparatus was compromised (Figure 1D,E). Furthermore, there was an increase in the Fo/Fm ratio of 54.54% and 86.66% when related to the same treatments (Figure 1F). However, when plants that were subjected to water deficit received pyruvate application (E2), there was a reduction in Fo (10.05%) and the Fo/Fm ratio (17.07%) and, consequently, an increase in Fm (7.08%), Fv (20.63%), and the Fv/Fm ratio (10.61%), which was statistically different to plants that did not receive calcium pyruvate (E3).

Table 1. F-test for chlorophyll *a* fluorescence and photosynthetic pigments in sugarcane genotypes under management strategies, 211 days after regrowth (DAR).

Sources of Variation	**F-Test**									
	Fo	Fm	Fv	Fv/Fm	Fv/Fo	Fo/Fm	F′	Fm′	ETR	Y
Strategies (E)	**	**	**	**	**	**	**	**	**	**
Genotype (G)	*	**	*	ns	ns	ns	*	**	**	ns
G × E	ns	ns	ns	ns	ns	ns	ns	ns	ns	ns
Block	*	**	ns	ns	ns	ns	ns	ns	ns	ns
CV (%)	9.85	4.16	6.53	4.36	16.62	9.03	15.90	12.06	9.85	17.98
	NPQ	QCN	YNO	YNPQ	Chl *a*	Chl *b*	Car			
Strategies (E)	**	**	**	**	**	**	**			
Genotype (G)	**	**	*	ns	**	**	**			
G × E	ns	ns	ns	ns	ns	ns	ns			
Block	ns	ns	ns	ns	ns	ns	ns			
CV (%)	14.96	2.96	18.69	19.92	18.50	17.91	13.83			

*, **, significant at 1% and 5%, respectively. ns—not significant. CV (%)—coefficient of variation.

The Fo increase in plants under water deficit may be the result of damage to the PSII reaction center or a reduced capacity to transfer the excitation energy from the antenna to the reaction center [20,21]. According to Rosseau et al. [22], Fo represents the status of the plant when the PSII reaction center is oxidized. An increase in this parameter directly reflects on Fm and Fv, which affects the flow of electrons between photosystems, thus decreasing the plant's ability to transfer energy to the formation of NADPH and ATP [10,21].

The maximum quantum efficiency of PSII (Fv/Fm ratio) expresses the capture efficiency of the excitation energy by the open reaction centers of PSII [23,24]. Under normal conditions, the Fv/Fm ratio can vary from 0.75 to 0.85, with drops in this ratio indicating structural damage to the thylakoids, affecting photochemical efficiency, CO_2 assimilation, and, above all, electron transport [23,25]. Values below 0.75 quantum^{-1} electrons indicate that damage is occurring to the plant's photosynthetic system. Reductions in the Fv/Fm ratio under water deficit conditions were recorded by Silva et al. [26] when studying the physiological parameters of four sugarcane genotypes (RB72454, RB72910, RB92579, and RB867515) under greenhouse conditions, with the authors finding a mean value of 0.71 electrons quantum^{-1} for the Fv/Fm ratio.

The lowest Fv/Fo ratio value in the E3 management strategy occurred due to a reduction in the variable fluorescence (Fv) or an increase in the initial fluorescence (Fo), indicating a change in the rate of electron transport from PSII to the primary electron acceptors [27]. Furthermore, the increase in the Fo/Fm ratio and Fo associated with the reduction in the Fv/Fm and Fv/Fo ratios suggests the occurrence of photoinhibition in plants cultivated under the E3 management strategy. According to Silva et al. [24], photoinhibition causes a slow reduction in photosynthesis, indicating that the plant was subjected to a stressful environment.

Therefore, maintaining high Fv/Fm and Fv/Fo ratios and a low Fo/Fm ratio improves the efficiency of radiation use and facilitates CO_2 assimilation [28]. In this context, although the plants that received pyruvate application obtained only 0.66 electrons quantum^{-1}, not reaching the ideal values of Fv/Fm (0.75 to 0.84 electrons quantum^{-1}), the supplementation of this compound improved the light absorption system of PSII compared with plants that did not receive this product (0.59 electrons quantum^{-1}), suggesting that pyruvate alleviated the deleterious effects of water deficit.

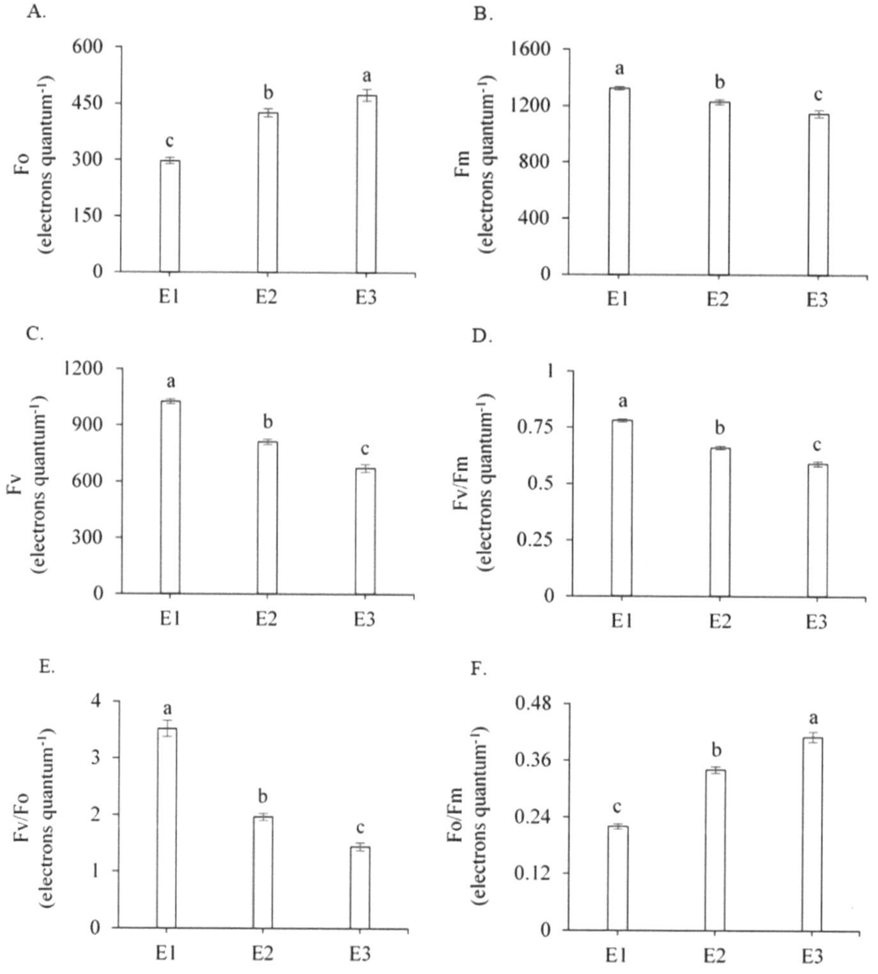

Figure 1. Mean values for initial fluorescence (Fo) (**A**), maximum fluorescence (Fm) (**B**), variable fluorescence (Fv) (**C**), maximum quantum efficiency of PSII (Fv/Fm) (**D**), photochemical efficiency (Fv/Fo) (**E**), and basal quantum efficiency of the non-photochemical process (Fo/Fm) (**F**) of sugarcane as a function of the three management strategies (E), 211 days after regrowth. Different lowercase letters indicate a statistical difference using the Tukey test ($p \leq 0.05$). Vertical bars represent the standard error of the mean (n = 3). E1—full irrigation, E2—water deficit plus 30 mM of calcium pyruvate, E3—water deficit without calcium pyruvate.

When studying the fluorescence parameters, an increase of 42.48% was observed in the F′ parameter in sugarcane plants under water deficit and without pyruvate application (E3) in relation to treatment E1, and 14.86% in relation to treatment E2 (Figure 2A). In the Fm′, ETR, and Y parameters in plants subjected to treatment E3, the reductions were 30.44%, 69.15%, and 46.97%, respectively, in relation to treatment E1, and 20.26%, 46.23%, and 32.69%, respectively, in relation to treatment E2 (Figure 2B–D).

These results indicate that, without the application of calcium pyruvate (E3), water deficit stress negatively influenced all the chlorophyll fluorescence parameters. These results corroborate those obtained by Souza et al. [13], Leanasawat et al. [15], and Verma et al. [29] in sugarcane under drought conditions, which recorded reductions in fluorescence emission levels.

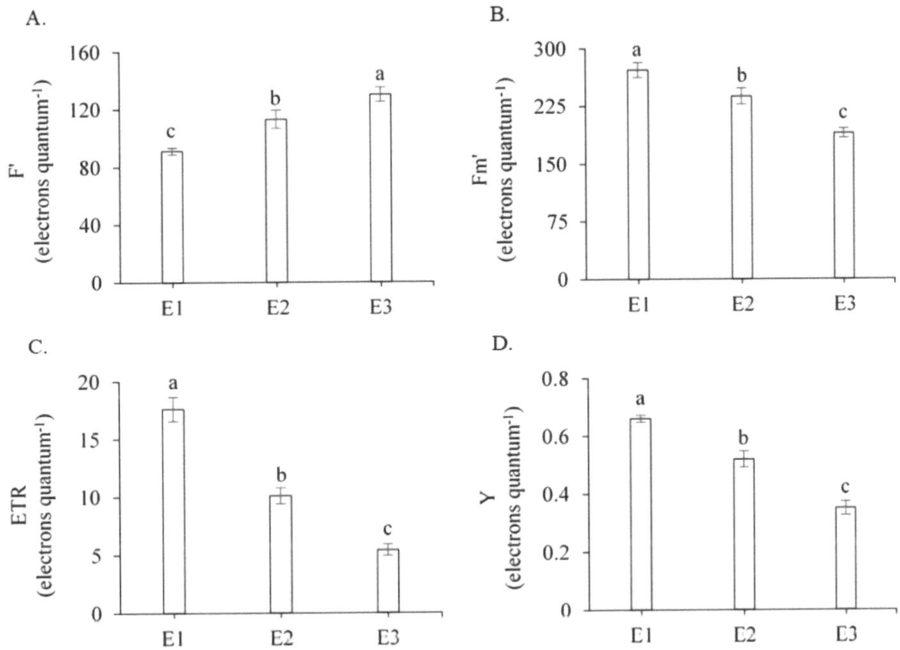

Figure 2. Mean values of initial fluorescence before the saturation pulse (F′) (**A**), maximum fluorescence after adaptation to saturating light (Fm′) (**B**), electron transport rate (ETR) (**C**), and quantum efficiency of PSII (Y) (**D**) of sugarcane as a function of the three management strategies (E), 211 days after regrowth. Different lowercase letters indicate statistical difference using the Tukey test ($p \leq 0.05$). Vertical bars represent the standard error of the mean (n = 3). E1—full irrigation, E2—water deficit plus 30 mM of calcium pyruvate, E3—water deficit without calcium pyruvate.

In addition to the Fv/Fm ratio, the ETR is also considered an important parameter to determine the efficiency of the photosynthetic apparatus [20]. In this research, a negative effect of water deficit on the electron transport rate was observed (Figure 2C), notably in plants that did not receive the foliar application of calcium pyruvate, indicating a reduced capacity for electron transport and compromised production of ATP and NADPH during the continuity of the photosynthetic process.

Through the analysis of the quenching coefficients, it was observed that water deficit during tillering and stalk elongation in sugarcane affected the photosynthetic system, with a significant increase in the NPQ (39.48%), QCN (5.06%), YNO (67.65%), and YNPQ (118.51%) parameters compared with management strategy E1 (Figure 3A–D). However, plants subjected to strategy E2 showed significant reductions ($p \leq 0.05$) in the NPQ (19.74%), QCN (2.41%), YNO (18.42%), and YNPQ (33.89%) parameters compared with plants subjected to water deficit without calcium pyruvate (Figure 3A–D).

Chlorophylls are excited when plants receive sunlight, making these pigments highly reactive. If there is no attenuation, they will become an element that generates oxidative stress [30]. This occurs because chlorophylls do not stop absorbing light; therefore, all excess light received must be dissipated. If this dissipation is carried out in the form of heat (NPQ), there is an increase in the QCN and YNO parameters, consequently increasing the YNPQ parameter, which corroborates the results found in this study, as the plants subjected to irrigation strategy E3 had reduced Fv/Fm ratio, Y, and ETR parameters.

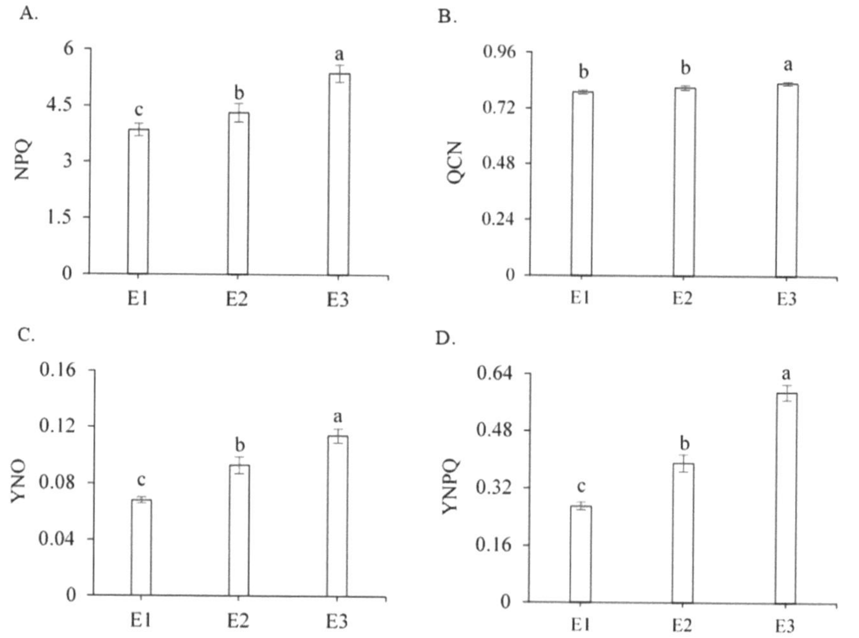

Figure 3. Mean values for Stern–Volmer non-photochemical quenching (NPQ) (**A**), complete non-photochemical quenching of chlorophyll fluorescence (QCN) (**B**), quantum yield of non-regulated photochemical quenching (YNO) (**C**), and quantum yield of regulated photochemical quenching (YNPQ) (**D**) in sugarcane as a function of the three management strategies (E), 211 days after regrowth. Different lowercase letters indicate a statistical difference using the Tukey test ($p \leq 0.05$). Vertical bars represent the standard error of the mean (n = 3). E1—full irrigation, E2—water deficit plus 30 mM of calcium pyruvate, E3—water deficit without calcium pyruvate.

Such reductions resulted in increased values of the Stern–Volmer non-photochemical quenching (NPQ), complete non-photochemical quenching of chlorophyll fluorescence (QCN), and quantum yield of non-regulated photochemical quenching (YNO) parameters, possibly triggering oxidative processes in the PSII of plants. In this situation, plants need to dissipate the excess absorbed light energy as heat since it exceeds their normal use for driving photosynthesis and electron transfer, thus increasing the quantum yield of regulated photochemical quenching (YNPQ) parameter. However, it appears that foliar supplementation with calcium pyruvate in plants under water deficit (E2) significantly improved all the studied chlorophyll fluorescence parameters.

Photosynthetic pigments play a fundamental role in the growth and development of plants, being responsible for the transmission of light energy for the production of photoassimilates [31]. In the present study, the levels of chlorophyll *a* (Chl *a*), chlorophyll *b* (Chl *b*), and carotenoids in sugarcane leaves were significantly reduced when plants were cultivated under water deficit in the tillering and stalk elongation phases (E2 and E3) compared with plants under full irrigation (E1) (Figure 4). However, the plants that received calcium pyruvate had increased levels of chlorophyll *a* (Figure 4A), *b* (Figure 4B), and carotenoids (Figure 4C), by 55.40%, 43.80%, and 55.70%, respectively, compared with the treatment without calcium pyruvate under water deficit (E3), expressing greater tolerance to this stress condition through the maintenance of photosynthetic pigments.

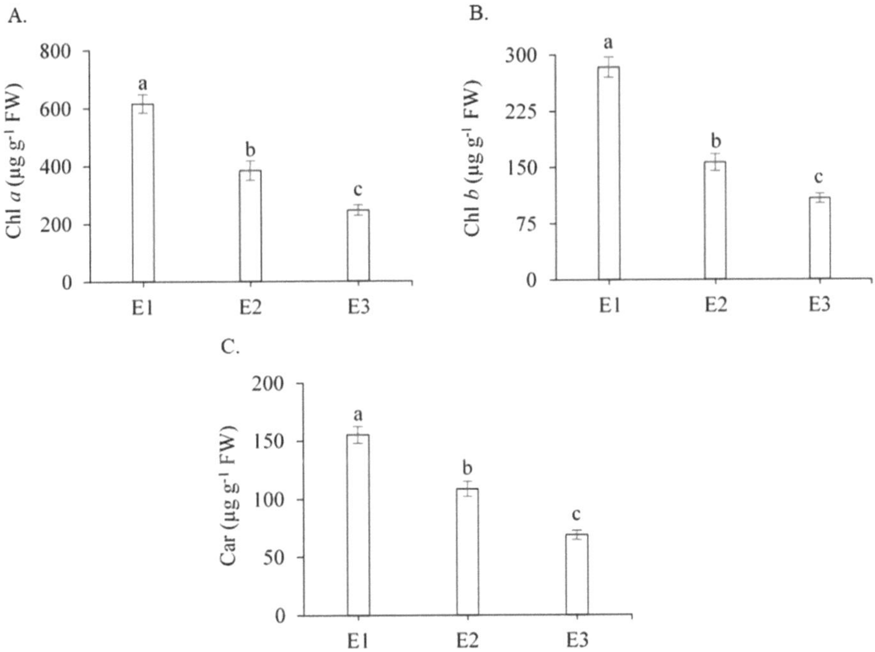

Figure 4. Mean values for chlorophyll *a* (**A**), chlorophyll *b* (**B**), and carotenoids (**C**) of sugarcane according to the three management strategies (E), 211 days after regrowth. Different lowercase letters indicate statistical difference using the Tukey test ($p \leq 0.05$). Vertical bars represent the standard error of the mean (n = 3). E1—full irrigation, E2—water deficit plus 30 mM of calcium pyruvate, E3—water deficit without calcium pyruvate.

The degradation in the levels of chlorophyll *a*, chlorophyll *b*, and carotenoids may be due to the reduced biosynthesis of chlorophyll resulting from photochemical disturbances caused by excess light in the PSII reaction centers [32]. Carotenoids are non-enzymatic antioxidants that target the excessive accumulation of reactive oxygen species [33]. Possibly, the low carotenoid content contributed to reducing the photosynthetic activity during the period of water deficit and increasing photodegradation, which resulted in a low Fv/Fm ratio. Souza et al. [13] also found reductions in fluorescence emission levels and photosynthetic pigment content in genotypes RB855536 and RB93509 that were subjected to water restriction during early growth.

In addition to sugarcane, there are also reports in the literature about the degradation of photosynthetic pigments in several crops under water deficit, e.g., wheat [34], sorghum [35], and maize [36]. Furthermore, although research focuses only on the harmful effects of water deficit, there are reports that organic substances can reduce the harmful effects of drought on plants. For example, Verma et al. [29] found that the foliar application of silicon, in addition to reducing the harmful effects of water deficit in sugarcane, also improves the plant's antioxidant defense system, by favoring the synthesis of photosynthetic pigments and the maximum quantum efficiency of photosystem II. In another study, Maia Júnior et al. [37] observed that, in sugarcane plants, the foliar application of glycine betaine mitigates the harmful effects of water deficit on the PSII photochemical apparatus.

In addition to these studies, Dias et al. [38] found that the effects of water deficit on the tillering and stalk elongation phases in sugarcane are alleviated by the exogenous application of 30 mM calcium pyruvate.

2.1. Multivariate Data Analysis

To corroborate the results presented in the univariate analysis, a multivariate analysis of the data was carried out using a cluster analysis and a principal component analysis.

2.1.1. Cluster Analysis

In order to verify the relationship between the genotypes studied, a cluster analysis was carried out by applying the Euclidean Distance (ED) as a measure of dissimilarity. As a subjective criterion for visual inspection, a cutoff was established between 4.0 and 5.0. For this analysis, only the variables with a significant difference between the genotypes were used, as observed in the F-test (Table 1). Three groups were formed: group 1 was formed by genotypes RB041443 and RB962962, group 2 by RB021754 and RB92579, and group 3 by RB863129 (Figure 5). These groups were characterized by greater homogeneity between the genotypes of each group and greater heterogeneity between the groups in relation to the parameters analyzed.

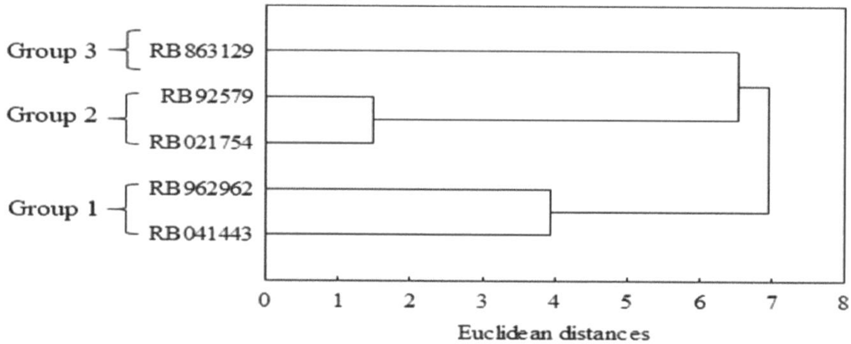

Figure 5. Clustering of sugarcane genotypes assembled with variables according to the isolated effect for genotypes.

2.1.2. Principal Component Analysis

The first two principal components explained 91.66% of the variance contained in the original variables. Furthermore, factor loadings with an absolute value greater than 0.60 were considered relevant. The first principal component (PC1) contributed 79.65% of the total variance, and the second component (PC2) contributed 12.01% of the remaining variance (Table 2).

The two-dimensional projections of the combinations of the sugarcane genotypes and irrigation management strategies on the first two PCs are illustrated in Figure 6A,B. The first principal component was formed from the combination of 15 physiological parameters, which was divided into two groups: group 1, formed by the parameters Fo, Fo/Fm ratio, F′, YNPQ, and YNO; and group 2, formed by the parameters Fm, Fv, Fv/Fm ratio, Fv/Fo ratio, Fm′, Y, ETR, Chl *a*, Chl *b*, and carotenoids. The groups formed in this component presented variations in opposite directions, i.e., as the parameters of group 1 increased, the parameters of group 2 decreased the physiological activity of the sugarcane genotypes. These changes were more significant in plants subjected to the E3 irrigation management strategy. However, when receiving calcium pyruvate application (E2), reductions were observed in the Fo, Fo/Fm ratio, F′, YNPQ, and YNO parameters and increases in were observed in the Fm, Fv, Fv/Fm ratio, Fv/Fo ratio, Fm′ Y, ETR, Chl *a*, Chl *b*, and carotenoids parameters in all genotypes in relation to the pyruvate-free strategy (E3). The second component is represented by the non-photochemical parameters, QCN and NPQ, which correlated proportionally.

Table 2. Eigenvalues, percentage of total variance explained, and correlation coefficients (r) between the original variables and the principal components.

	Principal Components	
	PC1	**PC2**
Eigenvalues (λ)	13.54	2.02
Percentage of Total Variance (S^2%)	79.65	12.01

PCs	Correlation coefficient (r)								
	Fo	Fm	Fv	Fv/Fm	Fv/Fo	Fo/Fm	F′	Fm′	Y
PC1	0.94 *	−0.80 *	−0.96 *	−0.97 *	−0.96 *	0.97 *	0.76 *	−0.88 *	−0.97 *
PC2	−0.11	−0.40	−0.18	−0.04	−0.10	−0.04	0.50	0.44	0.10
	ETR	QCN	NPQ	YNPQ	YNO	Chl *a*	Chl *b*	Car	
PC1	−0.98 *	0.68	0.69	0.95 *	0.81 *	−0.94 *	−0.90 *	−0.93 *	
PC2	−0.03	−0.72 *	−0.71 *	−0.18	0.54	−0.11	−0.01	−0.11	

* Variables considered in PCA: r = 0.10–0.39 (weak), 0.40–0.69 (moderate), and 0.70–1.00 (strong).

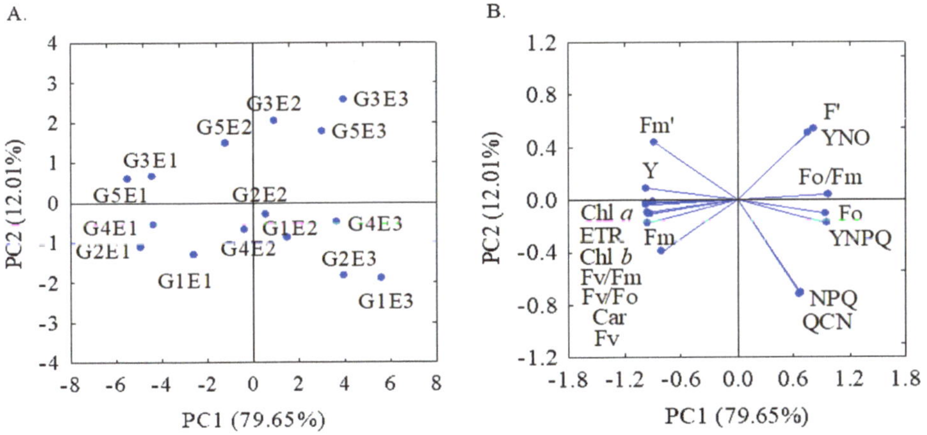

Figure 6. Two-dimensional projection of the principal component scores for the genotypes and treatments (**A**) and the sugarcane variables analyzed (**B**) in the first two principal components (PC1 and PC2). Initial fluorescence (Fo), maximum fluorescence (Fm), variable fluorescence (Fv), maximum quantum efficiency of photosystem II (Fv/Fm), maximum primary efficiency of the photochemical process in PSII (Fv/Fo), basal quantum efficiency of the non-photochemical process in PSII (Fo/Fm), initial fluorescence before the saturation pulse (F′), maximum fluorescence after adaptation to saturating light (Fm′), quantum efficiency of photosystem II (Y), electron transport rate (ETR), complete non-photochemical quenching of chlorophyll fluorescence (QCN), quantum yield of regulated photochemical quenching (YNPQ), quantum yield of non-regulated photochemical quenching (YNO), and content of chlorophyll *a* (Chl *a*), chlorophyll *b* (Chl *b*), and carotenoids (Car) in sugarcane plants grown under water deficit and calcium pyruvate application at 211 DAR. G1—RB863129, G2—RB92579, G3—RB962962, G4—RB021754, and G5—RB041443). E1—full irrigation, E2—water deficit plus 30 mM of calcium pyruvate, E3—water deficit without calcium pyruvate.

Based on the physiological parameters of sugarcane, the highest values of the Fo, Fo/Fm ratio, and F′ parameters were observed when plants were subjected to strategy E3, influencing the increase of the photochemical quenching parameters QCN, NPQ, YNPQ, and YNO. This increase suggests that there was damage to the plant's photosynthetic apparatus, since the high initial fluorescence indicates the dissipation of energy lost by the plant. The damage caused to the plant's photosynthetic apparatus contributed to the reduction in the Fm, Fv, Fv/Fm ratio, Fm′, Y, Chl *a*, Chl *b*, and carotenoids parameters.

3. Materials and Methods

3.1. Location of This Experiment

This experiment was conducted in a plant nursery belonging to the Agricultural Engineering Academic Unit of the Federal University of Campina Grande (UAEA-UFCG), located in the city of Campina Grande, PB, at the geographic coordinates 7°15′18″ S, 35 °52′28″ W, at a mean elevation of 550 m a.s.l (meters above sea level).

3.2. Treatments and Experimental Design

The treatments were obtained from three management strategies (E1—full irrigation throughout the crop cycle, E2—water deficit with the application of 30 mM calcium pyruvate, and E3—water deficit without the application of calcium pyruvate) and five commercial sugarcane genotypes (G1—RB863129, G2—RB92579, G3—RB962962, G4—RB021754, and G5—RB041443), distributed in randomized blocks in a 3 × 5 factorial design with three replications.

The pyruvate concentration (30 mM) was established based on a study developed by Shen et al. [39] with *Arabidopsis thaliana*, in which the authors used the exogenous application of pyruvate. Adjustments were made to the pyruvate concentration since the original research was carried out with a sample of incubated leaves from an uncultivated species (Arabidopsis). The adjustment was also based on the study developed by Barbosa et al. [16], who used 50 mM pyruvate in peanut plants under drought stress. The calcium pyruvate used in this research was chosen because it is considered a low-cost product and is easily found commercially, when compared with the other sources of pyruvate.

3.3. Description of the Experiment

Plastic containers with a capacity of 45 L adapted as drainage lysimeters were used in the experiment, receiving a 2.0 cm layer of crushed stone and a non-woven geotextile fabric (Bidim OP 30) at the bottom, and distributed in a 1.0 × 1.5 m spatial arrangement. Two 10.0-mm wide hoses were connected to each lysimeter, which were coupled to two containers with a volumetric capacity of 2.0 L to collect the drained water (Figure 7).

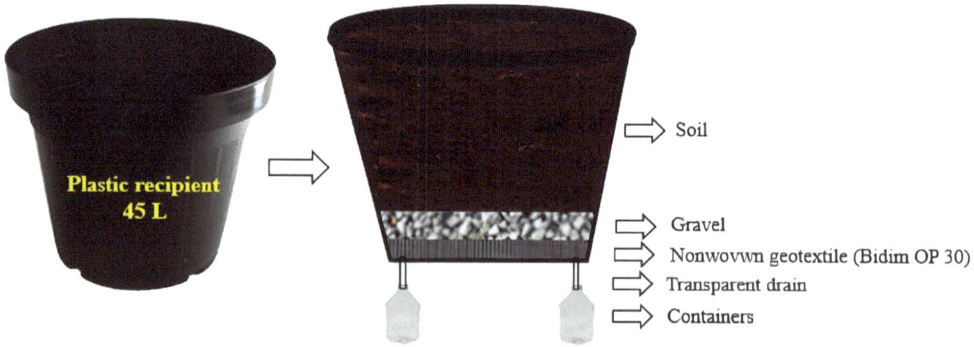

Figure 7. Illustration of the preparation of the drainage lysimeters for soil filling.

The pots were filled with soil that was classified as Regolithic Neosol (Entisol) with a sandy loam texture. The soil came from the municipality of Lagoa Seca, Paraíba, Brazil, and its physical and chemical attributes were determined according to the methodology described by Teixeira et al. [40] (Table 3).

Table 3. Chemical and physical attributes of the soil used in the experiment.

Chemical Attributes			Physical Attributes	
			Sand	63.48%
pH	6.50	-	Silt	25.14%
P	79.0	mg dm^{-3}	Clay	11.38%
K^+	0.24	cmolc dm^{-3}	Soil density	1.13 g cm^{-3}
Ca^{2+}	9.50	cmolc dm^{-3}	Particle density	2.72 g cm^{-3}
Na^+	0.51	cmolc dm^{-3}	Porosity	58.45%
Mg^{2+}	5.40	cmolc dm^{-3}	Sandy loam	
Al^{3+}	0.00	cmolc dm^{-3}	Matric potential (kPa)	Moisture (%)
H^+	0.90	cmolc dm^{-3}	Natural	0.55
SB	15.65	cmolc dm^{-3}	−10	24.86
CEC	16.55	cmolc dm^{-3}	−33	17.05
V	94.56	%	−100	12.57
M	0.00	%	−500	9.01
OM	8.10	g dm^{-3}	−1000	8.91
			−1500	8.84

pH (H_2O)—potential of hydrogen; SB—sum of bases; CEC—cation exchange capacity at pH 7.0; Mehlich (P, K, Na); Potassium chloride (KCl) 1N (Ca, Mg, and Al); Calcium acetate at pH 7.0 (H + Al); OM—organic matter; V—base saturation, and M—aluminum saturation.

The soil moisture values at the tensions of −10, −33, −100, −500, −1000, and −1500 kPa were used to adjust the soil water retention curve (Figure 8). In order to obtain the adjustment parameters, the volumetric moisture values (θ) corresponding to the applied matrix potentials (Ψm) were modeled using the RETC v.6 software and the non-linear model proposed by Van Genuchten [41].

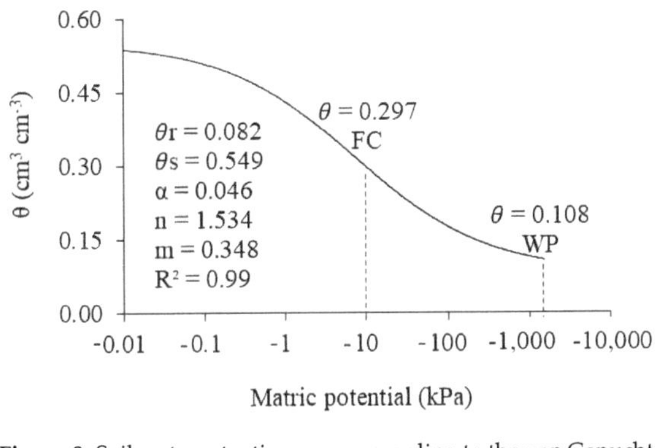

Figure 8. Soil water retention curve according to the van Genuchten model. θr: residual soil moisture; θs: soil moisture at saturation; α, n, and m are the model adjustment parameters; FC: field capacity and WP: wilting point.

After harvest (the first production of culms after planting), the second cultivation cycle began. From this point, the pots were irrigated regularly, close to the level corresponding to field capacity, until the moment of implementation of the treatments.

Fertilization with nitrogen, potassium, and phosphorus was applied weekly via irrigation water, with a total of 47.67 g of urea (45% N), 36.28 g of monoammonium phosphate (51% P_2O_5, 11% N), and 71.25 g of potassium chloride (60% K_2O). The micronutrients were applied at 15-day intervals to avoid nutritional deficiency by applying 1.0 g L^{-1} of Quimifol® (composition: Mg (5.0%); S (11.0%); B (3.5%); Cu (0.10%); Fe (0.20%); Mn (1.0%); Mo (0.10%); and Zn (6.0%)) with the aid of a knapsack sprayer.

Calcium pyruvate was purchased from Natusvita®. The solution was obtained by dissolving calcium pyruvate in distilled water minutes before spraying and applied with the aid of a Jacto XP knapsack sprayer with a capacity of 12 L, a working pressure (maximum) of 88 psi (6 bar), and a JD 12P Nozzle. Spraying was carried out at 5:00 p.m. on all the leaves of the plant. Between 100 and 200 mL of the solution was applied per experimental unit. Table 4 describes the period of the application of treatments.

Table 4. Application of treatments in sugarcane plants.

	Water Deficit		Applications		
	Period (DAR)	Total (Days)	Calcium Pyruvate	Water	Total
			Period (DAR)		
E1	-	-	-	-	-
E2	24 to 64 and 182 to 211	71	39 to 63 and 192 to 210	-	23
E3	24 to 64 and 182 to 211	71	-	39 to 63 and 192 to 210 *	23

* Plants under water deficit that did not receive calcium pyruvate (E3) were sprayed with distilled water plus an adhesive spreader. DAR—days after regrowth. Applications were carried out at two-day intervals between each application. E1—full irrigation, E2—water deficit plus 30 mM of calcium pyruvate, E3—water deficit without calcium pyruvate.

For better adhesion and absorption, an adhesive spreader was added to the solutions during spraying. Furthermore, to avoid drift caused by the wind during spraying, the plants in each pot were protected with plastic, and the soil was covered with an impermeable mantle to prevent runoff to the soil surface (Figure 9).

Figure 9. Protection of the sugarcane plants to prevent drift to other plants and runoff to the soil surface during spraying.

Irrigation was carried out daily, at 5:00 p.m., by applying the water volume corresponding to the demand of the plant undergoing each treatment. The volume applied to each pot, per irrigation event, was estimated individually using the water balance, according to Equation (1). A 20% leaching fraction was used monthly to remove excess salts from the ground [42].

$$VI = Va - Vd \tag{1}$$

where:

VI—the water volume to be used in the irrigation event (mL);
Va—the volume applied in the previous irrigation event (mL); and
Vd—the volume drained, quantified on the next morning (mL).

At 64 and 211 days after regrowth (DAR), soil samples were taken using a mini-auger, and the soil moisture content was determined using the standard oven method. Then, the soil samples were put in aluminum cans and weighed to obtain the wet mass. After this step, they were oven-dried at 105 °C for 72 h to obtain the dry mass, from which the soil moisture content was determined on a gravimetric basis [40], which is related to the soil matric potential (Table 5).

Table 5. Soil water moisture and matric potential (Ψm) recorded at the end of the water-deficit period in each treatment.

Genotypes	64 DAR					
	E1		E2		E3	
	Moisture (cm^3 cm^{-3})	Ψm (kPa)	Moisture (cm^3 cm^{-3})	Ψm (kPa)	Moisture (cm^3 cm^{-3})	Ψm (kPa)
RB863129	0.259	−18.6	0.149	−207.9	0.138	−305.8
RB92579	0.240	−25.9	0.143	−254.6	0.145	−237.5
RB962962	0.240	−25.9	0.142	−263.7	0.141	−273.3
RB021754	0.266	−16.5	0.145	−237.5	0.139	−294.1
RB041443	0.244	−24.1	0.149	−207.9	0.148	−214.8
	211 DAR					
	E1		E2		E3	
	Moisture (cm^3 cm^{-3})	Ψm (kPa)	Moisture (cm^3 cm^{-3})	Ψm (kPa)	Moisture (cm^3 cm^{-3})	Ψm (kPa)
RB863129	0.242	−25.0	0.126	−504.4	0.135	−342.9
RB92579	0.257	−19.2	0.133	−371.6	0.131	−403.9
RB962962	0.240	−25.9	0.133	−371.6	0.133	−371.6
RB021754	0.234	−28.8	0.127	−481.6	0.129	−440.9
RB041443	0.256	−19.6	0.130	−421.6	0.131	−403.9

Ψm—matric potential. DAR—days after regrowth. E1—full irrigation, E2—water deficit plus 30 mM of calcium pyruvate, E3—water deficit without calcium pyruvate.

3.4. Variables Analyzed

At the end of the water-deficit period (211 DAR), the initial fluorescence (Fo), the maximum fluorescence (Fm), and the variable fluorescence (Fv) were measured using a modulated pulse fluorometer: model OS5p from Opti-Science ((Hudson, NH, USA) Saturation flash intensity: 11,250 µmol and saturating pulse width: 0.8 s). From these results, the maximum quantum efficiency of photosystem II (Fv/Fm), the photochemical efficiency (Fv/Fo), and the basal quantum efficiency of the non-photochemical process (Fo/Fm) were determined. The evaluations were carried out on leaves with blades that were adapted to the dark for 30 min, using a clip to ensure that all the primary acceptors were fully oxidized [43].

After the fluorescence evaluations with adaptation to the dark, the evaluations were carried out under lighting conditions using the 'Yield' protocol (Saturation flash intensity: 11,250 µmol, saturating pulse width: 0.8 s, and default PAR: 120 µE) to determine the initial fluorescence before the saturation pulse (F′), the maximum fluorescence after adaptation to saturating light (Fm′), the electron transport rate (ETR), and the quantum efficiency of photosystem II (Y). From these results, the Stern–Volmer non-photochemical quenching coefficient (NPQ), the complete non-photochemical quenching coefficient of chlorophyll fluorescence (QCN), the quantum yield of regulated photochemical quenching (YNPQ), and the quantum yield of non-regulated photochemical quenching (YNO) were determined [44,45].

The content of the photosynthetic pigments (chlorophyll *a*, chlorophyll *b*, and carotenoids) was determined at 211 DAR, according to Arnon [46] and Lichtenthäler [47], by removing a 6-mm disc from the +2 leaf of each treatment. Each sample received 6.0 mL of 80% acetone P.A. Subsequently, the supernatants containing the extracted pigments were collected, and absorbance readings were taken on a spectrophotometer (model UV/VIS-UV1720, AKSON®, São Leopoldo, RS, Brazil) at the absorbance wavelengths (ABS) of 470, 645, and 663 nm, calculated using Equations (2)–(4), with the values expressed in micrograms of pigment per gram of fresh mass (µg g^{-1} FW).

$$\text{Clh } a = \frac{((12.7 \times \text{ABS}_{663}) - (2.79 \times \text{ABS}_{647})) \times \text{V}}{\text{FW}} \quad (2)$$

$$\text{Clh } b = \frac{((22.9 \times \text{ABS}_{647}) - (5.10 \times \text{ABS}_{663})) \times \text{V}}{\text{FW}} \quad (3)$$

$$\text{Car} = \frac{(((1000 \times \text{ABS}_{470}) - (1.82 \times \text{Clh } a) - (85.02 \times \text{Clh } b))/198) \times \text{V}}{\text{FW}} \quad (4)$$

where:

ABS_{470}, ABS_{663}, and ABS_{647}—the absorbances at 480, 663, and 645 nm, respectively;
V—the volume of 80% acetone used in extraction (mL); and
FW—the fresh matter (g).

At the end of the analyses in all treatments, soil moisture was restored to close to the level corresponding to field capacity.

3.5. Statistical Analysis

Prior to the analysis of variance, the data were subjected to the normality test (Shapiro–Wilk). The analysis of variance and the F-test were carried out with the positive results obtained from the previous tests. Then, the test of means was applied using the Tukey test ($p \leq 0.05$), using the software Sisvar 5.8 [48]. A multivariate analysis using a principal component analysis (PCA) was also employed; to this end, the data were normalized to a zero mean (= 0.0), unit variance (σ^2) = 1.0, eigenvalues (λ) > 1.0, and a total variance (σ^2) > 10% [49]. Only variables with a Pearson correlation coefficient above 0.6 were kept in the composition of each principal component (PC) [50]. The analyses were processed using the software Statistica 7.0 [51].

4. Conclusions

There is a dissimilarity in the fluorescence parameters and photosynthetic pigments of genotype RB863129 in relation to those of genotypes RB041443, RB96262, RB021754, and RB92579.

Foliar application of calcium pyruvate alleviates the deleterious effects of water deficit in sugarcane on the fluorescence parameters of chlorophyll *a* and photosynthetic pigments, without interaction with the genotypes. However, subsequent validation tests will be necessary to test and validate the adoption of this technology under field conditions.

Author Contributions: Conceptualization, M.d.S.D., F.d.A.d.S. and P.D.F.; methodology, M.d.S.D., F.d.A.d.S., M.d.F.C.d.S., A.M.d.L. and V.R.d.N.L.; software, M.d.S.D., F.d.A.d.S., A.A.R.d.S. and T.F.d.L.A.; validation, M.d.S.D., P.D.F., C.H.d.A.F. and L.S.R.; formal analysis, L.S.R. and F.d.A.d.S.; investigation, P.D.F., C.H.d.A.F. and M.d.S.D.; resources, P.D.F. and C.H.d.A.F.; data collection, M.d.S.D., F.d.A.d.S., M.d.F.C.d.S., V.R.d.N.L., C.N.d.L., A.M.d.L., R.F.d.L. and W.B.B.d.S.; writing—original draft preparation, M.d.S.D., F.d.A.d.S., M.d.F.C.d.S., V.R.d.N.L. and A.M.d.L.; writing—review and editing, M.d.S.D., P.D.F., F.d.A.d.S. and L.S.R.; supervision, P.D.F. and C.H.d.A.F.; project administration, P.D.F. All authors have read and agreed to the published version of the manuscript.

Funding: This research received no external funding.

Data Availability Statement: Data are contained within the article.

Acknowledgments: The authors thank the Graduate Program in Agricultural Engineering of the Federal University of Campina Grande, and the Coordination for the Improvement of Higher Education Personnel (CAPES) for the granting the scholarship to the first author.

Conflicts of Interest: The authors declare no conflict of interest.

References

1. Pan, S.; Zabed, H.M.; Wei, Y.; Qi, X. Technoeconomic and environmental perspectives of biofuel production from sugarcane bagasse: Current status, challenges and future outlook. *Ind. Crops Prod.* **2022**, *188*, 115684. [CrossRef]
2. Hoang, D.T.; Hiroo, T.; Yoshinobu, K. Nitrogen use efficiency and drought tolerant ability of various sugarcane varieties under drought stress at early growth stage. *Plant Prod. Sci.* **2019**, *22*, 250–261. [CrossRef]
3. Mohanraj, K.; Hemaprabha, G.; Vasantha, S. Biomass yield, dry matter partitioning and physiology of commercial and Erianthus introgressed sugarcane clones under contrasting water regimes. *Agric. Water Manag.* **2021**, *255*, 107035. [CrossRef]
4. Ebid, M.; Salim, B.; El-Gabry, Y. Evaluation of some sugarcane genotypes under drought condition. *Sci. J. Agric. Sci.* **2021**, *3*, 123–132. [CrossRef]
5. Feijó, A.R.; Viana, V.E.; Balbinot, A.; Fipke, M.V.; Souza, G.M.; do Amarante, L.; Avila, L.A. Vegetative stage induces tolerance to high temperature during anthesis in rice. *Plants* **2023**, *12*, 3133. [CrossRef] [PubMed]
6. Kumar, R.; Sagar, V.; Verma, V.C.; Kumari, M.; Gujjar, R.S.; Goswami, S.K.; Jha, S.K.; Pandey, H.; Dubey, A.K.; Srivastava, S.; et al. Drought and salinity stresses induced physio-biochemical changes in sugarcane: An overview of tolerance mechanism and mitigating approaches. *Front. Plant Sci.* **2023**, *14*, 1225234. [CrossRef] [PubMed]
7. Martinazzo, E.G.; Perboni, A.T.; Oliveira, P.V.; Bianchi, V.J.; Bacarin, M.A. Photosynthetic activity in japanese plum under water deficit and flooding. *Cienc. Rural* **2013**, *43*, 35–41. [CrossRef]
8. Sales, C.R.; Ribeiro, R.V.; Silveira, J.A.; Machado, E.C.; Martins, M.O.; Lagôa, A.M.M. Superoxide dismutase and ascorbate peroxidase improve the recovery of photosynthesis in sugarcane plants subjected to water deficit and low substrate temperature. *Plant Physiol. Biochem.* **2013**, *73*, 326–336. [CrossRef]
9. Yang, Y.; Nan, R.; Mi, T.; Song, Y.; Shi, F.; Liu, X.; Wang, Y.; Sun, F.; Xi, Y.; Zhang, C. Rapid and nondestructive evaluation of wheat chlorophyll under drought stress using hyperspectral imaging. *Int. J. Mol. Sci.* **2023**, *24*, 5825. [CrossRef]
10. Baker, N.R. Chlorophyll fluorescence: A probe of photosynthesis in vivo. *Annu. Rev. Plant Biol.* **2008**, *59*, 89–113. [CrossRef]
11. Li, N.; Pu, K.; Ding, D.; Yang, Y.; Niu, T.; Li, J.; Xie, J. Foliar Spraying of Glycine Betaine Alleviated Growth inhibition, photoinhibition, and oxidative stress in pepper (*Capsicum annuum* L.) seedlings under low temperatures combined with low light. *Plants* **2023**, *12*, 2563. [CrossRef] [PubMed]
12. Silva, M.A.; Santos, C.M.; Arantes, M.T.; Brunelli, M.C.; Holanda, L.A. Respostas fisiológicas de cultivares de cana-de-açúcar submetidas à deficiência hídrica e a reidratação. *Rev. Caatinga* **2013**, *26*, 28–35.
13. Souza, J.L.; Santos, R.B.; Nunes, V.V.; Torres, M.F.O.; Calazans, C.C.; Junior, L.F.G.O.; Silva-Mann, R. Déficit hídrico no desenvolvimento de cultivares de cana-de-açúcar. *Global Sci. Technol.* **2020**, *13*, 196–210.
14. Verma, K.K.; Anas, M.; Chen, Z.; Rajput, V.D.; Malviya, M.K.; Verma, C.L.; Singh, R.K.; Singh, P.; Song, X.P.; Li, Y.R. Silicon supply improves leaf gas exchange, antioxidant defense system and growth in *Saccharum officinarum* responsive to water limitation. *Plants* **2020**, *9*, 1032. [CrossRef] [PubMed]
15. Leanasawat, N.; Kosittrakun, M.; Lontom, W.; Songsri, P. Physiological and agronomic traits of certain sugarcane genotypes grown under field conditions as influenced by early drought stress. *Agronomy* **2021**, *11*, 2319. [CrossRef]
16. Barbosa, D.D.; Fernandes, P.D.; Marcelino, A.D.A.L.; Silva, F.A.; Dias, M.S.; Silva, C.R.C.; Santos, R.C. Exogenous pyruvate mitigates the detrimental effects of water stress in contrasting peanut genotypes. *Genet. Mol. Res.* **2021**, *20*, 1–14. [CrossRef]
17. Silva, F.A.; Dias, M.S.; Fernandes, P.D.; Marcelino, A.D.A.L.; Lima, A.M.; Pereira, R.F.; Barbosa, D.D.; Silva, M.F.C.; Silva, A.A.R.; Santos, R.C. Pyruvic acid as attenuator of water deficit in cotton plants varying the phenological stage. *Braz. J. Biol.* **2023**, *83*, e272003. [CrossRef]
18. Dias, M.S.; Fernandes, P.D.; Silva, F.A.; Marcelino, A.D.A.L.; Barbosa, D.D.; Santos, R.C.; Reis, L.S.; Lima, V.L.A. Pyruvate supplementation in cotton under water restriction varying the phenological phases. *Acta Sci. Agron.* **2023**, *45*, e61973. [CrossRef]
19. Taiz, L.; Zeiger, E.; Moller, I.M.; Murphy, A. *Fisiologia e Desenvolvimento Vegetal*, 6th ed.; Artmed: Porto Alegre, Brazil, 2017; 858p.
20. Baker, N.R.; Rosenqvist, E. Applications of chlorophyll fluorescence can improve crop production strategies: An examination of future possibilities. *J. Exp. Bot.* **2004**, *55*, 1607–1621. [CrossRef]
21. Lima, G.S.; Dias, A.S.; Soares, L.A.A.; Gheyi, H.R.; Nobre, R.G.; Silva, A.A.R. Eficiência fotoquímica, partição de fotoassimilados e produção do algodoeiro sob estresse salino e adubação nitrogenada. *Rev. Ciênc. Agrár.* **2019**, *43*, 214–225. [CrossRef]
22. Rousseau, C.; Belin, É.; Bove, E.; Rousseau, D.; Fabre, F.; Berruyer, R.; Guillaumès, J.; Manceau, C.; Jacques, M.; Boureau, T. High throughput quantitative phenotypingof plant resistance using chlorophyll fluorescence image analysis. *Plant Methods* **2013**, *9*, 17. [CrossRef] [PubMed]
23. Bolhàr-Nordenkampf, H.R.; Long, S.P.; Baker, N.R.; Öquist, G.; Schreider, U.; Lechner, E.G. Chlorophyll fluorescence as probe of the photosynthetic competence of leaves in the field: A review of current instrument. *Funct. Ecol.* **1989**, *3*, 497–514. [CrossRef]
24. Silva, M.A.; Jifon, J.L.; Silva, J.A.G.; Santos, C.M.; Sharma, V. Relationships between physiological traits and productivity of sugarcane in response to water deficit. *J. Agric. Sci.* **2014**, *152*, 104–118. [CrossRef]
25. García-Sánchez, F.; Simón-Grao, S.; Martínez-Nicolás, J.J.; Alfosea-Simón, M.; Liu, C.; Chatzissavvidis, C.; Pérez-Pérez, J.G.; Cámara-Zapata, J.M. Multiple stresses occurring with boron toxicity and deficiency in plants. *J. Hazard. Mater.* **2020**, *397*, e122713. [CrossRef] [PubMed]
26. Silva, P.P.; Soares, L.; Costa, J.G.; Silva Viana, L.; Andrade, J.C.F.; Gonçalves, E.R.; Santos, J.M.; Barbosa, G.V.S.; Nascimento, V.X.; Todaro, A.R.; et al. Path analysis for selection of drought tolerant sugarcane genotypes through physiological components. *Ind. Crops Prod.* **2012**, *37*, 11–19. [CrossRef]

27. Kumar, D.; Singh, H.; Raj, S.; Soni, V. Chlorophyll a fluorescence kinetics of mung bean (*Vigna radiata* L.) grown under artificial continuous light. *Biochem. Biophys. Rep.* **2020**, *24*, e100813. [CrossRef]
28. Herritt, M.T.; Pauli, D.; Mockler, T.C.; Thompson, A.L. Chlorophyll fluorescence imaging captures photochemical efficiency of grain sorghum (*Sorghum bicolor*) in a field setting. *Plant Methods* **2020**, *16*, 109. [CrossRef]
29. Verma, K.K.; Song, X.P.; Zeng, Y.; Guo, D.J.; Singh, M.; Rajput, V.D.; Malviya, M.K.; Wei, K.J.; Sharma, A.; Li, D.P.; et al. Foliar application of silicon boosts growth, photosynthetic leaf gas exchange, antioxidative response and resistance to limited water irrigation in sugarcane (*Saccharum officinarum* L.). *Plant Physiol. Biochem.* **2021**, *166*, 582–592. [CrossRef]
30. Ruban, A.V. Nonphotochemical chlorophyll fluorescence quenching: Mechanism and effectiveness in protecting plants from photodamage. *Plant Physiol.* **2016**, *170*, 1903–1916. [CrossRef]
31. Simkin, A.J.; Kapoor, L.; Doss, C.G.P.; Hofmann, T.A.; Lawson, T.; Ramamoorthy, S. The role of photosynthesis related pigments in light harvesting, photoprotection and enhancement of photosynthetic yield in planta. *Photosynth. Res.* **2022**, *152*, 23–42. [CrossRef]
32. Verma, K.K.; Song, X.P.; Lin, B.; Guo, D.J.; Singh, M.; Rajput, V.D.; Singh, R.K.; Singh, P.; Sharma, A.; Malviya, M.K.; et al. Silicon induced drought tolerance in crop plants: Physiological adaptation strategies. *Silicon* **2022**, *14*, 2473–2487. [CrossRef]
33. Spinoso-Castillo, J.L.; Moreno-Hernández, M.D.R.; Mancilla-Álvarez, E.; Sánchez-Segura, L.; Sánchez-Páez, R.; Bello-Bello, J.J. Arbuscular Mycorrhizal Symbiosis improves ex vitro acclimatization of sugarcane plantlets (*Saccharum* spp.) under drought stress conditions. *Plants* **2023**, *12*, 687. [CrossRef] [PubMed]
34. Hnilicka, F.; Lysytskyi, S.; Rygl, T.; Hnilickova, H.; Pecka, J. Effect of Short-Term Water Deficit on Selected Physiological Properties of Wheat (*Triticum aestivum* L.) Plants. *Agronomy* **2023**, *13*, 2892. [CrossRef]
35. Stefanov, M.; Rashkov, G.; Borisova, P.; Apostolova, E. Sensitivity of the Photosynthetic Apparatus in Maize and Sorghum under Different Drought Levels. *Plants* **2023**, *12*, 1863. [CrossRef] [PubMed]
36. Yousaf, M.I.; Riaz, M.W.; Shehzad, A.; Jamil, S.; Shahzad, R.; Kanwal, S.; Ghani, A.; Ali, F.; Abdullah, M.; Ashfaq, M.; et al. Responses of maize hybrids to water stress conditions at different developmental stages: Accumulation of reactive oxygen species, activity of enzymatic antioxidants and degradation in kernel quality traits. *PeerJ* **2013**, *11*, e14983. [CrossRef] [PubMed]
37. Maia Júnior, S.O.; Andrade, J.R.; Santos, C.M.; Santos, J.V.; Silva, L.K.S.; Clemente, P.R.A.; Ferreira, V.M.; Silva, J.V.; Endres, L. Foliar-applied glycine betaine minimizes drought stress-related impact to gas exchange and the photochemical efficiency of PSII in sugarcane. *Theor. Exp. Plant Physiol.* **2020**, *32*, 315–329. [CrossRef]
38. Dias, M.S.; Silva, F.A.; Fernandes, P.D.; Farias, C.H.A.; Silva, I.J.; Silva, M.F.C.; Lima, R.F.; Lacerda, C.N.; Lima, A.M.; Lima, V.R.N.; et al. Calcium pyruvate in the attenuation of the water deficit on the agro-industrial quality of ratoon sugarcane. *Braz. J. Biol.* **2023**, *83*, e275046. [CrossRef]
39. Shen, J.L.; Li, C.L.; Wang, M.; He, L.L.; Lin, M.Y.; Chen, D.H.; Zhang, W. Mitochondrial pyruvate carrier 1 mediates abscisic acid-regulated stomatal closure and the drought response by affecting cellular pyruvate content in *Arabidopsis thaliana*. *BMC Plant Biol.* **2017**, *17*, 217. [CrossRef]
40. Teixeira, P.C.; Donagemma, G.K.; Fontana, D.; Teixeira, W.G. *Manual de Métodos de Análise de Solo*, 3rd ed.; Embrapa Solos: Brasília, Brazil, 2017; 573p.
41. Van Genuchten, M.T. A closed-form equation for predicting the hydraulic conductivity on unsaturated soil. *Soil Sci. Soc. Am. J.* **1980**, *44*, 892–897. [CrossRef]
42. Soares, L.A.A.; Fernandes, P.D.; Lima, G.S.; Silva, S.S.; Moreira, R.C.; Medeiros, T.L. Phytomass and production components of colored cotton under salt stress in different phenological stages. *Rev. Bras. Eng. Agríc. Ambient.* **2021**, *25*, 132–138. [CrossRef]
43. Konrad, M.L.F.; Silva, J.A.B.; Furlani, P.R.; Machado, E.C. Trocas gasosas e fluorescência da clorofila em seis cultivares de cafeeiro sob estresse de alumínio. *Bragantia* **2005**, *64*, 339–347. [CrossRef]
44. Oxborough, K.; Baker, N.R. An instrument capable of image chlorophyll a fluorescence from intact leaves at very low irradiance and at the cellular and subcellular levels of organization. *Plant Cell Environ.* **1997**, *20*, 1473–1483. [CrossRef]
45. Kramer, D.M.; Johnson, G.; Kiirats, O.; Edwards, G.E. New fluorescence parameters for the determination of Q A redox state and excitation energy fluxes. *Photosynth. Res.* **2004**, *79*, 209. [CrossRef] [PubMed]
46. Arnon, D.I. Copper enzymes in isolated chloroplasts. *Polyphenoloxidase in Beta vulgaris. Plant Physiol.* **1949**, *24*, 1. [CrossRef]
47. Lichtenthaler, H.K. Chlorophylls and carotenoids: Pigments of photosynthetic biomembranes. In *Methods in Enzymology*; Packer, L., Douce, R., Eds.; Academic Press: Cambridge, MA, USA, 1987; Volume 148, pp. 350–382. [CrossRef]
48. Ferreira, D.F. Sisvar: A computer analysis system to fixed effects split plot type designs. *Braz. J. Biom.* **2019**, *37*, 529–535. [CrossRef]
49. Govaerts, B.; Sayre, K.D.; Lichter, K.; Dendooven, L.; Deckers, J. Influence of permanent planting in high bed and residue management on physical and chemical soil quality in rainfed corn/wheat systems. *Plant Soil* **2007**, *291*, 39–54. [CrossRef]
50. Hair, J.F.; Black, W.C.; Babin, B.J.; Anderson, R.E.; Tatham, R.L. *Análise Multivariada de Dados*, 6th ed.; Tradução Sant'Anna, A.S., Ed.; Bookman: Porto Alegre, Brazil, 2009; 688p.
51. Statsoft, Inc. *Programa Computacional Statistica*; 7.0; Statsoft: Hamburg, Germany, 2004; Available online: https://statsoft-academic.com.br (accessed on 10 September 2023).

plants

MDPI

Article

Water-Retaining Polymer and Planting Pit Size on Chlorophyll Index, Gas Exchange and Yield of Sour Passion Fruit with Deficit Irrigation

Antônio Gustavo de Luna Souto [1,*], Edinete Nunes de Melo [2], Lourival Ferreira Cavalcante [2], Ana Paula Pereira do Nascimento [2], Ítalo Herbert Lucena Cavalcante [2], Geovani Soares de Lima [3], Rafael Oliveira Batista [1], Hans Raj Gheyi [3], Reynaldo Teodoro de Fátima [3], Evandro Franklin de Mesquita [4], Gleyse Lopes Fernandes de Souza [2], Guilherme Romão Silva [2], Daniel Valadão Silva [1], Francisco de Oliveira Mesquita [5] and Palloma Vitória Carlos de Oliveira [1]

1 Postgraduate Program in Soil and Water Management, Universidade Federal do Semiárido, Mossoró 59625900, RN, Brazil; rafaelbatista@ufersa.edu.br (R.O.B.); daniel.valadao@ufersa.edu.br (D.V.S.); palloma.oliveira@alunos.ufersa.edu.br (P.V.C.d.O.)
2 Postgraduate Program in Agronomy, Universidade Federal da Paraíba, Areia 58397-000, PB, Brazil; enm@academico.ufpb.br (E.N.d.M.); lofeca@cca.ufpb.br (L.F.C.); ana.nascimento@academico.ufpb.br (A.P.P.d.N.); italo.cavalcante@univasf.edu.br (Í.H.d.L.C.); gleyse.souza@academico.ufpb.br (G.L.F.d.S.); grs@aluno.ueg.br (G.R.S.)
3 Postgraduate Program in Agricultural Engineering, Universidade Federal de Campina Grande, Campina Grande 58429900, PB, Brazil; geovani.soares@professor.ufcg.edu.br (G.S.d.L.); hans.gheyi@ufcg.edu.br (H.R.G.); reynaldo.teodoro@estudante.ufcg.edu.br (R.T.d.L.)
4 Department of Agrarian and Exact Sciences, State University of Paraíba, Catolé do Rocha 58429-500, PB, Brazil; evandrofranklin@servidor.uepb.edu.br
5 Postgraduate in Ecology and Conservation, State University of Paraiba, Campina Grande 58429-500, PB, Brazil; francisco.mesquita@servidor.uepb.edu.br
* Correspondence: gustavo.luna@ufersa.edu.br; Tel.: +55-83-99981-7898

Citation: de Luna Souto, A.G.; de Melo, E.N.; Cavalcante, L.F.; do Nascimento, A.P.P.; Cavalcante, Í.H.L.; de Lima, G.S.; Batista, R.O.; Gheyi, H.R.; de Fátima, R.T.; de Mesquita, E.F.; et al. Water-Retaining Polymer and Planting Pit Size on Chlorophyll Index, Gas Exchange and Yield of Sour Passion Fruit with Deficit Irrigation. *Plants* **2024**, *13*, 235. https://doi.org/10.3390/plants13020235

Academic Editor: Mingjun Li

Received: 18 November 2023
Revised: 31 December 2023
Accepted: 11 January 2024
Published: 15 January 2024

Abstract: Water availability is a limiting factor for the cultivation of sour passion fruit. Soil management techniques and the use of water-retaining polymers can increase soil water retention, reducing the frequency of irrigation in the crop. In this context, the objective of the research was to evaluate the gas exchange, the chlorophyll index, and the yield of the sour passion fruit cv. BRS GA1 as a function of irrigation depths, pit volumes, and doses of water-retaining polymer. The experiment was carried out in randomized blocks, in plots subdivided in a 2 × (2 × 5) arrangement, with irrigation depths of 70 and 100% of the crop evapotranspiration (ETc) as the main plot, the subplots with the volumes of pit of 64 and 128 dm^3, and doses of the water-retaining polymer of 0, 0.5, 1.0, 1.5, and 2.0 g dm^{-3}. The interaction of irrigation depths × pit volumes × doses of water-retaining polymer influences chlorophyll indexes, gas exchange, and water productivity, with positive impacts on yield of the sour passion fruit. The water depth of 70% of ETc increased the yield of sour passion fruit, in pits of 64 dm^3. The application of doses of up to 1.1 g dm^{-3} of the water-retaining polymer and irrigation with water of 70% of ETc is recommended, and a dose of 2.0 g dm^{-3} of the water-retaining polymer in a pit volume of 128 dm^3, associated with an irrigation depth of 100% ETc causes stress in sour passion fruit plants due to excess water.

Keywords: *Passiflora edulis* Sims; water stress; pit volume; hydrogel; physiology; fruit harvest

1. Introduction

The sour passion fruit (*Passiflora edulis* Sims) is a species belonging to the Passifloraceae family, cultivated in tropical and subtropical countries, being one of the fruit species in greater evidence in Brazil [1]. In 2022, it had a national average yield of 15.26 t ha^{-1} [2]; however, this is still far below the genetic and productive potential of the crop, which can exceed 50 t ha^{-1} [3]. Brazil is the largest producer and consumer of passion fruit in the world, with the states of Bahia and Ceará as the largest producers of the species with 207,488 and 177,291 t, respectively [2].

Passiculture has great socioeconomic importance for the semiarid region of the Brazilian Northeast; however, water in this region is the most limiting factor for obtaining high yields, due to high evaporative rates and irregular distribution of rainfall, making it necessary to use irrigation [4,5]. However, the use of water by agricultural crops is considered very low, as approximately 60% of the applied water, depending on the irrigation method, is lost by evaporation [6].

The limitation of water availability in sour passion fruit can cause oxidative stress [7], negatively affecting several physiological processes, such as a reduction in chlorophyll levels [4], gas exchange [5], assimilation of carbon dioxide [8,9], and in fruit yield [4,10]. In this regard, the use of irrigation associated with techniques that allow for greater storage and retention of water in the soil close to the absorbent roots can increase irrigation efficiency with reflections on greater water availability to the plant.

One of these techniques is the use of water-retaining polymers in the soil, which can store and act as regulators of water availability for crops, providing a reduction in nutrient losses by leaching, increasing yield and minimizing production costs [11–14]. For the sour passion fruit, the use of the water-retaining polymer provided satisfactory results in the formation of seedlings [15,16], an increase in photosynthetic rates, and fruit yield [4,5]. In other fruit species, beneficial effects of the water-retaining polymer were observed under water deficit conditions. In plum, *Prunus salicina* trees, the application of water-retaining polymers increased fruit yield and water use efficiency of plants under water deficit [17]. Alshallash et al. (2022) [18] highlight that the application of water-retaining polymers increases water retention in the soil, causing a positive impact on the yield and quality of mango fruit (*Mangifera indica* L.) cv. Shelly in arid conditions.

In the implantation of the sour passion fruit orchard, in addition to irrigation, one of the factors that also contributes to the development of the plants is the preparation of the soil. Efficient soil management techniques, such as preparation of pits, must be implemented to promote ideal conditions for root growth and better yields of the crop [19,20]. Adequate pit volume and soil structural quality provide a better soil–water–plant relation, optimizing the efficient use of water by plants [21,22].

Thus, the objective of this study was to evaluate the chlorophyll index, gas exchange, and the yield of the sour passion fruit cv. BRS GA1 as a function of irrigation depths, pit volumes, and doses of water-retaining polymer.

2. Materials and Methods

2.1. Characterization of the Experimental Area

The experiment was conducted from September 2018 to January 2020, at the Macaquinhos Farm, located in the municipality of Remígio, in the State of Paraíba, Brazil. The municipality is georeferenced by geographic coordinates: 7°00′1.95″ latitude South, 35°47′55″ longitude West of the Greenwich Meridian and altitude of 562 m, physiographically inserted in the 'Agreste Paraibano Mesoregion' and 'Curimataú Microregion' western. The climate of the region, according to the Köppen's classification [23], was classified as As′, which means hot and humid summer and rainy season concentrated between March and June.

Daily rainfall and evaporation data were recorded through readings in a rain gauge and Class "A" tank, respectively, and the values of temperature and relative air humidity were obtained with a Datalogger, model HT-70, installed close to the experiment (Figure 1a,b).

The soil in the experimental area was classified as Entisols Psamment [24]. Before the implementation of the experiment, six soil samples were collected per block in the 0–0.40 m layer, for the purpose of physical and chemical characterization regarding soil fertility (Table 1), according to methodologies described in [25].

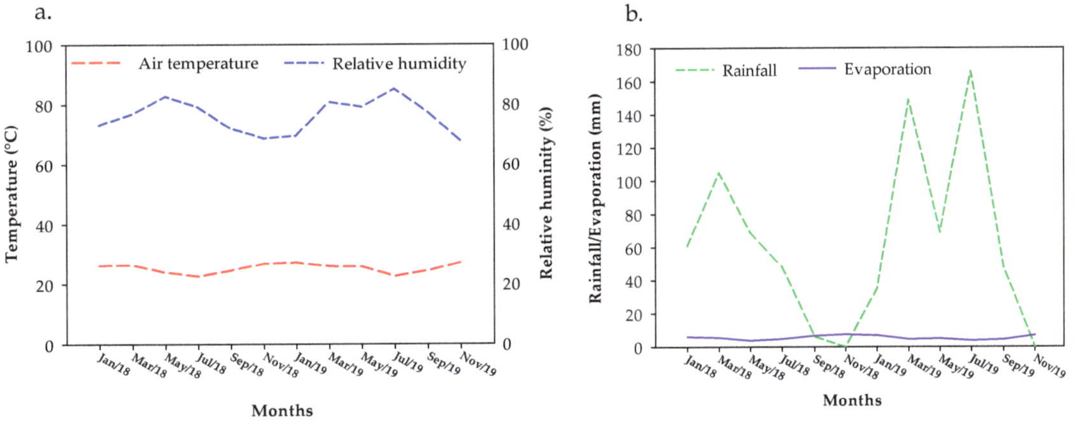

Figure 1. Mean monthly values of temperature and relative humidity of air (**a**) and rainfall and evaporation (**b**) during the experiment (1 January 2018 to 30 November 2019).

Table 1. Physical and chemical attributes of soil fertility in the depth of 0-0.40 m in the experimental area before passion fruit cultivation.

Physical Attributes		Chemical Attributes	
Coarse Sand (g kg^{-1})	602	pH in water (1:2.5)	5.37
Fine Sand (g kg^{-1})	194	SOM (g dm^{-3})	4
Silt (g kg^{-1})	132	P–Rem (mg dm^{-3})	45.6
Clay (g kg^{-1})	72	P (mg dm^{-3})	5.5
Dispersed clay (g kg^{-1})	10.5	S (mg dm^{-3})	6.55
Bulk density (kg dm^{-3})	1.54	K^+ (cmol$_c$ dm^{-3})	0.13
Particle density (kg dm^{-3})	2.77	Ca^{2+} (cmol$_c$ dm^{-3})	1.18
Total porosity (%)	44.5	Mg^{2+}(cmol$_c$ dm^{-3})	0.37
Degree of flocculation (%)	86	Na^+ (cmol$_c$ dm^{-3})	Dash
Dispersion index (%)	14	SB (cmol$_c$ dm^{-3})	1.68
Flocculation/dispersion ratio	7.75	$H^+ + Al^{3+}$ (cmol$_c$ dm^{-3})	1.8
Field capacity (%-v) *	8	Al^{3+} (cmol$_c$ dm^{-3})	0.18
Wilting point (%-v) **	3.75	CEC (cmol$_c$ dm^{-3})	3.48
Available water (%-v)	4.25	V (%)	47.85
Textural class	Sandy loam	Fertility class	Eutrophic

* Water retention at 0.01 MPa; ** water retention at 1.50 MPa; SOM = soil organic matter; SB = sum of exchangeable soil bases (SB = $K^+ + Ca^{2+} + Mg^{2+}$); CEC = cation exchange capacity of soil (SB + ($H^+ + Al^{3+}$); V(%) = saturation by exchangeable soil bases [(SB/CEC) × 100]. Extractors: SOM—Walkley-Black; P-Rem = Remaining phosphorus—$CaCl_2$—0.01 mol L^{-1}; P, K^+, Na^+—Mehlich 1; Ca^{2+}, Mg^{2+}, e Al^{3+}—KCl—1 mol L^{-1}; S = Sulfur—Acetic Acid—2 mol L^{-1}; $H^+ + Al^{3+}$= potential acidity—calcium acetate 0.5 mol L^{-1}—pH 7.0.

2.2. Experimental Design and Plant Material

The experiment was carried out in randomized blocks and split plot 2 × (2 × 5), with four replications and three plants per plot (Figure 2), in which the main plot was related to irrigation depths of 70 and 100% of the crop evapotranspiration (ETc), the subplots to pit volumes of 64 dm^3 (traditional) and 128 dm^3 (proposed), and doses of 0; 0.5, 1.0, 1.5, and 2.0 g of water-retaining polymer per dm^3 of soil.

Pit volumes were based on [19] and the doses of the water-retaining polymer according to the methodology contained in [5]. The water-retaining polymer used was Hydroplan®-EB/HyA. The plant material under study was the sour passion fruit (*Passiflora edulis* Sims), commercial cv. cv. BRS GA1, earlier known as Gigante Amarelo, propagated via seeds and obtained from a commercial nursery accredited by Embrapa and the Brazilian Ministry of Agriculture, Livestock and Supply (MAPA). The seedlings were transplanted when they presented 35 cm in height, 4 mm in diameter, and four pairs of fully expanded leaves.

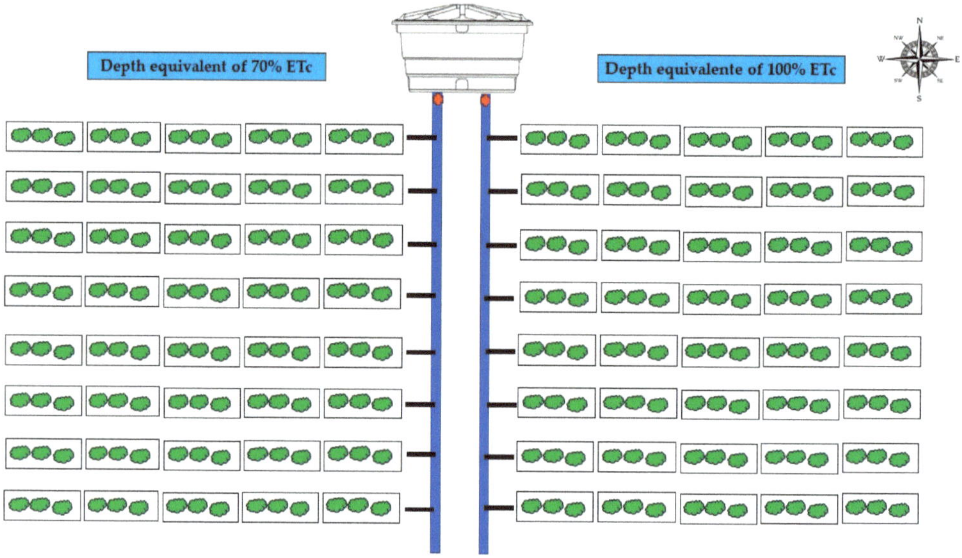

Figure 2. Layout of the experimental area.

2.3. Conducting the Experiment

The plants were distributed at a spacing of 2 m between plants and 3 m between rows in pits 0.4 m in depth with diameters of 0.45 and 0.64 m, corresponding to volumes of 64 dm^3 and 128 dm^3, respectively, to which 0.26% of cattle manure was added before transplanting. The amount of cattle manure applied was to raise the initial soil organic matter content from 0.4 to 2.0%. Before application, the cattle manure was characterized according to the pH and the contents of nutrients [25], as shown in Table 2.

Table 2. Chemical characterization of cattle manure used in the experiment.

Attribute	Value	Attribute	Value
pH (H_2O)	8.81	Mg^{2+} (g kg^{-1})	5.0
C (g kg^{-1})	159.0	S (g kg^{-1})	1.8
C/N	19:1	B (mg kg^{-1})	21.3
Na^+ (mg kg^{-1})	790	Cu (mg kg^{-1})	8.0
N (g kg^{-1})	8.3	Fe (mg kg^{-1})	991
P (g kg^{-1})	2.8	Mn (mg kg^{-1})	250
K (g kg^{-1})	10.4	Zn (mg kg^{-1})	58
Ca^{2+} (g kg^{-1})	8.2		

C = carbon oxidized by potassium dichromate and determined by colorimetry; C/N = carbon/nitrogen ratio; B and Fe—spectrophotometer UV-vis at the wavelengths of 460 and 508 nm, respectively; N—Kjeldahl by wet digestion; P—Mehlich-1 and photocolorimeter, 660 nm; S—spectrophotometry at 420 nm; K^+ and Na^+—flame photometer; Ca^{2+} and Mg^{2+}—atomic absorption spectrophotometer at 422.7 and 285.2 nm, respectively; Cu—atomic absorption spectrophotometer at 324.7 nm; Mn and Zn—atomic absorption spectrometry at 231.9 and 279.5 nm, respectively, with flame-acetylene air.

In the respective pits of 64 dm^3 and 128 dm^3, 50 and 100 g of FTE-BR12 (3.9% S, 1.8% B, 0.85% Cu, 2.0% Mn and 9.0% Zn), 45 and 90 g pit^{-1} of dolomitic limestone (PRNT = 80% and 28% CaO) to raise the base saturation of the soil from 48% to 70% [26] and 9 and 18 g of potassium chloride (KCl—60% of K_2O) to raise the potassium content of the soil from 60 to 90 mg dm^{-3}.

At 30 days after transplanting (DAT), cover fertilization with nitrogen (urea, 45% N) and potassium (KCl) were applied monthly, applying 15 g and 20 g, respectively, during the transplanting period in the vegetative growth phase. During the flowering phase, the values were increased to 24 g and 30 g, and at the end of flowering, which comprises the end of the harvest, applications of 33 g and 60 g were carried out, totaling 231 and 350 g per year^{-1}, respectively, of urea and potassium chloride (Figure 3).

Figure 3. Development stages of the sour passion fruit (initial growth—vegetative (**a**), beginning of flower bud emission (**b**) and fruit production (**c**)).

Fertilization with phosphorus was carried out from 60 DAT, starting with the joint application of N and K, with simple superphosphate (18% P_2O_5, 16% Ca and 8% S), applying 50 g every three months and 100 g at the end of the fruit harvest, totaling 250 g plant^{-1} year^{-1} in 4 applications of simple superphosphate, as recommended by [27].

Irrigations were carried out with water of low salinity (without any restrictions of use) for agriculture (ECw = 0.5 dS m^{-1} and SAR = 2.2 (mmol L^{-1})$^{1/2}$), using the micro-sprinkler irrigation system, with one emitter per plant and a flow of 60 L·h^{-1}, with a pressure of 0.2 MPa. Application of the daily volume (100% of ETc) was carried out based on the crop evapotranspiration (ETc) through ET_0 and the cultivation coefficient (Kc) of the sour passion fruit, with values of 0.43 in the vegetative phase, 0.94 in the flower in phase, and 1.04 in the fruiting phase, as recommended by [28]. The water depth equivalent to 70% of the ETc of the sour passion fruit was obtained through the product of the daily evapotranspiration requirement (100% of ETc) by a factor of 0.7.

2.4. Traits Analyzed

At 120 DAT, when the sour passion fruit plant was in full flower bud emission, the third pair of leaves from the branches located in the middle third of the plants was selected for reading of the chlorophyll index, chlorophyll fluorescence, and gas exchange. The evaluation of chlorophyll a (Chl a), b (Chl b), and total (Chl total) indexes were performed in the morning, using a portable ChlorofiLOG meter from FalKer®.

Chlorophyll fluorescence parameters referring to initial fluorescence (F_0), maximum fluorescence (Fm), and variable fluorescence (Fv), were measured after dark adaptation of leaves for 30 min with clamps, using a modulated Plant Efficiency Analyzer—PEA II® fluorometer. Gas exchange: net photosynthesis rates (A—µmol CO_2 m^{-2} s^{-1}), stomatal conductance (gs—mol H_2O m^{-2} s^{-1}), internal CO_2 concentration (Ci—µmol CO_2 m^{-2} s^{-1}), and leaf transpiration (E—mmol H_2O m^{-2} s^{-1}) were performed with a portable infrared carbon dioxide analyzer (IRGA), model LCpro-SD from BioScientific®, with the temperature set at 25 °C, irradiation of 1200 μmol m^{-2} s^{-1}, and air flow of 200 mL min^{-1} [29].

From the data obtained, water productivity (WP) was calculated, relating net photosynthesis to transpiration (A/E) [(μmol CO_2 m^{-2} s^{-1})(mmol H_2O m^{-2} s^{-1})$^{-1}$]. Total yield (t ha^{-1}) corresponds to the 2018 to 2019 harvests and was obtained by the product of production per plant and the density of plants per hectare (1667 plants ha^{-1}).

2.5. Statistical Analysis

The collected data were subjected to the normality of distribution test (Shapiro–Wilk test) at a probability level of 0.05. Data were submitted to analysis of variance using the F test ($p \leq 0.05$). The values referring to irrigation depths and pit volumes were compared using the F test ($p \leq 0.05$) and the doses of water-retaining polymer by linear (polynomial regression) and non-linear (sigmoidal and peak) models. For data analysis, the statistical software SISVAR version 5.6 [30] was used. A Pearson correlation analysis was performed between the variables and a correlation matrix was created using the "GGally" statistical package from R Studio [31].

3. Results

3.1. Chlorophyll Indices and Fluorescence

The leaf chlorophyll indices and the initial fluorescence of the sour passion fruit cv. BRS GA1 were affected by the interaction among irrigation depths, pit volume, and doses of water-retaining polymer. The irrigation depth × pit volumes interaction had a significant effect on the Fv and Fm of the sour passion fruit cv. BRS GA1, as seen below in Table 3.

Table 3. Summary of variance analysis for leaf chlorophyll indices and chlorophyll fluorescence of the sour passion fruit plants as a function of irrigation depths (ID), pit volume (PV), and doses of water-retaining polymer (WRP).

SV	DF	Mean Squares					
		Chl a	Chl b	Chl Total	F_0	Fm	Fv
Blocks	2	3.09 ^{ns}	0.11 ^{ns}	4.37 ^{ns}	13,952.06 ^{ns}	384,103.26 ^{ns}	378,388.86 ^{ns}
Irrigation depths (ID)	1	293.22 *	0.71 ^{ns}	322.75 *	640.26 ^{ns}	417,333.60 ^{ns}	449,800.41 ^{ns}
Error 1	2	4.64	0.42	0.003	17,617.86	63,411.80	49,449.26
Pit volume (PV)	1	3.66 ^{ns}	3.89 *	5.50	2912.06 ^{ns}	203,933.40 ^{ns}	157,388.81 ^{ns}
Water-retaining polymer (WRP)	4	18.09 **	5.28 **	41.03 **	1069.93 ^{ns}	42,117.66 ^{ns}	33,767.85 ^{ns}
ID × PV	1	1.47 ^{ns}	22.81 **	12.69 *	1288.06 ^{ns}	363,793.06 *	409,200.41 *
ID × WRP	4	49.89 **	1.24 ^{ns}	63.32 **	8125.26 *	50,025.68 ^{ns}	38,505.37 ^{ns}
PV × WRP	4	22.62 **	9.95 **	47.32 **	1398.40 ^{ns}	31,285.06 ^{ns}	31,029.52 ^{ns}
ID × PV × WRP	4	44.44 **	4.04 **	64.84 **	9042.06 *	94,404.98 ^{ns}	74,372.04 ^{ns}
Error 2	36	1.67	0.84	3.02	3010.46	71,006.90	62,160.80
Total	59						
CV_1		5.42	5.56	4.56	29.50	10.76	11.77
CV_2		3.25	7.83	3.37	12.19	11.39	13.19
Mean		39.75	11.75	51.51	449,93	2339.83	1889.78

SV = sources variation; DF = degree of freedom; CV_1 = coefficient of variation of the main plot; Chl a = chlorophyll a index; Chl b = chlorophyll b index; Chl Total = chlorophyll total index; F_0 = initial fluorescence; Fm = maximum fluorescence; Fv = variable fluorescence; CV_2 = coefficient of variation of the subplots; *, **, ^{ns} = significant at 0.05, 0.01 probability and not significant, respectively, by the F test at 0.05 probability.

The Chl a index in the leaves of the sour passion fruit cultivated in the 64 dm^3 pits and irrigated with a water depth of 70% ETc did not fit any regression model, showing an average value of 37.63 (Figure 4a). In the same pit volume, the chlorophyll a index of plants irrigated with 100% ETc in relation to those with 70% of ETc was higher in the absence and at a dose of 1.5 g of the water-retaining polymer. In the pits of 128 dm^3, the plants irrigated with a depth of 70% of ETc presented the highest values of Chl a (41.25) in the absence of the polymer (Figure 4b). However, in plants irrigated with 100% ETc, the Chl a increased linearly from 37.15 to 45.97 with increments of the dose from 0 to 2.0 g dm^{-3} of soil water-retaining polymer, representing an increase of 23.74%.

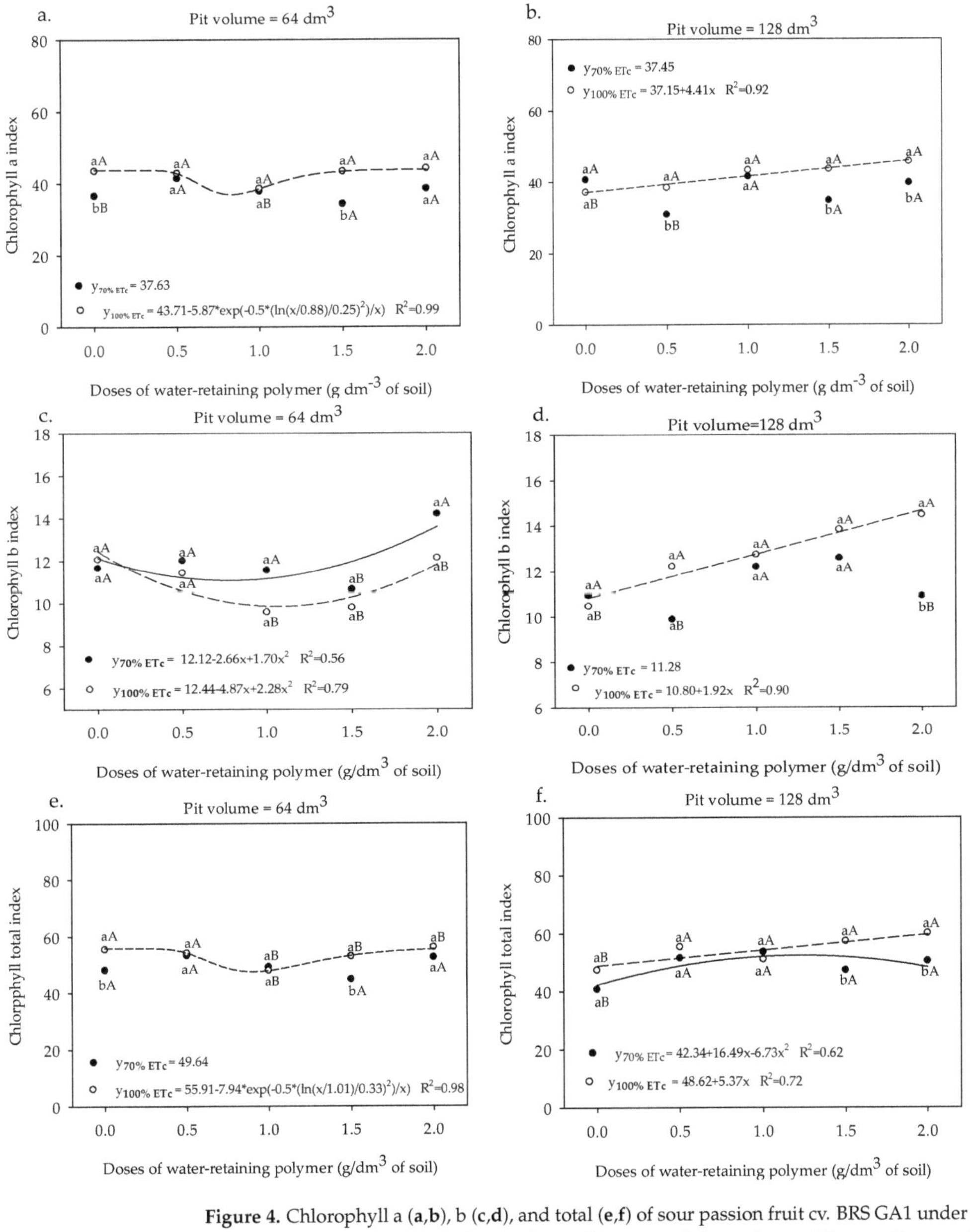

Figure 4. Chlorophyll a (**a**,**b**), b (**c**,**d**), and total (**e**,**f**) of sour passion fruit cv. BRS GA1 under irrigation depths, pit volumes of and, different doses of water-retaining polymer. Means with same lowercase letters do not differ in irrigation depths of 70% and 100% ETc within the same dose of the water-retaining polymer and the pit volume according to the Tukey test ($p > 0.05$), and means with same uppercase letters do not differ in pit volumes of 64 and 128 dm^3 within each dose of water-retaining polymer and irrigation depth according to the Tukey test at 0.05 probability.

In Chl b (Figure 4c,d), the indices were elevated with increasing doses of the water-retaining polymer from 0 to 2.0 g dm^{-3} of soil, except for plants grown in a pit of 128 dm^3 and irrigated with water depths of 70% of ETc, which showed an average value of 11.28. In the sour passion fruit cultivated in the 64 dm^3 pit irrigated with a 70% ETc depth, the Chl b leaf index increased from 11.58 to 14.14 when compared to the absence of the polymer for the dose of 2.0 g dm^{-3}; at 100% ETc depth in the same pit volume, the highest Chl b index was obtained in the absence of the polymer (12.48), with subsequent reduction. In the pit volume of 128 dm^3 and ETc depth of 100%, the Chl b index went from 10.80 to 14.76 between the lowest and highest doses of polymer, representing an increase of 36.66%. The association of the largest pit volume and polymer dose promoted a higher chlorophyll b index when the sour passion fruit plant was irrigated with a water depth of 100% ETc.

For the Chl total index, the plants grown in the 64 dm^3 pit irrigated with a depth of 70% ETc, there was no regression adjustment, represented by a mean of 49.64 (Figure 4e). In plants grown in the same pit volume, under irrigation with 100% ETc, the Chl total index did not vary between treatments without and with the highest dose of the polymer, obtaining an estimated value of 56.00. In plants grown in the 128 dm^3 pit and irrigated with 70% ETc (Figure 4f), it can be seen that increasing the polymer doses up to 1.12 g dm^{-3} provided the highest Chl total index (53.64), and in plants irrigated with 100% ETc, it increased from 48.92 to 59.67 between the pits without and with 2.0 g dm^{-3}.

The initial fluorescence values obtained under different doses of water-retaining polymer did not fit satisfactorily to any regression model when the sour passion fruit was cultivated in the 64 dm^3 pit and irrigated with a depth equivalent to 70% and 100% ETc, presenting mean values of 458.38 and 455.75, respectively (Figure 5a). On the other hand, passion fruit cultivated in a 128 dm^3 pit and irrigated with a water depth of 70% ETc showed reductions in F0 from a polymer dose of 1.0 g dm^{-3} of soil (Figure 5b). The highest F0 values were recorded in the sour passion fruit cultivated in the 128 dm^3 pit, under an irrigation depth of 100% ETc, which was increased from 400.49 (without polymer) to 500.41 at a dose of 2.0 g dm^{-3} of the polymer, an increase of 24.94%.

When evaluating the Fv of the sour passion fruit, it is observed that the pit volume of 128 dm^3 and the irrigation depth of 70% of ETc, provided the highest value of Fv, 2007.8; however, it did not differ statistically from the plants cultivated in the pit volume of 64 dm^3 under the depths of 70% and 100% ETc with values of 1945.2 and 1937.0, respectively. In the 128 dm^3 pit, increasing the irrigation depth from 70% to 100% of ETc reduced Fv by 338.5 in sour passion fruit plants (Figure 5c). Behavior similar to Fv was observed in Fm (Figure 5d), in which the plants of sour passion fruit cultivated in the pit of 128 dm^3 under irrigation with a water depth of 70% ETc presented the highest values of Fm (2442.8), but did not differ statistically from plants grown in a pit volume of 64 dm^3 and irrigated with 70% ETc (2403.5) and 100% ETc (2392.6). The lowest Fm values (2120.4) were recorded in plants grown in pits with a volume of 128 dm^3 and irrigated with 100% ETc.

3.2. *Gas Exchange and Yield*

The variables stomatal conductance, water productivity, and yield of the sour passion fruit cv. BRS GA1 were affected by the interaction of irrigation depths × pit volume × dose of water-retaining polymer (Table 4). The net photosynthesis was influenced by the interactions of PV × WRP and ID × WRP. The internal concentration of CO_2 responded to the interaction of ID × PV, and leaf transpiration responded to the interaction of ID × WRP.

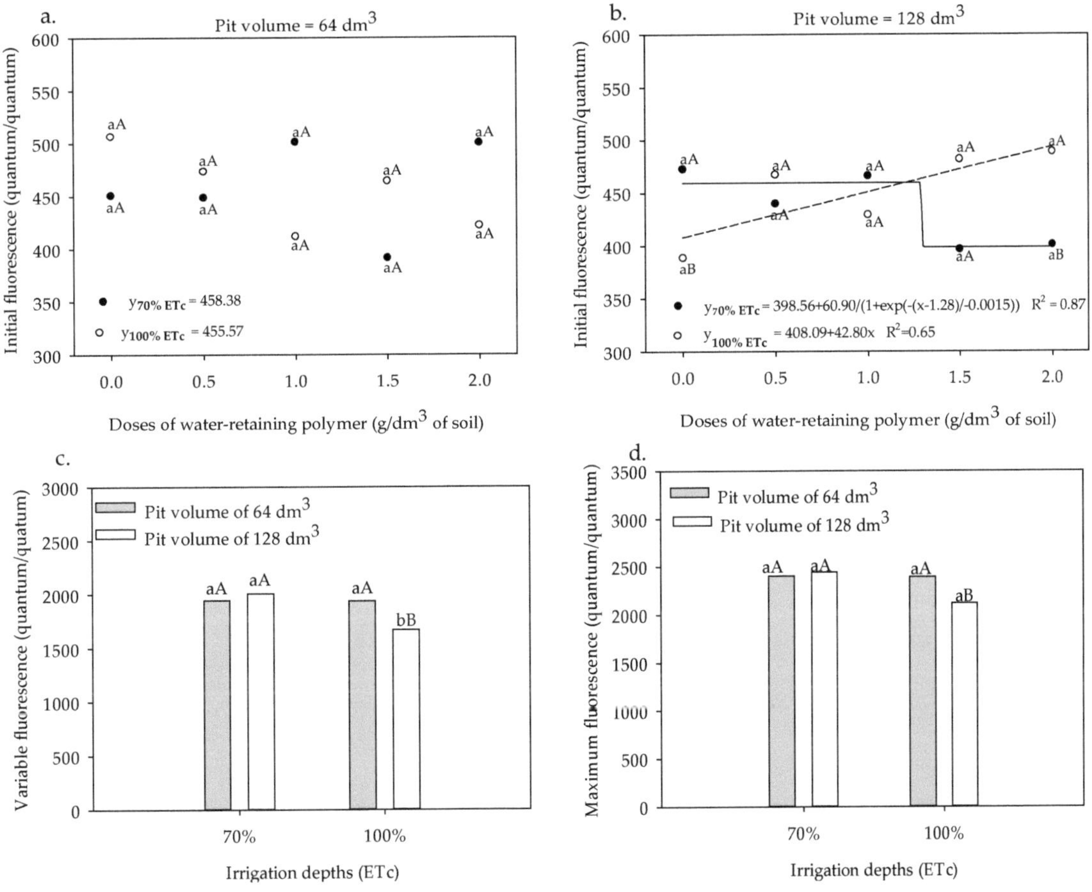

Figure 5. Initial fluorescence (**a**,**b**) in sour passion fruit cv. BRS GA1 under irrigation depths, pit volumes, and different doses of water-retaining polymer; variable fluorescence (**c**) and maximum fluorescence (**d**) under irrigation depths and pit volumes. Means with same lowercase letters do not differ in irrigation depths of 70% and 100% ETc within the same dose of the water-retaining polymer and pit volume according to the Tukey test at 0.05 probability, and means with same uppercase letters do not differ in pit volumes of 64 and 128 dm^3 within each dose of water-retaining polymer and irrigation depth according to the Tukey test ($p > 0.05$). (**a**,**b**). Means with same lowercase letters do not differ in pit volumes of 64 and 128 dm^3 within each irrigation depths according to the Tukey test at 0.05 probability, and means with equal capital letters not differ in irrigation depths of 70% and 100% within each pit volume according to the Tukey test at 0.05 probability (**c**,**d**).

The increase in the dose of the water-retaining polymer up to 1.21 g dm^{-3} of soil increased the *A* of the sour passion fruit in irrigation depths of 100% ETc, with a value of 11.06 μmol CO_2 m^{-2} s^{-1} (Figure 6a). In contrast, in plants irrigated with ETc equivalent to a depth of 70% ETc, the net photosynthesis rate was reduced from 15.19 to 11.27 μmol CO_2 m^{-2} s^{-1} with an increase in the dose from 0 to 2.0 g of the polymer. It was also noted that the net photosynthesis rate was greater in sour passion fruit irrigated with the lowest irrigation depth until the application of a dose of 0.5 g dm^{-3} of the water-retaining polymer, showing no statistical differences with higher doses for the depths of 100% ETc (Figure 6a). In the plants grown in a pit volume of 64 dm^3, the net photosynthesis rate presented an average value of 11.36 μmol CO_2 m^{-2} s^{-1}, which did not fit any tested regression model with increases in the doses of water-retaining polymer (Figure 6b). On the other hand, for those cultivated in a pit volume of 128 dm^3, increasing the dose of the water-retaining polymer up to 1.03 g dm^{-3} raised the *A* of sour passion fruit from 9.64 to 13.53 µmol CO_2 m^{-2} s^{-1}.

Table 4. Summary of analysis of variance for gas exchange and yield of the sour passion fruit plants as a function of irrigation depths (ID), pit volume (PV), and doses of water-retaining polymer (WRP).

SV	DF	Mean Squares					
		A	Gs	Ci	E	WP	Yield
Blocks	2	2.86 ^{ns}	0.0011 **	2908.71 **	0.0001 ^{ns}	5.00 ^{ns}	194.40 **
Irrigation depths (ID)	1	180.26 *	0.0325 **	317.40	6.01 *	1.66 ^{ns}	1.61 ^{ns}
Error 1	2	1.86	0.0001	1.55	0.06	0.46	0.65
Pit volumes (PV)	1	1.86 ^{ns}	0.0001 ^{ns}	1643.26 *	2.81 *	0.60 ^{ns}	248.06 **
Water-retaining polymer (WRP)	4	12.73 **	0.0010 ^{ns}	605.94 ^{ns}	1.77 *	0.62 ^{ns}	107.80 **
ID × PV	1	0.06 ^{ns}	0.0010 ^{ns}	1401.66 *	0.01 ^{ns}	0.06 ^{ns}	345.60 **
ID × WRP	4	17.85 **	0.0063 **	504.60 ^{ns}	2.89 **	0.70 ^{ns}	38.85 **
PV × WRP	4	14.50 **	0.0005 ^{ns}	348.97 ^{ns}	0.27 ^{ns}	0.55 ^{ns}	59.44 **
ID × PV × WRP	4	2.15 ^{ns}	0.0015 **	255.54 ^{ns}	1.39 ^{ns}	1.77 *	68.39 **
Error 2	36	2.29	0.0003	290.52	0.45	0.58	2.39
Total	59						
CV_1		11.85	2.48	0.53	6.54	19.52	4.34
CV_2		13.13	15.10	7.30	17.16	21.86	8.33
Mean		11.53	0.131	233.63	3.95	3.50	18.56

SV = sources variation; DF = degree of freedom; CV_1 = coefficient of variation of the main plot; A = net photosynthesis rates; gs = stomatal conductance; Ci = internal CO_2 concentration; E = leaf transpiration; WP = water productivity; CV_2 = coefficient of variation of the subplots; *, **, ^{ns} = significant at 0.05, 0.01 probability and not significant, respectively, according to the F test at 0.05 probability.

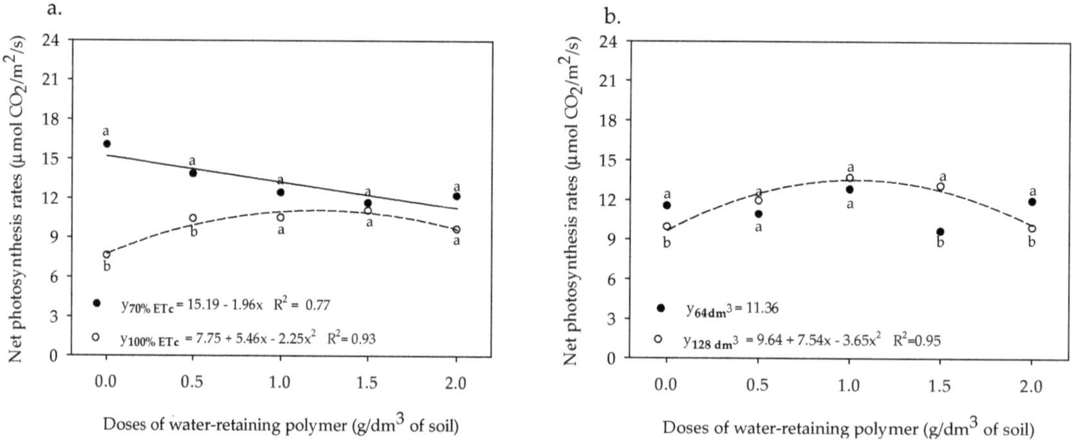

Figure 6. Net photosynthesis rate of sour passion fruit cv. BRS GA1 irrigated with water depths in the soil with different doses of water-retaining polymer (**a**) and cultivated in pit volumes in the soil with different doses of water-retaining polymer (**b**). Means with same lowercase letters in (**a**) do not differ in irrigation depths of 70% and 100% ETc within same dose of water-retaining polymer according to the Tukey test at 0.05 probability. Means with same lowercase letters in (**b**) do not differ in pit volumes of 64 dm^3 and 128 dm^3 within the same dose of water-retaining polymer according to the Tukey test at 0.05 probability.

The gs in the leaves of sour passion fruit cultivated in a pit volume of 64 dm^3 and irrigated with water depths of 70% and 100% ETc did not fit satisfactorily any regression model, presenting mean values of 0.16 and 0.10 mmol H_2O m^{-2} s^{-1}, respectively (Figure 7a). In the pit volume of 64 dm^3, only in the absence of the polymer and at a dose of 1.0 g dm^{-3} were there statistical differences in the stomatal conductance of sour passion fruit, with superiority in plants irrigated with 70% of ETc (Figure 7a). In the pit of 128 dm^3 and irrigation with 70% of ETc, the gs also did not fit any regression model, showing a mean value of 0.14 mmol m^{-2} s^{-1} (Figure 7b). At the 100% ETc depth, the gs was increased up to a dose of 1.06 g of the polymer dose, obtaining greater stomatal opening with a value of 0.1406 mmol CO_2 m^{-2} s^{-1}, an increase of 180.0% in relation to the absence of the polymer (Figure 7b).

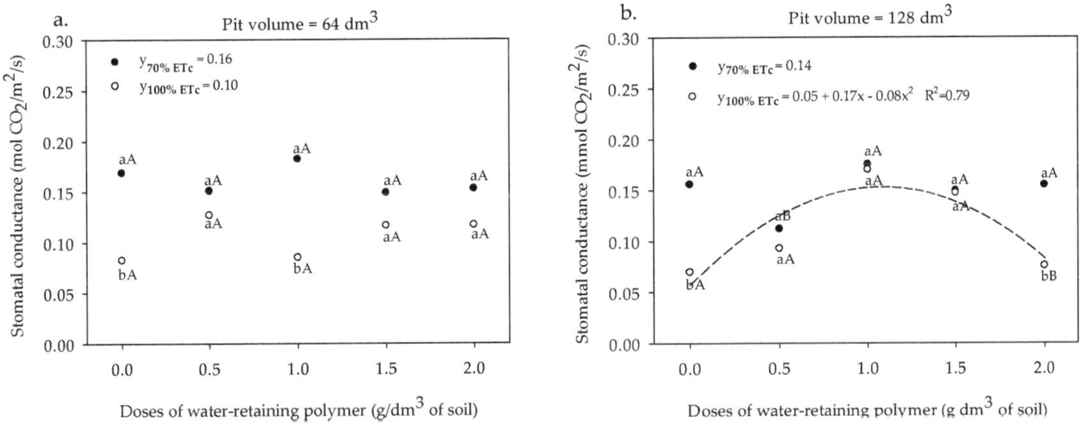

Figure 7. Stomatal conductance of the sour passion fruit cv. BRS GA1 irrigated with irrigation depths, pit volumes, and different doses of water-retaining polymer (**a**,**b**). Means with same lowercase letters do not differ in irrigation depths of 70% and 100% ETc within same dose of the water-retaining polymer and the pit volume according to the Tukey test at 0.05 probability, and means with same uppercase letters do not differ in pit volumes of 64 and 128 dm^3 within same dose of water-retaining polymer and irrigation depth according to the Tukey test at 0.05 probability.

Regarding the Ci of sour passion fruit in plants grown in the 64 dm^3 pit, a reduction of 2.20% was observed with an increase in irrigation depth. In the pit of 128 dm^3, the increase in irrigation depth from 70% to 100% ETc provided the highest leaf Ci (246.0 μmol CO_2 m^{-2} s^{-1}), showing an increase of 6.17%; however, statistically, irrigation depths did not differ from each other. At the 70% depth, increasing the pit volume from 64 to 128 dm^3 increased leaf Ci from 230.9 to 231.7 μmol CO_2 m^{-2} s^{-1} (Figure 8a). The leaf transpiration of sour passion fruit cv. BRS GA1 irrigated with 70% ETc showed a mean value of 4.21 mmol H_2O m^{-2} s^{-1} (Figure 8b). With 100% ETc, leaf *E* was increased to 4.28 mmol H_2O m^{-2} s^{-1} when the water-retaining polymer was applied at a dose of 1.12 g per dm^{-3} of soil.

The water productivity of the sour passion fruit cultivated in a pit volume of 64 dm^3 and irrigated with a depth of 70% ETc showed the highest PW of 4.04 [(mmol CO_2 m^{-2} s^{-1})(mmol H_2O m^{-2} s^{-1})$^{-1}$] in the absence of the polymer, and reduced by 16.13% when subjected to the application of 2.0 g dm^{-3} of the water-retaining polymer. At the irrigation depth of 100% ETc, the WP data did not fit satisfactorily any mathematical model, presenting a mean value of 3.38 [(mmol CO_2 m^{-2} s^{-1})(mmol H_2O m^{-2} s^{-1})$^{-1}$] (Figure 9a). The application of the doses of the water-retaining polymer up to a dose of 2.0 g dm^{-3} associated with a higher pit volume (128 dm^3) and irrigation with a depth of 70% ETc increased the WP of the sour passion fruit up to 4.85 [(mmol CO_2 m^{-2} s^{-1})(mmol H_2O m^{-2} s^{-1})$^{-1}$], showing an increase of 78.6% when compared to the absence of the polymer. In the sour passion fruit irrigated with ETc of 100% irrigation depth, the water productivity data did not fit satisfactorily any regression model and are represented by an average value of 3.38 [(mmol CO_2 m^{-2} s^{-1}) (mmol H_2O m^{-2} s^{-1})$^{-1}$] (Figure 9b).

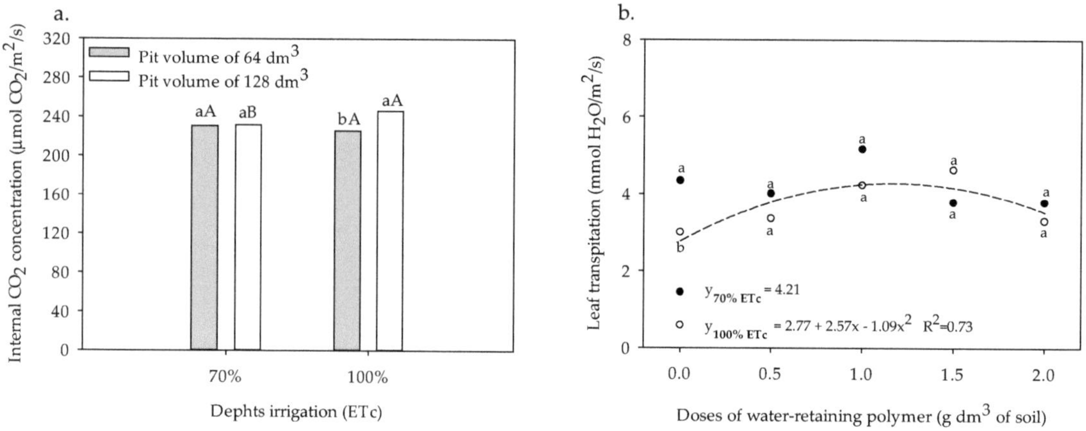

Figure 8. Internal CO_2 concentration in leaves of the sour passion fruit cv. BRS GA1 cultivated in pit volumes and irrigated with irrigation depths (**a**) and leaf transpiration of sour passion fruit cv. BRS GA1 irrigated with irrigation depths and different doses of water-retaining polymer (**b**). Means with the same lowercase letters do not differ in pit volumes of 64 and 128 dm^3 within each irrigation depth according to the Tukey test at 0.05 probability, and means with the same capital letters not differ in irrigation depths of 70% and 100% within each pit volume according to the Tukey test at 0.05 probability (**a**). Means with same lowercase letters do not differ in irrigation depths of 70% and 100% ETc within same dose of water-retaining polymer according to the Tukey test at 0.05 probability (**b**).

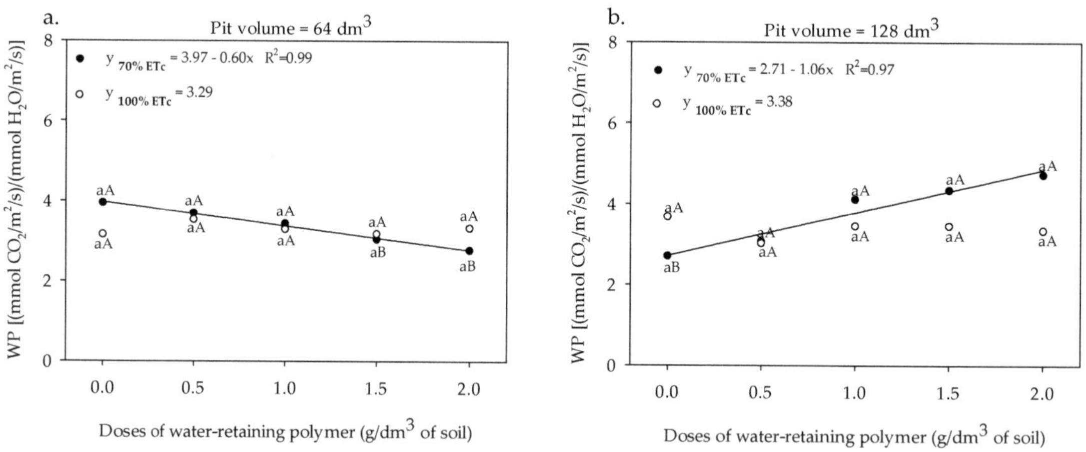

Figure 9. Water productivity (WP) of sour passion fruit under irrigation depth, pit volumes and, different doses of water-retaining polymer (**a**,**b**). Means with same lowercase letters do not differ in irrigation depths of 70% and 100% ETc within same dose of the water-retaining polymer and the pit volume according to the Tukey test at 0.05 probability and means with same uppercase letters not differ in pit volumes of 64 and 128 dm^3 within same dose of water-retaining polymer and irrigation depth according to the Tukey test at 0.05 probability.

The cultivation of sour passion fruit in a pit of 64 dm^3 and under 70% ETc provided a higher yield compared to that irrigated with 100% ETc, with an accumulated value of 31.5 t ha^{-1} in plants submitted to 1.1 g of the polymer (Figure 10a). In the same pit volume, increasing doses of the water-retaining polymer reduced fruit yield in sour passion fruit, resulting in losses of 29.45% when comparing the highest and lowest doses of the polymer (Figure 10a). The accumulated yield of the sour passion fruit cultivated in a pit of 128 dm^3 and irrigated with a depth of 70% ETc was increased up to a dose of 1.29 g polymer dm^{-3} of soil, with an estimated value of 18.8 t ha^{-1} of fruits (Figure 10b). In the sour passion fruit irrigated with an irrigation level of 100% ETc and cultivated in a pit volume of 128 dm^3, the yield was increased from 10.38 to 22.16 t ha^{-1}, representing gains of 113.48%. When comparing the maximum values of the accumulated yield obtained in each volume of the pit, it is verified that the volume of 64 dm^3 was superior in 40.82% in relation to the accumulated yield of the sour passion fruit cultivated in the pit of 128 dm^3 (Figure 10b).

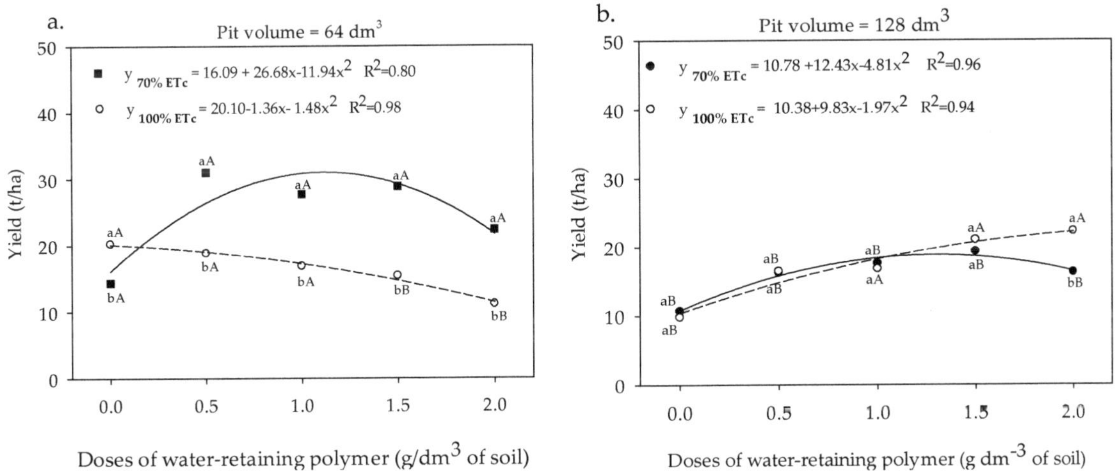

Figure 10. Yield of sour passion fruit under depths of irrigation, pit volumes, and different doses of water-retaining polymer (**a**,**b**). Means with same lowercase letters do not differ in irrigation depths of 70% and 100% ETc within same dose of the water-retaining polymer and the pit volume according to the Tukey test at 0.05 probability, and means with same uppercase letters do not differ in pit volumes of 64 and 128 dm^3 within same dose of water-retaining polymer and irrigation depth according to the Tukey test at 0.05 probability.

3.3. Correlation Matrix

Figure 11 presents the multiple correlation analysis between the different variables of chlorophyll index, fluorescence, gas exchange, and sour passion fruit yield. According to the positive Pearson correlation values, the increase in photosynthetic rate in sour passion fruit is associated with greater stomatal opening (0.649 ***) and leaf transpiration (0.493 ***). This greater stomatal conductance contributed to higher evaporation rates of the sour passion fruit (0.644 **). The increase in leaf chlorophyll indices in sour passion fruit increases the initial fluorescence of chlorophyll a, with values of 0.295 * (Chla), 0.266 * (Chlb), and 0.322 ** (total Chl), respectively; however, it reduced the water use—Chla (−0.373 **) and Chl total (−0.307 *). The variable fluorescence showed a strong relationship with the maximum fluorescence (0.976 ***), positively influencing the increase in the other variable. The Chlb index was the only variable that influenced the yield of passion fruit, causing a positive effect on fruit production (0.268 *).

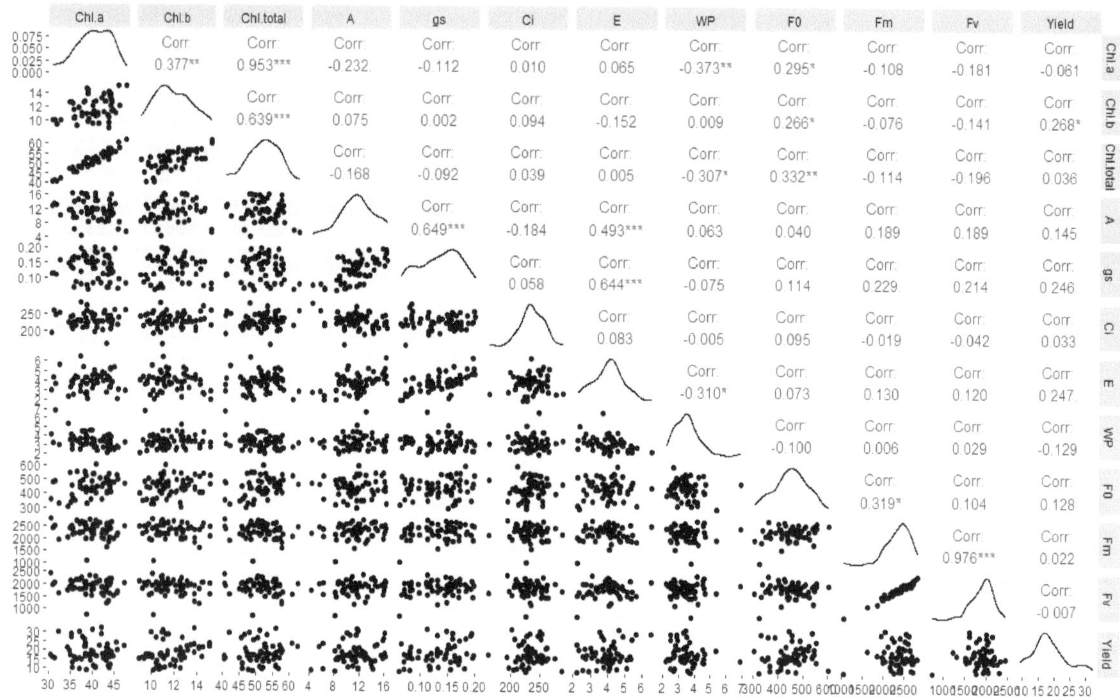

Figure 11. Pearson correlation matrix for the variables of chlorophyll index, fluorescence, gas exchange, and yield of the sour passion fruit under irrigation depths, pit volumes, and doses of water-retaining polymer. Chl a = chlorophyll a, Chl b = chlorophyll b, Chl total = chlorophyll total, A = net photosynthesis, gs = stomatal conductance, Ci = stomatal conductance, E = foliar transpiration, WP = water productivity, F0 = initial fluorescence, Fm = maximum fluorescence, and Fv = variable fluorescence, *, **, *** = correlation significant at 0.05, 0.01 and 0.001, respectively.

4. Discussion

The greater volume of the pits allows for greater root expansion and, consequently, greater use of water, and the application of the polymer associated with adequate water management promotes greater efficiency as a result of its soil conditioning action, in which it increases the capacity of soil water storage [32]. In this sense, the greater volume of the pit associated with greater water availability and greater water retention provided by the water-retaining polymer contributed to the higher chlorophyll indices in sour passion fruit (Figure 4). The linear increase observed in the Chl b index (Figure 4d) may be due to a negative adjustment of the enzymatic activity of Chl b reductase, responsible for the conversion of Chl b to Chl a [33], indicating that the photosynthetic apparatus is suffering damage in the chlorophyll fluorescence, which can be confirmed by observing the fluorescence data (Figure 5), which showed damage in the PSII. Behavior similar to our study was observed by [34] in eucalyptus (*Eucalyptus dunnii Maiden*) on the ground with a water-retaining polymer, which found increases in leaf chlorophyll levels depending on the irrigation frequency. The authors attributed to the water-retaining polymer a reduction in nutrient losses, mainly N, which is directly linked to the structuring composition of photosynthetic pigments in plants.

An increase in F0 may indicate that damage has occurred in the photosynthetic apparatus of plants, or a reduction in the ability to transfer excitation energy from the antenna to the photochemical reaction center [8,9]. According to Baker [35], an increase in F0 and a reduction in Fm are considered indicative of stress in plants, and according to the results observed (Figure 5d), the sour passion fruit cultivated in the pit of 128 dm^3 irrigated with a water depth of 100% ETc were under stress, probably caused by excess water present in the passion fruit root zone, which was aggravated by the increase in polymer doses.

The reduction in Fv in the pit of 128 dm^3 and irrigation with a depth of 100% ETc (Figure 5c) is related to the abiotic conditions to which the plants were submitted, in this case, the stress caused by excess water in the root zone of the crop, which caused damage to the photosynthetic apparatus, compromising the PSII [29]. The reduction in Fm in the greater volume of the pit with the increase in the irrigation depth (Figure 5d) demonstrates that high levels of water in the soil may have caused anoxia of the plant roots, initiating a stress condition, causing a reduction in the efficiency of photosystem II [36].

The increase in the net photosynthesis rate of the sour passion fruit at 70% ETc and an estimated dose of the water-retaining polymer of 0.5 g dm^{-3} of soil (Figure 6a) is related to the improvement in the maintenance of soil moisture close to the root system of the plants, provided by the retention of water in the polymer, made available by the respective water depth, increasing the availability of water and nutrients for plants, which contributes to increases in the net photosynthetic rate [4]. The highest net photosynthetic rates observed in Figure 6b are due to the fact that larger pit volumes allow for greater growth of lateral roots of the plants, providing greater absorption of irrigation water, since the highest concentrations of sour passion fruit roots are present at 0.60 cm from the stem, and the largest pit volume of the present study is 0.64 dm^3. This provided a greater contact area of the roots with water, due to the greater soil moisture provided by the application of the water-retaining polymer that acts by reducing water losses and increasing the availability of water, which can positively affect the net photosynthetic rate by supplying electrons from the photolysis of H_2O molecules [14,19,37].

The greater stomatal opening observed in the sour passion fruit irrigated with a depth of 70% ETc in the smallest pit volume and a water depth of 100% ETc in the pit with the largest volume with a dose of 1.06 g of the polymer (Figure 7) is related to the greater availability of water provided by the water-retaining polymer [4,13,14], that is, when water in the soil is not a limiting factor, the sap flow is high between the conducting vessels and the aerial part, stimulating stomatal opening and a greater frequency of water flow into the atmosphere [38,39].

The increase in Ci as a function of the greater irrigation depth and greater pit volume (Figure 8a) is related to the greater stomatal opening observed in (Figure 7b), and is indicative that, in fact, there was no restriction on the acquisition of CO_2 by the crop, since the greater the stomatal opening, the greater the diffusion of CO_2 in the substomatic chamber (Silva et al. 2015) [8]. However, in plants submitted to the lowest irrigation depth, the fixation process during the carboxylation phase was compromised, a fact observed in the present study, in which there was a reduction in the acquisition of CO_2 in the lowest water depth. The increase in the leaf transpiration rate of the sour passion fruit irrigated with a water depth of 100% ETc and a dose of 1.12 g dm^{-3} of the water-retaining polymer (Figure 8b) is related to the increase in the larger stomatal opening (Figure 7b), which was observed by [40] in passion fruit, in which stomatal conductance interfered with the evapotranspiration loss of leaves.

The application of the polymer resulted in considerable increases in photosynthetic rates (Figure 6a), which compensated for the higher transpiration rates (Figure 8b) and determined the increase in WP (Figure 9), which was reflected in fruit yield (Figure 10). The greater volume of the pit provides greater WP due to the greater presence of roots within a radius of 0.60 cm from the stem [19], which absorb and reduce water losses by lateral flow, optimizing the water productivity of the sour passion fruit (Figure 9). And according to [41], monitoring the WP plays an important role in the process of water losses through plant transpiration during gas exchange and, consequently, in the assimilation of CO_2.

The application of the polymer up to an estimated dose of 1.1 g dm^{-3} in the plants irrigated with 70% ETc (Figure 8a) probably reduced the losses of water and nutrients by leaching, contributing to the increase in yield of sour passion fruit, and according to [42–44], the polymer acts to maintain the physical and chemical properties of the soil, increasing the availability of nutrients, which is reflected in the increase in yield of the crops.

5. Conclusions

The interaction of irrigation depths × pit volumes × doses of water-retaining polymer influences chlorophyll indexes, initial fluorescence, gas exchange through stomatal conductance, leaf transpiration, and water productivity, with positive impacts on yield of the sour passion fruit. The irrigation depth of 70% ETc associated with 1.1 g of polymer dm^{-3} of soil is recommended, since it increased the yield of the sour passion fruit cultivated in a pit with a volume of 64 dm^3. Doses of 2.0 g dm^{-3} of the water-retaining polymer, in a pit volume of 128 dm^3 and associated with an irrigation depth of 100% ETc, probably cause stress in sour passion fruit plants due to excess water. As a future perspective, research should be conducted to understand the mechanism and interaction of soil–water-retaining polymer–plant in the acquisition of water and nutrients and the influence on the physiological and biochemical aspects and yield of fruit crops.

Author Contributions: Conceptualization, A.G.d.L.S., E.N.d.M. and L.F.C.; methodology, A.G.d.L.S., E.N.d.M. and A.P.P.d.N.; software, R.O.B., F.d.O.M. and E.F.d.M.; validation, L.F.C., Í.H.d.L.C. and H.R.G.; formal analysis, A.G.d.L.S.; G.S.d.L. and R.T.d.L.; investigation, G.L.F.d.S., G.R.S. and P.V.C.d.O.; resources, D.V.S., Í.H.d.L.C. and H.R.G.; data collection, A.G.d.L.S., E.N.d.M. and A.P.P.d.N.; writing—original draft preparation, A.G.d.L.S., E.N.d.M. and L.F.C.; writing—review and editing, H.R.G., D.V.S. and R.O.B.; visualization, E.F.d.M., F.d.O.M. and G.S.d.L.; supervision, L.F.C. and Í.H.d.L.C.; project administration, A.G.d.L.S. and L.F.C.; funding acquisition, A.G.d.L.S. All authors have read and agreed to the published version of the manuscript.

Funding: CNPq (National Council for Scientific and Technological Development), CAPES (Coordination for the Improvement of Higher Education Personnel), and UFPB (Universidade Federal da Paraíba).

Data Availability Statement: Data are contained within the article.

Acknowledgments: The authors would like to thank the Postgraduate Program in Agronomy (PPGAgro) of the Universidade Federal da Paraíba (UFPB), the National Council for Scientific and Technological Development (CNPq), and the Coordination for the Improvement of Higher Education Personnel (CAPES) for their support in carrying out this research.

Conflicts of Interest: The authors declare no conflicts of interest.

References

1. Thokchom, R.; Mandal, G. Production preference and importance of passion fruit (*Passiflora edulis*): A Review. *J. Agric. Eng. Food Technol.* **2017**, *4*, 27–30.
2. Instituto Brasileiro de Geografia e Estatística [IBGE]. *Brazilian Production of Passion Fruit*; IBGE: Rio de Janeiro, Brazil, 2021.
3. Freitas, G.C.d.O.; dos Santos, W.F.; da Paz, J.C.; da Silva, E.R. O cultivo do maracujazeiro no Centro—Oeste do Brasil. *Rev. Agrotecnol.* **2020**, *11*, 45–53.
4. Cavalcante, A.G.; Cavalcante, L.F.; Souto, A.G.d.L.; Cavalcante, A.C.P.; de Araújo, D.L.; do Nascimento, A.P.P.; Zanuncio, J.C. Physiology and production of yellow passion fruit with hydroabsorbent polymer and different irrigation depths. *Rev. Ceres* **2020**, *67*, 365–373. [CrossRef]
5. Araújo, D.L.; Souto, A.G.d.L.; Cavalcante, A.G.; Cavalcante, L.F.; Pereira, W.E.; de Melo, A.S. Physiological aspects of yellow passion fruit with use of hydrogel and mulching. *Rev. Caatinga* **2022**, *35*, 382–392. [CrossRef]
6. Lira, R.M.; dos Santos, A.N.; da Silva, J.S.; Barnabé, J.M.C.; de Barros, M.S.; Soares, H.R. A utilização de águas de qualidade inferior na agricultura irrigada. *Rev. Geama* **2015**, *1*, 341–362.
7. Qi, Y.; Ma, L.; Ghani, M.I.; Peng, Q.; Fan, R.; Hu, X.; Chen, X. Effects of Drought Stress Induced by Hypertonic Polyethylene Glycol (PEG-6000) on *Passiflora edulis* Sims Physiological Properties. *Plants* **2023**, *12*, 2296. [CrossRef]
8. da Silva, F.G.; Dutra, W.F.; Dutra, A.F.; de Oliveira, I.M.; Filgueiras, L.M.B.; de Melo, A.S. Gas exchange and chlorophyll fluorescence in eggplant plants under irrigation depths. *Rev. Bras. Eng. Agríc. Ambien.* **2015**, *10*, 946–952. [CrossRef]
9. da Silva, A.R.A.; Bezerra, F.M.L.; de Lacerda, C.F.; de Sousa, C.H.C.; Bezerra, M.A. Physiological responses of dwarf coconut plants under water deficit in salt-affected soils. *Rev. Caatinga* **2017**, *30*, 447–457. [CrossRef]

10. dos Anjos Veimrober Júnior, L.A.; da Silva, A.J.P.; Gheyi, H.R.; do Nascimento, F.A.L.; da Silva, M.G.; Vellame, L.M. Water productivity of passion fruit under different forms of propagation and soil-based irrigation management criteria. *Irrig. Sci.* **2022**, *40*, 423–433. [CrossRef]
11. Navroski, M.C.; Araújo, M.M.; de Oliveira Pereira, M.; Fior, C.S. Influência do polímero hidroretentor nas características do substrato comercial para produção de mudas florestais. *Interciência* **2016**, *41*, 357–361.
12. Pontes Filho, R.A.; Gondim, F.A.; Costa, M.C.G. Seedling growth of tree species under doses of hydrogel and two levels of luminosity. *Rev. Árvore* **2018**, *42*, e420112. [CrossRef]
13. Chang, L.; Xu, L.; Liu, Y.; Qiu, D. Superabsorbent polymers used for agricultural water retention. *Polym. Test.* **2021**, *94*, 107021. [CrossRef]
14. Malik, S.; Chaudhary, K.; Malik, A.; Punia, H.; Sewhag, M.; Berkesia, N.; Nagora, M.; Kalia, S.; Malik, K.; Kumar, D.; et al. Superabsorbent Polymers as a Soil Amendment for Increasing Agriculture Production with Reducing Water Losses under Water Stress Condition. *Polymers* **2023**, *15*, 161. [CrossRef]
15. Carvalho, R.P.; Cruz, M.d.C.M.; Martins, L.M. Irrigation frequency using hydro absorbent polymer in the production of yellow passion fruit seedlings. *Rev. Bras. Frut.* **2013**, *35*, 518–526. [CrossRef]
16. Fernandes, D.A.; Araujo, M.M.V.; Camili, E.C. Formation of yellow passion fruit seedlings under different irrigation depths and use of hydrogel. *Rev. Agric.* **2015**, *90*, 229–236. [CrossRef]
17. Khalil, G.A.; Ali, M.A.; Harhash, M.M.; ELSegieny, A.M. Impact of super absorbent polymers (hydrogels) on water use parameters of plum trees under water stress conditions. *J. Adv. Agric. Res.* **2022**, *27*, 791–803. [CrossRef]
18. Alshallash, K.S.; Sharaf, M.; Hmdy, A.E.; Khalifa, S.M.; Abdel-Aziz, H.F.; Sharaf, A.; Ibrahim, M.T.S.; Alharbi, K.; Elkelish, A. Hydrogel improved growth and productive performance of mango trees under semi-arid condition. *Gels* **2022**, *8*, 602. [CrossRef]
19. Lucas, A.A.T.; Frizzone, J.Á.; Coelho Filho, M.A. Characteristics of the root distribution of passion fruit under irrigation. *Irriga* **2012**, *17*, 245–250. [CrossRef]
20. Junghans, T.G.; de Jesus, O.N. *Maracujá do Cultivo à Colheita*, 1st ed.; Embrapa: Brasília, Brazil, 2017.
21. Tavares, O.C.H.; Lima, E.; Zonta, E. Growth and productivity of sugarcane plant grown in different soil preparation and harvesting systems. *Acta Sci. Agron.* **2010**, *32*, 61–68. [CrossRef]
22. Moraes, M.T.; Debiasi, H.; Franchini, J.; da Silva, V.R. Benefits of cover crops on the physical properties of the soil. In *Soil and Water Management and Conservation in Small Rural Properties in Southern Brazil: Alternative Management Practices Aimed at Soil and Water Conservation*; Tiecher, T., Ed.; UFRGS: Porto Alegre, Brazil, 2016; pp. 34–48.
23. Alvares, C.A.; Stape, J.L.; Sentelhas, P.C.; Gonçalves, J.L.M.; Sparovek, G. Köppen's climate classification map for Brazil. *Meteorol. Z.* **2013**, *22*, 711–728. [CrossRef]
24. US Soil Survey Staff. *Keys to Soil Taxonomy*; United States Department of Agriculture and Natural Resources Conservation Service: Lincoln, NE, USA, 2014; 332p.
25. Teixeira, P.C.; Donagemma, G.K.; Fontana, A. *Soil Analysis Methods Manual*, 3rd ed.; Embrapa: Brasília, Brazil, 2017; 573p.
26. Borges, A.L.; Coelho, E.F. Fertigation in Tropical Fruit Trees. In *Portuguese with English Summary*, 2nd ed.; Embrapa: Cruzdas Almas, Brazil, 2009; 179p.
27. Souza, J.T.A.; Nunes, J.C.; Cavalcante, L.F.; Nunes, J.A.d.S.; Pereira, W.E.; Freire, J.L.d.O. Effects of water salinity and organomineral fertilization on leaf composition and production in *Passiflora edulis*. *Rev. Bras. Eng. Agríc. Ambien.* **2018**, *22*, 235–240. [CrossRef]
28. Nogueira, E.; Gomes, E.R.; de Sousa, V.F.; da Silva, L.R.A.; Broetto, F. Crop coefficient and irrigation depths of yellow passion fruit under semi-arid conditions. In Proceedings of the II Inovagri International Meeting, Fortaleza, Brazil, 13–16 April 2014.
29. Freire, J.L.d.O.; Dias, T.J.; Cavalcantee, L.F.; Fernandes, P.D.; Lima Neto, A.J.d. Rendimento quântico e trocas gasosas em maracujazeiro amarelo sob salinidade hídrica, biofertilização e cobertura morta. *Rev. Ciênc. Agron.* **2014**, *45*, 82–91. [CrossRef]
30. Ferreira, D.F. Sisvar: A computer analysis system to fixed effects split plots type designs. *Rev. Bras. Biom.* **2019**, *37*, 529–535. [CrossRef]
31. R Core Team. *R: A Language and Environment for Statistical Computing*; R Foundation for Statistical Computing: Vienna, Austria, 2020.
32. de Carvalho, J.M.; Caldeira, D.S.A.; Vieira, C.L.; da Silva, G.V.B.; Rocha, R.R.; de Campos, I. Crescimento inicial de mudas de *Sapindus saponária* com uso de hidrogel e lâminas de água. *Sci. Elec. Arch.* **2022**, *15*, 7–12. [CrossRef]
33. Gomes Do Ó, L.M.; Cova, A.M.W.; Silva, P.C.C.; Gheyi, H.R.; de Azevedo Neto, A.D.; Ribas, R.F. Aspectos bioquímicos e fluorescência da clorofila a em plantas de minimelancia hidropônica sob estresse salino. *Irriga* **2021**, *26*, 221–239. [CrossRef]
34. Felippe, D.; Navroski, M.C.; Pereira, M.d.O.; Baptista, K.R.S.d. Hydrogel in the seedling growth of *Eucalyptus dunnii Maiden* under different irrigation management. *Rev. Ambient. Água* **2021**, *16*, e2582. [CrossRef]
35. Baker, N.R. Chlorophyll fluorescence: A probe of photosynthesis in vivo. *Annu. Rev. Plant Biol.* **2008**, *59*, 89–113. [CrossRef]
36. Cremon, T.; Scalon, S.d.P.Q.; Rosa, D.B.C.J.; Dresch, D.M. How does *Copaifera langsdorffii* respond to flooding under different irradiance levels? *Plant Biosyst.* **2022**, *156*, 68–78. [CrossRef]
37. Mendonça, T.G.; Querido, D.M.; Souza, C.F. Eficiência do polímero hidroabsorvente na manutenção da umidade do solo no cultivo de alface. *Rev. Bras. Agri. Irri.* **2015**, *9*, 239–245. [CrossRef]
38. Azizi, P.; Rafii, M.Y.; Maziah, M.; Abdullah, S.N.A.; Hanafi, M.M.; Latif, M.A.; Rashid, A.A.; Sahebi, M. Understanding the shoot apical meristem regulation: A study of the phytohormones, auxin and cytokinin, in rice. *Mech Dev.* **2015**, *135*, 1–15. [CrossRef]

39. Taiz, L.; Zeiger, E.; Møller, I.M.; Murphy, A. *Abiotic Stress*, 6th ed.; Artmed: Porto Alegre, Brazil, 2017; pp. 732–761.
40. Gomes, M.M.A.; Ramos, M.J.M.; Torres Netto, A.; Rosa, R.C.C.; Campostrini, E. Water relations, photosynthetic capacity, and growth in passion fruit (*Passiflora edulis* Sims f. flavicarpa Deg). *Rev. Ceres* **2018**, *65*, 135–143. [CrossRef]
41. Cajazeira, J.P.; Correa, M.C.M.; Almeida, E.I.B.; Queiroz, R.F.; Mesquita, R.O. Growth and gas exchange in white pitaya under different concentrations of potassium and calcium. *Rev. Ciênc. Agron.* **2018**, *49*, 112–121. [CrossRef]
42. Elshafie, H.S.; Camele, I. Applications of Absorbent Polymers for Sustainable Plant Protection and Crop Yield. *Sustainability* **2021**, *13*, 3253. [CrossRef]
43. Mansour, H.A.; Jiandong, H.; Hongjuan, R.; Kheiry, A.N.O.; Abd-Elmabod, S.K. Influence of using automatic irrigation system and organic fertilizer treatments on faba bean water productivity. *Int. J. GEOMATE* **2019**, *17*, 250–259. [CrossRef]
44. Mansour, H.A.; Abdel-Hady, M.; Eldardiry, E.I.; Bralts, V.F. Performance of automatic control different localized irrigation systems and lateral lengths for:1-Emitters clogging and maize (*Zea mays* L.) growth and yield. *Int. J. GEOMATE* **2015**, *9*, 1545–1552. [CrossRef]

plants

Article

Exploring the Differential Impact of Salt Stress on Root Colonization Adaptation Mechanisms in Plant Growth-Promoting Rhizobacteria

Lorena del Rosario Cappellari, Pablo Cesar Bogino, Fiorela Nievas, Walter Giordano and Erika Banchio *

INBIAS Instituto de Biotecnología Ambiental y Salud, CONICET—Universidad Nacional de Río Cuarto, Campus Universitario, Río Cuarto 5800, Argentina; lcappellari@exa.unrc.edu.ar (L.d.R.C.); pbogino@exa.unrc.edu.ar (P.C.B.); wgiordano@exa.unrc.edu.ar (W.G.)
* Correspondence: ebanchio@exa.unrc.edu.ar

Abstract: Salinity inhibits plant growth by affecting physiological processes, but soil microorganisms like plant growth-promoting rhizobacteria (PGPR) can alleviate abiotic stress and enhance crop productivity. However, it should be noted that rhizobacteria employ different approaches to deal with salt stress conditions and successfully colonize roots. The objective of this study was to investigate the effect of salt stress on bacterial survival mechanisms such as mobility, biofilm formation, and the autoaggregation capacity of three plant growth-promoting strains: *Pseudomonas putida* SJ04, *Pseudomonas simiae* WCS417r, and *Bacillus amyloliquefaciens* GB03. These strains were grown in diluted LB medium supplemented with 0, 100, 200, or 300 mM NaCl. Swimming and swarming mobility were evaluated in media supplemented with 0.3 and 0.5% agar, respectively. Biofilm formation capacity was quantified using the crystal violet method, and the autoaggregation capacity was measured spectrophotometrically. In addition, we evaluated in vitro the capacity of the strains to ameliorate the effects of saline stress in *Mentha piperita*. The study found that the GB03 strain exhibited enhanced swarming mobility when the salt concentration in the medium increased, resulting in a two-fold increase in the halo diameter at 300 mM. However, high concentrations of NaCl did not affect the swimming mobility. In contrast, swimming motility was reduced in WCS417r and SJ04 under salt stress. On the other hand, exposure to 300 mM NaCl resulted in a 180% increase in biofilm formation and a 30% rise in the percentage of autoaggregation in WCS417r. Conversely, the autoaggregation percentage of the strains SJ04 and GB03 remained unaffected by saline stress. However, for GB03, biofilm formation decreased by 80% at 300 mM. Simultaneously, inoculation with the three evaluated strains alleviated the detrimental effects of salinity on plant growth. Under 150 mM salt stress, all strains showed increased fresh weight, with GB03 and WCS417r improving by 40% and SJ04 exhibiting the most remarkable effect with a 70% rise compared to non-inoculated plants. Despite their different strategies for mitigating salt stress, the application of these strains presents a promising strategy for effectively mitigating the negative consequences of salt stress on plant cultivation.

Keywords: PGPR; salt stress; biofilm; motility; autoaggregation capacity; *Mentha piperita*

Citation: Cappellari, L.d.R.; Bogino, P.C.; Nievas, F.; Giordano, W.; Banchio, E. Exploring the Differential Impact of Salt Stress on Root Colonization Adaptation Mechanisms in Plant Growth-Promoting Rhizobacteria. *Plants* **2023**, *12*, 4059. https://doi.org/10.3390/plants12234059

Academic Editors: Geovani Soares de Lima, Lauriane Almeida dos Anjos Soares and Francisco Vanies Da Silva Sá

Received: 13 November 2023
Revised: 29 November 2023
Accepted: 1 December 2023
Published: 3 December 2023

1. Introduction

Excessive salt in soil inhibits plant growth and development, leading to decreased crop yield [1]. The accumulation of sodium (Na^+) in the root zone of plants can cause several deleterious effects, including alterations to soil physical–chemical properties. Soil salinity reduces soil organic matter content, soil water-holding capacity, and water infiltration and weakens soil structure and disrupts soil aggregate stability. Other common negative impacts of salinity on soil properties include increased soil pH, exchangeable sodium percentage (ESP), and sodium adsorption ratio (SAR), as well as reduced cation exchange capacity (CEC) and soil microbial community [2]. These effects restrict the absorption of water and usually cause ionic imbalances in plants [3,4]. Salt stress can also alter the levels of growth regulators, decrease protein synthesis, and impair photosynthetic processes in plants [5,6]. Additionally, salt stress can cause oxidative damage due to the accumulation of reactive oxygen species (ROS), which can be detrimental to cells at high concentrations [7–9]. If the intracellular concentration of ROS is not controlled, it can lead to damage to the cell structure by lipid peroxidation, protein oxidation, nucleic acid damage, and enzymatic inhibition, ultimately leading to cell death [10–12].

To mitigate salt stress, beneficial microorganisms known as plant growth-promoting rhizobacteria (PGPR) can play an important role. Rhizospheric bacteria can effectively colonize plant roots and maintain soil fertility by offering a favorable reinforcement to inorganic fertilizers and pesticides [13]. The promotion of growth in plants produced by PGPR can be diverse, resulting, for example, from the solubilization of phosphates, production of siderophores, biological fixation of nitrogen, production of the enzyme 1-aminocyclopropane 1-carboxylic acid (ACC) deaminase, production and regulation of phytohormones, biocontrol activity, production of volatile organic compounds (VOCs), and activation of induced systemic resistance (ISR) [14]. In the last decade, bacteria belonging to different genera (*Achromobacter, Azospirillum, Bacillus, Burkholderia, Enterobacter, Methylobacterium, Microbacterium, Paenibacillus, Pantoea, Pseudomonas, Rhizobium*, and *Variovorax*) have been reported to promote tolerance to various types of abiotic stress in plants, such as water, salt, and heat stress, as well as nutrient deficiency and heavy metals [15,16]. Several studies have shown that PGPR inoculation can improve plant growth and development under salt stress conditions [17–19]. The mechanisms behind these phenomena include osmotic adjustment, modulation of the plant antioxidant system, ion homeostasis, modulation of the phytohormonal balance, and increase in nutrient uptake [20]. For instance, our previous research showed that the microbial volatile organic compounds (mVOCs) produced by *Bacillus amyloliquefaciens* GB03 improve antioxidant status and ameliorate the effects of salt stress in *Mentha piperita* [21,22]. Similar effects were observed with mVOCs emitted by native *Pseudomonas putida* SJ46 and SJ04 isolated from the *Mentha piperita* rhizosphere [23,24]. Additionally, direct inoculation with *Pseudomonas simiae* WCS417r or *B. amyloliquefaciens* GB03 improved antioxidant status, leading to an enhancement of *M. piperita* plant growth under drought stress conditions [25]. However, the information available regarding the strategies employed by rhizobacteria to mitigate the negative effects of salt stress is limited [26].

Every living organism, whether it is unicellular or multicellular, possesses the inherent ability to respond to its environment [27]. Through the process of evolution, unicellular organisms have developed various mechanisms to effectively adapt and respond to changing environmental conditions [28]. Microbial motility is one of the most important features of microbial ecology and physiology [29,30]. The motility behaviors of bacteria enable them to search for nutrients but also represent an adaptive strategy to survive in harsh environmental conditions [27], such as saline soils. Flagella are necessary for both swimming and swarming motility [31]. Swimming is the typical motility in liquid media, while swarming occurs on semi-solid surfaces [31]. Bacteria within the genus *Pseudomonas* are able to move individually or collectively by swimming or swarming as a result of one or more polar flagella [32], while *Bacillus subtilis* possesses peritrichous flagella, which facilitate its motility [33]. Swarming is distinct from swimming since it requires secretions of a surfactant, an increase in flagellar density, and perhaps additional factors. The surfactant acts to reduce surface tension and create a thin layer of water within which to swarm [34,35]. The reason that cells require an increase in flagellar density is unknown, but it may be necessary to increase the total amount of thrust generated by the cell [34,35]. Moreover, biofilm formation and EPS production by PGPR are processes that help microorganisms endure adverse environments [36]. In saline environments, in particular, the high osmotic pressure usually causes cytoplasmic lysis and cell death. Hence, microorganisms develop biofilms, which protect them from such abiotic stress [37]. PGPR with EPS and biofilm production are better at colonizing plant root surfaces under unfavorable conditions [38].

In fact, biofilm formation is the most common strategy for bacterial life and survival in terrestrial habitats [39].

Biofilm formation and bacterial autoaggregation are two closely related processes that play crucial roles in bacterial adhesion and colonization on various surfaces, including biotic and abiotic surfaces [40]. Bacterial aggregations are highly specific processes that involve an interaction between surface molecules of microorganisms, which act as adhesins, and complementary receptors, including proteins and carbohydrates [41]. On the other hand, biofilm formation is a complex process that involves the attachment of bacteria to a surface, followed by the formation of a three-dimensional structure [42]. Autoaggregation plays a crucial role in the initial stages of biofilm formation by promoting bacterial adhesion and facilitating the formation of microcolonies [43]. Biofilms are multicellular communities embedded in a self-secreted extracellular matrix mainly composed of exopolysaccharides (EPSs) [44].

Epiphytic PGPR typically form biofilms once established in the root epidermis. In the case of *Bacillus subtilis*, biofilm formation is essential for root colonization, with the plant playing a crucial role in this interaction [45,46]. Particularly in *B. subtilis*, biofilm formation is facilitated by the extracellular matrix, which consists of two main components: an EPS (extracellular polymeric substance) and the protein TasA [45]. However, in the case of some *Pseudomonas* species, biofilm formation is not essential. Instead, they colonize the root as microcolonies, occupying the junctions between epidermal cells and are surrounded by plant-derived mucilage [47]. Therefore, different bacterial strains have different modes of arrangement in the roots, which can even vary depending on the type of plant being colonized [48].

As previously mentioned, microorganisms play a vital role in amending soil properties affected by saline soil and contribute to improving crop yield [49]. However, it should be noted that rhizobacteria employ different approaches to deal with salt stress conditions and successfully colonize roots. In this regard, the aim of this work was to evaluate different strategies of rhizobacteria to mitigate salt stress. Specifically, we analyzed the impact of salinity on the autoaggregation capacity, biofilm formation, and motility of three PGPR strains: *Pseudomonas simiae* WCS417r, *Pseudomonas putida* SJ04, and *Bacillus amyloliquefaciens* GB03. Additionally, we evaluated the capacity of plant growth promoted by these strains in salt-stressed *Mentha piperita* plants. These strains were evaluated based on their demonstrated capacity to enhance growth in various aromatic plant species under abiotic stress conditions by direct inoculation, as well as by the emission of VOCs [21,25,50–52]. However, it is not clear which strategies these strains use to adapt to adverse environments and promote plant growth.

2. Results

2.1. Effect of Salinity on Motility: Swimming and Swarming

After a 48 h incubation period, we assessed the motility of the different PGPR strains under salt conditions. All evaluated PGPR strains exhibited swimming motility, but only the WCS417r and SJ04 strains displayed a significant decrease when exposed to salt stress ($p \leq 0.05$) (Table 1; Figure 1). This decrease became more pronounced with higher salt concentrations. In contrast, GB03 did not show any variation in swimming motility across the different salt concentrations evaluated ($p > 0.05$). However, for strain GB03, it was observed that as the concentration of salt in the medium increased, swarming-type mobility also increased. On the other hand, WCS417r and SJ04 presented low swarming mobility under the different salt concentrations evaluated (Table 1; Figure 1).

2.2. Autoaggregation Assay

Bacterial autoaggregation was quantified in different salt concentrations. All strains evaluated showed high levels of autoaggregation under control conditions, with 60% for WCS417r and SJ04 and 80% for GB03 (Figure 2). Under salt stress conditions, the autoaggregation values did not differ from the control for the SJ04 and GB03 strains. In contrast, WCS417r showed a 30% increase ($p < 0.05$) under the most stressful conditions (200 and 300 mM), reaching values of around 90%.

Table 1. Diameters of swimming and swarming of the strains *P. simiae* WCS417r, *P. putida* SJ04, and *B. amyloliquefaciens* GB03 grown under different salt concentrations (means ± SEs). Means followed by the same letter within a column are not significantly different according to Fisher's LSD test ($p \leq 0.05$).

Strain	NaCl	Swimming (cm)	Swarming (cm)
WCS417r	0mM	5.05 ± 0.12 a	0.70 ± 0.06 a
	100 mM	5.50 ± 0.25 a	0.71 ± 0.08 a
	200 mM	3.86 ± 0.32 b	0.74 ± 0.06 a
	300 mM	1.56 ± 0.23 c	0.70 ± 0.07 a
SJ04	0 mM	7.84 ± 0.21 a	0.48 ± 0.06 a
	100 mM	6.58 ± 0.23 b	0.52 ± 0.08 a
	200 mM	4.83 ± 0.32 c	0.52 ± 0.04 a
	300 mM	3.12 ± 0.17 d	0.48 ± 0.07 a
GB03	0 mM	6.99 ± 0.32 a	2.78 ± 0.14 a
	100 mM	7.38 ± 0.19 a	3.49 ± 0.17 b
	200 mM	6.34 ± 0.34 a	3.95 ± 0.18 b
	300 mM	6.48 ± 0.68 a	6.07 ± 0.24 c

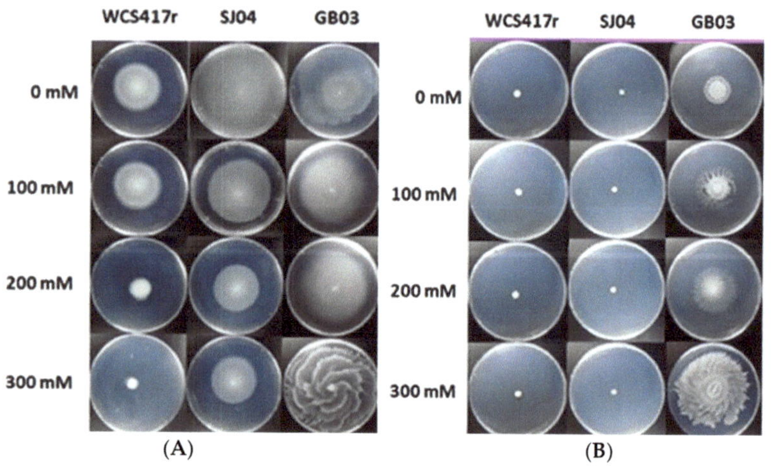

Figure 1. Effect of exposure to different concentrations of NaCl on the swimming (**A**) and swarming (**B**) motility of the strains *P. simiae* WCS417r, *P. putida* SJ04, and *B. amyloliquefaciens* GB03.

2.3. Effect of Salinity on Biofilm Formation

The ability to form biofilms has been observed to be influenced by exposure to salinity conditions in microorganisms. Therefore, the impact of salt on the formation of biofilms by different PGPR strains was explored. In Figure 3A, it can be observed that the growth of the WCS417r strain decreased considerably when it was exposed to high concentrations of NaCl (200 or 300 mM) ($p \leq 0.05$). Although the growth of SJ04 decreased in response to salt stress, the decrease was not as abrupt as for the WCS417r strain. GB03 growth was not affected by exposure to low concentrations of NaCl ($p > 0.05$), and only GB03 growth was greater when the strain was exposed to 300 mM NaCl ($p \leq 0.05$).

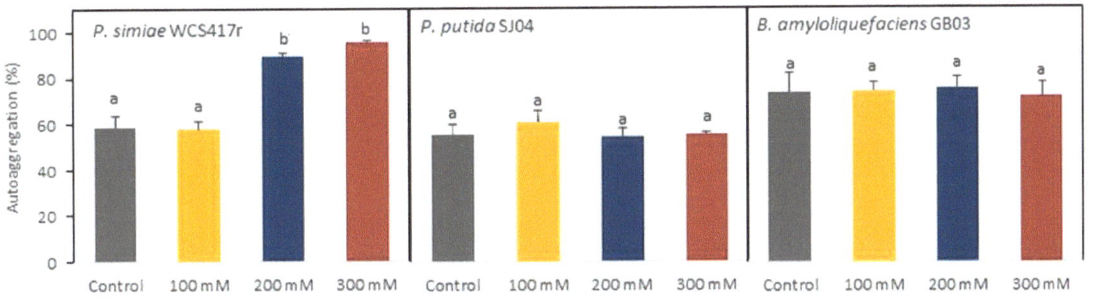

Figure 2. Percentage of autoaggregation of the strains *P. simiae* WCS417r, *P. putida* SJ04, and *B. amyloliquefaciens* GB03 under salt stress conditions (0 mM, 100 mM, 200 mM, and 300 mM). Values are means ± standard errors (SEs) of three independent replicates ($n = 3$). Means followed by the same letter in a given column are not significantly different according to Fisher's LSD test ($p \leq 0.05$).

Figure 3B shows the values of absorbance at 570 nm as a measure of optical density (OD), which reflects the capacity of biofilm formation of the different PGPR strains evaluated in diluted (1:10) LB medium supplemented with different concentrations of NaCl (0, 100, 200, and 300 mM). The OD values for *P. simiae* WCS417r were significantly higher under salt stress conditions in the medium without added NaCl (0 mM) ($p \leq 0.05$). This increase was almost 180% in relation to control conditions, and the values were similar regardless of the salt concentrations applied (Figure 3B). For SJ04, although the biofilm formation capacity increased when the strain was exposed to NaCl, the difference with respect to the control was not statistically significant ($p > 0.05$). In contrast, for GB03, this ability decreased significantly under saline stress as the treatment became more severe, being 40%, 76%, and 81% under 100, 200, and 300 mM, respectively (Figure 3B).

In spite of these observations about the capacity of biofilm formation, the biofilm/growth (B/G) (OD570/OD620) ratio can lead to more accurate conclusions about biofilm development since it is a normalized parameter for biofilm formation when the method of determination is based on CV measurements. The B/G ratio can be interpreted as a parameter of bacterial lifestyle under certain conditions. In the absence of stress, this ratio was higher for GB03 than for SJ04 and WCS417r (Figure 3C). Under saline stress, for GB03, the decrease in biofilm formation produced low B/G ratios; the higher the salt concentration, the lower the B/G ratio ($p \leq 0.05$). In contrast, the B/G ratio for both *Pseudomonas* strains, the WCS417r and SJ04 strains, increased with higher values corresponding to higher salt concentrations ($p < 0.05$). The highest OD 570/OD620 ratio was observed for the WCS417r strain at 300 mM (ca. 34).

2.4. Effect of PGPR Inoculation on Plant Growth under Salinity Conditions

To determine the effects of different strain inoculations on peppermint plants subjected to saline stress, saline concentrations of 0, 100, and 150 mM were used. Preliminary trials were also conducted with higher concentrations (200 and 300 mM), and it was observed that the explants failed to develop at these concentrations.

Peppermint plants subjected to salt stress exhibited a reduction in their shoot fresh weight, as expected, resulting in a 25% decrease compared to the control plants ($p < 0.05$).

The inoculation with WCS417r, SJ04, and GB03, under non-salt stress conditions, significantly enhanced the fresh weight by approximately 36%, 29%, and 50%, respectively, compared to non-inoculated plants (Figure 4). When subjected to 100 mM salt stress, plants treated with WCS417r or SJ04 displayed a significant rise of 67–79% in comparison to non-inoculated stressed plants ($p \leq 0.05$), while GB03 did not show any change in relation to the controls ($p > 0.05$). Under 150 mM salt stress, all tested strains exhibited an increase in fresh weight, with GB03 and WCS417r showing a 40% improvement, while the most remarkable effect was observed with SJ04, resulting in a 70% rise compared to non-inoculated plants ($p \leq 0.05$).

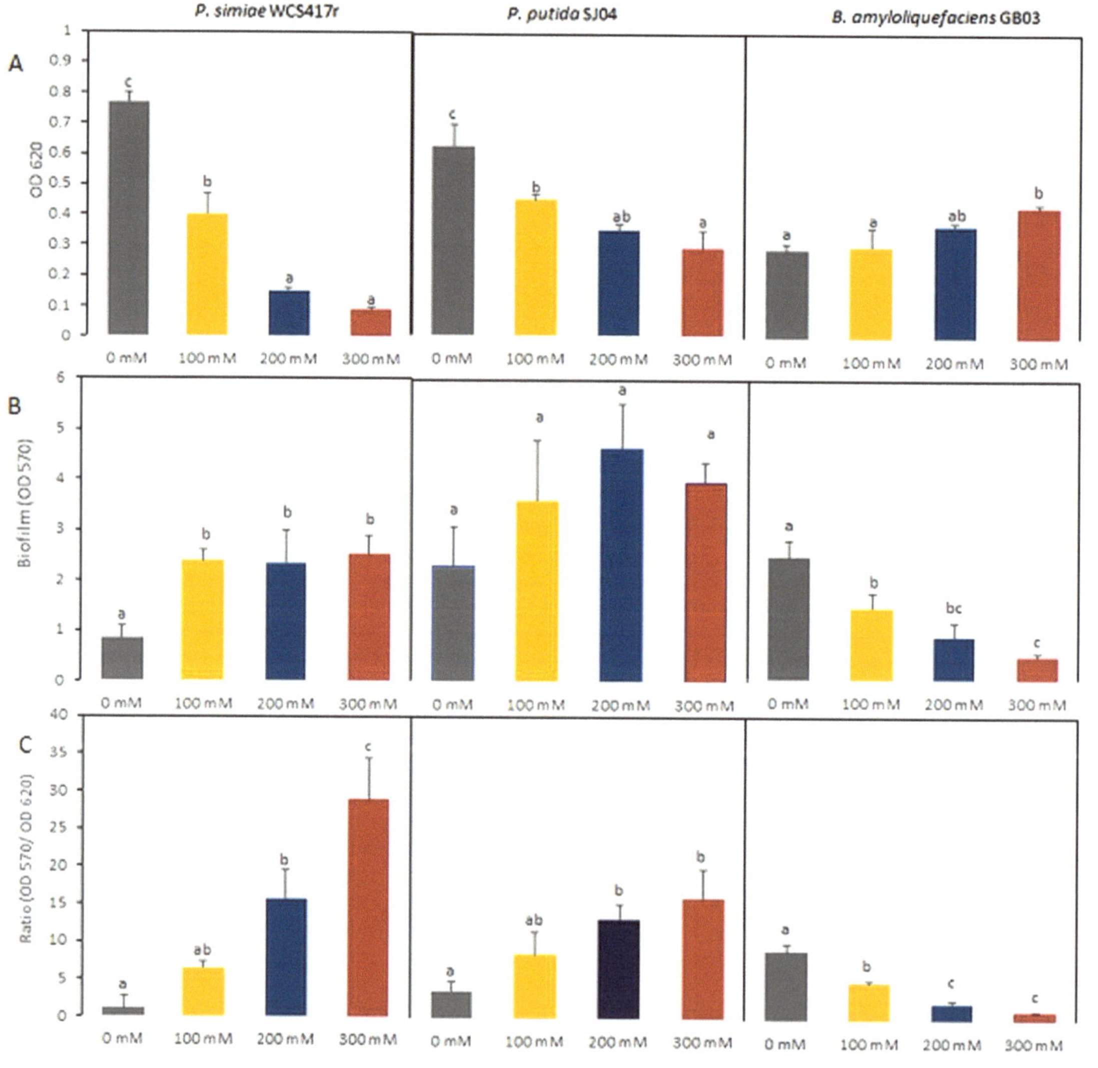

Figure 3. Biofilm formation of the strains *P. simiae* WCS417r, *P. putida* SJ04, and *B. amyloliquefaciens* GB03 under saline conditions (0, 100, 200, and 300 mM). (**A**) Growth of strains under different salt concentrations (OD). (**B**) Biofilm development on polystyrene surfaces shown as an average of OD measurements at 570 nm. (**C**) The relation between biofilm and growth is shown as the ratio between ODs at 570 and 620 nm. Values are means ± standard errors (SEs) of three independent replicates (n = 3). Means followed by the same letter in a given column are not significantly different according to Fisher's LSD test ($p \leq 0.05$).

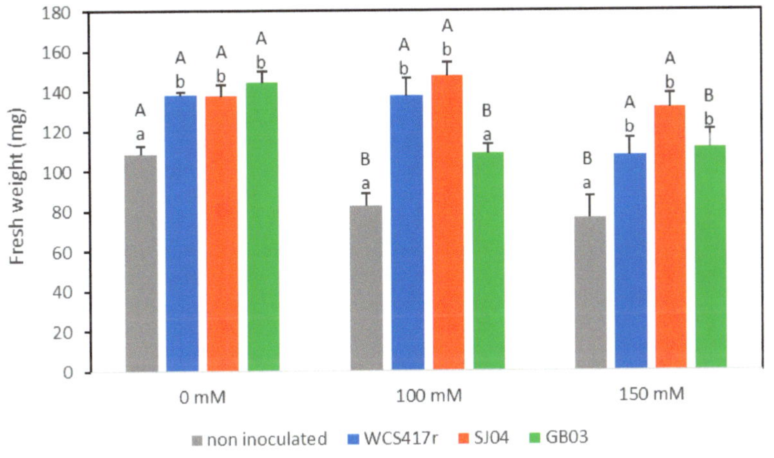

Figure 4. Effects of inoculation with different PGPR strains (*P. simiae* WCS417r, *P. putida* SJ04, and *B. amyloliquefaciens* GB03) on fresh weight of *Mentha piperita* plants grown in Murashige–Skoog medium with 0, 100, and 1500 mM NaCl. Values are means ± standard errors (SEs) of three independent replicates (n = 3). Values with different letters (capital letters for the same strains at different NaCl concentrations and lowercase letters for different strains at the same NaCl concentration) denote significant differences among treatment groups according to Fisher's LSD test ($p \leq 0.05$).

3. Discussion

Soil salinity is a significant environmental stress that affects plant growth [5]. High concentrations of salts cause lowered water potential in the soil, which results in a wide range of physiological and biochemical alterations in plants. Salinity affects almost all aspects of plant development, including germination, vegetative growth, and reproductive development [15,53]. It imposes ion toxicity, osmotic stress, nutrient deficiency, and oxidative stress on plants, which collectively lead to impaired physiological processes, abnormal biochemical mechanisms, nutrient imbalance, abridged photosynthesis, altered enzymatic activities, loss of cell turgor, and accumulation of ROS [12]. These effects ultimately lead to reduced crop productivity and huge economic losses [15,54,55]. Not only does salt affect plant productivity, but it also affects rhizobacteria, modifying their phenotypic characteristics related to root colonization [56]. However, nature has provided microorganisms with mechanisms to sense their environments and respond accordingly. In bacteria, these responses can be characterized as sensory adaptations [57].

Swimming and swarming represent individual and collective cell motility [58,59]. It has been suggested that motility plays a significant role in the search for nutrients, interaction with the environment, and survival in adverse environmental conditions [60]. Moreover, motility appears to be crucial for biofilm formation [61]. This is because motility, which depends on flagella, fimbriae, and pili, allows cells to move and adhere to appropriate host surfaces. In turn, this facilitates colonization [62]. In the present study, all tested strains showed some degree of swimming and swarming ability under control conditions. However, when salt concentration increased, the swimming-type mobility decreased in *Pseudomonas* strains, and in *B. amyloliquefaciens* GB03, the swarming-type mobility increased. It has been reported that all strains that could form biofilm were also positive in terms of motility [63,64], suggesting that motility is critical for biofilm formation [65]. However, our results revealed that under salt stress conditions, strains that have weak biofilm ability showed high swimming motility and also some degree of swarming, while other strains with high biofilm formation ability showed poor motility. Papadopoulou et al. [66] observed similar results in *P. putida* strains. The strain that revealed weak biofilm ability showed high swimming motility and some degree of swarming, while the strain that exhibited a high degree of swarming and swimming motility was not able to form biofilm in vitro. These results, in agreement with our research, support the idea that the intensity of motility does not correlate with the capacity for biofilm formation and that other factors may be involved. For instance, the chemical composition, viscosity, and pH of the culture medium, as well as the incubation temperature, have been identified as potential contributing factors [67]. Additionally, attachment and subsequent biofilm formation can be influenced by various surface structures. Specifically, surface proteins have been identified to play a crucial role in the initial adherence to surfaces [68]. Nevertheless, despite the fact that individual motility was reduced, this was not an impediment to forming biofilm, which reinforces the idea that biofilm formation constitutes an important protection mechanism.

Autoaggregation is a mechanism by which bacteria can form aggregates or clumps, which can affect the capacity of root colonization [68].. Autoaggregation can promote bacterial adhesion to surfaces, including plant roots, and facilitate biofilm formation, playing a crucial role in the initial stages of biofilm formation [43]. According to the present study, the percentage of autoaggregation was only increased in *P. simiae* WCS471r at high salt concentrations. This suggests that the effect of salinity on autoaggregation may depend on the bacterial species and environmental conditions [69]. A positive correlation was found between autoaggregation and biofilm formation in some bacterial species, including *Sinorhizobium meliloti*, but the correlation may not always hold true for other bacterial species [42]. In the present study, we observed a correlation between autoaggregation and biofilm formation only in WCS417r when exposed to 200 and 300 mM NaCl. Unfavorable growth conditions or low metabolic activity have been found to induce aggregative behavior in bacteria that normally grow in a dispersed, non-aggregated manner. In this context, autoaggregation may reflect a survival strategy that is triggered under hostile environmental conditions [68].

Biofilm-forming PGPR have been found to exhibit a number of beneficial features, such as facilitating resource acquisition and higher resistance to antibiotics and adverse environmental stresses (e.g., high temperature, extreme pH, salinity, and drought), and as a result, their chances of survival in competitive soil environments are enhanced [69–72]. Biofilm production is a significant characteristic that enables PGPR to withstand various environmental stresses and maintain a high cellular presence attached to plant roots. This allows them to exert their beneficial interactions in the rhizosphere [73]. Furthermore, biofilm formation not only protects PGPR from stressful conditions but also directly benefits plants [74]. Biofilm formation is a multistep process that requires the integration of various bacterial physiological processes, including quorum sensing [75,76], motility [64,77], autoaggregation [42], and EPS production [78,79]. The results of the present study revealed that biofilm formation was variable for the evaluated strains. Both *Pseudomonas* strains showed higher biofilm formation under salinity stress conditions, with similar results being registered for *P. stutzeri* MJL19, a bacterial strain isolated from the rhizosphere of halophyte plants [80]. It is important to note that in WCS417r and SJ04, biofilm production was promoted under high concentrations of NaCl (200 and 300 mM), despite the fact that this concentration inhibited bacterial growth, especially in the case of WCS417r (Figure 3A). This result supports the idea that biofilm can be associated with a protection mechanism, allowing bacteria to survive and thrive in environments containing high salt concentrations. Biofilm development by rhizobacteria is expected to enhance rhizosphere competence both under normal and stress conditions [69,81]. Biofilm-forming rhizobacteria are expected to colonize plant roots more stably and can remain active more consistently. Moreover, exopolysaccharides have water-retaining and adhesive characteristics and play a significant role in the stabilization of soil aggregates and the regulation of nutrients and water flow across plant roots [82]. *B. amyloliquefaciens* GB03 exhibited a decrease in biofilm formation as the salt concentration in the medium increased. However, its growth remained unaffected by exposure to NaCl. Related to this, other studies have demonstrated that salt-tolerant bacteria can survive in various salt concentrations and counteract the effects of salt through different mechanisms, including the accumulation of osmoprotectants, compatible solutes, and specialized extracellular polymeric substances (EPSs), which can contribute to improved plant growth in saline soil [83,84]. In agreement with this, Ansari et al. [56] found that *Bacillus subtilis* FAB10 displayed improved biofilm development when exposed to lower concentrations of NaCl treatment. However, at higher concentrations, there was a notable decrease in biofilm formation compared to the control.

Previous research has reported that increased NaCl concentration can reduce bacterial EPSs and biofilm formation [56]. However, plant growth-promoting rhizobacteria with EPS-producing characters can chelate different cations, including Na^+, and bind with Na^+ ions through the secretion of EPSs, which reduce their toxicity in the soil. Therefore, a large population of EPS-producing bacteria in the rhizosphere zone can reduce the concentration of available Na^+ for plant uptake and alleviate salt stress effects on plants in a saline environment. This suggests that halotolerant biofilm-producing rhizobacteria can enhance crop productivity under saline agro-ecosystems by decreasing the accumulation of harmful Na^+. EPSs under salinity stress can bind sodium ions and alleviate their toxic effect in the soil [85,86]. Biofilm formation in moderately halophilic bacteria is also believed to be helpful for plant growth improvement in saline soil [13,87], as we observed in inoculated *M. piperita* plants grown under 100 or 150 mM NaCl. When the plants were exposed to 100 or 150 mM, the fresh weight of peppermint plants that were not inoculated decreased, while, when plants were inoculated indistinctly with WCS417r, SJ04, or GB03, a significant increase in the fresh weight of the aerial parts was observed compared to their corresponding controls. Remarkably, both the GB03 and the *Pseudomonas* strains exhibited an ability to mitigate the negative effect of salinity, despite having distinct mechanisms of adaptation to saline environments. Similar results were observed in *M. piperita* plants inoculated with *P. simiae* WCS417r under drought stress [25]. In previous studies, we were able to determine that *M. piperita* plants grown in a saline medium and exposed to GB03 VOCs had significantly better morphological characteristics and higher total chlorophyll contents compared to control plants. In addition, they showed a reduction in malondialdehyde (MDA) levels, which are considered to be an indicator of lipid peroxidation and membrane damage, and had an increased antioxidant capacity in relation to plants cultivated under salt stress but not treated with mVOCs. Through chromatographic analysis, it was determined that the percentage of acetoin emission increased when the bacterium was exposed to NaCl [21]. These results demonstrate the potential of GB03 mVOCs to diminish the adverse effects of salt stress [22]. In a related study, Vaishnav et al. [88] observed an up-regulation of salt-tolerant gene expression in soybean plants that were inoculated with the biofilm-producing PGPR *P. simiae* strain AU under salinity stress conditions. Thus, biofilm-producing bacteria may possess advantages over other types and be able to thrive in adverse environments and deliver beneficial effects to plants.

4. Materials and Methods

4.1. Bacterial Strains and Culture Conditions

The three strains, previously reported as PGPR, *Pseudomonas simiae* WCS417r (formerly known as *P. fluorescens* WCS417r [89]), *Bacillus amyloliquefaciens* GB03 (originally described as *B. subtilis* GB03 [90]), and *Pseudomonas putida* SJ04 (GenBank KF312464.1), were isolated from the rhizosphere of *M. piperita* from a commercial plantation in the Villa Dolores Region, Córdoba, and used for the bacterial cultures [23]. Bacteria were grown on Luria–Bertani (LB) medium [91] (10 g/L tryptone, 5 g/L yeast extract, 5 g/L NaCl) for routine use and maintained in nutrient broth with 15% glycerol at −80 °C for long-term storage. For the different assays, the strains were grown in diluted (1:10) LB medium and supplemented with 0, 100, 200, or 300 mM NaCl.

4.2. Motility: Swarming and Swimming

The swimming and swarming of the WCS417r, GB03, and SJ04 strains were evaluated in the presence of different concentrations of NaCl (0, 100, 200, and 300 mM). Swimming motility was determined in plates containing diluted (1:10) LB medium supplemented with corresponding concentrations of NaCl (0, 100, 200, or 300 mM) and 0.3% *w*/*v* agar. A bacterial suspension was inoculated by puncture in the center of the plate. Subsequently, the plates were inverted and incubated for 48 h at 28 ± 2 °C, and the colony diameters were measured in centimeters (cm) [92].

For the swarming assay, different concentrations of NaCl were added to diluted (1:10) LB medium containing 0.5% *w*/*v* of agar. A quantity of 2 µL of an inoculum grown in LB medium for 24 h at 28 ± 2 °C was placed at the center of the agar surface and incubated for 48 h at 28 ± 2 °C. The swarming extent was determined by measuring the swarming diameters (cm) and comparing them with the corresponding controls [93]. This assay was repeated three times, and each experiment contained ten replicates.

4.3. Autoaggregation Assay

The autoaggregation capacity of the strains was determined as described by Sorroche et al. [42]. The bacteria were grown in LB medium overnight at 30 °C with rotation at 120 rpm until the stationary phase and were later centrifuged at 10,000 rpm for 10 min at 25 ± 2 °C and then resuspended in diluted (1:10) LB medium supplemented with the corresponding concentration of NaCl to attain an optical density OD620 nm of 1. Bacterial suspensions (5 mL) were then transferred to a glass tube (10 × 70 mm) and allowed to settle for 24 h at 4 °C. A 200 µL aliquot of the upper portion of the suspension was transferred to a microtiter plate, and the OD600 was measured (OD final) with a MicroELISA Auto Reader (Series 700 Microplate Reader; Cambridge Technology). A control tube was vortexed for 30 s, and the OD600 was determined (OD initial). The autoaggregation percentage was calculated as 100× (OD final/OD initial). This assay was repeated three times, and each experiment contained ten biological replicates.

4.4. Biofilm Formation Assay

Biofilm formation was determined macroscopically by a quantitative assay using 96-well microtiter dishes made of polyvinyl chloride. Biofilms were stained with crystal violet (CV) based on the method of O'Toole and Kolter [94], with modifications. The bacteria were grown in LB medium overnight at 30 °C with rotation at 120 rpm until the stationary phase and were later centrifuged at 10,000 rpm for 10 min at 25 °C and then re-suspended in an appropriate volume of diluted (1:10) LB medium supplemented with the corresponding concentration of NaCl (0, 100, 200, or 300 mM) to achieve an initial OD 620nm of 0.1. A quantity of 150 µL of each bacterial suspension was added to each well and incubated for 24 h at 28 ± 2 °C. Bacterial growth was quantified by measuring the absorbance of planktonic cells in each well at OD620 with a MicroELISA Auto Reader (Series 700 Microplate Reader; Cambridge Technology). Planktonic cells were gently aspirated with a pipette, each well was washed 3 times with saline solution (0.9% *w*/*v* NaCl), and cells adhered to the polystyrene support were stained with 180 µL of a CV aqueous solution (0.1% *w*/*v*) for 15 min. The wells were rinsed repeatedly with distilled water, and biofilm formation was assayed by the addition of 150 µL of 95% *v*/*v* ethanol. The OD570 of solubilized CV was measured with a MicroELISA Auto Reader, as described above. In parallel, sterile control cultures were established in diluted (1:10) LB medium supplemented with the corresponding concentration of NaCl. Relative BFA (biofilm-forming ability) was calculated as OD570/OD620 (biofilm quantified by staining with CV relative to planktonic growth measurement) [68]. This assay was repeated three times, and each experiment contained ten biological replicates for each treatment and control.

4.5. Plant Micropropagation

Young shoots from *Mentha piperita* plants grown in the Traslasierra Valley (Cordoba Province, Argentina) were surface disinfected by soaking for 1 min in 17% (*v*/*v*) sodium hypochlorite solution and then rinsed three times in sterile distilled water. The disinfected shoots were cultured in 100 mL Murashige and Skoog culture medium (MS) [95] containing macro- and micronutrients, vitamins, 0.66 mg/L indolebutyric acid (IBA), 0.025 mg/L naphthalene acetic acid (NAA), 0.7% (w/v) agar, and 3% (w/v) sucrose). The pH was adjusted to 5.6–5.8. After 30 days, apical meristems with foliar primordia that did not show contamination were aseptically removed from the terminal buds, and new explants were cultured in test tubes in 40 mL MS medium [23]. Plantlets obtained from tips were multiplied by single-node culture. Explants were grown in a growth chamber under controlled conditions of light (light/dark cycle: 16/8 h), temperature (22 ± 2 °C), and relative humidity (ca. 70%).

4.6. Inoculation Assay

The bacterial culture was grown in LB medium overnight at 30 °C with rotation at 120 rpm until the stationary phase and later centrifuged at 10,000 rpm, washed twice in 0.9% w/v NaCl by centrifugation (10,000 rpm, 10 min, 25 ± 2 °C), resuspended in sterile water, and adjusted to a final concentration of ca.10^9 colony-forming units cfu/mL for use as an inoculum.

Glass bottles (250 mL) coated with MS (0.5% % w/v agar) were inoculated with 100 µL bacterial suspension (10^8 cfu/bottle) and with sterile water for the treatment control. After solidification of the medium, one node from an aseptically cultured plant was placed in the center of the bottle. Distilled water was used as a negative control.

The glass bottles containing plants and bacteria were covered with aluminum foil, sealed with parafilm to avoid contamination, and placed in a growth chamber under controlled conditions of light (16 h/8 h light/dark cycle), temperature (22 ± 2 °C), and relative humidity (ca. 70%).

Salt stress was generated by the addition of NaCl. Different concentrations of salt were added to the MS media (plant growth media) during its preparation, including 0 mM, 100 mM, and 150 mM NaCl. A solution of 1000 mM NaCl was used to achieve the desired final NaCl concentration. The medium was then sterilized by autoclaving [96]. Previously, a test was carried out to determine the maximum concentration of salt that the plants could withstand. This assay was repeated three times, and each experiment contained ten biological replicates (10 bottles per treatment; 1 plant/bottle). After 30 days, the plants were harvested, each plant was removed from its glass bottle, and the shoot fresh weight was measured.

4.7. Statistical Analyses

The data were subjected to an analysis of variance (one-way ANOVA) followed by a comparison of multiple treatment levels with controls using Fisher's post hoc LSD (least significant difference) test. Differences between means were considered significant for p values ≤ 0.05. The Infostat software program, version 2020 (Group Infostat, Universidad Nacional de Córdoba, Argentina), was used for all the statistical analyses.

5. Conclusions

The present study revealed that the three evaluated PGPR strains have different strategies for mitigating salt stress. Mobility plays an important role in the survival of microorganisms under adverse conditions. The three strains studied exhibited swimming and swarming motility under control conditions. Under saline stress, only *B. amyloliquefaciens* GB03 increased swarming motility as a strategy to colonize new areas in search of more favorable conditions. *Pseudomonas* may utilize alternative adaptive mechanisms. On the other hand, autoaggregation capacity and biofilm formation represent strategies associated with the generation of protective structures in stress situations. In this context, when subjected to salt stress conditions, both the *P. simiae* WCS417r and *P. putida* SJ04 strains exhibited an enhancement in biofilm formation, while only the *P. simiae* WCS417r strain demonstrated an increase in the percentage of autoaggregation at high NaCl concentrations. Notably, a correlation between autoaggregation and biofilm formation was observed specifically for the *P. simiae* WCS417r strain at high concentrations of NaCl in the medium. However, all the evaluated PGPR strains were able to adapt to salt stress through the different mechanisms of biofilm formation, autoaggregation, and mobility, suggesting that these strains could be useful in promoting the survival, and therefore the colonization, of plants grown under salt stress conditions, which was reflected in the observed plant growth promotion of *M. piperita*. Thus, inoculation with these strains could represent a strategy for mitigating the effects of salt stress in plants grown under these conditions. However, with the aim of increasing agricultural productivity, future research should now focus on exploring the mechanisms of cross-talk between PGPR and plants in salinity stress environments in order to obtain a better understanding of the mechanisms by which PGPR and plants interact in adverse conditions, and also on investigating the importance of several adaptive mechanisms in plants toward salt stress.

Author Contributions: Conceptualization, E.B.; methodology, L.d.R.C., P.C.B. and F.N.; validation, P.C.B., F.N. and E.B.; formal analysis, E.B.; investigation, L.d.R.C.; resources, W.G., L.d.R.C. and E.B.; data curation, L.d.R.C.; writing—original draft preparation, L.d.R.C.; writing—review and editing, E.B. All authors have read and agreed to the published version of the manuscript.

Funding: This research was funded by the Secretaría De Ciencia y Técnica De La Universidad Nacional De Río Cuarto (grant number: PPI 2020-2022: C529-1), the Consejo Nacional De Investigaciones Científicas Y Técnicas (CONICET), and the Agencia Nacional De Promoción Científica y Tecnológica (ANPCYT).

Data Availability Statement: The data presented in this study are available on request from the corresponding author. The data are not publicly available due to privacy restrictions.

Acknowledgments: The authors are grateful to Paul Hobson, a native speaker, for editorial assistance.

Conflicts of Interest: The authors declare no conflict of interest.

References

1. Atkinson, N.J.; Urwin, P.E. The Interaction of Plant Biotic and Abiotic Stresses: From Genes to the Field. *J. Exp. Bot.* **2012**, *63*, 3523–3543. [CrossRef] [PubMed]
2. Bello, S.K.; Alayafi, A.H.; AL-Solaimani, S.G.; Abo-Elyousr, K.A.M. Mitigating Soil Salinity Stress with Gypsum and Bio-Organic Amendments: A Review. *Agronomy* **2021**, *11*, 1735. [CrossRef]
3. Patil, N.M.; Datir, S.S.; Shah, P.V. Salt-Induced Physiological and Biochemical Changes in Two Varieties of *Linum usitatissimum* L. *Int. J. Curr. Microbiol. App. Sci.* **2015**, *4*, 296–304.
4. Joshi, S.; Nath, J.; Singh, A.K.; Pareek, A.; Joshi, R. Ion Transporters and Their Regulatory Signal Transduction Mechanisms for Salinity Tolerance in Plants. *Physiol. Plant.* **2022**, *174*, e13702. [CrossRef]
5. Zhao, S.; Zhang, Q.; Liu, M.; Zhou, H.; Ma, C.; Wang, P. Regulation of Plant Responses to Salt Stress. *Int. J. Mol. Sci.* **2021**, *22*, 4609. [CrossRef] [PubMed]
6. Verma, S.; Negi, N.P.; Pareek, S.; Mudgal, G.; Kumar, D. Auxin Response Factors in Plant Adaptation to Drought and Salinity Stress. *Physiol. Plant.* **2022**, *174*, e13714. [CrossRef] [PubMed]
7. Bharti, N.; Barnawal, D. Chapter Five—Amelioration of Salinity Stress by PGPR: ACC Deaminase and ROS Scavenging Enzymes Activity. In *PGPR Amelioration in Sustainable Agriculture*; Singh, A.K., Kumar, A., Singh, P.K., Eds.; Woodhead Publishing: Sawston, UK, 2019; pp. 85–106. [CrossRef]
8. Devireddy, A.R.; Zandalinas, S.I.; Fichman, Y.; Mittler, R. Integration of Reactive Oxygen Species and Hormone Signaling During Abiotic Stress. *Plant J.* **2021**, *105*, 459–476. [CrossRef]
9. Sunera; Amna; Saqib, S.; Uddin, S.; Zaman, W.; Ullah, F.; Ayaz, A.; Asghar, M.; Rehman, S.U.; Munis, M.F.H.; et al. Characterization and Phytostimulatory Activity of Bacteria Isolated from Tomato (*Lycopersicon esculentum* Mill.) Rhizosphere. *Microb. Pathog.* **2020**, *140*, 103966. [CrossRef]
10. Das, K.; Roychoudhury, A. Reactive Oxygen Species (ROS) and Response of Antioxidants as ROS-Scavengers during Environmental Stress in Plants. *Front. Environ. Sci.* **2014**, *2*, 53. [CrossRef]
11. Huang, H.; Ullah, F.; Zhou, D.X.; Yi, M.; Zhao, Y. Mechanisms of ROS Regulation of Plant Development and Stress Responses. *Front. Plant Sci.* **2019**, *10*, 800. [CrossRef]
12. Hasanuzzaman, M.; Bhuyan, M.H.M.B.; Zulfiqar, F.; Raza, A.; Mohsin, S.M.; Mahmud, J.A.; Fujita, M.; Fotopoulos, V. Reactive Oxygen Species and Antioxidant Defense in Plants under Abiotic Stress: Revisiting the Crucial Role of a Universal Defense Regulator. *Antioxidants* **2020**, *9*, 681. [CrossRef]
13. Shultana, R.; Kee Zuan, A.T.; Yusop, M.R.; Saud, H.M. Characterization of Salt-Tolerant Plant Growth-Promoting Rhizobacteria and the Effect on Growth and Yield of Saline-Affected Rice. *PLoS ONE* **2020**, *15*, e0238537. [CrossRef]

14. Bhattacharyya, P.N.; Jha, D.K. Plant Growth-Promoting Rhizobacteria (PGPR): Emergence in Agriculture. *World J. Microbiol. Biotechnol.* **2012**, *28*, 1327–1350. [CrossRef]
15. Shrivastava, P.; Kumar, R. Soil salinity: A Serious Environmental Issue and Plant Growth Promoting Bacteria as One of the Tools for its Alleviation. *Saudi J. Biol. Sci.* **2015**, *22*, 123–131. [CrossRef] [PubMed]
16. Grover, M.; Ali, S.Z.; Sandhya, V.; Rasul, A.; Venkateswarlu, B. Role of Microorganisms in Adaptation of Agriculture Crops to Abiotic Stresses. *World J. Microbiol. Biotechnol.* **2011**, *27*, 1231–1240. [CrossRef]
17. Zhang, H.; Kim, M.S.; Sun, Y.; Dowd, S.E.; Shi, H.; Paré, P.W. Soil Bacteria Confer Plant Salt Tolerance by Tissue-Specific Regulation of the Sodium Transporter *HKT1*. *Mol. Plant Microbe Interact.* **2008**, *21*, 737–744. [CrossRef] [PubMed]
18. Farag, M.A.; Zhang, H.; Ryu, C.M. Dynamic Chemical Communication Between Plants and Bacteria through Airborne Signals: Induced Resistance by Bacterial Volatiles. *J. Chem. Ecol.* **2013**, *39*, 1007–1018. [CrossRef] [PubMed]
19. Fincheira, P.; Quiroz, A. Microbial Volatiles as Plant Growth Inducers. *Microbiol. Res.* **2018**, *208*, 63–75. [CrossRef]
20. Giannelli, G.; Potestio, S.; Visioli, G. The Contribution of PGPR in Salt Stress Tolerance in Crops: Unravelling the Molecular Mechanisms of Cross-Talk between Plant and Bacteria. *Plants* **2023**, *12*, 2197. [CrossRef] [PubMed]
21. Cappellari, L.R.; Banchio, E. Microbial Volatile Organic Compounds Produced by *Bacillus amyloliquefaciens* GB03 Ameliorate the Effects of Salt Stress in *Mentha piperita* Principally through Acetoin Emission. *J. Plant Growth Regul.* **2019**, *39*, 764–775. [CrossRef]
22. Cappellari, L.R.; Chiappero, J.; Palermo, T.B.; Giordano, W.; Banchio, E. Volatile Organic Compounds from Rhizobacteria Increase the Biosynthesis of Secondary Metabolites and Improve the Antioxidant Status in *Mentha piperita* L. Grown under Salt Stress. *Agronomy* **2020**, *10*, 1094. [CrossRef]
23. Santoro, M.V.; Bogino, P.C.; Nocelli, N.; Cappellari, L.R.; Giordano, W.F.; Banchio, E. Analysis of Plant Growth-Promoting Effects of Fluorescent *Pseudomonas* Strains Isolated from *Mentha piperita* Rhizosphere and Effects of Their Volatile Organic Compounds on Essential Oil Composition. *Front. Microbiol.* **2016**, *7*, 1085. [CrossRef] [PubMed]
24. Gil, S.S.; Cappellari, L.R.; Giordano, W.; Banchio, E. Antifungal Activity and Alleviation of Salt Stress by Volatile Organic Compounds of Native *Pseudomonas* Obtained from *Mentha piperita*. *Plants* **2023**, *12*, 1488. [CrossRef] [PubMed]
25. Chiappero, J.; Cappellari, L.R.; Sosa Alderete, L.G.; Palermo, T.B.; Banchio, E. Plant Growth Promoting Rhizobacteria Improve the Antioxidant Status in *Mentha piperita* Grown under Drought Stress Leading to an Enhancement of Plant Growth and Total Phenolic Content. *Ind. Crops Prod.* **2019**, *139*, 111553. [CrossRef]
26. Ma, X.; Pan, J.; Xue, X.; Zhang, J.; Guo, Q. A Bibliometric Review of Plant Growth-Promoting Rhizobacteria in Salt-Affected Soils. *Agronomy* **2022**, *12*, 2304. [CrossRef]
27. Shamim, S.; Rehman, A.; Qazi, M.H. Swimming, Swarming, Twitching, and Chemotactic Responses of *Cupriavidus metallidurans* CH34 and *Pseudomonas putida* mt2 in the Presence of Cadmium. *Arch. Environ. Contam. Toxicol.* **2014**, *66*, 407–414. [CrossRef] [PubMed]
28. Hoch, J.A. Two-Component and Phosphorelay Signal Transduction. *Curr. Opin. Microbiol.* **2000**, *3*, 165–170. [CrossRef]
29. Rashid, M.H.; Kornberg, A. Inorganic Polyphosphate is Needed for Swimming, Swarming, and Twitching Motilities of *Pseudomonas aeruginosa*. *Proc. Natl. Acad. Sci. USA* **2000**, *97*, 4885–4890. [CrossRef]
30. Barrionuevo, M.R.; Vullo, D.L. Bacterial Swimming, Swarming and Chemotactic Response to Heavy Metal Presence: Which Could be the Influence on Wastewater Biotreatment Efficiency? *World J. Microbiol. Biotechnol.* **2012**, *28*, 2813–2825. [CrossRef]
31. Mordini, S.; Osera, C.; Marini, S.; Scavone, F.; Bellazzi, R.; Galizzi, A.; Calvio, C. The Role of SwrA, DegU and PD3 in *fla/che* Expression in *B. subtilis*. *PLoS ONE* **2013**, *8*, e85065. [CrossRef]
32. Bouteiller, M.; Dupont, C.; Bourigault, Y.; Latour, X.; Barbey, C.; Konto-Ghiorghi, Y.; Merieau, A. *Pseudomonas flagella*: Generalities and Specificities. *Int. J. Mol. Sci.* **2021**, *22*, 3337. [CrossRef] [PubMed]
33. Guttenplan, S.B.; Shaw, S.; Kearns, D.B. The Cell Biology of Peritrichous Flagella in *Bacillus subtilis*. *Mol. Microbiol.* **2013**, *87*, 211–229. [CrossRef]
34. Kearns, D.B.; Losick, R. Swarming motility in undomesticated *Bacillus subtilis*. *Mol. Microbiol.* **2003**, *49*, 581–590. [CrossRef] [PubMed]
35. Hall, A.N.; Subramanian, S.; Oshiro, R.T.; Canzoneri, A.K.; Kearns, D.B. SwrD (YlzI) Promotes Swarming in *Bacillus subtilis* by Increasing Power to Flagellar Motors. *J. Bacteriol.* **2017**, *200*, e00529-17. [CrossRef]
36. Gontia-Mishra, I.; Sapre, S.; Deshmukh, R.; Sikdar, D.; Tiwari, S. Biofilm formation and EPS production by plant growth-promoting rhizobacteria under abiotic stress conditions. In *Plant Growth Promoting Rhizobacteria for Sustainable Agriculture*; Springer: Singapore, 2020; pp. 119–136.
37. Yin, K.; Li, Y.; Zhao, F.; Zhang, K. The role of microbial biofilms in the salt tolerance of plants. In *Microbial Biofilms*; Springer: Singapore, 2019; pp. 1–19.
38. Steele, A.; Franklin, M.J.; Underwood, A.J. The role of bacterial biofilms in the persistence of marine organisms. In *Marine Animal Forests*; Springer: Dordrecht, The Netherlands, 2014; pp. 177–198.
39. Fujishige, N.A.; Kapadia, N.N.; Hirsch, A.M. A Feeling for the Micro-organism: Structure on a Small Scale. Biofilms on Plant Roots. *Bot. J. Linn. Soc.* **2006**, *150*, 79–88. [CrossRef]
40. Trunk, T.; Khalil, H.S.; Leo, J.C. Bacterial Autoaggregation. *AIMS Microbiol.* **2018**, *4*, 140. [CrossRef]
41. Simoes, M. Antimicrobial Strategies Effective Against Infectious Bacterial Biofilms. *Curr. Med. Chem.* **2011**, *18*, 2129–2145. [CrossRef]

42. Sorroche, F.G.; Spesia, M.B.; Zorreguieta, A.; Giordano, W. A Positive Correlation Between Bacterial Autoaggregation and Biofilm Formation in Native *Sinorhizobium meliloti* Isolates from Argentina. *Appl. Environ. Microbiol.* **2012**, *78*, 4092–4101. [CrossRef]
43. Karygianni, L.; Ren, Z.; Koo, H.; Thurnheer, T. Biofilm Matrixome: Extracellular Components in Structured Microbial Communities. *Trends Microbiol.* **2020**, *28*, 668–681. [CrossRef]
44. Wei, Q.; Ma, L.Z. Biofilm Matrix and Its Regulation in *Pseudomonas aeruginosa*. *Int. J. Mol. Sci.* **2013**, *14*, 20983–21005. [CrossRef]
45. Beauregard, P.B.; Chai, Y.; Vlamakis, H.; Losick, R.; Kolter, R. *Bacillus subtilis* Biofilm Induction by Plant Polysaccharides. *Proc. Natl. Acad. Sci. USA* **2013**, *110*, E1621–E1630. [CrossRef]
46. Allard-Massicotte, R.; Tessier, L.; Lécuyer, F.; Lakshmanan, V.; Lucier, J.F.; Garneau, D.; Caudwell, L.; Vlamakis, H.; Bais, H.P.; Beauregard, P.B. *Bacillus subtilis* early colonization of *Arabidopsis thaliana* roots involves multiple chemotaxis receptors. *MBio* **2016**, *7*, 1110–1128. [CrossRef]
47. Barahona, E.; Navazo, A.; Yousef-Coronado, F.; Aguirre de Cárcer, D.; Martínez-Granero, F.; Espinosa-Urgel, M.; Martín, M.; Rivilla, R. Efficient Rhizosphere Colonization by *Pseudomonas fluorescens* f113 Mutants Unable to Form Biofilms on Abiotic Surfaces. *Environ. Microbiol.* **2010**, *12*, 3185–3195. [CrossRef]
48. Poonguzhali, S.; Madhaiyan, M.; Yim, W.J.; Kim, K.A.; Sa, T.M. Colonization Pattern of Plant Root and Leaf Surfaces Visualized by Use of green-fluorescent-marked Strain of *Methylobacterium suomiense* and its Persistence in Rhizosphere. *Appl. Microbiol. Biotechnol.* **2008**, *78*, 1033–1043. [CrossRef]
49. Kumar, R.; Swapnil, P.; Meena, M.; Selpair, S.; Yadav, B.G. Plant Growth-Promoting Rhizobacteria (PGPR): Approaches to Alleviate Abiotic Stresses for Enhancement of Growth and Development of Medicinal Plants. *Sustainability* **2022**, *14*, 15514. [CrossRef]
50. Banchio, E.; Bogino, P.; Santoro, M.V.; Torres, L.; Zygadlo, J.; Giordano, W. Systemic Induction of Monoterpene Biosynthesis in *Origanum* × *majoricum* by Soil Bacteria. *J. Agric. Food Chem.* **2010**, *58*, 650–654. [CrossRef] [PubMed]
51. Cappellari, L.R.; Santoro, M.V.; Nievas, F.; Giordano, W.; Banchio, E. Increase of Secondary Metabolite Content in Marigold by Inoculation with Plant Growth-Promoting Rhizobacteria. *Appl. Soil Ecol.* **2013**, *70*, 16–22. [CrossRef]
52. Cappellari, L.R.; Santoro, M.V.; Reinoso, H.; Travaglia, C.; Giordano, W.; Banchio, E. Anatomical, Morphological, and Phytochemical Effects of Inoculation with Plant Growth-Promoting Rhizobacteria on Peppermint (*Mentha piperita*). *J. Chem. Ecol.* **2015**, *41*, 149–158. [CrossRef] [PubMed]
53. Zörb, C.; Geilfus, C.M.; Dietz, K.J. Salinity and Crop Yield. *Plant Biol.* **2019**, *21*, 31–38. [CrossRef] [PubMed]
54. Khan, M.A.; Sahile, A.A.; Jan, R.; Asaf, S.; Hamayun, M.; Imran, M.; Adhikari, A.; Kang, S.-M.; Kim, K.-M. Halotolerant Bacteria Mitigate the Effects of Salinity Stress on Soybean Growth by Regulating Secondary Metabolites and Molecular Responses. *BMC Plant Biol.* **2021**, *21*, 176. [CrossRef]
55. Kamran, M.; Parveen, A.; Ahmar, S.; Malik, Z.; Hussain, S.; Chattha, M.S.; Saleem, M.H.; Adil, M.; Heidari, P.; Chen, J.T. An Overview of Hazardous Impacts of Soil Salinity in Crops, Tolerance Mechanisms, and Amelioration Through Selenium Supplementation. *Int. J. Mol. Sci.* **2020**, *21*, 148. [CrossRef]
56. Ansari, F.A.; Ahmad, I.; Pichtel, J. Growth Stimulation and Alleviation of Salinity Stress to Wheat by the Biofilm Forming *Bacillus pumilus* Strain FAB10. *Appl. Soil Ecol.* **2019**, *143*, 45–54. [CrossRef]
57. Raberg, M.; Reinecke, F.; Reichelt, R.; Malkus, U.; Ko, S.; Po, M.; Fricke, W.F. *Ralstonia eutropha* H16 Flagellation Changes According to Nutrient Supply and State of Poly (3-hydroxybutyrate) Accumulation. *Appl. Environ. Microbiol.* **2008**, *74*, 4477–4490. [CrossRef]
58. Packiavathy, I.A.S.V.; Priya, S.; Pandian, S.K.; Ravi, A.V. Inhibition of Biofilm Development of Uropathogens by Curcumin—An Anti-Quorum Sensing Agent from *Curcuma longa*. *Food Chem.* **2014**, *148*, 453–460. [CrossRef]
59. Ariel, G.; Rabani, A.; Benisty, S.; Partridge, J.D.; Harshey, R.M.; Be'er, A. Swarming Bacteria Migrate by Lévy Walk. *Nat. Commun.* **2015**, *6*, 8396. [CrossRef]
60. Nies, D.H. Microbial Heavy-Metal Resistance. *Appl. Microbiol. Biotechnol.* **1999**, *51*, 730–750. [CrossRef]
61. Drenkard, E. Antimicrobial Resistance of *Pseudomonas aeruginosa* Biofilms. *Microbes Infect.* **2003**, *5*, 1213–1219. [CrossRef]
62. Tan, M.S.; White, A.P.; Rahman, S.; Dykes, G.A. Role of Fimbriae, Flagella and Cellulose on the Attachment of *Salmonella typhimurium* ATCC 14028 to Plant Cell Wall Models. *PLoS ONE* **2016**, *11*, e0158311. [CrossRef]
63. Karatan, E.; Watnick, P. Signals, Regulatory Networks, and Materials that Build and Break Bacterial Biofilms. *Microbiol. Mol. Biol. Rev.* **2009**, *73*, 310–347. [CrossRef]
64. Verstraeten, N.; Braeken, K.; Debkumari, B.; Fauvart, M.; Fransaer, J.; Vermant, J.; Michiels, J. Living on a Surface: Swarming and Biofilm Formation. *Trends Microbiol.* **2008**, *16*, 496–506. [CrossRef]
65. Lemon, K.P.; Higgins, D.E.; Kolter, R. Flagella-Mediated Motility is Critical for *Listeria monocytogenes* Biofilm Formation. *J. Bacteriol.* **2007**, *189*, 4418–4424. [CrossRef]
66. Papadopoulou, A.; Matsi, T.; Kamou, N.; Avdouli, D.; Mellidou, I.; Karamanoli, K. Decoding the Potential of a New *Pseudomonas putida* Strain for Inducing Drought Tolerance of Tomato (*Solanum lycopersicum*) Plants through Seed Biopriming. *J. Plant Physiol.* **2022**, *271*, 153658. [CrossRef]
67. Tokárová, V.; Sudalaiyadum Perumal, A.; Nayak, M.; Shum, H.; Kašpar, O.; Rajendran, K.; Mohammadi, M.; Tremblay, C.; Gaffney, E.A.; Martel, S.; et al. Patterns of bacterial motility in microfluidics-confining environments. *Proc. Natl. Acad. Sci. USA* **2021**, *118*, e2013925118. [CrossRef]

68. Bogino, P.C.; Oliva, M.D.l.M.; Sorroche, F.G.; Giordano, W. The Role of Bacterial Biofilms and Surface Components in Plant-Bacterial Associations. *Int. J. Mol. Sci.* **2013**, *14*, 15838–15859. [CrossRef]
69. Bogino, P.; Abod, A.; Nievas, F.; Giordano, W. Water-Limiting Conditions Alter the Structure and Biofilm-Forming Ability of Bacterial Multispecies Communities in the Alfalfa Rhizosphere. *PLoS ONE* **2013**, *8*, e79614. [CrossRef]
70. Mah, T.F.; Pitts, B.; Pellock, B.; Walker, G.C.; Stewart, P.S.; O'Toole, G.A. A Genetic Basis for *Pseudomonas aeruginosa* Biofilm Antibiotic Resistance. *Nature* **2003**, *426*, 306–310. [CrossRef]
71. Ahemad, M.; Kibret, M. Mechanisms and Applications of Plant Growth Promoting Rhizobacteria: Current Perspective. *J. King Saud Univ. Sci.* **2014**, *26*, 1–20. [CrossRef]
72. Kimani, V.N.; Chen, L.; Liu, Y.; Raza, W.; Zhang, N.; Mungai, L.K.; Shen, Q.; Zhang, R. Characterization of Extracellular Polymeric Substances of *Bacillus amyloliquefaciens* SQR9 Induced by Root Exudates of Cucumber. *J. Basic Microbiol.* **2016**, *56*, 1183–1193. [CrossRef]
73. Santoyo, G.; Urtis-Flores, C.A.; Loeza-Lara, P.D.; Orozco-Mosqueda, M.d.C.; Glick, B.R. Rhizosphere Colonization Determinants by Plant Growth-Promoting Rhizobacteria (PGPR). *Biology* **2021**, *10*, 475. [CrossRef]
74. Ansari, F.A.; Ahmad, I. Alleviating Drought Stress of Crops Through PGPR: Mechanism and Application. In *Microbial Interventions in Agriculture and Environment*; Singh, D., Gupta, V., Prabha, R., Eds.; Springer: Singapore, 2019. [CrossRef]
75. Parsek, M.R.; Greenberg, E.P. Sociomicrobiology: The Connections Between Quorum Sensing and Biofilms. *Trends Microbiol.* **2005**, *13*, 27–33. [CrossRef]
76. Hooshangi, S.; Bentley, W.E. From Unicellular Properties to Multicellular Behavior: Bacteria Quorum Sensing Circuitry and Applications. *Curr. Opin. Biotechnol.* **2008**, *19*, 550–555. [CrossRef]
77. Klausen, M.; Heydorn, A.; Ragas, P.; Lambertsen, L.; Aaes-Jørgensen, A.; Molin, S.; Tolker-Nielsen, T. Biofilm Formation by *Pseudomonas aeruginosa* Wild Type, Flagella and Type IV Pili Mutants. *Mol. Microbiol.* **2003**, *48*, 1511–1524. [CrossRef]
78. Sutherland, I.W. Biofilm Exopolysaccharides: A Strong and Sticky Framework. *Microbiol.* **2001**, *147*, 3–9. [CrossRef]
79. Rinaudi, L.V.; González, J.E. The Low-Molecular-Weight Fraction of Exopolysaccharide II from *Sinorhizobium meliloti* is a Crucial Determinant of Biofilm Formation. *J. Bacteriol.* **2009**, *191*, 7216–7224. [CrossRef]
80. Lami, M.J.; Adler, C.; Caram-Di Santo, M.C.; Zenoff, A.M.; de Cristóbal, R.E.; Espinosa-Urgel, M.; Vincent, P.A. *Pseudomonas stutzeri* MJL19, a Rhizosphere-Colonizing Bacterium that Promotes Plant Growth Under Saline Stress. *J. Appl. Microbiol.* **2020**, *129*, 1321–1336. [CrossRef]
81. Redmile-Gordon, M.; Gregory, A.S.; White, R.P.; Watts, C.W. Soil Organic Carbon, Extracellular Polymeric Substances (EPS), and Soil Structural Stability as Affected by Previous and Current Land-Use. *Geoderma* **2020**, *363*, 114143. [CrossRef]
82. Mohd, M.A.; Iqbal, A. In vitro Biofilm Development and Enhanced Rhizosphere Colonization of *Triticum aestivum* by *Fluorescent Pseudomonas* sp. *J. Pure Appl. Microbiol.* **2019**, *13*, 1441–1449. [CrossRef]
83. Kumar, A.; Singh, S.; Gaurav, A.K.; Srivastava, S.; Verma, J.P. Plant Growth-Promoting Bacteria: Biological Tools for the Mitigation of Salinity Stress in Plants. *Front. Microbiol.* **2020**, *11*, 1216. [CrossRef]
84. Ramasamy, K.P.; Mahawar, L. Coping with salt stress-interaction of halotolerant bacteria in crop plants: A mini review. *Front. Microbiol.* **2023**, *14*, 1077561. [CrossRef]
85. Bhagat, N.; Raghav, M.; Dubey, S.; Bedi, N. Bacterial Exopolysaccharides: Insight into Their Role in Plant Abiotic Stress Tolerance. *J. Microbiol. Biotechnol.* **2021**, *31*, 1045–1059. [CrossRef]
86. Morcillo, R.J.L.; Manzanera, M. The Effects of Plant-Associated Bacterial Exopolysaccharides on Plant Abiotic Stress Tolerance. *Metabolites* **2021**, *11*, 337. [CrossRef]
87. Haque, M.M.; Biswas, M.S.; Mosharaf, M.K.; Haque, M.A.; Islam, M.S.; Nahar, K.; Islam, M.; Shozib, H.B.; Islam, M. Halotolerant Biofilm-Producing Rhizobacteria Mitigate Seawater-Induced Salt Stress and Promote Growth of Tomato. *Sci. Rep.* **2022**, *12*, 5599. [CrossRef]
88. Vaishnav, A.; Kumari, S.; Jain, S.; Varma, A.; Tuteja, N.; Choudhary, D.K. PGPR-Mediated Expression of Salt Tolerance Gene in Soybean through Volatiles under Sodium Nitroprusside. *J. Basic Microbiol.* **2016**, *56*, 1274–1288. [CrossRef]
89. Berendsen, R.L.; van Verk, M.C.; Stringlis, I.A.; Zamioudis, C.; Tommassen, J.; Pieterse, C.M.; Bakker, P.A. Unearthing the Genomes of Plant-Beneficial *Pseudomonas* Model Strains WCS358, WCS374 and WCS417. *BMC Genom.* **2015**, *16*, 539. [CrossRef]
90. Choi, S.K.; Jeong, H.; Kloepper, J.W.; Ryu, C.M. Genome Sequence of *Bacillus amyloliquefaciens* GB03, an Active Ingredient of the First Commercial Biological Control Product. *Genome Announc.* **2014**, *2*, e01092-14. [CrossRef]
91. Bertani, G. Studies on lysogenesis. *I. The Mode of Phage Liberation by Lysogenic Escherichia coli. J. Bacteriol.* **1951**, *62*, 293–300. [CrossRef]
92. Sorroche, F.; Bogino, P.; Russo, D.M.; Zorreguieta, A.; Nievas, F.; Morales, G.M.; Hirsch, A.M.; Giordano, W. Cell Autoaggregation, Biofilm Formation, and Plant Attachment in a *Sinorhizobium meliloti* lpsB Mutant. *Mol. Plant Microbe Interact.* **2018**, *31*, 1075–1082. [CrossRef]
93. Vezza, M.E.; Olmos Nicotra, M.F.; Agostini, E.; Talano, M.A. Biochemical and molecular characterization of arsenic response from *Azospirillum brasilense* Cd, a bacterial strain used as plant inoculant. *Environ. Sci. Pollut. Res.* **2020**, *27*, 2287–2300. [CrossRef]
94. O'Toole, G.A.; Kolter, R. Initiation of biofilm formation in *Pseudomonas fluorescens* WCS365 proceeds via multiple, convergent signalling pathways: A genetic analysis. *Mol. Microbiol.* **1998**, *28*, 449–461. [CrossRef]

95. Murashige, T.; Skoog, F. A Revised Medium for Rapid Growth and Bio Assays with Tobacco Tissue Cultures. *Physiol. Plant* **1962**, *15*, 473–497. [CrossRef]
96. Cappellari, L.R.; Gil, S.S.; Palermo, T.B.; Palermo, J.S.; Meneguzzi, R.; Giordano, W.; Banchio, E. A Practical Method to Investigate the Effect of Volatile Organic Compounds Emitted by Rhizobacteria on Plant Growth Under Conditions of Salt Stress. *MethodsX* **2023**, *10*, 102099. [CrossRef] [PubMed]

plants

MDPI

Article

Foliar Application of Melatonin Positively Affects the Physio-Biochemical Characteristics of Cotton (*Gossypium hirsutum* L.) under the Combined Effects of Low Temperature and Salinity Stress

Yuanyuan Fu [1,2,†], Lang Xin [1,†], Abdoul Kader Mounkaila Hamani [3,*], Weihao Sun [2], Hongbo Wang [1], Abubakar Sunusi Amin [2], Xingpeng Wang [1], Anzhen Qin [2] and Yang Gao [2,4,*]

1 College of Water Conservancy and Architecture Engineering, Tarim University, Alar 843300, China; fyy2016060105@163.com (Y.F.); 18942777099@163.com (L.X.); 120200047@taru.edu.cn (H.W.); 13999068354@163.com (X.W.)
2 Key Laboratory of Crop Water Use and Regulation, Ministry of Agriculture and Rural Affairs, Institute of Farmland Irrigation, Chinese Academy of Agriculture Sciences, Xinxiang 453002, China; sunweihao1996@163.com (W.S.); sunusiabubakar@yahoo.com (A.S.A.); 2020y90100004@caas.cn (A.Q.)
3 College of Tropical Crops, Hainan University, Haikou 570100, China
4 Western Agricultural Research Center, Chinese Academy of Agricultural Sciences, Changji 831100, China
* Correspondence: m_abdoulkader@yahoo.com (A.K.M.H.); gaoyang@caas.cn (Y.G.)
† These authors contributed equally to this work.

Abstract: Low temperature and soil salinization during cotton sowing and seedling adversely affect cotton productivity. Exogenous melatonin (MT) can alleviate the damage caused to plants under non-biological stress; thus, applying MT is a means to improve the growth condition of crops under stress. However, achieving this goal requires a thorough understanding of the physiological regulatory mechanisms of MT on cotton seedlings under low temperature and salinity stress. This study could bring new knowledge on physio-biochemical mechanisms that improve the tolerance of cotton seedlings to combined effects of low temperature and salt stress using an exogenous foliar application of MT. The phytotron experiment comprised two temperature levels of cold stress and control and five MT treatments of 0, 50, 100, 150, and 200 μM and two salinity levels of 0 and 150 mM NaCl. Compared with the control treatments (non-salinity stress under cold stress and control), the combined stress of salt and low temperature reduced cotton seedlings' biomass and net photosynthetic rate (P_n), aggravated the membrane damage, reduced the potassium (K^+) content, and increased the sodium (Na^+) accumulation in the leaves and roots. Under NaCl stress, exogenously sprayed 50–150 μM MT increased the biomass and gas exchange parameters of cotton seedlings under salt and low temperature combined with salt stress, reduced the degree of membrane damage, and regulated the antioxidant enzyme, ion homeostasis, transport, and absorption of cotton seedlings. The pairwise correlation analysis of each parameter using MT shows that the parameters with higher correlation with MT at cold stress are mainly malondialdehyde (MDA), peroxidase (POD), and catalase (CAT). The highest correlation coefficient at 25 °C is observed between the K^+ and Na^+ content in cotton seedlings. The conclusion indicates that under salt and low-temperature stress conditions, exogenous application of MT primarily regulates the levels of P_n, superoxide dismutase (SOD), andPOD in cotton seedlings, reduces Na^+ and MDA content, alleviates damage to cotton seedlings. Moreover, the most significant effect was observed when an exogenous application of 50–150 μM of MT was administered under these conditions. The current study's findings could serve as a scientific foundation for salinity and low-temperature stress alleviation during the seedling stage of cotton growth.

Keywords: cotton seedlings; ion homeostasis; foliar melatonin; membrane damage; low temperature; salt stress

check for updates

Citation: Fu, Y.; Xin, L.; Mounkaila Hamani, A.K.; Sun, W.; Wang, H.; Amin, A.S.; Wang, X.; Qin, A.; Gao, Y. Foliar Application of Melatonin Positively Affects the Physio-Biochemical Characteristics of Cotton (*Gossypium hirsutum* L.) under the Combined Effects of Low Temperature and Salinity Stress. *Plants* **2023**, *12*, 3730. https://doi.org/10.3390/plants12213730

Academic Editor: Juan Barceló

Received: 15 September 2023
Revised: 20 October 2023
Accepted: 26 October 2023
Published: 31 October 2023

1. Introduction

Plants are often exposed to various environmental stresses, including drought, heavy-metal contamination, heat stress, low temperature, and soil salinization. Among these stresses, salinity and low temperature are important abiotic factors that are consequently associated with limited plant growth and productivity [1–3]. Moreover, over 50% of the world's arable land is predicted to be affected by salinity by 2050 [4,5], and 15% of the suitable agricultural areas are affected by temperature stress worldwide [6]. Soil salinization is often accompanied by temperature stress because changes in ambient temperature are more frequent than changes in other abiotic factors. On the other hand, changes in ambient temperature rapidly aggravate other environmental stresses, including salinity [1,6].

Salinity is a worldwide concern that reduces agricultural output and imposes substantial revenue losses [7]. Salinity caused by NaCl has become a focus of environmental research investigations. NaCl-associated salinity leads to a wide range of changes in plant metabolism [8,9]. An increasing NaCl concentration affects plants in several ways. It causes nutrient deficiencies, osmotic stress, and specific ion toxicity, affecting several physiological mechanisms involved in plant metabolism [10]. Salt stress affects a range of important mechanisms, including photosynthesis, energy and lipid metabolism, and protein synthesis [11,12]. Low-temperature stress generally affects plant growth and induces reactive oxygen species production, damaging cell membrane structure [13]. Previous studies reported that low-temperature stress rapidly increases antioxidant activities, such as superoxide dismutase (SOD), peroxidase (POD), and catalase (CAT) [14,15].

Melatonin (N-acetyl-5-methoxytryptamine; MT), which was initially identified and isolated from the pineal glands of cows, is an important multifunctional hormone that is involved in modulating a wide range of animal physiological processes, such as sleep [16], immunity [17], circadian rhythm and antioxidant activity [18]. Melatonin was discovered for the first time in 1995 in vascular plants as an indoleamine hormone [19]. MT plays a key role in plant growth and development by exerting an auxin-like function [20,21]. The first plant phytomelatonin receptor (CAND2/PMTR1) was recently detected in *Arabidopsis thaliana*, allocating to melatonin the concept of phytohormone [22,23]. A new phytomelatonin receptor (named *ZmPMTR1*) that plays a key role in plant osmotic and drought stress tolerance was also identified in *Zea mays* [24]. Several studies revealed the importance of MT in mitigating biotic and abiotic stresses [25–33]. Known as an antioxidant and free radical scavenger, exogenously applied MT improved plants' tolerance to biotic and abiotic stresses, which confers to plant stress resistance by enhancing photosynthesis, ion homeostasis, and antioxidant enzyme activities [34,35]. Exogenous foliar application of MT is involved in numerous physiological processes to improve plant resistance to salinity stress [36]. A recent study demonstrated that exogenously applied MT could sustain a high photosynthetic rate in tea plants, which enhances salt tolerance via its effects on antioxidant response against environmental stress [37]. Exogenous application of MT under salt stress showed alleviating effects on horticultural crops, such as an increase in primary root length and antioxidant activity in sunflower [38], an increase in net photosynthetic rate and stomatal conductance in tomato [39], and improving photosynthetic efficiency, endogenous melatonin content, and ion homeostasis in upland cotton [40]. On the other hand, exogenous application of melatonin was found to improve the resistance capacity of horticultural crops to low-temperature stress by increasing photosynthetic rate, antioxidant activity, and decreasing malondialdehyde (MDA) content in tea plants [41], increasing endogenous MT content, antioxidant activity, and decreasing MDA content in bermudagrass [42]; and increasing photosynthetic carbon assimilation, activities of antioxidant enzymes, levels of non-enzymatic antioxidants, anddecreasing the impact of cold on tomato membrane damage and ROS accumulation [39].

Cotton is the world's most important natural textile fiber with an impressive economic value and is considered the backbone of the textile industry. Despite the capacity of cotton plants to tolerate environmental stress, exposure to stress conditions could negatively affect cotton growth, development, and yield [43,44]. When cotton is grown in arid/semiarid regions, seedlings are often exposed to both low-temperature and salinity stresses. Therefore, several investigations have aroused keen interest in improving cotton resilience. Previous studies revealed the mechanism of exogenous melatonin on crop resilience improvement under a single stress of either salinity or low temperature. Understanding the significance of exogenous MT on cotton seedlings' physio-biochemical characteristics under the combined stress of salt and low temperature would have great theoretical and practical meaning. Therefore, the main purpose of this study was to determine the physiological and biochemical responses of low-temperature and NaCl-stressed cotton seedlings sprayed with exogenous melatonin.

2. Results

2.1. The Effect of MT on Cotton Seedling Biomass under Low Temperature and Salt Stress

Compared with the control, low temperature and salt stress, as well as their combined stress, significantly reduced the shoot dry weight, while root difference did not significantly affect it. Compared with the control (CK) at 15 °C, the coupled stress of salt and low temperature insignificantly affected the dry weights of the shoot and root of cotton seedlings (Figure 1). Under the salt stress condition at 25 °C, although the aboveground dry weight of cotton seedlings treated with exogenous MT increased slightly (compared with 0 μM MT), it was not significant, and there was no significant difference between different MT treatments. Under 25 °C, when compared to the 0 μM MT-saline treatment, exogenous MT significantly increased root dry weight, except for 200 μM MT; the highest value was obtained with 150 μM MT.

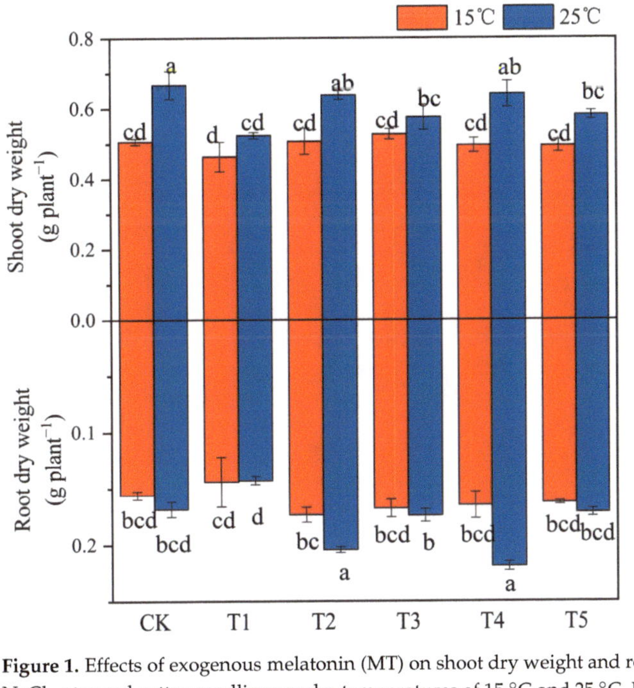

Figure 1. Effects of exogenous melatonin (MT) on shoot dry weight and root dry weight of 150 mM NaCl−stressed cotton seedlings under temperatures of 15 °C and 25 °C. Note: CK = Control (0 μM salt0 μM MT), T1 = 150 mM NaCl+0 μM (MT), T2 = 150 mM NaCl+50 μM (MT), T3 = 150 mM NaCl+100 μM (MT), T4 = 150 mM NaCl+150 μM (MT), T5 = 150 mM NaCl+200 μM (MT). Values are means ± standard deviation (n = 3). Different letters represent significant differences at $p < 0.05$ level between the experimental treatments.

2.2. Effects of MT on Gas Exchange Parameters of Cotton Seedling Leaves under Low Temperature and Salt Stress

Leaf gas exchange parameters, including the net photosynthetic rate (P_n), stomatal conductance (g_s), intercellular carbon dioxide concentration (C_i), and transpiration rate (T_r), were measured for evaluating the physiological responses of cotton seedlings to MT under the combined stress. Salinity (0 µM at 25 °C) or low temperature (control at 15 °C) stress alone have significantly reduced P_n, g_s, T_r, and C_i of cotton seedlings, in comparison with the control treatment at 25 °C. Compared with control at 15 °C, the combined stress reduced the P_n, g_s, T_r, and C_i of cotton seedlings (Figure 2). Under the salt stress condition at 25 °C, with the increase in exogenous MT concentration, P_n, g_s, T_r, and C_i all showed a trend of increase firstly and then decreased in the 200 µM MT treatment compared with the non-MT treated saline treatment. Except for C_i, the highest values of gas exchange parameters appeared in the 150 µM MT treatment. Under the combined stress of low temperature and salinity when compared to the MT-untreated saline, the application of different concentrations of exogenous MT insignificantly increased P_n, g_s, and T_r of cotton seedlings, and with the increase in MT concentration, P_n, g_s, and T_r also showed a trend of increasing and then decreasing. The highest value of P_n was obtained with the 100 µM MT treatment.

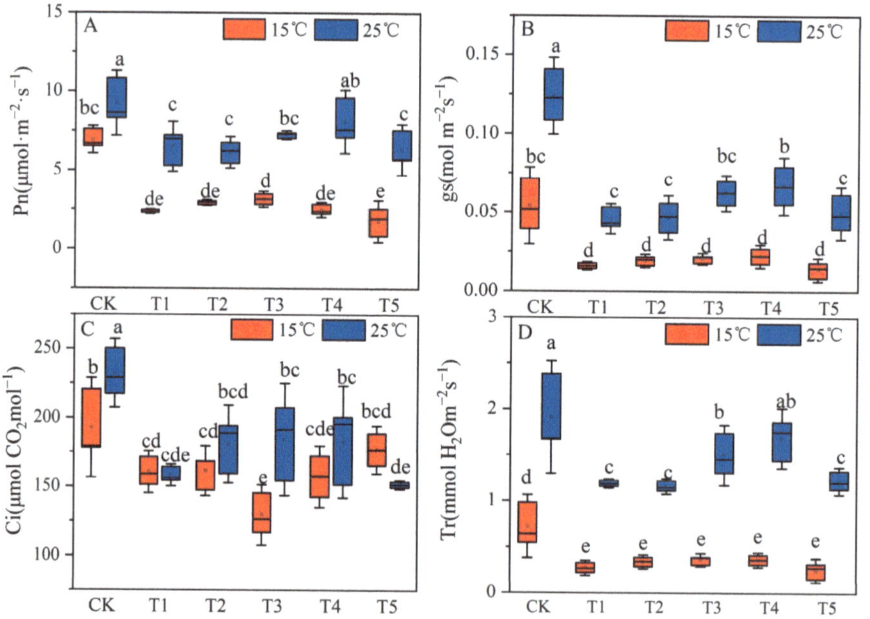

Figure 2. Effects of exogenous melatonin (MT) on (**A**) net photosynthetic (P_n), (**B**) stomatal conductance(g_s), (**C**) transpiration rate (T_r), and (**D**) intracellular CO_2 concentration (C_i) of cotton seedlings under the condition of 150 mM NaCl and temperatures of 15 °C and 25 °C. Note: CK = Control (0 µM salt+0 µM MT), T1 = 150 mM NaCl+0 µM (MT), T2 = 150 mM NaCl+50 µM (MT), T3 = 150 mM NaCl+100 µM (MT), T4 = 150 mM NaCl+150 µM (MT),T5 = 150 mM NaCl+200 µM (MT). Values are means ± standard deviation (n = 3). Different letters represent significant differences at $p < 0.05$.

2.3. The Effect of MT on Cotton Seedling Lipid Peroxidationunder Low Temperature and Salt Stress

Compared with control at 15 °C, low temperature plus salt stress did not significantly affect lipid peroxidation, measured as MDA content, nor the production rate of superoxide anion in leaves (Figure 3). All the concentrations of exogenous MT significantly decreased the MDA content under the combined stress of salt plus low temperature in comparison with the saline treatment in the absence of MT. Only the concentration of 100 µM MT significantly decreased the superoxide anion production rate under the combined stress of salt plus low temperature in comparison with the MT-untreated saline. At 25 °C, the exogenously applied MT (150 µM) significantly reduced both the superoxide anion production rate and MDA content in cotton seedling leaves under salt stress. At the same time (under 25 °C), compared with the saline treatment in the absence of MT, the superoxide anion production rate was insignificantly reduced by 29.6–4.4% under the 50–100 µM MT treatment, and the MDA content was significantly reduced by 43.3–67.7% under the 100–200 µM MT treatments, respectively. Under the combined stress of low temperature and saline conditions, spraying different concentrations of MT, compared with saline treatment in the absence of MT, the superoxide anion production rate significantly dropped to the value of 0.0034 µmol min^{-1} mg^{-1} port under the 150 µM MT treatment. With the increase in exogenous MT concentration, the content of MDA in leaves gradually decreased under the combined stress of low temperature and salt, and it significantly decreased to become stable when treated with 100–200 µM MT.

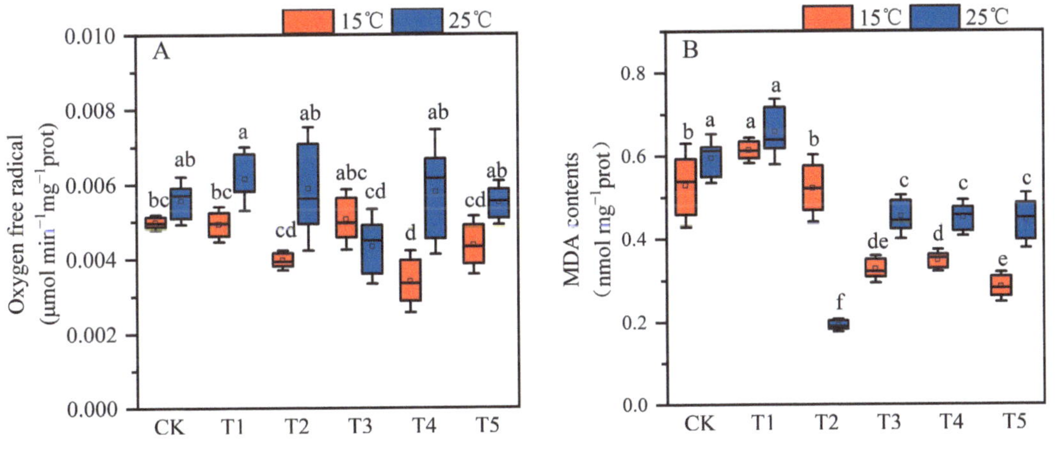

Figure 3. Effects of exogenous melatonin (MT) on (**A**) Superoxide anion content and (**B**) malondialdehyde (MDA) content of cotton seedlings under 150 mM NaCl-stressed and temperatures of 15 °C and 25 °C. Note: CK = Control (0 µM salt+0 µM MT), T1= 150 mM NaCl+0 µM (MT), T2 = 150 mM NaCl+50 µM (MT), T3 = 150 mM NaCl+100 µM (MT), T4 = 150 mM NaCl+150 µM (MT),T5 = 150 mM NaCl+200 µM (MT). Values are means ± standard deviation (n = 3). Different letters represent significant differences at $p < 0.05$.

2.4. Effect of MT on Antioxidant Enzyme Activities under Low Temperature and Salt Stress

Table 1 shows the effect of exogenously applied MT on the antioxidant enzymes of cotton under salt stress alone or combined. Compared with controls at 15 °C, combined stress significantly reduced the contents of POD and CAT in leaves, and salt stress alone reduced SOD and POD activities. The combined stress of low temperature and salt did not affect the SOD activity in comparison with the control treatment. Compared with the saline treatment in the absence of MT at 25 °C, exogenously sprayed MT significantly reduced the SOD and POD activity and did not affect CAT activity in leaves. Compared with the saline treatment in the absence of MT under the combined stress of low temperature and salt, spraying 50–100 µM MT increased SOD and POD activity and decreased that of CAT; salt stress alone showed the same trend. Using an interactive analysis between the treatments, it was found that different temperature conditions had a significant effect on the SOD and CAT activities, but the POD of leaves under salt stress did not significantly affect them. Different MT concentrations significantly impacted POD and CAT while having an insignificant effect on the SOD of cotton seedlings under salt stress. The POD and CAT activities in cotton seedling leaves were affected by the interaction of temperature and MT.

Table 1. Response of antioxidant enzymes activities to MT in cotton seedlings under temperatures of 15 °C and 25 °C and salt stress. Values are means ± standard deviation.

Treatment	Antioxidant Enzyme Activities in Leaf Tissues					
	SOD/U × g^{-1} × FW		POD/U × g^{-1} × FW		CAT/U × g^{-1} × FW	
	15 °C	25 °C	15 °C	25 °C	15 °C	25 °C
CK	67.32 ^{cd}	108.32 ^{b}	13.32 ^{a}	7.38 ^{b}	0.46 ^{a}	0.01 ^{b}
T1	69.27 ^{cd}	78.52 ^{c}	4.38 d ^{e}	5.09 ^{cd}	0.11 ^{b}	0.02 ^{b}
T2	147.18 ^{a}	45.32 ^{f}	5.17 ^{cd}	5.95 ^{c}	0.03 ^{b}	0.07 ^{b}
T3	137.57 ^{a}	51.96 ^{ef}	4.46 ^{de}	3.55 ^{ef}	0.04 ^{b}	0.02 ^{b}
T4	140.41 ^{a}	58.09 d ^{ef}	4.45 ^{de}	3.79 ^{ef}	0.04 ^{b}	0.02 ^{b}
T5	101.91 ^{b}	63.78 ^{de}	3.14 ^{f}	3.95 ^{ef}	0.03 ^{b}	0.04 ^{b}
Temp	11.79 **		1.43		29.48 **	
MT	2.60		1.46 **		20.29 **	
Temp× MT	1.35		9.90 **		33.43 **	

Note: CK = Control (0 µM salt + 0 µM MT), T1 = 150 mM NaCl + 0 µM (MT), T2 = 150 mM NaCl + 50 µM (MT), T3 = 150 mM NaCl + 100 µM (MT), T4 = 150 mM NaCl + 150 µM (MT), T5 = 150 mM NaCl + 200 µM (MT). Treatments: 15 °C, 25 °C; CK, control. Temp is temperature; MT is melatonin. Values are means ± standard deviation (n = 3). Different letters represent significant differences at $p < 0.05$. ** $p < 0.01$.

2.5. Effect of MT on Ion Homeostasis and Absorption under Different Temperatures and Salt Stress

Results (Figure 4) show that compared with the control treatment at 25 °C, salt-stressed treatment alone significantly reduced the potassium (K^+) ion content in the leaves and roots by 15% and 50% and increased the sodium (Na^+) ion by 4.2 and 2.5 times-, respectively. The combined stress reduced the K^+ content in the roots and significantly increased the accumulation of Na^+ in the leaves and roots. After foliar spraying of different concentrations of MT, the K^+ content in the roots was significantly increased when compared with the saline treatment in the absence of MT at 25 °C, and the K^+ content in the leaves significantly increased with the exogenous foliar application of 150 µM when compared with the saline treatment in the absence of MT at 15 °C. The Na^+ content in the leaves and roots was significantly reduced, but the difference between different concentrations of MT was insignificant. Under the combined stress conditions, compared with the saline treatment in the absence of MT, the application of exogenous MT to the K^+ content in the leaves and roots of cotton seedlings did not significantly affect. Different concentrations of exogenous MT significantly reduced the content of Na^+ in the leaves and roots of cotton seedlings when compared with the saline treatment in the absence of MT at 25 °C.

Compared with the control treatment, the coupled stress of low temperature and salt significantly increased the Na^+/K^+ ratio in the leaves and roots of cotton seedlings (5.4 and 2.9 times, respectively) (Figure 5A,B), and significantly increased the Na^+ uptake roots (5.9 times) (Figure 5D). Exogenous foliar spraying of different MT concentrations significantly reduced the Na^+/K^+ ratio in the leaves and roots under the salt stress at 25 °C and correspondingly reduced the Na^+ absorption. Under the combined stress, the exogenous foliar supplementation with MT, the Na^+/K^+ ratio in the leaves did not significantly affect. The Na^+/K^+ ratio in the roots significantly decreased with the foliar application of MT at 200 µM under the combined stress. On the other hand, compared with the control, the combined stress significantly increased the transfer of K^+ ions from roots to shoots and then from shoots to leaves. After exogenously foliar spraying of different concentrations of MT, the low temperature plus salt stress significantly increased K^+ transport from roots to aboveground parts. Compared with the saline treatment in the absence of MT, different concentrations of MT slightly reduced the transport of K^+ from roots to shoots.

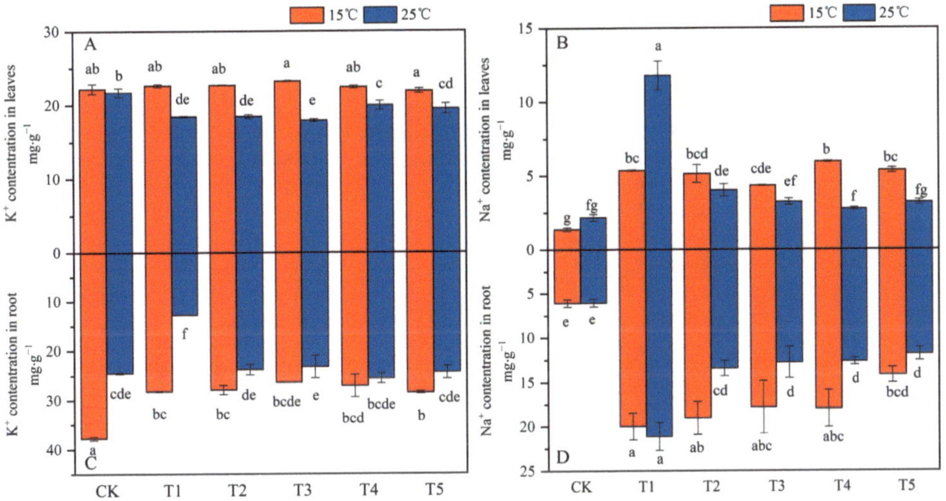

Figure 4. Effects of exogenous melatonin (MT) on (**A**) K^+ concentration in the leaves, (**B**) Na^+ concentration in the leaves, (**C**) K^+ concentration in the roots, and (**D**) Na^+ concentration in the roots of cotton seedlings under the 150 mM NaCl-stressed and temperatures of 15 °C and 25 °C. Note: CK = Control (0 μM salt+0 μM MT), T1 = 150 mM NaCl+0 μM (MT), T2 = 150 mM NaCl+50 μM (MT), T3 = 150 mM NaCl+100 μM (MT), T4 = 150 mM NaCl+150 μM (MT),T5 = 150 mM NaCl+200 μM (MT). Values are means ± standard deviation (n = 3). Different letters represent significant differences at $p < 0.05$.

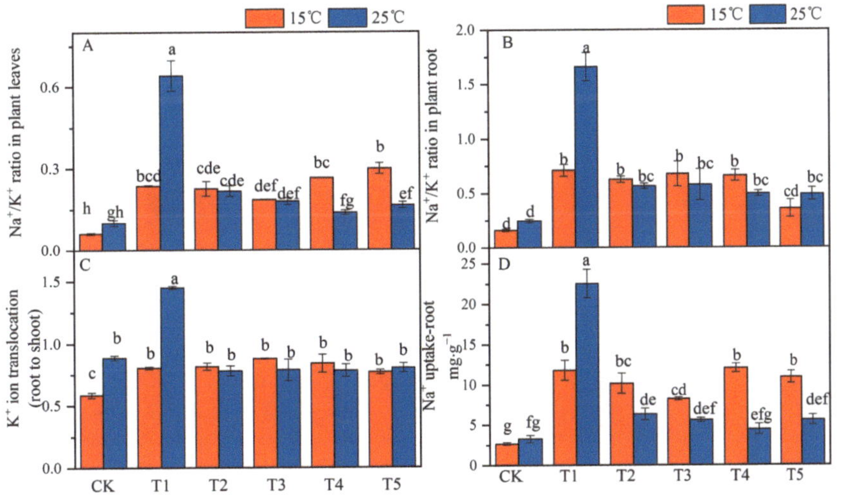

Figure 5. Effects of exogenous Melatonin (MT) on (**A**) Na^+/K^+ ratio in plant leaves, (**B**) Na^+/K^+ ratio in plant roots, (**C**) K^+ ion translocation (root to shoot), and (**D**) Na^+ uptake in-root of cotton seedlings under the 150 mM NaCl-stressed and temperatures of 15 °C and 25 °C. Note: CK = Control (0 μM salt+0 μM MT), 0 μM = 150 mM NaCl+0 μM (MT), 50 μM = 150 mM NaCl+50 μM (MT), 100 μM = 150 mM NaCl+100 μM (MT), 150 μM = 150 mM NaCl+150 μM (MT), 200 μM = 150 mM NaCl+200 μM (MT).Values are means ± standard deviation (n = 3). Different letters represent significant differences at $p < 0.05$.

3. Discussion

Abiotic stresses such as salinity and low temperature severely limit the growth and development of plants. Moreover, nearly 50% of the annual yield loss of major crops worldwide is related to abiotic stresses [45]. The combined effect of low temperature and salt stress reduces leaf area, relative water content, water potential, transpiration rate, fresh weight, and dry weight of plant stems and roots [46]. To gain new knowledge of cotton seedling responses to exogenously applied MT under combined low temperature and salt stress, seedlings in a phytotron were foliar sprayed with MT to identify their physiological and biochemical responses. In this study, the combined effect of low temperature and salt stress significantly and negatively affected seedlings' physiological and biochemical mechanisms (Table 1, Figures 1–6). In agreement with our findings, previous studies have concluded that the effects of combined stress on crop growth and productivity may be devastating [47,48]. For example, drought and high temperature, salinity and high temperature, ozone and salinity, ozone and high temperature, nutrient stress and drought, nutrient stress and salinity, ultraviolet light and high temperature, ultraviolet light and drought, strong light and heat, drought or low temperature and other stress interactions showed significant negatives impacts on crop yield [48,49]. Literature reported It has been reported that combined stress such as drought plus high-temperature stress has a more serious negative impact on the number of wheat tillers, chlorophyll content, yield, and harvest index than a single stress of drought or high temperature [50,51]. In this study, the combination of low temperature plus salinity stress has a greater impact on cotton seedling biomass when compared to the control treatment (Figure 1). This is mainly related to the senescence and dehydration of plant cells under abiotic stress, and salinity limits cell elongation and division, which leads to a decrease in the growth rate of roots and shoots and a decrease in dry matter accumulation [52]. After exogenous foliar supplementation with MT under the different stresses of low temperature and salinity plus low temperature, the aboveground and underground biomass remained unaffected, while the underground biomass was significantly increased by the foliar supplementation with MT (50, 100 and 150 μM) under salt stress at 25 °C in comparison with the saline treatment in the absence of MT (Figure 1). This is related to the ability of MT to regulate the physical process of cell wall extension to induce plant root growth and promote the accumulation of dry matter [53].

Photosynthesis is the basis of plant growth and development, but it is susceptible to environmental stresses (particularly sensitive to low-temperature stress). The main components of photosynthesis (electron transport, Calvin cycle, stomatal conductance) are negatively affected under low-temperature stress [46]. In a previous study, we observed that sustained leaf gas exchange parameters, including P_n, g_s, T_r, and C_i are essential for cotton seedlings' survival under salt stress conditions [43]. Regulation of leaf gas exchange parameters is reported to be an important aspect of improving crop resilience to different biotic and abiotic stress conditions [54]. In this study, g_s and T_r decreased in response to salinity alone, low temperature alone, and salinity plus low temperature, and plants reduced water loss by closing their stomata. The closing of stomata leads to insufficient CO_2 and inhibits photosynthesis (Figure 2). This is consistent with the research results of Chatrath et al. [55] who reported that plants may close their stomata under saline conditions to reduce water loss. As shown in Figure 7B, under the combined stress of salinity plus low temperature the antioxidant enzyme activity of cotton seedlings unilaterally affects photosynthetic parameters. Biomass is not directly affected by photosynthetic parameters; membrane damage and other parameters have no mutual influence. After the foliar application of MT (100 and 150 μM) under salt stress at 25 °C, the g_s and T_r of cotton seedlings were increased, thereby increasing their P_n, but the foliar application of MT did not alter photosynthesis parameters under the combined effect of salinity plus low temperature (Figure 2). This is in contradiction with the research results of Irshad et al. [56] and Zhang et al. [33], who observed that exogenous MT significantly improved P_n under low-temperature stress and salt stress, respectively. This difference between our findings and those of previous studies is probably due to (1) the different species used as plant material (*Medicago truncatula* in a previous study and *Gossypium hirsutum* L. in the current study), (2) the MT application mode (previous studies applied MT by seed priming and or hand MT application one week before cold stress induction, while foliar MT treatment and low-temperature stress were conducted simultaneously in this study).

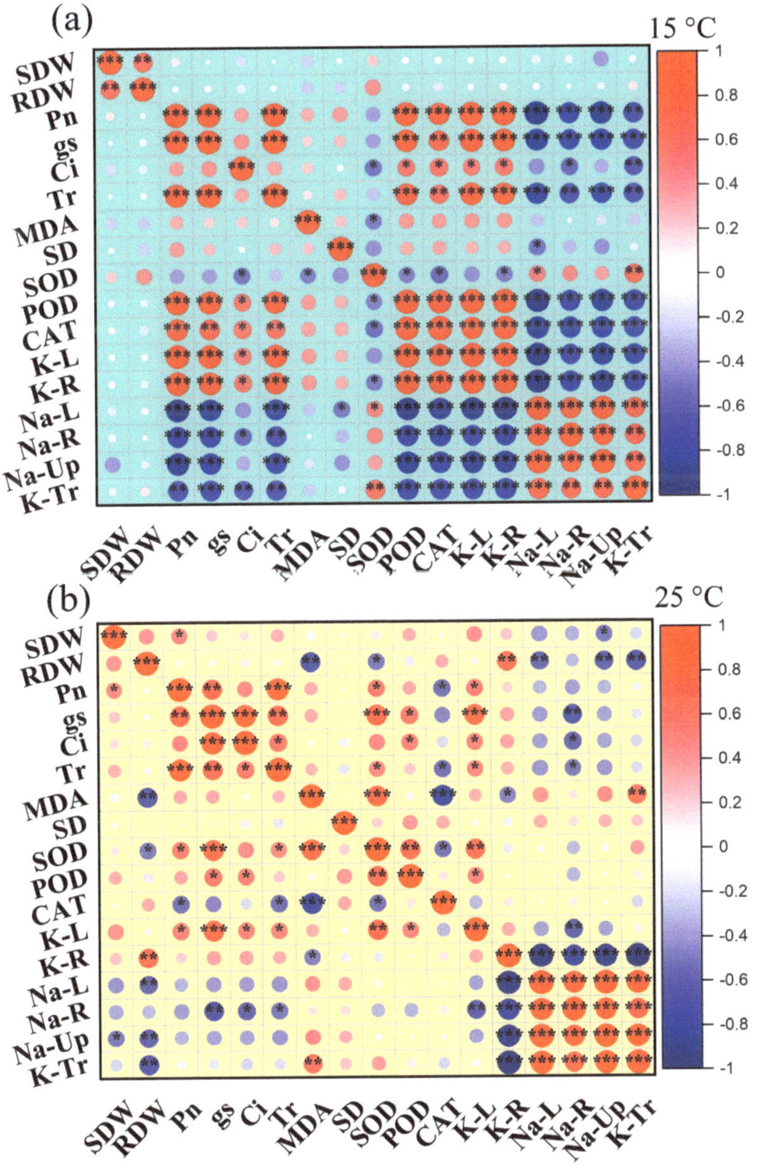

Figure 6. Correlation analysis of exogenous MT on various parameters of salt-stressed cotton seedlings under (**a**) low temperature (15 °C) and (**b**) 25 °C. SDW, shoot dry weight; RDW, root dry weight; *Pn*, net photosynthetic rate; g_s, stomatal conductance; C_i, intracellular CO_2 concentration; T_r, transpiration rate; K-L, leaf K^+ content; K-R, root K^+ content; Na-L, leaf Na^+ content; Na-R, root Na^+ content; Na^+ uptake; K^+ translocation. * $p < 0.05$, ** $p < 0.01$ and *** $p < 0.001$.

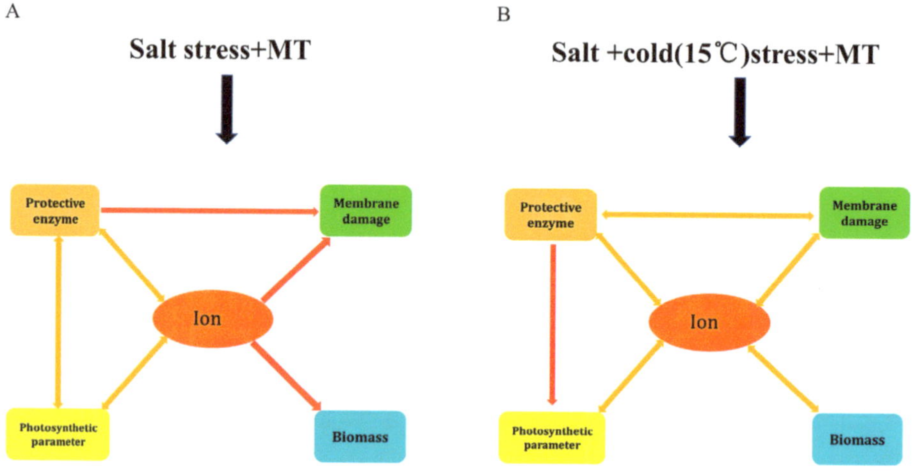

Figure 7. Effects of exogenous MT on the physiological mechanism of cotton seedlings under (**A**) salt stress and (**B**) under coupled low temperature and salt stress. The yellow double-orientation arrow indicates that the two affect each other; the red one-way arrow indicates unilateral influence.

The translocation of substances and ions in plants mainly depends on the integrity of the structure and function of the cellular membranes. When the plant is subjected to abiotic stress, the plant's cellular membrane system is the first to be impacted and damaged. One of the damages caused by abiotic stress on cell membranes is lipid peroxidation. The final product of peroxidation damage to cell membranes by reactive oxygen species is MDA. The production of reactive oxygen species (ROS) such as H_2O_2, superoxide anion, hydroxyl anion, and MDA under salt conditions severely destroys chlorophyll content, destroys lipids and mitochondria, and causes plant necrosis [33]. In this study, under the combined stress, cotton seedlings' MDA content increased compared with the control, although the superoxide anion production rate remained unchanged compared with the control, which led to membrane lipid peroxidation (Figure 3). The research results are consistent with those of Parveen et al. [57]. Abiotic stress causes an increase in the level of active oxygen in the plant body and simultaneously activates the defense system to reduce the harm of active oxygen. SOD, CAT, and POD are the key enzymes to remove active oxygen in cells, and the ability to remove active oxygen depends on the coordination of these enzymes [58]. As the first line of defense for ROS removal, SOD can disproportionate superoxide anions into O_2 and H_2O_2, and POD and CAT can remove H_2O_2 in peroxisomes [59]. In this experiment, compared with the control at 15 °C, salt stress alone as well as the coupled stress of salt plus low temperature decreased the CAT and POD activities. Furthermore, the low-temperature stress increased the POD and CAT activities and decreased the SOD activity when compared to the control treatment at 25 °C (Table 1). MT can increase the activity of antioxidant enzymes to alleviate the damage caused by abiotic stress on plants. Studies have shown that the application of exogenous MT can increase the SOD, POD, and CAT activities in cucumber under salt-stress environments [60] and increase the SOD activity in wheat [61] and cucumber [62] under low-temperature stress. In addition, under salt and heat stress, applying MT increased tomato CAT activity and reduced SOD activity [63]. In this experiment under the combined stress of salinity plus low temperature, after applying the exogenous MT, compared with the salt treatment without the exogenous MT, the exogenous spraying of MT at a concentration of 200 μM decreased the POD activity of cotton seedlings. In contrast, under the combined stress of salinity plus low temperature, spraying of different concentrations of MT significantly increased the SOD activity of cotton seedlings (Table 1). Using interactive analysis, it is known that temperature has a very significant impact on SOD and CAT, and MT has a very significant impact on POD and CAT. At the same time, POD and CAT are affected by the interaction of temperature and MT.

Under salt stress, the excessive accumulation of Na^+ in plant cells leads to the production of ROS, which breaks the dynamic balance of the scavenging system and initiates membrane lipid peroxidation and degreasing. The membrane structure is destroyed, and the MDA content increases [58]. In this study, after MT under salt stress, the opposite conclusion was reached and found that the Na^+ content in cotton seedlings was reduced, the activity of antioxidant enzymes, the degree of lipid peroxidation was increased, the degree of lipid peroxidation damage was reduced, and the photosynthetic parameters and dry matter content were increased. Previous studies reported that the high accumulation of Na^+ in the apoplasts of roots and stems easily triggers osmotic stress [64]. The higher accumulation of Na^+ leads to an increase in the ratio of Na^+/K^+ ratio and promotes ion imbalance, thereby hindering metabolic functions and destroying plant activities [33]. For example, photosynthesis declines, and dry matter accumulation decreases, which ultimately leads to reduced crop growth [53], production of reactive oxygen species [65], and plant damage. The results of this study showed that, compared with the control at 25 °C, salinity reduced the K^+ accumulation in the leaves and roots of cotton seedlings and significantly increased the accumulation of Na^+, resulting in a significant increase in the ratio of Na^+/K^+ and at the same time increased the absorption of Na^+ (Figures 4 and 5). However, after spraying exogenous MT at 25 °C under saline conditions, Na^+ accumulation in leaves and roots was significantly reduced. The K^+ accumulation in the leaves was increased, and the Na^+/K^+ ratio was reduced. Since the K^+ dynamic balance is affected by the accumulation of Na^+, an appropriate Na^+/K^+ should be maintained to improve the salt tolerance of crops [66]. Studies have found that under abiotic stress, plants, by closing stomata and reducing transpiration, may limit the transport of Na^+ to plant tissues [67]. The results of this study indicate that compared to salt stress, the stomatal conductance is reduced under the combined stress of low temperature and salt, and the Na^+/K^+ ratio is lower than that of salt stress. In the current study, using interactive analysis, temperature was shown to significantly affect K^+, Na^+, Na^+/K^+ ratio, Na^+ absorption, and K^+ transport in roots and leaves, whereas Na^+ content in leaves was not affected. The influence of exogenous MT on K^+ content is insignificant. The Na^+ content in root removal is not affected by the interaction of temperature and MT, and the others are all affected by the interaction of temperature and MT.

Studies have shown that low-temperature stress can enhance the antioxidant defense ability by regulating the metabolic and photosynthetic reactions of tea plants, wheat, Arabidopsis, and other plants, thereby improving the cold tolerance of plants [37,68]. The results of this study showed that when exogenous MT is applied under low temperature plus salt stress, the ion content directly affects the protective enzymes, photosynthetic parameters, membrane damage, and biomass, while the protective enzymes and photosynthetic parameters negatively affect the ion content. Protective enzymes affect membrane damage in one direction and have a two-way effect on photosynthesis (Figure 7). In summary, using interactive analysis, we have shown that temperature has a significant effect on the shoots and root biomasses under salt stress conditions. Exogenous MT successfully increased the biomass and physiological and biochemical indexes of cotton seedlings under the combined stresses. In addition, reducing the absorption of ROS and Na^+ in plants is the focus of this study, indicating that exogenous MT has the potential to restore crop growth under salt stress. The relationships between various parameters of cotton seedlings at 15 °C and 25 °C under salt stress conditions are shown in Figure 6 and Table 2. It was observed that improving antioxidant enzyme activities with exogenous MT results in an improvement in leaf gas exchange parameters and K^+ uptake while decreasing Na^+ uptake.

Table 2. Stepwise regression of parameters of salt-stressed cotton seedlings at 15 °C and 25 °C under exogenous melatonin.

Independent Variable		Low Temperature (15 °C) + Salt Stress (150 mM)			25 °C + Salt Stress (150 mM)		
		Stepwise Regression Equation	R^2	*p*	Stepwise Regression Equation	R^2	*p*
X1	shoot dry weight	y = 1.443x2 + 0.266	0.40	0.01	y = 0.20x12 + 0.192x14 − 0.101x16 + 0.147	0.87	0.00
X2	root dry weight	y = 0.278x1 + 0.023	0.40	0.01	y = −0.070x17 + 0.244	0.47	0.00
X3	P_n	y = −0.014x5 + 3.090x6 + 0.398x10 + 2.046	0.97	0.00	y = 3.558x6 + 2.126	0.72	0.00
X4	g_s	y = −0.002x12 + 0.001x10 + 0.061x6 + 0.034	1.00	0.00	y = 0.002x13 + 0.001x9 + 0.000x5−0.108	0.92	0.00
X5	C_i	y = −135.623x17 − 269.164	0.39	0.06	y = 891.435x4 + 122.900	0.72	0.00
X6	T_r	y = 0.025x12 − 0.016x10 + 15.698x4−0.468	0.99	0.00	y = 0.203x3−0.033	0.72	0.00
X7	MDA	y = −0.002x9 + 0.635	0.23	0.04	y = −0.070x10 + 0.007x9 + 0.209x17 + 0.120	0.92	0.00
X8	Superoxide anion	y = 0.000x14 − 0.005	0.26	0.03	/	/	/
X9	SOD	y = 205.005x17 − 49.711	0.40	0.01	y = −1.211x15 + 7.685x10 + 101.543x7 − 1.916	0.96	0.00
X10	POD	y = 0.125x15 + 130.185x4 + 2.404x11 − 0.422	0.98	0.00	y = −4.530x7 + 32.189x11 + 0.083x9 + 0.531	0.85	0.00
X11	CAT	y = −0.052x15 − 2.979x6 + 0.329x10 − 0.498	0.96	0.00	y = 0.002x14 − 0.114x7 + 0.073	0.73	0.00
X12	K^+-L	y = −0.057x16 + 0.042x15 − 35.334x4 + 0.604x13 + 22.223x17 − 11.876	0.96	0.00	y = 37.733x4 + 16.815	0.57	0.00
X13	K^+-R	y = 0.42x11 + 48.308x4 + 1.020x12 − 33.413x17 + 31.137	1.00	0.00	y = 0.237x16 + 1.078x12−23.081x17 + 20.755	0.99	0.00
X14	Na^+-L	y = 4.983x2 + 9.996x1 + 0.492x16 − 5.787	0.99	0.00	y = −0.896x7 + 4.021x1 + 0.534x16 − 1.744	1.00	0.00
X15	Na^+-R	y = 0.758x13−4.227x11 + 4.298x12 + 2.025x14 − 110.436	0.87	0.00	y = 17.741x17 − 0.150x9 + 6.932	0.91	0.00
X16	Na^+-uptake	y = −0.993x2 − 20.381x1 + 2.027x14 + 11.795	1.00	0.00	y = 1.713x7 − 7.595x1 + 1.869x14 + 3.308	1.00	0.00
X17	K^+ translocation	y = 0.011x11 + 1.427x4 + 0.031x12 − 0.030x13 + 0.912	1.00	0.00	y = 0.119x7 + 0.012x16 + 0.040x12 − 0.039x13 + 0.856	0.99	0.00

Note: SDW, shoot dry weight; RDW, root dry weight; P_n, net photosynthetic rate; g_s, stomatal conductance; C_i, intracellular CO_2 concentration; T_r, transpiration rate; K-L, leaf K^+ content; K-R, root K^+ content; Na-L, leaf Na^+ content; Na-R, root Na^+ content; Na^+ uptake; K^+ translocation.

Expression of the phythomelatonin receptor (*ZmPMTR1*) induced using abiotic stresses, including low temperature and salt, suggests its potential function in maize responses to various abiotic stresses and may be useful for crop genetic improvement [24]. Recently, various studies aimed to investigate the function of melatonin as a signaling molecule that is involved in plant growth regulation and stress responses, mainly the identification and analysis of its receptors in various plant species [28,69]. More in-depth studies on cotton are required in the future perspectives on the MT-receptor interaction (identification, location, regulation, etc.). Exogenous MT has different regulatory mechanisms on the ion content, photosynthetic parameters, membrane damage, protective enzyme activities, and biomass accumulation of cotton seedlings under coupled salt and low-temperature stress (Figure 7). Protecting enzymes under salt stress affects photosynthesis in two directions while affecting photosynthesis in one direction under low temperatures plus salt stress. The antioxidant enzyme activity under salt stress affects membrane damage one way, and ion content one-way affects membrane damage and biomass, while under the combined stress of low temperature plus salt, the antioxidant enzymeand membrane damage, ion content, and membrane damage and biomass have two-way effects. Under double coercion, MT's adjustment of the above parameters is more complex.

4. Materials and Methods

4.1. Plant Material and Experimental Design

The cotton variety, Xinluzhong37, was purchased from The Seed Industry Co., Ltd., Alar City, China. It is a mid-early maturing upland cotton variety with a growth period of about 140 days. Melatonin (MT), sodium chloride (NaCl), sodium sulfate (Na_2SO_4), and other chemical reagents are all analytically pure, purchased from Beijing Soleibao and Sinopharm Chemical Reagent Co., Ltd. (Shanghai, China).

The experiment was carried out in a phytotron in the Xinxiang Comprehensive Experimental Station of the Chinese Academy of Agricultural Sciences (35.09° N, 113.48° E, and altitude 81 m). The humidity of the climate room is 40–50%, the photoperiod is 12 h, the light intensity is 600 μmol m^{-2} s^{-1} (lighting is provided using LED lights), and the temperature is controlled according to the experimental design. The experimental design is shown in Table 3.

Table 3. The detailed experimental treatments.

Temperature	Treatment Label	NaCl Dose (mM)	MT Dose (μM)
15 °C	CK	0	0
	T1	150	0
	T2	150	50
	T3	150	100
	T4	150	150
	T5	150	200
25 °C	CK	0	0
	T1	150	0
	T2	150	50
	T3	150	100
	T4	150	150
	T5	150	200

Uniform size and plump cotton seeds were selected, disinfected with 1000 times diluted solution of carbendazim, and rinsed with deionized water after 30 min. The sterilized seeds were sown at a depth of 2 cm in PVC pots (diameter 6 cm, height 24 cm) containing 780 g of fine sand (the sand was sterilized at high temperature). After sowing, the PVC pots were covered with an opaque shading board to keep the surface of the sand moist and facilitate the germination of seeds. For germination, the PVC pots were placed in a phytotron with a day/night temperature of 25/20 °C. After the seeds germinated, the shading was removed from the pots. Seven (7) plants per treatment with one plant/pot were used in this experiment, with a total number of 70 plants, and three (3) replications per treatment were used for all the measurements. When the seedlings grew to one true leaf, seedlings were watered with 1/2 Hoagland nutrient solution [9]. The pH of the nutrient solution was between 5.8 and 6.2; each pot was watered every five days with about 80 mL of the 1/2 Hoagland solution. When the seedlings were grown to two leaves and one heart leaf, they were randomly divided into groups to start the salt and temperature treatments. For room temperature (25 °C) and low temperature (15 °C) treatments, seedlings were placed in phytotrons with day/night temperatures of 25/18 °C and 15/8 °C, respectively. On the 1st, 4th, and 9th day under the temperature treatment, seedlings were irrigated with 150 mmol L^{-1} salt water (NaCl: Na_2SO_4 = 9:1) in the morning. Every day after turning off the lights, different concentrations of MT (0, 50, 100, 150, 200 μmol L^{-1}) were sprayed on the upper surface of the leaves of salt-stressed cotton seedlings at 15 and 25 °C. On the 10th day, the fully expanded leaves were sampled for the measurement of various indicators.

4.2. Measurements

4.2.1. Determination of Cotton Seeding Biomass and Gas Exchange Parameters

Three plants were randomly selected for each treatment, the shoots were separated from the roots, and the fresh weights were, respectively weighed and put into sample bags. The samples were put into a sample bag at 105 °C for 30 min and oven-dried at 75 °C for 48 h; the dry weight was then measured.

The LI-6400XT photosynthesis measuring system (Licor, Lincoln, NE, USA) was used to measure the net photosynthetic rate (P_n), stomatal conductance (g_s) intercellular carbon dioxide concentration (C_i), and transpiration rate (T_r) of the upper fully expanded leaves of cotton seedlings between 9:00–11:00 in the morning with photosynthetically active radiation (PAR) at 1000 $\mu mol \cdot m^{-2} \cdot s^{-1}$.

4.2.2. Determination of MDA, Superoxide Anion and Antioxidant Enzyme Activities

About 0.3 g of fresh leaf tissue was ground using a pestle and mortar. An amount of 3.0 mL of 0.05 M precooled phosphate buffer (pH 7.8) was added to the homogenate, which was then centrifuged at 15,000× g at 4 °C for 20 min. Lipid peroxidation was measured spectrophotometrically from MDA content using a thiobarbituric acid (TBA) reaction following the method described by Heath and Packer [70]. At the same time, the same supernatant was used to determine the leaves' superoxide anion production rate [71].

About 0.3 g of fresh leaf tissue was ground using a pestle and mortar. An amount of 2.7 mL of 0.05 M precooled phosphate buffer (pH 7.8) was added to the homogenate, which was then centrifuged at 15,000× g at 4 °C for 20 min. After obtaining the supernatant, refer to the following method to determine the protective enzyme activity. SOD activity was determined spectrophotometrically from the inhibition of the photochemical reduction of nitroblue tetrazolium (NBT) at 560 nm [72]. POD activity was measured using the determination of guaiacol oxidation by H_2O_2 at 470 nm [73]. CAT activity was measured by monitoring the disappearance of H_2O_2 at 240 nm [74].

4.2.3. Determination of Seeding Biomass and Ion Content

On the 10th day after the treatment, three cotton seedling leaves and roots were randomly selected for each treatment, oven-dried at 105 °C for 30 min, and then continued drying at 75 °C for 48 h; the dry weight was then measured. The dried leaves and roots were ground into powder. Each sample, weighing 0.15 g, was digested with H_2SO_4-H_2O_2 and diluted in a 100-mL volumetric flask. The filtered supernatant was used to determine the Na^+ and K^+ ion contents in the leaves and roots of cotton seedlings using a flame photometer. Na^+ uptake at the seedlings' root surface and K^+ translocation from root to shoot were computed by following the equations described by Shabala and Mackay [75,76], respectively.

$$\text{Na ion uptake at root} = \frac{\text{Sum of Na concentration in tissues}}{\text{Total root dry wright of}} \tag{1}$$

$$\text{Translocation factor} = \frac{\text{K concentration in leaves}}{\text{K concentrantion in roots}} \tag{2}$$

4.3. Statistical Analysis

The data were statistically analyzed using one-way ANOVA using EXCEL 2020 and DPS V13.5 software. All experimental data were expressed as means ± standard deviation. All treatment means (n = 3) were compared for any significant differences using Duncan's multiple range tests at $p < 0.05$. Data fitting and graphical presentation were carried out in Origin-Pro 2021a (Origin Lab, Northampton, MA, USA). A general linear regression model was used to fit the relationships between parameters.

5. Conclusions

The current study demonstrated that exogenous foliar application of MT significantly affects the physiological and biochemical characteristics of cotton subjected to combined stress of salinity and low temperature. Under salt stress, the spraying of 50 to 150 μM MT on the leaf surface increased the biomass, P_n, and SOD and POD activities while reducing membrane damage of cotton seedling leaves. The exogenously supplied MT reduced the accumulation of Na^+ and MDA under the coupled stress of low temperature and salt. The regulatory mechanism of exogenously applied MT remains different under salt stress alone as compared to the combined salinity and low-temperature stress. The regulation of the mechanism under the combined stress of low temperature and salinity is more complex. Further study is necessary to investigate the molecular mechanisms of MT-regulated salinity and low-temperature stress tolerance in cotton.

Author Contributions: Conceptualization, X.W. and Y.G.; methodology, Y.F., L.X. and A.K.M.H.; software, L.X. and A.S.A.; validation, X.W. and Y.G.; formal analysis, Y.F. and L.X.; investigation, W.S. and H.W.; data curation, Y.F. and L.X.; writing—original draft preparation, L.X., A.K.M.H. and Y.F.; writing—review and editing, A.K.M.H., A.Q. and Y.G.; funding acquisition, X.W. and Y.G. All authors have read and agreed to the published version of the manuscript.

Funding: This work was supported by the National Natural Science Foundation of China (No. 51879267, 51790534), China Agriculture Research System of MOF and MARA (CARS-15-13), XPCC finance science and technology project (2021AA003, 2022BC009), and the Agricultural Science and Technology Innovation Program (ASTIP), Chinese Academy of Agricultural Sciences.

Data Availability Statement: Not applicable.

Acknowledgments: We would like to thank all the lab mates, the Key Laboratory of Crop Water Use and Regulation manager, the Ministry of Agriculture and Rural Affairs, Xinxiang, Henan, and the China Scholarship Council (CSC).

Conflicts of Interest: The authors declare that they have no known competing financial interest or personal relationships that could have appeared to influence the work reported in this paper.

References

1. Ashraf, M.; Foolad, M.R. Roles of glycine betaine and proline in improving plant abiotic stress resistance. *Environ. Exp. Bot.* **2007**, *59*, 206–216. [CrossRef]
2. Mishra, S.; Kumar, S.; Saha, B.; Awasthi, J.; Dey, M.; Panda, S.K.; Sahoo, L. Crosstalk between Salt, Drought, and Cold Stress in Plants: Toward Genetic Engineering for Stress Tolerance. In *Abiotic Stress Response in Plants*; Springer: New York, NY, USA, 2016; pp. 57–88.
3. Naing, A.H.; Kim, C.K. Abiotic stress-induced anthocyanins in plants: Their role in tolerance to abiotic stresses. *Physiol. Plant.* **2021**, *172*, 1711–1723. [CrossRef] [PubMed]

4. Demiral, T.; Türkan, I. Exogenous glycinebetaine affects growth and proline accumulation and retards senescence in two rice cultivars under NaCl stress. *Environ. Exp. Bot.* **2006**, *56*, 72–79. [CrossRef]
5. Saeed, N.A.; Ahmad, M.; Mukhtar, Z. GM Cotton for Stress Environments. In *Cotton Precision Breeding*; Springer: Berlin/Heidelberg, Germany, 2021; pp. 257–280.
6. Çiçek, N.; Çakırlar, H. Effects of salt stress on some physiological and photosynthetic parameters at three different temperatures in six soya bean (*Glycine max* L. Merr.) cultivars. *J. Agron. Crop Sci.* **2008**, *194*, 34–46. [CrossRef]
7. Prajapati, P.; Gupta, P.; Kharwar, R.N.; Seth, C.S. Nitric oxide mediated regulation of ascorbate-glutathione pathway alleviates mitotic aberrations and DNA damage in *Allium cepa* L. under salinity stress. *Int. J. Phytoremediation* **2023**, *25*, 403–414. [CrossRef] [PubMed]
8. Dong, Y.; Jinc, S.; Liu, S.; Xu, L.; Kong, J. Effects of exogenous nitric oxide on growth of cotton seedlings under NaCl stress. *J. Soil Sci. Plant Nutr.* **2014**, *14*, 1–13. [CrossRef]
9. Hamani, A.K.M.; Chen, J.; Soothar, M.K.; Wang, G.; Shen, X.; Gao, Y.; Qiu, R. Application of Exogenous Protectants Mitigates Salt-Induced Na(+) Toxicity and Sustains Cotton (*Gossypium hirsutum* L.) Seedling Growth: Comparison of Glycine Betaine and Salicylic Acid. *Plants* **2021**, *10*, 380. [CrossRef]
10. Munns, R. Comparative physiology of salt and water stress. *Plant Cell Environ.* **2002**, *25*, 239–250. [CrossRef]
11. Parida, A.K.; Das, A.B. Salt tolerance and salinity effects on plants: A review. *Ecotoxicol. Environ. Saf.* **2005**, *60*, 324–349. [CrossRef]
12. Mariyam, S.; Bhardwaj, R.; Khan, N.A.; Sahi, S.V.; Seth, C.S. Review on nitric oxide at the forefront of rapid systemic signaling in mitigation of salinity stress in plants: Crosstalk with calcium and hydrogen peroxide. *Plant Sci.* **2023**, *336*, 111835. [CrossRef]
13. Yan, P.; Li, G.; Sun, H.; Zhang, Z.; Yang, R.; Sun, J. Can arbuscular mycorrhizal fungi and biochar enhance plant resistance to low-temperature stress? *Agron. J.* **2021**, *113*, 1457–1466. [CrossRef]
14. Airaki, M.; Leterrier, M.; Mateos, R.M.; Valderrama, R.; Chaki, M.; Barroso, J.B.; Del Rio, L.A.; Palma, J.M.; Corpas, F.J. Metabolism of reactive oxygen species and reactive nitrogen species in pepper (*Capsicum annuum* L.) plants under low temperature stress. *Plant Cell Environ.* **2012**, *35*, 281–295. [CrossRef]
15. Sun, B.; Liu, G.-L.; Phan, T.T.; Yang, L.-T.; Li, Y.-R.; Xing, Y.-X. Effects of cold stress on root growth and physiological metabolisms in seedlings of different sugarcane varieties. *Sugar Tech.* **2017**, *19*, 165–175. [CrossRef]
16. Srinivasan, V.; Pandi-Perumal, S.R.; Trahkt, I.; Spence, D.W.; Poeggeler, B.; Hardeland, R.; Cardinali, D.P. Melatonin and Melatonergic Drugs on Sleep: Possible Mechanisms of Action. *Int. J. Neurosci.* **2009**, *119*, 821–846. [CrossRef] [PubMed]
17. Calvo, J.R.; Gonzalez-Yanes, C.; Maldonado, M. The role of melatonin in the cells of the innate immunity: A review. *J. Pineal Res.* **2013**, *55*, 103–120. [CrossRef]
18. Zhan, Y.; Wu, T.; Zhao, X.; Wang, Z.; Chen, Y. Comparative physiological and full-length transcriptome analyses reveal the molecular mechanism of melatonin-mediated salt tolerance in okra (*Abelmoschus esculentus* L.). *BMC Plant Biol.* **2021**, *21*, 1–16.
19. Hattori, A.; Migitaka, H.; Iigo, M.; Itoh, M.; Yamamoto, K.; Ohtani-Kaneko, R.; Hara, M.; Suzuki, T.; Reiter, R.J. Identification of melatonin in plants and its effects on plasma melatonin levels and binding to melatonin receptors in vertebrates. *Biochem. Mol. Biol. Int.* **1995**, *35*, 627–634. [PubMed]
20. Byeon, Y.; Back, K. An increase in melatonin in transgenic rice causes pleiotropic phenotypes, including enhanced seedling growth, delayed flowering, and low grain yield. *J. Pineal Res.* **2014**, *56*, 408–414. [CrossRef] [PubMed]
21. Hardeland, R. Melatonin in plants–diversity of levels and multiplicity of functions. *Front. Plant Sci.* **2016**, *7*, 198. [CrossRef]
22. Wei, J.; Li, D.X.; Zhang, J.R.; Shan, C.; Rengel, Z.; Song, Z.B.; Chen, Q. Phytomelatonin receptor PMTR 1-mediated signaling regulates stomatal closure in Arabidopsis thaliana. *J. Pineal Res.* **2018**, *65*, e12500. [CrossRef]
23. Back, K. Melatonin metabolism, signaling and possible roles in plants. *Plant J.* **2021**, *105*, 376–391. [CrossRef] [PubMed]
24. Wang, L.-F.; Lu, K.-K.; Li, T.-T.; Zhang, Y.; Guo, J.-X.; Song, R.-F.; Liu, W.-C. Maize Phytomelatonin Receptor1 functions in plant tolerance to osmotic and drought stress. *J. Exp. Bot.* **2021**, *73*, 5961–5973. [CrossRef] [PubMed]
25. Liang, C.; Zheng, G.; Li, W.; Wang, Y.; Hu, B.; Wang, H.; Wu, H.; Qian, Y.; Zhu, X.G.; Tan, D.X. Melatonin delays leaf senescence and enhances salt stress tolerance in rice. *J. Pineal Res.* **2015**, *59*, 91–101. [CrossRef] [PubMed]
26. Sharma, A.; Wang, J.; Xu, D.; Tao, S.; Chong, S.; Yan, D.; Li, Z.; Yuan, H.; Zheng, B. Melatonin regulates the functional components of photosynthesis, antioxidant system, gene expression, and metabolic pathways to induce drought resistance in grafted Carya cathayensis plants. *Sci. Total Environ.* **2020**, *713*, 136675. [CrossRef] [PubMed]
27. Shi, H.; Jiang, C.; Ye, T.; Tan, D.-X.; Reiter, R.J.; Zhang, H.; Liu, R.; Chan, Z. Comparative physiological, metabolomic, and transcriptomic analyses reveal mechanisms of improved abiotic stress resistance in bermudagrass [*Cynodon dactylon* (L). Pers.] by exogenous melatonin. *J. Exp. Bot.* **2015**, *66*, 681–694. [CrossRef]
28. Arnao, M.B.; Hernández-Ruiz, J. Melatonin: A new plant hormone and/or a plant master regulator? *Trends Plant Sci.* **2019**, *24*, 38–48. [CrossRef]
29. Moustafa-Farag, M.; Almoneafy, A.; Mahmoud, A.; Elkelish, A.; Arnao, M.B.; Li, L.; Ai, S. Melatonin and its protective role against biotic stress impacts on plants. *Biomolecules* **2020**, *10*, 54. [CrossRef]
30. Turk, H.; Genisel, M. Melatonin-related mitochondrial respiration responses are associated with growth promotion and cold tolerance in plants. *Cryobiology* **2020**, *92*, 76–85. [CrossRef]
31. Zhan, H.; Nie, X.; Zhang, T.; Li, S.; Wang, X.; Du, X.; Tong, W.; Song, W. Melatonin: A small molecule but important for salt stress tolerance in plants. *Int. J. Mol. Sci.* **2019**, *20*, 709. [CrossRef]
32. Fan, J.; Xie, Y.; Zhang, Z.; Chen, L. Melatonin: A multifunctional factor in plants. *Int. J. Mol. Sci.* **2018**, *19*, 1528. [CrossRef]

33. Zhang, Y.; Zhou, X.; Dong, Y.; Zhang, F.; He, Q.; Chen, J.; Zhu, S.; Zhao, T. Seed priming with melatonin improves salt tolerance in cotton through regulating photosynthesis, scavenging reactive oxygen species and coordinating with phytohormone signal pathways. *Ind. Crops Prod.* **2021**, *169*, 113671. [CrossRef]
34. Bajwa, V.S.; Shukla, M.R.; Sherif, S.M.; Murch, S.J.; Saxena, P.K. Role of melatonin in alleviating cold stress in A rabidopsis thaliana. *J. Pineal Res.* **2014**, *56*, 238–245. [CrossRef] [PubMed]
35. Zhang, H.M.; Zhang, Y. Melatonin: A well-documented antioxidant with conditional pro-oxidant actions. *J. Pineal Res.* **2014**, *57*, 131–146. [CrossRef] [PubMed]
36. Kamiab, F. Exogenous melatonin mitigates the salinity damages and improves the growth of pistachio under salinity stress. *J. Plant Nutr.* **2020**, *43*, 1468–1484. [CrossRef]
37. Li, J.; Yang, Y.; Sun, K.; Chen, Y.; Chen, X.; Li, X. Exogenous melatonin enhances cold, salt and drought stress tolerance by improving antioxidant defense in tea plant (*Camellia sinensis* (L.) O. Kuntze). *Molecules* **2019**, *24*, 1826. [CrossRef]
38. Mukherjee, S.; David, A.; Yadav, S.; Baluška, F.; Bhatla, S.C. Salt stress-induced seedling growth inhibition coincides with differential distribution of serotonin and melatonin in sunflower seedling roots and cotyledons. *Physiol. Plant.* **2014**, *152*, 714–728. [CrossRef]
39. Ding, F.; Liu, B.; Zhang, S. Exogenous melatonin ameliorates cold-induced damage in tomato plants. *Sci. Hortic.* **2017**, *219*, 264–271. [CrossRef]
40. Shen, J.; Chen, D.; Zhang, X.; Song, L.; Dong, J.; Xu, Q.; Hu, M.; Cheng, Y.; Shen, F.; Wang, W. Mitigation of salt stress response in upland cotton (*Gossypium hirsutum*) by exogenous melatonin. *J. Plant Res.* **2021**, *134*, 857–871. [CrossRef]
41. Li, J.; Arkorful, E.; Cheng, S.; Zhou, Q.; Li, H.; Chen, X.; Sun, K.; Li, X. Alleviation of cold damage by exogenous application of melatonin in vegetatively propagated tea plant (*Camellia sinensis* (L.) O. Kuntze). *Sci. Hortic.* **2018**, *238*, 356–362. [CrossRef]
42. Fan, J.; Hu, Z.; Xie, Y.; Chan, Z.; Chen, K.; Amombo, E.; Chen, L.; Fu, J. Alleviation of cold damage to photosystem II and metabolisms by melatonin in Bermudagrass. *Front. Plant Sci.* **2015**, *6*, 925. [CrossRef]
43. Hamani, A.K.M.; Wang, G.; Soothar, M.K.; Shen, X.; Gao, Y.; Qiu, R.; Mehmood, F. Responses of leaf gas exchange attributes, photosynthetic pigments and antioxidant enzymes in NaCl-stressed cotton (*Gossypium hirsutum* L.) seedlings to exogenous glycine betaine and salicylic acid. *BMC Plant Biol.* **2020**, *20*, 1–14. [CrossRef] [PubMed]
44. Xie, F.; Wang, Q.; Sun, R.; Zhang, B. Deep sequencing reveals important roles of microRNAs in response to drought and salinity stress in cotton. *J. Exp. Bot.* **2015**, *66*, 789–804. [CrossRef] [PubMed]
45. Valliyodan, B.; Nguyen, H.T. Understanding regulatory networks and engineering for enhanced drought tolerance in plants. *Curr. Opin. Plant Biol.* **2006**, *9*, 189–195. [CrossRef] [PubMed]
46. Hu, Z.; Fan, J.; Xie, Y.; Amombo, E.; Liu, A.; Gitau, M.M.; Khaldun, A.; Chen, L.; Fu, J. Comparative photosynthetic and metabolic analyses reveal mechanism of improved cold stress tolerance in bermudagrass by exogenous melatonin. *Plant Physiol. Biochem.* **2016**, *100*, 94–104. [CrossRef]
47. Mittler, R. Abiotic stress, the field environment and stress combination. *Trends Plant Sci.* **2006**, *11*, 15–19. [CrossRef]
48. Mittler, R.; Blumwald, E. Genetic engineering for modern agriculture: Challenges and perspectives. *Annu. Rev. Plant Biol.* **2010**, *61*, 443–462. [CrossRef]
49. Suzuki, N.; Rivero, R.M.; Shulaev, V.; Blumwald, E.; Mittler, R. Abiotic and biotic stress combinations. *N. Phytol.* **2014**, *203*, 32–43. [CrossRef]
50. Zandalinas, S.I.; Balfagón, D.; Arbona, V.; Gómez-Cadenas, A.; Inupakutika, M.A.; Mittler, R. ABA is required for the accumulation of APX1 and MBF1c during a combination of water deficit and heat stress. *J. Exp. Bot.* **2016**, *67*, 5381–5390. [CrossRef]
51. Zandalinas, S.I.; Rivero, R.M.; Martínez, V.; Gómez-Cadenas, A.; Arbona, V. Tolerance of citrus plants to the combination of high temperatures and drought is associated to the increase in transpiration modulated by a reduction in abscisic acid levels. *BMC Plant Biol.* **2016**, *16*, 1–16. [CrossRef]
52. Khalil, R.; Yusuf, M.; Bassuony, F.; Haroun, S.; Gamal, A. Alpha-tocopherol reinforce selenium efficiency to ameliorates salt stress in maize plants through carbon metabolism, enhanced photosynthetic pigments and ion uptake. *South Afr. J. Bot.* **2022**, *144*, 1–9. [CrossRef]
53. Sarropoulou, V.N.; Therios, I.N.; Dimassi-Theriou, K.N. Melatonin promotes adventitious root regeneration in in vitro shoot tip explants of the commercial sweet cherry rootstocks CAB-6P (*Prunus Cerasus* L.), Gisela 6 (*P. cerasus* × *P. canescens*), and MxM 60 (*P. avium* × *P. mahaleb*). *J. Pineal Res.* **2012**, *52*, 38–46. [CrossRef] [PubMed]
54. Chaves, M.M.; Flexas, J.; Pinheiro, C. Photosynthesis under drought and salt stress: Regulation mechanisms from whole plant to cell. *Ann. Bot.* **2009**, *103*, 551–560. [CrossRef] [PubMed]
55. Chatrath, A.; Mandal, P.; Anuradha, M. Effect of secondary salinization on photosynthesis in fodder oat (*Avena sativa* L.) genotypes. *J. Agron. Crop Sci.* **2000**, *184*, 13–16. [CrossRef]
56. Irshad, A.; Rehman, R.N.U.; Kareem, H.A.; Yang, P.; Hu, T. Addressing the challenge of cold stress resilience with the synergistic effect of Rhizobium inoculation and exogenous melatonin application in Medicago truncatula. *Ecotoxicol. Environ. Saf.* **2021**, *226*, 112816. [CrossRef]
57. Parveen, A.; Ahmar, S.; Kamran, M.; Malik, Z.; Ali, A.; Riaz, M.; Abbasi, G.H.; Khan, M.; Sohail, A.B.; Rizwan, M. Abscisic acid signaling reduced transpiration flow, regulated Na+ ion homeostasis and antioxidant enzyme activities to induce salinity tolerance in wheat (*Triticum aestivum* L.) seedlings. *Environ. Technol. Innov.* **2021**, *24*, 101808. [CrossRef]

58. Hernandez, J.A.; Jiménez, A.; Mullineaux, P.; Sevilia, F. Tolerance of pea (*Pisum sativum* L.) to long-term salt stress is associated with induction of antioxidant defences. *Plant Cell Environ.* **2000**, *23*, 853–862. [CrossRef]
59. Anjum, S.A.; Ashraf, U.; Tanveer, M.; Khan, I.; Hussain, S.; Shahzad, B.; Zohaib, A.; Abbas, F.; Saleem, M.F.; Ali, I. Drought induced changes in growth, osmolyte accumulation and antioxidant metabolism of three maize hybrids. *Front. Plant Sci.* **2017**, *8*, 69. [CrossRef]
60. Wang, L.; Liu, J.; Wang, W.; Sun, Y. Exogenous melatonin improves growth and photosynthetic capacity of cucumber under salinity-induced stress. *Photosynthetica* **2016**, *54*, 19–27. [CrossRef]
61. Turk, H.; Erdal, S.; Genisel, M.; Atici, O.; Demir, Y.; Yanmis, D. The regulatory effect of melatonin on physiological, biochemical and molecular parameters in cold-stressed wheat seedlings. *Plant Growth Regul.* **2014**, *74*, 139–152. [CrossRef]
62. Marta, B.; Szafrańska, K.; Posmyk, M.M. Exogenous melatonin improves antioxidant defense in cucumber seeds (*Cucumis sativus* L.) germinated under chilling stress. *Front. Plant Sci.* **2016**, *7*, 575. [CrossRef]
63. Martinez, V.; Nieves-Cordones, M.; Lopez-Delacalle, M.; Rodenas, R.; Mestre, T.C.; Garcia-Sanchez, F.; Rubio, F.; Nortes, P.A.; Mittler, R.; Rivero, R.M. Tolerance to stress combination in tomato plants: New insights in the protective role of melatonin. *Molecules* **2018**, *23*, 535. [CrossRef]
64. Zhao, C.; Zhang, H.; Song, C.; Zhu, J.-K.; Shabala, S. Mechanisms of plant responses and adaptation to soil salinity. *Innovation* **2020**, *1*, 100017. [CrossRef]
65. Shahzad, R.; Khan, A.L.; Bilal, S.; Waqas, M.; Kang, S.-M.; Lee, I.-J. Inoculation of abscisic acid-producing endophytic bacteria enhances salinity stress tolerance in Oryza sativa. *Environ. Exp. Bot.* **2017**, *136*, 68–77. [CrossRef]
66. Rasouli, F.; Kiani-Pouya, A.; Tahir, A.; Shabala, L.; Chen, Z.; Shabala, S. A comparative analysis of stomatal traits and photosynthetic responses in closely related halophytic and glycophytic species under saline conditions. *Environ. Exp. Bot.* **2021**, *181*, 104300. [CrossRef]
67. Perin, E.C.; da Silva Messias, R.; Borowski, J.M.; Crizel, R.L.; Schott, I.B.; Carvalho, I.R.; Rombaldi, C.V.; Galli, V. ABA-dependent salt and drought stress improve strawberry fruit quality. *Food Chem.* **2019**, *271*, 516–526. [CrossRef] [PubMed]
68. Yu, Y.; Lv, Y.; Shi, Y.; Li, T.; Chen, Y.; Zhao, D.; Zhao, Z. The role of phyto-melatonin and related metabolites in response to stress. *Molecules* **2018**, *23*, 1887. [CrossRef]
69. Arnao, M.B.; Hernández-Ruiz, J. Is phytomelatonin a new plant hormone? *Agronomy* **2020**, *10*, 95. [CrossRef]
70. Heath, R.L.; Packer, L. Photoperoxidation in isolated chloroplasts: I. Kinetics and stoichiometry of fatty acid peroxidation. *Arch. Biochem. Biophys.* **1968**, *125*, 189–198. [CrossRef]
71. Tian, M.; Gu, Q.; Zhu, M. The involvement of hydrogen peroxide and antioxidant enzymes in the process of shoot organogenesis of strawberry callus. *Plant Sci.* **2003**, *165*, 701–707. [CrossRef]
72. Dhindsa, R.S.; Plumb-Dhindsa, P.; Thorpe, T.A. Leaf senescence: Correlated with increased levels of membrane permeability and lipid peroxidation, and decreased levels of superoxide dismutase and catalase. *J. Exp. Bot.* **1981**, *32*, 93–101. [CrossRef]
73. Lacan, D.; Baccou, J.-C. High levels of antioxidant enzymes correlate with delayed senescence in nonnetted muskmelon fruits. *Planta* **1998**, *204*, 377–382. [CrossRef]
74. Jiang, Y.; Huang, B. Effects of calcium on antioxidant activities and water relations associated with heat tolerance in two cool-season grasses. *J. Exp. Bot.* **2001**, *52*, 341–349. [CrossRef] [PubMed]
75. Shabala, S.; Mackay, A. Ion transport in halophytes. *Adv. Bot. Res.* **2011**, *57*, 151–199.
76. Malik, R.N.; Husain, S.Z.; Nazir, I. Heavy metal contamination and accumulation in soil and wild plant species from industrial area of Islamabad, Pakistan. *Pak. J. Bot.* **2010**, *42*, 291–301.

MDPI AG
Grosspeteranlage 5
4052 Basel
Switzerland
Tel.: +41 61 683 77 34

Plants Editorial Office
E-mail: plants@mdpi.com
www.mdpi.com/journal/plants

www.ingramcontent.com/pod-product-compliance
Lightning Source LLC
Chambersburg PA
CBHW041430210926
51850CB00013B/3391
9783725827343